Lecture Notes in Computer Science   12730

More information about this subseries at http://www.springer.com/series/7407

Rahul Santhanam · Daniil Musatov (Eds.)

Computer Science – Theory and Applications

16th International Computer Science Symposium in Russia, CSR 2021
Sochi, Russia, June 28 – July 2, 2021
Proceedings

 Springer

Editors
Rahul Santhanam
University of Oxford
Oxford, UK

Daniil Musatov 🆔
Innovations & HighTech
Moscow Institute of Physics and Technology
Moscow, Russia

ISSN 0302-9743   ISSN 1611-3349 (electronic)
Lecture Notes in Computer Science
ISBN 978-3-030-79415-6   ISBN 978-3-030-79416-3 (eBook)
https://doi.org/10.1007/978-3-030-79416-3

LNCS Sublibrary: SL1 – Theoretical Computer Science and General Issues

This Springer imprint is published by the registered company Springer Nature Switzerland AG
The registered company address is: Gewerbestrasse 11, 6330 Cham, Switzerland

Preface

This volume contains extended abstracts of the papers selected for presentation at CSR 2021: the 16th International Computer Science Symposium in Russia. The meeting was held during June 28 – July 2, 2021, in Sochi. Due to the COVID-19 crisis, the meeting was held in a hybrid format, with most international attendees participating online.

There were 68 submissions. Each submission was reviewed by at least 2, and on the average 3.1, Program Committee members. The committee decided to accept 28 papers. The reviewing process was smoothly run using the Easychair conference system.

This was the 16th edition of the annual series of meetings. Previous editions were held in St. Petersburg (2006), Ekaterinburg (2007), Moscow (2008), Novosibirsk (2009), Kazan (2010), St. Petersburg (2011), Nizhny Novgorod (2012), Ekaterinburg (2013), Moscow (2014), Listvyanka (2015), St. Petersburg (2016), Kazan (2017), Moscow (2018), Novosibirsk (2019), and online (2020; scheduled to held in Ekaterinburg). The symposium covers a broad range of topics in Theoretical Computer Science.

The distinguished CSR 2021 keynote lecture was given by Tim Roughgarden (Columbia University, USA). The eight other invited speakers were, in alphabetical order, Amin Coja-Oghlan (Goethe University, Germany), Edith Elkind (University of Oxford, UK), Hugo Gimbert (CNRS, France), Ugo dal Lago (University of Bologna, Italy), Merav Parter (Weizmann Institute of Science, Israel), Toniann Pitassi (University of Toronto, Canada; Columbia University, USA), Joel Spencer (New York University, USA), and Jens Vygen (University of Bonn, Germany). The Program Committee included 23 scientists from 14 countries and was chaired by Rahul Santhanam (University of Oxford, UK).

The Program Committee selected the paper "Real tau-Conjecture for sum-of-squares: a unified approach to lower bounds and derandomization" by Pranjal Dutta (Chennai Mathematical Institute, India) to received both the Best Paper award and the Best Student Paper award. Both awards were sponsored by Springer.

We would like to thank the invited speakers for agreeing to give talks, the Programme Committee members for all their effort in helping to finalize the program, the external reviewers for their expert opinions, and the publishers Springer for offering financial support for the event. We thank the Sirius Mathematical Center in Sochi for hosting the event and for offering financial support for travel and accommodation. We also thank the Caucasus Mathematical Center at Adyghe State University in Maykop and the Moscow Institute of Physics and Technology for organizational help.

May 2021

Rahul Santhanam
Daniil Musatov

Organization

Program Committee

Elena Arseneva	St. Petersburg University, Russia
Aleksandrs Belovs	University of Latvia, Latvia
Simina Branzei	Purdue University, USA
Andrei Bulatov	Simon Fraser University, Canada
Anupam Das	University of Birmingham, UK
Laure Daviaud	City, University of London, UK
Susanna F. de Rezende	Institute of Mathematics of the Czech Academy of Sciences, Czech Republic
Laurent Doyen	CNRS and ENS Paris-Saclay, France
Piotr Faliszewski	AGH University of Science and Technology, Poland
Pawel Gawrychowski	University of Wroclaw, Poland
Heng Guo	University of Edinburgh, UK
Siyao Guo	New York University, USA
Shuichi Hirahara	National Institute of Informatics, Japan
Michael Kapralov	Ecole Polytechnique Fédérale de Lausanne, Switzerland
Telikepalli Kavitha	Tata Institute of Fundamental Research, India
Jesper Nederlof	Eindhoven University of Technology, the Netherlands
Alexander Okhotin	St. Petersburg State University, Russia
Sofya Raskhodnikova	Boston University, USA
Alexander Razborov	University of Chicago, USA
Laura Sanita	University of Waterloo, Canada
Rahul Santhanam	University of Oxford, UK
Amir Yehudayoff	Institute for Advanced Study, USA
Meirav Zehavi	Ben-Gurion University, Israel

Additional Reviewers

Anshu, Anurag
Barloy, Corentin
Bauer, Johann
Ben-Sasson, Eli
Bhaskar, Siddharth
Bodlaender, Hans L.
Boehmer, Niclas
Bonacina, Ilario
Borovitskiy, Viacheslav
Bredereck, Robert
Byrka, Jaroslaw

Calamoneri, Tiziana
Chen, Lijie
Courcelle, Bruno
Cucker, Felipe
Downey, Rod
Duchon, Philippe
Dudek, Bartlomiej
Duparc, Jacques
Elias, Marek
Enright, Jessica
Faenza, Yuri

Fernau, Henning
Fici, Gabriele
Fluschnik, Till
Fogarty, Seth
Golovach, Petr
Golovnev, Alexander
Gupta, Sushmita
Har-Peled, Sariel
Huang, Shenwei
Ilango, Rahul
Inamdar, Tanmay
Jain, Pallavi
Janczewski, Wojciech
Jugé, Vincent
Kallaugher, John
Kawamura, Akitoshi
Keldenich, Phillip
Klute, Fabian
Kobayashi, Yusuke
Koivisto, Mikko
Kokkinis, Ioannis
Kumar, Akash
Kuperberg, Denis
Laekhanukit, Bundit
Lauria, Massimo
Li, Yinan
Lovett, Shachar
Makarychev, Konstantin
Manabe, Yoshifumi
Manzoni, Luca
Miltzow, Till
Misra, Pranabendu
Miyahara, Daiki
Mozes, Shay
Myrisiotis, Dimitrios

Narayanaswamy, N. S.
Norman, Gethin
Pandit, Supantha
Paradise, Orr
Penelle, Vincent
Pous, Damien
Pradic, Pierre
Purser, David
Qian, Li
Rampersad, Narad
Rao, Shravas
Ryzhikov, Andrew
Sahlot, Vibha
Saurabh, Nitin
Seiller, Thomas
Seshadhri, C.
Shen, Alexander
Shimizu, Nobutaka
Shinagawa, Kazumasa
Shur, Arseny
Skopalik, Alexander
Staals, Frank
Stephan, Frank
Takaoka, Asahi
Talmon, Nimrod
Tawari, Anuj
Traub, Vera
Tzameret, Iddo
Vitanyi, Paul
Volkovich, Ilya
Wlodarczyk, Michal
Wu, Kewen
Xiao, Mingyu
Xie, Zhiye

Contents

Computational Complexity of Multi-player Evolutionarily Stable Strategies

Manon Blanc[1] and Kristoffer Arnsfelt Hansen[2]($\boxtimes$) (iD)

[1] ENS Paris-Saclay, Gif-Sur-Yvette, France
manonblanc@free.fr
[2] Aarhus University, Aarhus, Denmark
arnsfelt@cs.au.dk

Abstract. In this paper we study the computational complexity of computing an evolutionary stable strategy (ESS) in multi-player symmetric games. For two-player games, deciding existence of an ESS is complete for Σ_2^p, the second level of the polynomial time hierarchy. We show that deciding existence of an ESS of a multi-player game is closely connected to the second level of the *real* polynomial time hierarchy. Namely, we show that the problem is hard for a complexity class we denote as $\exists^D \cdot \forall \mathbb{R}$ and is a member of $\exists \forall \mathbb{R}$, where the former class restrict the latter by having the existentially quantified variables be Boolean rather then real-valued. As a special case of our results it follows that deciding whether a given strategy is an ESS is complete for $\forall \mathbb{R}$.

1 Introduction

First introduced by Maynard Smith and Price in [27,28], a central concept emerging from evolutionary game theory is that of an evolutionary stable strategy (ESS) in a symmetric two-player game in strategic form. Each pure strategy of the game is viewed as a type of possible individuals of a population. A mixed strategy of the game then corresponds to describing the proportion of each type of individual of the population, which as a simplifying assumption is considered to be infinite. The population is engaged in a pairwise conflict where two individuals are selected at random and receive payoffs depending on their respective types. The population is expected to evolve in a way where strategies that achieve a higher payoff than others will spread in the population. A strategy σ is an ESS if it outperforms any "mutant" strategy $\tau \neq \sigma$ adopted by a small fraction of the population. Otherwise we say that σ may be invaded. An ESS is in particular a symmetric Nash equilibrium (SNE), but, unlike a SNE, it is not guaranteed to exist.

The Hawk-Dove game [28], presented with concrete payoffs in Fig. 1, is a classic example where an ESS may explain the proportion of the population tending to engage in aggressive behavior. The game has a unique SNE σ, where

The second author is supported by the Independent Research Fund Denmark under grant no. 9040-00433B.

R. Santhanam and D. Musatov (Eds.): CSR 2021, LNCS 12730, pp. 1–17, 2021.
https://doi.org/10.1007/978-3-030-79416-3_1

	Hawk Dove	
	Hawk	Dove
Hawk	-1,-1	2,0
Dove	0,2	1,1

Fig. 1. Hawk-Dove game

the players choose Hawk with probability $\frac{1}{2}$, and this is in fact an ESS. Note first that $u(\sigma,\sigma) = (-1)\left(\frac{1}{2}\right)^2 + 2\left(\frac{1}{2}\right)^2 + 0\left(\frac{1}{2}\right)^2 + 1\left(\frac{1}{2}\right)^2 = \frac{1}{2}$. Consider now any strategy profile τ that chooses Hawk with probability p. Then $u(\tau,\sigma) = (-1+2)p/2 + (1+0)(1-p)/2 = \frac{1}{2}$ as well. However, $u(\sigma,\tau) = \frac{3}{2} - 2p$ and $u(\tau,\tau) = 1 - 2p^2$, and thus $u(\sigma,\tau) - u(\tau,\tau) = 2(p-\frac{1}{2})^2$, which means that σ outperforms τ if $p \neq \frac{1}{2}$.

While the two-player setting is the typical setting to study ESS, the concept may in a natural way be generalized to the setting of multi-player games, as established by Palm [33] and Broom, Cannings, and Vickers [11]. This allows one to model populations that engage in conflicts involving more than two individuals. Many of the two-player games typically studied in the context of ESS readily generalize to multi-player games, including the Hawk-Dove and Stag Hunt games (cf. [12]). For a naturally occurring example, Broom and Rychtář [12, Example 9.1] argue that the cooperative hunting method of carousel feeding by killer whales may be modeled as a multi-player Stag Hunt game.

The computational complexity of computing an ESS was first studied by Etessami and Lochbihler [19]. We shall denote the problem of deciding whether a given symmetric game in strategic form has an ESS as ∃ESS and similarly the problem of deciding whether a given strategy is an ESS of the given game as IsESS. Previous work has been concerned only with two-player symmetric games in strategic form. Etessami and Lochbihler proved that ∃ESS is hard both for NP and coNP and is contained in Σ_2^p. Nisan [32] showed that ∃ESS is hard for the class coDP, which is the class of unions of languages from NP and coNP. From both works it also follows that the problem IsESS is coNP-complete. Finally Conitzer [14] showed Σ_2^p-completeness for ∃ESS. The direct but important consequence of these results is that any algorithm for computing an ESS in a general game can be used to solve Σ_2^p-complete problems. For instance, we cannot expect to be able to compute an ESS in a simple way using a SAT solver.

One may observe that the above hardness results for two-player games also generalize to apply to m-player games, for any fixed $m \geq 3$. Note that, since a reduction showing Σ_2^p-hardness must produce an m-player symmetric game, this is not a trivial observation (in particular adding "dummy" players, each having a single strategy, to a nontrivial symmetric game would result in a non-symmetric game). One would however suspect that the problems ∃ESS and IsESS become significantly harder for m-player games, when $m \geq 3$. Namely, starting with the work of Schaefer and Štefankovič [36], several works have shown that many natural decision problems concerning Nash equilibrium (NE) in 3-player strategic form games are ∃ℝ-complete [4–6,20,24]. These results stand in contrast to the two-player setting, where the same decision problems are NP-complete [15,21].

The class $\exists\mathbb{R}$ is the complexity class that captures the decision problem for the existential theory of the reals [36], or alternatively, is the constant free Boolean part of the real analogue $\mathrm{NP}_\mathbb{R}$ in the Blum-Shub-Smale model of computation [10]. Clearly we have $\mathrm{NP} \subseteq \exists\mathbb{R}$, and from the decision procedure for the existential theory of the reals by Canny [13] it follows that $\exists\mathbb{R} \subseteq \mathrm{PSPACE}$. We consider it likely that NP is a strict subset of $\exists\mathbb{R}$, which would mean that the above mentioned decision problems concerning NE become strictly harder as the number of players increase beyond two.

We confirm that the problems $\exists$ESS and IsESS indeed are likely to become harder for multi-player games by proving hardness of the problems for discrete complexity classes defined in terms of real complexity classes that we consider likely to be stronger than Σ_2^p and NP. Our results are perhaps most easily stated in terms of the decision problem for the first order theory of the reals $\mathrm{Th}(\mathbb{R})$. Just like the class $\exists\mathbb{R}$ corresponds to existential fragment $\mathrm{Th}_\exists(\mathbb{R})$ of $\mathrm{Th}(\mathbb{R})$, we can consider classes $\forall\mathbb{R}$ and $\exists\forall\mathbb{R}$ corresponding to the universal fragment $\mathrm{Th}_\forall(\mathbb{R})$ and the existential-universal fragment $\mathrm{Th}_{\exists\forall}(\mathbb{R})$ of $\mathrm{Th}(\mathbb{R})$, respectively. It is easy to see that the problem $\exists$ESS belongs to $\exists\forall\mathbb{R}$ and that IsESS belongs to $\forall\mathbb{R}$. We show that for 5-player games, the problem $\exists$ESS is hard for the subclass of $\exists\forall\mathbb{R}$ where the block of universal quantifiers is restricted to range over Boolean variables. For the problem IsESS we completely characterize its complexity for 5-player games by proving that the problem is also hard for $\forall\mathbb{R}$. Our hardness results thus imply that any algorithm for computing an ESS in a 5-player game can be used to solve quite general problems involving real polynomials. In particular it indicates that computing an ESS is significantly more difficult than deciding if a system of real polynomials has no solution, which is a basic problem complete for $\forall\mathbb{R}$.

Our proof of hardness for $\exists$ESS combines ideas of the Π_2^p-completeness proof of the problem MINMAXCLIQUE by Ko and Lin [26], the reduction from the complement of MINMAXCLIQUE to $\exists$ESS for two-player games by Conitzer [14], and the direct translation of solutions of a polynomial system to strategies of a game by Hansen [24], in addition to new ideas.

We leave the problem of determining the precise computational complexity of $\exists$ESS as an interesting open problem. The class $\exists\forall\mathbb{R}$ is the natural real complexity class generalization of Σ_2^p. Together with Σ_2^p-completeness of $\exists$ESS for the setting of two-player games, this might lead one to expect that $\exists$ESS should be $\exists\forall\mathbb{R}$-hard for multi-player games. However, a basic property of the set of evolutionary stable strategies is that any ESS is an isolated point in the space of strategies [2, Proposition 3], which means that the set of evolutionary stable strategies is always a *discrete* set. Expressing $\exists$ESS in $\mathrm{Th}_{\exists\forall}(\mathbb{R})$, the universal quantifier range over all potential ESS and the existential quantifier over potential invading strategies. The fact that the set of ESS is a discrete set could possibly mean that the universal quantifier could be made discrete as well. We also note that we do not even know whether $\exists$ESS is hard for $\exists\mathbb{R}$, which is clearly a prerequisite for $\exists\forall\mathbb{R}$-hardness.

1.1 Other Related Work

Starting with the universality theorem of Mnëv [30], which in particular implies that deciding whether an arrangement of pseudolines is *stretchable* is complete for $\exists\mathbb{R}$, a large number of problems are by now known to be complete for $\exists\mathbb{R}$. A crucial insight used for the first $\exists\mathbb{R}$-completeness result concerning games by Schaefer and Štefankovič [36] was that the $\exists\mathbb{R}$-complete QUAD remains complete when asking for a solution of the polynomial system in the unit ball. This was also used by Schaefer [35] to prove that deciding rigidity of linkages is $\forall\mathbb{R}$-complete, and similar insights were used by Abrahamsen, Adamaszek, and Miltzow in their proof of $\exists\mathbb{R}$-completeness of the classic art gallery problem [1].

So far much fewer results are known concerning larger fragments of the first-order theory of the reals. Bürgisser and Cucker [10] study decision problems about general semialgebraic sets and show that the problem of deciding whether such a set contains an isolated point is hard for $\forall\mathbb{R}$ and contained in $\exists\forall\mathbb{R}$. Dobbins, Kleist, Miltzow, and Rząʒewski [18] prove $\forall\exists\mathbb{R}$-completeness for certain problems concerned with embedding graphs in the plane. For problems concerning games, Gimbert, Paul, and Srivathsan [22] show that deciding whether in a two-player extensive form game with imperfect recall a player has a behavior strategy with positive payoff is hard both for $\exists\mathbb{R}$ and $\forall\mathbb{R}$ while being contained in $\exists\forall\mathbb{R}$.

2 Preliminaries

2.1 Strategic Form Games

We present here basic definitions concerning strategic form games, mainly to establish our notations. A finite m-player strategic form game $\mathcal{G}$ is given by finite sets $S_1, \ldots, S_m$ of actions (*pure strategies*) together with *utility functions* $u_1, \ldots, u_m : S_1 \times \cdots \times S_m \to \mathbb{R}$. A choice of an action $a_i \in S_i$ for each player together forms a pure strategy profile $a = (a_1, \ldots, a_m)$. Let $\Delta(S_i)$ denote the set of probability distributions on S_i. A *(mixed) strategy* for player i is then an element $x_i \in \Delta(S_i)$. We may conveniently identify an action a_i with the strategy that assigns probability 1 to a_i. A strategy x_i for each player i together form a strategy profile $x = (x_1, \ldots, x_m)$. For fixed i we denote by x_{-i} the *partial* strategy profile $(x_1, \ldots, x_{i-1}, x_{i+1}, \ldots, x_m)$ for all players except player i, and if $x_i' \in \Delta(S_i)$ we denote by $(x_i'; x_{-i})$ the strategy profile $(x_1, \ldots, x_{i-1}, x_i', x_{i+1}, \ldots, x_m)$. The utility functions extend to strategy profiles by letting $u_i(x) = \mathrm{E}_{a \sim x} u_i(a_1, \ldots, a_m)$. We shall also refer to $u_i(x)$ as the *payoff* of player i. A strategy profile x is a Nash equilibrium (NE) if $u_i(x) \geq u_i(x_i'; x_{-i})$ for all i and all $x_i' \in \Delta(S_i)$. Every finite strategic form game $\mathcal{G}$ has an NE [31].

In this paper we shall only consider *symmetric* games. The game $\mathcal{G}$ is symmetric if all players have the same set S of actions and where the utility function of a given player depends only on the action of that player (and not the identity of the player) together with the *multiset* of actions of the other players. More precisely we say that $\mathcal{G}$ is *symmetric* if there is a finite set S such that $S_i = S$, for every $i \in [m]$, and such that for every permutation π on $[m]$, every $i \in [m]$ and every $(a_1, \ldots, a_m) \in S^m$ it holds that $u_i(a_1, \ldots, a_m) = u_{\pi^{-1}(i)}(a_{\pi(1)}, \ldots, a_{\pi(m)})$. It follows that a symmetric game $\mathcal{G}$ is fully specified by S and u_1; for notational simplicity we let $u = u_1$. A strategy profile $x = (x_1, \ldots, x_m)$ is symmetric if $x_1 = \cdots = x_m$. If a symmetric strategy profile x is an NE it is called a symmetric NE (SNE). Every finite strategic form symmetric game $\mathcal{G}$ has a SNE [31].

A single strategy $\sigma \in \Delta(S)$ defines the symmetric strategy profile σ^m. More generally, given $\sigma, \sigma_1, \ldots, \sigma_r \in \Delta(S)$ and $m_1, \ldots, m_r \geq 1$ with $m_1 + \cdots + m_r = m - 1$, we denote by $(\sigma; \sigma_1^{m_1}, \ldots, \sigma_r^{m_r})$ a strategy profile where player 1 is playing using strategy σ and m_i of the remaining players are playing using strategy σ_i, for $i = 1, \ldots, r$. By the assumptions of symmetry, the payoff $u(\sigma; \sigma_1^{m_1}, \ldots, \sigma_r^{m_r})$ is well defined.

2.2   Evolutionary Stable Strategies

Our main object of study is the notion of evolutionary stable strategies as defined by Maynard Smith and Price [28] for 2-player games and generalized to multi-player games by Palm [33] and Broom, Cannings, and Vickers [11]. We follow below the definition given by Broom et al.

Definition 1. *Let $\mathcal{G}$ be a symmetric game given by S and u. Let $\sigma, \tau \in \Delta(S)$. We say that σ is evolutionary stable (ES) against τ if there is $\varepsilon_\tau > 0$ such that for all $0 < \varepsilon < \varepsilon_\tau$ we have*

$$u(\sigma; \tau_\varepsilon^{m-1}) > u(\tau; \tau_\varepsilon^{m-1}) \,, \tag{1}$$

where $\tau_\varepsilon = \varepsilon\tau + (1 - \varepsilon)\sigma$ is the strategy that plays according to τ with probability ε and according to σ with probability $1 - \varepsilon$. We say that σ is an evolutionary stable strategy (ESS) if σ is ES against every $\tau \neq \sigma$. If σ is not ES against τ we also say that τ invades σ.

The supremum over ε_τ for which Eq. 1 holds is called the *invasion barrier* for τ. If σ is an ESS and there exists $\varepsilon_\sigma > 0$ such that for all $\tau \neq \sigma$ the invasion barrier ε_τ for τ satisfies $\varepsilon_\tau \geq \varepsilon_\sigma$, we say that σ is an ESS with *uniform* invasion barrier ε_σ. For 2-player games any ESS has a uniform invasion barrier [25]. Milchtaich [29] give a simple example of an ESS in a 4-player game without a uniform invasion barrier.

The following simple lemma due to Broom et al. [11] provides a useful alternative characterization of an ESS.

Lemma 1. *A strategy σ is ES against τ if and only if there exists $0 \leq j < m$ such that $u(\sigma; \tau^j, \sigma^{m-1-j}) > u(\tau; \tau^j, \sigma^{m-1-j})$ and that for all $0 \leq i < j$, $u(\sigma; \tau^i, \sigma^{m-1-i}) = u(\tau; \tau^i, \sigma^{m-1-i})$.*

For the case of 2-player games, this alternative characterization is actually the original definition of an ESS given by Maynard Smith and Price [28], and the definition of an ESS we use was stated for the case of 2-player games by Taylor and Jonker [37]. A straightforward corollary of the characterization is that if σ is an ESS then σ^m is a SNE.

By the *support* of an ESS σ, $\text{Supp}(\sigma)$, we refer to the set of pure strategies i such that are played with non-zero probability under the strategy σ.

2.3 Real Computational Complexity

While we are mainly interested in the computational complexity of discrete problems, it is useful to discuss a model of computation operating on real-valued input. We use this to define the complexity class $\exists^D \cdot \forall \mathbb{R}$, used to formulate our main result. Alternatively we may simply define this class in terms of a restriction of the decision problem for the first-order theory of the reals, as explained in the next subsection. The reader may thus defer reading this subsection.

A standard model for studying computational complexity in the setting of reals is that of Blum-Shub-Smale (BSS) machines [7]. A BSS machine takes a vector $x \in \mathbb{R}^n$ as an input and performs arithmetic operations and comparisons at unit cost. In addition the machine may be equipped with a finite set of real-valued *machine constants*. In this way a BSS machine accepts a *real language* $L \subseteq \mathbb{R}^\infty$, where $\mathbb{R}^\infty = \bigcup_{n \geq 0} \mathbb{R}^n$. Imposing polynomial time bounds we obtain the complexity classes $\text{P}_\mathbb{R}$ and $\text{NP}_\mathbb{R}$ for deterministic and nondeterministic BSS machines, respectively, forming real-valued analogues of P and NP. Cucker [16] defined the real analogue $\text{PH}_\mathbb{R}$ of the polynomial time hierarchy formed by the classes $\Sigma_k^\mathbb{R}$ and $\Pi_k^\mathbb{R}$, for $k \geq 1$. The class $\Sigma_{k+1}^\mathbb{R}$ may be defined as real languages accepted by a nondeterministic *oracle* BSS machine in polynomial time using an oracle language from $\Sigma_k^\mathbb{R}$ with $\Sigma_1^\mathbb{R} = \text{NP}_\mathbb{R}$, and $\Pi_k^\mathbb{R}$ is simply the class of complements of languages of $\Sigma_k^\mathbb{R}$. For natural problems such as TSP or KNAPSACK with real-valued input the search space remains discrete. Goode [23] introduced the notion of *digital nondeterminism* (cf. [17]) restricting nondeterministic guesses to the set $\{0, 1\}$, which when imposing polynomial time bounds define the class $\text{DNP}_\mathbb{R}$. One may also define a polynomial hierarchy based on digital nondeterminism giving rise to classes $\text{D}\Sigma_k^\mathbb{R}$ and $\text{D}\Pi_k^\mathbb{R}$, for $k \geq 1$.

Another convenient way to define the classes described above is by means of complexity class *operators* (cf. [9,38]). Here we shall consider existential or universal quantifiers over either real-valued or Boolean variables whose number is bounded by a polynomial. For a real complexity class $\mathcal{C}$, define $\exists^\mathbb{R} \cdot \mathcal{C}$ as the class of real languages L for which there exists $L' \in \mathcal{C}$ and a polynomial p such that $x \in L$ if and only if $\exists y \in \mathbb{R}^{\leq p(|x|)} : \langle x, y \rangle \in L'$. For a real (or discrete) complexity class $\mathcal{C}$, define $\exists^D \cdot \mathcal{C}$ as the class of real (or discrete) languages L for which there exists $L' \in \mathcal{C}$ and a polynomial p such that $x \in L$ if and only if $\exists y \in \{0, 1\}^{\leq p(|x|)} : \langle x, y \rangle \in L'$. Replacing existential quantifiers with universal quantifiers we analogously obtain definitions of classes $\forall^\mathbb{R} \cdot \mathcal{C}$ and $\forall^D \cdot \mathcal{C}$. We now have that $\Sigma_{k+1}^\mathbb{R} = \exists^\mathbb{R} \cdot \Pi_k^\mathbb{R}$, $\text{D}\Sigma_{k+1}^\mathbb{R} = \exists^D \cdot \text{D}\Pi_k^\mathbb{R}$, as well as $\Sigma_{k+1}^P = \exists^D \cdot \Pi_k^P$, for $k \geq 1$. We shall also consider mixing real and discrete operators. In such cases one may not always have an equivalent definition in terms of oracle machines. For instance, while $\exists^\mathbb{R} \cdot \text{coDNP} = \text{NP}_\mathbb{R}^{\text{DNP}_\mathbb{R}}$ we can only prove the inclusion $\exists^D \cdot \text{coNP}_\mathbb{R} \subseteq \text{DNP}_\mathbb{R}^{\text{NP}_\mathbb{R}}$ and in particular we do not know if $\text{NP}_\mathbb{R} \subseteq \exists^D \cdot \text{coNP}_\mathbb{R}$.

To study discrete problems we define the *Boolean part* of a real language $L \subseteq \mathbb{R}^\infty$ as $\mathrm{BP}(L) = L \cap \{0,1\}^*$ and of real complexity classes $\mathcal{C}$ as $\mathrm{BP}(\mathcal{C}) = \{\mathrm{BP}(L) \mid L \in \mathcal{C}\}$. The Boolean part of a real complexity class is thus a discrete complexity class and may be compared with other discrete complexity classes defined for instance using Turing machines. Furthermore, since we are interested in *uniform* discrete complexity we shall disallow machine constants. Indeed, a single real number may encode an infinite sequence of discrete advice strings, which for instance implies that $\mathrm{P/poly} \subseteq \mathrm{BP}(\mathrm{P}_\mathbb{R})$. For a class $\mathcal{C}$ defined above we denote by $\mathcal{C}^0$ the analogously defined class without machine constants. Several classes given by Boolean parts of constant free real complexity are defined specifically in the literature. Most prominently is the class $\mathrm{BP}(\mathrm{NP}_\mathbb{R}^0)$ which also captures the complexity of the existential theory of the reals. It has been named $\exists\mathbb{R}$ by Schaefer and Štefankovič [36] as well as NPR by Bürgisser and Cucker [10]; we shall use the former notation $\exists\mathbb{R}$. We further let $\forall\mathbb{R} = \mathrm{BP}(\mathrm{coNP}_\mathbb{R}^0)$ as well as $\exists\forall\mathbb{R} = \mathrm{BP}(\Sigma_2^{\mathbb{R},0}) = \exists^\mathbb{R} \cdot \forall\mathbb{R}$ and $\forall\exists\mathbb{R} = \mathrm{BP}(\Pi_2^{\mathbb{R},0}) = \forall^\mathbb{R} \cdot \exists\mathbb{R}$. We shall in particular be interested in the class $\exists^\mathrm{D} \cdot \forall\mathbb{R}$. Clearly, from the definitions above we have that this class contains both the familiar classes $\forall\mathbb{R}$ and Σ_2^p and is itself contained in $\exists\forall\mathbb{R}$. In fact $\exists^\mathrm{D} \cdot \forall\mathbb{R}$ contains the class $(\Sigma_2^\mathrm{p})^{\mathrm{PosSLP}}$, where PosSLP is the problem of deciding whether an integer given by a division free arithmetic circuit is positive, as introduced by Allender et al. [3]. This follows since $\mathrm{P}^{\mathrm{PosSLP}} = \mathrm{BP}(\mathrm{P}_\mathbb{R}^0)$ [3, Proposition 1.1], and thus

$$(\Sigma_2^\mathrm{p})^{\mathrm{PosSLP}} = \exists^\mathrm{D} \cdot \forall^\mathrm{D} \cdot \mathrm{P}^{\mathrm{PosSLP}} = \exists^\mathrm{D} \cdot \forall^\mathrm{D} \cdot \mathrm{BP}(\mathrm{P}_\mathbb{R}^0)$$
$$\subseteq \exists^\mathrm{D} \cdot \mathrm{BP}(\forall^\mathbb{R} \cdot \mathrm{P}_\mathbb{R}^0) = \exists^\mathrm{D} \cdot \mathrm{BP}(\mathrm{coNP}_\mathbb{R}^0) = \exists^\mathrm{D} \cdot \forall\mathbb{R} \ .$$

2.4   The First-Order Theory of the Reals

The discrete complexity classes $\mathrm{BP}(\Sigma_k^{\mathbb{R},0})$ and $\mathrm{BP}(\Pi_k^{\mathbb{R},0})$ may alternatively be characterized using the decision problem for the first-order theory of the reals. We denote by $\mathrm{Th}(\mathbb{R})$ the set of all true first-order sentences over the reals. We shall consider the restriction to sentences in *prenex normal form*

$$(Q_1 x_1 \in \mathbb{R}^{n_1}) \cdots (Q_k x_k \in \mathbb{R}^{n_k}) \varphi(x_1, \ldots, x_k) \ , \tag{2}$$

where φ is a quantifier free Boolean formula of equalities and inequalities of polynomials with integer coefficients, where each Q_i is one of the quantifiers $\exists$ or $\forall$, typically alternating, and gives rise to k blocks of quantified variables. The restriction of $\mathrm{Th}(\mathbb{R})$ to formulas in prenex normal form with k being a fixed constant and also $Q_1 = \exists$ is complete for $\mathrm{BP}(\Sigma_k^{\mathbb{R},0})$; when instead $Q_1 = \forall$ it is complete for $\mathrm{BP}(\Pi_k^{\mathbb{R},0})$. In particular, the *existential theory of the reals* $\mathrm{Th}_\exists(\mathbb{R})$, where $k = 1$ and $Q_1 = \exists$, is complete for $\exists\mathbb{R}$. Similarly $\mathrm{Th}_{\forall\exists}(\mathbb{R})$ where $k = 2$ and $Q_1 = \forall$ is complete for $\forall\exists\mathbb{R}$; when we furthermore restrict the first quantifier block to Boolean variables the problem becomes complete for $\exists^\mathrm{D} \cdot \forall\mathbb{R}$.

2.5   Real Polynomials with Discrete Quantification

In this section we shall prove that the following problem, $\forall^D \text{HOM4FEAS}(\Delta)$, is complete for the complexity class $\forall^D \cdot \exists \mathbb{R}$. In Sect. 3 we use the complement of this problem to prove our main result of $\exists^D \cdot \forall \mathbb{R}$-hardness of $\exists \text{ESS}$.

Denote by $\Delta^n \subseteq \mathbb{R}^{n+1}$ the n-simplex $\{x \in \mathbb{R}^{n+1} \mid x \geq 0 \wedge \sum_{i=1}^{n+1} x_i = 1\}$ and similarly by $\Delta_c^n \subseteq \mathbb{R}^n$ the corner n-simplex $\{x \in \mathbb{R}^n \mid x \geq 0 \wedge \sum_{i=1}^{n} x_i \leq 1\}$.

Definition 2 ($\forall^D \text{HOM4FEAS}(\Delta)$). *For the problem $\forall^D \text{HOM4FEAS}(\Delta)$ we are given as input rational coefficients $a_{i,\alpha}$, where $i \in \{0,\dots,n\}$ and $\alpha \in [m]^4$ forming the polynomial*

$$F(y,z) = F_0(z) + \sum_{i=1}^{n} y_i F_i(z) \ ,$$

where

$$F_i(z) = \sum_{\alpha \in [m]^4} a_{i,\alpha} \prod_{j=1}^{4} z_{\alpha_j} \ , \text{ for } i = 0,\dots,n \ .$$

We are to decide whether for all $y \in \{0,1\}^n$ there exists $z \in \Delta^{m-1}$ such that $F(y,z) = 0$.

The proof of $\forall^D \cdot \exists \mathbb{R}$-hardness of $\forall^D \text{HOM4FEAS}(\Delta)$ given below is mainly a combination of existing ideas and proofs, and the reader may thus defer reading it.

Theorem 1. *The problem $\forall^D \text{HOM4FEAS}(\Delta)$ is complete for $\forall^D \cdot \exists \mathbb{R}$, and remains $\forall^D \cdot \exists \mathbb{R}$-hard even with the promise that for all $y \in \{0,1\}^n$ and $z \in \mathbb{R}^m$ it holds that $F(y,z) \geq 0$.*

Proof. We shall prove hardness of $\forall^D \text{HOM4FEAS}(\Delta)$ by describing a general reduction from a language L in $\forall^D \cdot \exists \mathbb{R}$ in several steps making use of reductions that proves several problems involving real polynomials $\exists \mathbb{R}$-hard. Consider first the standard complete problem QUAD for $\exists \mathbb{R}$ which is that of deciding if a system of multivariate quadratic polynomials have a common root [8,36]. The general reduction from a language L in $\exists \mathbb{R}$ to QUAD works by treating the input x as variables and computes, based *only* on $|x|$ and not the actual value of x, a system of quadratic polynomials $q_i(x,y)$, $i = 1,\dots,\ell$, where $y \in \mathbb{R}^{p(|x|)}$ for some polynomial p. The system has the property that for all x it holds that $x \in L$ if and only if there exists y such that $q_i(x,y) = 0$, for all i.

Suppose now that $L \in \forall^D \cdot \exists \mathbb{R}$. Then there is L' in $\exists \mathbb{R}$ and a polynomial p such that $x \in L$ if and only if $\forall y \in \{0,1\}^{p(|x|)} : \langle x,y \rangle \in L'$. On input x we may apply the reduction from L' to QUAD and in this way obtain a system of quadratic equations $q_i(x,y,z)$, $i = 1,\dots,\ell_1$ where $z \in \mathbb{R}^{p_1(|x|)}$ such that $\langle x,y \rangle \in L'$ if and only if there exists $z \in \mathbb{R}^{p_1(|x|)}$ such that $q_i(x,y,z) = 0$ for all i. At this point we may just treat x as fixed constants, and we view the system as polynomials in variables (y,z), suppressing the dependence on

x in the notation. Define $n = p(|x|)$. We next introduce additional existentially quantified variables $w \in \mathbb{R}^n$, substitute w_i for y_i in all polynomials, and then add new polynomials $w_i - y_i$, for $i \in [n]$. Renaming polynomials and bundling the existentially quantified variables we now have a system of polynomials $q_i(y, z)$, $i \in [\ell_2]$ where $z \in \mathbb{R}^{m_2}$, where $m_2 \leq p_2(|x|)$ for some polynomial p_2, such that $x \in L$ if and only if

$$\forall y \in \{0,1\}^n \exists z \in \mathbb{R}^{m_2} \forall i \in [\ell_2] : q_i(y, z) = 0 \ ,$$

and where each polynomial q_i depends on at most 1 coordinate of y.

For the next step we use that QUAD remains $\exists\mathbb{R}$-hard when asking for a solution in the unit ball [34], or analogously in the corner simplex [24]. Applying the reduction of [24, Proposition 2] we first rewrite each variable z_i as a difference $z_i = z_i^+ - z_i^-$ of two non-negative real variables z_i^+ and z_i^- and then introduce additional existentially quantified variables $w_0, \ldots, w_t$ for suitable $t = O(\log \tau + m_2)$, where τ is the maximum bitlength of the coefficients of the given system. Then polynomials are added that together implement t steps of repeated squaring of $\frac{1}{2}$, i.e. we add polynomials $w_t - \frac{1}{2}$, and $w_{j-1} - w_j$, for $j \in [t]$, which means that any solution must then have $w_0 = 2^{-2^t}$. In the given polynomial system we now substitute z_i by $(z_i^+ - z_i^-)/w_0$ in each of the polynomials and then multiply them by w_0^2 to clear w_0 from the denominators. For suitable t this means that if for fixed y, the given system of polynomials has a solution $z \in \mathbb{R}^{m_2}$, then the transformed system has a solution (z^+, z^-, w) in $\Delta_c^{2m_2+t+1}$. Note also, that since the variables x_i are not divided by w_0, multiplying by w_0^2 causes an increase in the degree of the polynomials, but the degree in the other variables remains at most 2. Again, renaming polynomials and bundling the existentially quantified variables we now have a system of polynomials $q_i(y, z)$, $i \in [\ell_3]$ where $z \in \mathbb{R}^{m_3}$, where $m_3 \leq p_3(|x|)$ for some polynomial p_3, such that $x \in L$ if and only if

$$\forall y \in \{0,1\}^n \exists z \in \Delta_c^{m_3} \forall i \in [\ell_3] : q_i(y, z) = 0 \ ,$$

and where each polynomial q_i depends on at most 1 coordinate of y.

The next step simply consists of homogenizing the polynomials in the existentially quantified variables z. For this we simply introduce a *slack variable* $z_{m_3+1} = 1 - \sum_{i=1}^{m_3} z_i$ and homogenize by multiplying terms by $\sum_{i=1}^{m_3+1} z_i$ or $\sum_{i=1}^{m_3+1} \sum_{j=1}^{m_3+1} z_i z_j$ as needed. Letting q_i' be the homogenization of q_i we now have that $x \in L$ if and only if

$$\forall y \in \{0,1\}^n \exists z \in \Delta^{m_3} \forall i \in [\ell_3] : q_i'(y, z) = 0 \ ,$$

and where each polynomial q_i' depends on at most 1 coordinate of y and are homogeneous of degree 2 in the variables z.

For the final step we reuse the idea of the reduction from QUAD to 4FEAS, which merely takes the sum of the squares of every given polynomial. Thus we let

$$F(y, z) = \sum_{i=1}^{\ell_3} (q'(y, z))^2 \ .$$

We note that $(q'(y,z))^2 \geq 0$ for all y and z and is homogeneous of degree 4 in the variables z. Further, since $y_j^2 = y_j$ for any $y_j \in \{0,1\}$ we may replace all occurrences of y_j^2 by y_j thereby obtaining an equivalent polynomial (when $y \in \{0,1\}^n$) of the form of Definition 2. We have that for every fixed $y \in \{0,1\}^n$ and all $z \in \mathbb{R}^m$ that $F(y,z) = 0$ if and only if $q_i(y,z) = 0$ for all i. Thus $x \in L$ if and only if

$$\forall y \in \{0,1\}^n \exists z \in \Delta^{m_3} F(y,z) = 0 \ ,$$

which completes the proof of hardness. Let us also note that the definition of F guarantees that $F(y,z) \geq 0$ for all $y \in \{0,1\}^n$ and $z \in \mathbb{R}^m$ it holds that $F(y,z) \geq 0$. Since on the other hand clearly $\forall^D \text{HOM4FEAS}(\Delta) \in \forall^D \cdot \exists\mathbb{R}$ the result follows.   □

As a special case, (when there are no universally quantified variables) the proof gives a reduction from the $\exists\mathbb{R}$-complete problem QUAD to the problem HOM4FEAS(Δ), where we are given as input a homogeneous degree 4 polynomial $F(z)$ in m variables with rational coefficients and are to decide whether there exists $z \in \Delta^{m-1}$ such that $F(z) = 0$. Also, we clearly have that HOM4FEAS(Δ) is a member of $\exists\mathbb{R}$ and therefore have the following result.

Theorem 2. *The problem* HOM4FEAS(Δ) *is complete for* $\exists\mathbb{R}$, *and remains* $\exists\mathbb{R}$-*hard even when assuming that for all* $z \in \mathbb{R}^m$ *it holds that* $F(z) \geq 0$.

3   Complexity of ESS

In this section we shall prove our results for deciding existence of an ESS. In the proof we will re-use a trick used by Conitzer [14] for the case of 2-player games, where by duplicating a subset of the actions of a game we ensure that no ESS can be supported by any of the duplicated actions as shown in the following lemma. Here, by duplicating an action we mean that the utilities assigned to any pure strategy profile involving the duplicated action is defined to be equal to the utility for the pure strategy profile obtained by replacing occurrences of the duplicated action by the original action. The precise property is as follows.

Lemma 2. *Let* $\mathcal{G}$ *be an* m-*player symmetric game given by* S *and* u. *Suppose that* $s, s' \in S$ *are such that for all strategies* τ *we have* $u(s; \tau^{m-1}) = u(s'; \tau^{m-1})$. *Then* s *can not be in the support of an ESS* σ.

Proof. Suppose σ is a strategy with $s \in \text{Supp}(\sigma)$. Let σ' be obtained from σ by moving the probability mass of s to s'. From our assumption we then have $u(\sigma; \tau^{m-1}) = u(\sigma'; \tau^{m-1})$ for all τ. In particular we have $u(\sigma; \sigma_\varepsilon^{m-1}) = u(\sigma'; \sigma_\varepsilon^{m-1})$, for all $\varepsilon > 0$, where we have $\sigma_\varepsilon = \varepsilon\sigma' + (1-\varepsilon)\sigma$. This means that σ' invades σ and σ is therefore not an ESS.   □

We now state and prove the main result of this paper.

Theorem 3. $\exists$ESS *is* $\exists^D \cdot \forall\mathbb{R}$-*hard for 5-player games.*

Proof. We prove our result by giving a reduction from the *complement* of the problem $\forall^D\mathrm{Hom4Feas}(\Delta)$ to $\exists\mathrm{ESS}$. It follows from Theorem 1 that the former problem is complete for $\exists^D \cdot \forall\mathbb{R}$. Thus let $a_{i,\alpha}$ be given rational coefficients, with $i = 0, \ldots, n$ and $\alpha \in [m]^4$, forming the polynomials $F(y, z)$ and $F_i(z)$, for $i = 0, \ldots, n$ as in Definition 2. We may assume that for all $y \in \{0,1\}^n$ and all $z \in \mathbb{R}^m$ it holds that $F(y, z) \geq 0$. Also, without loss of generality we may assume that each F_i is *symmetrized*, i.e. that for all i and α, if π is a permutation on $[4]$, then defining $\pi \cdot \alpha \in [m]^4$ by $(\pi \cdot \alpha)_i = \alpha_{\pi(i)}$, we have that $a_{i,\alpha} = a_{i,\pi \cdot \alpha}$. Namely, we may simply replace each coefficient $a_{i,\alpha}$ by the average of all coefficients of the form $a_{i,\pi \cdot \alpha}$. This leaves the *functions* given by the expressions for F_i unchanged, but crucially ensures that the game defined below is symmetric.

We next define a 5-player game $\mathcal{G}$ based on F. The strategy set is naturally divided in three parts $S = S_1 \cup S_2 \cup S_3$. These are defined as follows.

$$
\begin{aligned}
S_1 &= \{(i, \alpha, b) \mid i \in \{0, \ldots, n\},\ \alpha \in [m]^4,\ b \in \{0, 1\}\} \\
S_2 &= \{\gamma\} \\
S_3 &= \{1, \ldots, m\}
\end{aligned}
\tag{3}
$$

An action (i, α, b) of S_1 thus identifies a term of F_i together with $b \in \{0, 1\}$, which is supposed to be equal to y_i. When convenient we may describe the actions of S_1 by pairs (t, b), where $t = (i, \alpha)$ for some i and α. The single action γ is used for rewarding inconsistencies in the choices of b among strategies of S_1. Finally, a probability distribution on S_3 will define an input z. Let $M = (n + 1)m^4$ be the total number of terms of F. Thus $|S_1| = 2M$.

We shall *duplicate* all actions of $S_2 \cup S_3$ and let duplicates behave exactly the same regarding the utility function defined below. By Lemma 2 it then follows that any ESS σ of $\mathcal{G}$ must have $\mathrm{Supp}(\sigma) \subseteq S_1$. For simplicity we describe the utilities of $\mathcal{G}$ without the duplicated actions.

When all players are playing an action of S_1 we define

$$
u((t_1, b_1), \ldots, (t_5, b_5)) = \begin{cases} 2 & \text{if } t_1 \notin \{t_2, \ldots, t_5\} \\ 1 & \text{if } t_1 \in \{t_2, \ldots, t_5\} \text{ and } t_1 = t_j \Rightarrow b_1 = b_j \\ 0 & \text{otherwise} \end{cases} . \tag{4}
$$

Before defining the remaining utilities, we consider the payoff of strategies that play uniformly on the set of terms and according to a fixed assignment y. Define the number T by

$$
T = 2 - \frac{4}{M} + \frac{6}{M^2} - \frac{4}{M^3} + \frac{1}{M^4} .
\tag{5}
$$

Lemma 3. *Let $y \in \{0,1\}^n$, let $y_0 \in \{0,1\}$ be arbitrary, and define σ_y to be the strategy that plays (i, α, y_i) with probability $\frac{1}{M}$ for all α, and the remaining strategies with probability 0. Then $u(\sigma_y^5) = T$.*

Proof. Note that $2 - u(\sigma_y^5)$ is precisely the probability of the union of the events $t_1 = t_j$, where $j = 2, \ldots, 5$, and t_j is the term chosen by player j. For fixed

t_1, these events are independent and each occurs with probability $\frac{1}{M}$. By the principle of inclusion-exclusion we thus have

$$2 - u(\sigma_y^5) = \frac{4}{M} - \frac{6}{M^2} + \frac{4}{M^3} - \frac{1}{M^4} .$$

We will construct the game $\mathcal{G}$ in such a way that any ESS σ will have $u(\sigma^5) = T$. Making use of Lemma 3, we now define utilities when at least one player is playing the action γ. In case at least two players are playing γ, these players receive utility 0 while the remaining players receive utility T. In case exactly one player is playing γ, the player receives utility $T + 1$ in case there are two players that play actions (i, α, b) and (i, α', b') with $b \neq b'$; otherwise the player receives utility T. In either case, when exactly one player is playing γ, the remaining players receive utility T.

We finally define utilities when one player is playing an action from S_1 and the remaining four players are playing an action from S_3. Suppose for simplicity of notation that player j is playing action $\beta_j \in S_3$, for $j = 1, \ldots, 4$, while player 5 is playing action (i, α, b). We let player 5 receive utility T. In case $\alpha = \beta$ the first four players receive utility $T - a_{i,\alpha}$; otherwise they receive utility T. Here we use that $a_{i,\alpha} = a_{i,\pi \cdot \alpha}$ for any permutation π on $[4]$, to ensure that $\mathcal{G}$ is symmetric.

At this point we have only partially specified the utilities of the game $\mathcal{G}$; we simply let all remaining unspecified utilities equal T, thereby completing the definition of $\mathcal{G}$.

We are now ready to prove that $\mathcal{G}$ has an ESS if and only if there exists $y \in \{0, 1\}^n$ such that $F(y, z) > 0$ for all $z \in \Delta^{m-1}$. Suppose first that $y \in \{0, 1\}^n$ exists such that $F(y, z) > 0$ for all $z \in \Delta^{m-1}$. We define $\sigma = \sigma_y$ as in Lemma 3 and show that any $\tau \neq \sigma$ satisfies the conditions of Lemma 1 thereby proving that σ is an ESS of $\mathcal{G}$. Suppose that $\tau \neq \sigma$ invades σ. Consider first playing τ against σ^4. From the proof of Lemma 3 it follows that playing a strategy of form (i, α, b) against σ^4 gives payoff T if $b = y_i$ and otherwise payoff strictly below T. The strategies of $S_2 \cup S_3$ all give payoff T against σ^4. It follows that to invade σ, τ can only play strategies from S_1 contained in $\text{Supp}(\sigma)$. Let us write $\tau = \delta_1 \tau_1 + \delta_2 \tau_2 + \delta_3 \tau_3$ as a convex combination of strategies τ_j with $\text{Supp}(\tau_j) \subseteq S_j$, for $j = 1, 2, 3$. We shall consider playing τ against (τ, σ^3) and argue that $\tau_1 = \sigma$ if $\delta_1 > 0$ and that $\delta_2 = 0$. Note first that if a strategy of S_3 is played, all players receive utility T, so we may focus on the case when all players play using strategies from $S_1 \cup S_2$. Suppose that $\delta_1 > 0$ and let $p_t = \text{Pr}_{\tau_1}[t]$, where t is a term of F. Using the principle of inclusion-exclusion we have

$$2 - u(\tau_1; \tau_1, \sigma^3) = \sum_t p_t \left[\frac{3}{M} - \frac{3}{M^2} + \frac{1}{M^3} + p_t \left(1 - \frac{3}{M} + \frac{3}{M^2} - \frac{1}{M^3} \right) \right]$$

$$= \frac{3}{M} - \frac{3}{M^2} + \frac{1}{M^3} + \left(1 - \frac{3}{M} + \frac{3}{M^2} - \frac{1}{M^3} \right) \sum_t p_t^2 .$$

By Jensen's inequality, $\sum_t p_t^2 \geq M \left(\sum_t p_t / M \right)^2 = \frac{1}{M}$, with equality if and only if $p_t = \frac{1}{M}$ for all t. This means that $u(\tau_1; \tau_1, \sigma^3) \leq T$, with equality if and only if $p_t = \frac{1}{M}$ for all t. Thus if $\tau_1 \neq \sigma$, then $u(\tau_1; \tau_1, \sigma^3) < u(\sigma; \tau_1, \sigma^3) = T$, where the last equality may be derived again using the principle of inclusion-exclusion. Now, since $\text{Supp}(\tau_1) \subseteq \text{Supp}(\sigma)$ when $\delta_1 > 0$, playing γ can give utility at most T but gives utility 0 in case another player plays γ as well.

Combining these observations it follows that unless $\delta_2 = 0$ and $\tau_1 = \sigma$ when $\delta_1 > 0$ we have $u(\sigma; \tau, \sigma^3) > u(\tau; \tau, \sigma^3)$. Thus we may now assume that this is the case, i.e., that $\tau = \delta_1 \sigma + \delta_3 \tau_3$. From the definition of $\mathcal{G}$ we now have that $u(\tau; \tau^j, \sigma^{4-j}) \leq u(\sigma; \tau^j, \sigma^{4-j}) = T$, for $j = 1, 2, 3$. For τ to invade σ it is thus required that $u(\tau; \tau^4) \geq u(\sigma; \tau^4)$, and it follows from the definition of $\mathcal{G}$ that this is equivalent to $u(\tau_3; \tau_3^3, \sigma) \geq T$. Now $\tau_3 \in \Delta(S_3) = \Delta^{m-1}$ and by assumption we have $F(y, \tau_3) > 0$. Furthermore we have $u(\tau_3; \tau_3^3, \sigma) = T - F(y, \tau_3)/M$ and thus $u(\tau_3; \tau_3^3, \sigma) < T$, which means σ is actually ES against τ.

Suppose now on the other hand that σ is an ESS of $\mathcal{G}$. First, since we duplicated the actions of $S_2 \cup S_3$, it follows from Lemma 2 that $\text{Supp}(\sigma) \subseteq S_1$. We next show that for all terms t, if $\sigma(t, b) > 0$, then unless $\sigma(t, 1 - b) = 0$, σ can be invaded. Suppose that t is a term of F, let $p_0 = \sigma(t, 0)$ and $p_1 = \sigma(t, 1)$, and suppose that $p_0 > 0$ and $p_1 > 0$. Suppose without loss of generality that $p_0 \geq p_1$. Note now that

$$u((t, 0); \sigma^4) - u((t, 1); \sigma^4) = (1 - p_1)^4 - (1 - p_0)^4 \geq 0 \ ,$$

which can be seen by noting that the left hand side of the equality does not change when replacing all utilities of 2 by 1. Similarly

$$u((t, 0); (t, 0), \sigma^3) - u((t, 1); (t, 1), \sigma^3) = (1 - p_1)^3 - (1 - p_0)^3 \geq 0 \ .$$

Define the strategy σ' from σ by playing the strategy $(t, 0)$ with probability $p = p_0 + p_1$, the strategy $(t, 1)$ with probability 0, and otherwise according to σ. Then

$$u(\sigma'; \sigma^4) - u(\sigma^5) = (p_0 + p_1)u((t, 0); \sigma^4) - p_0 u((t, 0); \sigma^4) - p_1 u((t, 1); \sigma^4)$$
$$= p_1(u((t, 0); \sigma^4) - u((t, 1); \sigma^4)) \geq 0 \ ,$$

and by definition, $u((t, 0); (t, 1), \sigma_3) = u((t, 1); (t, 0), \sigma_3) = 0$. Thus we have

$$u(\sigma'; \sigma', \sigma^3) - u(\sigma; \sigma', \sigma^3)$$
$$= (p_0 + p_1)^2 u((t, 0); (t, 0), \sigma^3) - p_0(p_0 + p_1)u((t, 0); (t, 0), \sigma^3)$$
$$= (p_0 p_1 + p_1^2)u((t, 0); (t, 1), \sigma^3) > 0.$$

which means that σ' invades σ. Since σ is an ESS, this means that for each term t there is $b_t \in \{0, 1\}$ such that σ plays $(t, 1 - b_t)$ with probability 0. Let $p_t = \sigma(t, b_t)$ and define the function $h : \mathbb{R} \to \mathbb{R}$ by $h(p) = 4p - 6p^2 + 4p^3 - p^4$, and note that $\frac{d}{dp}h(p) = 4(1 - p)^3$. By the principle of inclusion-exclusion we have

$$2 - u(\sigma^5) = \sum_t p_t h(p_t) \ .$$

Suppose now there exists terms t and t' such that $p_t < p_{t'}$. Since h is strictly increasing on $[0, 1]$ we also have $h(p_t) < h(p_{t'})$, and therefore $p_t h(p_t) + p_{t'} h(p_{t'}) >$

$p_{t'}h(p_t) + p_t h(p_{t'})$. Define σ' to play t with probability $p_{t'}$, t' with probability p_t, and otherwise according to σ. We then have

$$(2 - u(\sigma^5)) - (2 - u(\sigma'; \sigma^4)) = p_t(h(p_t) - h(p_{t'})) + p_{t'}(h(p_{t'}) - h(p_t)) > 0 ,$$

and therefore $u(\sigma'; \sigma^4) > u(\sigma^5)$, which means that σ' invades σ. Since σ is an ESS this means that $p_t = \frac{1}{M}$ for all t. From the proof of Lemma 3 it then follows that $u(\sigma^5) = T$.

Suppose now that there exists $i \in [n]$ and α, α' such that $b_{(i,\alpha)} \neq b_{(i,\alpha')}$. But then $u(\gamma; \sigma^4) > T = u(\sigma^5)$, which means that γ invades σ. Since σ is an ESS there must exist $y \in \{0, 1\}^n$ (and some $y_0 \in \{0, 1\}$) such that $\sigma = \sigma_y$, using the notation of Lemma 3.

Finally, let $z \in \Delta^{m-1} = \Delta(S_3)$. By definition of u we have $u(z; z^j, \sigma^{4-j}) = T = u(\sigma; z^j, \sigma^{4-j})$, for all $j \in \{0, 1, 2\}$. Next $u(z; z^3, \sigma) = T - F(y, z)/M$ while we have $u(\sigma; z^3, \sigma) = T$. For σ to be ES against z we must thus have $F(y, z) > 0$, and this concludes the proof.   □

The best upper bound on the complexity of ∃ESS we know is membership of ∀∃ℝ which easily follows from definitions. For the simpler problem IsESS of determining whether a given strategy is an ESS we can fully characterize its complexity.

Theorem 4. IsESS *is ∀ℝ-complete for 5-player games.*

Proof. Clearly IsESS belongs to ∀ℝ. To show ∀ℝ-hardness we reduce from the *complement* of the problem Hom4Feas(Δ) to IsESS. It follows from Theorem 2 that the former problem is complete for ∀ℝ. From F we construct the game $\mathcal{G}$ as in the proof of Theorem 3 letting $n = 0$. We let σ be the uniform distribution on the set of actions $(0, \alpha, 0)$, where $\alpha \in [m]^4$. It then follows from the proof of Theorem 3 that σ is an ESS of $\mathcal{G}$ if and only if $F(z) > 0$ for all $z \in \Delta^{m-1}$. Since we may assume that $F(z) \geq 0$ for all $z \in \mathbb{R}^m$ this completes the proof.   □

4   Conclusion

We have shown the problem ∃ESS to be hard for $\exists^D \cdot \forall\mathbb{R}$ and member of $\exists\forall\mathbb{R}$. The main open problem is to characterize the precise complexity of ∃ESS, perhaps by improving the upper bound. Another point is that our hardness proofs construct 5-player games, whereas the recent and related ∃ℝ-completeness results for decision problems about NE in multi-player games holds already for 3-player games. This leads to the question about the complexity of ∃ESS and IsESS in 3-player and 4-player games. The reason that we end up with 5-player games is that we construct a degree 4 polynomial in the reduction, rather than (a system of) degree 2 polynomials as used in the related ∃ℝ-completeness results. In both cases a number of players equal to the degree is used to simulate evaluation of a monomial and a last player is used to select the monomial. For our proof we critically use that the degree 4 polynomial involved in the reduction may be assumed to be non-negative.

References

1. Abrahamsen, M., Adamaszek, A., Miltzow, T.: The art gallery problem is ∃ℝ-complete. In: STOC 2018, pp. 65–73. ACM (2018). https://doi.org/10.1145/3188745.3188868
2. Accinelli, E., Martins, F., Oviedo, J.: Evolutionary game theory: a generalization of the ESS definition. Int. Game Theory Rev. **21**(4), 1950005 (19 pages) (2019). https://doi.org/10.1142/S0219198919500051
3. Allender, E., Bürgisser, P., Kjeldgaard-Pedersen, J., Miltersen, P.B.: On the complexity of numerical analysis. SIAM J. Comput. **38**(5), 1987–2006 (2009). https://doi.org/10.1137/070697926
4. Berthelsen, M.L.T., Hansen, K.A.: On the computational complexity of decision problems about multi-player Nash equilibria. In: Fotakis, D., Markakis, E. (eds.) SAGT 2019. LNCS, vol. 11801, pp. 153–167. Springer, Cham (2019). https://doi.org/10.1007/978-3-030-30473-7_11
5. Bilò, V., Mavronicolas, M.: A catalog of ∃ℝ-complete decision problems about Nash equilibria in multi-player games. In: Ollinger, N., Vollmer, H. (eds.) STACS 2016. LIPIcs, vol. 47, pp. 17:1–17:13. Schloss Dagstuhl - Leibniz-Zentrum für Informatik (2016). https://doi.org/10.4230/LIPIcs.STACS.2016.17
6. Bilò, V., Mavronicolas, M.: ∃ℝ-complete decision problemsabout symmetric Nash equilibria in symmetric multi-player games. In: Vollmer, H., Vallé, B. (eds.) STACS 2017. LIPIcs, vol. 66, pp.13:1–13:14. Schloss Dagstuhl–Leibniz-Zentrum für Informatik (2017). https://doi.org/10.4230/LIPIcs.STACS.2017.13
7. Blum, L., Schub, M., Smale, S.: On a theory of computation and complexity over the real numbers: Np-completeness, recursive functions and universal machines. Bull. Amer. Math. Soc. **21**, 1–46 (1989). https://doi.org/10.1090/S0273-0979-1989-15750-9
8. Complexity and Real Computation. Springer, New York (1998). https://doi.org/10.1007/978-1-4612-0701-6
9. Borchert, B., Silvestri, R.: Dot operators. Theor. Comput. Sci. **262**(1), 501–523 (2001). https://doi.org/10.1016/S0304-3975(00)00323-6
10. Bürgisser, P., Cucker, F.: Exotic quantifiers, complexity classes, and complete problems. Foundations Comput. Math. **9**, 135–170 (2009). https://doi.org/10.1007/s10208-007-9006-9
11. Broom, M., Cannings, C., Vickers, G.T.: Multi-player matrix games. Bltn Mathcal Biology **59**, 931–952 (1997). https://doi.org/10.1007/BF02460000
12. Broom, M., Rychtář, J.: Game-Theoretical Models in Biology. Chapman and Hall (2013)
13. Canny, J.: Some algebraic and geometric computations in PSPACE. In: Proceedings of the Annual ACM Symposium on Theory of Computing, pp. 460–467 (1988). https://doi.org/10.1145/62212.62257
14. Conitzer, V.: The exact computational complexity of evolutionarily stable strategies. Math. Oper. Res. **44**(3), 783–792 (2019). https://doi.org/10.1287/moor.2018.0945
15. Conitzer, V., Sandholm, T.: New complexity results about Nash equilibria. Games Econ. Behav. **63**(2), 621–641 (2008). https://doi.org/10.1016/j.geb.2008.02.015
16. Cucker, F.: On the complexity of quantifier elimination: the structural approach. Comput. J. **36**(5), 400–408 (1993). https://doi.org/10.1093/comjnl/36.5.400
17. Cucker, F., Matamala, M.: On digital nondeterminism. Math. Syst. Theory **29**(6), 635–647 (1996). https://doi.org/10.1007/BF01301968

18. Dobbins, M., Kleist, L., Miltzow, T., Rzążewski, P.: ∀∃ℝ-completeness and area-universality. In: WG 2018. pp. 164–175 (2018). https://doi.org/10.1007/978-3-030-00256-5_14
19. Etessami, K., Lochbihler, A.: The computational complexity of evolutionary stable strategies. Int. J. Game Theor. **37**, 93–113 (2007). https://doi.org/10.1007/s00182-007-0095-0
20. Garg, J., Mehta, R., Vazirani, V.V., Yazdanbod, S.: ∃ℝ-completeness for decision versions of multi-player (symmetric) Nash equilibria. ACM Trans. Econ. Comput. **6**(1), 1:1–1:23 (2018). https://doi.org/10.1145/3175494
21. Gilboa, I., Zemel, E.: Nash and correlated equilibria: some complexity considerations. Games Economic Behav. **1**(1), 80–93 (1989). https://doi.org/10.1016/0899-8256(89)90006-7
22. Gimbert, H., Paul, S., Srivathsan, B.: A bridge between polynomial optimization and games with imperfect recall. In: Seghrouchni, A.E.F., Sukthankar, G., An, B., Yorke-Smith, N. (eds.) AAMAS 2020, pp. 456–464. International Foundation for Autonomous Agents and Multiagent Systems (2020). https://doi.org/10.5555/3398761.3398818
23. Goode, J.B.: Accessible telephone directories. J. Symbolic Logic **59**(1), 92–105 (1994). https://doi.org/10.2307/2275252
24. Hansen, K.A.: The real computational complexity of minmax value and equilibrium refinements in multi-player games. Theory Comput. Syst. **63**(7), 1554–1571 (2018). https://doi.org/10.1007/s00224-018-9887-9
25. Hofbauer, J., Schuster, P., Sigmund, K.: A note on evolutionary stable strategies and game dynamics. Journal of Theoretical Biology **81**(3), 609–612 (1979). https://doi.org/10.1016/0022-5193(79)90058-4
26. Ko, K.I., Lin, C.L.: On the complexity of min-max optimization problems and their approximation. In: AYDU, D.Z., Pardalos, P.M. (eds.) Minimax and Applications, pp. 219–239. Springer, Boston (1995). https://doi.org/10.1007/978-1-4613-3557-3_15
27. Maynard Smith, J.: The theory of games and the evolution of animal conflicts. J. Theor. Biol. **47**(1), 209–221 (1974). https://doi.org/10.1016/0022-5193(74)90110-6
28. Maynard Smith, J., Price, G.: The logic of animal conflict. Nature **246**, 15–18 (1973)
29. Milchtaich, I.: Static Stability in Games. Working Papers 2008–04, Bar-Ilan University, Department of Economics (2008). https://ideas.repec.org/p/biu/wpaper/2008-04.html
30. Mnev, N.E.: The universality theorems on the classification problem of configuration varieties and convex polytopes varieties. In: Viro, O.Y., Vershik, A.M. (eds.) Topology and Geometry — Rohlin Seminar. LNM, vol. 1346, pp. 527–543. Springer, Heidelberg (1988). https://doi.org/10.1007/BFb0082792
31. Nash, J.: Non-cooperative games. Ann. Math. **2**(54), 286–295 (1951). https://doi.org/10.2307/1969529
32. Nisan, N.: A note on the computational hardness of evolutionary stable strategies. iN: Electronic Colloquium on Computational Complexity (ECCC), vol. 13, no. 076 (2006). http://eccc.hpi-web.de/eccc-reports/2006/TR06-076/index.html
33. Palm, G.: Evolutionary stable strategies and game dynamics for n-person games. J. Mathe. Biol. **19**, 329–334 (1984). https://doi.org/10.1007/BF00277103
34. Schaefer, M.: Complexity of some geometric and topological problems. In: Eppstein, D., Gansner, E.R. (eds.) GD 2009. LNCS, vol. 5849, pp. 334–344. Springer, Heidelberg (2010). https://doi.org/10.1007/978-3-642-11805-0_32

35. Schaefer, M.: Realizability of graphs and linkages. In: Pach, J. (ed.) Thirty Essays on Geometric Graph Theory, pp. 461–482. Springer, New York (2013)
36. Schaefer, M., Štefankovič, D.: Fixed points, nash equilibria, and the existential theory of the reals. Theory Comput. Syst. **60**(2), 172–193 (2015). https://doi.org/10.1007/s00224-015-9662-0
37. Taylor, P.D., Jonker, L.B.: Evolutionary stable strategies and game dynamics. Math. Biosci. **40**(1), 145–156 (1978). https://doi.org/10.1016/0025-5564(78)90077-9
38. Zachos, S.: Probabilistic quantifiers, adversaries, and complexity classes: an overview. In: Selman, A.L. (ed.) Structure in Complexity Theory. LNCS, vol. 223, pp. 383–400. Springer, Heidelberg (1986). https://doi.org/10.1007/3-540-16486-3_112

Injective Colouring for H-Free Graphs

Jan Bok[1], Nikola Jedličková[1], Barnaby Martin[2], Daniël Paulusma[2], and Siani Smith[2(✉)]

[1] Faculty of Mathematics and Physics, Charles University, Prague, Czech Republic
bok@iuuk.mff.cuni.cz, jedlickova@kam.mff.cuni.cz
[2] Department of Computer Science, Durham University, Durham, UK
{barnaby.d.martin,daniel.paulusma,siani.smith}@durham.ac.uk

Abstract. A function $c : V(G) \rightarrow \{1, 2, \ldots, k\}$ is a k-colouring of a graph G if $c(u) \neq c(v)$ whenever u and v are adjacent. If any two colour classes induce the disjoint union of vertices and edges, then c is called injective. Injective colourings are also known as $L(1,1)$-labellings and distance 2-colourings. The corresponding decision problem is denoted INJECTIVE COLOURING. A graph is H-free if it does not contain H as an induced subgraph. We prove a dichotomy for INJECTIVE COLOURING for graphs with bounded independence number. Then, by combining known with further new results, we determine the complexity of INJECTIVE COLOURING on H-free graphs for every H except for one missing case.

1 Introduction

Graph colouring is a well-studied topic in Computer Science and Discrete Mathematics, both for theoretical and practical reasons. The classical variant is to give each vertex of a graph a colour in such a way that two adjacent vertices are not coloured alike while using as few colours as possible. Formally, a *colouring* of a graph $G = (V, E)$ is a mapping $c : V \rightarrow \{1, 2, \ldots\}$ such that $c(u) \neq c(v)$ for each pair of vertices u, v with $uv \in E$. If $c(V) \subseteq \{1, \ldots, k\}$, then c is also called a k-*colouring*. The integer $c(u)$ is the *colour* of u, and the set of all vertices of the same colour i is a *colour class* of c. The problem COLOURING is to decide if a given graph G has a k-colouring for some given integer k. It is well-known that COLOURING is NP-complete even if $k = 3$.

We prove new results for a well-studied variant of graph colouring. We denote the disjoint union two graphs F and G by $F + G = (V(F) \cup V(G), E(F) \cup E(G))$; the disjoint union of s copies of G by sG and the n-vertex path by P_n. A colouring is *injective* if the union of any two colour classes of c induces an $sP_1 + tP_2$ for some integers $s \geq 0$ and $t \geq 0$. Injective colourings are also known as $L(1,1)$-*labellings* or *distance*-2 *colourings*. An equivalent formulation is to say that a colouring of a graph G is injective if all the neighbours of every vertex of G are coloured differently. The problem INJECTIVE COLOURING is to decide whether a given

The research in this paper was supported by GAUK 1580119, SVV–2020–260578 and the Leverhulme Trust (RPG-2016-258).

R. Santhanam and D. Musatov (Eds.): CSR 2021, LNCS 12730, pp. 18–30, 2021.
https://doi.org/10.1007/978-3-030-79416-3_2

graph has an injective k-colouring for some given integer k. In the literature (see, for example, [6, 8, 10]) it is sometimes allowed that injective colourings give adjacent vertices the same colour. This is *in contrast to our paper:* we emphasize that all colourings in our paper are proper (as can be seen from their definitions).

It is known that INJECTIVE COLOURING is NP-complete for split graphs [1], unit disk graphs [15] and planar graphs [13], respectively. Moreover, INJECTIVE COLOURING is NP-complete for line graphs of bipartite graphs of girth at least g for any fixed integer $g \geq 3$ [14] and cubic graphs [7] even when $k = 4$. On the positive side, INJECTIVE COLOURING is polynomial-time solvable for graphs of bounded treewidth [16]. Injective colourings, viewed as $L(1,1)$-labellings, belong to the distance constrained labelling framework; we refer to the $L(h,k)$-labelling survey of Calamoneri [3] for more algorithmic and structural results.

In the above results of [1, 7, 13–16] the input is restricted to some special graph class. Restricting the input is also the focus in our paper. Our goal is to obtain *complexity dichotomies*, which we compare with similar dichotomies for COLOURING. In particular we consider classes of H-*free* graphs, that is, graphs that do not contain some fixed graph H as an induced subgraph, or equivalently, cannot be modified into H by a sequence of vertex deletions. On a side note, the difference between the *chromatic number* χ and *injective chromatic number* χ_i of a graph can be arbitrarily large as illustrated by the $(n+1)$-vertex star $K_{1,n}$, which has chromatic number 2 but injective chromatic number $n + 1$.

1.1 Dichotomies for Colouring

Král' et al. [12] completely classified the complexity of COLOURING for H-free graphs. We write $H_1 \subseteq_i H_2$ to say that H_1 is an induced subgraph of H_2.

Theorem 1 ([12]). *Let H be a graph. Then* COLOURING *for H-free graphs is polynomial-time solvable if $H \subseteq_i P_1 + P_3$ or $H \subseteq_i P_4$ and* NP-*complete otherwise.*

The complexity of k-COLOURING, the variant of COLOURING where k is *fixed* (in other words, not part of the input) has not been classified yet; in particular there are infinite families of open cases when H is a *linear forest* (disjoint union of paths); for example, the complexities of 3-COLOURING for P_t-free graphs for $t \geq 8$ and k-COLOURING for sP_3-free graphs for $s \geq 2$ and $k \geq 4$ are still open; see also [4, 5, 11].

A set of vertices in a graph G is *independent* if each pair of vertices is non-adjacent. The *independence number* $\alpha(G)$ of G is the size of a largest independent set of G. The class of sP_1-free graphs coincides with the class of graphs with $\alpha \leq s - 1$. Hence, Theorem 1 immediately implies the following dichotomy.

Theorem 2. *Let $s \geq 1$ be an integer. Then* COLOURING *for graphs with $\alpha \leq s$ is polynomial-time solvable if $s \leq 2$ but* NP-*complete if $s \geq 3$.*

The *complement* of a graph G is the graph $\overline{G}$ with vertex set $V(G)$ and an edge between two distinct vertices u and v if and only if $uv \notin E(G)$. A k-colouring of G is a partition of $V(G)$ into at most k independent sets. Hence, a

(k-)colouring of G corresponds to a *(k-)clique-covering* of $\overline{G}$, which is a partition of $V(\overline{G}) = V(G)$ into (at most k) cliques. The *clique covering number* $\overline{\chi}(G)$ of G is the smallest number of cliques in a clique-covering of G. Note that $\overline{\chi}(G) = \chi(\overline{G})$. The following theorem follows from a known result of [9] and a standard trick (see Sect. 2).

Theorem 3. *Let $s \geq 1$ be an integer. Then* COLOURING *for graphs with $\overline{\chi} \leq s$ is polynomial-time solvable if $s \leq 2$ but* NP-*complete if $s \geq 3$.*

Note that Theorems 2 and 3 are incomparable: for $p \geq 2$, the class of graphs with $\alpha \leq p$ is a proper superclass of the class of graphs with $\overline{\chi} \leq p$. For example, as the class of K_3-free graphs properly contains the class of bipartite graphs, the class of $3P_1$-free graphs, that is, graphs with $\alpha \leq 2$, properly contains the class of co-bipartite graphs, that is, graphs with $\overline{\chi} \leq 2$. Hence, the polynomial part of Theorem 2 is stronger than the polynomial part of Theorem 3, and the reverse statement holds for the hardness parts of these two dichotomies.

1.2   Dichotomies for Injective Colouring

In contrast to the situation for k-COLOURING, the variant of INJECTIVE COLOURING where the number k of colours is fixed has been completely settled: in [2] we proved the following dichotomy for INJECTIVE k-COLOURING; note that INJECTIVE 3-COLOURING is polynomial-time solvable, as the only yes-instances are graphs of maximum degree 2 (disjoint unions of cycles and paths).

Theorem 4. *Let H be a graph and $k \geq 4$ be an integer. Then* INJECTIVE k-COLOURING *for H-free graphs is polynomial-time solvable if H is a linear forest and* NP-*complete otherwise.*

Our first result, proven in Sect. 2, is a similar dichotomy as Theorem 2.

Theorem 5. *Let $s \geq 1$ be an integer. Then* INJECTIVE COLOURING *for graphs with $\alpha \leq s$ is polynomial-time solvable if $s \leq 3$ but* NP-*complete if $s \geq 4$.*

Comparing Theorem 2 with Theorem 5, we see that the jump in complexity for COLOURING happens when $s = 3$ instead of $s = 4$. This has the following consequence. It is well known and easy to see that an injective colouring of a graph G is a colouring of the *square G^2* of G. The latter graph is obtained from G by adding an edge between every pair of non-adjacent vertices in G that have a common neighbour in G. If G has $\alpha \leq 3$, then G^2 has $\alpha \leq 3$ as well. However, COLOURING for graphs with $\alpha \leq 3$ is NP-complete by Theorem 2. Hence, we do not have any polynomial-time algorithm for COLOURING that we can use. Instead of this, we will develop a direct approach to obtain polynomial-time solvability of INJECTIVE COLOURING for $4P_1$-free graphs (on a side note, it can be observed that the hard instances with $\alpha = 3$ of COLOURING are not squares of graphs).

The hardness part of Theorem 5 follows from the hardness part of the following analogue of Theorem 3 for INJECTIVE COLOURING, which we prove in Sect. 2.

Theorem 6. *Let $s \geq 1$ be an integer. Then* INJECTIVE COLOURING *for graphs with $\overline{\chi} \leq s$ is polynomial-time solvable if $s \leq 3$ but* NP-*complete if $s \geq 4$.*

In [2] we combined known results from [1,14] with some new results to obtain a partial dichotomy for INJECTIVE COLOURING where we left open ten cases of linear forests H, including the cases $H = 4P_1$ and $H = 5P_1$, which we now have solved (these casess correspond to graphs with $\alpha \leq 3$ and $\alpha \leq 4$, respectively, so follow from Theorem 5). In this paper we solve a total of nine open cases. In Sect. 3 we use our result for $4P_1$-free graphs from Theorem 5 to obtain new polynomial-time algorithms for two superclasses of $4P_1$-free graphs, namely when $H = 2P_1 + P_3$ and $H = 3P_1 + P_2$. In the same section we also prove polynomial-time solvability if $H = P_1 + P_4$. As a consequence of our results we can update the partial dichotomy of [2] as follows, leaving only one missing case.

Theorem 7. *Let $H \neq 2P_1 + P_4$ be a graph. Then* INJECTIVE COLOURING *for H-free graphs is polynomial-time solvable if $H \subseteq_i P_1 + P_4$ or $H \subseteq_i 2P_1 + P_3$ or $H \subseteq_i 3P_1 + P_2$ and* NP-*complete otherwise.*

Compared to Theorem 1 we notice that INJECTIVE COLOURING is polynomial-time solvable for larger classes of H-free graphs than COLOURING. We prove Theorem 7 in Sect. 3 as well and state some relevant open problems in Sect. 4.

2   The Proofs of Theorems 3, 5 and 6

An injective colouring c of a graph G is *optimal* if G has no injective colouring using fewer colours than c. An injective colouring c is *ℓ-injective* if every colour class of c has size at most ℓ. An ℓ-injective colouring c of a graph G is *ℓ-optimal* if G has no ℓ-injective colouring that uses fewer colours than c. We start with a useful lemma for the case where $\ell = 2$ that we will also use in the next section.

Lemma 1. *An optimal 2-injective colouring of a graph G can be found in polynomial time.*

Proof. Let c be a 2-injective colouring of G. Then each colour class of size 2 in G corresponds to a *dominating* edge of $\overline{G}$ (an edge uv of a graph is dominating if every other vertex in the graph is adjacent to at least one of u, v). Hence, the end-vertices of every non-dominating edge in $\overline{G}$ have different colours in G. Algorithmically, this means we may delete every non-dominating edge of $\overline{G}$ from $\overline{G}$; note that we do not delete the end-vertices of such an edge.

Let μ^* be the size of a maximum matching in the graph obtained from $\overline{G}$ after deleting all non-dominating edges of $\overline{G}$. The edges in such a matching will form exactly the colour classes of size 2 of an optimal 2-injective colouring of G. Hence, the injective chromatic number of G is equal to $\mu^* + (|V(G)| - 2\mu^*)$. It remains to observe that we can find a maximum matching in a graph in polynomial time by using a standard algorithm. $\square$

We are now ready to prove the following result.

Lemma 2. INJECTIVE COLOURING *is polynomial-time solvable for* $4P_1$*-free graphs, or equivalently, graphs with* $\alpha \leq 3$*, and thus for graphs with* $\overline{\chi} \leq 3$*.*

Proof. Let $G = (V, E)$ be a $4P_1$-free graph on n vertices. We first analyze the structure of injective colourings of G. Let c be an optimal injective colouring of G. As G is $4P_1$-free, every colour class of c has size at most 3. From all optimal injective colourings, we choose c such that the number of size-3 colour classes is as small as possible. We say that c is *class-3-optimal.*

Suppose c contains a colour class of size 3, say colour 1 appears on three distinct vertices u_1, u_2 and u_3 of G. As G is $4P_1$-free, $\{u_1, u_2, u_3\}$ dominates G. As c is injective, this means that every vertex in $G - \{u_1, u_2, u_3\}$ is adjacent to exactly one vertex of $\{u_1, u_2, u_3\}$. Hence, we can partition $V \setminus \{u_1, u_2, u_3\}$ into three sets T_1, T_2 and T_3, such that for $i \in \{1, 2, 3\}$, every vertex of T_i is adjacent to u_i and not to any other vertex of $\{u_1, u_2, u_3\}$. If two vertices t, t' in the same T_i, say T_1, are non-adjacent, then $\{t, t', u_2, u_3\}$ induces a $4P_1$, a contradiction. Hence, we partitioned V into three cliques $T_i \cup \{u_i\}$. We call the cliques T_1, T_2, T_3, the *T-cliques* of the triple $\{u_1, u_2, u_3\}$.

Let $t \in T_i$ for some $i \in \{1, 2, 3\}$. For $i \in \{0, 1, 2\}$ we say that t is *i-clique-adjacent* if t has a neighbour in zero, one or two cliques of $\{T_1, T_2, T_3\} \setminus T_i$, respectively. By the definition of an injective colouring and the fact that every T_i is a clique, a 1-clique-adjacent vertex of $T_1 \cup T_2 \cup T_3$ belongs to a colour class of size at most 2, and a 2-clique-adjacent vertex of $T_1 \cup T_2 \cup T_3$ belongs to a colour class of size 1. Hence, all the vertices that belong to a colour class of size 3 are 0-clique-adjacent. The partition of $V(G)$ is illustrated in Fig. 1.

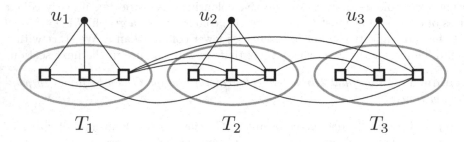

Fig. 1. The partition of $V(G)$ from Lemma 2. The squares inside each T_i, $i \in \{1, 2, 3\}$, represent the sets of 0-clique-adjacent, 1-clique-adjacent and 2-clique-adjacent vertices in T_i, respectively.

We now use the fact that c is class-3-optimal. Let $t \in V \setminus \{u_1, u_2, u_3\}$, say $t \in T_1$, be i-clique-adjacent for $i = 0$ or $i = 1$. Then we may assume without loss of generality that t has no neighbours in T_2. If t belongs to a colour class of size 1, then we can set $c(u_2) := c(t)$ to obtain an optimal injective colouring with fewer size-3 colour classes, contradicting our choice of c.

We now consider the 0-clique-adjacent vertices again. Recall that these are the only vertices, other than u_1, u_2 and u_3, that may belong to a colour class of size 3. As every T_i is a clique, every colour class of size 3 (other than $\{u_1, u_2, u_3\}$) has exactly one vertex of each T_i. Let $\{w_1, w_2, w_3\}$ be another colour class of size 3 with $w_i \in T_i$ for every $i \in \{1, 2, 3\}$. Let $x \in T_1 \setminus \{w_1\}$ be another 0-clique-adjacent vertex. Then swapping the colours of w_1 and x yields another class-3-optimal injective colouring of G. Hence, we derived the following claim, which summarizes the discussion above and where statement (iv) follows from (i)–(iii).

Claim. *Let c be a class-3-optimal injective colouring of G with $c(u_1) = c(u_2) = c(u_3)$ for three distinct vertices u_1, u_2, u_3 and with $p \geq 0$ other colour classes of size 3. Then the following four statements hold:*

(i) *All 0-clique-adjacent and 1-clique-adjacent vertices belong to a colour class of size at least 2.*

(ii) *Let $S = \{y_1, \ldots, y_s\}$ be the set of 2-clique-adjacent vertices. Then $\{y_1\}, \ldots, \{y_s\}$ are exactly the size-1 colour classes.*

(iii) *For $i \in \{1, 2, 3\}$, let $x_1^i, \ldots, x_{q_i}^i$ be the 0-clique-adjacent vertices of T_i and assume without loss of generality that $q_1 \leq q_2 \leq q_3$. Then $p \leq q_1$ and if $p \geq 1$, we may assume without loss of generality that the size-3 classes, other than $\{u_1, u_2, u_3\}$, are $\{x_1^1, x_1^2, x_1^3\}, \ldots, \{x_p^1, x_p^2, x_p^3\}$.*

(iv) *The number of colours used by c, or equivalently, the number of colour classes of c is equal to $1 + s + p + \frac{1}{2}(n - s - 3(p+1)) = \frac{1}{2}n + \frac{1}{2}s - \frac{1}{2}p - \frac{1}{2}$.*

We are now ready to present our algorithm. We first find, in polynomial time, an optimal 2-injective colouring of G by Lemma 1. We remember the number of colours used. Recall that the colour classes of every injective colouring of G have size at most 3. So, it remains to compute an optimal injective colouring for which at least one colour class has size 3.

We consider each triple u_1, u_2, u_3 of vertices of G and check if $\{u_1, u_2, u_3\}$ can be a colour class. That is, we check if $\{u_1, u_2, u_3\}$ is an independent set and has corresponding T-cliques T_1, T_2, T_3. This takes polynomial time. If not, then we discard $\{u_1, u_2, u_3\}$. Otherwise we continue as follows. Let $S = \{y_1, \ldots, y_s\}$ be the set of 2-clique adjacent vertices in $T_1 \cup T_2 \cup T_3$. Exactly the vertices of S will form the size-1 colour classes by Claim (ii). For $i \in \{1, 2, 3\}$, let $x_1^i, \ldots, x_{q_i}^i$ be the 0-clique-adjacent vertices of T_i, where we assume without loss of generality that $q_1 \leq q_2 \leq q_3$. By Claim (iii), any injective colouring of G which has $\{u_1, u_2, u_3\}$ as one of its colour classes has at most q_1 other colour classes of size 3 besides $\{u_1, u_2, u_3\}$. As can be seen from Claim (iv), the value $\frac{1}{2}n + \frac{1}{2}s - \frac{1}{2}p - \frac{1}{2}$ is minimized if the number p of size-3 colour classes is maximum.

From the above we can now do as follows. For $p = q_1, \ldots, 1$, we check if G has an injective colouring with exactly p colour classes of size 3. We stop as soon as we find a yes-answer or if p is set to 0. We first set $\{x_1^1, x_1^2, x_1^3\}, \ldots, \{x_p^1, x_p^2, x_p^3\}$ as the colour classes of size 3 by Claim (iii). Let Z be the set of remaining 0-clique-adjacent and 1-clique-adjacent vertices. We use Lemma 1 to check in polynomial time if the subgraph of G induced by $S \cup Z$ has an injective colouring that uses $s + \frac{1}{2}(n - s - 3(p+1))$ colours (which is the minimum number of colours possible). If so, then we stop and note that after adding the size-3 colour classes we obtained an injective colouring of G that uses $\frac{1}{2}n + \frac{1}{2}s - \frac{1}{2}p - \frac{1}{2}$ colours, which we remember. Otherwise we repeat this step after first setting $p := p - 1$.

As the above procedure for a triple u_1, u_2, u_3 takes polynomial time and the number of triples we must check is $O(n^3)$, our algorithm runs in polynomial time. We take the 3-injective colouring that uses the smallest number of colours and compare it with the number of colours used by the optimal 2-injective colouring that we computed at the start. Our algorithm then returns a colouring with the smallest of these two values as its output. □

We now show two hardness results.

Lemma 3. COLOURING *is* NP-*complete for graphs with* $\overline{\chi} \leq 3$.

Proof. The LIST COLOURING problem takes as input a graph G and a *list assignment* L that assigns each vertex $u \in V(G)$ a list $L(u) \subseteq \{1, 2, \ldots\}$. The question is whether G admits a colouring c with $c(u) \in L(u)$ for every $u \in V(G)$. Jansen [9] proved that LIST COLOURING is NP-complete for unions of two complete graphs. This is the problem we reduce from.

Let G be a graph with a list assignment L and assume that $V(G)$ can be split into two (not necessarily disjoint) cliques K and K'. We set $A_1 := K$ and $A_2 := K \setminus K'$. As both A_1 and A_2 are cliques, we have that $\overline{\chi}(G) \leq 2$. We may assume without loss of generality that the union of all the lists $L(u)$ is $\{1, \ldots, k\}$ for some integer k. We now extend G by adding a clique A_3 of k new vertices $v_1, \ldots, v_k$ and by adding an edge between a vertex x_ℓ and a vertex $u \in V(G)$ if and only if $\ell \notin L(u)$. This yields a new graph G' with $\overline{\chi}(G') \leq 3$. It is readily seen that G has a colouring c with $c(u) \in L(u)$ for every $u \in V(G)$ if and only if G' has a k-colouring. □

We use Lemma 3 to prove the next lemma, which uses the same arguments as the proof of our NP-hardness result for $6P_1$-free graphs [2]. The only difference is that we now reduce from COLOURING for graphs with $\overline{\chi} \leq 3$ instead of COLOURING for graphs with $\overline{\chi} \leq 4$ as we did in [2].

Lemma 4. INJECTIVE COLOURING *is* NP-*complete for graphs with* $\overline{\chi} \leq 4$, *and thus for* $5P_1$-*free graphs, or equivalently, graphs with* $\alpha \leq 4$.

We are now ready to prove three theorems.

Proofs of Theorems 3, 5 and 6. Theorem 3 follows from combining Theorem 1, after observing that graphs with $\overline{\chi} \leq 2$ are $3P_1$-free and thus $(P_1 + P_3)$-free, with Lemma 3, whereas Theorems 5 and 6 follow from combining Lemmas 2 and 4.

3   The Proof of Theorem 7

In this section we prove our final theorem. We start by showing three new polynomial results. We shall use Lemma 2, on $4P_1$-free graphs, in the proofs of Lemmas 6 and 7.

Lemma 5. INJECTIVE COLOURING *is polynomial-time solvable for* $(P_1 + P_4)$-*free graphs.*

Proof. Let G be a $(P_1 + P_4)$-free graph. Since connected P_4-free graphs have diameter at most 2, no two vertices can be coloured alike in an injective colouring. Hence, the injective chromatic number of a P_4-free graph is equal to the number of its vertices. Consequently, INJECTIVE COLOURING is polynomial-time solvable for P_4-free graphs. From now on, we assume that G is not P_4-free.

We first show that any colour class in any injective colouring of G has size at most 2. For contradiction, assume that c is an injective colouring of G such that there exists some colour, say colour 1, that has a colour class of size at least 3. Let $P = x_1 x_2 x_3 x_4$ be some induced P_4 of G.

We first consider the case where colour 1 appears at least twice on P. As no vertex has two neighbours coloured with the same colour, the only way in which this can happen is when $c(x_1) = c(x_4) = 1$. By our assumption, $G - P$ contains a vertex u with $c(u) = 1$. As G is $(P_1 + P_4)$-free, u has a neighbour on P. As every colour class is an independent set, this means that u must be adjacent to at least one of x_2 and x_3. Consequently, either x_2 or x_3 has two neighbours with colour 1, a contradiction.

Now we consider the case where colour 1 appears exactly once on P, say $c(x_h) = 1$ for some $h \in \{1, 2, 3, 4\}$. Then, by our assumption, $G - P$ contains two vertices u_1 and u_2 with colour 1. As G is $(P_1 + P_4)$-free, both u_1 and u_2 must be adjacent to at least one vertex of P, say u_1 is adjacent to x_i and u_2 is adjacent to x_j. Then $x_i \neq x_j$, as otherwise G has a vertex with two neighbours coloured 1. As every colour class is an independent set, we have that $x_h \notin \{x_i, x_j\}$, and hence, x_h, x_i, x_j are distinct vertices. Moreover, x_h is not a neighbour of x_i or x_j, as otherwise x_i or x_j has two neighbours coloured 1. Hence, we may assume without loss of generality that $h = 1$, $i = 3$ and $j = 4$. As every colour class is an independent set, u_1 and u_2 are non-adjacent. However, now $\{x_1, u_1, x_3, x_4, u_2\}$ induces a $P_1 + P_4$, a contradiction.

Finally, we consider the case where colour 1 does not appear on P. Let u_1, u_2, u_3 be three vertices of $G - P$ coloured 1. As before, $\{u_1, u_2, u_3\}$ is an independent set and each u_i has a different neighbour on P. We first consider the case where x_1 or x_4, say x_4 is not adjacent to any u_i. Then we may assume without loss of generality that $u_1 x_1$ and $u_2 x_2$ are edges. However, now $\{x_4, u_1, x_1, x_2, u_2\}$ induces a $P_1 + P_4$, which is not possible. Hence, we may assume without loss of generality that $u_1 x_1$, $u_2 x_2$ and $u_4 x_4$ are edges of G. Again we find that $\{x_4, u_1, x_1, x_2, u_2\}$ induces a $P_1 + P_4$, a contradiction.

From the above, we find that each colour class in an injective colouring of G has size at most 2. This means we can use Lemma 1.   □

Lemma 6. INJECTIVE COLOURING *is polynomial-time solvable for* $(2P_1 + P_3)$-*free graphs.*

Proof. Let $G = (V, E)$ be a $(2P_1 + P_3)$-free graph. We may assume without loss of generality that G is connected and by Lemma 2 that G has an induced $4P_1$. We first show that any colour class in any injective colouring of G has size at most 2. For contradiction, assume that c is an injective colouring of G such that there exists some colour, say colour 1, that has a colour class of size at least 3. Let $U = \{u_1, \ldots, u_p\}$ for some $p \geq 3$ be the set of vertices of G with $c(u_i) = 1$ for $i \in \{1, \ldots, p\}$.

As c is injective, every vertex in $G - U$ has at most one neighbour in U. Hence, we can partition $G - U$ into (possibly empty) sets $T_0, \ldots, T_p$, where T_0 is the set of vertices with no neighbour in U and for $i \in \{1, \ldots, p\}$, T_i is the set of vertices of $G - U$ adjacent to u_i.

We first claim that T_0 is empty. For contradiction, assume $v \in T_0$. As G is connected, we may assume without loss of generality that v is adjacent to some vertex $t \in T_1$. Then $\{u_2, u_3, u_1, t, v\}$ induces a $2P_1 + P_3$, a contradiction. Hence, $T_0 = \emptyset$.

We now prove that every T_i is a clique. For contradiction, assume that t and t' are non-adjacent vertices of T_1. Then $\{u_2, u_3.t, u_1, t'\}$ induces a $2P_1 + P_3$, a contradiction. Hence, every T_i and thus every $T_i \cup \{u_i\}$ is a clique.

We now claim that $p = 3$. For contradiction, assume that $p \geq 4$. As G is connected and U is an independent set, we may assume without of generality that there exist vertices $t_1 \in T_1$ and $t_2 \in T_2$ with $t_1 t_2 \in E$. Then $\{u_3, u_4, u_1, t_1, t_2\}$ induces a $2P_1 + P_3$, a contradiction. Hence, $p = 3$.

Now we know that V can be partitioned into three cliques $T_1 \cup \{u_1\}$, $T_2 \cup \{u_2\}$ and $T_3 \cup \{u_3\}$. However, then G is $4P_1$-free, a contradiction. We conclude that every colour class of every injective colouring of G has size at most 2. This means we can use Lemma 1. $\qquad\square$

Lemma 7. INJECTIVE COLOURING *is polynomial-time solvable for* $(3P_1 + P_2)$-*free graphs.*

Proof. Let G be a $(3P_1 + P_2)$-free graph on n vertices. We may assume without loss of generality that G is connected and by Lemma 2 that G has an induced $4P_1$. As before, we will first analyze the structure of injective colourings of G. We will then exploit the properties found algorithmically.

Let c be an injective colouring of G that has a colour class U of size at least 3. So let $U = \{u_1, \ldots, u_p\}$ for some $p \geq 3$ be the set of vertices of G with, say colour 1. As c is injective, every vertex in $G - U$ has at most one neighbour in U. Hence, we can partition $G - U$ into (possibly empty) sets $T_0, \ldots, T_p$, where T_0 is the set of vertices with no neighbour in U and for $i \in \{1, \ldots, p\}$, T_i is the set of vertices of $G - U$ adjacent to u_i.

Assume that $p \geq 4$. As G is connected, there exists a vertex $v \notin U$ but that has a neighbour in U, say $v \in T_1$. Then $\{u_2, u_3, u_4, u_1, v\}$ induces a $3P_1 + P_2$, a contradiction. Hence, we have shown the following claim.

Claim 1. Every injective colouring of G is ℓ-injective for some $\ell \in \{1, 2, 3\}$.

We continue as follows. As $p = 3$ by Claim 1, we have $V(G) = U \cup T_0 \cup T_1 \cup T_2 \cup T_3$. Suppose T_0 contains two adjacent vertices x and y. Then $\{u_1, u_2, u_3, x, y\}$ induces a $3P_1 + P_2$, a contradiction. Hence, T_0 is an independent set. As G is connected, this means each vertex in T_0 has a neighbour in $T_1 \cup T_2 \cup T_3$.

Suppose T_0 contains two vertices x and y with the same colour, say $c(x) = c(y) = 2$. Let $v \in T_1 \cup T_2 \cup T_3$, say $v \in T_1$ be a neighbour of x. Then, as $c(x) = c(y)$ and c is injective, v is not adjacent to y. As T_0 is independent, x and y are not adjacent. However, now $\{u_2, u_3, y, x, v\}$ induces a $3P_1 + P_2$, a contradiction. Hence, every vertex in T_0 has a unique colour. Suppose T_0 contains a vertex x and $T_1 \cup T_2 \cup T_3$ contains a vertex v such that $c(x) = c(v)$. We may assume without loss of generality that $v \in T_1$. Then $\{u_2, u_3, x, v, u_1\}$ induces a $3P_1 + P_2$, a contradiction.

Finally, suppose that $T_1 \cup T_2 \cup T_3$ contain two distinct vertices v and v' with $c(v) = c(v')$. Let $x \in T_0$. Then x is not adjacent to at least one of v, v', say $xv \notin E$ and also assume that $v \in T_1$. Then $\{u_2, u_3, x, v, u_1\}$ induces a $3P_1 + P_2$. Hence, we have shown the following claim.

Claim 2. If c is 3-injective and U is a size-3 colour class such that G has a vertex not adjacent to any vertex of U, then all colour classes not equal to U have size 1.

We note that the injective colouring c in Claim 2 uses $n - 2$ distinct colours.

We continue as follows. From now on we assume that $T_0 = \emptyset$. Every T_i is ($P_1 + P_2$)-free, as otherwise, if say T_1 contains an induced $P_1 + P_2$, then this $P_1 + P_2$, together with u_2 and u_3, forms an induced $3P_1 + P_2$, which is not possible. Hence, each T_i induces a complete r_i-partite graph for some integer r_i (that is, the complement of a disjoint union of r_i complete graphs). Hence, we can partition each T_i into r_i independent sets $T_i^1, \ldots, T_i^{r_i}$ such that there exists an edge between every vertex in T_i^a and every vertex in T_i^b if $a \neq b$. See also Fig. 2.

Suppose G contains another colour class of size 3, say v_1, v_2 and v_3 are three distinct vertices coloured 2. If two of these vertices, say v_1 and v_2, belong to the same T_i, say T_1, then u_1 has two neighbours with the same colour. This is not possible, as c is injective. Hence, we may assume without loss of generality that $v_i \in T_i^1$ for $i \in \{1, 2, 3\}$.

Suppose that T_1^2 contains two vertices s and t. Then, as s and t are adjacent to v_1, both of them are not adjacent to v_2 (recall that $c(v_1) = c(v_2)$ and c is injective). Hence, $\{s, t, u_3, v_2, u_2\}$ induces a $3P_1 + P_2$ (see Fig. 2). We conclude that for every $i \in \{1, 2, 3\}$, the sets $T_i^2, \ldots, T_i^{r_i}$ have size 1.

We will now make use of the fact that G contains an induced $4P_1$. We note that each $T_i \cup \{u_i\}$ is a clique, unless $|T_i^1| \geq 2$. As $V(G) = T_1 \cup T_2 \cup T_3 \cup \{u_1, u_2, u_3\}$ and G contains an induced $4P_1$, we may assume without loss of generality that T_1^1 has size at least 2. Recall that $v_1 \in T_1^1$. Let $z \neq v_1$ be some further vertex of T_1^1. If z is not adjacent to v_2, then $\{z, v_1, u_3, v_2, u_2\}$ induces a $3P_1 + P_2$, which is not possible. Hence, z is adjacent to v_2. For the same reason, z is adjacent to v_3. This is not possible, as c is injective and v_2 and v_3 both have colour 2. Hence, we have proven the following claim.

Claim 3. If c is 3-injective and U is a size-3 colour class such that each vertex of $G - U$ is adjacent to a vertex of U, then c has no other colour class of size 3.

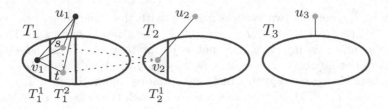

Fig. 2. The situation in Lemma 7 where T_1^2 contains two vertices s and t. We show that this situation cannot happen, as it would lead to a forbidden induced $3P_1 + P_2$. Note that each u_i is adjacent to all vertices of T_i and not to any vertices of T_j for $j \neq i$. There may exist edges between vertices of different sets, but these are not drawn.

We are now ready to present our polynomial-time algorithm. We first use Lemma 1 to find in polynomial time an optimal 2-injective colouring of G. We remember the number of colours it uses.

By Claim 1, it remains to find an optimal 3-injective colouring with at least one colour class of size 3. We now consider each set $\{u_1, u_2, u_3\}$ of three vertices. We discard our choice if u_1, u_2, u_3 do not form an independent set or if $V(G) \setminus \{u_1, u_2, u_3\}$ cannot be partitioned into sets $T_0, \ldots, T_4$ as described above. Suppose we have not discarded our choice of vertices u_1, u_2, u_3. We continue as follows.

If $T_0 \neq \emptyset$, then by Claim 2 the only 3-injective colouring of G (subject to colour permutation) with colour class $\{u_1, u_2, u_3\}$ is the colouring that gives each u_i the same colour and a unique colour to all the other vertices of G. This colouring uses $n - 2$ colours and we remember this number of colours.

Now suppose $T_0 = \emptyset$. By Claim 3, we find that $\{u_1, u_2, u_3\}$ is the only colour class of size 3. Recall that no vertex in $G - \{u_1, u_2, u_3\} = T_1 \cup T_2 \cup T_3$ is adjacent to more than one vertex of $\{u_1, u_2, u_3\}$. Hence, we can apply Lemma 1 on $G - \{u_1, u_2, u_3\}$. This yields an optimal 2-injective colouring of $G - \{u_1, u_2, u_3\}$. We colour u_1, u_2, u_3 with the same colour and choose a colour that is not used in the colouring of $G - \{u_1, u_2, u_3\}$. This yields a 3-injective colouring of G that is optimal over all 3-injective colourings with colour class $\{u_1, u_2, u_3\}$. We remember the number of colours.

As the above procedure takes polynomial time and there are $O(n^3)$ triples to consider, we find in polynomial time an optimal 3-injective colouring of G that has at least one colour class of size 3 (should it exist). We compare the number of colours used with the number of colours of the optimal 2-injective colouring of G that we found earlier. Our algorithm returns the minimum of the two values as the output. Since both colourings are found in polynomial time, we conclude that our algorithm runs in polynomial time. □

We are now ready to prove Theorem 7.

Theorem 7 (restated). *Let $H \neq 2P_1 + P_4$ be a graph. Then* INJECTIVE COLOURING *for H-free graphs is polynomial-time solvable if $H \subseteq_i P_1 + P_4$ or $H \subseteq_i 2P_1 + P_3$ or $H \subseteq_i 3P_1 + P_2$ and* NP-*complete otherwise.*

Proof. The problem is readily seen to be in NP. Let C_n denote the n-vertex cycle. In [2], we proved that INJECTIVE 4-COLOURING is NP-complete for C_3-free graphs (in fact, the proof of [2] shows NP-completeness for bipartite graphs). Hence, if $C_3 \subseteq_i H$, then INJECTIVE COLOURING is NP-complete for H-free graphs. If $C_p \subseteq_i H$ for some $p \geq 4$ or $K_{1,3} \subseteq_i H$, then we use a result of Mahdian [14], namely that for every $g \geq 4$ and $k \geq 4$, INJECTIVE k-COLOURING is NP-complete for line graphs of bipartite graphs of girth at least g: by setting $g = p + 1$ we obtain a class of $(C_p, K_{1,3})$-free graphs.

In the remaining case H is a linear forest. If $5P_1 \subseteq_i H$, then we use Lemma 4. If $2P_2 \subseteq_i H$, then we obtain a superclass of $(2P_2, C_4, C_5)$-free graphs (split graphs) for which Bodlaender et al. [1] proved NP-completeness. In all other cases, $H \subseteq_i P_1 + P_4$, $H \subseteq_i 2P_1 + P_3$ or $H \subseteq_i 3P_1 + P_2$, and we use Lemma 5, 6 or 7, respectively. □

4   Conclusions

We extended the partial classification of INJECTIVE COLOURING for H-free graphs in [2] to a new classification that leaves open only the case where $H = 2P_1 + P_4$. Note that each of the graphs $P_1 + P_4$, $2P_1 + P_3$ and $3P_1 + P_2$ is an induced subgraph of $2P_1 + P_4$. Our strategy for solving the cases where $H = 3P_1 + P_2$ or $H = 2P_1 + P_3$ is based on the presence of an induced $4P_1$; as otherwise we can use our algorithm for $4P_1$-free graphs from Lemma 2. We then argue in each of the two cases that the number of size-3 colour classes is small. However, such an approach no longer works for the unknown case where $H = 2P_1 + P_4$, as there exist connected $(2P_1 + P_4)$-free graphs with an induced $4P_1$ that have an arbitrarily large number of size-3 colour classes. On the other hand, the algorithms for the cases where $H \in \{P_1 + P_4, 2P_1 + P_3, 3P_1 + P_2\}$ might be useful for a polynomial-time algorithm for the case $H = 2P_1 + P_4$ (should such an algorithm exist). However, we first need to understand the structure of these graphs better and leave the case where $H = 2P_1 + P_4$ for future work.

Open Problem 1. *Determine the complexity of* INJECTIVE COLOURING *for* $(2P_1 + P_4)$-*free graphs.*

We recall that INJECTIVE 3-COLOURING is polynomial-time solvable for general graphs and that in [2] we proved that, for every $k \geq 4$, INJECTIVE k-COLOURING is NP-complete for bipartite graphs and thus for graphs of girth at least 4. Mahdian [14] proved that for every $g \geq 4$ and $k \geq 4$, INJECTIVE EDGE k-COLOURING is NP-complete for bipartite graphs of girth at least g. For the vertex variant, such a result does not seem to be known (except when $g = 4$). Hence, we pose the following challenging but highly interesting open problem (see also [2]).

Open Problem 2. *For every* $g \geq 5$, *determine the complexity of* INJECTIVE COLOURING *and* INJECTIVE k-COLOURING *($k \geq 4$) for graphs of girth at least* g.

References

1. Bodlaender, H.L., Kloks, T., Tan, R.B., van Leeuwen, J.: Approximations for lambda-colorings of graphs. Comput. J. **47**, 193–204 (2004)
2. Bok, J., Jedličková, N., Martin, B., Paulusma, D., Smith, S.: Acyclic colouring, star colouring and injective colouring for H-free graphs. In: Proceedings ESA 2020, LIPIcs, vol. 173, pp. 22:1–22:22 (2020)
3. Calamoneri, T.: The $L(h, k)$-labelling problem: an updated survey and annotated bibliography. Comput. J. **54**, 1344–1371 (2011)
4. Chudnovsky, M., Huang, S., Spirkl, S., Zhong, M.: List-three-coloring graphs with no induced $P_6 + rP_3$. CoRR, abs/1806.11196 (2018)
5. Golovach, P.A., Johnson, M., Paulusma, D., Song, J.: A survey on the computational complexity of colouring graphs with forbidden subgraphs. J. Graph Theory **84**, 331–363 (2017)
6. Hahn, G., Kratochvíl, J., Širáň, J., Sotteau, D.: On the injective chromatic number of graphs. Discret. Math. **256**, 179–192 (2002)
7. Heggernes, P., Telle, J.A.: Partitioning graphs into generalized dominating sets. Nordic J. Comput. **5**, 128–142 (1998)
8. Hell, P., Raspaud, A., Stacho, J.: On injective colourings of chordal graphs. In: Laber, E.S., Bornstein, C., Nogueira, L.T., Faria, L. (eds.) LATIN 2008. LNCS, vol. 4957, pp. 520–530. Springer, Heidelberg (2008). https://doi.org/10.1007/978-3-540-78773-0_45
9. Jansen, K.: Complexity Results for the Optimum Cost Chromatic Partition Problem, Universität Trier, Mathematik/Informatik, Forschungsbericht, pp. 96–41 (1996)
10. Jin, J., Baogang, X., Zhang, X.: On the complexity of injective colorings and its generalizations. Theoret. Comput. Sci. **491**, 119–126 (2013)
11. Klimošová, T., Malík, J., Masařík, T., Novotná, J., Paulusma, D., Slívová, V.: Colouring $(P_r + P_s)$-free graphs. In: Proceedings ISAAC 2018, LIPIcs, pp. 123:5:1–5:13 (2018)
12. Král', D., Kratochvíl, J., Tuza, Z., Woeginger, G.J.: Complexity of coloring graphs without forbidden induced subgraphs. In: Brandstädt, A., Le, V.B. (eds.) WG 2001. LNCS, vol. 2204, pp. 254–262. Springer, Heidelberg (2001). https://doi.org/10.1007/3-540-45477-2_23
13. Lloyd, E.L., Ramanathan, S.: On the complexity of distance-2 coloring. Proc. ICCI **1992**, 71–74 (1992)
14. Mahdian, M.: On the computational complexity of strong edge coloring. Discret. Appl. Math. **118**, 239–248 (2002)
15. Sen, A., Huson, M.L.: A new model for scheduling packet radio networks. Wireless Netw. **3**, 71–82 (1997)
16. Zhou, X., Kanari, Y., Nishizeki, T.: Generalized vertex-coloring of partial k-trees. IEICE Trans. Fundamentals Electron. Commun. Comput. Sci. **E83-A**, 671–678 (2000)

Variants of the Determinant Polynomial and the **VP**-Completeness

Prasad Chaugule$^{(\boxtimes)}$, Nutan Limaye, and Shourya Pandey

Indian Institute of Technology, Bombay, India
{prasad,nutan,shouryap}@cse.iitb.ac.in

Abstract. The determinant is a canonical VBP-complete polynomial in the algebraic complexity setting. In this work, we introduce two variants of the determinant polynomial which we call $\texttt{StackDet}_n(X)$ and $\texttt{CountDet}_n(X)$ and show that they are VP and VNP complete respectively under p-projections. The definitions of the polynomials are inspired by a combinatorial characterisation of the determinant developed by Mahajan and Vinay (SODA 1997). We extend the combinatorial object in their work, namely *clow sequences*, by introducing additional edge labels on the edges of the underlying graph. The idea of using edge labels is inspired by the work of Mengel (MFCS 2013).

Keywords: Algebraic circuits · VP-Completeness · Determinant family

1 Introduction

In an influential paper of Valiant [12], a complexity theoretic view of algebraic computation was presented. This work led to a classification of polynomials based on the ease of computing them. Consequently, complexity classes such as VF, VBP, VP and VNP were defined and investigated in many follow-up papers. These algebraic classes were designed with the intention of mimicking Boolean complexity classes. It was believed that they would give rise to equally interesting, but potentially easier to resolve questions. For example, the question of separating the classes VP and VNP turned out to be very interesting, like its Boolean counterpart, namely the famous question of separating NP from P.

While there are many parallels between these two worlds, over the years, many crucial differences between them have also surfaced. Specifically, in the Boolean world, many naturally occurring problems have been found to be *complete* for the classes NP and P[1]. Although many naturally occurring polynomials

---

[1] A problem P is said to be complete for a Boolean complexity class $\mathcal{C}$ if $P \in \mathcal{C}$ and any problem P' in $\mathcal{C}$ *reduces* to P in polynomial time.

N. Limaye—Funded by SERB Project no. MTR/2017/000909.

R. Santhanam and D. Musatov (Eds.): CSR 2021, LNCS 12730, pp. 31–55, 2021.
https://doi.org/10.1007/978-3-030-79416-3_3

are known to be complete[2] for VNP, until very recently no natural polynomial was known to be complete for VP.

The process of finding many complete problems for a complexity class is crucial in many ways. For one, each complete problem presents a potentially different way of understanding the class. It also makes the complexity class rich and robust. In this work, we contribute to the class of VP-complete polynomials.

Until as recently as 2014, hardly any natural VP-complete polynomials were known. In Durand et al. [2] and Mahajan et al. [5], many interesting and fairly natural families of polynomials were shown to be VP-complete. In [1], a few more polynomials complete for VP were presented. All these polynomials were based on counting graph homomorphisms[3].

In this work we define two fairly simple to state variants of the determinant polynomial and show that they are VP and VNP complete. As the determinant is known to be complete for the class VBP[4] (a class known to be contained in VP), this gives a satisfactory way of using the same base polynomial, namely the determinant polynomial, whose generalisations capture the class VP and VNP.

The determinant polynomial is a central object of study in algebraic complexity theory. Classically, the determinant has been studied for many centuries by mathematicians, physicists, numerical analysts and computer scientists.

The determinant is known to be *easy to compute*. In this respect, it enjoys a rather rare place in computation; it is an extremely useful quantity which is also efficiently computable. The classical efficient algorithms for the determinant are typically variants of the Guassian elimination method. In last three to four decades, other approaches for computing the determinant have also been proposed. One such example is an innovative approach proposed by Mahajan and Vinay [6], which gave the first combinatorial characterization of the determinant that yielded an efficient algorithm.

In this work, we take our inspiration from this combinatorial characterization of the determinant polynomial and define two variants of the determinant which we call $\texttt{StackDet}_n$ and $\texttt{CountDet}_n$. We show that they are complete for the classes VP and VNP, respectively.

The main proof idea comes from a paper of Mengel [8], which introduces characterisations of VP and VNP using *Algebraic Branching Programs (ABPs)*

---

[2] A polynomial $P_n(X)$ is said to be complete for an algebraic complexity class $\mathcal{A}$ if $P_n(X)$ can be computed in $\mathcal{A}$ and any polynomial $P'_m(Y)$ can be obtained from $P_n(X)$ by setting the variables in X to variables in Y or field constants. For formal definitions see Sect. 2.

[3] See also [3] for interesting variants of homomorphism polynomials.

[4] An algebraic branching program (ABP) is a directed layered acyclic graph with a source s and a sink t. The edges are labelled with formal variables or field constants. The weight of an s to t path π is the product of the weights on the edges of π. The polynomial computed by the ABP is the sum of weights of all the s to t paths. A family f_n with $s(n)$ number of variables and degree $d(n)$ where both $s(n)$ and $d(n)$ are polynomially bounded in n is said to be in VBP iff there exist algebraic branching program of size polynomially bounded in n which computes f_n. For more details see [11].

with memory. In that work, informally speaking, it is shown that when ABPs are appended with *stack-like* memory, then they capture the class VP, and when they are appended with *counter-like* memory, they characterise the class VNP. We use these ideas and combine them with the combinatorial characterisation of the determinant to define our polynomial families.

The proof that shows that the determinant polynomial is complete for VBP can be adapted in a very straightforward way along with the *ABP with memory characterisations* of VP and VNP from the work of [8], to obtain polynomial families that are hard for these classes. However, like many other classes of polynomials (see for instance polynomial families from [7,10]), they are circuit-description dependent. From the work started by Durand et al. the quest has been to find circuit-description independent polynomial families complete for VP. We are able to achieve that here. The polynomial families we obtain here are circuit-description independent as desired and are variants of the determinant polynomial, which make them substantially different from the previous works [1,2,5].

1.1 Our Results

Before going into the details of our results, we recall the combinatorial characterisation of the determinant. Let Y be an $m \times m$ matrix, with (i,j)th entry equal to $y_{i,j}$. It is known that the determinant of Y is sum of signed cycle covers of the directed graph represented by Y. This is one of the many combinatorial definitions of the determinant, but as is, it is not known to give rise to an efficient computational procedure. Mahajan and Vinay generalized cycle covers using a notion of *clow sequences* and proved that the sum of signed clow sequences also equals the determinant. They then proved that the signed sum of clow sequences is efficiently computable.

StackDet$_m$ **and** CountDet$_m$. We also use sum of signed clow sequences to define our polynomial. In our case, the graph has some additional edge labels. For StackDet$_m$ (for CountDet$_m$), the labels come from a *stack alphabet* (*counter alphabet*, resp.). Based on these labels, we get two types of clow sequences; those which are *stack-realizable* (*counter-realizable*) and those which are not. The polynomial sums only the prior clow sequences. We show that StackDet$_m$ is VP-complete and that CountDet$_m$ is VNP-complete. We state our main theorem.

Theorem 1. StackDet$_n$$(X)$ *is VP-complete and* CountDet$_n$$(X)$ *is VNP-complete over any field under p-projections.*

2 Preliminaries

In this paper, a graph is always a directed graph unless specified otherwise.

Let $G = (V, E)$ be a graph where $V = [n]$. A walk $(u_1, u_2, \ldots, u_{k+1})$ in G is called a closed walk, or a clow, if $u_1 = u_{k+1}$, u_1 is the least numbered vertex in the walk and for any $2 \leq i \leq k$, $u_i \neq u_1$. The vertex u_1 is called *the head of the clow*. We use $\deg(\mathcal{C})$ to denote the number of edges in $\mathcal{C}$ (counted with multiplicity), i.e. in this case k.

Definition 1 (A clow sequence [6]). *A clow sequence $\widehat{\mathcal{C}} = \langle \mathcal{C}_1, \ldots \mathcal{C}_\ell \rangle$ in a graph $G = (V, E)$ is an ordered tuple of clows such that $\text{Head}(\mathcal{C}_1) < \text{Head}(\mathcal{C}_2) < \text{Head}(\mathcal{C}_3) < \ldots < \text{Head}(\mathcal{C}_\ell)$ and $\deg(\widehat{\mathcal{C}}) = \sum_{i=1}^{\ell} \deg(\mathcal{C}_i) = n$, where $V = [n]$. The sign of a clow sequence, $\text{sign}(\widehat{\mathcal{C}})$, is $(-1)^{n+\ell}$.*

Definition 2 (Stack graphs and Counter graphs). *A stack graph is a directed graph $G = (V, E, \Sigma, \phi)$, where V is a set of vertices, E is a set of edges. The set Σ is a symbol set. The function ϕ labels every edge of the graph with either $\text{Push}(s)$, $\text{Pop}(s)$ for some $s \in \Sigma$ or with No-op[5]. A counter graph $G = (V, E, \Sigma, \phi)$ is very similar to the stack graph except in this case, the function ϕ labels every edge of the graph with either $\text{Read}(s)$, $\text{Write}(s)$ for some $s \in \Sigma$ or with No-op.*

We call $\text{Push}(s)$, $\text{Pop}(s)$ as the stack operations whereas $\text{Read}(s)$, $\text{Write}(s)$ as the counter operations. No-op is both a stack as well as a counter operation. Let $s_1 = [a_1, a_2, \ldots, a_m]$ and $s_2 = [b_1, b_2, \ldots, b_n]$ be two sequences of stack operations (or counter operations) then concatenation of s_1 followed by s_2 (denoted as $s_1 \square s_2$) is the ordered sequence $[a_1, a_2, \ldots, a_m, b_1, b_2, \ldots, b_n]$. It is easy to extend this definition of concatenation of two sequences to any number of sequences. Let $\mathcal{W} = (u_1, \ldots, u_{k+1})$ be a walk of length k in a stack graph (or counter graph) G. We define $Seq[\mathcal{W}]$ to be the sequence of stack operations (counter operations, respectively) along the edges in this walk, i.e. $[\phi(u_1, u_2), \phi(u_2, u_3), \ldots, \phi(u_k, u_{k+1})]$.

Definition 3 (Stack-realizable sequence). *A stack-realizable sequence of operations is a sequence of stack operations which can be inductively formed using the following rules :*

- *The empty sequence is a stack-realizable sequence.*
- *If P is a stack-realizable sequence then $\text{Push}(s) \square P \square \text{Pop}(s)$ is stack-realizable $\forall s \in \Sigma$.*
- *If P is a stack-realizable sequence then $\text{No-op} \square P$ and $P \square \text{No-op}$ are also stack-realizable.*
- *If P and Q are stack-realizable sequences then $P \square Q$ is a stack-realizable sequence.*

For example, $[\text{Push}(a), \text{Push}(b), \text{Pop}(b), \text{Push}(c), \text{No-op}, \text{Pop}(c), \text{Pop}(a), \text{No-op}]$ is a stack-realizable sequence, whereas $[\text{Push}(a), \text{Pop}(b)]$ is not.

Definition 4 (Counter-realizable sequence). *A sequence of counter operations P is said to be counter-realizable if for every $s \in \Sigma$, $\text{Write}(s)$ and $\text{Read}(s)$ occur equal number of times in P and for every prefix P' of P, the number of times $\text{Write}(s)$ occurs in P' is at least as much as the number of times $\text{Read}(s)$ appears in P'.*

---

[5] No-op stands for No-operation.

A directed walk $\mathcal{W}$ in a stack graph (or counter graph) G is called *stack-realizable walk* (or counter-realizable walk, respectively) if and only if $Seq[\mathcal{W}]$ is stack-realizable (or counter-realizable, respectively).

Definition 5 (A realizable clow sequence). *A clow sequence* $\widehat{\mathcal{C}} = \langle \mathcal{C}_1, \ldots, \mathcal{C}_\ell \rangle$ *of a stack graph (or counter graph) G is called stack-realizable (or counter-realizable, respectively) if and only if* $Seq[\mathcal{C}_1] \square Seq[\mathcal{C}_2] \square \ldots \square Seq[\mathcal{C}_\ell]$ *is a stack-realizable sequence (or counter-realizable, respectively).*

Let X be a set of variables. Consider a stack graph (or a counter graph) $G = (V, E, \Sigma, \phi)$ with an edge-labeling function $\mathcal{L} : E \to X \cup \mathbb{F}$. For some $1 \le j \le \ell$ and clow $\mathcal{C}_j = (u_1, u_2, \ldots, u_{k+1})$, $mon(\mathcal{C}_j)$ denotes monomial formed by multiplying the labels of the edges in $\mathcal{C}_j$, i.e. $mon(\mathcal{C}_j) = \prod_{i=1}^{k} \mathcal{L}\left((u_i, u_{i+1})\right)$ and $mon(\widehat{\mathcal{C}}) = \prod_{i=1}^{\ell} mon(\mathcal{C}_i)$.

Before going into the details of the definition of the Stack Determinant, we recall the definition of a determinant of a graph G as stated in [6].

Definition 6 (Determinant). *Consider a directed graph $G = (V, E)$. Let* $\mathcal{L} : E \longrightarrow X \cup \mathbb{F}$, *where X is the set of variables. The determinant polynomial* $\mathtt{Det}(X)$ *is defined as follows :* $\mathtt{Det}(X) = \sum_{\substack{\text{All clow} \\ \text{sequences } \widehat{\mathcal{C}} \text{ of degree } |V|}} sign(\widehat{\mathcal{C}}) mon(\widehat{\mathcal{C}}).$

Definition 7 (General Stack Determinant). *Consider a stack graph $G = (V, E, \Sigma, \phi)$. Let* $\mathcal{L} : E \longrightarrow X \cup \mathbb{F}$, *where X is the set of variables. Let $\phi : E \longrightarrow \bigcup_{a \in \Sigma} \{\mathtt{Push(a)}, \mathtt{Pop(a)}\} \cup \{\mathtt{No\text{-}op}\}$ be any edge map. The general stack determinant polynomial* $\mathtt{GenStackDet}(X)$ *is defined as follows :*

$$\mathtt{GenStackDet}(X) = \sum_{\substack{\text{All stack realizable clow} \\ \text{sequences } \widehat{\mathcal{C}} \text{ of degree } |V|}} sign(\widehat{\mathcal{C}}) mon(\widehat{\mathcal{C}}).$$

In Definition 7, we now fix our graph family G_n and instantiate the function ϕ to Φ and define the stack determinant family $\mathtt{StackDet_n}(X)$.

Definition 8 (Stack Determinant). *Let $\Sigma = \{s_1, \ldots, s_n\}$. Consider a stack graph $G_n = (V, E, \Sigma, \Phi)$ with $V = [4n]$ and $E = \{e_{i,j} = (i,j) | 1 \le i, j \le 4n\}$. Let $\mathcal{L}((i,j)) = x_{i,j}$. We define the function $\Phi : E \longrightarrow \bigcup_{a \in \Sigma} \{\mathtt{Push(a)}, \mathtt{Pop(a)}\} \cup \{\mathtt{No\text{-}op}\}$ such that for all $i \in [n]$, $\Phi(e_{k=4(i-1)+1, k+1}) = \mathtt{Push(s_i)}$ and $\Phi(e_{k+2, k+3}) = \mathtt{Pop(s_i)}$. All the other remaining edges are mapped to $\mathtt{No\text{-}op}$. The stack determinant polynomial $\mathtt{StackDet_n}(X)$ is defined as follows :*

$$\mathtt{StackDet_n}(X) = \sum_{\substack{\text{All stack realizable clow} \\ \text{sequences } \widehat{\mathcal{C}} \text{ of degree } |V|}} sign(\widehat{\mathcal{C}}) mon(\widehat{\mathcal{C}})$$

Definition 9 (General Counter Determinant). *Consider a counter graph $G = (V, E, \Sigma, \phi)$. Let* $\mathcal{L} : E \longrightarrow X \cup \mathbb{F}$, *where X is the set of variables. Let $\phi : E \longrightarrow \bigcup_{a \in \Sigma} \{\mathtt{Write(a)}, \mathtt{Read(a)}\} \cup \{\mathtt{No\text{-}op}\}$ be any edge map. The general counter determinant polynomial* $\mathtt{GenCountDet}(X)$ *is defined as follows :*

$$\mathtt{GenCountDet}(X) = \sum_{\substack{\text{All counter realizable clow} \\ \text{sequences } \widehat{\mathcal{C}} \text{ of degree } |V|}} sign(\widehat{\mathcal{C}}) mon(\widehat{\mathcal{C}}).$$

In Definition 9, we now instantiate the function ϕ to Φ and redefine the counter determinant (with respect to Φ).

Definition 10 (Counter Determinant). *Let $\Sigma = \{s_1, \ldots, s_n\}$. Consider a counter graph $G_n = (V, E, \Sigma, \Phi)$ with $V = [4n]$ and $E = \{e_{i,j} = (i, j) | 1 \leq i, j \leq 4n\}$. Let the function $\Phi : E \longrightarrow \bigcup_{a \in \Sigma} \{\mathtt{Write}(a), \mathtt{Read}(a)\} \cup \{\mathtt{No\text{-}op}\}$ be such that for all $i \in [n]$, $\Phi(e_{k=4(i-1)+1, k+1}) = \mathtt{Write}(s_i)$ and $\Phi(e_{k+2, k+3}) = \mathtt{Read}(s_i)$. All the other remaining edges are mapped to $\mathtt{No\text{-}op}$. Let $\mathcal{L}((i, j)) = x_{i,j}$. The counter determinant polynomial $\mathtt{CountDet_n}(X)$ is defined as follows :*

$$\mathtt{CountDet_n}(X) = \sum_{\substack{\text{All counter\_realizable clow} \\ \text{sequences } \widehat{\mathcal{C}} \text{ of degree } |V|.}} sign(\widehat{\mathcal{C}}) mon(\widehat{\mathcal{C}}).$$

Definition 11. *A polynomial family $\{f_n\}$ is said to be a projection of a family $\{g_n\}$, denoted as $\{f_n\} \leq \{g_n\}$, if for every f_n (where $n \in \mathbb{N}$), there exist some $m \in \mathbb{N}$ where f_n can be computed by g_m by setting the variables of g_m to either the variables of f_n or the field constants. If m is polynomially bounded in n, it is said to be a p-projection, denoted by $\{f_n\} \leq_p \{g_n\}$.*

Definition 12. *A p-bounded family $\{f_n\}$ is complete for class $\mathcal{C}$, if $f_n \in \mathcal{C}$ and for every $\{g_n\} \in \mathcal{C}$, $\{g_n\} \leq_p \{f_n\}$.*

3   Upper Bounds for Variants of Determinant Family

In this section, we discuss the upperbounds of $\mathtt{StackDet_n}(X)$ and $\mathtt{CountDet_n}(X)$. Before going into the details of the upper bound proof, we recall the definition of the stack branching program (SBP) and the definition of the random access branching program (RABP) and the characterization of the classes VP and VNP using SBP and RABP respectively [8].

Definition 13 (SBP [8]). *A stack branching program $G = (V, E)$ (over Σ) is an algebraic branching program with a function $\phi : E \longrightarrow \bigcup_{a \in \Sigma} \{\mathtt{Push}(a), \mathtt{Pop}(a)\} \cup \{\mathtt{No\text{-}op}\}$. The polynomial computed by G is $f_G = \sum_{\mathcal{P}} mon(\mathcal{P})$, where the sum is over all the stack-realizable s-t paths in G. The size of a stack branching program G is the number of vertices in it, i.e., $|V|$*

Definition 14 (RABP [8]). *A random access branching program $G = (V, E)$ (over Σ) is an algebraic branching program with an additional function $\phi : E \longrightarrow \bigcup_{a \in \Sigma} \{\mathtt{Write}(a), \mathtt{Read}(a)\} \cup \{\mathtt{No\text{-}op}\}$. The polynomial computed by G is $f_G = \sum_{\mathcal{P}} mon(\mathcal{P})$, where the sum is over all the counter-realizable s-t paths in G. The size of a random access branching program G is the number of vertices in it.*

Lemma 1 ([8]). *A family $\{f_n\}$ is in VP if and only if there exist a stack branching program family $\mathcal{S}_n$ of size $\mathrm{poly}(n)$ to compute $\{f_n\}$. A family $\{f_n\}$ is in VNP if and only if there exist a random access branching program family $\mathcal{R}_n$ of size $\mathrm{poly}(n)$ to compute $\{f_n\}$.*

The upper bound proofs is motivated by the ABP upper bound for the Determinant polynomial proved by [6]. The determinant is known to be equal to the sum of signed clow sequences. This combinatorial definition of the determinant was used in [6] to obtain an ABP upper bound. Those familiar with the proof of [6] may notice that the definitions of $\texttt{StackDet}_n(X)$ and $\texttt{CountDet}_n(X)$ are inspired by this definition of the determinant. We observe that, just like the combinatorial definition of the determinant is used to obtain an ABP upper bound in [6], our definition of $\texttt{StackDet}_n(X)$ and $\texttt{CountDet}_n(X)$ allow us to compute them using an SBP and RABP, respectively.

Construction of an SBP Computing $\texttt{StackDet}_n(X)$: Let $G_n = (V, E, \Sigma, \Phi)$ and $\mathcal{L}$ be as in the definition of $\texttt{StackDet}_n(X)$. Consider the complete directed graph $G'_n = (V, E)$, i.e. G_n without the stack symbols and labels. Let A_n denote the adjacency matrix of this graph under the labelling $\mathcal{L}$, i.e. $A_n[i, j] = x_{i,j}$. From the result of [6], we get an ABP, say $\mathcal{B}_n$ (of size poly(n)), that computes the determinant of A_n. From $\mathcal{B}_n$ we obtain an SBP $\mathcal{S}_n$, by simply defining the function ϕ. We inherit ϕ from the Φ defined in the stack graph G_n as follows. Let B_n be the graph underlying the ABP $\mathcal{B}_n$. In B_n some edges are labelled with X variables, while some other edges are labelled with field constants. The function ϕ for all edges which are labelled with field constants is set to No-op. Consider any edge (p, q) in B_n that is labelled with an X variable. Suppose the edge is labelled $x_{i,j}$, then we let $\phi((p, q)) = \Phi((i, j))$.

The following statement can now be proved in a straightforward way, which finishes the proof of the upper bound.

Claim. Let $\widehat{C}$ be any clow sequence in G_n. The SBP $\mathcal{S}_n$ has a stack-realizable path from s to t with weight $sign(\widehat{C}) \cdot mon(\widehat{C})$ if and only if $\widehat{C}$ is a stack-realizable clow sequence of degree $|V|$.

Proof. Let us start by recalling the construction of an ABP for the determinant polynomial from [6]. First recall that the determinant polynomial $\texttt{Det}_n$ is defined as follows in [6].

$$\texttt{Det}_n(G'_n) = \sum_{\widehat{C} \text{ a clow sequence of degree } |V|} sign(\widehat{C}) mon(\widehat{C})$$

It was shown that there exist an algebraic branching program $\mathcal{B}_n$ (with s as the source vertex and t as the sink vertex and two special nodes t^+ and t^-) of size $\mathcal{O}(n^3)$ which computes $\texttt{Det}_n(G)$. The ABP $\mathcal{B}_n$ has the following properties.

- For every clow sequence $\widehat{C} = \langle C_1, C_2, \ldots, C_k \rangle$ of degree $|V|$ and positive signature, there exists a unique $s - t$ path $\mathcal{P}$ in $\mathcal{B}_n$ such that path $\mathcal{P}$ is obtained by unwinding the clows in the clow sequence $\widehat{C} = \langle C_1, C_2, \ldots, C_k \rangle$ into paths, $\mathcal{P}_1, \mathcal{P}_2, \ldots, \mathcal{P}_k$, respectively and then stitching these paths together in order $\mathcal{P}_1$ followed by $\mathcal{P}_2$ and so on upto $\mathcal{P}_k$ and then followed by a single edge $\hat{e}$ labelled by $+1$ from t^+ to t. For negative signature, it is similar; except the last edge is labelled -1 and is from t^- to t.

– The variable labels on the edges (except the last edge) in $s - t$ path $\mathcal{P}$ in $\mathcal{B}_n$ are consistent with the variable labels on the edges in the closed walks in the clow sequence $\mathcal{C}$ of G_n.
– There are no $s - t$ paths in $\mathcal{B}_n$ other than the kind of paths stated above.

As the SBP $\mathcal{S}_n$ has the same underlying graph as $\mathcal{B}_n$, i.e. $\mathcal{B}_n$, ignoring Φ, we get a bijection between clow sequences of the stack graph G_n and s to t paths in $\mathcal{B}_n$.

Stack graph $\mathcal{S}_n$ is obtained by specifying ϕ along with $\mathcal{B}_n$. Note that the set of s to t paths in $\mathcal{S}_n$ and $\mathcal{B}_n$ continue to be the same. In $\mathcal{S}_n$ some paths become stack-realizable under the function ϕ. Consider a stack-realizable path $\mathcal{P}$ in $\mathcal{S}_n$. It has a corresponding clow sequence $\widehat{\mathcal{C}}$ associated with it in G_n. As the labels of $\mathcal{P}$ are consistent with those on $\widehat{\mathcal{C}}$, we get that $\widehat{\mathcal{C}}$ is a stack-realizable clow sequence.

Conversely, if we start with a stack-realizable clow sequence of G_n, we will find an s to t stack-realizable path in $\mathcal{S}_n$. This finishes the proof. □

Remark 1. To show that $\texttt{CountDet}_n(\mathrm{X})$ is in VNP, we can construct RABP $\mathcal{R}_n$ which computes $\texttt{CountDet}_n(\mathrm{X})$ using arguments very similar as in the case of the construction of SBP $\mathcal{S}_n$ computing $\texttt{StackDet}_n(\mathrm{X})$.

4 $\texttt{StackDet}_n(\mathrm{X})$ Is Hard for VP

In this section we prove that $\texttt{StackDet}_n(X)$ is VP-hard. We start by proposing two simple approaches for proving the hardness and discuss why they do not seem to work directly.

– The first way is to mimic the construction used to show that the determinant polynomial is VBP hard. Start with a stack branching program P computing f. P has designated nodes s and t. Add an extra vertex, say α, and add edges from t to α and from α to s. Also add self-loops on all the vertices of P except α. Then do the following.
  (a) Firstly observe that the stack-realizable clow sequences of this graph can be partitioned into two sets, say $\mathcal{G}$ and $\mathcal{B}$. Prove that the clow sequences in $\mathcal{B}$ pairwise cancel each other and their weights add up to zero.
  (b) Moreover, show that the signed clow sequences in $\mathcal{G}$ are in one-to-one correspondence with the monomials of f.
  (c) Finally prove that the sum of signed clow sequences in $\mathcal{G}$ is equal to $\texttt{StackDet}_n$.
  While (a) and (b) above can be proved, (c) does not seem to be true. This is because we do not have any control over the map ϕ used in P. Note that in the definition of $\texttt{StackDet}_n$ Φ is a fixed map, whereas, in P, ϕ depends on the polynomial f. For instance, it is possible that stack symbols repeat themselves several times in ϕ, while in Φ they do not as per the definition. To obtain a graph *along with the* Φ as defined in $\texttt{StackDet}_n$ does not seem feasible in this straightforward proof idea.

– A possible fix to the above problem is to update the definition of $\mathtt{StackDet}_n$ so that it allows for a ϕ that arises from the underlying stack branching program P that computes f. Unfortunately, that leads to polynomial families that are circuit-description dependent.

We will work on the first approach above. Our proof steps consist of the additional effort required to make this approach work. We state the main three steps in the proof outline.

Step 1: Let $\mathcal{U}_m$ be a universal circuit [2,10,11] of size $\mathrm{poly}(m)$ computing an m-variate, degree $\mathrm{poly}(m)$ polynomial $f_m(Y) \in \mathsf{VP}$. We obtain *a universal block circuit* $\tilde{\mathcal{U}}_m$, which has some more structure than $\mathcal{U}_m$ and computes $f_m(Y)$.

Step 2: We take the directed graph underlying the circuit $\tilde{\mathcal{U}}_m$ and transform it into another graph G_N with N vertices, where $N = \mathrm{poly}(m)$ and $N = 4n$ for some parameter n. The graph G_N has the following properties.

– All the cycle covers of G_N have the same sign (say +ve sign w.l.o.g.).
– All the cycle covers can be classified into two categories: *good cycle covers*, say $\mathcal{G}$, and *bad cycle covers*, say $\mathcal{B}$; and the sum of weights of the good cycle covers equals $f_m(Y)$. (We will define these notions formally below.)

Step 3: From G_N we obtain a stack graph H_N with the following properties. The set of stack-realizable clow sequences in H_N which are cycle covers, equals $\mathcal{G}$. Moreover, the sum of signed weights of stack-realizable cycle covers in H_N equals the sum of signed weights of cycle covers in $\mathcal{G}$, i.e. equal to $f_m(Y)$. and the sum of signed weights of stack-realizable clow sequences that are not cycle covers equals 0. Overall, the sum of signed weights of stack-realizable clow sequences of H_N equals $f_m(Y)$.

We can now interpret H_N as a complete graph, where $\mathcal{L}((i,j)) = 0$ if (i,j) is not an edge in H_N. We will show that the polynomial $\mathtt{StackDet}_n$ defined with respect to H_N equals $f_m(Y)$.

The Step 1 and 2 above are obtained using the ideas of Block Trees from [1]. Step 3 above uses the cancellation trick from [6], but now in the context of stack-realizable clow sequences (instead of clow sequences) and with respect to an SBP (instead of an ABP).

4.1 VP-Hardness of $\mathtt{StackDet}_n(X)$ Step 1

Recall that from the constructions in [2,10,11], we can assume the following properties about the universal circuit. The circuit $\mathcal{U}_m$ has m variables, size $s(m)$ and each even layer is a $+$ gate, while each odd layer is a $\times$ gate. The output gate is a $\times$ gate. The $\times$ gates are multiplicatively disjoint[6] and have fan-in bounded by 2. The input gates have fanin 0, fanout 1. The total depth[7] of the circuit

---

[6] A multiplication gate α with children gates α_ℓ and α_r in an arithmetic circuit C is called multiplicatively disjoint if the subcircuits rooted at α_ℓ and α_r are disjoint.

[7] The depth of the circuit is the length of the longest input gate to output gate path.

is $2c\lceil \log m \rceil + 1$, where c is some fixed constant. Say it computes a polynomial $f_m(Y)$ of degree $\text{poly}(m)$.[8]

We now create a circuit $\tilde{\mathcal{U}}_m$, which will have the same depth, each even layer will again consist of $+$ gates and each odd layer of $\times$ gates. It will continue to be multiplicatively disjoint and its size will be $\text{poly}(s(m))$. It is created as follows:

Block Structure. In the jth layer of $\tilde{\mathcal{U}}_m$ we create $t(j) = 2^{\lfloor \frac{j}{2} \rfloor}$ many blocks. The blocks on the j layer are denoted by $B_1^{(j)}, B_2^{(j)}, \ldots, B_{t(j)}^{(j)}$.

Gates. If j is odd - Let $g_1, \ldots, g_r$ be the $\times$ gates appearing in $\mathcal{U}_m$ in layer j. In $\tilde{\mathcal{U}}_m$, each block B has one copy of $g_1, \ldots g_r$.

If j is even - Let $g_1, \ldots, g_r$ be the $+$ gates appearing in $\mathcal{U}_m$ in layer j. Each block B in jth layer in $\tilde{\mathcal{U}}_m$ has $s(m)$ sub-blocks. Each sub-block has one copy of $g_1, \ldots, g_r$. (That is, there are $s(m)$ copies of each gate in each block and there are $t(j)$ many blocks. So each gate is copied $t(j) \cdot s(m)$ times. Note that this is polynomially bounded in $\text{poly}(m)$.)

Wires: Let g be a $+$ gate in layer j with children $g_1, g_2, \ldots g_r$ in $\mathcal{U}_m$. Then the copy of g in $B_i^{(j)}$ has copies of $g_1, \ldots, g_r$ from block $B_i^{(j+1)}$ as its children for each $i \in t(j)$. Let g be a $\times$ gate in layer j with children $g_{\text{left}}, g_{\text{right}}$ in $\mathcal{U}_m$. Also among the different gates that use g_{left}, let g be the kth such gate. Then (the unique) copy of g in $B_i^{(j)}$ has kth copy of g_{left} from block $B_{2i-1}^{(j+1)}$ as its child. Similarly, among the gates that use g_{right}, let g be the k'th such gate. Then the copy of g in $B_i^{(j)}$ has k'th copy of g_{right} from block $B_{2i}^{(j+1)}$ as its child. Finally, we only keep the minimal circuit, i.e. we remove gates that eventually do not feed into the output gate.

This completes the description of $\tilde{\mathcal{U}}_m$. The construction is exactly the same as the construction of D'_n in [1]. We call this the universal block circuit.

Claim. The polynomial computed by $\tilde{\mathcal{U}}_m$ is $f_m(Y)$ and the size of the circuit is polynomial in $s(m)$, say $p(m)$, which in turn is polynomial in m.

We skip the proof details (See [9] for proof details) (Fig. 1).

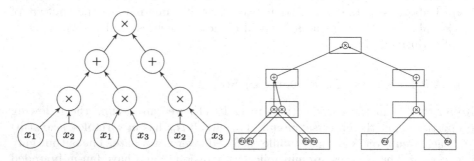

Fig. 1. $\mathcal{U}_m$ computing $(x_1 x_2 + x_1 x_3)(x_2 x_3)$ and the corresponding $\tilde{\mathcal{U}}_m$

---

[8] This description is slightly different as compared to the one in [2], but it is easy to see that we can get this form for a universal circuit using ideas from [10].

4.2   VP-Hardness of StackDet$_n(X)$ Step 2

We now consider the graph underlying the universal block circuit created in Step 1. We direct all the edges in this graph from top (i.e. from the output gate) to bottom (to the input gates). The top-most layer has a single vertex, which is the output gate. Each layer j has $t(j)$-many blocks. We denote this directed graph by $V_{p(m)}$, where $p(m)$ is the number of vertices in this graph. We take two views of this underlying graph; a *coarse* view and a *fine* view. The fine view is simply the whole graph $V_{p(m)}$, while the coarse view is the graph formed by the block structure.

Block Tree. For the coarse view, we think of each block of $V_{p(m)}$ as a vertex. We call these *block vertices*. Two blocks vertices B, B' are said to be connected if and only if $\exists u \in B$ and $v \in B'$ such that there is an edge between u and v in $V_{p(m)}$. We refer to (B, B') as a *block edge*. By observing the connections in $V_{p(m)}$, it is easy to see that the coarse view results into a tree. We call this tree $T_{\Delta(m)}$, where $\Delta(m)$ denotes the number of leaf nodes in the tree. Let B be a vertex in T_Δ. If B is on an even layer, then it has only one child. We call these the *unary blocks*. If it is on an odd layer then it has two children. We call these blocks *binary blocks*. A path formed by block edges is called a *block path*. When m is clear from the context, we use V_p and T_Δ to talk about these two graphs.

Construction of G$_N$.

- For any binary block B and any vertex $u \in B$, we do the following. Let B_ℓ and B_r be the two children of B in T_Δ. Let $u_\ell \in B_\ell$ and $u_r \in B_r$ such that (u, u_ℓ) and (u, u_r) are edges in V_p. We sub-divide the edge (u, u_r) into (u, z_u) and (z_u, u_r). We delete the edge (u, z_u) from the graph, but retain the edge (z_u, u_r). For any node u in a binary block, we use Couple(u) to denote the pair of edges $\{(u, u_\ell), (z_u, u_r)\}$. (Couple$(u)$ is not defined for a u in a unary block.)

  Note that this creates a new graph which is disconnected. If we look at the coarse view of this new graph then it is a collection of Δ block paths, let us call them $\mathcal{P}_1, \ldots, \mathcal{P}_\Delta$. Each block path contains exactly one leaf node of T_Δ. We will assume that the block paths are numbered such that the ith leaf node of T_Δ belongs to $\mathcal{P}_i$.

- We add two more vertices for each block path. We add a source vertex s_i and a sink vertex t_i for each $i \in [\Delta]$. We also add edges from s_i to all the vertices in the first block in the block path $\mathcal{P}_i$. The vertices in the last block in any block path are vertices corresponding to input gates in $\tilde{\mathcal{U}}_m$ and hence are labelled with input variables Y. Let u be a vertex in the leaf block of the path $\mathcal{P}_i$ labelled $y \in Y$. We add a directed edge (u, t_i) and label it with y. (We do this for each vertex in every leaf block of all block paths.) The graphs thus obtained are called $\mathcal{R}_1, \ldots, \mathcal{R}_\Delta$.

- We now identify t_i with s_{i+1} for $1 \le i \le \Delta - 1$. We use $\mathcal{R}$ to denote the graph thus formed and θ_i to denote the vertex formed by identifying t_i with s_{i+1} for $1 \le i \le \Delta - 1$. Addtionally, we want to ensure that the number of vertices in the resultant graph is a multiple of 4 (This will help in defining a stack graph in the next step). To ensure this, we add three[9] additional vertices $\alpha_1, \alpha_2, \alpha_3$ and the following directed edges to obtain a graph D_N: $(t_\Delta, \alpha_3), (\alpha_3, \alpha_2), (\alpha_2, \alpha_1), (\alpha_1, s_1)$. We add self-loops on all the vertices except on α_1, α_2 and α_3. The edges which are not labelled with variables from Y are labelled 1 (Fig. 2).

The graph thus obtained is denoted by G_N, where N is the number of vertices in it. It is easy to note that $N = \text{poly}(p(m))$ which is $\text{poly}(m)$. We have also ensured that $N = 4n$ for some parameter n.

Definition 15. *We say that a cycle cover $\widehat{C} = \langle C_1, \ldots, C_k \rangle$ of G_N is a good cycle cover if for any vertex u appearing in $\widehat{C}$ for which $\texttt{Couple}(u)$ is defined, either both the edges in $\texttt{Couple}(u)$ are present in $\widehat{C}$ or neither is. All the other cycle covers are called bad cycle covers. Let $\mathcal{G}$ denote the set of all good cycle covers of G_N and $\mathcal{B}$ denote the set of all the bad cycle covers.*

Claim. All the cycle covers of G_N have the same sign. Moreover, the sum of weights of good cycle covers equals $f_m(Y)$.

Proof. Recall the graphs $\mathcal{R}_1, \ldots, \mathcal{R}_\Delta$ that we created from $\mathcal{P}_1, \ldots, \mathcal{P}_\Delta$.

Consider any path π from s_i to t_i in $\mathcal{R}_i$. The first edge of π must be from s_i to a vertex belonging to the first block, and the last edge of π must be from a vertex belonging to the last block to the vertex t_i. All intermediate edges must connect adjacent blocks. So, the number of edges in π is one more than the number of blocks in $\mathcal{R}_i$. Therefore all paths from s_i to t_i in $\mathcal{R}_i$ have the same number of edges, say p_i.

Consider any path Π from s_1 to t_Δ. For any $2 \le i \le \Delta$, the vertex s_i must belong to Π (because deleting s_i disconnects the graph into two components, where s_1 and t_Δ belong to different components). This means Π can be viewed as a composition of the paths $\pi_1, \pi_2, \ldots, \pi_\Delta$, where π_i is a path from s_i to t_i for all $1 \le i \le \Delta$. This path π_i is also a path in $\mathcal{R}_i$, so it has length p_i. Therefore the path Π has length $p_1 + p_2 + \cdots + p_\Delta$, which we call q, say. In all, any path from s_1 to t_Δ has the same length q.

Let $\widehat{C} = \langle C_1, C_2, \cdots, C_k \rangle$ be a cycle cover of G_N, and consider a cycle of the cycle cover $\widehat{C}$ that α_1 belongs to, say C_1. The only incoming edge to α_1 is via t_Δ, and the only edge outgoing from α_1 is to α_2. This means the edges (t_Δ, α_1) and (α_1, α_2) belong to C_1. The only outgoing edge from α_2 is to α_3, and the only outgoing edge from α_3 is to s_1. Therefore, the edges (α_2, α_3) and (α_3, s_1) also belong to C_1. So, C_1 contains a path from t_Δ to s_1 via α_1, α_2 and α_3. The remaining part of C_1 is a path from s_1 to t_Δ. This path does not use the vertices α_1, α_2, and α_3, so it is also a path in $\mathcal{R}$. As shown before, any such path from s_1 to t_Δ has length $q = p_1 + p_2 + \cdots + p_\Delta$. Therefore C_1 is a cycle of length $q + 4$.

Consider a cycle $C_j \ne C_1$ in the cycle cover $\widehat{C}$. This cycle cannot use the vertices α_1, α_2 and α_3. Furthermore, if C_j is not a loop, then it is a cycle in $\mathcal{R}$,

---

[9] As Δ is a power of 2, it is easy to note that adding three new vertices will always make the total number of vertices of graph G_N a multiple of 4.

which contradicts the fact that $\mathcal{R}$ is a DAG. Therefore $\mathcal{C}_j$ is a loop. In all, the cycle cover $\widehat{\mathcal{C}}$ has exactly one cycle $\mathcal{C}_1$ of length $q+4$ passing through α_1, α_2, and α_3, and $N-q-4$ loops covering the vertices not present in the cycle $\mathcal{C}_1$. Either way, the sign of any cycle cover $\widehat{\mathcal{C}}$ is fixed. It is also easy to see from the above discussion that there is a one-to-one correspondence between a path Π from s_1 to t_Δ in $\mathcal{R}$ and cycles covers of G_N.

We will now show that the good cycle covers of G_N have a one-to-one correspondence with the parse trees of $\widetilde{\mathcal{U}}_m$.

Let $\mathcal{T}$ be any parse tree of $\widetilde{\mathcal{U}}_m$. For any vertex u corresponding to a $\times$ gate of $\widetilde{\mathcal{U}}_m$, such that u_l is the left child and u_r is the right child of u in $\mathcal{T}$, split the edge (u, u_r) into (u, z_u) and (z_u, u_r) and delete edge (u, z_u). This splits $\mathcal{T}$ into Δ paths $Q_1, Q_2, \cdots, Q_\Delta$, where Q_i belongs to $\mathcal{R}_i$ for each $i \in [\Delta]$. These Δ paths (when concatenated appropriately) trace out a path Π in D_N. This path can be completed into a cycle $\mathcal{C}_1$ in G_N. This cycle $\mathcal{C}_1$ along with self-loops on all the other vertices outside of $\mathcal{C}_1$, forms a cycle cover $\widehat{\mathcal{C}}$ of G_N. Note that, the way this cycle cover was created, for each u in a binary block of V_p, either both edges of Couple(u) are present in $\widehat{\mathcal{C}}$ or neither edge of Couple(u) is present in $\widehat{\mathcal{C}}$. Therefore $\widehat{\mathcal{C}}$ is a good cycle cover. It is easy to see that the cycle cover has weight equal to the monomial computed by $\mathcal{T}$ in $\widetilde{\mathcal{U}}_m$.

For the converse, we show that a good cycle cover of G_N can be traced back to a unique parse tree of $\widetilde{\mathcal{U}}_m$. Let $\widehat{\mathcal{C}} = \langle \mathcal{C}_1, \mathcal{C}_2, \cdots, \mathcal{C}_k \rangle$ be a good cycle cover of G_N. Let $\mathcal{C}_1$ be the big cycle and the rest of the cycles in the cover be self-loops. Let E_1 denote the edges that $\mathcal{C}_1$ shares with graphs $\mathcal{P}_1, \mathcal{P}_2, \ldots, \mathcal{P}_\Delta$. As this is a good cycle cover, for each vertex u in $\mathcal{C}_1$ for which Couple(u) is defined, edges (u, u_ℓ) and (z_u, u_r) are both present in $\mathcal{C}_1$. We will identify z_u with u for all such vertices. This will give rise to a unique parse tree of $\mathcal{U}_m$. $\qquad \square$

Remark 2. To be able to sum over only the good cycle covers, we need a mechanism to ensure that either both or none of the edges in Couple(u) are selected in any cycle cover $\widehat{\mathcal{C}}$. In Valiant's work [12] for instance, this is ensured by using an *iff graph gadget*. If we can come up with such a gadget (in the determinant setting) then we will be able to show that Det$_n$ is VP-complete, thereby showing VP= VBP. We ensure *coupling* using the stack symbols.

4.3 VP-Hardness of StackDet$_n(X)$ Step 3

We would like to modify the graph G_N so that we filter out good cycle covers, while killing all the bad cycle covers. We achieve this using stack symbols. Specifically, we create a stack graph H_N from G_N to achieve this.

Construction of H_N. For a vertex u for which Couple(u) is defined, we set $\phi((u, u_\ell)) = $ Push(s_u) and $\phi((z_u, u_r)) = $ Pop(s_u). For all the other edges, ϕ is set to No-op.

Claim. Consider the stack graph H_N constructed as above.

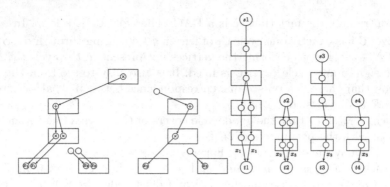

Fig. 2. Graphs $\mathcal{P}_1, \ldots, \mathcal{P}_\Delta$ and $\mathcal{R}_1, \ldots, \mathcal{R}_\Delta$.

- The sum of signed weights of stack-realizable cycle covers in H_N equals the signed sum of weights of cycle covers in $\mathcal{G}$, i.e. equal to $f_m(Y)$.
- The sum of signed weights of stack-realizable clow sequences that are not cycle covers equals 0.

Proof. **Part 1.** From the proof of Claim in Sect. 4.2, we have that there is a bijection between parse trees of $\tilde{\mathcal{U}}_m$ and good cycle covers of G_N. To prove the first part of the claim, we will show that there is a bijective map from a good cycle covers of G_N to stack-realizable cycle covers of H_N.

We start with some notations. Let $\widehat{\mathcal{C}}$ be a good cycle cover in G_N. Let $\mathcal{T}_{\widehat{\mathcal{C}}}$ be the unique parse tree corresponding to $\widehat{\mathcal{C}}$. Let $\widehat{\mathcal{C}} = \langle \mathcal{C}_1, \ldots, \mathcal{C}_k \rangle$ and $\mathcal{C}_1$ be the long cycle, while all other $\mathcal{C}_i$s be self-loops. (Any good cycle cover has this structure as we established in the proof of Claim in Sect. 4.2.) Let $U_{\widehat{\mathcal{C}}} = \{u_1, \ldots, u_\tau\}$ be the subset of vertices in $\mathcal{C}_1$ for which **Couple** is defined. Note that the output gate, let us call it u^*, of $\mathcal{T}_{\widehat{\mathcal{C}}}$ belongs to $U_{\widehat{\mathcal{C}}}$.

We say that a vertex $u \in U_{\widehat{\mathcal{C}}}$ has rank k, denoted as $\mathtt{rank}(u)$, if it appears at distance $2k - 1$ from the leaves in $\mathcal{T}_{\widehat{\mathcal{C}}}$. (Note that, vertices in $U_{\widehat{\mathcal{C}}}$ appear at only odd distance from the leaves in $\mathcal{T}_{\widehat{\mathcal{C}}}$.)

For $u \in U_{\widehat{\mathcal{C}}}$ such that $\mathtt{rank}(u) = 1$, u_ℓ and u_r are leaves, i.e. nodes corresponding to input gates. For a vertex u in $U_{\widehat{\mathcal{C}}}$ such that $\mathtt{rank}(u) > 1$, let u_ℓ and u_r be its two children in $\mathcal{T}_{\widehat{\mathcal{C}}}$. Let u' be u_ℓ's unique child in $\mathcal{T}_{\widehat{\mathcal{C}}}$ and let u'' be the unique child of u_r in $\mathcal{T}_{\widehat{\mathcal{C}}}$. Note that $u', u'' \in U_{\widehat{\mathcal{C}}}$ and $\mathtt{rank}(u') = \mathtt{rank}(u'') = \mathtt{rank}(u) - 1$.

Let $\Pi_{\widehat{\mathcal{C}}}$ be the unique path traced out by $\mathcal{C}_1$ in $\mathcal{R}$. (Recall, $\mathcal{R}$ is the graph obtained by concatenating $\mathcal{R}_i$ for $i \in [\Delta]$ as described in the construction.)

For a vertex $u \in U_{\widehat{\mathcal{C}}}$, such that $\mathtt{rank}(u) = 1$, we use $\Pi_{[u]}$ to denote the subpath of $\Pi_{\widehat{\mathcal{C}}}$ from u to u_r. Given the structure of the subtree rooted at u in $\mathcal{T}_{\widehat{\mathcal{C}}}$, and assuming that u_r appears in $\mathcal{R}_{i+1}$ for some $i \in [\Delta - 1]$, we get that $\Pi_{[u]} = (u, u_\ell) \cdot (u_\ell, \theta_i) \cdot (\theta_i, z_u) \cdot (z_u, u_r)$. (Recall that θ_i was the vertex obtained by identifying t_i of $\mathcal{R}_i$ with s_{i+1} of $\mathcal{R}_{i+1}$ for $i \in [\Delta - 1]$.)

On the other hand, for $u \in U_{\widehat{\mathcal{C}}}$ and $\mathtt{rank}(u) > 1$ such that u_r appears in $\mathcal{R}_{i+1}$ for some $i \in [\Delta - 1]$, we use $\Pi_{[u]}$ to denote the subpath of Π corresponding to the entire subtree rooted at u in $\mathcal{T}_{\widehat{\mathcal{C}}}$. Specifically, for the given the structure of the subtree rooted at u in $\mathcal{T}_{\widehat{C}}$, $\Pi_{[u]} = (u, u_\ell) \cdot (u_\ell, u') \cdot \Pi_{[u']} \cdot (\theta_i, z_u) \cdot (z_u, u_r) \cdot (u_r, u'') \cdot \Pi_{[u'']}$. We will now prove the following statement.

$$\text{For any } u \in U_{\widehat{C}}, Seq[\Pi_{[u]}] \text{ is stack-realizable in } H_N. \tag{1}$$

If we are able to show this, then in particular for $u^* \in U_{\widehat{C}}$ we will get that $\Pi_{[u^*]}$ is stack-realizable. This will then imply that $(s_1, u^*) \cdot \Pi_{[u^*]} \cdot (\theta_\Delta, t_\Delta)$ is also stack-realizable, because both (s_1, u^*) and $(\theta_\Delta, t_\Delta)$ are No-op edges.

We prove (1) by induction on $\text{rank}(u)$. Suppose $\text{rank}(u) = 1$ and say $u_r \in \mathcal{R}_{i+1}$, then as noted above, $\Pi_{[u]} = (u, u_\ell) \cdot (u_\ell, \theta_i) \cdot (\theta_i, z_u) \cdot (z_u, u_r)$. From our function ϕ defined for H_N, we see that $Seq[\Pi_{[u]}] = \text{Push}(s_u)\square\text{No-op}\square\text{No-op}\square\text{Pop}(s_u)$. Therefore it is stack-realizable.

Suppose $\text{rank}(u) = k > 1$ and say that $u_r \in \mathcal{R}_{i+1}$. In this case, as noted above, we have $\Pi_{[u]} = (u, u_\ell) \cdot (u_\ell, u') \cdot \Pi_{[u']} \cdot (\theta_i, z_u) \cdot (z_u, u_r) \cdot (u_r, u'') \cdot \Pi_{[u'']}$. From this, we see that $Seq[\Pi_{[u]}] = \text{Push}(s_u)\square\text{No-op}\square Seq[\Pi_{[u']}] \square\text{No-op}\square\text{Pop}(s_u)\square\text{No-op}\square Seq[\Pi_{[u'']}]$. As $\text{rank}(u'), \text{rank}(u'') < k$, by induction hypothesis we have that $Seq[\Pi_{[u']}]$ and $Seq[\Pi_{[u'']}]$ are stack-realizable. Therefore, we get that $Seq[\Pi_{[u]}]$ is also stack-realizable.

It is not hard to argue that bad cycle covers of G_N get mapped to cycle covers of H_N, which are not stack-realizable by a similar argument.

Part 2. Recall that in the proof of Claim in Sect. 4.2 we showed that any cycle cover of G_N consists of one big cycle and a collection of self-loops. Similarly, it is easy to see that in H_N any clow sequence has a certain structure: except for one clow, which will be of length $\geq p + 4$, all other clows in the clow sequence are self-loops.

We first note that this unique long clow will contain the vertex α_1. Suppose it does not contain α_1, then no other clow in the clow sequence can cover α_1 (as all other clows are self-loops and α_1 does not have a self-loop on it). But suppose α_1 is not covered by any clow in the clow sequence, then the degree of such a clow sequence is strictly less than $|V|$.

Under the ordering in which vertex α_1 gets the lowest number, say 1, the long clow will be the first clow in the sequence, say $\mathcal{C}_1$ and α_1 will be its head.

We will adopt ideas from [6] in order to argue that the sum of weights of stack-realizable clow sequences which are not cycle covers is 0 in H_N. Like in [6], we define an involution on the signed clow sequences. (Recall that an involution is a bijective map ψ such that ψ^2 is identity.) The map ψ will have the property that any stack-realizable clow sequence $\widehat{C}$ which is not a cycle cover, is paired off with another stack-realizable clow sequence $\widehat{C}'$ which is again not a cycle cover and the monomials corresponding to $\widehat{C}$ and $\widehat{C}'$ are the same, but $sign(\widehat{C}') = -sign(\widehat{C})$. For clow sequence $\widehat{C}$ which is a cycle cover, the map ψ maps it to itself, i.e. it is identity for cycle covers.

Let $\widehat{C}$ be a stack-realizable clow sequence, which is not a cycle cover. We start walking along the edges of $\mathcal{C}_1$ starting from the head. One of the following two cases will happen first.

- **Case 1.** Either we will encounter a vertex v in $\mathcal{C}_1$ such that there exists a $\mathcal{C}_i \in \widehat{C}$ for $i > 1$, such that $\mathcal{C}_i$ is a self-loop at vertex v.

– **Case 2.** Or we will encounter a vertex u that has $\beta \geq 1$ self-loops in C_1.

First note that, if $\widehat{C}$ is not a cycle cover then one of the two cases must occur.

Suppose Case 1 occurs. In this case, consider $\widehat{C}'$ obtained from $\widehat{C}$ by merging cycle C_i with C_1, by attaching it at v in C_1. We will define ψ of $\widehat{C}$ to be this $\widehat{C}'$. It is easy to see that if $\widehat{C}$ is a stack-realizable clow sequence, then so is $\widehat{C}'$. Both have the same set of edges. And $\widehat{C}'$ has one less component than $\widehat{C}$, i.e. their signs are opposite.

On the other hand, suppose Case 2 occurs. In this case, consider $\widehat{C}'$ obtained from $\widehat{C}$ by detaching one of the β-many self-loops from u and adding that as a separate cycle in $\widehat{C}'$. To observe that $\widehat{C}'$ thus obtained is a stack-realizable clow sequence, we first note that there is no other clow in $\widehat{C}'$ with the same head as this newly added self-loop. This is easy to see, because if say there was already a clow in $\widehat{C}'$ with u as its head, then we would have been in Case 1 above.

We also observe that if $\widehat{C}$ is stack-realizable, then detaching a self-loop, which is a No-op edge, will ensure that $\widehat{C}'$ is also stack-realizable. Here again, $\widehat{C}$ and $\widehat{C}'$ have the same set of edges and $\widehat{C}$ has one less component than $\widehat{C}'$, i.e. they have opposite signs.

Note that in both the cases above, if $\psi(\widehat{C}) = \widehat{C}'$ then $\psi(\widehat{C}') = \widehat{C}$. Hence, we have the desired involution.   □

If we now show an ordering of the vertices of graph H_N such that the function ϕ in graph H_N can be exactly mapped to the function Φ as defined in Definition 8 then the VP-hardness immediately follows. It is easy to show that such an ordering always exist.

Ordering of the Vertices of Graph H_N. We will now show that there is an ordering of the vertices of H_N, which gives a graph $G_n = (V, E, \Sigma, \Phi)$ as defined in Definition 8 and a labelling function $\mathcal{L}$ as defined in Definition 8, such that $f_m(Y)$ can be obtained as a projection of $\texttt{StackDet}_n(X)$ defined with respect to G_n, which finishes the proof.

We now come up with such an ordering. We start by ordering vertices $\theta_1, \ldots, \theta_{\Delta-1}$ and α_1, α_2 and α_3. Note that these vertices must appear in any cycle, which is not a self-loop. If we start traversing any such cycle from α_1, then we will visit these vertices in the following order $\langle \alpha_1, s_1, \theta_1, \ldots, \theta_{\Delta-1}, t_\Delta, \alpha_3, \alpha_2, \alpha_1 \rangle$. We number these vertices in the reverse order, i.e. α_1 gets numbered 1, α_2 gets 2, α_3 is numbered 3, t_Δ is numbered 4 and so on till s_1 is numbered $\Delta + 4$. This numbering ensures that all the edges that appear between these vertices get No-op label on them.

Now, let $u_1, u_2 \ldots, u_\tau$ be the vertices for which Couple is defined. Let $\texttt{Couple}(u_i) = \{(u_i, u_{i\ell}), (z_{u_i}, u_{ir})\}$. For every $i \in [\tau]$, let the four vertices $u_i, u_{i\ell}, z_{u_i}, u_{ir}$ be numbered as $4(i-1) + 1 + (\Delta + 4)$, $4(i-1) + 2 + (\Delta + 4)$, $4(i-1) + 3 + (\Delta + 4)$ and $4(i-1) + 4 + (\Delta + 4)$ respectively[10]. It is easy to check that such an ordering always gives distinct numbers to all the vertices of the graph and this ordering is consistent with Φ from Definition 8.

---

[10] It is not too hard to see that $\Delta + 4$ is a multiple of 4. (As Δ is a power of 2.).

The labelling function $\mathcal{L}$ retains the labels of all the edges of H_N as they are. For any two vertices u, v in H_N, such that there is no edge in H_N between u and v, we add such an edge in G_n, but set $\mathcal{L}((u, v)) = 0$. This labelling function now ensures that when we consider the $\texttt{StackDet}_n$ polynomial with respect to G_n we obtain $f_m(Y)$.

5 VNP-Hardness of the $\texttt{CountDet}_n(X)$

In this section we first show that $\texttt{CountDet}_n(X)$ is hard for VNP. We show the VNP-hardness in two cases[11]. We will first show that the permanent polynomial[12] can be computed as a projection of $\texttt{CountDet}_n(X)$. This will prove that $\texttt{CountDet}_n(X)$ is VNP-hard over fields of characteristic $\neq 2$. To show it's hardness over fields of characteristic 2, we will show that it can compute another polynomial, namely $\texttt{EC}_m^*$, as a projection, where $n = \text{poly}(m)$. This polynomial was shown to be VNP-complete over fields of characteristic 2 in [4].

5.1 Details Regarding VNP-Hardness of $\texttt{CountDet}_n(X)$ When Char $\neq 2$

We will first show that the Permanent polynomial can be computed as a projection of $\texttt{CountDet}_n(X)$. This will prove that $\texttt{CountDet}_n(X)$ is VNP-hard over fields of char $\neq 2$.

Let $Y = \{y_{1,1}, y_{1,2}, \ldots, y_{m,m}\}$. We will show that $\texttt{Perm}_m(Y)$ can be obtained as a projection of $\texttt{CountDet}_n(X)$, where $n = \text{poly}(m)$. To prove this, we create a counter graph H_N, such that $N = \text{poly}(m)$ and the following properties hold.

– All the counter-realizable cycle covers in H_N have the same sign.
– Moreover, the sum of the weights of the counter-realizable clow sequences which are cycle covers, equals $\texttt{Perm}_m$ and the sum of the signed weights of the counter-realizable clow sequences which are not cycle covers $= 0$.

Then by simple re-ordering of the vertices of H_N and adding edges to make it a complete graph G_n, as in the definition of $\texttt{CountDet}_n(X)$, we get that $\texttt{Perm}_m(Y)$ can be obtained as a projection of $\texttt{CountDet}_n(X)$.

In order to describe the construction of H_N, we first create $2m$ smaller counter graphs, $W_1, \ldots, W_m$ and $R_1, \ldots, R_m$. For each $i \in [m]$, $W_i = (V_i^w, E_i^w, \Sigma_i^w, \phi_i^w)$ is as follows.

---

[11] Note that the VNP-hardness can also possibly be shown in a single step via constructing a RABP $\mathcal{R}_n$ which computes a polynomial family $\mathcal{P}_n$ (known to be hard for VNP for all fields) and converting it to a counter graph G_n such that $\texttt{CountDet}_n$ defined over G_n computes $\mathcal{P}_n$. However, constructing an RABP $\mathcal{R}_n$ (and converting it then to a counter graph G_n) such that the function ϕ in G_n getting exactly mapped to the function Φ defined in Definition 10 is not immediate. Therefore, we show the hardness in two steps.

[12] Recall that $\texttt{Perm}_m(Y) = \sum_{\sigma:\text{ permutation of } [m]} \prod_{i \in [m]} y_{i,\sigma(i)}$.

- $V_i^w = \{s_i^w, t_i^w\} \cup \{u_{i,1}, \ldots, u_{i,m}\} \cup \{v_{i,1}, \ldots, v_{i,m}\}$. $E_i^w = \bigcup_{j \in [m]} \{(s_i^w, u_{i,j})\} \cup \bigcup_{j \in [m]} \{(v_{i,j}, t_i^w)\} \cup \bigcup_{j \in [m]} \{(u_{i,j}, v_{i,j})\}$. $\Sigma_i^w = \{\alpha_{i,1}, \alpha_{i,2}, \ldots, \alpha_{i,m}\}$.
- For each $j \in [m]$, $\phi_i^w((u_{i,j}, v_{i,j})) = \texttt{Write}(\alpha_{i,j})$. ϕ_i^w is No-op for all other edges in E_i^w.

Similarly, for each $i \in [m]$, $R_i = (V_i^r, E_i^r, \Sigma_i^r, \phi_i^r)$ can be described as follows.

- $V_i^r = \{s_i^r, t_i^r\} \cup \{a_{i,1}, \ldots, a_{i,m}\} \cup \{b_{i,1}, \ldots, b_{i,m}\}$. $E_i^r = \bigcup_{j \in [m]} \{(s_i^r, a_{i,j})\} \cup \bigcup_{j \in [m]} \{(b_{i,j}, t_i^r)\} \cup \bigcup_{j \in [m]} \{(a_{i,j}, b_{i,j})\}$. $\Sigma_i^r = \{\alpha_{1,i}, \alpha_{2,i}, \ldots, \alpha_{m,i}\}$. For each $j \in [m]$, $\phi_i^r((a_{i,j}, b_{i,j})) = \texttt{Read}(\alpha_{j,i})$. ϕ_i^r is No-op for all other edges in E_i^r.

Let H_N' be the graph formed by identifying t_i^w with s_{i+1}^w for $1 \le i \le m-1$ and by identifying t_m^w with s_1^r and also identifying t_i^r with s_{i+1}^r for $1 \le i \le m-1$. We also add labels on the edges of H_N'. We define $\mathcal{L}((u_{i,j}, v_{i,j})) = y_{i,j}$ for $i, j \in [m]$. For all other edges, $\mathcal{L}$ is set to 1. We first make the following observation about H_N'.

Claim. For each monomial in $\mathcal{M}$ in $\texttt{Perm}_m(Y)$, there is a unique counter-realizable path π from s_1^w to t_m^r in H_N' such that $\prod_{e \in \pi} \mathcal{L}(e) = \mathcal{M}$.
For any counter-realizable path π from s_1^w to t_m^r in H_N', $\prod_{e \in \pi} \mathcal{L}(e)$ corresponds to a unique monomial of $\texttt{Perm}_m(Y)$.

Proof. From our construction of H_N', it is easy to see that the vertices $t_1^w, t_2^w, \ldots, t_m^w$ and the vertices $t_1^r, t_2^r, \ldots, t_{m-1}^r$ are all cut-vertices in H_N', and deleting any one of them disconnects the vertices s_1^w and t_m^r. This means any path π from s_1^w to t_m^r passes through the vertices t_i^w for $1 \le i \le m$ and t_i^r for $1 \le i \le m-1$. Therefore, π can be viewed as a composition of the $2m$ paths $\pi_1^w, \pi_2^w, \ldots, \pi_m^w, \pi_1^r, \pi_2^r, \ldots, \pi_m^r$ in that order, where π_i^w is the subpath of π between s_i^w and t_i^w for $1 \le i \le m$, and π_i^r is the subpath of π between s_i^r and t_i^r for $1 \le i \le m$. In fact, for any such $2m$ paths, their composition (in that order) is a path from s_1^w to t_m^r in H_N'.

We now proceed with the proof of the claim. Any monomial $\mathcal{M}$ in $\texttt{Perm}_m(Y)$ is of the form $\prod_{i=1}^m y_{i,\sigma(i)}$, where σ is a permutation of $[m]$. The path π is constructed as follows: take π_i^w to be the path $s_i^w, u_{i,\sigma(i)}, v_{i,\sigma(i)}, t_i^w$, and π_i^r to be the path $s_i^r, u_{i,\sigma^{-1}(i)}^r, v_{i,\sigma^{-1}(i)}^r, t_i^r$, both for $1 \le i \le m$. Consider the sequence of counter operations along π, other than the No-op operations. The only edges that have such operations are the edges $(u_{i,\sigma(i)}^w, v_{i,\sigma(i)}^w)$ for $1 \le i \le m$ and $(u_{i,\sigma^{-1}(i)}^r, v_{i,\sigma^{-1}(i)}^r)$ for $1 \le i \le m$. This implies that the counter operations encountered in π are $\texttt{Write}(\alpha_{1,\sigma(1)}), \texttt{Write}(\alpha_{2,\sigma(2)}), \ldots, \texttt{Write}(\alpha_{m,\sigma(m)})$, $\texttt{Read}(\alpha_{\sigma^{-1}(1),1}), \texttt{Read}(\alpha_{\sigma^{-1}(2),2}), \ldots, \texttt{Read}(\alpha_{\sigma^{-1}(m),m})$ in that order. Now, σ is a permutation of $[m]$, so the pairs $(\sigma^{-1}(j), j)$ for $1 \le j \le m$ are a permutation of the pairs $(j, \sigma(j))$ for $1 \le j \le m$. Therefore, the m symbols read are exactly the m symbols written, possibly in a different order. Since the write operations all come before the read operations, this sequence of counter operations is indeed a counter-realizable sequence. Moreover, the only edges that have labels other than 1 are the edges $(u_{i,\sigma(i)}^w, v_{i,\sigma(i)}^w)$ for $1 \le i \le m$, and these edges have labels $y_{i,\sigma(i)}$. Therefore, π is a counter-realizable path from s_1^w to t_m^r in H_N' computing the monomial $\prod_{e \in \pi} \mathcal{L}(e) = \prod_{i=1}^m y_{i,\sigma(i)} = \mathcal{M}$.

Conversely, let π be a counter-realizable path from s_1^w to t_m^r in H_N'. For each $1 \leq i \leq m$, the path π_i^w is a path from s_i^w to t_i^w. Any such path clearly is of the form $s_i^w, u_{i,f_i}^w, v_{i,f_i}^w, t_i^w$ for some $1 \leq f_i \leq m$. Similarly, for each $1 \leq i \leq m$, the path π_i^r is a path from s_i^r to t_i^r of the form $s_i^r, u_{g_i,i}^r, v_{g_i,i}^r, t_i^r$ for some $1 \leq g_i \leq m$. We represent the j_is and k_is using two functions $f, g : [m] \rightarrow [m]$ defined as $f(i) = j_i$ for all $i \in [m]$ and $g(i) = k_i$ for all $i \in [m]$. From the previous paragraph, the sequence of counter operations along π other than No-op is $\texttt{Write}(\alpha_{1,f(1)}), \texttt{Write}(\alpha_{2,f(2)}), \ldots, \texttt{Write}(\alpha_{m,f(m)}),$ $\texttt{Read}(\alpha_{g(1),1}), \texttt{Read}(\alpha_{g(2),2}), \ldots, \texttt{Read}(\alpha_{g(m),m})$ in that order. This sequence is counter-realizable, because π is a counter-realizable path.

For each $1 \leq j \leq m$, the operation $\texttt{Read}(\alpha_{g(j),j})$ appears in the sequence of counter operations. This means $\texttt{Write}(\alpha_{g(j),j})$ is an operation earlier in the sequence. The only such write operation appearing in the sequence is $\texttt{Write}(\alpha_{g(j),f(g(j))})$, so $f(g(j)) = j$. Similarly, for each $1 \leq j \leq m$, the operation $\texttt{Write}(\alpha_{j,f(j)})$ appears in the sequence of counter operations, which means $\texttt{Read}(\alpha_{j,f(j)})$ appears later in the sequence. The only such read operation appearing in the sequence is $\texttt{Read}(\alpha_{g(f(j)),f(j)})$, so $g(f(j)) = j$. Therefore $f(g(j)) = g(f(j)) = j$ for all $j \in [m]$, so f and g are both permutations of $[m]$ and are inverses of each other. We rewrite f as σ and g as σ^{-1}. The monomial computed by π is, therefore, $\prod_{e \in \pi} \mathcal{L}(e) = \prod_{i=1}^{m} y_{i,\sigma(i)}$, which is a monomial of $\texttt{Perm}_m(Y)$. $\qquad\square$

We now construct the graph H_N from graph H_N'. If m is odd, then we add a vertex α and edges (α, s_1^w) and (t_m^r, α) and if m is even, then we add three vertices $\alpha_1, \alpha_2, \alpha_3$ and edges (α_1, s_1^w) (α_2, α_1), (α_3, α_2) and (t_m^r, α_3). This ensures that, $N = 4n$ for some parameter n, where N is the number of vertices in H_N. We set the weights of all the extra added edges as 1 and label it with No-op. We now add self-loops on all the vertices with weight 1 except the α vertices. All the self-loop edges have the label of No-op on it. Consider the counter graph H_N constructed as above, we will argue that the sum of weights of counter-realizable cycle covers in H_N equals the $\texttt{Perm}_m(Y)$ and the sum of signed weights of counter-realizable clow sequences that are not cycle covers equals 0. Without loss of generality, we assume that m is odd. Similarly, we can extend our arguments for even m.

We already know from our claim that there exists a bijection between the set of monomials in $\texttt{Perm}_m(Y)$ and the set of counter-realizable paths between s_1^w to t_m^r in graph H_N'. It is therefore sufficient to show a bijection between the set of counter-realizable paths between s_1^w to t_m^r in graph H_N' and the set of all counter-realizable cycle covers of graph H_N. We also argue that the sign of every cycle cover of graph H_N is same (w.l.o.g., we assume it to be positive). Since, α is a vertex in H_N without any self loop, any cycle cover $\widehat{\mathcal{C}} = \langle \mathcal{C}_1, \mathcal{C}_2, \mathcal{C}_3, \ldots, \mathcal{C}_k \rangle$ of H_N must cover α with some cycle, w.l.o.g., we call it $\mathcal{C}_1$ which have both the edges (α, s_1^w) and (t_m^r, α), and all other cycles in $\widehat{\mathcal{C}}$ are self-loops on all the vertices which are not covered in cycle $\mathcal{C}_1$. It is easy to observe that the length of any cycle in H_N which uses vertex α is always equal to $6m + 2$. Therefore, the total number of vertices of graph H_N which are not covered in this long cycle and which will get covered by self loops in any cycle cover is $N - 6m - 2$. It immediately follows that the sign of every cycle cover of graph H_N is same.

We now show a bijection between the set of all counter-realizable cycle covers of graph H_N and the set of counter-realizable paths between s_1^w to t_m^r in graph H'_N. It is easy to see that a cycle cover $\widehat{\mathcal{C}} = \langle \mathcal{C}_1, \mathcal{C}_2, \mathcal{C}_3, \ldots, \mathcal{C}_k \rangle$ of H_N is counter-realizable iff the long cycle which uses the vertex α is counter-realizable, w.l.o.g., we call the long cycle $\mathcal{C}_1$. It is easy to note that, for any counter-realizable cycle cover $\widehat{\mathcal{C}} = \langle \mathcal{C}_1, \mathcal{C}_2, \mathcal{C}_3, \ldots, \mathcal{C}_k \rangle$, the cycle $\mathcal{C}_1$ must be formed by an edge (α, s_1^w), followed by a unique counter-realizable directed path $\mathcal{P}$ between s_1^w to t_m^r (of graph H'_N), followed by an edge (t_m^r, α). Also, for every counter-realizable directed path $\mathcal{P}$ from s_1^w to t_m^r (of graph H'_N), one can form a unique counter-realizable long cycle (and therefore a counter-realizable cycle cover) in H_N where the long cycle is (α, s_1^w), followed by directed path $\mathcal{P}$ between s_1^w to t_m^r, followed by (t_m^r, α). This finishes the first part of our argument.

We now argue that the signed sum of all counter-realizable clow sequences of graph H_N which are not cycle covers is equal to 0. We now argue that there exist no clow sequence in graph H_N which does not contain the vertex α in any of its clow. Suppose there exist some clow which does not contain α then we consider the graph formed by deleting the vertex α in H_N, in such a graph, the only closed walks possible are single-loops on each vertex of such a graph. But the total degree of such a clow sequence can never be equal to N, therefore, such a clow sequence is not a valid clow sequence. Let us assume that α is the least numbered vertex in H_N, say, numbered with 1. Since, α is a vertex without a self-loop, any clow involving α must have edges (α, s_1^w) and (t_m^r, α). It is easy to see that in H_N, any counter-realizable clow sequence, say, $\widehat{\mathcal{C}} = \langle \mathcal{C}_1, \mathcal{C}_2, \ldots, \mathcal{C}_k \rangle$ satisfies the property that except the first clow $\mathcal{C}_1$ (which involves the vertex α), all other clows in the clow sequence $\widehat{\mathcal{C}}$ are self-loops. α_1 will be the head of $\mathcal{C}_1$. It is crucial to note that since all self–loops in graph H_N are labelled with No-op, the clow sequence $\widehat{\mathcal{C}}$ is counter-realizable iff the first clow $\mathcal{C}_1$ is counter-realizable.

We can now use similar ideas discussed in part 2 of the previous claim to show that there always exists an involution ψ on the set of counter-realizable clow sequences of graph H_N such that ψ will map any counter-realizable cycle cover to itself and for any counter-realizable clow sequence which is not a cycle cover, say $\widehat{\mathcal{C}}$, there exist another counter-realizable clow sequence which is not a cycle cover, $\widehat{\mathcal{C}'}$ such that $\psi(\widehat{\mathcal{C}}) = \widehat{\mathcal{C}'}$ and $\psi(\widehat{\mathcal{C}'}) = \widehat{\mathcal{C}}$ and the monomials associated with both $\widehat{\mathcal{C}}$ and $\widehat{\mathcal{C}'}$ are same but their signatures are opposite.

Obtaining G_n from H_N. To obtain G_n from this H_N, we need to give an ordering on the vertices that is consistent with Definition 10 and ensure that G_n is a complete graph. Ensuring the latter is easy. We add all the missing edges and set $\mathcal{L}$ value for them to 0.

To describe the ordering, let us first assume that m is odd. When m is even, the ordering can be worked out similarly. We first introduce some notation. For $1 \leq i \leq m-1$ let us denote the vertex obtained by fusing t_i^w with s_{i+1}^w by θ_i. Let us denote the vertex obtained by fusing t_m^w with s_1^r by θ_m. Also for $1 \leq i \leq m-1$, let us denote the vertex obtained by fusing t_i^r with s_{i+1}^r by θ'_i.

The ordering can now be described as follows. Vertex α is set to 1 and the vertex t_m^r is set to 2. The vertices θ_1' to θ_{m-1}' are numbered in reverse order, i.e. θ_{m-1}' is set to 3, θ_{m-2}' to 4 and so on up to θ_1' is set to $m+1$. We also number the vertices θ_1 to θ_m in reverse order starting from $2m+1$ down to $m+2$. We number the vertex s_1^w as $2m+2^{13}$

Now, let us assume that the symbols in Σ are ordered in some arbitrary order, say $a_1,\ldots,a_{m^2}$. In H_N, let $\mathcal{E}$ be defined as $\{e \mid \phi(e) \neq \text{No-op}\}$. From our construction of H_N, no two edges in $\mathcal{E}$ share any endpoints. Also for each $a_i \in \Sigma$, there is a unique edge with $\text{Write}(a_i)$ on it and a unique edge with $\text{Read}(a_i)$ on it. We now fix the following ordering: the tail of the edge with $\text{Write}(a_i)$ on it is assigned $4(i-1)+1+(2m+2)$ and its head is assigned $4(i-1)+2+(2m+2)$, the tail of the edge with $\text{Read}(a_i)$ on it is assigned $4(i-1)+3+(2m+2)$ and finally, its head is assigned $4(i-1)+4+(2m+2)$.

5.2 Details Regarding VNP-Hardness of $\text{CountDet}_n(X)$ When Char = 2

Over characteristic 2, Perm_m is known to be easy. Therefore, to prove VNP-hardness over characteristic 2, we use a different VNP-hard polynomial, which was shown to be VNP-hard over characteristic 2 fields in a work of Hrubes [4]. The polynomial is based on the algebraic variant of the well-known Edge Cover problem. We start by defining the polynomial.

Definition 16. *Let* $m = \binom{\tau}{2}$ *for some parameter* τ. *Let* $G = (V,E)$ *be a complete undirected graph on* τ *vertices, i.e.* $V = [\tau]$ *and* $E = \{(i,j) \mid 1 \leq i < j \leq \tau\}$ *and let edge* $e = (i,j)$ *be labelled with* $y_{i,j}$.
$\text{EC}_m^*(Y) = \sum_{E' \subseteq E, E' \text{ is an edge cover}} \prod_{(i,j) \in E', i<j} y_{i,j}$

In [4], the above polynomial was shown to be VNP-hard. We will show that we can write $\text{EC}_m^*(Y)$ as a projection of $\text{CountDet}_n(X)$, where $n = \text{poly}(m)$.

For this, we will define $m+1$ counter graphs, $W, R_1,\ldots,R_m$, which when interconnected appropriately will give us another counter graph H_N, where $N = \text{poly}(m)$ and it has the following two properties.

- All the counter-realizable clow sequences in H_N have the same sign.
- Moreover, the sum of the weights of the counter-realizable clow sequences which are cycle covers equals EC_m^* and the sum of the weights of the counter-realizable clow sequences which are not cycle covers = 0.

Construction of W. For each edge $(i,j) \in E$ such that $1 \leq i < j \leq \tau$, we add a directed path $\rho_{i,j} = \langle (s_{i,j},i),(i,j),(j,t_{i,j}) \rangle$ in W. We will call $s_{i,j}$ the source of $\rho_{i,j}$ and $t_{i,j}$ the sink of $\rho_{i,j}$. We arrange these paths in a linear order $\rho_{1,2},\ldots,\rho_{1,\tau},\rho_{2,3},\ldots,\rho_{2,\tau},\ldots\rho_{\tau-1,\tau}$ one after the other. Addtionally, we do the following. We rename $s_{1,2}$ as s_1 and $t_{\tau-1,\tau}$ as t_τ

---

[13] For an odd m, note that $2m+2$ is always a multiple of 4.

- Add edges from s_1 to all the other sources, i.e. $\forall 1 \leq i < j \leq \tau$, add edge $(s_1, s_{i,j})$.
- Add edges from the sink of all the paths to t_τ, i.e. $\forall 1 \leq i < j \leq \tau$, add $(t_{i,j}, t_\tau)$.
- Also add edges from sink of a path to the sources of all the paths that come after it in the above order.
- We define $\phi((s_{i,j}, i)) = \texttt{Write}(\alpha_{i,j})$ and $\phi((j, t_{i,j})) = \texttt{Write}(\alpha_{j,i})$ for all $i \leq 1 < j \leq \tau$. We also define $\phi((s_1, 1)) = \texttt{Write}(\alpha_{1,2})$ and $\phi((\tau, t_\tau)) = \texttt{Write}(\alpha_{\tau,\tau-1})$ For all the other edges we set ϕ to be No-op. We also assign $\mathcal{L}((i,j)) = y_{i,j}$.

Let π be any path from s_1 to t_τ in W. We will say that an edge (i,j) of G is traversed in π if $\rho_{i,j}$ is in π, i.e. all the three edges in $\rho_{i,j}$ are traversed in π. It is easy to see the following property holds.

Observation 1. *Let $S \subseteq E$, then there is a path π_S in W such that it traverses exactly the set of edges in S. Moreover, if S is an edge cover then for each vertex $i \in [\tau]$ we would have at least one edge $(i,j) \in S$ such that upon traversing the path π_S we would have done $\texttt{Write}(\alpha_{i,j})$ and $\texttt{Write}(\alpha_{j,i})$ for that edge.*

Construction of $R_1, \ldots, R_\tau$. We have one graph R_i for each vertex $i \in [\tau]$. This graph will allow for reading the symbols $\alpha_{i,j}$ for all $j \neq i$. For each $i \in [\tau]$, we describe $R_i = (V_i, E_i, \Sigma_i, \phi_i)$. Here, $V_i = \bigcup_{j \in [\tau] \setminus \{i\}} \{a_{i,j}\} \cup \bigcup_{j \in [\tau] \setminus \{i\}} \{b_{i,j}\}$. Let min and max denotes the minimum and maximum number in $[\tau] \setminus \{i\}$. $E_i = \{(a_{i,j}, b_{i,j}) \mid j \in [\tau] \setminus \{i\}\} \cup \{(a_{i,min}, a_{i,j'}) \mid j' \in [\tau] \setminus \{i\} \text{ and } j' > min\}$ $\cup \{(b_{i,j}, b_{i,max}) \mid j \in [\tau] \setminus \{i\} \text{ and } j < max\} \cup \bigcup_{k \in [\tau] \setminus \{i\}} \{(b_{i,k}, a_{i,j}) \mid j > k\}$. $\phi((a_{i,j}, b_{i,j})) = \texttt{Read}(\alpha_{i,j})$, where $j \neq i$ and ϕ for all the other edges is No-op.

We relabel $a_{i,min}$ as a_i^* and we relabel $b_{i,max}$ as b_i^*. We observe the following about R_i.

Observation 2. – *Let π be any path from a_i^* to b_i^*. There exists at least one $j \in [\tau], j \neq i$ such that we encounter $\texttt{Read}(\alpha_{i,j})$ along π.*
 – *Let $S \subseteq [\tau] \setminus \{i\}$, there exists a path from a_i^* to b_i^*, say π_S, that encounters exactly the set $\{\texttt{Read}(\alpha_{i,j}) \mid j \in S\}$ along it.*

Construction of H_N. We now interconnect W and $R_1, \ldots, R_\tau$ to create the graph H_N as follows. We add an edge from t_τ to a_1^*. We also add edges from b_i^* to a_{i+1}^* for $1 \leq i \leq \tau - 1$. Finally, we add an edge from b_τ^* to s_1 and self-loops on all nodes other than b_τ^* and s_1. We first observe the following properties about H_N.

Claim. Let Π be any counter-realizable path from s_1 to b_τ^* in H_N. The product of the Y variables along Π, corresponds to a unique monomial in $\texttt{EC}_m^*(Y)$.

Conversely, if $\mathcal{M}$ is a monomial in $\texttt{EC}_m^*$ then there is a unique counter-realizable path Π in H_N from s_1 to b_τ^* such that the product of Y variables along Π equals $\mathcal{M}$.

Proof. Let Π be a counter-realizable path from s_1 to b_τ^* in H_N. Clearly, Π is obtained by concatenating the following paths and edges in this order: $\pi_0 \cdot (t_\tau, a_1^*) \cdot \pi_1 \cdot (b_1^*, a_2^*)\pi_2 \ldots \cdot (b_{\tau-1}^*, a_\tau^*) \cdot \pi_\tau$, where π_0 is a directed path from s_1 to t_τ in W and π_i is a directed path from a_i^* to b_i^* in R_i for $1 \le i \le \tau$.

By part 1 of Observation 2, we know that for each $i \in [\tau]$, π_i must encounter $\mathtt{Read}(\alpha_{i,j})$ for at least one $j \ne i$. As the path is counter-realizable, there will not be any read operation that does not have a corresponding write operation before it. Let S be a subset of edges of G that are traversed in π_0.

As all reads must find a corresponding write along π_0, we have that for each vertex i, there must be at least one edge (i, j) in S, which results into $\mathtt{Write}(\alpha_{i,j})$ and $\mathtt{Write}(\alpha_{j,i})$ along π_0. Therefore, S must be an edge cover. Hence the monomial obtained by taking product of the Y variables along the path gives rise to a monomial $\mathtt{EC}_m^*$.

Conversely, let $\mathcal{M}$ be a monomial in $\mathtt{EC}_m^*$. Then $\mathcal{M}$ corresponds to a subset of edges S of E that forms an edge cover of G. By Observation 1 we know that there exists a unique path in W that traverses exactly the set of edges in S. Let us call this path π_S. As S is an edge cover, we know that for each i in the vertex set of G, there exists at least one $j \ne i$ such that $\mathtt{Write}(\alpha_{i,j})$ occurs in π_S. For $i \in [\tau]$, let $U_i = \{\alpha_{i,j} \mid j \in [\tau] \setminus \{i\}$ and $\mathtt{Write}(\alpha_{i,j})$ occured along $\pi_S\}$. We know that $|U_i| \ge 1$ for $i \in [\tau]$. From the second part of Observation 2 we know that we can uniquely append π_S with paths $\pi_{U_1}, \pi_{U_2}, \ldots \pi_{U_\tau}$, where π_{U_i} is the unique path between a_i^* and b_i^* that traverses the set U_i. Therefore the path $\pi_S \cdot (t_\tau, a_1^*) \cdot \pi_{U_1} \cdot (b_1^*, a_2^*)\pi_{U_2} \ldots \cdot (b_{\tau-1}^*, a_\tau^*) \cdot \pi_{U_\tau}$ is a counter-realizable path and the product of the Y variables along it gives rise to the monomial $\mathcal{M}$. $\square$

Consider the counter graph H_N as defined in Sect. 5.2. We will argue that the sum of weights of counter-realizable cycle covers in H_N equals the $\mathtt{EC}_m^*(Y)$ and the sum of signed weights of counter-realizable clow sequences which are not cycle covers equals 0. We already know from Claim 5.2, that there exists a bijection between the set of monomials in $\mathtt{EC}_m^*(Y)$ and the set of counter-realizable paths between s_1 to b_τ^* in graph H_N. It is therefore sufficient to show that there exists a bijection between the set of counter-realizable paths between s_1 to b_τ^* in graph H_N and the set consisting of all counter-realizable cycle covers of graph H_N. Since, we are working on fields of characteristic 2, sign of every cycle cover of graph H_N is positive.

We now show a bijection between the set of all counter-realizable cycle covers of graph H_N and the set of counter-realizable paths between s_1 to b_τ^* in graph H_N. It is easy to note that any cycle cover $\widehat{\mathcal{C}} = \langle \mathcal{C}_1, \mathcal{C}_2, \mathcal{C}_3, \ldots, \mathcal{C}_k \rangle$ of H_N must use the edge (b_τ^*, s_1), this is because, if it does not use this edge, then there is no way to cover vertices s_1 and b_τ^* by any other cycle in cycle cover $\widehat{\mathcal{C}}$ in graph H_N. We assume that $\mathcal{C}_1$ is the cycle in cycle cover $\widehat{\mathcal{C}}$ which uses the edge (b_τ^*, s_1). It is also easy to note that all the vertices which are not covered in $\mathcal{C}_1$ must be covered by self-loops on each of them in cycle cover $\widehat{\mathcal{C}}$, that is, in other words, all cycles in the cycle cover $\widehat{\mathcal{C}}$, except $\mathcal{C}_1$ are all self-loops. This is because, after deleting vertices b_τ^* and s_1, the only cycles left in graph H_N are self-loops. It is easy to see that for any counter-realizable cycle cover $\widehat{\mathcal{C}} = \langle \mathcal{C}_1, \mathcal{C}_2, \mathcal{C}_3, \ldots, \mathcal{C}_k \rangle$, the cycle

$\mathcal{C}_1$ must be formed by an edge (b_τ^*, s_1), followed by a unique counter-realizable directed path $\mathcal{P}$ between s_1 to b_τ^* of graph H_N. Also, for every counter-realizable directed path $\mathcal{P}$ between s_1 to b_τ^*, one can form a unique counter-realizable long cycle (and therefore a counter-realizable cycle cover) in H_N where the long cycle is formed by an edge (b_τ^*, s_1), followed by a unique counter-realizable directed path $\mathcal{P}$ between s_1 to b_τ^* of graph H_N. This finishes the first part of our argument.

We now prove that the sum of signed weights of all counter-realizable clow sequences of graph H_N which are not cycle covers is equal to 0. We first show that there exist no clow sequence in graph H_N without using the vertex s_1 in any of its clow. For the sake of contradiction, let us assume that there exists a clow sequence $\widehat{\mathcal{C}'} = \langle \mathcal{C}_1', \mathcal{C}_2', \mathcal{C}_3', \dots, \mathcal{C}_k' \rangle$ which does not use vertex s_1. We now consider the graph formed by deleting the vertex s_1 in H_N, in such a graph, the only closed walks possible are single-loops on each vertex of such a graph. It is easy to see that the total degree of $\widehat{\mathcal{C}'}$ is always less than N, therefore, $\widehat{\mathcal{C}'}$ is not a valid clow sequence. Let us assume that s_1 is the least numbered vertex in H_N, say, numbered with 1.

Since, s_1 is a vertex without a self-loop, any clow which uses s_1 must have the edge (b_τ^*, s_1). It is easy to see that any counter-realizable clow sequence, say, $\widehat{\mathcal{C}} = \langle \mathcal{C}_1, \mathcal{C}_2, \dots, \mathcal{C}_k \rangle$ consists of the big cycle $\mathcal{C}_1$ (which uses the vertex s_1) and all other clows in $\widehat{\mathcal{C}}$ are self-loops. s_1 will be the head of $\mathcal{C}_1$. It is crucial to note that since all self–loops in graph H_N are labelled with No-op, the clow sequence $\widehat{\mathcal{C}}$ is counter-realizable iff the clow $\mathcal{C}_1$ is counter-realizable.

Using similar ideas discussed above, it can be shown that there exists an involution ψ defined on the set of all counter-realizable clow sequences of graph H_N, such that the function ψ maps every counter-realizable clow sequence which is also a cycle cover to itself and for any counter-realizable clow sequence $\widehat{\mathcal{C}}$, which is not a cycle cover, there exists a counter-realizable clow sequence $\widehat{\mathcal{C}'}$ which is also not a cycle cover such that the monomials associated with both $\widehat{\mathcal{C}}$ and $\widehat{\mathcal{C}'}$ are same but with opposite signatures, also, $\psi(\widehat{\mathcal{C}}) = \widehat{\mathcal{C}'}$ and $\psi(\widehat{\mathcal{C}'}) = \widehat{\mathcal{C}}$.

Ordering of the vertices of H_N. To finish the proof we also show that we can give a complete ordering of the vertices of H_N consistent with Definition 10 and turn it into a complete graph to obtain G_n to fit the description of G_n as in Definition 10. To make it a complete graph, simply add all the missing edges and assign $\mathcal{L}$ for the newly added edges to 0. To get the ordering, fix any arbitrary ordering for the set Σ, the stack alphabet of H_N, say $a_1, a_2, \dots, a_{|\Sigma|}$. In H_N, let $\mathcal{E}$ be defined as $\{e \mid \phi(e) \neq \text{No-op}\}$. From our construction of H_N, no two edges in $\mathcal{E}$ share any endpoints. Also for each $a_i \in \Sigma$, there is a unique edge with $\text{Write}(a_i)$ on it and a unique edge with $\text{Read}(a_i)$ on it. We now fix the following ordering: the tail of the edge with $\text{Write}(a_i)$ on it is assigned $4(i-1)+1$ and its head is assigned $4(i-1)+2$, the tail of the edge with $\text{Read}(a_i)$ on it is assigned $4(i-1)+3$ and finally, its head is assigned $4(i-1)+4$.

Observation 3. *It is worth noting that another plausible (and probably more natural) way to define the polynomial family in Definition 8 (and Definition 10, respectively) is to consider only the stack-realizable cycle covers* *(counter-realizable cycle covers, respectively) instead of the stack-realizable clow sequences (counter-realizable clow sequences, respectively) in the overall summation. In case of such a variant of* $\text{StackDet}_n(X)$, *the VP-hardness immediately follows whereas the VP-upperbound is not known. However, in case of such a variant of* $\text{CountDet}_n(X)$, *both VNP-upperbound and VNP-hardness follows.*

References

1. Chaugule, P., Limaye, N., Varre, A.: Variants of homomorphism polynomials complete for algebraic complexity classes. In: Du, D.-Z., Duan, Z., Tian, C. (eds.) COCOON 2019. LNCS, vol. 11653, pp. 90–102. Springer, Cham (2019). https://doi.org/10.1007/978-3-030-26176-4_8
2. Durand, A., Mahajan, M., Malod, G., de Rugy-Altherre, N., Saurabh, N.: Homomorphism polynomials complete for VP. In: LIPIcs-Leibniz International Proceedings in Informatics, vol. 29 (2014)
3. Engels, C.: Dichotomy theorems for homomorphism polynomials of graph classes. J. Graph Algorithms Appl. **20**(1), 3–22 (2016)
4. Hrubes, P.: On hardness of multilinearization, and VNP completeness in characteristics two. In: Electronic Colloquium on Computational Complexity (ECCC), vol. 22, p. 67 (2015). http://eccc.hpi-web.de/report/2015/067
5. Mahajan, M., Saurabh, N.: Some complete and intermediate polynomials in algebraic complexity theory. Theory Comput. Syst. **62**(3), 622–652 (2018)
6. Mahajan, M., Vinay, V.: Determinant: combinatorics, algorithms, and complexity. Technical report (1997)
7. Mengel, S.: Characterizing arithmetic circuit classes by constraint satisfaction problems. In: Aceto, L., Henzinger, M., Sgall, J. (eds.) ICALP 2011. LNCS, vol. 6755, pp. 700–711. Springer, Heidelberg (2011). https://doi.org/10.1007/978-3-642-22006-7_59
8. Mengel, S.: Arithmetic branching programs with memory. In: Chatterjee, K., Sgall, J. (eds.) MFCS 2013. LNCS, vol. 8087, pp. 667–678. Springer, Heidelberg (2013). https://doi.org/10.1007/978-3-642-40313-2_59
9. Chaugule, P., Limaye, N., Pandey, S.: Variants of the determinant polynomial and the VP-completeness, October 2020. https://eccc.weizmann.ac.il/report/2020/152/. Posted 07 Oct 2020
10. Raz, R.: Elusive functions and lower bounds for arithmetic circuits. In: Proceedings of the Fortieth Annual ACM Symposium on Theory of Computing, pp. 711–720. ACM (2008)
11. Shpilka, A., Yehudayoff, A.: Arithmetic circuits: a survey of recent results and open questions. Found. Trends® Theor. Comput. Sci. **5**(3–4), 207–388 (2010)
12. Valiant, L.G.: Completeness classes in algebra. In: Proceedings of the Eleventh Annual ACM Symposium on Theory of Computing, STOC 1979, pp. 249–261 (1979)

Dynamic Complexity of Expansion

Samir Datta[1,4], Anuj Tawari[1,3], and Yadu Vasudev[2(✉)]

[1] Chennai Mathematical Institute, Chennai, India
{sdatta,atawari}@cmi.ac.in
[2] Indian Institute of Technology, Madras, India
yadu@cse.iitm.ac.in
[3] Dhirubhai Ambani Institute of Information and Communication Technology,
Gandhinagar, India
[4] UMI ReLaX, Chennai, India

Abstract. Dynamic Complexity was introduced by Immerman and Patnaik [25] (see also [14]). It has seen a resurgence of interest in the recent past, see [2,4,8–12,24,26,30,31] for some representative examples. Use of linear algebra has been a notable feature of some of these papers. We extend this theme to show that the gap version of spectral expansion in bounded degree graphs can be maintained in the class $\mathsf{DynAC^0}$ (also known as DynFO, for domain independent queries) under batch changes (insertions and deletions) of $O(\frac{\log n}{\log \log n})$ many edges.

The spectral graph theoretic material of this work is based on the paper by Kale-Seshadri [23]. Our primary technical contribution is to maintain up to logarithmic powers of the transition matrix of a bounded degree undirected graph in $\mathsf{DynAC^0}$.

Keywords: Dynamic complexity · Expansion · Eigenvalues · Random walks

1 Introduction

Computational complexity conventionally deals with problems in which the entire input is given to begin with and does not change with time. However, in practice, the input is not always static and may undergo frequent changes with time. For instance, one may want to efficiently update the result of a query under insertion or deletion of tuples into a database. In such a scenario, recomputing the solution from scratch after every update may be unnecessarily computationally intensive. In this work, we deal with problems whose solution can be maintained by one of the simplest possible models of computation: polynomial size boolean circuits of bounded depth. The resulting complexity class $\mathsf{DynAC^0}$ is equivalent to Pure SQL in computational power when we think of graphs (and other structures) encoded as a relational database. It is also surprisingly powerful as witnessed by the result showing that $\mathsf{DynAC^0}$ is strong enough to maintain transitive closure in directed graphs [9]. The primary idea in that paper was to reformulate the problem in terms of linear algebra. We follow the same theme to show that expansion in bounded degree graphs can be maintained in $\mathsf{DynAC^0}$.

© Springer Nature Switzerland AG 2021
R. Santhanam and D. Musatov (Eds.): CSR 2021, LNCS 12730, pp. 56–77, 2021.
https://doi.org/10.1007/978-3-030-79416-3_4

1.1   The Model of Dynamic Complexity

In the *dynamic* (graph) model we start with an empty graph on a fixed set of vertices. The graph evolves by the insertion/deletion of a single edge in every time step and some property which can be periodically queried, has to be maintained by an algorithm. The dynamic complexity of the algorithm is the static complexity for each step. If the updates and the queries can be executed in a static class $\mathcal{C}$ the dynamic problem is said to belong to $\mathsf{Dyn}\mathcal{C}$. In this paper, $\mathcal{C}$ is often a complexity class defined in terms of bounded depth circuits[1] such as $\mathsf{AC}^0, \mathsf{TC}^0$, where AC^0 is the class of polynomial size constant depth circuits with AND and OR gates of unbounded fan-in; TC^0 circuits may additionally have MAJORITY gates. We encourage the reader to refer to any textbook (e.g. Vollmer [29]) for precise definitions of the standard circuit complexity classes. The model was first introduced by Immerman and Patnaik [25] (see also Dong, Su, and Topor [14]) who defined the complexity class DynFO.

There are several roughly equivalent ways to view the complexity class DynFO as capturing: (i) The dynamic complexity of maintaining a Pure SQL database under fixed (first order) updates and queries (the original formulation from [25]). (ii) The circuit dynamic complexity of maintaining a property where the updates and queries use uniform AC^0 circuits (see [3] for the equivalence of uniform AC^0 and FO) (iii) The parallel dynamic complexity of maintaining a property where the updates and queries use constant time on a CRCW PRAM (for precise definition of Concurrent RAM see [22])

The first characterization is popular in the Logic and Database community while the second is common in more complexity theoretic contexts. The third one is useful to compare and contrast this class with dynamic algorithms which essentially classify dynamic problems in terms of the sequential time for updates and queries. Operationally, our procedure is easiest to view in terms of the second or even the third viewpoint. We would like to emphasize that modulo finer variations based on built in predicates (like arithmetic and order) in the first variation, uniformity in the second one and built in predicates (like shift) in the third one, the three viewpoints are entirely equivalent. Because it seems necessary to include arithmetic and order in the structure to take care of multiple updates, we abuse notation and write DynFO for what is more precisely DynFO($<$, $+$, $\times$) referred to in previous papers like [10,13].

The archetypal example of a dynamic problem is maintaining reachability ("is there a directed path from s to t"), in a digraph. This problem has recently [9] been shown to be maintainable in the class DynAC^0 - a class where the reachability query can be maintained after edge insertions and deletions using AC^0 circuits. This answers an open question from [25]. Even more recently this result has been extended to batch changes of size $O(\frac{\log n}{\log \log n})$ (see [12]). In this work, we study expansion testing under batch changes of size similar to above.

---

[1] Occasionally we will refer to the (dlogtime-)uniform versions of these circuit classes and we adopt the convention that, whenever unspecified, we mean the uniform version.

1.2   Expansion Testing in Dynamic Graphs

In this paper we study the dynamic complexity of checking the expansion of a bounded-degree graph under edge updates. For a degree-bounded graph G, let λ_G denote the second largest eigenvalue of the normalized adjacency matrix of G. We are interested in the following problem, which we call *Expansion Testing*:

Definition 1. *(Expansion Testing) Given a graph G, degree bound d, and a parameter α, decide whether $\lambda_G \leq \alpha$ or $\lambda_G \geq \alpha'$ where $\alpha' = 1 - (1 - \alpha)^2/5000$.*

A bounded-degree graph G is an expander if λ_G, is bounded away from 1. Expanders are a very useful class of graphs with a variety of applications in algorithms and computational complexity, for instance in derandomization. This arises due to the many useful properties of an expander such as the fact that an expander has no small cuts and that random walks on an expander mix well.

Our aim is to dynamically maintain an approximation of the second largest eigenvalue of a dynamically changing graph in DynAC^0. The second largest eigenvalue is also closely related to the conductance of the graph. A related problem that is frequently studied in the context of approximation is a promise problem, where we are interested whether a graph is an expander with parameter λ or its expansion is small. Such promise problems for algebraic problems were studied in the context of locally-testable codes and the PCP theorem. Proving a lower bound for the gap version implies a lower bound for the original approximation version.

We take a first step towards the problem of approximating the expansion in DynAC^0 in this paper. In particular, we show that for a graph G, we can answer if the second largest eigenvalue of the graph is less than or equal to a parameter α (meaning that G is a good expander) or if λ_G is greater than or equal to α' where α' is polynomially related to α. The study of a related promise problem of testing expansion was initiated in the sparse model of property testing by Goldreich and Ron [15], and testers for spectral expansion by Kale and Seshadhri [23] and vertex expansion by Czumaj and Sohler [7] are known.

Our algorithm is borrowed from the property testing algorithm of [23] where it is shown that if $\lambda_G \leq \alpha$, then random walks of logarithmic length from every vertex in G will converge to the uniform distribution. On the contrary, if $\lambda_G \geq \alpha'$ then this is not the case for at least one vertex in G. The key technical contribution in the paper is a method to maintain the logarithmic powers of the normalized adjacency matrix of a dynamic graph when there are few edge modifications.

1.3   Overview of the Algorithm

The Kale-Seshadhri algorithm [23] estimates the collision probability of several logarithmically long random walks using the lazy transition matrix from a small set of randomly chosen vertices. It uses these to give a probabilistically robust test for the gap version of conductance. We would like to extend this test to the dynamic setting where the graph evolves slowly by insertion/deletion of small number of edges. Moreover, in our dynamic complexity setting the metric to measure the algorithm is not the sequential time but parallel time using polynomially many processors. Thus it suffices to maintain the collision probabilities in constant parallel time with polynomially many processors to be able to solve the gap version of conductance in DynAC^0.

This brings us to our main result:

Theorem 1. *(Dynamic Expansion test)* *Given the promise that the graph remains bounded degree (degree at most d) after every round of updates,* **Expansion testing**[2] *can be maintained in* DynAC^0 *under* $O(\frac{\log n}{\log \log n})$ *changes.*

To prove the theorem we show how to maintain the generating function of at most logarithmic length walks of a transition matrix when the matrix is changed by almost logarithmically many edges[3] in one step.

The algorithm is based on a series of reductions, from the above problem ultimately to two problems – integer determinant of an almost logarithmic matrix modulo a small prime and interpolation of a rational polynomial of polylogarithmic degree. Each reduction is in the class AC^0. Moreover if there are errors in the original data the errors do not increase after a step. On the other hand the entries themselves lengthen in terms of number of bits. To keep them in control we have to truncate the entries at every step increasing the error. We can continue to use these values for a number of steps before the error grows too large. Then we use a matrix that contains the required generating function computed from scratch. Unfortunately this "from scratch" computation takes logarithmically many steps. But by this time $O(\frac{\log^2 n}{\log \log n})$ changes have accumulated. Since we deal with almost logarithmic many changes in logarithmically many steps by working at twice the speed in other words the AC^0 circuits constructed will clear off two batches in one step and thus are of twice the height. Using this, we catch up with the current change in logarithmically many steps. Hence, we spawn a new circuit at every time step which will be useful logarithmically many steps later.

The crucial reductions are as follows:

- (Lemma 9) Dynamically maintaining the aforesaid generating function reduces to powering an almost logarithmic matrix of univariate polynomials to logarithmic powers by adapting (the proof of) a method by Hesse [19].
- (Lemma 12) Logarithmically powering an almost logarithmic sized matrix reduces to powering a collection of similar sized matrices but to only an almost logarithmic power using the Cayley-Hamilton theorem along. This further requires the computation of the characteristic polynomial via an almost logarithmic sized determinant and interpolation.
- (Lemma 13) To compute M^i for i smaller than the size of M, we consider the power series $(I - zM)^{-1}$ and show that we can use interpolation and small

---

[2] See Sect. 3 for a formal definition.

[3] $O(\frac{\log n}{\log \log n})$ to be precise – for us the term "almost logarithmic" is a shorthand for this.

determinants (of triangular matrices) to read off the small powers of M from it.
- (Lemma 14) We reduce rational determinant to integer determinant modulo p. We invoke a known result from [12] to place this in AC^0.

Since interpolation of polylogarithmic degree polynomials is in AC^0, this rounds off the reductions and the outline of the proof of:

Theorem 2. *(Main Technical Result: Informal) Let T be an $n \times n$ dynamic transition matrix, in which, there are at most $O(\frac{\log n}{\log \log n})$ changes in a step. Then we can maintain in DynAC^0, a matrix $\tilde{T}$ such that $\max_{i,j} |\tilde{T}[i,j] - T^{\log n}[i,j]| < \frac{1}{n^{\omega(1)}}$.*

The complexity theoretic motivation of the problem is based on lower bounds from [5] that show that detecting if there is an s,t-path of length $\Omega(\log n)$ requires superpolynomial sized (size $n^{\Omega(\frac{\log^{1/d} n}{d})}$) for depth d) AC^0 circuits. In contrast we show that we can not only maintain if there exists such a path in DynAC^0 but also accurately approximate the number of such short paths. In the parallel algorithm we have traded off work (though only to a large polynomial) for a constant parallel time. This qualitatively overshadows the trivial static doubling-of-exponents algorithm that has linear work, though a logarithmic parallel time. Notice that it is not clear how to achieve a DynAC^0 bound even for single edge changes without using our techniques. We highlight bulk changes (almost logarithmic) over single edge changes as popular in recent literature (see e.g. [28]).

1.4   Related Work

The conductance of a graph, also referred to as the uniform sparsest cut in many works, is an important metric of the graph. Several works have addressed the question of maintaining an approximate value of the conductance in a dynamic graph subject to edge changes.

Goranci [16] describes a sequential dynamic incremental algorithm (only edge insertions allowed) with polylogarithmic approximation and sublinear worst-case update time. In [17], the authors give a fully dynamic algorithm (both edge insertions and deletions allowed) with slightly sublinear approximation and polylogarithmic amortized update time. On the other hand, our work gives a fully dynamic algorithm in a parallel setting. Another difference is that the algorithm in [16,17] outputs an approximate value of the conductance while our algorithm only solves the gap version.

It is interesting to note that in [17], dynamic (approximate) algorithms for many natural graph problems are obtained via an algorithm which maintains the expander hierarchy. They show that many problems like $s-t$ maximum flow, $s-t$ minimum cut, connectivity, have an easy solution when we are promised that the graph remains an expander after every round of updates. So, testing expansion of a graph is a natural first step towards designing dynamic algorithms for these problems in the parallel setting.

There has also been significant related work on investigating the dynamic complexity of problems like rank, reachability and matching under single edge changes [8,9,19] and under batch changes [10,12].

2 Preliminaries

We start by putting down a convention we have already been using. We refer by *almost logarithmic* (in n) to a function that grows like $O(\frac{\log n}{\log \log n})$.

The primary circuit complexity class we will deal with is AC^0, consisting of languages recognisable by a Boolean circuit family with $\wedge$, $\vee$-gates of unbounded fan-in along with $\neg$-gates of fan-in one where the size of the circuit is a polynomial in the length of the input and crucially the depth of the circuit is a constant (independent of the input). Occasionally, we will deal with the circuit class AC^1 in which the depth of the circuit is logarithmic in the size of the input. Since we are more interested in providing AC^0-upper bounds, our circuits will be Dlogtime-uniform. There is a close connection between uniform AC^0 and the formal logic class FO – to the extent that [3] shows that the version $\mathsf{FO}(<, +, \times)$ is essentially identical to AC^0. We will henceforth not distinguish between the two.

The goal of a dynamic program is to answer a given query on an input graph under changes that insert or delete edges. We assume that the number of vertices in the graph is fixed and initially the number of edges in the graph is zero. Of course, we assume an encoding for the graph as a string and any natural encoding works.

The complexity of the dynamic program is measured by a complexity class $\mathcal{C}$ (such as AC^0) and those queries which can be answered constitute the class $\mathsf{Dyn}\mathcal{C}$. In other words the (circuit) class $\mathcal{C}$ can handle each update given some polynomially many stored bits.

Traditionally the changes under which this can be done was fixed to one but recently [10,12] this has been extended to batch changes. In this work we will allow nonconstantly many batch changes (of cardinality $O(\frac{\log n}{\log \log n})$).

One technique which has proved important for dealing with batch changes is a form of pipelining suited for circuit/logic classes called "muddling" [11,27]. Suppose we have a static parallel circuit $\mathcal{A}$ of non-constant depth that can process the input to a form from where the query is answerable easily and in addition we have a dynamic program $\mathcal{P}$ consisting of constant depth circuits that can maintain the query but the correctness of the results is guaranteed for only a small number of batches. This situation may arise if e.g. the dynamic program uses an approximation in computing the result and the ensuing errors add up across several steps making the results useless after a while. On the other hand the static circuit does precise computation always but takes too much depth.

We have in our particular case (an adapted and modified version of the "muddling" lemmas from [11,12]) the following lemma.

Lemma 1. *Let M be a matrix with $b = O(\log^2 n)$-bit rational entries. Suppose we have two routines available:*

- An algorithm $\mathcal{A}$ that can compute $M^{\log n}$ by an AC^1 circuit.
- A dynamic program $\mathcal{P}$ specified by an AC^0-circuit that can approximately maintain [4] $M^{\log n}$ under batch changes of size $l = O(\frac{\log n}{\log \log n})$ for $\Omega(\log n)$ batches.

Then, we have an AC^0-circuit that will approximately maintain $M^{\log n}$ under batch changes of size l for arbitrarily many batches.

Proof. Suppose the circuit for $\mathcal{A}$ has depth $c_{\mathcal{A}} \log n$ and the circuit for $\mathcal{P}$ has depth $c_{\mathcal{P}}$. Then we will show how to construct a circuit C_t at time t of depth $d = (c_{\mathcal{A}} + 2c_{\mathcal{P}}) \log n = c \log n$ that will compute the value $M^{\log n}$ which is correct at the time $t + \log n$. At a time there are $\log n$ circuits extant viz. $C_{t-\log n+1}, C_{t-\log n+2}, \ldots, C_t$ which will deliver the correct value of $M^{\log n}$ at times $t + 1, t + 2, \ldots, t + \log n$ respectively. Since the size of each circuit C_i is polynomial in n so is the size $s_c(n)$ of c layers of C_i. Thus, we can think that each layer of the overall circuit consists of c layers of each of $C_{t-\log n+1}, \ldots, C_t$ of total size $s_c(n) \log n$ per layer i.e. it is an AC^0 circuit.

We prove following lemma about the complexity of powering matrices with rational entries.

Lemma 2. *Let A be an $n \times n$ matrix with rational entries. Each entry is represented with a precision of n bits. Then computing A^ℓ, where $\ell = O(\log n)$, is in AC^1.*

Proof. We describe the algorithm $\mathcal{A}$ that works in AC^1. Notice that two $n \times n$ matrices with entries that are rationals with at most polynomial in n bits each can be multiplied in TC^0 (see e.g. [20,29]). Hence raising a matrix A to the $\log n$-th power can be done by TC-circuits of depth $O(\log \log n)$ (by repeated squaring). We also know that TC^0 is a subset of NC^1 [29] which in turn has AC-circuits of depth $O(\frac{\log n}{\log \log n})$ (just cut up the circuit into NC-circuits of depth $\log \log n$ and expand each subcircuit into a DNF-formula of size $2^{2^{\log \log N}} = n$ – thus overall we get a depth reduction by a factor of $\log \log n$ at the expense of a linear blowup in size). Now by substituting these AC-circuit in the TC-circuits of depth $\log \log n$ we get an AC^1 circuit.

Next we have some preliminaries related to error analysis. A rational $r \in [-1, 1]$ is said to have a B-bit approximation $\tilde{r}$ if $|r - \tilde{r}| \leq 2^{-B}$.

Lemma 3. *Let $f(z)$ be a polynomial of degree d with entries that are rationals not necessarily smaller than 1. Suppose, $z_i = \frac{i}{(3d)^2}$ for $i \in \{0, \ldots, d\}$ are $d + 1$ values. If we know B-bit approximations $\tilde{f}_i$ to the values $f(z_i)$, then the interpolant of these values is a function $\tilde{f}$ whose coefficients are at least B-bit approximations of the corresponding coefficients of f.*

---

[4] By approximately maintaining a matrix we mean that we maintain an $\frac{1}{n^{\omega(1)}}$- entrywise additive approximation to the matrix after every update.

Proof. (Lagrange) interpolation can be viewed as computing $V^{-1}F$ where V is a $(d+1) \times (d+1)$ Vandermonde matrix [5], such that $V_{ij} = z_i^j$ while $F_i = f(z_i)$ are entries of a column vector. The determinant of the Vandermonde matrix is $\prod_{0 \leq j < i \leq d}(z_i - z_j)$. This equals $\prod_{0 \leq j < i \leq d} \frac{i-j}{(3d)^2} = \left(\prod_{i=1}^d i!\right) \frac{1}{(3d)^{d(d-1)}}$. On the other hand, the various co-factors are upper bounded in magnitude by $d! \prod_{i=1}^d z_i^{d-i+1} = d! \prod_{i=1}^d \frac{i^{d-i+1}}{(3d)^{2(d-i+1)}} = d! \frac{1^d 2^{d-1} \ldots (d-1)^2 d}{(3d)^2 \sum_{i=1}^d i} = d! \frac{\prod_{i=1}^d i!}{(3d)^{d(d+1)}}$ by considering the monomial with the largest magnitude. Thus an entry of the inverse i.e. the ratio of a co-factor and the determinant is upper bounded by: $\frac{d!(3d)^{d(d-1)}}{(3d)^{d(d+1)}} = \frac{d!}{(3d)^{2d}} < 1$.

Hence the coefficients of $V^{-1}(F - \tilde{f})$ (where $\tilde{f}$ is the column vector with entries $\tilde{f}_i$) are bounded by 2^{-B} completing the proof. Notice that we do not use the magnitude of $f(z_i)$ or of $\tilde{f}_i$ in the proof but only that their difference is small.

Lemma 4. *Let A be a $k \times k$ matrix with entries that are b-bit rationals smaller than k^{-1}. Let $\tilde{A}$ be a $k \times k$ matrix each of whose entries is a B-bit approximation to the corresponding entry of A. Then assuming $B = \Omega(k^2)$, $\det(\tilde{A})$ is a B-bit approximation to $\det(A)$.*

Proof. Difference between corresponding monomials in the two determinants is easily seen to be upper bounded by $2^{-B} k^{-(k-1)}$ in magnitude. Notice that the assumption $B = \Omega(k^2)$ tacitly implies that we can neglect all monomials that contain more than one term of magnitude 2^{-B} and just need to consider the terms that consist of exactly one 2^{-B} and the rest being the actual entries. Hence the (signed) sum over all monomial (differences) is upper bounded by 2^{-B} in magnitude.

Finally, we present some basic results about logarithmic space computations which will be useful to us. First, we show that reachability in graphs can be decided by bounded depth boolean circuits of subexponential size.

Lemma 5. *(Folklore) Given an input graph G with $|V(G)| = n$ and two fixed vertices s and t, there is a circuit of depth $2d$ and size $n^{n^{1/d}}$ which can decide if there is a path from s to t in G.*

See [5], pg. 613 for a proof.

In the following we denote by $A^{\leq l}$ the words in the language A that are of length at most l.

Lemma 6. *Suppose $A \in L$ is a language. Then for constant $c > 0$, $A^{\leq \log^c n}$ has an AC^0 circuit of depth $O(1)$ and size $n^{O(1)}$.*

Proof. Undirected reachability is L-hard under first order reductions by Cook and McKenzie [6]. Hence A reduces to undirected reachability by first order

---

[5] We assume that the indices of the Vandermonde matrix run in $\{0, \ldots, d\}$ and that $0^0 = 1$ for convenience.

reductions. Thus given N, we can construct an undirected graph G_N of size some N^k, and two vertices s, t thereof using first order formulas such that for all w of length at most N, $w \in A$ iff s, t are connected in G_N. But Lemma 5 tells us that there exists a (very-uniform) AC-circuit of size $N^{kN^{k/d}}$ and depth $2d$ that determines connectivity in G_N. Taking $N = \log^c n$, the size of the circuit becomes $2^{kc \log \log n \log^{kc/d} n}$. Now pick $d = kc + 1$ then the size becomes sublinear in n (because the exponent is sublogarithmic).

3   Maintaining Expansion in Bounded Degree Graphs

In this section, we aim to prove Theorem 1. Our algorithm is based on Kale and Seshadhri's work on testing expansion in the property testing model [23]. Our algorithm differs from theirs in that we are working on a dynamic graph, and the major technical challenge is an efficient way to maintain the powers of the normalized adjacency matrix. In this section, we will describe the algorithm and its correctness. In the subsequent sections, we will detail the method to update the power of the normalized adjacency matrix when a small number of entries change.

To prove the theorem, we will first look at the conductance of a graph G. For a vertex cut $(S, \overline{S})$ with $|S| \leq n/2$, the conductance of the cut (denoted by $\Phi_G(S)$) is the probability that one step of the lazy random walk leaves the set S. A *lazy random walk* on a graph from a vertex v, chooses a neighbor uniformly at random with probability $1/2d$ and chooses to stay at v with probability $1 - d(v)/2d$. The conductance of the graph Φ_G is the minimum of $\Phi_G(S)$ over all vertex cuts $(S, \overline{S})$. For a d-degree-bounded graph G, we will think of G as a $2d$-regular graph where each vertex $v \in V$ has $2d - d(v)$ self-loops. The main idea behind the algorithm in [23] is to perform many lazy random walks of length $k = O(\log n)$ from a fixed vertex s and count the number of pairwise collisions between the endpoints of these walks. We can compute exactly the probability that two different random walks starting at s collide at their endpoints by computing $S_s = \sum_{u \in [n]} T^k[s][u] \cdot T^k[s][u]$, where T is a transition matrix of the graph. Since T is symmetric, the matrix T^k must be a symmetric matrix. Then S_s is equal to the (s, s) entry of the matrix T^{2k}. Hence, it suffices to maintain the (s, s) entry of the matrix T^{2k}.

To analyze the lazy random walks in our setting, we will look at transition matrices T such that $T[v, v] = 1 - d(v)/2d$ for every $v \in V$ and $T[u, v] = d(u)/2d$ for every edge $(u, v) \in G$. Notice that it is equivalent to a random-walk on a $2d$-regular graph, where each vertex u with degree $d(u)$ has $2d - d(u)$ self-loops, and therefore we can use the lemma stated above on the graph.

For a vertex $v \in G$, let π_v^k denote the distribution over V of lazy random walks of length k starting from v. The distance of this distribution from the stationary distribution (which is uniform in this case), denoted by $D_k(v)$ is given by $D_k(v)^2 = \sum_{u \in V} \left(\pi_v^k(u) - \frac{1}{n} \right)^2 = \sum_{u \in V} \pi_v^k(u)^2 - \frac{1}{n}$.

Observe that $\sum_{u \in V} \pi_v^k(u)^2 = T^{2k}[v,v]$ as shown in the lemma above. We now state a technical lemma about the existence of vertex v such that $D_k(v)$ is high if the graph has low conductance.

Lemma 7 ([23]). *For a graph $G(V,E)$, let $S \subset V$ be a set of size $s \leq n/2$ such that the cut $(S, \overline{S})$ has conductance less than δ. Then, for any integer $l > 0$, there exists a vertex $v \in S$ such that $D_l(v) > \frac{1}{2\sqrt{s}}(1 - 4\delta)^l$.*

We defer the description of our algorithm and its proof of correctness to Sect. 4.7.

4   Maintaining the Logarithmic Power of a Matrix

We are given an $n \times n$ lazy-transition matrix T that varies dynamically with the batch insertion/deletion of almost logarithmically $(O(\frac{\log n}{\log \log n}))$ many edges per time step. We want to maintain each entry of sum of powers: $\sum_{i=0}^{\log n}(xT)^i$. Notice that the exponent $\log n$ arises from the Kale-Seshadhri expansion-testing algorithm which needs the probabilities of walks of length $\log n$. On the other hand, the almost logarithmic bound on the small number of changes is a consequence of the reductions described below from the dynamic problem above to ultimately, determinants of small matrices and interpolation of small degree polynomials. Here interpolation can be done for degrees up to polylogarithmic but known techniques [12] permit determinants of at most almost logarithmic size in AC^0 yielding this bottleneck. Another way to view this bottleneck is: while from Lemmata 5, 6, polylogarithmic length inputs of languages in L (or even NL: see [12]) can be decided in AC^0, such bounds are not known for languages reducible to determinants. A b-bit rational is a pair consisting of an integer α a natural number β such that $|\alpha| < \beta \leq 2^b$. Its value is $\frac{\alpha}{\beta}$. By a mild abuse of notation we conflate the pair (α, β) with its value $\frac{\alpha}{\beta}$.

Remark 1. First, notice that every b-bit rational is smaller than 1 by definition. Second, a B-bit approximation $\tilde{r}$ to a rational r may itself be a b-bit rational for some $b \neq B$. This is because the two statements $|r - \tilde{r}| \leq 2^{-B}$ and $\tilde{r} = \frac{\alpha}{\beta}$ where $|\alpha| < \beta \leq 2^b$ are independent.

We need some definitions and begin with the definition of a dynamic matrix and the associated problems of maintaining dynamic matrix powers. Let $l \in \mathbb{N}$. A matrix $A \in \mathbb{Q}^{n \times n}$ is said to be (n, b, l)-dynamic if:

- each entry is a b-bit rational
- at every step there is a change in the entries of some $l \times l$ submatrix of A to yield a new matrix A'. The change matrix $\Delta A = A' - A$ has entries bounded by $\frac{1}{2}$ in magnitude.

DynMatPow(n, b, k, l) is the problem of maintaining the value of each entry of $\sum_{i=0}^k (xA)^i$ for an (n, b, l)-dynamic matrix.

Let **DynBipMatPow**(n, b, k, l) be the special case of
DynMatPow(n, b, k, l) where the change matrix ΔA has a support that is a
bipartite graph with all edges from one bipartition to another.

Next, we define problems to which the dynamic problems will be reduced to.
We begin with polynomial matrix powering. The last condition in the following
bounding the constant term of entries of the powered matrix is a technical one
for controlling the error.

Let **MatPow**(n, d, b, k) be the problem of determining $\sum_{i=0}^{k}(xA)^i$ for a
matrix $A \in \mathbb{Q}^{n \times n}[x]$ where all the following hold:

- the degree of the polynomials is upper bounded by d
- each coefficient is a b-bit rational
- the constant term of each non-diagonal polynomial entry is upper bounded
  by $(3n)^{-1}$

The next group of definitions involve the problems we ultimately reduce the
intermediate matrix powering algorithm to. These include various determinant
problems, polynomial interpolation and polynomial division. Let **Det**(n, b, v) be
the problem of computing the value of the determinant of an $n \times n$ matrix with
entries that are b-bit rationals bounded by $v < 1$ in magnitude.

Let **Det**$_p(n)$ be the problem of computing the value of the determinant of
an $n \times n$ matrix with entries that are from $\mathbb{Z}_p$ for a prime p.

Let **DetPoly**(n, d, b) be the problem of computing the value of the determinant of an $n \times n$ matrix with entries that are degree d polynomials of b-bit
rational coefficients.

Let **Interpolate**(d, b) be the problem of computing the coefficients of a univariate polynomial of degree d, where the coefficients are rationals (not necessarily smaller than one) and where $d + 1$ evaluations of the polynomial on b-bit
rationals are given.

Let **Div**(n, m, b) be the problem of computing the quotient of a univariate
polynomial $g(x)$ of degree n when it is divided by a polynomial $f(x)$ of degree
m where both polynomials are monic with other entries being b-bit rationals.

In the rest of this section, we will use the following variables consistently:
n number of nodes in the graph, $l = O(\frac{\log n}{\log \log n})$, the number of changes in
one batch, $k = O(\log n)$, the exponent to which we want to raise the transition
matrix, $b = \log^{O(1)} n$, the number of bits in the representation and $d \leq \log^{O(1)} n$,
the degree of a polynomial.

4.1   Preprocessing

We need the following technical lemma to preprocess the matrix into a form
where application of the next lemma becomes possible.

Lemma 8. **DynMatPow**(n, b, k, l) *reduces to*
DynBipMatPow$(2n, b, 2k, l)$ *via a local[6] AC^0-reduction.*

---

[6] That is, changing a "small" submatrix of the input dynamic matrix results in a
"small" submatrix change in the output dynamic matrix of the reduction. The notion
of smallness being almost logarithmic.

Proof. Let A be an (n, b, l)-dynamic matrix. Let B be the following 2×2 block matrix with entries from $\mathbb{Q}^{n \times n}$:

$$\begin{pmatrix} 0_n & A \\ I_n & 0_n \end{pmatrix}.$$

Here $0_n, I_n$ are respectively the $n \times n$ all zeroes, identity matrices. Then clearly,

$$B^{2k} = \begin{pmatrix} A^k & 0_n \\ 0_n & A^k \end{pmatrix}$$

Notice that:

$$B' - B = \begin{pmatrix} 0_n & A' - A \\ 0_n & 0_n \end{pmatrix},$$

is a directed bipartite graph with all edges from the first partition of n vertices to the second partition of n vertices, completing the proof.

4.2 Generalising Hesse's Construction

Let G be a weighted directed graph with a weight function $w : E \to \mathbb{R}^+$ and weighted adjacency matrix A. Let $H = H_G^{(k)}(x)$ denote the weighted graph with weighted adjacency matrix $A_H = \sum_{i=0}^{k} (xA)^i$ where k is an integer and x is a formal (scalar) variable. Let G' be a graph on the vertices of G differing from G in a "few" edges and A' be its adjacency matrix.

Lemma 9. *Suppose there exists a partition of the affected vertices into two sets U_i, U_o such that all inserted and deleted edges are from a vertex in U_i to a vertex in U_o. Consider the matrices Δ_σ, for $\sigma \in \{+, -\}$, of dimension $|U| + 2$, viewed as a weighted adjacency matrix of a graph on $U \cup \{s, t\}$ where $s, t \in V(G) \setminus U$, and whose entries are defined as below:*

$$\Delta_\sigma[u, v] = \begin{cases} \sigma w_{uv} x & \text{if } u \in U_i \text{ and } v \in U_o \\ A_H[u, v] & \text{if } u \in U_o \text{ and } v \in U_i \\ A_H[s, v] & \text{if } u = s \text{ and } v \in U_i \\ A_H[u, t] & \text{if } v = t \text{ and } u \in U_o \\ 0 & \text{otherwise} \end{cases}$$

the number of s, t walks in G' of length $k \leq \ell$ are given by the coefficient of x^k in $A_H[s, t] + \Delta_b^k[s, t]$.

Proof. We will prove this separately for the cases $\sigma = +$ and $\sigma = -$. The proof follows the general strategy of Hesse's proof in [19].

When $\sigma = +$, we insert edges into the graph G. When new edges are added to G, the total number of walks from s to t is the sum of the number of walks that are already present and the new walks due to the insertion of the new edges. Observe that all the $s - t$ walks in the graph on $U \cup \{s, t\}$ must pass through the new edges and every such walk of length l is counted exactly once in $\Delta_+^k[s, t]$.

The more interesting case is when $\sigma = -$, and edges are deleted from G. In this case we $A_H[s,t]$ contains all walks from s to t including the deleted edges, and we need to delete only those walks that contain at least one edge that is deleted. The proof follows along the same lines as Hesse's proof when a single edge is deleted. The idea is to show that every walk from s to t of length l is counted exactly once in $A_H[s,t] + \Delta^k_-[s,t]$.

Let P be any $s-t$ walk in G that contains edges that are deleted. Firstly, P is counted exactly once in $A_H[s,t]$. Suppose that r of the deleted edges occur in P and the ith edge occurs k_i times. Among the k_i occurrences of the ith edge we can choose l_i occurrences, for each i, and this gives a walk where these are the edges from U_i to U_o that we choose in graph on $U \cup \{s,t\}$, and the remaining are counted in the walk from s to U_i, U_i to U_o and U_o to t. There are $\prod_{i=1}^{r} \binom{k_i}{l_i}$ such choices, and for each choice the corresponding summand for the walk in $\Delta^k_-[s,t]$ is $(-1)^{l_1+l_2+\cdots+l_r}$. When $l_1 = l_2 = \ldots = l_r = 0$, the walk is counted in $A_H[s,t]$. Therefore, the contribution of the walk P to the sum is given by

$$\sum_{l_1=0}^{k_1} \sum_{l_2=0}^{k_2} \cdots \sum_{l_r=0}^{k_r} (-1)^{l_1+l_2+\cdots+l_r} \prod_{i=1}^{r} \binom{k_i}{l_i} = \sum_{l_1=0}^{k_1} \sum_{l_2=0}^{k_2} \cdots \sum_{l_r=0}^{k_r} \prod_{i=1}^{r} (-1)^{l_i} \binom{k_i}{l_i}$$

$$= \prod_{i=1}^{r} \left(\sum_{l_i=0}^{k_i} (-1)^{l_i} \binom{k_i}{r_i} \right) = 0.$$

Since the walks that do not pass through the deleted edges never appear in the new graph on $U \cup \{s,t\}$ that we created and are hence counted in $A_H[s,t]$, this completes the proof for the case $\sigma = -$.

From the lemma above, we can conclude the following reduction.

Lemma 10. DynBipMatPow$(2n,b,k,l)$ *reduces to* **MatPow**(l,k,b,k) *via an* AC^0*-reduction.*

Proof. Let A denote the $(2n,b,l)$-dynamic matrix such that the support of the changes is a bipartite graph. Let $A_H = \sum_{i=0}^{k} (xA)^i$. From Lemma 9, we know that if l entries of A change, then the new sum $A_{H'}[s,t]$ can be computed in two steps, first by computing $A_H[s,t] + \Delta^k_-[s,t]$ to obtain $A_{H''}[s,t]$ and then computing $A_{H'}[s,t]$ as $A_{H''}[s,t] + \Delta^k_+[s,t]$. The lemma follows from these observations. $\square$

Lemma 11. *Let* $\tilde{A}_H$ *be a* b*-bit approximation of the matrix* A_H, *then the corresponding matrix* $\tilde{A}_{H'}$ *obtained from* $\tilde{A}_H$ *is a* $b-1$*-bit approximation of* $A_{H'}$.

Proof. Let $\tilde{A}_H[s,t] = A_H[s,t] - E[s,t]$ where E denotes an error-matrix with each entry a polynomial with coefficients upper-bounded by $1/2^b$. Each entry in $\tilde{A}_H$ is represented by a k-degree polynomial with b-bit rational coefficents.

We can compute $\tilde{A}_{H'}[s,t] = \tilde{A}_H[s,t] + \tilde{\Delta}^k_\sigma[s,t]$, where $\tilde{\Delta}^k_\sigma$ can be constructed from $\tilde{A}_{H'}$. We can write $\tilde{\Delta}^k_\sigma = \Delta^k_\sigma - E'$, where E' is an error matrix consisting of polynomials of degree at most k. We will now show that the coefficients of these polynomials are upper-bounded by $1/2^b$. First observe that every entry of Δ_σ is a degree k polynomial with coefficients at most $1/2$. We will bound the term corresponding to Δ_σ^k and the remainder separately. Since each entry of Δ_σ is at most $1/2$, we can bound the first term by $\frac{k}{2^k 2^b}$. The remainder of the sum can be upper bounded by $\frac{k 2^k}{2^2 2^{2b}}$. Therefore, each coefficient of the polynomials of this matrix is bounded by $\frac{k}{2^k 2^b} + \frac{k 2^k}{2^2 2^{2b}} \le 1/2^b$.

Therefore, we can write $\tilde{A}_{H'} = \tilde{A}_H + \tilde{\Delta}_\sigma^k = A_H + \Delta_\sigma^k - E''$ where E'' is an error matrix with each entry bounded by $1/2^{b-1}$.

4.3   From Powering to Small Determinants

We first need to reduce the exponent from logarithmic to almost logarithmic. The following lemma in fact reduces it from polylogarithmic to almost logarithmic.

Lemma 12. $\mathbf{MatPow}(l, d, b, k)$ AC^0-*reduces to the conjunction of the following:* $\mathbf{MatPow}(l, 0, lb, l)$, $\mathbf{DetPoly}(l, 1, lb + 2k^2 d)$, $\mathbf{Interpolate}(dk, kl^2 b + k^3 ld)$ *and* $\mathbf{Div}(k, l, l^2 b + 2k^2 ld)$

We use a trick (see e.g. [1,18]; notice that the treatment is similar but not identical to that in [1] because there we had to power only constant sized matrices) to reduce large exponents to exponents bounded by the size of the matrix for any matrix powering problem via the Cayley-Hamilton theorem (see e.g. Theorem 4, Sect. 6.3 in Hoffman-Kunze[21]).

Proof. Given a matrix $M \in \mathbb{Q}^{l \times l}[x]$, let M_i be the value of the polynomial matrix M with rationals x_i substituted instead of x, for $i \in \{0, \ldots, dk\}$. Here $x_0, \ldots, x_{dk}$ are $dk + 1$ distinct, sufficiently small rationals (say $x_i = \frac{i}{(3dk)^2}$). Let $\chi_{M_i}(z)$ denote its characteristic polynomial $det(zI - M_i)$. We write $z^k = q_i(z)\chi_{M_i}(z) + r_i(z)$ for unique polynomials q_i, r_i such that $deg(r_i) < deg(\chi_{M_i}) = l$. Now, $M_i{}^k = q_i(M_i)\chi_{M_i}(M_i) + r_i(M_i) = r_i(M_i)$. Here, the last equality follows from the Cayley-Hamilton theorem that asserts that $\chi_{M_i}(M_i) = 0_l$. But $r_i(z)$ is a polynomial of degree strictly less than the dimension of M_i and each monomial in this involves powering M_i to an exponent bounded by $l-1$. Finally computing M^k reduces to interpolating each entry from the corresponding entries of M_i^k.

Now we analyse this algorithm. First we evaluate the matrix at $dk + 1$ points $x_0, x_1, \ldots, x_{dk}$ where $x_i = \frac{i}{(3dk)^2}$. This yields a matrix M_i whose entries are bounded by $\frac{1}{3k} + \sum_{j=1}^{dk} i^j (3dk)^{-2j} < \frac{1}{3k} + \sum_{j=1}^{d} (3k)^{-j} < \frac{1}{3k} + \frac{1}{3k-1} < k^{-1} < (3l)^{-1}$ in magnitude. We then compute the characteristic polynomial of M_i. Notice that the value of $det(zI - M_i)$ is a monic polynomial with coefficient of z^{l-j} bounded by $j!\binom{l}{j}(3l)^{-j} < 1$ for $j > 0$. Suppose, the denominator of an entry of M is bounded by β. Then the denominator of M_i is bounded by $\beta(3dk)^{2dk}$. Moreover, the denominator of this coefficient is further bloated to at most $\beta^l(3dk)^{2ldk} \le 2^{lb+2k^2 d}$ (where we use that $l \log 3dk \approx k$). Thus this corresponds to an instance of $\mathbf{DetPoly}(l, 1, lb + 2k^2 d)$.

In the next step, we divide z^k by the characteristic polynomial of M_i, $\chi_{M_i}(z)$. This corresponds to an instance of $\mathbf{Div}(k, l, l^2 b + 2k^2 ld)$.

For computing the evaluation of the remainder polynomial on an M_i, we need to power an $l \times l$ matrix M_i with lb-bit rational entries to exponents bounded by at most $l - 1$. This can be accomplished by $\mathbf{MatPow}(l, 0, lb, l)$ by recalling that the each entry of M_i is bounded by $(3l)^{-1}$.

Finally, we obtain M^k by interpolation. Every entry of M^k is a polynomial of degree at most dk. Every coefficient of this polynomial is an kb-bit rational and moreover the evaluation on entries of $r_i(M_i^l)$ are given which are $kl^2 b + 2k^3 ld$-bit i.e. via $\mathbf{Interpolate}(dk, kl^2 b + 2k^3 ld)$.

4.4 Reducing Small Matrix Powering to Small Matrix Determinant

Next, we reduce almost logarithmic powers of almost logarithmic sized matrices to almost logarithmic sized determinants of polynomials.

Lemma 13. $\mathbf{MatPow}(l, d, b, l)$ AC^0-reduces to the conjunction of the following: $\mathbf{Det}(l + 1, lb, \frac{1}{l+1})$, $\mathbf{DetPoly}(l, 1, lb)$ and $\mathbf{Interpolate}(dl, lb)$.

Proof. Let $A^{(j)}$ be the univariate polynomial matrix $A = A(x)$ evaluated at point $x = x_j$ where $x_0, \ldots, x_{dl}$ are $dl + 1$ distinct rationals say $x_j = \frac{j}{(3dl)^2}$. Consider the infinite power series $p^{(s,t,j)}(z) = (I - zA^{(j)})^{-1}[s, t]$. $(I - zA^{(j)})^{-1} = \sum_{i=0}^{\infty} z^i (A^{(j)})^i$. Thus $p^{(s,t,j)}(z)$ is the generating function of $(A^{(j)})^i[s, t]$ parameterised on i. $p^{(s,t,j)}(z)$ can be also be written, by Cramer's rule (See, for example, Sect. 5.4, p. 161, Hoffman-Kunze [21]), as the ratio of two determinants – the numerator being the determinant of the (t, s)-th minor of $(I - zA^{(j)})$, say $D^{(s,t,j)}(z)$ and the denominator being the determinant $D^{(j)}(z)$ of $I - zA^{(j)}$. Thus $p^{(s,t,j)}(z) = \frac{D^{(s,t,j)}(z)}{D^{(j)}(z)}$. In other words, $D^{(s,t,j)}(z) = p^{(s,t,j)}(z) D^{(j)}(z)$. Now let us compare the coefficients of z^i on both sides[7]:

$$D_i^{(s,t,j)} = \sum_{k=0}^{i} p_k^{(s,t,j)} D_{i-k}^{(j)}$$

Letting, i run from 0 to degree of $D^{(j)}$ which is $l = $ dimension of A, we get $l + 1$ equations in the $l + 1$ unknowns $p_i^{(s,t,j)}$ for $i \in \{0, \ldots, l\}, j \in \{0, \ldots, d\}$. Equivalently, this can be written as the matrix equation: $M^{(j)} \pi = d^{(j)}$ where $M^{(j)}$ is an $(l + 1) \times (l + 1)$ matrix with entries $M_{ik}^{(j)} = D_{i-k}^{(j)}$, for $0 \leq k \leq i \leq l$ and zero for all other values of i, k lying in $\{0, \ldots, l\}$. Similarly, $d^{(j)}$ is a vector with entries $d_k^{(j)} = D_k^{(s,t,j)}$ and π the vector with $l + 1$ unknowns $\pi_k = p_k^{(s,t,j)}$ again for $i, k \in \{0, \ldots, l\}$. Notice that specifically in this argument, indices of matrices/vectors start at 0 instead of 1 for convenience.

Next, we show that the matrix M is invertible. We make the trivial but crucial observation from which the subsequent proposition is immediate.

---

[7] Here we use the convention that a_i denotes the coefficient of z^i in $a(z)$, where $a(z)$ is a power series (or in particular, a polynomial).

Observation 3 *The constant term in $D^{(j)}(z) = det(I - zA^{(j)})$ is 1.*

Proposition 1. $M^{(j)}$ *is a lower triangular matrix with all principal diagonal entries equal to 1 hence has determinant 1.*

Next we can interpolate the values of $A^i[s,t]$ from the values of $(A^{(j)})^i[s,t] = [z^i]p^{(s,t,j)}$ for $i \in \{0, \ldots, l\}, j \in \{0, \ldots, \frac{dl}{(3dl)^2}\}$.

Now we analyse this algorithm. First, we evaluate the matrix at $dl+1$ distinct rationals. Each entry of the j-th matrix is now bounded by $\frac{1}{3l} + \sum_{i=1}^{dl} j^i (3dl)^{-2i} < (l+1)^{-1}$. We then compute determinant of the matrix $I - zA^{(j)}$ where $A \in \mathbb{Q}^{l \times l}$. The number of bits in the denominators are less than 2^{bl} and moreover the values of coefficient of z^j is less than $j!\binom{l}{j}(l+1)^{-j} < 1$ if $j > 0$. Thus the coefficients are lb-bit rationals. Hence, computing the determinant of $I - zA^{(j)}$ corresponds to an instance of **DetPoly**$(l, 1, lb)$.

The next step is to compute the inverse of a $(l+1) \times (l+1)$ matrix $M^{(j)}$ above. The (a,b) entry of $M^{(j)}$ is either zero or equals the coefficient of z^{a-b} in a cofactor of $(I - zA^{(j)})$. By a logic similar to that used in the proof of Lemma 12 these are all lb-bit rationals. Further, a crude upper bound on their values is $(l+1)^{-1}$ as above. Thus we get an instance of **Det**$(l+1, lb, (l+1)^{-1})$.

Finally, we find the matrix A^i from the values of $(A^{(j)})^i$. This corresponds to an instance of **Interpolate**(dl, lb).

First we reduce the problem of computing almost logarithmic sized determinants of small polynomials to computing almost logarithmic sized determinants over small rationals.

Lemma 14. **DetPoly**(l, d, b) AC^0-*reduces to the conjunction of the following:* **Interpolate**(dl, lb) *and* **Det**$(l, lb, (l+1)^{-1})$.

Proof. Here, we need to compute the determinant of an $l \times l$ matrix with each entry a degree $\leq d$ polynomial with b bit coefficients. Clearly, the determinant is a degree at most ld polynomial. So, we plug in $ld+1$ different values $x_0, x_1, \ldots, x_{ld}$, where $x_i = \frac{i}{(3ld)^2}$ into the determinant polynomial. This yields a determinant with entries bounded by $(l+1)^{-1}$ in magnitude as in the proof of Lemma 12. The next step is to interpolate a degree at most ld polynomial. The coefficient of x^m where $m \leq ld$ is bounded by 1 as in the previous lemmas, and the number of bits is at most lb as well.

4.5   Computing Small Determinants

Then we show how to compute small rational determinants by using Chinese Remaindering and the computation of determinants over small fields.

Lemma 15. *For every $c > 0$,* **Det**$(\frac{\log n}{\log \log n}, \log^c n, v) \in \mathsf{AC}^0$

Proof. The basic idea is to use the Chinese Remainder Theorem, CRT (see e.g. [20]) with prime moduli that are of magnitude at most $\log^{c+1} n$ to obtain the determinant of $2^b A$ which is an integer matrix (since the entries are b-bit rationals of magnitude at most 2^b). For $n^{O(1)}$ primes this problem is solvable in TC^0 by [20] and hence in L. Thus, by Lemma 6 it is in AC^0 for primes of magnitude polylogarithmic. We of course need to compute the determinants modulo the small primes for which we crucially use Lemma 16 below.

Lemma 16 ([12, **Theorem 8**]). *If $p \in n^{O(1)}$ is a prime then,* $\mathbf{Det}_p(\frac{\log n}{\log \log n}) \in \mathsf{AC}^0$.

4.6   The Complexity of Polynomial Division

We use a slight modification of the Kung-Sieveking algorithm as described in [1,18]. The algorithm in [1] worked over finite fields while here we apply it to divide polynomials of small heights and degrees over rationals. The algorithm and its proof of correctness follow Lemma 7 from [1] in verbatim. We reproduce the relevant part for completeness (with minor emendments to accommodate for the characteristic).

Lemma 17. *Let $g(x)$ of degree n and $f(x)$ of degree m be monic univariate polynomials over $\mathbb{Q}[x]$, such that $g(x) = q(x)f(x)+r(x)$ for some polynomials $q(x)$ of degree $(n-m)$ and $r(x)$ of degree $(m-1)$. Then, given the coefficients of g and f, the coefficients of r can be computed in TC^0. In other words $\mathbf{Div}(n, m, b) \in \mathsf{TC}^0$ if $m < n$ and $b = n^{O(1)}$.*

Proof. Let $f(x) = \sum_{i=0}^{m} a_i x^i$, $g(x) = \sum_{i=0}^{n} b_i x^i$, $r(x) = \sum_{i=0}^{m-1} r_i x^i$ and $q(x) = \sum_{i=0}^{n-m} q_i x^i$. Since f, g are monic, we have $a_m = b_n = 1$. Denote by $f_R(x)$, $g_R(x)$, $r_R(x)$ and $q_R(x)$ respectively the polynomial with the i-th coefficient a_{m-i}, b_{n-i}, r_{m-i-1} and q_{n-m-i} respectively. Then note that $x^m f(1/x) = f_R(x)$, $x^n g(1/x) = g_R(x)$, $x^{n-m} q(1/x) = q_R(x)$ and $x^{m-1} r(1/x) = r_R(x)$. We use the Kung-Sieveking algorithm (as implemented in [1]). The algorithm is as follows: First, compute $\tilde{f}_R(x) = \sum_{i=0}^{n-m}(1 - f_R(x))^i$ via interpolation. Then compute $h(x) = \tilde{f}_R(x)g_R(x) = c_0 + c_1 x + \ldots + c_{d(n-m)+n}x^{m(n-m)+n}$ from which the coefficients of $q(x)$ can be obtained as $q_i = c_{m(n-m)+n-i}$. Finally compute $r(x) = g(x) - q(x)f(x)$. The proof of correctness of the algorithm is identical to that in [1]. The proof of the lemma is immediate because polynomial product is in TC^0 from [20]. $\qquad\square$

Lemma 18. $\mathbf{Div}(k, l, b)$ AC^0*-reduces to* $\mathbf{Interpolate}(kl, kb)$

Proof. In the first step of the algorithm from the proof of Lemma 17, we need to interpolate a polynomial of degree at most $(k - l)l$. Also, the coefficients of the polynomial are rationals with rationals that are $(k - l)b$ bits long. Notice that we do not require the coefficients of the polynomial to be smaller than 1.

4.7   Putting It Together with Error Analysis

We now reach the main theorem of this Section:

Theorem 4. *(Main Technical Result: Formal Statement of Theorem 2)*
Let T be an $(n, \log n, \log^2 n, \frac{\log n}{\log \log n})$-dynamic transition matrix. Then we can maintain in DynAC^0, a matrix $\tilde{T}$ such that $\max_{i,j} |\tilde{T}[i,j] - T^{\log n}[i,j]| < \frac{1}{n^{\omega(1)}}$.

Proof.

$\textbf{DynMatPow}(n, b, k, l)$

$\qquad (Lemma\ 8)\quad \leq^{\mathsf{AC}^0} \textbf{DynBipMatPow}(2n, b, 2k, l)$

$\qquad (Lemma\ 10)\quad \leq^{\mathsf{AC}^0} \textbf{MatPow}(l, d, b, 2k)$

$\qquad (Lemma\ 12)\quad \leq^{\mathsf{AC}^0} \textbf{MatPow}(l, 0, lb, l) \wedge$
$\qquad\qquad\qquad\qquad \textbf{DetPoly}(l, 1, lb + 8k^2 d) \wedge$
$\qquad\qquad\qquad\qquad \textbf{Interpolate}(2dk, 2kl^2 b + 8k^3 ld) \wedge$
$\qquad\qquad\qquad\qquad \textbf{Div}(2k, l, l^2 b + 8k^2 ld)$

$\qquad (Lemma\ 13)\quad \leq^{\mathsf{AC}^0} \textbf{DetPoly}(l, 1, lb) \wedge$
$\qquad\qquad\qquad\qquad \textbf{Det}(l + 1, lb, (l + 1)^{-1}) \wedge$
$\qquad\qquad\qquad\qquad \textbf{Interpolate}(0, lb) \wedge$
$\qquad\qquad\qquad\qquad \textbf{DetPoly}(l, 1, lb + 8k^2 d) \wedge$
$\qquad\qquad\qquad\qquad \textbf{Interpolate}(2dk, 2kl^2 b + 8k^3 ld) \wedge$
$\qquad\qquad\qquad\qquad \textbf{Div}(2k, l, l^2 b + 8k^2 ld)$

$\qquad (Lemma\ 14)\quad \leq^{\mathsf{AC}^0} \textbf{Interpolate}(dl, ldb) \wedge$
$\qquad\qquad\qquad\qquad \textbf{Det}(l, ldb, (l + 1)^{-1}) \wedge$
$\qquad\qquad\qquad\qquad \textbf{Interpolate}(2dk, 2kl^2 b + 8k^3 ld) \wedge$

$\qquad (Lemma\ 18)\qquad \textbf{Interpolate}(2kl, 2kl^2 b + 8k^3 ld)$

$\qquad\qquad\qquad \equiv\ \textbf{Interpolate}(2 \log^2 n, \dfrac{18 \log^5 n}{\log \log n}) \wedge$

$\qquad\qquad\qquad\quad \textbf{Det}(\dfrac{\log n}{\log \log n}, \dfrac{\log^4 n}{(\log \log n)}, (\dfrac{\log n}{\log \log n} + 1)^{-1})$

Each **DynMatPow** call[8] boils down to a number of **Det, Interpolate** calls as above. By Lemmata 3, 4, 11 there is a single bit of loss of precision.

However, the length of the bit representation grows by a factor upper bounded by $18\frac{\log^3 n}{\log \log n}$ at every batch. Thus to keep the number of bits under control we need to truncate the matrix at $\log^2 n$-bits again so that now the powered matrix is now a $\log^2 n - 1$-bit approximation. This -1 will deteriorate at

---

[8] For ease of readability we use d for $2k$ in the second reduction from **DynBipMatPow** to **MatPow** above.

every step by one so that we can afford to perform at least $\Omega(\log n)$ steps before we recompute the results from scratch i.e. do muddling.

We can do muddling by invoking Lemma 1 where we pick $\mathcal{A}$ to be the algorithm from Lemma 2 and with the above sequence of reductions as the dynamic program $\mathcal{P}$ for handling a batch (or actually two batches – one old and one new) of changes.

Proof. (of Theorem 1) We can now describe our algorithm for testing expansion. After each update, we use Theorem 4 to obtain the matrix $\tilde{T}$ such that $|\tilde{T} - T^k| \leq 1/n^3$, where $k = \log n/\Phi^2$. Therefore for each v, we $\tilde{T}(v,v)$ such that $|\tilde{T}[v,v] - \sum_{u \in V} \pi_v^k(u)^2| \leq 1/n^3$. We now test if $\tilde{T}[v,v] \leq \frac{1}{n}\left(1 + \frac{2}{n}\right)$ for each $v \in G$, and reject if this is not the case even for one $v \in G$. The correctness of this algorithm follows from the two claims below.

Claim. If $\lambda_G \leq \alpha$, then $\tilde{T}[v,v] \leq \frac{1}{n}\left(1 + \frac{2}{n}\right)$ for every $v \in G$.

Proof. If $\lambda_G \leq \alpha$, let $\eta = 1 - \alpha$. Now,

$$D_l(v)^2 = \|\pi_v^l - \frac{1}{n}\|_2^2$$
$$= \lambda_G^{2l} \leq (1 - \eta)^{2l}$$
$$\leq \frac{1}{n^2}, \text{ for } l = \frac{\log n}{1 - \alpha}.$$

Therefore, $T^{2l}[v,v] = \sum_{u \in V} \pi_v^\ell(u)^2 \leq \frac{1}{n}\left(1 + \frac{1}{n}\right)$. Since $|\tilde{T}[v,v] - T^{2\ell}[v,v]| \leq 1/n^3$, we have $\tilde{T}[v,v] \leq \frac{1}{n}\left(1 + \frac{2}{n}\right)$ for every $v \in V$.

Claim. If $\lambda_G \geq \alpha'$, then there exists a vertex $v \in G$ such that $\tilde{T}[v,v] > \frac{1}{n}\left(1 + \frac{2}{n}\right)$.

Proof. If $\lambda_G \geq \alpha'$, then we know that $\Phi_G \leq (1-\alpha)/50$. Therefore, there exists a vertex cut $(S, \overline{S})$ such that $\Phi_G(S) \leq (1-\alpha)/50$. From Lemma 7 we can conclude that there exists a vertex v such that $D_l(v)^2 > \frac{1}{4s}(1 - (1-\alpha)/12)^{2l} \geq \frac{1}{2n}(1 - (1-\alpha)/12)^{2l}$. For $l = \ln n/(1-\alpha)$, we have $D_\ell(v)^2 > \frac{1}{2n^{1+\epsilon}}$ for a small constant $\epsilon > 0$. The collision probability $\sum_{u \in V} \pi_v^\ell(u)^2$ is therefore at least $\frac{1}{n}\left(1 + \frac{1}{n^\epsilon}\right)$. From Theorem 4, we know that $\tilde{T}[v,v] \geq \frac{1}{n}\left(1 + \frac{1}{2n^\epsilon}\right) > \frac{1}{n}\left(1 + \frac{2}{n}\right)$.

5   Conclusion

In this paper we solve a gap version of the expansion testing problem, wherein we want to test if the expansion is greater than α or less than a α'. The dependence of α' on α in this paper is due to the approach of using random walks and testing the conductance. It is natural to ask if there is an alternate method leading to a better dependence. A more natural question is whether we can maintain an approximation of the second largest eigenvalue of a dynamic graph.

An alternative direction of work would be to improve on the number of updates allowed per round fromn almost logarithmic to, say, polylogarithmic.

Acknowledgements. SD would like to thank Anish Mukherjee, Nils Vortmeier and Thomas Zeume for many interesting and illuminating conversations over the years and in particular for discussions that ultimately crystallized into Lemma 9. We would like to thank Eric Allender for clarification regarding previous work. SD was partially funded by a grant from Infosys foundation and SERB-MATRICS grant MTR/2017/000480. AT was partially funded by a grant from Infosys foundation.

References

1. Allender, E., Balaji, N., Datta, S.: Low-depth uniform threshold circuits and the bit-complexity of straight line programs. In: Csuhaj-Varjú, E., Dietzfelbinger, M., Ésik, Z. (eds.) MFCS 2014. LNCS, vol. 8635, pp. 13–24. Springer, Heidelberg (2014). https://doi.org/10.1007/978-3-662-44465-8_2
2. Barceló, P., Romero, M., Zeume, T.: A more general theory of static approximations for conjunctive queries. In: 21st International Conference on Database Theory, ICDT 2018, Vienna, Austria, March 26–29, 2018), pp. 7:1–7:22 (2018). https://doi.org/10.4230/LIPIcs.ICDT.2018.7
3. Barrington, D.A.M., Immerman, N., Straubing, H.: On uniformity within nc^1. J. Comput. Syst. Sci. **41**(3), 274–306 (1990). https://doi.org/10.1016/0022-0000(90)90022-D
4. Bouyer, P., Jugé, V.: Dynamic complexity of the dyck reachability. In: Esparza, J., Murawski, A.S. (eds.) FoSSaCS 2017. LNCS, vol. 10203, pp. 265–280. Springer, Heidelberg (2017). https://doi.org/10.1007/978-3-662-54458-7_16
5. Chen, X., Oliveira, I.C., Servedio, R.A., Tan, L.: Near-optimal small-depth lower bounds for small distance connectivity. In: Wichs, D., Mansour, Y. (eds.) Proceedings of the 48th Annual ACM SIGACT Symposium on Theory of Computing, STOC 2016, Cambridge, MA, USA, June 18–21, 2016, pp. 612–625. ACM (2016). https://doi.org/10.1145/2897518.2897534
6. Cook, S.A., McKenzie, P.: Problems complete for deterministic logarithmic space. J. Algorithms **8**(3), 385–394 (1987). https://doi.org/10.1016/0196-6774(87)90018-6
7. Czumaj, A., Sohler, C.: Testing expansion in bounded-degree graphs. Comb. Probab. Comput. **19**(5–6), 693–709 (2010). https://doi.org/10.1017/S096354831000012X
8. Datta, S., Hesse, W., Kulkarni, R.: Dynamic complexity of directed reachability and other problems. In: Esparza, J., Fraigniaud, P., Husfeldt, T., Koutsoupias, E. (eds.) ICALP 2014. LNCS, vol. 8572, pp. 356–367. Springer, Heidelberg (2014). https://doi.org/10.1007/978-3-662-43948-7_30
9. Datta, S., Kulkarni, R., Mukherjee, A., Schwentick, T., Zeume, T.: Reachability is in dynfo. J. ACM **65**(5), 33:1–33:24 (2018). https://doi.org/10.1145/3212685
10. Datta, S., Kumar, P., Mukherjee, A., Tawari, A., Vortmeier, N., Zeume, T.: Dynamic complexity of reachability: How many changes can we handle? In: Czumaj, A., Dawar, A., Merelli, E. (eds.) In: 47th International Colloquium on Automata, Languages, and Programming, ICALP 2020, Saarbrücken, Germany, July 8–11, 2020, (Virtual Conference). LIPIcs, vol. 168, pp. 122:1–122:19. Schloss Dagstuhl - Leibniz-Zentrum für Informatik (2020). https://doi.org/10.4230/LIPIcs.ICALP.2020.122
11. Datta, S., Mukherjee, A., Schwentick, T., Vortmeier, N., Zeume, T.: A strategy for dynamic programs: start over and muddle through. Log. Methods Comput. Sci. **15**(2), 1–14 (2019). https://lmcs.episciences.org/5442

12. Datta, S., Mukherjee, A., Vortmeier, N., Zeume, T.: Reachability and distances under multiple changes. In: Chatzigiannakis, I., Kaklamanis, C., Marx, D., Sannella, D. (eds.) In: 45th International Colloquium on Automata, Languages, and Programming, ICALP 2018, Prague, Czech Republic, July 9–13, 2018. LIPIcs, vol. 107. Schloss Dagstuhl - Leibniz-Zentrum fuer Informatik (2018). https://doi.org/10.4230/LIPIcs.ICALP.2018.120

13. Datta, S., Mukherjee, A., Vortmeier, N., Zeume, T.: Reachability and distances under multiple changes. In: 45th International Colloquium on Automata, Languages, and Programming, ICALP 2018, Prague, Czech Republic, July 9–13, 2018, pp. 120:1–120:14 (2018)

14. Dong, G., Su, J., Topor, R.W.: Nonrecursive incremental evaluation of datalog queries. Ann. Math. Artif. Intell. **14**(2–4), 187–223 (1995). https://doi.org/10.1007/BF01530820

15. Goldreich, O., Ron, D.: On testing expansion in bounded-degree graphs. In: Goldreich, O. (ed.) Studies in Complexity and Cryptography. Miscellanea on the Interplay between Randomness and Computation. LNCS, vol. 6650, pp. 68–75. Springer, Heidelberg (2011). https://doi.org/10.1007/978-3-642-22670-0_9

16. Goranci, G.: Dynamic graph algorithms and graph sparsification: new techniques and connections. CoRR abs/1909.06413 (2019). http://arxiv.org/abs/1909.06413

17. Goranci, G., Räcke, H., Saranurak, T., Tan, Z.: The expander hierarchy and its applications to dynamic graph algorithms. CoRR abs/2005.02369 (2020). https://arxiv.org/abs/2005.02369

18. Healy, A., Viola, E.: Constant-depth circuits for arithmetic in finite fields of characteristic two. In: Durand, B., Thomas, W. (eds.) STACS 2006. LNCS, vol. 3884, pp. 672–683. Springer, Heidelberg (2006). https://doi.org/10.1007/11672142_55

19. Hesse, W.: The dynamic complexity of transitive closure is in dyntc^0. Theor. Comput. Sci. **296**(3), 473–485 (2003). https://doi.org/10.1016/S0304-3975(02)00740-5

20. Hesse, W., Allender, E., Barrington, D.A.M.: Uniform constant-depth threshold circuits for division and iterated multiplication. J. Comput. Syst. Sci. **65**(4), 695–716 (2002). https://doi.org/10.1016/S0022-0000(02)00025-9

21. Hoffman, K., Kunze, R.: Linear algebra. Englewood Cliffs, New Jersey (1971)

22. Immerman, N.: Expressibility and parallel complexity. SIAM J. Comput. **18**(3), 625–638 (1989). https://doi.org/10.1137/0218043

23. Kale, S., Seshadhri, C.: An expansion tester for bounded degree graphs. SIAM J. Comput. **40**(3), 709–720 (2011). https://doi.org/10.1137/100802980

24. Muñoz, P., Vortmeier, N., Zeume, T.: Dynamic graph queries. In: 19th International Conference on Database Theory, ICDT 2016, Bordeaux, France (March 15–18, 2016), pp. 14:1–14:18 (2016). https://doi.org/10.4230/LIPIcs.ICDT.2016.14

25. Patnaik, S., Immerman, N.: Dyn-FO: A parallel, dynamic complexity class. J. Comput. Syst. Sci. **55**(2), 199–209 (1997). https://doi.org/10.1006/jcss.1997.1520

26. Schmidt, J., Schwentick, T., Vortmeier, N., Zeume, T., Kokkinis, I.: Dynamic complexity meets parameterised algorithms. In: 28th EACSL Annual Conference on Computer Science Logic, CSL 2020, Barcelona, Spain, January 13–16, 2020, pp. 36:1–36:17 (2020). https://doi.org/10.4230/LIPIcs.CSL.2020.36

27. Schwentick, T., Vortmeier, N., Zeume, T.: Dynamic complexity under definable changes. ACM Trans. Database Syst. **43**(3), 12:1–12:38 (2018). https://doi.org/10.1145/3241040

28. Schwentick, T., Vortmeier, N., Zeume, T.: Sketches of dynamic complexity. SIGMOD Rec. **49**(2), 18–29 (2020). https://doi.org/10.1145/3442322.3442325

29. Vollmer, H.: Introduction to Circuit Complexity - A Uniform Approach. Texts in Theoretical Computer Science. An EATCS Series. Springer, Berlin, Heidelberg (1999). https://doi.org/10.1007/978-3-662-03927-4
30. Zeume, T.: The dynamic descriptive complexity of k-clique. Inf. Comput. **256**, 9–22 (2017). https://doi.org/10.1016/j.ic.2017.04.005
31. Zeume, T., Schwentick, T.: On the quantifier-free dynamic complexity of reachability. Inf. Comput. **240**, 108–129 (2015). https://doi.org/10.1016/j.ic.2014.09.011

Real τ-Conjecture for Sum-of-Squares: A Unified Approach to Lower Bound and Derandomization

Pranjal Dutta$^{(\boxtimes)}$

Chennai Mathematical Institute, Chennai, India
pranjal@cmi.ac.in

Abstract. Koiran's real τ-conjecture asserts that if a non-zero real univariate polynomial f can be written as $\sum_{i=1}^{k} \prod_{j=1}^{m} f_{ij}$, where each f_{ij} contains at most t monomials, then the number of distinct real roots of f is polynomially bounded in kmt. Assuming the conjecture with parameter $m = \omega(1)$, one can show that VP $\neq$ VNP (i.e. symbolic permanent requires superpolynomial-size circuit). In this paper, we propose a τ-conjecture for sum-of-squares (SOS) model (equivalently, $m = 2$).

For a univariate polynomial f, we study the *sum-of-squares* representation (SOS), i.e. $f = \sum_{i \in [s]} c_i f_i^2$, where c_i are field elements and the f_i's are univariate polynomials. The size of the representation is the number of monomials that appear across the f_i's. Its minimum is the *support-sum* $S(f)$ of f. We *conjecture* that any real univariate f can have at most $O(S(f))$-many real roots. A random polynomial satisfies this property. We connect this conjecture with two central open questions in algebraic complexity– matrix rigidity and VP vs. VNP.

The concept of matrix rigidity was introduced by Valiant (MFCS 1977) and independently by Grigoriev (1976) in the context of computing linear transformations. A matrix is rigid if it is far (in terms of Hamming distance) from any low rank matrix. We know that rigid matrices exist, yet their explicit construction is still a major open question. Here, we show that SOS-τ-conjecture implies construction of such matrices. Moreover, the conjecture also implies the famous Valiant's hypothesis (Valiant, STOC 1979) that VNP is exponentially harder than VP. Thus, this new conjecture implies both the fundamental problems by Valiant.

Furthermore, strengthening the conjecture to sum-of-cubes (SOC) implies that blackbox-PIT (Polynomial Identity Testing) is in P. This is the first time a τ-conjecture has been shown to give a polynomial-time PIT. We also establish some special cases of this conjecture, and prove tight lower bounds for restricted depth-2 models.

Keywords: τ-conjecture · Matrix rigidity · Real root · VP · VNP · PIT

1 Introduction

An *algebraic circuit* over an underlying field $\mathbb{F}$ is a natural model that represents a polynomial compactly. It is a layered directed acyclic graph with the leaf nodes

© Springer Nature Switzerland AG 2021
R. Santhanam and D. Musatov (Eds.): CSR 2021, LNCS 12730, pp. 78–101, 2021.
https://doi.org/10.1007/978-3-030-79416-3_5

as the input variables $x_1, \ldots, x_n$, and constants from $\mathbb{F}$. All the other nodes are labeled as $+$ and $\times$ gates. The output of the root node is the polynomial computed by the circuit. Two important complexity parameters of a circuit are: 1) the *size*, the number of edges and nodes, 2) the *depth*, the number of layers.

The famous Shub-Smale τ-conjecture [48] is a conjecture in algebraic complexity, asserting that a univariate polynomial which is computable by a small algebraic circuit has a small number of integer roots. It was established in [48] that the τ-conjecture implies $\mathsf{P}_\mathbb{C} \neq \mathsf{NP}_\mathbb{C}$, for the Blum–Shub–Smale (BSS) model of computation over the complex numbers [6,7]. Bürgisser [9] obtained a similar result for the algebraic version of P vs. NP, namely, VP vs. VNP (informally defined below), which was originally proposed by Valiant [53].

The class $\mathsf{VP}_\mathbb{F}$ contains the families of n-variate polynomials of degree $\mathsf{poly}(n)$ over $\mathbb{F}$, computed by circuits of $\mathsf{poly}(n)$-size. The class $\mathsf{VNP}_\mathbb{F}$ can be seen as a non-deterministic analog of the class $\mathsf{VP}_\mathbb{F}$[1]. Informally, it contains the families of n-variate polynomials that can be written as an exponential sum of polynomials in VP; for formal definitions, see Sect. 2. VP is contained in VNP, and it is believed that this containment is *strict* (Valiant's Hypothesis [53]). For more details, see [11,31,47]. Unless specified otherwise, we consider field $\mathbb{F} = \mathbb{R}$, the field of real numbers, or $\mathbb{C}$, the field of complex numbers.

One possible disadvantage of the τ-conjecture is the reference of *integer roots*. As a natural approach to the τ-conjecture, one can try to bound the number of *real roots* instead of integer roots. However, a mere replacement of "integer roots" by "real roots" fails miserably as the number of real roots of a univariate polynomial can be *exponential* in its circuit size; for e.g. Chebyshev polynomials [49]. Interestingly, Koiran [24] came up with the following τ-conjecture for the *restricted* (depth-4) circuits. It states that if $f \in \mathbb{R}[x]$ is a polynomial of the form $f = \sum_{i=1}^{k} \prod_{j=1}^{m} f_{ij}$, where each $f_{ij} \in \mathbb{R}[x]$ is t-sparse [2], then the number of *distinct* real roots of f can at most be $\mathsf{poly}(kmt)$. Note that, the conjecture is true for $m = 1$, by Descartes' rule of signs (Lemma 5). Using the celebrated depth-4 reduction [2,25], it was established that real τ-conjecture with $m = \omega(1)$, yields a strong separation in the *constant-free* settings, i.e. $\mathsf{VP}_0 \neq \mathsf{VNP}_0$. Later, it was shown to imply $\mathsf{VP} \neq \mathsf{VNP}$, see [16,50].

Before trying to prove a lower bound in the general settings, we would like to remark that one of the major open problems in algebraic complexity is to prove any *super-linear* lower bound for *linear circuits*. These are simple circuits where we are only allowed to use addition and multiplication by a scalar. By definition, they can only compute linear (affine) functions. In fact, any algebraic circuit, computing a set of linear functions, can be converted into a linear circuit with only a constant blow-up in size, see [11, Theorem 13.1]. Clearly, every set of n linear functions on n variables can be represented by a matrix in $\mathbb{F}^{n \times n}$, which can be computed by a linear circuit of size $O(n^2)$.

Given the ubiquitous role linear transformations play in computing, understanding the inherent complexity of explicit linear transformations is impor-

---

[1] We will drop the subscript $\mathbb{F}$ whenever $\mathbb{F}$ is implicitly clear or does not matter.

[2] $f := \sum_{i=0}^{n} a_i x^i$ is t-sparse if at most t of the coefficients $a_0, \ldots, a_n$ are non-zero.

tant. Using dimension argument/counting, it can be shown that a random matrix requires $\Omega(n^2)$-size circuit. However, showing the same for an explicit $A_n \in \mathbb{F}^{n \times n}$, still remains open. The standard notion of explicitness is that there is a deterministic algorithm which outputs the matrix A_n in $\text{poly}(n)$-time. Weak super-linear lower bounds are known for constant-depth linear circuits, using superconcentrators and their minimal size, see [5,37,39,51]. It is also known that this technique alone is *insufficient* for proving lower bounds for logarithmic depth.

The quest for showing super-linear lower bound for logarithmic-depth lead to the notion of *matrix rigidity*, a pseudorandom property of matrices, introduced by Valiant [52], and independently by Grigoriev [17].

Definition 1 (Matrix rigidity). *A matrix A over $\mathbb{F}$ is (r, s)-rigid, if one needs to change $> s$ entries in A to obtain a matrix of rank $\leq r$. That is, one cannot decompose A into $A = R + S$, where $\text{rank}(R) \leq r$ and $\text{sp}(S) \leq s$, where $\text{sp}(S)$ is the sparsity of S, i.e., the number of nonzero entries in S.*

Valiant [52] showed that an explicit construction of a $(\epsilon \cdot n, n^{1+\delta})$-rigid matrix, for some $\epsilon, \delta > 0$, will imply a *super-linear* lower bound for linear circuits of depth $O(\log n)$; for a simple proof, see [47, Theorem 3.22]. Pudlak [38] observed that similar rigidity parameters will imply even *stronger* lower bounds for constant depth circuits. Here, we remark that a random matrix is $(r, (n - r)^2)$-rigid, but the best explicit constructions have rigidity $(r, n^2/r \cdot \log(n/r))$ [15,46], which is *insufficient* for proving lower bounds. For recent works, we refer to [4,14,41].

The interplay between proving lower bounds and derandomization is one of the central themes in complexity theory [34]. In algebraic complexity, the central derandomization question is to design an efficient deterministic algorithm for *Polynomial Identity Testing* (PIT) in the *blackbox* model, i.e. to test the zeroness of a given algebraic circuit via query access. Though the celebrated *Polynomial Identity Lemma* [12,35,45,55]) gives a randomized polynomial-time algorithm for blackbox-PIT, finding a deterministic polynomial-time algorithm has been a long-standing open question. The problem also naturally appears in the geometric approaches to the $\text{P} \neq \text{NP}$ question, e.g. [18,32,33].

Efficient blackbox-PIT and circuit lower bounds are strongly intertwined [1,19,23]. However, any connection between constructing explicit rigid matrix and proving $\text{VP} \neq \text{VNP}$ (or $\text{PIT} \in \text{P}$) is not *clear*. Further, to the best of our knowledge, we do not know any *singular* problem that solves these *central* problems. Towards that, we conjecture 1, the SOS-τ-conjecture, and prove Theorems 1,2, & 5, below stated:

> *If the number of real roots of any $f \in \mathbb{R}[x]$ is bounded by a constant multiple of the sum of the sparsity of the univariates when written as sum-of-squares, then there exists an explicit $(\epsilon \cdot n, n^{1+\delta})$-rigid matrix for $\epsilon, \delta > 0$. It also implies exponential separation between $\text{VP}_{\mathbb{C}}$ and $\text{VNP}_{\mathbb{C}}$. Further, strengthening the requirement to sum-of-cubes, puts blackbox-PIT in P.*

The novelty of this work is to introduce the τ-conjecture in the SOS (or SOC)-model and link with the three main problems in algebraic complexity. The sum-of-squares representation (SOS) is one of the most fundamental in number theory and algebra [36,40]; it has found many applications in approximation & optimization, see [29,30,43]. Intuitively, the analytic nature of SOS makes this conjecture a *viable* path to resolve the long standing questions.

1.1   Sum-of-Squares (SOS) Model and a τ-Conjecture

We say that a univariate polynomial $f(x) \in R[x]$ over a ring R is computed as a *sum-of-squares* (SOS) if

$$f = \sum_{i=1}^{s} c_i f_i^2 , \tag{1}$$

for some *top-fanin* s, where $f_i(x) \in R[x]$ and $c_i \in R$.

Definition 2 (Support-sum size $S_R(f)$, [13]). *The* size *of the representation of f in (1) is the* support-sum, *the sum of the sparsity (or support size) of the polynomials f_i. The support-sum size of f, denoted by $S_R(f)$ [3], is defined as the minimal support-sum of f.*

Let $|f|_0$ denote the sparsity of f. For any field $R = \mathbb{F}$ of characteristic $\neq 2$, we have $\sqrt{|f|_0} \leq S_{\mathbb{F}}(f) \leq 2|f|_0 + 2$. The lower bound can be shown by counting monomials. The upper bound follows from the identity: $f = (f+1)^2/4 - (f-1)^2/4$. In particular, the SOS-model is *complete* for any field of characteristic $\neq 2$. Further, a standard geometric-dimension argument implies that $S_{\mathbb{F}}(f) = \Theta(d)$, for *most* univariate polynomials f of degree d, since $|f|_0 = \Theta(d)$, for a random f.

Any #P/poly-explicit f_d (for definition, see Sect. 2), which satisfies $S(f_d) \geq \Omega(d^{1/2+\varepsilon})$, for some sub-constant function $\varepsilon(d)$, implies VP $\neq$ VNP [13, Theorem 6]. However, in this work, we are interested in the number of real roots of f in terms of $S(f)$. Since the sparsity of f can be at most $S(f)^2$, f can have at most $S(f)^2$-many real roots by Descartes' rule of signs (Lemma 5). Further, it can be shown that a random polynomial f can have at most $O(S(f))$-many real roots, similar to [8, Theorem 1.1 with $k = 2$].

Moreover, if f and g have sparsity s, is the number of real roots of $fg + 1$ linear in s? This question was originally designed by Arkadev Chattopadhyay, as the simplest case of Koiran's tau-conjecture; unfortunately we do not yet understand it. Motivated thus, we conjecture the following. Motivated thus, we conjecture the following.

Conjecture 1 (SOS-τ-conjecture). Consider any non-zero polynomial $f(x) \in \mathbb{R}[x]$. Then, there exists a positive constant $c > 0$ such that the number of distinct real roots of f is at most $c \cdot S_{\mathbb{R}}(f)$.

Remark. 1. One can show that Conjecture 1 implies $S_{\mathbb{C}}((x+1)^d) \geq \Omega(d)$; see Lemma 9. This is almost identical to [21,22], where strong distribution property of complex roots with multiplicities were shown to be implied, from the real τ-conjecture.

---

[3] We will write $S(f)$ whenever the underlying ring is clear or does not matter.

2. In the Eq. (1), we could restrict the degrees of f_i to be $O(d \log d)$. This might help us proving the conjecture; for details, see Sect. 3.2 (& remark).
3. The $fg + 1$ case happens to be a special case of this new conjecture for 3 squares. For the two squares,i.e. any $f \in \mathbb{R}[x]$ of the form $c_1 f_1^2 + c_2 f_2^2$, can have at most $O(|f_1|_0 + |f_2|_0)$-many real roots; for details see Theorem 9 in Sect. 6.1.

1.2 Main Results

The leitmotif of this paper is the interplay between the SOS-τ-conjecture and derandomization/hardness questions in algebraic complexity. We start by asking: can the τ-conjecture in the SOS-model imply the existence of explicit rigid matrices? We evince a positive answer. This is the first time a τ-conjecture has been shown to solve the long-awaited matrix rigidity problem.

Theorem 1. *If Conjecture 1 holds, then there exist $\epsilon > 0$ and a "very"-explicit family of real matrices $(A_n)_n$ such that A_n is $(\epsilon \cdot n, n^{1+\delta})$-rigid, for any $\delta < 1$.*

Remark. 1. The matrix A_n is not only $\mathsf{poly}(n)$-explicit, it is 'very' explicit in the good common sense: one could consider as simple as binomial coefficients, recorded one row at a time. This is quite interesting given the recent dramatic developments that have killed virtually all known candidates.
2. Our proof requires the upper bound of $O(S(f))$ in Conjecture 1; any weaker upper bound *does not* yield the same rigidity parameter.

Theorem 1 implies *super-linear* lower bound for linear circuits [52]. Interestingly, Conjecture 1 is robust enough to show strong lower bounds in the *general* circuit settings as well.

Theorem 2. *Conjecture 1 implies that $\mathsf{VNP}_\mathbb{C}$ is exponentially harder than $\mathsf{VP}_\mathbb{C}$.*

Remark. 1. One could directly obtain that Conjecture 1 implies $S_\mathbb{R}(f_d) \geq \Omega(d)$, where $f_d := \prod_{i=1}^{d}(x - i)$. However, to separate VP and VNP, using the proof techniques of [13] (with $\varepsilon = 1$) *require* GRH (Generalized Riemann Hypothesis).
2. To show an unconditional lower bound, we work with $f_d := \sum_{i=0}^{d} 2^{2i(d-i)} \cdot x^i$ (a similar family was considered in [16, Eq. 8]). However, the hardness proof is completely different from [16], due to disparate settings and parameters.
3. The hardness proof presented here does not require any *fine-grained* decomposition, as required in [13],i.e. via algebraic branching programs (ABP) or circuit-depth boosting techniques. For the simplicity and completeness, we present a self-contained proof in Sect. 3.2.
4. The exponential separation puts blackbox-PIT $\in$ QP (*Quasi*poly) [23].

We introduce a τ-conjecture for the sum-of-cubes (SOC-model) in Sect. 4, and show that sufficient strengthening (of the measure) gives a *polynomial*-time PIT for general circuits (Theorem 5). This is the *first* time a τ-conjecture has been shown to derandomize PIT *completely*.

We also show lower bounds for symmetric depth-2 circuits and invertible depth-2 circuits in Sect. 5, by studying $S(f_d)$, for different explicit polynomials f_d (in the restricted sense); for details see Theorem 6.

1.3 Proof Ideas

The τ-conjecture for $\sum^k \prod^m \sum^t \prod$-model (with $m = \omega(1)$) has been shown to imply circuit hardness [16,24,26]. This work is more about remodeling the τ-conjecture in the *simplest* format possible and viewing this as the pivotal problem of interest. Although the proofs of the theorems are standard and obtained from clever maneuvering of the existing techniques, the implications are quite far-reaching. Moreover, the τ-conjecture for $m = 1$ is true using Descartes' rule (Lemma 5) while $m = 2$ (*equivalent* to SOS-model) *solves* almost everything. This makes the whole regime tantalisingly close to being realisable.

Proof idea of Theorem 1. If A is not $(\epsilon n, n^{1+\delta})$-rigid, then one can show that A can be written as BC, where 'sparse' matrices B and C can have at most $2cn^2 + 2n^{1+\delta}$ non-zero entries (item 1–4 in Sect. 3.1). Now, the idea is to use $f_d := \prod_{i \in [d]}(x - i)$, to construct matrices A_n that cannot be factored thus.

Define $d := n^2 - 1$, $[x_1]_n := \begin{bmatrix} 1 & x_1 & \cdots & x_1^{n-1} \end{bmatrix}$, and similarly $[x_2]_n$. Define polynomial $g_n(x_1, x_2)$ such that after Kronecker substitution: $g_n(x_1, x_2) \mapsto g_n(x, x^n) = f_d$. Finally, define matrix A_n such that $[x_2]_n A_n [x_1]_n^T = g_n(x_1, x_2)$. Note that A_n is $\mathsf{poly}(n)$-explicit.

Assume Conjecture 1. Then, $S_{\mathbb{R}}(f_d) > \delta' \cdot d$ for some $\delta' > 0$. We show that any layered linear circuit of depth-2 computing A_n has size $> (\delta'/2) \cdot d$. Suppose not, i.e. $A_n = BC$, with $B \in \mathbb{R}^{n \times t}$, $C \in \mathbb{R}^{t \times n}$ such that $t \leq (\delta'/2) \cdot d$. Then, $[x_2]_n B \, C \, [x_1]_n^T = g_n(x_1, x_2)$. We deduce that $f_d = \sum_{i \in [t]} \ell_i(x) \, \tilde{\ell}_i(x^n)$, where $([x_2]_n B)_i =: \tilde{\ell}_i(x_2)$ and $(C \, [x]_n^T)_i =: \ell_i(x_1)$. Note that, f_d can easily be written as sum of $2t$ squares with sum of sparsity $\leq \delta' \cdot d$, a contradiction.

This ensures that the number of nonzero entries in B and C is 'large'. Therefore, choosing ϵ and δ carefully, we have A_n *is rigid*. We remark that we can also work with $f_d = (x + 1)^d$, but one needs to use Lemma 9.

Proof idea of Theorem 2. We work with $f_d := \sum_{i=0}^d 2^{2i(d-i)} \cdot x^i$. Note that, f_d satisfies the Kurtz condition (Theorem 3). Thus, assuming Conjecture 1, $S_{\mathbb{R}}(f_d) \geq \Omega(d)$, which further implies $S_{\mathbb{C}}(f_d) \geq \Omega(d)$ (Lemma 10). This implies exponential hardness of VP, by [13, Theorem 6] (with $\varepsilon = 1$); however in this setting, the hardness proof is much *simpler* due to stronger hardness assumption and a classical decomposition (Lemma 1), instead of a convoluted one, *via* ABPs, required in [13] (for general ε).

The overall idea is to convert f_d to an *exponentially hard* multivariate polynomial. Usually, (inverse) Kronecker type substitution is used for univariate to multivariate conversion ([16,24]); here we *do not* use the Kronecker due to a technical barrier [4]. Instead, we use a *multilinear* map ϕ [13] that sends x^i

---

[4] Kronecker would give a *naive* bound of $\binom{k+kn/2}{k} > (n+1)^k > d$; which is useless.

to $\phi(x^i) := \prod_{j\in[n],\,\ell\in[0\ldots k-1]} y_{j,\ell}$, where $\ell \cdot k^{j-1}$ contributes to the $\mathrm{base}_k(i)$-*representation* in the j-th position; here $n = O(\log d)$ and k is a large constant.

Define, $\phi(f_d) =: P_{n,k}$, by linear extension. By construction, $P_{n,k}$ is a kn-variate degree-n multilinear polynomial. For large constant k, we show that $\mathrm{size}(P_{n,k}) > d^{1/7} = 2^{\Omega(kn)}$. The proof goes via contradiction. If the size is smaller, then using a classical decomposition (Lemma 1), $P_{n,k}$ can be written as sum of $\le O(d^{1/7} \cdot n^2)$-many Q_i^2's; where the intermediate polynomial Q_i (kn-variate) has degree at most $2n/3$. Thus, a naive upper bound on the support-sum $S_{\mathbb{C}}(f_d)$ is $O(d^{1/7} \cdot n^2 \cdot \binom{kn+2n/3}{2n/3})) = O(d^{6/7} \log^2 d) = o(d)$, a contradiction.

As the coefficients are *easy* to compute, $(P_{n,k})_n \in \mathsf{VNP}$, by Valiant's criterion (Theorem 4). Therefore, the conclusion follows.

1.4 Comparison with Prior Works

Technically, our SOS-τ-conjecture is incomparable to the earlier τ-conjectures as all of the previous works [16, 24, 26] used the standard depth-reduction results [1, 2, 20, 25], hence, they were concerned with the sum-of unbounded-powers $\sum \prod^m \sum \prod$, with $m = \omega(1)$, while we work with $m = 2$. As mentioned earlier, this is the *first* time we are showing connections to matrix-rigidity and PIT; these were perhaps always desired of, nonetheless *never achieved*.

Moreover, the measure $S(f)$ in the τ-conjecture is different from the usual circuit-size. If we consider the expression in (1) as a $\sum \bigwedge^2 \sum \prod$-formula, then the support-sum is the number of $\prod$-operations directly above the input level. However, the usual measure is the size of the depth-4 circuit $\sum^k \prod^m \sum^t \prod$. Even if we substitute $m = 2$, there is *no* direct dependence of t (individual sparsity of the intermediate polynomials f_i) on $S(f)$, which implies that the sparsity of some f_i could be *large*. However, the upper bound requirement in [24] is $\mathrm{poly}(kmt)$ while the SOS-τ-conjecture *demands* a linear (*stronger*) dependence on $S(f)$.

Further, the polynomial family and the proof used in [16, 24] are *different* from those in Theorem 2, as it relies on depth-4 reduction and the usual Kronecker map while our proof relies on multilinearization ([13]) and a folklore decomposition (Lemma 1), see [42, 44, 47]. In [13], a *fine-grained* decomposition (using algebraic branching programs) was used which made the parameters and the hardness proof more intricate; this is *not at all* required in this work.

2 Preliminaries

Basic notation. Let $[n] = \{1, \ldots, n\}$. In general, for $a < b$, $[a, b]$ denotes all integers $a \le i \le b$. For $i \in \mathbb{N}$ and $b \ge 2$, we denote by $\mathrm{base}_b(i)$ the unique k-tuple $(i_1, \ldots, i_k)$ such that $i = \sum_{j=1}^{k} i_j\, b^{j-1}$. In the special case $b = 2$, we define $\mathrm{bin}(i) := \mathrm{base}_2(i)$. $\mathbb{R}$ denotes the real field while $\mathbb{C}$ denotes the complex field.

For a matrix S, $\mathrm{sp}(S)$, the sparsity of S, is the number of non-zero entries.

Binomial Inequality. We use the following standard bound on binomial coefficients for $1 \leq k \leq n$,

$$\binom{n}{k} \leq \left(\frac{en}{k}\right)^k. \qquad (2)$$

Polynomials and Real Roots. For a polynomial $p \in \mathbb{F}[x]$, supp(p), the *support of p* is the set of nonzero monomials in p. The *sparsity* or *support size* of p is $|p|_0 = |\text{supp}(p)|$. By coef(p) we denote the *coefficient vector* of p (in some order).

For an exponent vector $\boldsymbol{e} = (e_1, \ldots, e_k)$, $\boldsymbol{x}^{\boldsymbol{e}}$ denotes the monomial $x_1^{e_1} \ldots x_k^{e_k}$.

When are all the roots of a univariate polynomial *real* and *distinct*? Kurtz [28] came up with the following *tight* and *sufficient* condition.

Theorem 3 ([28]). *Let f be a real polynomial of degree $n \geq 2$ with positive coefficients. If*

$$a_i^2 > 4a_{i-1}a_{i+1}, \ \forall i \in [n-1],$$

then all the roots of f are real and distinct.

Primer on Algebraic Complexity. The *algebraic circuit complexity* of a polynomial f, denoted by size(f), is the size of the smallest circuit computing f. A family of n-variate polynomials $(f_n)_n$ over $\mathbb{F}$ is in VNP if there exists a family of polynomials $(g_n)_n$ in VP such that for every $\boldsymbol{x} = (x_1, \ldots, x_n)$ one can write $f_n(\boldsymbol{x}) = \sum_{w \in \{0,1\}^{t(n)}} g_n(\boldsymbol{x}, w)$, for some polynomial $t(n)$ which is called the *witness size.*

VP and VNP have several closure properties. In particular, they are closed under substitution. That is, for a polynomial $f(\boldsymbol{x}, \boldsymbol{y}) \in$ VP (or VNP), also $f(\boldsymbol{x}, \boldsymbol{y}_0) \in$ VP (respectively VNP), for any values $\boldsymbol{y}_0$ from $\mathbb{F}$ assigned to the variables in $\boldsymbol{y}$.

The explicitness of the family plays a major role in its usefulness in algebraic complexity.

Definition 3 (Explicit functions). *Let $(f_d)_d$ be a polynomial family, where $f_d(x)$ is of degree d. The family is* explicit, *if its coefficient-function is computable in time* poly log(d) *and each coefficient can be at most* poly(d)*-bits long. The coefficient-function gets input (j, i, d) and outputs the j-th bit of the coefficient of x^i in f_d.*

Valiant [53] gave a useful *sufficient* condition for the explicitness of a polynomial family $f_n(\boldsymbol{x})$ so that it is in VNP. For a proof, see [10, Proposition 2.20].

Theorem 4 (Valiant's criterion, [53]). *Let function $\phi : \{0,1\}^n \to \{0,1\}^n$ be computable in* P/poly. *Then, the family of polynomials defined by $f_n(\boldsymbol{x}) := \sum_{e \in \{0,1\}^n} \phi(\boldsymbol{e}) \cdot \boldsymbol{x}^{\boldsymbol{e}}$, is in* VNP.

The following lemma is a classical decomposition lemma using frontiers, for details see [3,42,54]. For a proof, see [44, Lemma 5.12] (with frontier $m = d/3$).

Lemma 1 (Sum of product-of-2). *Let $f(\boldsymbol{x})$ be an n-variate, homogeneous, degree d polynomial computed by a homogeneous circuit Φ of size s. Then, there exist polynomials $f_{ij} \in \mathbb{F}[\boldsymbol{x}]$ s.t.*

$$f(\boldsymbol{x}) = \sum_{i=1}^{s} f_{i1} \cdot f_{i2} , \quad \text{with the following properties:} \tag{3}$$

1) $d/3 \leq \deg(f_{i1}), \deg(f_{i2}) \leq 2d/3, \forall i \in [s],$ *and* **2)** $\deg(f_{i1}) + \deg(f_{i2}) = d, \forall i \in [s].$

Remark. It is well known that each homogeneous part can be computed by a homogeneous circuit of size $O(sd^2)$. Thus, for non-homogeneous polynomials, s can be replaced by $O(sd^2)$.

3   Proof of the Main Results

In this section, we prove the main theorems, namely Theorem 1–2.

3.1   SOS-τ-Conjecture to Matrix Rigidity: Proof of Theorem 1

We argue via *linear circuits* which we have defined in Sect. 1. Linear circuits can compute linear functions (see [27, Sec.1.2]). As a graph, the nodes of a linear circuits are either input nodes or addition nodes, and the edges are labeled by scalars. If an edge from u to v is labeled by $c \in \mathbb{F}$, then the output of u is multiplied by c and then given as input to v.

We eventually establish that any matrix $A \in \mathbb{R}^{n \times n}$ which is not (r, s)-rigid, for some r, s, can be computed by a depth-2 circuit of size $2rn + s + n$; see item 1–4 below. This will be crucial in the proof of Theorem 1. We give this bound over any general field $\mathbb{F}$.

1. Let $\boldsymbol{a} = (a_1, \ldots, a_n)$ be a vector. Consider $\boldsymbol{a}$ as a linear function $\mathbb{F}^n \to \mathbb{F}$. It can be computed by a linear circuit of depth 1 with n inputs and one addition-gate as output gate. The edge from the i-th input is labeled by a_i. The size of the circuit is n. However, we omit edges labeled by 0. Hence, the size of the circuit is actually $\mathrm{sp}(\boldsymbol{a}) \leq n$, the sparsity of $\boldsymbol{a}$.
   Similarly, we consider an $n \times n$ matrix A as a linear transformation $\mathbb{F}^n \to \mathbb{F}^n$. For each row vector of A we get a linear circuit as described above. Hence we represent A by circuit of depth 1 with n output gates and size $\mathrm{sp}(A) \leq n^2$.
2. The model gets already interesting for linear circuits of depth 2. Suppose $A = BC$, where B is a $n \times r$ matrix and C is a $r \times n$ matrix. Then we can take the depth-1 circuit for C at the bottom as in item 1 and combine it with the depth-1 circuit for B on top. The resulting depth-2 circuit is *layered*: all edges go either from the bottom to the middle layer, or from the middle to the top layer. The size of the circuit is $\mathrm{sp}(B) + \mathrm{sp}(C) \leq 2rn$.

In particular, there is a representation $A = BC$ with $r = \text{rank}(A)$. Hence the rank of A is involved in the circuit size bound for A. Also note that r is bounded by the size of the circuit because be omit all zero-edges.

Note that any layered linear circuit of depth 2 in turn gives a factorization of A as a product of 2 matrices, $A = BC$, where the top edges define B and the bottom edges C.

3. Let $A = BC + D$, where B, C are as above and D is a $n \times n$ matrix. Then we can represent A by a depth-2 circuit for BC as in item 2 plus edges from the inputs directly to the output nodes to represent D as in item 1. The resulting circuit has depth 2 and size $\text{sp}(B) + \text{sp}(C) + \text{sp}(D) \leq 2rn + n^2$, but it would not be layered. We can transform it into a layered circuit by writing A as $A = BC + ID$, where I is the $n \times n$ identity matrix. Then we get a depth-2 circuit for ID similar to BC and can combine the two circuits into one. The size increases by $\leq n$ edges for I.

4. Now consider matrix A that is not (r, s)-rigid, for some r, s. Hence, we can write A as $A = R + S$, where $\text{rank}(R) = r$ and $\text{sp}(S) = s$. Then R can be written as as $R = BC$, where B is a $n \times r$ matrix and C is a $r \times n$ matrix. From item 3, we have that $A = BC + S$ has a layered linear circuit of depth 2 of size $\leq 2rn + s + n$.

Proof (Proof of Theorem 1). Consider the polynomial family $f_d := \prod_{i \in [d]}(x - i)$. Let $d =: n^2 - 1$, for some $n \in \mathbb{N}$. Conjecture 1 implies that $S_{\mathbb{R}}(f_d) > \delta' \cdot d$, for some $\delta' > 0$. Note that $\delta' \leq 2$, as $S_{\mathbb{R}}(f_d) \leq 2d + 2$, from the upper bound (see Sect. 1.1). This δ' will play a crucial role in the proof.

Define the bivariate polynomial $g_n \in \mathbb{R}[x_1, x_2]$ from f_d such that after the Kronecker substitution, $g_n(x, x^n) = f_d$. It is easy to construct g_n from a given d; just convert every x^e, for $e \in [0, d]$ to $x_1^{e_1} \cdot x_2^{e_2}$, where $e =: e_1 + e_2 \cdot n$, and $0 \leq e_i \leq n - 1$. Thus, the individual degree of each x_i in g_n is at most $n - 1$.

Let $g_n(x_1, x_2) = \sum_{1 \leq i, j \leq n} a_{i,j} x_1^{i-1} x_2^{j-1}$. By the definition of f_d, $a_{i,j} = \text{coef}_{x^{(i-1) + (j-1)n}}(f_d)$. Define the $n \times n$ matrix $A_n = (a_{i,j})_{1 \leq i,j \leq n}$ and vectors

$$[x_1]_n = \left(1 \ x_1 \ \cdots \ x_1^{n-1}\right) , \quad [x_2]_n = \left(1 \ x_2 \ \cdots \ x_2^{n-1}\right) .$$

Thus, $g_n(x_1, x_2) = [x_1]_n A_n [x_2]_n^T$. Further, $a_{i,j}$ is $\text{poly}(n)$-computable implies A_n is $\text{poly}(n)$-explicit. Next we show a lower bound on the linear circuit size of A_n.

Lemma 2. *Conjecture 1 $\implies$ any layered linear circuit of depth 2 that computes A_n, has size $> (\delta'/2) \cdot d$.*

Proof of Lemma 2. Conjecture 1 implies that $S_{\mathbb{R}}(f_d) > \delta' \cdot d$, for some $\delta' > 0$. We show that, size of the linear circuit computing A_n has size $> (\delta'/2) \cdot d$.

Assume that this is false. Then we can write $A_n = BC$, where $B \in \mathbb{R}^{n \times t}$, $C \in \mathbb{R}^{t \times n}$, such that $t \leq \text{sp}(B) + \text{sp}(C) \leq (\delta'/2) \cdot d$.

Denote

$$[x_1]_n B = \left(\ell_1(x_1) \ \ell_2(x_1) \ \cdots \ \ell_t(x_1)\right) \quad \text{and} \quad C [x_2]_n^T = \left(\tilde{\ell}_1(x_2) \ \tilde{\ell}_2(x_2) \ \cdots \ \tilde{\ell}_t(x_2)\right)^T .$$

Then

$$g_n(x_1, x_2) = [x_1]_n A_n [x_2]_n^T = [x_1]_n B C [x_2]_n^T = \sum_{i=1}^{t} \ell_i(x_1) \tilde{\ell}_i(x_2).$$

Since $\mathrm{sp}(B) + \mathrm{sp}(C) \leq (\delta/2) \cdot d$, we have $\sum_{i=1}^{t} (|\ell_i|_0 + |\tilde{\ell}_i|_0) \leq (\delta'/2) \cdot d$. Substituting $x_1 = x$ and $x_2 = x^n$, we get

$$f_d(x) = g(x, x^n) = \sum_{i=1}^{t} \ell_i(x) \tilde{\ell}_i(x^n) = \sum_{i=1}^{t} \left(\frac{\ell_i(x) + \tilde{\ell}_i(x^n)}{2} \right)^2 - \sum_{i=1}^{t} \left(\frac{\ell_i(x) - \tilde{\ell}_i(x^n)}{2} \right)^2.$$

Thus, we have a representation of f_d as $\leq 2t \leq \delta' \cdot d$ sum of squares. Note that, this means

$$S_{\mathbb{R}}(f_d) \leq \sum_{i=1}^{t} 2 \cdot \left(|\ell_i|_0 + |\tilde{\ell}_i|_0 \right) \leq \delta' \cdot d, \tag{4}$$

contradicting the assumption on the hardness of f_d. This proves Lemma 2.

We now show that A_n is $((\delta'/8) \cdot n, n^{1+\delta})$-rigid, for any $\delta < 1$. For the sake of contradiction, assume that this is false. Then there is a $\delta < 1$, and a decomposition $A_n = R + S$, where $\mathrm{rank}(R) = r = (\delta'/8) \cdot n$, and $\mathrm{sp}(S) = s = n^{1+\delta}$. By item 4 above, A_n has a layered linear circuit C_n of depth 2 of size

$$\mathrm{size}(C_n) \leq 2rn + s + n \leq \frac{\delta' \cdot n^2}{4} + 2n^{1+\delta}. \tag{5}$$

Recall that δ' is a constant and $\delta < 1$. Hence, for large enough n, we have $2n^{1+\delta} \leq \delta' \cdot (\frac{n^2-2}{4})$. Note: $\delta = 1$ is not achievable as $\delta' \leq 2$. Now, we can continue the inequalities in (5) by

$$\mathrm{size}(C_n) \leq \delta' \cdot (\frac{n^2 - 1}{2}) = (\delta'/2) \cdot d. \tag{6}$$

For the last equation, recall that $d = n^2 - 1$. The bound in (6) contradicts Lemma 2. Therefore we conclude that A_n is $(\epsilon \cdot n, n^{1+\delta})$-rigid for any $\delta < 1$, where $\epsilon := \delta'/8$ (remember δ' was fixed at the beginning).

Remark. The same proof holds over $\mathbb{C}$, using Lemma 10.

3.2 SOS-τ-Conjecture to Exponential Hardness: Proof of Theorem 2

Proof. We will construct an explicit (multivariate) polynomial family from the univariate $f_d := \sum_{i=0}^{d} 2^{2i(d-i)} \cdot x^i$, and show that it requires *exponential size* circuit (assuming Conjecture 1). Moreover, we show that the family is in VNP, and the conclusion would directly follow.

Kurtz Condition. We show that the coefficients $a_i := 2^{2i(d-i)}$ satisfies the Kurtz condition (Theorem 3). For that, it suffices to check that

$$4i(d-i) > 2 + 2(i-1)(d-i+1) + 2(i+1)(d-i-1),$$

which is true since LHS - RHS=2. Therefore, roots of f_d are all distinct and real.

Construction. We will construct $(P_{n,k})_n$ from f_d, where $P_{n,k}$ is a multilinear degree-n and kn-variate polynomial, where k is a fixed constant (to be fixed in Lemma 4), and $n = O(\log d)$; thus $kn = O(\log d)$.

The basic relation between d, n and k is that $k^n \geq d+1 > (k-1)^n$. Introduce kn many new variables $y_{j,\ell}$, where $1 \leq j \leq n$ and $0 \leq \ell \leq k-1$. Let $\phi_{n,k}$ be the map,

$$\phi_{n,k} : x^i \mapsto \prod_{j=1}^{n} y_{j,i_j} \text{ ,where } i =: \sum_{j=1}^{n} i_j \cdot k^{j-1}, \ 0 \leq i_j \leq k-1.$$

For $i \in [0, d]$, $\phi_{n,k}$ maps x^i uniquely to a multilinear monomial of degree n. By linear extension, define $\phi_{n,k}(f_d) =: P_{n,k}$. By construction, $P_{n,k}$ is n-degree, kn-variate multilinear polynomial. Let $\psi_{n,k}$ be the homomorphism that maps any degree-n multilinear monomial, defined on variables $y_{j,\ell}$, such that $y_{j,\ell} \mapsto x^{\ell \cdot k^{j-1}}$. Trivially, $\psi_{n,k} \circ \phi_{n,k}(f) = f$, for any degree $\leq d$ polynomial $f \in \mathbb{C}[x]$.

Lemma 3. $(P_{n,k})_n \in \mathsf{VNP}$.

Proof. By construction, $P_{n,k}$ is a kn-variate, individual degree-n multilinear polynomial. Hence,

$$P_{n,k} = \sum_{e \in \{0,1\}^{kn}} \gamma(e) \cdot y^e.$$

Here, y denotes the kn variables $y_{j,\ell}$ where $1 \leq j \leq n$ and $0 \leq \ell \leq k-1$, and e denotes the exponent-vector. As each x^e in supp(f_d) maps to a monomial y^e *uniquely*; given e, one can easily compute $e := \sum_{j=1}^{n} e_j \cdot k^{j-1}$, and thus $\gamma(e) = \text{coef}_{x^e}(f_d) = 2^{2e(d-e)}$. Note that, $\gamma(e) < 2^{d^2}$, for all e. We also remark that each bit of $\gamma(e)$ is computable in poly($\log d$) = poly(kn)-time.

Write each $\gamma(e)$ in binary, i.e. $\gamma(e) =: \sum_{j=0}^{d^2-1} \gamma_j(e) \cdot 2^j$, where $\gamma_j(e) \in \{0,1\}$ is computable in P. As $d^2 - 1 < k^{2n}$, introduce new variables $z = (z_1, \ldots, z_m)$, where $m := 2n \log k = O(n)$ [so that, $d^2 - 1 \leq 2^m - 1$]; and consider the auxiliary polynomial $\tilde{\gamma}(e, z) := \sum_{j \in \{0,1\}^m} \gamma_j(e) \cdot z^{\text{bin}(j)}$. Here, we identify $j \in [0, 2^m - 1]$ as a unique $j \in \{0,1\}^m$, via bin(j),i.e. $\gamma_j = \gamma_j$. Let $z_0 := (2^{2^0}, \ldots, 2^{2^{m-1}})$. Note that, $\tilde{\gamma}(e, z_0) = \gamma(e)$. Finally, consider the $(m + kn)$-variate (where $m + kn = O(n)$) auxiliary polynomial $h_{n,k}(y, z)$ as:

$$h_{n,k}(y, z) := \sum_{e \in \{0,1\}^{kn}} \tilde{\gamma}(e, z) \cdot y^e = \sum_{e \in \{0,1\}^{kn}} \sum_{j \in \{0,1\}^m} \gamma_j(e) \cdot z^j \cdot y^e.$$

Then, we have $h_{n,k}(y, z_0) = P_{n,k}(y)$. Since each bit $\gamma_j(e)$ is computable in P, thus by Valiant's criterion (Theorem 4), we have $(h_{n,k}(y, z))_n \in \mathsf{VNP}$. As VNP is *closed* under substitution, it follows that $(P_{n,k}(y))_n \in \mathsf{VNP}$.

Next we show that $P_{n,k}$ is exponentially hard assuming Conjecture 1.

Lemma 4. *Conjecture 1 implies $P_{n,k}$ requires exponential-size circuit.*

Proof. We show that over $\mathbb{C}$, size of the minimal circuit computing $P_{n,k}$, namely $\text{size}(P_{n,k}) > d^{1/7} = 2^{\Omega(kn)}$. If not, then apply Lemma 1 to conclude that

$$P_{n,k} = \sum_{i=1}^{s} c_i \cdot Q_i^2 \implies f_d = \sum_{i=1}^{s} c_i \cdot \psi_{n,k}(Q_i)^2,$$

where, $\deg(Q_i) \leq 2n/3$, and $s = O(d^{1/7} \cdot n^2)$. Above equation implies: $S_{\mathbb{C}}(f_d) \leq s \cdot \binom{kn+2n/3}{2n/3}$. We want to show that $S_{\mathbb{C}}(f_d) \leq o(d)$, this will contradict Conjecture 1. This is because the coefficients of f_d satisfies the Kurtz condition implying f_d has all distinct real roots, then Conjecture 1 implies that $S_{\mathbb{R}}(f_d) \geq \Omega(d) \implies S_{\mathbb{C}}(f_d) \geq \Omega(d)$, from Lemma 10.

By assumption, $s \leq O(d^{1/7} \cdot \log^2 d)$. It suffices to show that $\binom{kn+2n/3}{2n/3} \leq d^{5/7}$, so that $S(f_d) \leq O(d^{6/7} \cdot \log^2 d) = o(d)$, the desired contradiction. Use Equation (2) to show the upper bound on the binomial:

$$\binom{kn + 2n/3}{2n/3} \leq (e + 3ek/2)^{2n/3} \leq (5(k-1))^{2n/3} \leq (k-1)^{5n/7} \leq d^{5/7}.$$

The second inequality holds for $e + 3ek/2, \leq 5(k-1)$; so $k \geq 9$ suffices. For the third inequality to be true, $(k-1)^{5/7} \geq (5(k-1))^{2/3}$ suffices; this holds true for $(k-1)^{1/21} \geq 5^{2/3} \iff k \geq 5^{14} + 1$. We also used $d \geq (k-1)^n$ (by assumption).

Both the above Lemma 3-4 imply the desired conclusion.

Remark. As $\deg(Q_i) \leq 2n/3$, we have $\deg(\psi_{n,k}(Q_i)) \leq 2n/3 \cdot (k-1) \cdot k^{n-1} < n.k^n = O(nd) = O(d \log d)$. Thus, it is enough to consider the restricted-degree SOS representation, and prove the conjecture.

4   A τ-Conjecture for Sum-of-Cubes and Derandomization

It was shown in [13] that a strong lower bound in the sum-of-cubes model leads to a *complete* derandomization of blackbox-PIT. We say that a univariate polynomial $f(x) \in R[x]$ over a ring R is computed as a *sum-of-cubes* (SOC), if

$$f = \sum_{i=1}^{s} c_i f_i^3, \tag{7}$$

for some top-fanin s, where $f_i(x) \in R[x]$ and $c_i \in R$.

Definition 4 (Support-union size $U_R(f,s)$, [13]). *The size of the representation of f in (7) is the size of the support-union, namely the number of distinct monomials in the representation, $|\bigcup_{i=1}^{s} \text{supp}(f_i)|$, where support $\text{supp}(f_i)$ denotes the set of monomials with a nonzero coefficient in the polynomial $f_i(x)$. The support-union size of f with respect to s, denoted $U_R(f,s)$, is defined as the minimal support-union size when f is written as in (7).*

If we consider the expression in (7) as a $\sum \bigwedge^3 \sum \prod$-circuit, then the support-union size is the number of $\prod$-operations directly above the input level (unlike $\sum \bigwedge^2 \sum \prod$-*formula* in Definition (2)).

The two measures– support-union and support-sum –are largely incomparable, since $U(\cdot)$ has the extra argument s.

Here, we remark that for any polynomial f, we have $|f|_0^{1/3} \le U_{\mathbb{F}}(f, s) \le |f|_0 + 1$, where the upper bound is for $s \ge 3$, and for fields $R = \mathbb{F}$ of characteristic $\ne 2, 3$. The upper bound follows from the identity: $f = (f + 2)^3/24 + (f - 2)^3/24 - f^3/12$. Hence, the SOC-model is *complete* for any field of characteristic $\ne 2, 3$. The lower bound can be shown by counting monomials.

It is unclear whether an SOC representation with support-union $= o(d)$ exists for a very small fanin s over $\mathbb{F} = \mathbb{Q}$ [5]. This trade-off between the measure U, and the top-fanin s lead to the definition of hardness in the SOC-model.

Definition 5 (SOC-hardness, [13]). *A* poly(d)-*time explicit univariate polynomial family* $(f_d)_d$ *is SOC-hard, if there exists a positive constant* $\varepsilon < 1/2$, *such that* $U_{\mathbb{F}}(f_d, d^\varepsilon) = \Omega(d)$.

The existence of an SOC-hard family implies blackbox-PIT $\in$ P [13, Theorem 11]. Owing the same tenable approach to connect PIT with the number of real roots, we conjecture the following.

Conjecture 2 (SOC-τ-conjecture). Consider any non-zero polynomial $f \in \mathbb{R}[x]$. Then, there exist positive constants $\varepsilon < 1/2$, and c such that the number of distinct real roots of f is at most $c \cdot U_{\mathbb{R}}(f, d^\varepsilon)$.

Remark. We show that $f = c_1 f_1^3 + c_2 f_2^3$, has at most $O(\mathrm{supp}(f_1) \cup \mathrm{supp}(f_2))$-many real roots (see Theorem 10).

Now, we show a complete derandomization of blackbox-PIT assuming Conjecture 2.

Theorem 5 (Derandomization). *If SOC-τ-conjecture holds, then blackbox-PIT $\in$ P.*

Proof sketch. Consider the polynomial family $f_d := \prod_{i \in [d]} (x - i)$. It is trivial to see that $(f_d)_d$ is poly(d)-explicit. Moreover, if the SOC-τ-conjecture is true, then there exists a $\varepsilon < 1/2$, such that $U_{\mathbb{R}}(f_d, d^\varepsilon) = \Omega(d)$ implying $(f_d)_d$ is an explicit SOC-hard family. Invoking [13, Theorem 11], the conclusion follows.

5   Lower Bound for Restricted Models

Kumar and Volk [27] showed a strong connection between matrix rigidity and depth-2 linear circuit lower bound. They argued (similarly done in [38] in a different language) that depth-2 $\Omega(n^2)$ lower bound for an explicit matrix is *necessary* and *sufficient* for proving *super*-linear lower bound for general $O(\log n)$-depth circuits (or matrix rigidity).

---

[5] For large $s = \Omega(d^{1/2})$, $U_{\mathbb{C}}(f, s)$ is small [13, Corollary 10].

Symmetric Depth- 2 Circuit. Over $\mathbb{R}$, it is a circuit of the form $B^T \cdot B$, for some $B \in \mathbb{R}^{m \times n}$. [Over $\mathbb{C}$, one should take the conjugate-transpose B^* instead of B^T.] Symmetric circuits are a natural computational model for computing a positive semi-definite (*PSD*) matrix.

Invertible Depth- 2 Circuit. It is a circuit $B \cdot C$, where at least one of the matrices B, C is invertible. We stress that invertible circuits can compute non-invertible matrices. Invertible circuits generalize many of the common matrix decompositions, such as QR decomposition, eigen decomposition, singular value decomposition (SVD), and LUP decomposition.

[27, Thms.1.3 & 1.5] prove asymptotically optimal lower bounds for both the models.

Theorem 6. [27] *There exists an explicit family of real $n \times n$ PSD matrices $(A_n)_{n \in \mathbb{N}}$ such that every symmetric circuit (respectively invertible circuits) computing A_n (over $\mathbb{R}$) has size $\Omega(n^2)$.*

We present a simple, *alternative* proof of Theorem 6 using lower bounds on the SOS representation (with restriction) of two different explicit families f_d over $\mathbb{R}$. For details, see Theorems 7, and 8, in Sect. 5.

Before going into details, we state a classical lemma due to Descartes which will be used throughout the paper.

Lemma 5 (Descartes' rule of signs). *Let $p(x) \in \mathbb{R}[x]$ be a polynomial with t many monomials. Then, number of distinct positive roots in $p(x)$ can be at most $t - 1$.*

Remark. An s-sparse polynomial $f \in \mathbb{C}[x]$ can have at most $2(s - 1)$-many real roots. A real root a of f must be a real root of both the real part $\Re(f)$ and the imaginary part $\Im(f)$. By above, there can be at most $s - 1$ many positive roots. The same bound holds for negative roots by $x \mapsto -x$.

5.1   Lower Bound for Symmetric Circuits Over $\mathbb{R}$: Proof of the First Part of Theorem 6

We state a lemma from classical mathematics for the study of fewnomials and give a simple proof. This would be critical to prove explicit lower bounds.

Lemma 6 (Hajós Lemma). *Suppose $f(x) \in \mathbb{C}[x]$ be a univariate polynomial with $t \geq 1$ monomials. Let α be a non-zero root of $f(x)$. Then, the multiplicity of α in f can be at most $t - 1$.*

Proof. We will prove this by induction on t. When $t = 1$, $f(x) = a_m x^m$ for some m. It has no non-zero roots and we are trivially done. Assume that, it is true upto t. We want to prove the claim for $t + 1$.

Suppose $|f|_0 = t + 1$. There exists $m \geq 0$ such that $f(x) = x^m \cdot g(x)$, with $|g|_0 = t + 1$ and $g(0) \neq 0$. It suffices to prove the claim for g. Let, α be a non-zero root of $g(x)$. Suppose, $g(x) = (x - \alpha)^s \cdot h(x)$, for some $s \geq 1$ and $h(\alpha) \neq 0$.

Observe that, multiplicity of α in g' is $s - 1$. As $g(0) \neq 0$, $|g'|_0 = t$. Therefore by induction hypothesis, $s - 1 \leq t - 1 \implies s \leq t$. Hence, multiplicity of α in g (thus in f) can be at most t. This finishes the induction step.

Corollary 1. *Suppose $f(x) = (x+\alpha)^t \cdot g(x)$, for some non-zero α and $g(\cdot)$, then we must have $|f|_0 \geq t + 1$.*

We prove an important lower bound on SOS representation for a non-zero multiple of $(x + 1)^d$; it will be important to prove the first part of Theorem 6.

Lemma 7. *Let $f(x)$ be a non-zero polynomial in $\mathbb{R}[x]$. Suppose, there exist non-zero $\ell_i \in \mathbb{R}[x]$, for $i \in [m]$ such that $(x + 1)^d \cdot f(x) = \sum_{i=1}^{m} \ell_i^2$. Then, $\sum_{i \in [m]} |\ell_i|_0 \geq m \cdot (\lfloor d/2 \rfloor + 1)$.*

Proof. Denote $g(x) := \gcd(\ell_1, \ldots, \ell_m)$. We will prove that $(x + 1)^t \mid g(x)$ where $t := \lfloor d/2 \rfloor$. Suppose not, assume that $(x + 1)^k \| g(x)$ (i.e. $(x + 1)^{k+1} \nmid g(x)$) such that $k < t$ (and thus $d - 2k > 0$). Then, $g(x) = h(x) \cdot (x + 1)^k$ where $h(x) \in \mathbb{R}[x]$ with $h(-1) \neq 0$. Define $\tilde{\ell}_i := \ell_i / (x + 1)^k$. By assumption, $(x + 1) \nmid \gcd(\tilde{\ell}_1, \ldots, \tilde{\ell}_m) =: h(x)$. Thus,

$$\sum_{i=1}^{k} \ell_i(x)^2 = (x+1)^d \cdot f(x) \implies \sum_{i=1}^{m} \tilde{\ell}_i(x)^2 = (x+1)^{d-2k} \cdot f(x)$$

$$\implies \sum_{i=1}^{m} \tilde{\ell}_i(-1)^2 = 0$$

$$\implies \tilde{\ell}_i(-1) = 0, \quad \forall i \in [1, m]$$

$$\implies (x+1) \mid \tilde{\ell}_i(x), \quad \forall i \in [1, m]$$

$$\implies (x+1) \mid \gcd(\tilde{\ell}_1, \ldots, \tilde{\ell}_m) = h(x)$$

which is a contradiction. Thus, $k \geq t$.

   This implies, each ℓ_i is non-zero polynomial multiple of $(x + 1)^t$. Since Corollary 1 implies that $|\ell_i|_0 \geq t + 1$, for all $i \in [m]$; the lemma follows.

Recall that a *symmetric* depth-2 circuit (over $\mathbb{R}$) is a circuit of the form $B^T \cdot B$ for some $B \in \mathbb{R}^{m \times n}$. We prove the *first* part of Theorem 6.

Theorem 7 (Reproving Theorem 1.3 of [27]). *There exists an explicit family of real $n \times n$ PSD matrices $\{A_n\}_{n \in \mathbb{N}}$ such that every symmetric circuit computing A_n (over $\mathbb{R}$) has size $\Omega(n^2)$.*

Proof. Denote $[x]_n := [1 \, x \ldots x^{n-1}]$. Denote $k := \lfloor n/2 \rfloor$. Define $g_i(x) := (x + 1)^k \cdot x^{\lfloor (i-1)/2 \rfloor}$, for $i \in [n]$. Note that, $\deg(g_i) = k + \lfloor (i-1)/2 \rfloor \leq k + \lfloor (n-1)/2 \rfloor = n - 1$. Define $n \times n$ matrix M_n such that

$$M_n \cdot [x]_n^T := \begin{bmatrix} g_1(x) \\ g_2(x) \\ \vdots \\ g_n(x) \end{bmatrix} .$$

It is easy to see that $g_1, g_3, g_5, \ldots$ are linearly independent over $\mathbb{R}$. Therefore, $\text{rank}(M_n) = \text{rank}_{\mathbb{R}}(g_1(x), \ldots, g_n(x)) = \lfloor (n-1)/2 \rfloor + 1 = \lfloor (n+1)/2 \rfloor$.

Define $A_n := M_n^T \cdot M_n$. By definition, A_n is PSD and $\text{rank}(A_n) = \lfloor (n+1)/2 \rfloor$. This follows from the classical fact that for any matrix A over $\mathbb{R}$, $\text{rank}(A^T A) = \text{rank}(A)$. Also A_n is *explicit* (entries are P-computable from definition). Now, assume there is some $m \times n$ matrix B such that $A_n = B^T \cdot B$. Then, denote $B[x]_n := \begin{bmatrix} \ell_1 & \ell_2 & \ldots & \ell_m \end{bmatrix}^T$, where $\ell_i \in \mathbb{R}[x]$ are univariate polynomials of degree at most $n-1$. Observe that number of *non-zero* entries in B is precisely $\sum_{i \in [m]} |\ell_i|_0$. Thus, it suffices to show that $\sum_{i \in [m]} |\ell_i|_0 \geq \Omega(n^2)$.

As $\text{rank}(B) = \text{rank}(B^T B) = \text{rank}(A_n) = \lfloor (n+1)/2 \rfloor$, we must have $m \geq \lfloor (n+1)/2 \rfloor$. Thus,

$$A_n = B^T \cdot B \implies [x]_n M_n^T \cdot M_n [x]_n^T = [x]_n B^T \cdot B [x]_n^T$$

$$\iff \sum_{i=1}^{n} g_i(x)^2 = \sum_{i=1}^{m} \ell_i^2$$

$$\iff (x+1)^{2k} \cdot f(x) = \sum_{i=1}^{m} \ell_i^2, \text{ where} f(x) := \sum_{i=1}^{n} x^{2 \cdot \lfloor (i-1)/2 \rfloor}$$

$$\implies \sum_{i=1}^{m} |\ell_i|_0 \geq (\lfloor (n+1)/2 \rfloor) \cdot (k+1) \geq \frac{n^2}{4} \qquad \text{by Lemma 7.}$$

5.2   Lower Bound for Invertible Circuits Over $\mathbb{R}$: Proof of the Second Part of Theorem 6

This subsection is devoted to proving the *second* part of Theorem 6. This proof uses SOS lower bound for a bivariate polynomial. Investigating sum of product of two polynomials is similar to looking at the SOS; as, one can write $f \cdot g = ((f+g)/2)^2 - ((f-g)/2)^2$. The summand fan-in at most doubles. Thus, proving lower bound for sum of product of two polynomials is 'same' as proving SOS lower bound. The following lemma proves certain lower bound on sum of sparsity when a specific *bivariate* polynomial is written as sum of product of two polynomials (with certain restrictions).

Lemma 8. *Let $f_d := f_{d,t}(x,y) := \left(\prod_{i \in [d]} (x-i)(y-i) \right) \cdot p(x,y)$, for some polynomial $p \in \mathbb{R}[x,y]$ such that $\deg_x(p) = \deg_y(p) = t$. Suppose, $f_d = \sum_{i \in [d+t+1]} \ell_i(x) \cdot \tilde{\ell}_i(y)$, where $\ell_i, \tilde{\ell}_i$'s are polynomials of degree at most $d+t$; with the additional property that $\tilde{\ell}_1, \ldots, \tilde{\ell}_{d+t+1}$ are $\mathbb{R}$-linearly independent.*

Then, $\sum_{i=1}^{d+t+1} |\ell_i|_0 \geq m \cdot (d+1)$, where m is the number of non-zero ℓ_i.

Proof. Suppose, $g(x) := \gcd(\ell_1, \ldots, \ell_{d+t+1})$. We claim that $\prod_{i=1}^{d} (x-i) \mid g(x)$. Note that, it suffices to prove the claim; as, $\prod_{i=1}^{d} (x-i) \mid \ell_i(x)$ for each non-zero ℓ_i implies $|\ell_i|_0 \geq d+1$ by Lemma 5.

We prove the claim by contradiction. Suppose, there exists $j \in [d]$ such that $x - j \nmid g(x)$, so $g(j) \neq 0$. Fix this j. Hence, there exists i such that $\ell_i(j) \neq 0$.

In particular, $v := \begin{bmatrix} \ell_1(j) \; \ell_2(j) \; \ldots \; \ell_{d+t+1}(j) \end{bmatrix}^T \neq \mathbf{0}$. Define the $(d+t+1) \times (d+t+1)$ matrix A as

$$[y]_{d+t+1} \cdot A := \begin{bmatrix} \tilde{\ell}_1 \; \tilde{\ell}_2 \ldots \tilde{\ell}_{d+t+1} \end{bmatrix}, \text{where} [y]_{d+t+1} := \begin{bmatrix} 1 \; y \; \ldots \; y^{d+t} \end{bmatrix}.$$

Observe: $\mathrm{rank}_{\mathbb{R}}(\tilde{\ell}_1, \ldots, \tilde{\ell}_{d+t+1}) = d+t+1 \iff A$ is invertible. But,

$$v \neq \mathbf{0} \text{ and } A \text{ is invertible} \implies A \cdot v \neq \mathbf{0}$$
$$\implies [y]_{d+t+1} \cdot Av \neq 0$$
$$\implies \sum_{i=1}^{d+t+1} \tilde{\ell}_i(y) \cdot \ell_i(j) \neq 0$$
$$\implies f_{d,t}(j, y) \neq 0$$

which is a contradiction! Therefore, $\prod_{i=1}^{d}(x-i) \mid g(x)$ and so we are done.

Recall that an *invertible* depth-2 circuit computes a matrix A such that whenever $A = BC$, either B or C has to be invertible. We prove the *second* part of Theorem 6.

Theorem 8 (Reproving Theorem 1.5 of [27]). *There exists an explicit family of $n \times n$ PSD matrices $\{A_n\}_{n \in \mathbb{N}}$ such that every invertible circuit over $\mathbb{R}$ computing A_n has size $\Omega(n^2)$.*

Proof. Denote $k := \lfloor n/2 \rfloor$. Define $g_i(x) := \prod_{i=1}^{k}(x-i) \cdot x^{\lfloor (i-1)/2 \rfloor}$, for $i \in [n]$. Note that $\deg(g_i) = k + \lfloor (i-1)/2 \rfloor \leq k + \lfloor (n-1)/2 \rfloor = n - 1$. Define the $n \times n$ matrix M_n as

$$M_n \cdot [x]_n^T := \begin{bmatrix} g_1(x) \\ g_2(x) \\ \vdots \\ g_n(x) \end{bmatrix}.$$

It is easy to see that $g_1, g_3, g_5, \ldots$ are linearly independent over $\mathbb{R}$. Therefore, $\mathrm{rank}(M_n) = \mathrm{rank}_{\mathbb{R}}(g_1(x), \ldots, g_n(x)) = \lfloor (n-1)/2 \rfloor + 1 = \lfloor (n+1)/2 \rfloor$.

Define $A_n := M_n^T \cdot M_n$. By definition, A_n is PSD and $\mathrm{rank}(A_n) = \lfloor (n+1)/2 \rfloor$. This follows from the classical fact that for any matrix A, $\mathrm{rank}(A^T A) = \mathrm{rank}(A)$ over $\mathbb{R}$. Also A_n is *explicit* (entries are P-computable from definition).

Suppose, there exists $n \times n$ invertible matrix B and some $n \times n$ matrix C such that $A_n = B \cdot C$ (the other case where C is invertible is similar). Note that, from classical property of rank of matrices, $\mathrm{rank}(C) \geq \mathrm{rank}(A_n) = \lfloor (n+1)/2 \rfloor$. With the usual notation of $[x]_n$ and $[y]_n$ used before, denote

$$[y]_n \cdot B := \begin{bmatrix} \tilde{\ell}_1(y) \; \tilde{\ell}_2(y) \ldots \tilde{\ell}_n(y) \end{bmatrix} \text{ and } C \cdot [x]_n^T := \begin{bmatrix} \ell_1(x) \; \ell_2(x) \ldots \ell_n(x) \end{bmatrix}^T.$$

Note that the degree of each $\ell_i, \tilde{\ell}_i$ can be at most $n - 1$. Thus,

$$A_n = B \cdot C \implies [y]_n M_n^T \cdot M_n [x]_n^T = [y]_n \cdot B \cdot C \cdot [x]_n^T$$

$$\iff \sum_{i=1}^{n} g_i(x) \cdot g_i(y) = \sum_{i=1}^{n} \ell_i(x) \cdot \tilde{\ell}_i(y)$$

$$\iff \left(\prod_{i=1}^{k} (x - i)(y - i) \right) \cdot p(x, y) = \sum_{i=1}^{n} \ell_i(x) \cdot \tilde{\ell}_i(y)$$

where $p(x, y) := \sum_{i \in [n]} (xy)^{\lfloor (i-1)/2 \rfloor}$. The LHS is actually of the form $f_{k, \lfloor (n-1)/2 \rfloor}(x, y)$ as in Lemma 8. From the lower bound on rank of C, we know that there must be at least $\lfloor (n+1)/2 \rfloor$ many non-zero ℓ_i's. Therefore, Lemma 8 gives us $\sum_{i=1}^{n} |\ell_i|_0 \geq \lfloor (n+1)/2 \rfloor \cdot (k+1) \geq n^2/4$.

Remark. The defined matrix A_n in the above proof *also* works for the theorem 7. For that, one needs to replace the polynomial $\prod_{i=1}^{d} (x - i) \cdot f(x)$, in theorem 7, and prove similar lower bound on sum of sparsity. The proof details of theorem remains *almost* unchanged until at the very end, one has to use Descartes' rule (lemma 5) instead of Lemma 1.

6 τ-Conjectures for Top-Fanin 2 Hold True

In this section we show that both SOS-τ-conjecture and SOC-τ-conjecture hold true for top fanin-2.

6.1   SOS-τ-Conjecture for Sum of Two Squares

We show that when f is a sum of two squares, number of real roots is indeed linear in the support-sum.

Theorem 9. *If $f = \sum_{i=1}^{s} c_i \cdot f_i^2 \in \mathbb{R}[x]$, where $s \leq 2$, then f can have at most $O(\sum_{i=1}^{s} |f_i|_0)$-many real roots.*

Proof. There are two cases to consider:

Case I ($s = 1$): In this case, $f = c_1 \cdot f_1^2$. Thus, the real roots of f are precisely the roots of f_1. However, by Descartes' rule (Lemma 5), f_1 can have at most $2(|f_1|_0 - 1)$-many real roots.

Case II ($s = 2$): Without loss of generality, assume that c_1 and c_2 are of opposite signs ; otherwise, any real root of f must also be roots of f_1 and f_2, and trivially we are done by Lemma 5. When, the signs are opposite, note that, f has the following factoring over $\mathbb{R}[x]$:

$$f = c_1 \cdot (f_1 + \gamma \cdot f_2) \cdot (f_1 - \gamma \cdot f_2), \text{ where } \gamma := \sqrt{-c_2/c_1} \in \mathbb{R}.$$

It directly follows that $|f_1 \pm \gamma \cdot f_2|_0 \leq |f_1|_0 + |\gamma \cdot f_2|_0 = |f_1|_0 + |f_2|_0$. However, the real roots of f must also be real roots of $f_1 \pm \gamma \cdot f_2$. Each $f_1 \pm \gamma \cdot f_2$ can have at most $2(|f_1|_0 + |f_2|_0) - 2$ many real roots, by Descartes' rule (Lemma 5). Therefore, the conclusion follows.

Remark. We could strengthen the above theorem by replacing $O(|\bigcup_{i \in [2]} \text{supp}(f_i)|)$. Since, $|\text{supp}(f_1 \pm \gamma \cdot f_2)| \leq |\text{supp}(f_1) \bigcup \text{supp}(f_2)|$, using Lemma 5, the conclusion follows.

6.2 SOC-τ-Conjecture for Sum of Two Cubes

We show that when f is a sum of two squares, number of real roots is indeed linear in the support-union.

Theorem 10. *If $f = \sum_{i=1}^{s} c_i \cdot f_i^3 \in \mathbb{R}[x]$ where $s \leq 2$, then f can have at most $O(|\bigcup_{i=1}^{s} \text{supp}(f_i)|)$-many real roots.*

Proof. There are two cases to consider:

Case I ($s = 1$): In this case, $f = c_1 \cdot f_1^3$. Thus, the real roots of f are precisely the roots of f_1. However, by Descartes' rule (Lemma 5), f_1 can have at most $2(|f_1|_0 - 1)$-many real roots.

Case II ($s = 2$): Note that, f has the following factoring over $\mathbb{R}[x]$:

$$ f = c_1 \cdot (f_1 + \gamma \cdot f_2) \cdot (f_1^2 - \gamma \cdot f_1 f_2 + \gamma^2 \cdot f_2^2), \text{ where } \gamma := \sqrt[3]{c_2/c_1} \in \mathbb{R}. $$

However,

$$ f_1^2 - \gamma \cdot f_1 f_2 + \gamma^2 \cdot f_2^2 = (f_1 - \frac{\gamma}{2} \cdot f_2)^2 + (\frac{3\gamma^2}{4}) \cdot f_2^2, $$

which has $O(|\bigcup_{i=1}^{2} \text{supp}(f_i)|)$-many real roots by Theorem 9 (and its remark). Also $f_1 + \gamma \cdot f_2$ has at most $O(|\bigcup_{i=1}^{2} \text{supp}(f_i)|)$-many real roots by Descartes' rule (Lemma 5). Moreover, any real root of f must also be real roots of either $f_1 + \gamma \cdot f_2$ or $f_1^2 - \gamma \cdot f_1 f_2 + \gamma^2 \cdot f_2^2$. Therefore, the conclusion follows.

7 SOS-τ-Conjecture to SOS Lower Bound on $(x+1)^d$

Lemma 9. *If Conjecture 1 is true, then $S_{\mathbb{C}}(f_d) \geq \Omega(d)$, where $f_d := (x+a)^d$, for any $0 \neq a \in \mathbb{R}$.*

Before proving the above, we establish an interesting lemma. For $f \in \mathbb{C}[x]$, we denote $\Re(f)$ as the real part of f, and $\Im(f)$ as the imaginary part, i.e. $f = \Re(f) + \iota \cdot \Im(f)$. Note, $|\Re(f)|_0, |\Im(f)|_0 \leq |f|_0$.

Lemma 10. $S_{\mathbb{R}}(\Re(f)) \leq 2 \cdot S_{\mathbb{C}}(f)$, *for any $f \in \mathbb{C}[x]$.*

Proof. Suppose, $f(x) = \sum_{i=1}^{s} f_i^2$, for $f_i \in \mathbb{C}[x]$ is a *minimal* representation in SOS-model over $\mathbb{C}$ (we ignore the constants c_i as in Equation (1) as $\sqrt{c_i}$ can be taken inside), i.e. $S_{\mathbb{C}}(f) = \sum_{i=1}^{s} |f_i|_0$. Note that

$$\Re(f) = \sum_{i=1}^{s} \Re(f_i^2) = \sum_{i=1}^{s} \Re(\Re(f_i) + \iota \cdot \Im(f_i))^2$$

$$= \sum_{i=1}^{s} \left(\Re(f_i)^2 - \Im(f_i)^2 \right).$$

The last expression implies that

$$S_{\mathbb{R}}(\Re(f)) \leq \sum_{i=1}^{s} |\Re(f_i)|_0 + \sum_{i=1}^{s} |\Im(f_i)|_0 \leq \sum_{i=1}^{s} 2|f_i|_0 = 2 \cdot S_{\mathbb{C}}(f).$$

Proof of Lemma 9. It suffices to prove the bound for $f_d = (x + 1)^d$, as $S_{\mathbb{C}}((x + a)^d) = S_{\mathbb{C}}((x + 1)^d)$ [just by replacing $x \mapsto x/a$]. Consider the complex polynomial $g_d(x) := f_d(\iota x) + f_d(-\iota x)$. Its degree is either d, if d is even, or $d - 1$, if it is odd. The roots are of the form

$$\iota \cdot \frac{1 - \zeta}{1 + \zeta},$$

where ζ is d-th root of -1 ($\zeta \neq 1$). There are again either d or $d - 1$ such roots, depending on the parity of d. Further, they are all *distinct*. Since $|\zeta| = 1$, each root

$$\iota \cdot \frac{1 - \zeta}{1 + \zeta} = \iota \cdot \frac{(1 - \zeta)(1 + \overline{\zeta})}{(1 + \zeta)(1 + \overline{\zeta})} = \iota \cdot \frac{\overline{\zeta} - \zeta}{|1 + \zeta|^2} = \frac{2\Im(\zeta)}{|1 + \zeta|^2}$$

is real. Therefore, $g_d(x)$ must be a real polynomial with *distinct real* roots. Hence Conjecture 1 implies that $S_{\mathbb{R}}(g_d) = \Omega(d)$. Using Lemma 10, one can directly conclude that $S_{\mathbb{C}}(g_d) = \Omega(d)$. It is straightforward to see that $S_{\mathbb{C}}(f)$ remains unchanged under the map $x \mapsto c \cdot x$, for any $c \neq 0$. Therefore, in particular, $S_{\mathbb{C}}(f_d(\iota x)) = S_{\mathbb{C}}(f_d(-\iota x)) = S_{\mathbb{C}}(f_d)$. Finally, we must have

$$\Omega(d) = S_{\mathbb{C}}(g_d) \leq S_{\mathbb{C}}(f_d(\iota x)) + S_{\mathbb{C}}(f_d(-\iota x)) = 2 \cdot S_{\mathbb{C}}(f_d).$$

Hence, the conclusion follows.

8   Conclusion

This work effectively establishes that studying the number of real roots of univariate polynomials for sum-of-squares representation (respectively cubes) is fecund. In fact, proving a strong upper bound suffices to solve three major open problems in algebraic complexity.

Here are some immediate questions of interest which require rigorous investigation.

1. Does SOS-τ-conjecture solve PIT completely? The current proof technique fails to reduce from cubes to squares.
2. Prove the upper bound on the number of real roots for the *sum of constantly many squares*. Currently we only know it for $s = 2$ (Theorem 9).
3. Does SOC-τ-conjecture hold for a 'generic' polynomial f (over $\mathbb{Q}$)?
4. Can we *weaken* the requirement on the upper bound for matrix rigidity (Theorem 1)?

Acknowledgments. A part of this work was done when the author was a visiting scholar at IIT Kanpur. The author thanks Prof. Nitin Saxena and Prof. Thomas Thierauf for the initial discussions and encouragement to write the draft. The author is supported by Google Ph.D. Fellowship. The author would also like to thank Pranav Bisht, Bhargav CS and the anonymous reviewers for their important feedback.

References

1. Agrawal, M., Ghosh, S., Saxena, N.: Bootstrapping variables in algebraic circuits. Proc. Natl. Acad. Sci. **116**(17), 8107–8118 (2019). https://doi.org/10.1073/pnas.1901272116. Earlier in Symposium on Theory of Computing, 2018 (STOC 2018)
2. Agrawal, M., Vinay, V.: Arithmetic circuits: a chasm at depth four. In: Foundations of Computer Science. In: 49th Annual IEEE Symposium on Foundations of Computer Science, FOCS 2008, pp. 67–75. IEEE (2008)
3. Allender, E., Jiao, J., Mahajan, M., Vinay, V.: Non-commutative arithmetic circuits: depth reduction and size lower bounds. Theor. Comput. Sci. **209**(1–2), 47–86 (1998). https://doi.org/10.1016/S0304-3975(97)00227-2
4. Alman, J., Chen, L.: Efficient construction of rigid matrices using an NP oracle. In: 2019 IEEE 60th Annual Symposium on Foundations of Computer Science (FOCS), pp. 1034–1055. IEEE (2019)
5. Alon, N., Pudlak, P.: Superconcentrators of depths 2 and 3; odd levels help (rarely). J. Comput. Syst. Sci. **48**(1), 194–202 (1994)
6. Blum, L., Cucker, F., Shub, M., Smale, S.: Algebraic settings for the problem "P$\neq$NP". In: The Collected Papers of Stephen Smale, vol. 3, pp. 1540–1559. World Scientific (2000)
7. Blum, L., Shub, M., Smale, S.: On a theory of computation and complexity over the real numbers: NP-completeness, recursive functions and universal machines. Bull. New Ser. Am. Math. Soc. **21**(1), 1–46 (1989)
8. Briquel, I., Bürgisser, P.: The real tau-conjecture is true on average. Random Struct. Algorithms **57**(2), 279–303 (2020)
9. Bürgisser, P.: On defining integers and proving arithmetic circuit lower bounds. Comput. Complex. **18**(1), 81–103 (2009)
10. Bürgisser, P.: Completeness and Reduction in Algebraic Complexity Theory. Algorithms and computation in mathematics. Springer, Heidelberg (2013). https://doi.org/10.1007/978-3-662-04179-6
11. Bürgisser, P., Clausen, M.: Algebraic Complexity Theory. Grundlehren der mathematischen Wissenschaften. Springer, Berlin, Heidelberg (2013)
12. Demillo, R.A., Lipton, R.J.: A probabilistic remark on algebraic program testing. Inf. Proc. Lett. **7**(4), 193–195 (1978)

13. Dutta, P., Saxena, N., Thierauf, T.: A largish sum-of-squares implies circuit hardness and derandomization. In: 12th Innovations in Theoretical Computer Science (ITCS) (2021)
14. Dvir, Z., Golovnev, A., Weinstein, O.: Static data structure lower bounds imply rigidity. In: Proceedings of the 51st Annual ACM SIGACT Symposium on Theory of Computing, pp. 967–978 (2019)
15. Friedman, J.: A note on matrix rigidity. Combinatorica **13**(2), 235–239 (1993)
16. Garcia-Marco, I., Koiran, P., Tavenas, S.: Log-concavity and lower bounds for arithmetic circuits. In: Italiano, G.F., Pighizzini, G., Sannella, D.T. (eds.) MFCS 2015. LNCS, vol. 9235, pp. 361–371. Springer, Heidelberg (2015). https://doi.org/10.1007/978-3-662-48054-0_30
17. Grigoriev, D.Y.: Using the notions of seperability and independence for proving the lower bounds on the circuit complexity. notes of the leningrad branch of the steklov mathematical institute (1976)
18. Grochow, J.A.: Unifying known lower bounds via geometric complexity theory. Comput. Complex. **24**(2), 393–475 (2015)
19. Guo, Z., Kumar, M., Saptharishi, R., Solomon, N.: Derandomization from algebraic hardness: treading the borders. In: 60th IEEE Annual Symposium on Foundations of Computer Science, FOCS 2019, pp. 147–157 (2019). https://doi.org/10.1109/FOCS.2019.00018 online version: https://mrinalkr.bitbucket.io/papers/newprg.pdf
20. Gupta, A., Kamath, P., Kayal, N., Saptharishi, R.: Arithmetic circuits: A chasm at depth three. In: 2013 IEEE 54th Annual Symposium on Foundations of Computer Science, pp. 578–587. IEEE (2013)
21. Hrubes, P.: On the Real τ-Conjecture and the Distribution of Complex Roots. Theor. Comput. **9**(1), 403–411 (2013)
22. Hrubes, P.: On the distribution of runners on a circle. Eur. J. Comb. **89**, 103137 (2020). https://doi.org/10.1016/j.ejc.2020.103137
23. Kabanets, V., Impagliazzo, R.: Derandomizing polynomial identity tests means proving circuit lower bounds. Computat. Complex. **13**(1–2), 1–46 (2004)
24. Koiran, P.: Shallow circuits with high-powered inputs. In: Innovations in Computer Science - ICS (2011)
25. Koiran, P.: Arithmetic circuits: The chasm at depth four gets wider. Theor. Comput. Sci. **448**, 56–65 (2012)
26. Koiran, P., Portier, N., Tavenas, S., Thomassé, S.: A τ-Conjecture for newton polygons. Found. Comput. Math. **15**(1), 185–197 (2015)
27. Kumar, M., Volk, B.L.: Lower Bounds for Matrix Factorization. In: Proceedings of the 35th Computational Complexity Conference (CCC) (2020)
28. Kurtz, D.C.: A sufficient condition for all the roots of a polynomial to be real. Am. Math. Mon. **99**(3), 259–263 (1992)
29. Lasserre, J.B.: A sum of squares approximation of nonnegative polynomials. SIAM Rev. **49**(4), 651–669 (2007)
30. Laurent, M.: Sums of squares, moment matrices and optimization over polynomials. Emerging Applications of Algebraic Geometry, pp. 157–270. Springer, New York, NY (2009). https://doi.org/10.1007/978-0-387-09686-5_7
31. Mahajan, M.: Algebraic complexity classes. In: Agrawal, M., Arvind, V. (eds.) Perspectives in Computational Complexity. PCSAL, vol. 26, pp. 51–75. Springer, Cham (2014). https://doi.org/10.1007/978-3-319-05446-9_4
32. Mulmuley, K.: Geometric complexity theory V: efficient algorithms for noether normalization. J. Am. Math. Soc. **30**(1), 225–309 (2017)

33. Mulmuley, K.D.: Geometric complexity theory V: equivalence between blackbox derandomization of polynomial identity testing and derandomization of noether's normalization lemma. In: FOCS, pp. 629–638 (2012)
34. Nisan, N., Wigderson, A.: Hardness vs randomness. J. Comput. Syst. Sci. 49(2), 149–167 (1994)
35. Ore, O.: Über höhere kongruenzen. Nor. Mat. Foren. Skr. 1(7), 15 (1922)
36. Pfister, A.: Hilbert's seventeenth problem and related problems on definite forms. In: Mathematical Developments Arising from Hilbert Problems, Proc. Sympos. Pure Math, vol. 28, pp. 483–489 (1976)
37. Pippenger, N.: Superconcentrators. SIAM J. Comput. 6(2), 298–304 (1977)
38. Pudlak, P.: Communication in bounded depth circuits. Combinatorica 14(2), 203–216 (1994)
39. Radhakrishnan, J., Ta-Shma, A.: Bounds for dispersers, extractors, and depth-two superconcentrators. SIAM J. Discrete Math. 13(1), 2–24 (2000)
40. Ramanujan, S.: On the expression of a number in the form $ax^2 + by^2 + cz^2 + du^2$. Proc. Camb. Philos. Soc. 19, 11–21 (1917)
41. Ramya, C.: Recent progress on matrix rigidity–a survey. arXiv preprint arXiv:2009.09460 (2020)
42. Raz, R.: Elusive functions and lower bounds for arithmetic circuits. Theor. Comput. 6(1), 135–177 (2010). https://doi.org/10.4086/toc.2010.v006a007
43. Reznick, B.: Extremal PSD forms with few terms. Duke Math. J. 45(2), 363–374 (1978)
44. Saptharishi, R.: A survey of lower bounds in arithmetic circuit complexity. Github survey (2019)
45. Schwartz, J.T.: Fast probabilistic algorithms for verification of polynomial identities. J. ACM 27(4), 701–717 (1980)
46. Shokrollahi, M.A., Spielman, D.A., Stemann, V.: A remark on matrix rigidity. Inf. Proc. Lett. 64(6), 283–285 (1997)
47. Shpilka, A., Yehudayoff, A.: Arithmetic circuits: a survey of recent results and open questions. Found. Theor. Comput. Sci. 5(3–4), 207–388 (2010)
48. Shub, M., Smale, S.: On the intractability of Hilbert's Nullstellensatz and an algebraic version of "NP≠ P?". Duke Math. J. 81(1), 47–54 (1995)
49. Smale, S.: Mathematical problems for the next century. Math. Intell. 20(2), 7–15 (1998)
50. Tavenas, S.: Bornes inferieures et superieures dans les circuits arithmetiques. Ph.D. Thesis (2014)
51. Valiant, L.G.: On non-linear lower bounds in computational complexity. In: Proceedings of the Seventh Annual ACM Symposium on Theory of Computing, pp. 45–53 (1975)
52. Valiant, L.G.: Graph-theoretic arguments in low-level complexity. In: Gruska, J. (ed.) MFCS 1977. LNCS, vol. 53, pp. 162–176. Springer, Heidelberg (1977). https://doi.org/10.1007/3-540-08353-7_135
53. Valiant, L.G.: Completeness classes in algebra. In: Proceedings of the 11th Annual ACM symposium on Theory of computing, pp. 249–261. ACM (1979)
54. Valiant, L.G., Skyum, S., Berkowitz, S., Rackoff, C.: Fast parallel computation of polynomials using few processors. SIAM J. Comput. 12(4), 641–644 (1983). https://doi.org/10.1137/0212043
55. Zippel, R.: Probabilistic algorithms for sparse polynomials. In: Proceedings of the International Symposium on Symbolic and Algebraic Computation, pp. 216–226. EUROSAM 1979 (1979)

Dichotomy Result on 3-Regular Bipartite Non-negative Functions

Austen Z. Fan[✉][iD] and Jin-Yi Cai

University of Wisconsin-Madison, Madison, WI 53706, USA
{afan,jyc}@cs.wisc.edu

Abstract. We prove a complexity dichotomy theorem for a class of Holant problems on 3-regular bipartite graphs. Given an arbitrary non-negative weighted symmetric constraint function $f = [x_0, x_1, x_2, x_3]$, we prove that the bipartite Holant problem Holant $(f \mid (=_3))$ is *either* computable in polynomial time *or* #P-hard. The dichotomy criterion on f is explicit.

Keywords: Dichotomy theorem · Holant problem · Bipartite graph

1 Introduction

Holant problems are also called edge-coloring models. They can express a broad class of counting problems, such as counting matchings (#MATCHINGS), perfect matchings (#PM), edge-colorings, cycle coverings, and a host of counting orientation problems such as counting Eulerian orientations.

Given an input graph $G = (V, E)$, we identify each edge $e \in E$ as a variable over some finite domain D, and identify each vertex v as a constraint function f_v, also called a signature. Then the *partition function* is the following sum of product $\sum_{\sigma:E \to D} \prod_{v \in V} f_v(\sigma|_{E(v)})$, where $E(v)$ denotes the edges incident to v and $f_v(\sigma|_{E(v)})$ is the evaluation of f_v on the restriction of σ on $E(v)$. For example, #MATCHINGS and #PM are counting problems specified by the constraint function AT-MOST-ONE, respectively EXACT-ONE, which outputs value 1 if the input bits have *at most* one 1, respectively *exactly* one 1, and outputs 0 otherwise. Thus, every term $\prod_{v \in V} f_v(\sigma|_{E(v)})$ evaluates to 1 or 0, and is 1 iff the assignment σ is a mathching, respectively a perfect mathching. The framework of Holant problems is intimately related to Valiant's holographic algorithms [16]. Holant problems can encompass all counting constraint satisfaction problems (#CSP), in particular all vertex-coloring models from statistical physics. For #CSP, Bulatov [3] proved a sweeping complexity dichotomy. Dyer and Richerby [9] gave another proof of this dichotomy and also showed that the dichotomy is decidable. Cai, Chen, and Lu in [6] extended this to nonnegative weighted case. This was further generalized to complex weighted #CSP [5]. However, no full dichotomy has been proved for Holant problems.

Supported by NSF CCF-1714275.

R. Santhanam and D. Musatov (Eds.): CSR 2021, LNCS 12730, pp. 102–115, 2021.
https://doi.org/10.1007/978-3-030-79456-3_6

Every #CSP problem can be easily expressed as a Holant problem. On the other hand, Freedman, Lovász and Schrijver [10] proved that the prototypical Holant problem #PM cannot be expressed as a vertex-coloring model using any real constraint function, and this was extended to complex constraint functions [7]. Thus Holant problems are strictly more expressive.

In this paper we consider the Boolean domain $D = \{0, 1\}$. Significant knowledge has been gained about the complexity of Holant problems [1,2,4,8,13,14]. However, there has been very limited progress on bipartite Holant problems where the input graph $G = (U, V, E)$ is bipartite, and every constraint function on U and V comes from two separately specified sets of constraint functions. This is not an oversight; the reason is a serious technical obstacle. When the graph is bipartite and, say, r-regular, there is a curious number theoretic limitation as to what types of subgraph fragments, called gadgets, one can possibly construct. It turns out that every constructible gadget must have a rigid arity restriction; e.g., if the gadget represents a constraint function that can be used for a vertices in U or in V, the arity (the number of input variables) of the function must be congruent to 0 modulo r.

We initiate in this paper the study of Holant problems on bipartite graphs. To be specific, we classify Holant problems on 3-regular bipartite graphs where vertices of one side are labeled with a nonnegative weighted symmetric constraint function $f = [x_0, x_1, x_2, x_3]$, which takes value x_0, x_1, x_2, x_3 respectively when the input has Hamming weight 0, 1, 2, 3; the vertices of the other side are labeled by the ternary equality function $(=_3)$. Such graphs can be viewed as incidence graphs of hypergraphs. Thus one can interpret problems of this type as on 3-uniform (every hyperedge has size 3) 3-regular (every vertex appears in 3 hyperedges) hypergraphs. In this view, e.g., counting perfect matchings on 3-uniform hypergraphs (the number of subsets of hyperedges that cover every vertex exactly once) corresponds to the EXACT-ONE function $[0, 1, 0, 0]$. Alternatively one can think of them as set systems where every set has 3 elements and every element is in 3 subsets, and we count the number of exact-3-covers (#X3C). Denote the problem of computing the partition function in this case as Holant $(f \mid (=_3))$. We prove that for all f, this problem is *either* computable in polynomial time *or* #P-hard, depending on an explicit dichotomy criterion on x_0, x_1, x_2, x_3. Suppose $(X, \mathcal{S})$ is a 3-uniform set system where every $x \in X$ appears in 3 sets in $\mathcal{S}$. In the 0–1 case, this dichotomy completely classifies the complexity of counting the number of ways to choose $\mathcal{S}' \subseteq \mathcal{S}$ while satisfying some local constraint specified by f. In the more general nonnegative weighted case, this is to compute a weighted sum of products. The set of problems of the form Holant $(f \mid (=_3))$ is an infinite family of bipartite Holant problems that encompass all these concrete problems such as #X3C. This set of problems are also equivalent to counting CSP problems with one constraint f and each variable occurs exactly three times, sometimes called ternary CSP of degree 3. Thus, to give a complexity classification of all of them is interesting in its own right. However, what motivates us more is the fact that this serves as the most basic yet nontrivial setting for the classification program of the complexity of all bipartite Holant problems.

As mentioned earlier the main technical obstacle to this regular bipartite Holant dichotomy is the number theoretic arity restriction. We overcome this obstacle by considering *straddled* constraint functions, i.e., those functions which have some input variables that must be connected to one side of the bipartite graph while some other variables must be connected to the other side. Then we introduce a lemma that let's interpolate a *degenerate* constraint function by iterating the straddled function construction. Using a Vandermonde system we can succeed in this interpolation. Typically in proving a Holant dichotomy, getting a degenerate constraint function is useless and signifies failure. But here we turn the table, and transform this "failure" to a "success" by "peel" off a constraint function which is a tensor factor, whereby to break the number theoretic arity restriction.

This paper is a mere starting point for understanding bipartite Holant problems. Almost every generalization is an open problem at this point, including more than one constraint function on either side, other regularity parameter r, real or complex valued constraint functions which allow cancellations, etc. The bigger picture is to gain a systematic understanding of all such counting problems in a classification program, a theme seems second to none in its centrality to counting complexity theory, short of proving FP $\neq$ #P.

2 Preliminaries

In this paper we consider the following subclass of Holant problems. An input 3-regular bipartite graph $G = (U, V, E)$ is given, where each vertex on V is assigned the EQUALITY of arities 3 ($=_3$) and each vertex on U is assigned a ternary symmetric constraint function (also called a signature) f with nonnegative values. The problem is to compute the quantity

$$\text{Holant}(G) = \sum_{\sigma:E\to\{0,1\}} \prod_{u\in U} f\left(\sigma|_{E(u)}\right) \prod_{v\in V} (=_3)\left(\sigma|_{E(v)}\right)$$

Equivalently, this can be stated as a weighted counting constraint satisfaction problem on Boolean variables defined on 3-regular bipartite graphs: The input is a 3-regular bipartite $G = (U, V, E)$, where every $v \in V$ is a Boolean variable and every $u \in U$ represents the nonnegative valued constraint function f. An edge $(u, v) \in E$ indicates that v appears in the constraint at u. Being 3-regular means that every constraint has 3 variables and every variable appears in 3 constraints.

We adopt the notation $f = [x_0, x_1, x_2, x_3]$ to represent the ternary symmetric signature f where $f(0,0,0) = x_0$, $f(0,0,1) = f(0,1,0) = f(1,0,0) = x_1$, $f(0,1,1) = f(1,0,1) = f(1,1,0) = x_2$ and $f(1,1,1) = x_3$. The EQUALITY of arities 3 is $(=_3) = [1,0,0,1]$. For clarity, we shall call vertices in V are on the right hand side (RHS) and vertices in U are on the left hand side (LHS). We denote this problem Holant $(f \mid (=_3))$.

A gadget, such as those illustrated in Fig. 1, is a bipartite graph $G = (U, V, E_{in}, E_{out})$ with internal edges E_{in} and dangling edges E_{out}. There can be m dangling edges internally incident to vertices from U and n dangling edges internally incident to vertices from V. These $m + n$ dangling edges correspond to Boolean variables $x_1, \ldots, x_m, y_1, \ldots, y_n$ and the gadget defines a signature

$$f(x_1, \ldots, x_m, y_1, \ldots, y_n) = \sum_{\sigma: E_{in} \to \{0,1\}} \prod_{u \in U} f\left(\widehat{\sigma}|_{E(u)}\right) \prod_{v \in V} (=_3)\left(\widehat{\sigma}|_{E(v)}\right),$$

where $\widehat{\sigma}$ denotes the extension of σ by the assignment on the dangling edges.

To preserve the bipartite structure, we must be careful in any gadget construction how each external wire is to be connected, i.e., as an input variable whether it is on the LHS (like those of f which can be used to connect to $(=_3)$ on the RHS), or it is on the RHS (like those of $(=_3)$ which can be used to connect to f on the LHS). See illustrations in Fig. 1.

If a gadget construction produces a constraint function g such that all of its variables are on the LHS. We claim that its arity must be a multiple of 3. This follows from the following more general statement that if a constraint function has n input variables on the LHS and m input variables on the RHS, then $n \equiv m \bmod 3$. This can be easily proved by induction on the number of occurrences of f and $(=_3)$ in a gadget construction Γ: If either f or $(=_3)$ do not occur, then the function is a tensor product of $(=_3)$ or $(=_3)$, thus clearly of arity $n \equiv 0 \bmod 3$ or $m \equiv 0 \bmod 3$. Suppose f or $(=_3)$ both occur and let x be an external dangling edge. It is internally connected to a vertex v, which is labeled either f (if $v \in U$) or $(=_3)$ (if $v \in V$). Let v have exactly $k \in \{1, 2, 3\}$ incident edges that are dangling edges. Now we remove v and get a gadget Γ' with fewer occurrences of f and $(=_3)$. The induction hypothesis and a simple accounting of the arity for variables on the LHS and the RHS separately complete the induction.

One idea that is instrumental in this paper is to use gadgets that produce *straddled* signatures. For example the gadget G_1 in Fig. 1(a) has one variable on the LHS and one variable on the RHS. Such gadgets can be iterated while respecting the bipartite structure. The *signature matrix* of such straddled gadgets will adopt the notation that row indices denote the input from LHS and column indices denote the input from RHS. For example, the signature matrix for G_1 in Fig. 1(a) where we place $[x_0, x_1, x_2, x_3]$ on the square and $(=_3)$ on the circle will be $\begin{pmatrix} x_0 & x_2 \\ x_1 & x_3 \end{pmatrix}$. A binary signature is *degenerate* if it has determinant 0. We will use such constructions to interpolate a degenerate straddled signature so that we can "split" it to get unary signatures. To justify that we can indeed "split" a degenerate straddled signature, whenever we connect a unary signature, we have to "use up" another unary signature, for the pair was created by that degenerate straddled signature. Intuitively, we can "use all other edges up" by connecting them to form known positive global factors of the Holant value. This intuition is encoded into the following lemma. Effectively, we can split a degenerate binary straddled signature into unary signatures to be freely used.

Lemma 1. *Let f and g be two nonnegative valued signatures. If a degenerate nonnegative binary straddled signature $\begin{pmatrix} 1 & x \\ y & xy \end{pmatrix}$ can be obtained or interpolated in the problem* Holant $(f \mid g)$, *then*

$$\text{Holant} \, (f \mid \{g, [1, x]\}) \leq_T \text{Holant} \, (f \mid g). \tag{1}$$

A similar statement holds for adding the unary $[1, y]$ on the LHS.

Remark 1. The same proof applies for $\begin{pmatrix} y & xy \\ 1 & x \end{pmatrix}$ to get unary signatures $[1, x]$ on the RHS, or unary $[y, 1]$ on the LHS.

Proof. We prove (1). Let f and g have arity m and n respectively. We may assume that g is not a multiple of $[0, 1]^{\otimes n}$ (including identically 0), for otherwise Holant $(f \mid \{g, [1, x]\})$ can be computed in PF, since all signatures on the RHS are degenerate and can be applied directed as unary signatures on copies of f.

Let $k = \gcd(m, n)$, and $s = n/k \geq 1$. Consider any bipartite signature grid $\Omega = (G, \pi)$ for Holant $(f \mid \{g, [1, x]\})$. Let N_f, N_g, N_u be the numbers of occurrences of $f, g, [1, x]$ respectively. Then we have

$$mN_f = nN_g + N_u,$$

thus $N_u \equiv 0 \bmod k$. Let $t = N_u/k \geq 0$. We may assume $t \geq 1$, for otherwise $[1, x]$ does not occur and the reduction is trivial.

We will compute $(\text{Holant} \, (G))^s$, the s-th power of the value Holant (G), using an oracle for Holant $(f \mid g)$. Since the value Holant (G) is nongenative, we can obtain it from $(\text{Holant} \, (G))^s$.

In G, we replace each occurrence of $[1, x]$ with the binary straddled signature $\begin{pmatrix} 1 & x \\ y & xy \end{pmatrix}$, with one end connected to LHS, and leaving one edge yet to be connected to RHS. This creates a total of N_u such edges. This is equivalent to connecting N_u copies of the unary signature $[1, x]$ to LHS, and having N_u copies of the unary signature $[1, y]$ yet to be connected to RHS. Now in s disjoint copies of Ω there will be $sN_u = stk$ copies of $[1, y]$ to be connected, to which we create t copies of g. In other words we take $g^{\otimes t}$ with total arity stk and connect all stk unary signatures $[1, y]$ to it. Since $y \geq 0$ and g is not a multiple of $[0, 1]^{\otimes n}$, we get an easily computable positive factor.

The main result of this paper is the following:

Theorem 1. Holant $([x_0, x_1, x_2, x_3] \mid (=_3))$ *where $x_i \geq 0$ for $i = 0, 1, 2, 3$ is #P-hard except in the following cases, for which the problem is in* FP.

1. $[x_0, x_1, x_2, x_3]$ *is degenerate;*
2. $x_1 = x_2 = 0$;
3. $[(x_1 = x_3 = 0) \wedge (x_0 = x_2)]$ *or* $[(x_0 = x_2 = 0) \wedge (x_1 = x_3)]$.

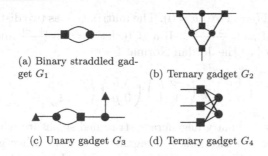

(a) Binary straddled gadget G_1

(b) Ternary gadget G_2

(c) Unary gadget G_3 (d) Ternary gadget G_4

Fig. 1. Some gadgets

In case 1 the signature $[x_0, x_1, x_2, x_3]$ decomposes into three unary signatures. In case 2 $[x_0, 0, 0, x_3]$ is a generalized equality. In case 3 the signature is in the affine class; see more details about these tractable classes in [4]. Therefore the Holant problem is in FP in cases 1–3. The main claim lies in that all other cases are #P-hard. When proving our main result, we shall apply the following dichotomy theorem on 2–3 Holant problem [12]:

Theorem 2. *Suppose $a, b \in \mathbb{C}$, and let $X = ab$, $Z = \left(\frac{a^3 + b^3}{2}\right)^2$. Then* Holant $([a, 1, b] \mid (=_3))$ *is #P-hard except in the following cases, for which the problem is in* P.

1. $X = 1$;
2. $X = Z = 0$;
3. $X = -1$ and $Z = 0$;
4. $X = -1$ and $Z = -1$.

In fact, since this paper mainly concerns with nonnegative valued functions, when establishing #P-hardness we usually only need to consider the exceptional case $X = 1$.

3 Interpolation from a Binary Straddled Signature

When x_0 and x_3 are not both 0, we can normalize the signature. If $x_0 \neq 0$ we divide the signature by x_0, and get the form $[1, a, b, c]$, with $a, b, c \geq 0$. If $x_0 = 0$, but $x_3 \neq 0$, we can flip all 0 and 1 inputs, which amounts to a reversal of the signature and get the above form. This does not change the complexity since the Holant value is only modified by a known nonzero factor.

Consider arguably the simplest possible gadget G_1 in Fig. 1. We have the signature matrix $G_1 = \begin{pmatrix} 1 & b \\ a & c \end{pmatrix}$. Let $\Delta = \sqrt{(1-c)^2 + 4ab}$ denote the nonnegative square root, and assume for now $\Delta \neq 0$, and we take the positive square root.

Note that $\Delta = 0$ iff $(c = 1) \wedge (ab = 0)$. The matrix G_1 has two distinct eigenvalues $\lambda = \frac{-\Delta+(1+c)}{2}$ and $\mu = \frac{\Delta+(1+c)}{2}$. If $a \neq 0$, let $x = \frac{\Delta-(1-c)}{2a}$ and $y = \frac{\Delta+(1-c)}{2a}$. The matrix for G_1 has the Jordan Normal Form

$$\begin{pmatrix} 1 & b \\ a & c \end{pmatrix} = \begin{pmatrix} -x & y \\ 1 & 1 \end{pmatrix} \begin{pmatrix} \lambda & 0 \\ 0 & \mu \end{pmatrix} \begin{pmatrix} -x & y \\ 1 & 1 \end{pmatrix}^{-1}. \tag{2}$$

We now interpolate a binary degenerate straddled signature which will be used several times later. To interpolate a signature h from other given signatures means that there is a reduction from any instance I where h occurs, to polynomially many instances $\{I_s\}$ where h does not occur, and from the Holant values of these polynomially many $\{I_s\}$ we can compute the Holant value of I in polynomial time.

Lemma 2. *Given the binary straddled signature $G_1 = \begin{pmatrix} 1 & b \\ a & c \end{pmatrix}$ with $a \neq 0$ and $\Delta > 0$, we can get unary signatures $[1, x]$ on RHS or $[y, 1]$ on LHS.*

Proof. If $c = ab$, we have $G_1 = \begin{pmatrix} 1 & b \\ a & ab \end{pmatrix}$ which is a nonzero degenerate nonnegative binary straddled signature. Otherwise, for $\Delta > 0$, we have $x + y = \Delta/a > 0$. Consider

$$D = \frac{1}{x+y} \begin{pmatrix} y & xy \\ 1 & x \end{pmatrix} = \begin{pmatrix} -x & y \\ 1 & 1 \end{pmatrix} \begin{pmatrix} 0 & 0 \\ 0 & 1 \end{pmatrix} \begin{pmatrix} -x & y \\ 1 & 1 \end{pmatrix}^{-1}.$$

Given any signature grid Ω where the binary degenerate straddled signature D appears n times, we form gadgets G_1^s where $0 \leq s \leq n$ by iterating the G_1 gadget s times and replacing each occurrence of D with G_1^s. (Here for $s = 0$ we simply replace each occurrence of D by an edge.) Denote the resulting signature grid as Ω_s. We stratify the assignments in the Holant sum for Ω according to assignments to $\begin{pmatrix} \lambda & 0 \\ 0 & \mu \end{pmatrix}$ as:

- $(0, 0)$ i times;
- $(1, 1)$ j times;

with $i + j = n$; all other assignments will contribute 0 in the Holant sum. Let $c_{i,j}$ be the sum over all such assignments of the products of evaluations (including the contributions from $\begin{pmatrix} -x & y \\ 1 & 1 \end{pmatrix}$ and its inverse). Then we have

$$\text{Holant}_{\Omega_s} = \sum_{i+j=n} (\lambda^i \mu^j)^s \cdot c_{i,j}$$

and $\text{Holant}_\Omega = c_{0,n}$. Since $\Delta > 0$, the coefficients form a full rank Vandermonde matrix. Thus we can interpolate D by solving the linear system of equations in polynomial time. Ignoring a nonzero factor, we may split D into unary signatures $[y, 1]$ on LHS or $[1, x]$ on RHS.

We have $\Delta \geq |1-c|$, and $x,y \geq 0$. Thus we can separate D and obtain unary signatures as specified in Lemma 1.

The following lemma lets us interpolate unary signatures on the RHS from a binary gadget with a straddled signature and a suitable unary signature s on the RHS. Mathematically, the proof is essentially the same as in [15], but technically Lemma 3 applies to binary straddled signatures.

Lemma 3. *Let $M \in \mathbb{R}^{2 \times 2}$ be the signature matrix for a binary straddled gadget which is diagonalizable with distinct non-zero eigenvalues, and let $s = [a,b]$ be a unary signature on RHS that is not a row eigenvector of M. Then $\{s \cdot M^j\}_{j \geq 0}$ can be used to interpolate any unary signature on RHS.*

4 A Basic Lemma

In this section, we prove Lemma 5 which will be invoked in the remaining cases.

When $\neg (x_0 = x_3 = 0)$, we normalize the signature $[x_0, x_1, x_2, x_3]$ to be $[1, a, b, c]$, where $a, b, c \geq 0$. We first deal with a special case when $a = b$ and $c = 1$, which will be used in the proof of Lemma 5. By a slight abuse of notation, we say $[1, a, b, c]$ is #P-hard or in FP if the problem Holant $([1, a, b, c] \mid (=_3))$ is #P-hard or in FP, respectively.

Lemma 4. *$[1, a, a, 1]$ is #P-hard unless $a = 0$ or $a = 1$ in which case the problem is in FP.*

Proof. When $a = 0$ or 1, the signature $[1, a, a, 1]$ is clearly in FP. Suppose $a \neq 0, 1$, we have $\Delta = 2a > 0$. By Lemma 2, we can interpolate the unary signature $[1, x] = [1, 1]$ on RHS. Connect $[1, 1]$ on RHS to $[1, a, a, 1]$ on LHS, we get the binary signature $[1+a, 2a, 1+a]$ on LHS. Invoke Theorem 2, as $a \neq 1$, the problem $[1+a, 2a, 1+a]$ is #P-hard, and thus $[1, a, a, 1]$ is #P-hard.

We are now ready to prove Lemma 5, the basic lemma.

Lemma 5. *When $ab \neq 0$, the problem $[1, a, b, c]$ is #P-hard unless it is degenerate in which case the problem is in FP.*

Proof. As $ab \neq 0$, we have $a \neq 0$, and $\Delta > 0$. By Lemma 2, we can interpolate the unary signature $[1, x]$ on RHS or $[y, 1]$ on LHS where $x = \frac{\Delta - (1-c)}{2a}$ and $y = \frac{\Delta + (1-c)}{2a}$. Consider the unary gadget G_3 in Fig. 1 where we place the signature $(=_3)$ at the circles, $[y, 1]$ at the triangles, and $[1, a, b, c]$ at the square. It has the unary signature $[y^2 + yb, ya + c]$ on the RHS. By Lemma 3 and using G_1 as the straddled signature, we can interpolate any unary gadget, in particular $\Delta_0 := [1, 0]$ and $\Delta_1 := [0, 1]$, on the RHS unless one of the eigenvalues equals to 0 or the signature of G_3 is a row eigenvector of G_1. Since $a, b, c \geq 0$, for the two eigenvalues λ and μ, only $\lambda = \frac{-\Delta + (1+c)}{2}$ can possibly be 0; but if $\lambda = 0$ we have $c = ab$, which we will handle shortly. Since $\Delta > 0$ we have $x + y > 0$, and Eq. (2) holds. The row eigenvectors of G_1 are the rows of $\begin{pmatrix} -x & y \\ 1 & 1 \end{pmatrix}^{-1}$, namely they are proportional to $[1, -y]$ and $[1, x]$. Since $y > 0$ by $ab \neq 0$, the only possibility that $[y^2 + yb, ya + c]$ is proportional to a row eigenvector of G_1 is $\frac{ya+c}{y^2+yb} = x$.

Observe that $xy = \frac{b}{a}$. Solve for y we get $(b - a^2) y = ac - b^2$. If $b = a^2$, then $ac = b^2$ and thus $[1, a, b, c]$ is degenerate. Otherwise, $y = \frac{ac-b^2}{b-a^2}$. Now plug into $y = \frac{\Delta+(1-c)}{2a}$, we have $\Delta + (1 - c) = \frac{2a(ac-b^2)}{b-a^2}$ which implies $(a^3 - b^3 - ab(1 - c))(ab - c) = 0$ (the high order multi-variable polynomial magically factors out).

We now divide our discussion into three cases: (1) $[(a^3 - b^3 - ab(1 - c) \neq 0) \wedge (ab - c \neq 0)]$, (2) $a^3 - b^3 - ab(1 - c) = 0$, and (3) $ab - c = 0$.

Case 1: $(a^3 - b^3 - ab(1 - c) \neq 0) \wedge (ab - c \neq 0)$

We can interpolate Δ_0 and Δ_1 on RHS. By connecting Δ_0 and Δ_1 on RHS to $[1, a, b, c]$ on LHS, we get the binary signatures $[1, a, b]$ and $[a, b, c]$ on LHS. That is we have

$$\text{Holant} ([1, a, b] \mid (=_3)) \leq_T \text{Holant} ([1, a, b, c] \mid (=_3))$$

and

$$\text{Holant} ([a, b, c] \mid (=_3)) \leq_T \text{Holant} ([1, a, b, c] \mid (=_3)).$$

By Theorem 2, the problem Holant $([1, a, b, c] \mid (=_3))$ is #P-hard unless $\frac{1}{a} \cdot \frac{b}{a} = 1$ and $\frac{a}{b} \cdot \frac{c}{b} = 1$, in which case the signature $[1, a, b, c]$ is degenerate and thus the problem is in FP.

Case 2: $a^3 - b^3 - ab(1 - c) = 0$

We have $1 - c = \frac{a^3-b^3}{ab}$ and thus $\Delta = \sqrt{(1 - c)^2 + 4ab} = \frac{a^3+b^3}{ab}$. Thus the unary signature interpolated by Lemma 2 on RHS is $[1, x] = [1, \frac{\Delta-(1-c)}{2a}] = [1, \frac{b^2}{a}]$. Connect $[1, x]$ on RHS to $[1, a, b, c]$ on LHS, we get the binary signature $[1 + \frac{b^2}{a}, a + \frac{b^3}{a^2}, b + \frac{b^2c}{a^2}]$ on LHS. By Theorem 2, the problem $[1 + \frac{b^2}{a}, a + \frac{b^3}{a^2}, b + \frac{b^2c}{a^2}]$ is #P-hard and thus the problem $[1, a, b, c]$ is #P-hard unless $(1 + \frac{b^2}{a})(b + \frac{b^2c}{a^2}) = (a + \frac{b^3}{a^2})^2$, which, after substituting c, simplifies to $(a^2 - b)(a^3 + ab + 2b^3) = 0$. Since $a, b > 0$, we have $a^3 + ab + 2b^3 \neq 0$ and thus $a^2 - b = 0$. This combined with $a^3 - b^3 = ab(1 - c)$ gives $c = a^3$, and thus $[1, a, b, c]$ is degenerate.

Case 3: $ab - c = 0$

Observe that under this case we have $\Delta = 1 + c$. The unary signature interpolated on RHS is $[1, x] = [1, \frac{\Delta-(1-c)}{2a}] = [1, \frac{c}{a}] = [1, b]$. We connect $[1, x]$ on RHS to $[1, a, b, c]$ on LHS to get binary signature $[1 + ab, a + b^2, b + bc]$ on LHS. That is we have

$$\text{Holant} ([1 + ab, a + b^2, b + bc] \mid (=_3)) \leq_T \text{Holant} ([1, a, b, c] \mid (=_3))$$

By Theorem 2, this problem is #P-hard unless $(1 + ab)(b + bc) = (a + b^2)^2$ which implies $(a^2 - b)(b^3 - 1) = 0$. If $a^2 - b = 0$, since we are under the case $ab - c = 0$, the signature $[1, a, b, c]$ is degenerate. If $b^3 - 1 = 0$, since b is a positive real number, we have $b = 1$. Thus the signature $[1, a, b, c]$ is simply $[1, a, 1, a]$.

Now consider the ternary gadget $G_4 = [2 + 2a^3, 2a + 2a^2, 2a + 2a^2, 2 + 2a^3]$ on LHS where we place the circles to be ($=_3$) and the squares to be $[1, a, 1, a]$. This signature by G_4 has the form in Lemma 4, and we see that it is #P-hard unless $2 + 2a^3 = 2a + 2a^2$. So the problem $[1, a, 1, a]$ is #P-hard and thus the problem $[1, a, b, c]$ is #P-hard unless $a^3 - a^2 - a + 1 = (a-1)^2(a+1) = 0$, which implies $a = 1$. When $a = 1$, the signature $[1, a, 1, a]$ is degenerate.

The proof is now complete.

We are now ready to prove a series of lemmas which lead to our main result, Theorem 1. The basic idea is to reduce cases to Lemma 5, and handle exceptional cases separately.

5   Case $\neg(x_0 = x_3 = 0)$

With the help of Lemma 5 we can quickly finish this case. We may normalize $[x_0, x_1, x_2, x_3]$ to $[1, a, b, c]$ by dividing a nonzero factor, and flipping all input 0's and 1's if necessary (i.e., taking the reversal of the signature). The exceptional cases in Lemma 5 for $[1, a, b, c]$ are $a = 0$ or $b = 0$. We first consider the case $a \neq 0$ but $b = 0$.

Lemma 6. *If $a \neq 0$ and $b = 0$, then $[1, a, b, c]$ is #P-hard.*

Proof. By gadget G_2 where we place the squares to be $[1, a, 0, c]$ and the circles to be ($=_3$), we have a ternary signature $[1 + 3a^3, a + a^4, a^2, a^3 + c^4]$ on LHS. Thus Lemma 5 applies (after normalizing) and $[1 + 3a^3, a + a^4, a^2, a^3 + c^4]$ is #P-hard unless it is degenerate, i.e., $(a + a^4)^2 = (1 + 3a^3)(a^2)$ and $(a^2)^2 = (a + a^4)(a^3 + c^4)$. The first equality forces $a = 1$ for positive a. Then we get $\frac{1}{2} = 1 + c^4 \geq 1$, which is a contradiction.

Now consider the case $a = 0$. We have the following subcases (1) $c = 1$, (2) $c = 0$, and (3) $(c \neq 1) \wedge (c \neq 0)$. These are handled by the following three lemmas.

Lemma 7. *If $a = 0$ and $c = 1$, then $[1, 0, b, 1]$ is #P-hard unless $b = 0$, in which case the problem is in* FP.

Proof. By flipping 0 and 1 in the input, we equivalently consider the signature $[1, b, 0, 1]$. This case is dealt with in Lemma 6 except when $b = 0$ in which case it is in FP.

Lemma 8. *If $a = 0$ and $c = 0$, then $[1, 0, b, 0]$ is #P-hard unless $b = 0$ or $b = 1$, in which case the problem is in* FP.

Proof. By gadget G_1 where we place $[1, 0, b, 0]$ at the squares and ($=_3$) at the circles, we have the binary degenerate straddled signature $\begin{pmatrix} 1 & b \\ 0 & 0 \end{pmatrix}$. We can thus split and get unary signature $[1, b]$ on RHS. Connect $[1, b]$ on RHS and $[1, 0, b, 0]$ on LHS, we get the binary signature $[1, b^2, b]$ on LHS. By Theorem 2, $[1, b^2, b]$

is #P-hard unless $b = 0$ or $b = 1$. When $b = 0$, the original signature $[1, 0, b, 0]$ becomes $[1, 0, 0, 0]$ which is degenerate and thus the problem is tractable. When $b = 1$, the signature $[1, 0, b, 0]$ becomes $[1, 0, 1, 0]$ which is an affine signature and thus the problem is also tractable ([4] p.70).

Lemma 9. *If $a = 0$, $c \neq 0$, and $c \neq 1$, then $[1, 0, b, c]$ is #P-hard unless $b = 0$, in which case the problem is in* FP.

Proof. By flipping 0 and 1 in the input, we equivalently consider the signature $[c, b, 0, 1]$. Thus Lemma 6 applies after normalizing, except when $b = 0$ in which case $[c, 0, 0, 1]$ is in FP. ∎

The discussion for the case $\neg(x_0 = x_3 = 0)$ is now complete.

6   Case $x_0 = x_3 = 0$

If both $x_1 = x_2 = 0$, the signature is identically zero and the problem is trivially in FP.

Suppose exactly one of x_1 and x_2 is 0. In this case, by normalizing and possibly flipping 0 and 1 in the input, it suffices to consider the ternary signature $[0, 1, 0, 0]$.

Lemma 10. *The problem* Holant $([0, 1, 0, 0] \mid (=_3))$ *is #P-hard.*

Proof. We begin with a reduction from Restricted Exact Cover by 3-Set (RX3C) [11]. This is a restricted version of the Set Exact Cover problem where every set has exactly 3 elements and every element is in exactly 3 sets. Given any instance for RX3C, construct the 3-regular bipartite graph whose vertices on LHS are elements of the set X and vertices on RHS are 3-element subsets of X. Connect an edge between one vertex v on LHS to one vertex C on RHS if and only if $v \in C$. Then every nonzero term in the Holant sum for Holant $([0, 1, 0, 0] \mid (=_3))$ exactly corresponds to one solution of RX3C. We observe that the reduction given in [11] (from the Exact Cover by 3 Set problem (X3C)) is parsimonious. And it is well known that SAT reduces to X3C parsimoniously via 3-Dimensional Matching (3DM). Thus we have the reduction chain:

$$\# \text{SAT} \leq_T \# \text{3DM} \leq_T \# \text{X3C} \leq_T \# \text{RX3C} \leq_T \text{Holant} ([0, 1, 0, 0] \mid (=_3))$$

We conclude that the problem $[0, 1, 0, 0]$ is #P-hard. ∎

It remains to consider the case when $x_1 \cdot x_2 \neq 0$. We normalize the signature to be $[0, 1, b, 0]$.

Lemma 11. *If $b \neq 0$, then the problem* Holant $([0, 1, b, 0] \mid (=_3))$ *is #P-hard.*

Proof. By gadget G_2 where we place the squares to be $[0, 1, b, 0]$ and the circles to be $(=_3)$, we have the ternary signature $[3b^2, 1 + 2b^3, 2b + b^4, 3b^2]$ on LHS. Lemma 5 applies (after normalizing) and the problem $[3b^2, 1 + 2b^3, 2b + b^4, 3b^2]$

is #P-hard unless it is degenerate. The condition for $[3b^2, 1 + 2b^3, 2b + b^4, 3b^2]$ being degenerate is

$$\begin{cases} \frac{3b^2}{1+2b^3} = \frac{1+2b^3}{2b+b^4} \\ \frac{1+2b^3}{2b+b^4} = \frac{2b+b^4}{3b^2} \end{cases}$$

The second equation simplifies to $(b^3 - 1)^2 = 0$. Thus when $b \neq 1$, we conclude that $[0, 1, b, 0]$ is #P-hard.

Proceed with the case $[0, 1, 1, 0]$. By gadget G_4, where we place the squares to be $[0, 1, 1, 0]$ and the circles to be $[1, 0, 0, 1]$, we get the (reversal invariant) ternary signature $[3, 2, 2, 3]$ on LHS. By Lemma 5 this problem is #P-hard.

We note that the problem Holant $([0, 1, 1, 0] \mid (=_3))$ when restricted to planar graphs is in FP.

Problem : Pl-#HyperGragh-Moderate-3-Cover
Input : A planar 3-uniform 3-regular hypergraph G.
Output : The number of subsets of hyperedges that cover every vertex with no vertex covered three times.

This is exactly the problem Pl-Holant$([0, 1, 1, 0] \mid (=_3))$, the restriction of Holant $([0, 1, 1, 0] \mid (=_3))$ to planar graphs. Its P-time tractability is seen by the following holographic reduction using the Hadamard matrix $H = \begin{pmatrix} 1 & 1 \\ 1 & -1 \end{pmatrix}$ to counting weighted perfect matchings. Under this transformation both signatures $[0, 1, 1, 0]$ and $(=_3)$ are transformed to matchgate signatures.

$$H^{\otimes 3}(=_3) = H^{\otimes 3}\left[\begin{pmatrix} 1 \\ 0 \end{pmatrix}^{\otimes 3} + \begin{pmatrix} 0 \\ 1 \end{pmatrix}^{\otimes 3}\right] = \begin{pmatrix} 1 \\ 1 \end{pmatrix}^{\otimes 3} + \begin{pmatrix} 1 \\ -1 \end{pmatrix}^{\otimes 3} = [2, 0, 2, 0],$$

and

$$[0, 1, 1, 0](H^{-1})^{\otimes 3} = \frac{1}{8}\left[(1, \ 1)^{\otimes 3} - (1, \ 0)^{\otimes 3} - (0, \ 1)^{\otimes 3}\right] H^{\otimes 3} = \frac{1}{4}[3, 0, -1, 0].$$

Since both these are matchgate signatures, the problem is reduced to counting weighted perfect matchings on planar graphs. Thus the planar problem Pl-Holant $([0, 1, 1, 0] \mid (=_3))$ can be computed in polynomial time using Kasteleyn's algorithm (see [16] and [4]). For readers unfamiliar with holographic algorithms, we will describe this algorithm for Pl-#HyperGragh-Moderate-3-Cover in more detail in an Appendix.

Acknowledgement. We sincerely thank Shuai Shao for many helpful discussions.

Appendix

The holographic algorithm for Pl-#HyperGragh-Moderate-3-Cover is achieved by a many-to-many reduction from this problem to the problem of counting (weighted) planar perfect matchings (Pl-#PM). Given planar weighted graph $G = (V, E, w)$ where $w : E \to \mathbb{R}$ is a weight function for the edge set, any perfect matching $M \subseteq E$ has weight $w(M) = \prod_{e \in M} w(e)$, and the partition function of the problem Pl-#PM is

$$\#PM(G) = \sum_{\text{perfect matchings } M} w(M).$$

Notice that when all weights $w(e) = 1$, i.e., in the unweighted case, $\#PM(G)$ simply counts the number of perfect matchings in G. Kasteleyn's algorithm can compute this partition function $\#PM(G)$ in polynomial time on planar graphs. It is important to note that Kasteleyn's algorithm can handle the case when $w(e)$ are both positive and negative real (even complex) numbers, which is important in this case for us.

The values of Holant $([0,1,1,0] \mid (=_3))$ and Holant $(\frac{1}{4}[3,0,-1,0] \mid [2,0,2,0])$ on the same instance graph G are exactly the same by Valiant's Holant Theorem [16], as

$$[0,1,1,0](H^{-1})^{\otimes 3} = \frac{1}{4}[3,0,-1,0] \quad \text{and} \quad H^{\otimes 3}(=_3) = [2,0,2,0].$$

Notice that this is a many-to-many transformation among solutions of one problem Holant $([0,1,1,0] \mid (=_3))$ (subsets of hyperedges that cover every vertex with no vertex covered three times) and another Holant $(\frac{1}{4}[3,0,-1,0] \mid [2,0,2,0])$.

Next we will use the following gadgets, called matchgates, to "implement" the constraint functions $\frac{1}{4}[3,0,-1,0]$ and $[2,0,2,0]$ by perfect matchings. Consider the following two matchgates.

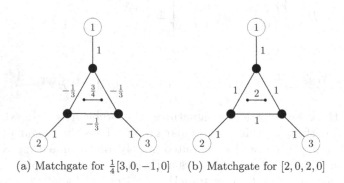

(a) Matchgate for $\frac{1}{4}[3,0,-1,0]$ (b) Matchgate for $[2,0,2,0]$

Fig. 2. Matchgates

For the matchgate in Fig. 2 (a), we consider its perfect matchings, when any subset $\emptyset \subseteq S \subseteq \{1,2,3\}$ of the external nodes labeled $1,2,3$ are removed. For $S = \emptyset$ there is a unique perfect matching M with weight $w(M) = 3/4$. If $|S| = 2$, we get $w(S) = -1/4$. If $|S|$ is odd, then $w(S) = 0$. In this sense the matchgate has the symmetric signature $\frac{1}{4}[3,0,-1,0]$.

A similar calculation shows that the matchgate in Fig. 2 (b), has the symmetric signature $[2,0,2,0]$.

Now we take a planar input bipartite graph G for Pl-Holant($[0,1,1,0] \mid (=_3)$), replace each vertex of degree three on the LHS (which has the label $[0,1,1,0]$) by the matchgate in Fig. 2 (a), replace each vertex of degree three on the RHS (which has the label $[1,0,0,1]$) by the matchgate in Fig. 2 (b), and for any edge in G add one edge with weight 1 between the corresponding external vertices from their respective matchgates. This creates a planar graph G', of size linearly bounded by that of G.

A moment reflection should convince the reader that the value $\#\mathrm{PM}(G')$ is exactly the same as the Holant value $\mathrm{Holant}\,(G)$.

References

1. Backens, M.: A complete dichotomy for complex-valued Holant^c. In: Chatzigiannakis, I., Kaklamanis, C., Marx, D., Sannella, D. (eds.) 45th International Colloquium on Automata, Languages, and Programming, ICALP 2018, Prague, Czech Republic, 9–13 July 2018. LIPIcs, vol. 107, pp. 12:1–12:14 (2018)
2. Backens, M., Goldberg, L.A.: Holant clones and the approximability of conservative Holant problems. ACM Trans. Algorithms **16**(2), 23:1–23:55 (2020)
3. Bulatov, A.A.: The complexity of the counting constraint satisfaction problem. J. ACM **60**(5) (2013)
4. Cai, J., Chen, X.: Complexity Dichotomies for Counting Problems: Volume 1, Boolean Domain. Cambridge University Press (2017)
5. Cai, J., Chen, X.: Complexity of counting CSP with complex weights. J. ACM **64**(3), 19:1–19:39 (2017)
6. Cai, J., Chen, X., Lu, P.: Nonnegative weighted #CSP: an effective complexity dichotomy. SIAM J. Comput. **45**(6), 2177–2198 (2016)
7. Cai, J., Govorov, A.: Perfect matchings, rank of connection tensors and graph homomorphisms. In: Chan, T.M. (ed.) Proceedings of the Thirtieth Annual ACM-SIAM Symposium on Discrete Algorithms, SODA 2019, San Diego, California, USA, 6–9 January 2019, pp. 476–495. SIAM (2019)
8. Cai, J., Guo, H., Williams, T.: A complete dichotomy rises from the capture of vanishing signatures. SIAM J. Comput. **45**(5), 1671–1728 (2016)
9. Dyer, M., Richerby, D.: An effective dichotomy for the counting constraint satisfaction problem. SIAM J. Comput. **42**(3), 1245–1274 (2013)
10. Freedman, M., Lovász, L., Schrijver, A.: Reflection positivity, rank connectivity, and homomorphism of graphs. J. Amer. Math. Soc. **20**(1), 37–51 (2007)
11. Gonzalez, T.F.: Clustering to minimize the maximum intercluster distance. Theoret. Comput. Sci. **38**, 293–306 (1985)
12. Kowalczyk, M., Cai, J.: Holant problems for 3-regular graphs with complex edge functions. Theory Comput. Syst. **59**(1), 133–158 (2016)
13. Lin, J., Wang, H.: The complexity of Boolean Holant problems with nonnegative weights. SIAM J. Comput. **47**(3), 798–828 (2018)
14. Shao, S., Cai, J.: A dichotomy for real boolean holant problems. In: Proceedings of the 61st IEEE Annual Symposium on Foundations of Computer Science (FOCS), pp. 1091–1102 (2020). https://doi.org/10.1109/FOCS46700.2020.00105
15. Vadhan, S.P.: The complexity of counting in sparse, regular, and planar graphs. SIAM J. Comput. **31**(2), 398–427 (2001)
16. Valiant, L.G.: Holographic algorithms. SIAM J. Comput. **37**(5), 1565–1594 (2008)

Upper Bounds on Communication in Terms of Approximate Rank

Anna Gál$^{(\boxtimes)}$ and Ridwan Syed

Department of Computer Science, University of Texas, Austin, TX, USA
{panni,ridwan}@cs.utexas.edu

Abstract. We show that any Boolean function with approximate rank r can be computed by bounded error quantum protocols without prior entanglement of complexity $O(\sqrt{r}\log r)$. In addition, we show that any Boolean function with approximate rank r and discrepancy δ can be computed by deterministic protocols of complexity $O(r)$, and private coin bounded error randomized protocols of complexity $O((\frac{1}{\delta})^2 + \log r)$. Our deterministic upper bound in terms of approximate rank is tight up to constant factors, and the dependence on discrepancy in our randomized upper bound is tight up to taking square-roots. Our results can be used to obtain lower bounds on approximate rank. We also obtain a strengthening of Newman's theorem with respect to approximate rank.

Keywords: Communication complexity · Approximate rank · Quantum communication

1 Introduction

The log-rank conjecture is one of the most intriguing open problems in communication complexity. Lovász and Saks [26] conjectured that the deterministic communication complexity $D(f)$ of a Boolean function f is upper bounded by $(\log rank(f))^k$ for some constant k, where $rank(f)$ denotes the rank over the reals of the communication matrix of the function f. It is a classical result in communication complexity by Mehlhorn and Schmidt [29] that $D(f) \geq \log rank(f)$. It is easy to see that $D(f) \leq rank(f) + 1$, but until a few years ago no one was able to obtain upper bounds sublinear in $rank(f)$. The current best upper bound was obtained by Lovett [27] who proved that $D(f) \leq O(\sqrt{rank(f)}\log rank(f))$. Considering separations, a series of results (see [33]) showed that the deterministic communication complexity can be superlinear in the logarithm of $rank(f)$. Currently the largest known separation is nearly quadratic: [14] shows that there are functions with $D(f) \geq \tilde{\Omega}((\log rank(f))^2)$.

Krause [19] extended the log-rank lower bound method to the context of randomized communication complexity, by considering the approximate rank of communication matrices. Different definitions of approximate rank have been

A. Gál—Part of this work was done while visiting the Simons Institute for the Theory of Computing in Berkeley.

R. Santhanam and D. Musatov (Eds.): CSR 2021, LNCS 12730, pp. 116–130, 2021.
https://doi.org/10.1007/978-3-030-79416-3_7

introduced in various contexts, see for example [3]. For $\alpha \geq 1$, the α-*approximate rank* of a matrix A with $+1, -1$ entries is the smallest possible rank of a matrix B that sign represents the matrix A (e.g. each entry of B has the same sign as the corresponding entry in A) with the additional condition that the absolute values of the entries are at least 1 and at most α. Krause [19] proved that the logarithm of α-approximate rank provides lower bounds on bounded error randomized communication complexity with private coins, allowing error at most $\epsilon = \frac{1}{2} - \frac{1}{2\alpha}$. The logarithm of approximate rank also lower bounds quantum communication [10]. We state the precise bounds in Sect. 2.

In light of these lower bounds, it is natural to consider the analogue of the log-rank conjecture for approximate rank in the context of randomized communication, and quantum communication. The "log-approximate-rank" conjecture and its quantum version were stated in the survey by Lee and Shraibman [21]. The conjecture states that $R_\epsilon(f) \leq (\log \mathrm{rk}^\alpha(f))^k$ for $\epsilon = \frac{1}{2} - \frac{1}{2\alpha}$ and some constant k, where R_ϵ denotes private coin randomized communication with error ϵ and rk^α denotes α-approximate rank.

If we relax the definition of approximate rank and allow arbitrary real entries in the matrix that sign represents the communication matrix, then we obtain the definition of "sign rank" (sometimes referred to as "dimension complexity"), which is a well studied measure. The appropriate communication model to consider is "unbounded error communication complexity", where it is only required that the error probability is less than $1/2$. Paturi and Simon [34] proved that the unbounded error communication complexity essentially equals to the logarithm of sign rank of the function, thus in this model the corresponding version of the log-rank conjecture holds.

In the case of randomized and quantum communication complexity, with error bounded by a constant $\epsilon < 1/2$, the conjecture has been recently refuted [5,11,38]. Until recently, the largest known separation between the logarithm of approximate rank and quantum communication was nearly quadratic [4]. A quadratic separation between the logarithm of approximate rank and randomized communication is witnessed by the disjointness function. The n-bit disjointness function can be computed by $O(\sqrt{n})$-qubit quantum protocols [1] and thus it has approximate rank $2^{O(\sqrt{n})}$ by the lower bound of [10] on quantum communication. On the other hand it requires $\Omega(n)$ randomized communication [17,35]. A breakthrough result of Chattopadhyay et al. [11] gave an example of a Boolean function with approximate rank r that requires $\Omega(r^{1/4})$ randomized communication. [5,38] showed that the same function requires $\Omega(r^{1/12})$ quantum communication.

1.1   Our Results

In this paper we consider upper bounds on communication in terms of approximate rank. Approximate rank can be significantly smaller than rank. For example the equality function gives an exponential separation: the approximate rank of the n-bit equality function is $\Theta(n)$ while its rank is 2^n. Thus, upper bounds on communication complexity in terms of approximate rank can potentially give much sharper upper bounds on communication complexity. Our randomized upper bounds also involve the inverse of discrepancy, which is upper bounded by the square root of approximate rank, and in turn can be arbitrarily small compared to approximate rank: the discrepancy of the equality function is constant.

Our main results are the following upper bounds: We show that any Boolean function with approximate rank r and discrepancy δ can be computed by

- deterministic protocols of complexity $O(r)$,
- private coin bounded error randomized protocols of $O((\frac{1}{\delta})^2 + \log r)$ complexity and
- bounded error quantum protocols without prior entanglement of complexity $O(\sqrt{r} \log r)$.

The example of the equality function shows that our deterministic upper bound in terms of approximate rank is tight up to constant factors, and that the additive $\log r$ term is necessary in our randomized upper bound. In addition, the disjointness function shows that the dependence on discrepancy in our randomized bound is tight up to taking square-roots.

While in the deterministic case the upper bound $D(f) \leq rank(f) + 1$ is immediate from bounding the number of different rows of the matrix, as far as we know, a linear upper bound on communication in terms of approximate rank alone has not been established before, considering deterministic or private coin bounded-error randomized communication. We show that any Boolean function with α-approximate rank r can be computed by deterministic protocols of complexity $O(r)$. This bound in turn can also be used to obtain lower bounds on approximate rank (see Sect. 3.1).

Lovett's result [27] implies that deterministic communication (and hence randomized communication) is upper bounded by square-root of approximate rank with an additional $(\log rank(f))^2$ factor. This follows since Lovett's proof gives that $D(f) \leq O(\frac{1}{disc(f)}(\log rank(f))^2)$, where $disc(f)$ is the discrepancy of f, and since $\frac{1}{disc(f)}$ is bounded above by the square-root of approximate rank [25]. However, when approximate rank is much smaller than $rank(f)$, this bound may be superlinear in approximate rank.

Results of Linial and Shraibman (Claim 2. in [25]), and Klauck [18] imply public coin bounded error randomized protocols with communication $O((\frac{1}{disc(f)})^2)$ and therefore with communication linear in approximate rank. It is well known by a theorem of Newman [31] that public coin protocols can be converted to private coin protocols (with slightly larger error) at the cost of additional $O(\log n + \log(1/\rho))$ bits where ρ is the (additive) increase in error. However, applying Newman's theorem as stated, does not give our claimed bound when the approximate rank is very small. We obtain our result on private coin protocols by giving a strengthening of Newman's theorem with respect to approximate rank.

Considering quantum protocols (without prior entanglement) we are able to match Lovett's bound in terms of approximate rank. We show that any Boolean function with α-approximate rank r can be computed by quantum protocols of complexity $O(\alpha^2 \sqrt{r} \log r)$. To obtain these bounds, we first show that any function with α-approximate rank r can be computed by $O(\log r)$ communication and error at most $\frac{1}{2} - \frac{1}{2\alpha\sqrt{r}}$ by private coin randomized (or quantum) protocols. Previously, private coin protocols with $O(\log r)$ communication but with error $\frac{1}{2} - \frac{1}{2\alpha r}$ were given in [12]. Even with our improved error bound, amplifying the correctness of this protocol classically would only give bounded error private coin protocols with $O(r \log r)$ communication. Using the amplitude amplification technique of [30] allows us to obtain bounded error quantum protocols with $O(\alpha^2 \sqrt{r} \log r)$ communication.

2 Preliminaries

Let $f : X \times Y \to \{-1, 1\}$ be a Boolean function. We write $D(f)$, $R_\epsilon(f)$, $R_\epsilon^{pub}(f)$, $U(f)$ to respectively denote the deterministic communication complexity, ϵ-error randomized communication complexity with private coins, ϵ-error randomized communication complexity with public coins, and unbounded error randomized communication complexity with private coins. For background in classical communication complexity we refer to [20]. For the quantum analogues of these measures we write $Q(f), Q_\epsilon(f), Q_\epsilon^*(f)$ to respectively denote the exact quantum communication complexity, ϵ-error quantum communication complexity without prior entanglement, and ϵ-error quantum communication complexity with prior entanglement. We postpone a brief review of the quantum communication model until Sect. 5.

We will often identify f with its communication matrix M with entries $M[x, y] = f(x, y)$. Our primary complexity measure of interest is the approximate rank.

Definition 1 *[19] (see also [21]). Let $f : X \times Y \to \{-1, 1\}$ and let M be the communication matrix of f. Fix some real $\alpha \geq 1$. The α-approximate rank of f is defined as*

$$rk^\alpha(f) := \min_{A \ :\ 1 \leq A_{i,j} \cdot M_{i,j} \leq \alpha} rk(A)$$

where rk is the usual rank over $\mathbb{R}$, and the entries of A are reals. We say such a matrix A is an approximating matrix for f. Note that when $\alpha = 1$, this measure coincides with the usual rank of f. When we do not bound the entries of A, this measure coincides with the sign rank of f, denoted $rk^\infty(f)$.

We say that a matrix B sign represents a communication matrix M if $B_{i,j} M_{i,j} > 0$ for all i, j. Note that an approximating matrix for f sign represents the communication matrix of f.

For a matrix A, we call a decomposition of the form $A = UV$ a *d-dimensional factorization* if the rows (resp. columns) of U (resp. V) are in $\mathbb{R}^d$. Recall that the rank of a matrix is the minimum d for which such a factorization is possible.

We will also be interested in bounds on the ℓ_2 norm of vectors in such decompositions, which is captured by factorization norms. Approximate factorization norm (γ_2 norm) was introduced by Linial and Shraibman [25], who showed that the logarithm of α-approximate γ_2 norm is a lower bound for quantum communication with entanglement and error $\epsilon = \frac{1}{2} - \frac{1}{2\alpha}$.

Definition 2 *[25]. Let $f : X \times Y \to \{-1, 1\}$ be a Boolean function, and let M be its communication matrix. Fix some real $\alpha \geq 1$. The α-approximate factorization norm of f is defined as*

$$\gamma_2^\alpha(f) = \min_{A=UV \; : \; 1 \leq A_{i,j} \cdot M_{i,j} \leq \alpha} \ell(U) \cdot \ell(V^T)$$

where $\ell(B)$ denotes the maximum ℓ_2 norm of a row of B. When $\alpha = 1$, we omit the superscript and simply write $\gamma_2(f)$. When we do not bound the entries of A, we write $\gamma_2^\infty(f)$.

The approximate γ_2 norm can be significantly smaller than the approximate rank. For example the γ_2 norm of the n-bit equality function is constant, while its approximate rank is $\Omega(n)$ [2,22].

Our results exploit the following relationship shown by [22] between γ_2^α and rk^α. See also [23,37] for more on relating factorization norm and rank.

Theorem 1. *For any $f : X \times Y \to \{-1, 1\}$ and $\alpha > 1$, we have*

$$\gamma_2^\alpha(f) \leq \alpha\sqrt{rk^\alpha(f)}$$

Moreover this bound can be witnessed by an approximating matrix $A = UV$ with a factorization of dimension $rk^\alpha(f)$ such that $\ell(U) \leq \sqrt{rk^\alpha(f)}$ and $\ell(V) \leq \alpha$.

For this factorization, we will find it convenient to explicitly refer to the rows (resp. columns) of U (resp. V), and we will additionally enforce uniformity in the respective ℓ_2 norms by slightly increasing dimension.

Lemma 1. *Let $f : X \times Y \to \{-1, 1\}$ be a Boolean function with $rk^\alpha(f) = r$. There exist factorization vectors $\{u_x\}_{x \in X}$ and $\{v_y\}_{y \in Y}$ in $\mathbb{R}^{r+2}$ such that for all x, y, $\|u_x\| = \sqrt{r}$, $\|v_y\| = \alpha$, and $1 \leq (u_x \cdot v_y)f(x, y) \leq \alpha$.*

Proof. Let $A = UV$ be the factorization guaranteed by Theorem 1. Write the row of U corresponding to $x \in X$ as u'_x. Let $s_x = r - \|u'_x\|_2^2$. Write the column of V corresponding to $y \in Y$ as v'_y. Let $s_y = \alpha^2 - \|v'_y\|_2^2$. Finally define $u_x = u'_x \oplus (s_x, 0)$ and $v_y = v'_y \oplus (0, s_y)$. It is straightforward to check that these vectors satisfy the claims of the lemma.

We record the lower bounds alluded to in the introduction. For $\frac{1}{2} > \epsilon > 0$ let $\alpha = \frac{1}{1-2\epsilon}$. We use log to denote $\log_2$ unless otherwise indicated.

- $D(f) \geq \log rk(f) \geq \log \gamma_2(f)$ [23,29]
- $R_\epsilon(f) \geq \log rk^\alpha(f) \geq 2 \log \gamma_2^\alpha(f) - 2 \log \alpha$ [19,25]
- $U(f) \geq \log rk^\infty(f)$ [34]

- $Q(f) \geq \log \mathrm{rk}(f)/2$ [9,10]
- $Q_\epsilon(f) \geq \log \mathrm{rk}^\alpha(f)/2$ [10]
- $Q_\epsilon^*(f) \geq \log \gamma_2^\alpha(f) - \log \alpha - 2$ [25]

Another important measure we consider is discrepancy.

Definition 3. *For a Boolean function* $f : X \times Y \to \{-1,1\}$ *its discrepancy is defined as*

$$disc(f) = \min_\mu \max_R | \sum_{(x,y)\in R} f(x,y)\mu(x,y)|$$

where μ *is an arbitrary distribution over* $X \times Y$ *and* $R = A \times B$ *with* $A \subseteq X$, $B \subseteq Y$ *is an arbitrary rectangle.*

It is known that $\frac{1}{disc(f)}$ is equivalent to $\gamma_2^\infty(f)$ up to constant factors [24]. Thus, by Theorem 1 for constant $\alpha \geq 1$ and any Boolean function f we have

$$\Theta(\frac{1}{disc(f)}) = \gamma_2^\infty(f) \leq \gamma_2^\alpha(f) \leq \alpha\sqrt{\mathrm{rk}^\alpha(f)} = O(\sqrt{\mathrm{rk}(f)}). \tag{1}$$

For more on communication lower bounds and the relationships between these and other measures we refer to an excellent survey by Lee and Shraibman [21].

For public coin protocols, the following upper bounds have been established in terms of discrepancy.

Lemma 2 *[25]. Let* $f : X \times Y \to \{1,-1\}$ *be a Boolean function. Then there is an* $O(1)$-*bit public coin randomized protocol for* f *with error at most* $\frac{1}{2} - \frac{1}{2K\gamma_2^\infty(f)}$, *where* $1.5 \leq K \leq 1.8$ *is the Grothendieck constant.*

Since $\frac{1}{disc(f)}$ is equivalent to $\gamma_2^\infty(f)$ up to constant factors [24], this also gives a bound with respect to discrepancy. A more direct argument in terms of discrepancy is given by Klauck [18].

Lemma 3 *[18]. Let* $f : X \times Y \to \{1,-1\}$ *be a Boolean function. Then there is an* $O(1)$-*bit public coin randomized protocol for* f *with error at most* $\frac{1}{2} - \frac{disc(f)}{2}$.

The following upper bound on public coin randomized protocols follows by standard amplification.

Theorem 2 *[18,25]. Let* $f : X \times Y \to \{1,-1\}$ *be a Boolean function. Then* $R_{1/3}^{pub}(f) = O((\frac{1}{disc(f)})^2)$.

The following private coin protocol with $O(\log \mathrm{rk}^\alpha(f))$ communication was observed in [12]. This protocol is similar to the proof of Paturi and Simon [34] in the unbounded error model.

Lemma 4 *[12]. Let* $f : X \times Y \to \{1,-1\}$ *be a Boolean function and* $\alpha > 1$. *Then there is a* $\log(4\mathrm{rk}^\alpha(f))$-*bit private coin randomized protocol with error at most* $\frac{1}{2} - \frac{1}{2\alpha \mathrm{rk}^\alpha(f)}$,

For functions f with $\mathrm{rk}^\alpha(f) = r$, amplifying this protocol gives private coin bounded error protocols with $O(\alpha^2 r^2 \log r)$ communication. We improve the error bounds of the protocol of Lemma 4 and we use the existence of this protocol as a starting point for our results.

3   Deterministic Communication

In this section we give upper bounds on the number of different rows and columns of Boolean matrices with given approximate rank. These estimates directly yield upper bounds on deterministic communication in terms of approximate rank.

We need the following theorem of Krause [19].

Theorem 3 *[19]. Let $f : X \times Y \to \{1, -1\}$ be a Boolean function, and let M be the communication matrix of f. If there is a private coin randomized protocol for f with c-bit communication and error at most $\frac{1}{2} - \frac{1}{s}$, then there is a matrix B with nonzero integer entries that sign represents M such that the absolute values of the entries of B are at most $t = 8s2^c$, and B has rank at most 2^c.*

Note that the lower bound of Krause stated in the Introduction and Sect. 2 is a simplified version of this theorem, without requiring that the approximating matrix has integer elements.

Theorem 3 allows us to estimate the number of different rows and columns in the communication matrix of a function with given approximate rank.

Lemma 5. *Let $f : X \times Y \to \{1, -1\}$ be a Boolean function and $\alpha \geq 1$. Let M be the communication matrix of the function f. If $\mathrm{rk}^\alpha(f) = r$, then M has at most $(2t)^{4r}$ distinct rows and columns, where $t = 64\alpha r^2$.*

Proof. Applying Theorem 3 to the protocol from Lemma 4 we know that there is a matrix B with nonzero integer entries that sign represents M such that the absolute values of the entries of B are at most $t = 64\alpha r^2$, and B has rank at most $4r$. Thus the number of different rows and columns of B is at most $(2t)^{4r}$. But since B sign represents M, the number of different rows (and columns) of M cannot be larger than the number of different rows (and columns) of B.

Shachar Lovett [28] pointed out to us that our bound can be improved as follows.

Lemma 6 *[28]. Let $f : X \times Y \to \{1, -1\}$ be a Boolean function and $\alpha \geq 1$. Let M be the communication matrix of the function f. If $\mathrm{rk}^\alpha(f) = r$, then M has at most $(\alpha + 2)^r$ distinct rows and columns.*

Proof Let $A \in R^{|X| \times |Y|}$ be an approximating matrix for f. Let V denote the linear span of the rows of A. Consider a maximal set of pairwise different rows of M, and denote the corresponding rows of A by $v_1, \ldots, v_N$.

We will use the notation $W = V \cap [-\alpha, \alpha]^{|Y|}$. By definition, $v_i \in W$ for each $i \in [N]$. Moreover, for $i_1 \neq i_2$ there is an index $1 \leq j \leq |Y|$ such that $|v_{i_1}(j) - v_{i_2}(j)| \geq 2$, where $v_i(j)$ denotes the j-th coordinate of the vector v_i.

We also use the notation $\beta W = \{\beta w | w \in W\}$ for constant $\beta > 0$, and we will consider $v + \beta W = \{v + u | u \in \beta W\}$.

Let us fix $\beta = \frac{1}{\alpha+1}$. With this notation, we have that the sets $v_i + \beta W$ are pairwise disjoint for $i \in [N]$. Thus, $\sum_{i \in [N]} \mathrm{Vol}(v_i + \beta W) = N\mathrm{Vol}(\beta W) \leq \mathrm{Vol}((1+\beta)W)$. This implies that $N \leq (\frac{1+\beta}{\beta})^r = (\alpha+2)^r$.

The above bounds on the number of different rows and columns imply the following upper bound on deterministic communication in terms of approximate rank. The proof follows by the simple observation, that for any Boolean function f, $D(F)$ is at most $1+$ the logarithm of the number of different rows (or columns) of the communication matrix representing the function f.

Theorem 4. *Let $f : X \times Y \to \{1, -1\}$ be a Boolean function and $\alpha \geq 1$. Then $D(f) \leq 1 + c_\alpha rk^\alpha(f)$, where $c_\alpha = \log(\alpha+2)$.*

Note that this bound is tight up to constant factors, as demonstrated by the equality function.

3.1   Lower Bounds on Approximate Rank

Theorem 4 can be used to derive lower bounds on approximate rank. First we note that it implies that for constant α the separation between approximate rank and rank is at most exponential. This also holds with respect to the nonnegative rank $rk^+(f)$ of the corresponding $0/1$ matrix, since the log-rank lower bound of Mehlhorn and Schmidt [29] also gives $D(f) \geq \log rk^+(f)$.

Corollary 1. *Let $f : X \times Y \to \{1, -1\}$ be a Boolean function and let $\alpha > 1$ be a constant. Then $rk^\alpha(f) \geq \Omega(\log rk^+(f)) \geq \Omega(\log rk(f))$.*

Next we note that by Theorem 4 any lower bound on the deterministic communication complexity of a function yields lower bounds on its approximate rank. Previous lower bounds on approximate rank are usually based on measures like discrepancy, factorization norm and trace norm. Our method will not give larger than n lower bounds for communication problems involving n-bit inputs for the players, but it can give interesting lower bounds for functions where the previously used measures are not very large.

For the n-bit Greater Than function GT_n Braverman and Weinstein [7] proved that $\frac{1}{disc(GT_n)} \geq \sqrt{n}$. This implies using (1) that for constant α the approximate rank of GT_n is $\Omega(n)$. Theorem 4 provides another proof of the $\Omega(n)$ lower bound on the approximate rank of GT_n, with improved dependence on α.

Corollary 2. *For $\alpha > 1$, $rk^\alpha(GT_n) \geq \frac{n}{\log(\alpha+2)}$.*

We also obtain a new proof of the $\Omega(n)$ lower bound of Alon [2] on the approximate rank of the n-bit Equality function EQ_n.

Corollary 3. *For $\alpha > 1$, $rk^\alpha(EQ_n) \geq \frac{n}{\log(\alpha+2)}$.*

4   Randomized Communication

Using Lemma 5 we obtain the following strengthening of Newman's theorem [31] with respect to approximate rank.

Theorem 5. *Let $f : X \times Y \to \{1, -1\}$ be a Boolean function and $\alpha \geq 1$. For every $\rho > 0$ and every $\epsilon > 0$, $R_{\epsilon+\rho}(f) \leq R_\epsilon^{pub}(f) + O(\log rk^\alpha(f) + \log(\alpha + 2) + \log \rho^{-1})$.*

Proof. Let M be the communication matrix of the function. If the number of different rows of the matrix M is at most N_1 and the number of different columns of M is at most N_2 then the players can determine the value of f by running a protocol for a function $f' : X' \times Y' \to \{1, -1\}$ with $|X'| = N_1$ and $|Y'| = N_2$. Newman's theorem [31] gives that $R_{\epsilon+\rho}(f) \leq R_\epsilon^{pub}(f) + O(\log \log(N_1 N_2) + \log \rho^{-1})$, where N_1 is the number of different rows of M and N_2 is the number of different columns of M. The statement follows by Lemma 5.

Theorem 5 allows us to simulate public coin protocols by private coin protocols efficiently with respect to approximate rank. We obtain the following bounds on private coin protocols.

Theorem 6. *Let $f : X \times Y \to \{1, -1\}$ be a Boolean function and $\alpha \geq 1$. There is a private coin protocol computing f with $O(\log rk^\alpha(f) + \log(\alpha + 2))$ communication and error at most $\frac{1}{2} - \frac{disc(f)}{4}$.*

Proof. Follows by applying Theorem 5 with $\rho = \frac{disc(f)}{4}$ to the protocol given by Lemma 3, and using that $\frac{1}{disc(f)} = O(\alpha \sqrt{rk^\alpha(f)})$ by (1).

Theorem 7. *Let $f : X \times Y \to \{1, -1\}$ be a Boolean function and $\alpha \geq 1$. Then $R_{1/3}(f) = O((\frac{1}{disc(f)})^2 + \log rk^\alpha(f) + \log(\alpha + 2))$.*

Proof. Follows by applying Theorem 5 to Theorem 2.

The equality function shows that the additive $\log rk^\alpha(f)$ term is necessary in Theorem 7, since $\frac{1}{disc(EQ_n)} = O(\gamma_2(EQ_n)) = O(1)$. The disjointness function shows that the dependence on $\frac{1}{disc(f)}$ in Theorem 7 is tight up to taking square-roots: it is known that $\frac{1}{disc(DISJ_n)} = O(n)$ [20], and $\log rk^\alpha(DISJ_n) = O(\sqrt{n})$ by combining the existence of $O(\sqrt{n})$-qubit quantum protocols [1] for $DISJ_n$ and the lower bound of [10] on quantum protocols as noted in the introduction. On the other hand $DISJ_n$ requires $\Omega(n)$ randomized communication [17,35].

5   Quantum Communication

5.1   Quantum Communication Model

We assume basic familiarity with quantum information, and refer to [32] for more background. The state space of a quantum communication protocol is comprised of three registers: Alice's private register, Bob's private register, and a shared register. Each of these registers may be of some arbitrary fixed size which may depend on the function (but not the inputs) being computed. We assume that Alice and Bob are given their inputs x, y encoded as computational basis states $|x\rangle, |y\rangle$, and that the initial state of the protocol is

$$|\text{init}\rangle = |x\rangle |\mathbf{0}\rangle |y\rangle$$

Alice and Bob then alternate sending each other messages across the channel. More precisely, when it is Alice's ith turn to speak she applies some unitary U^{A_i} to her register and the shared register, and when it is Bob's jth turn to speak he applies some unitary U^{B_j} to his register and the shared register. At the end of the protocol, one of the players measures the qubits in the channel and as a function of the result outputs a result for the protocol. Recall that we write $Q(f), Q_\epsilon(f)$ to respectively denote the exact quantum communication complexity and ϵ-error quantum communication complexity and ϵ-error quantum communication.

5.2   Root Approximate Rank Upper Bound

Let $\text{rk}^\alpha(f) = r$ and assume as well that $\alpha > 1$ is a constant. Our main protocol combines two standard techniques: quantum fingerprinting[1] and amplitude amplification. The first, is for Alice and Bob to associate their (potentially long) inputs x and y with significantly shorter quantum states or *fingerprints* [8]. In particular, we will have Alice and Bob associate their inputs with quantum states which encode the factorization vectors $u_x, v_y \in \mathbb{R}^{\mathcal{O}(r)}$ given by Lemma 1. In particular, such vectors (up to normalization) can be encoded with $O(\log r)$ qubits. Alice and Bob can then perform a distributed variant of the Hadamard test on their encoded states, which will allow them to estimate the inner product of their respective fingerprints, and in turn compute $f(x, y)$ with bias $1/O(\sqrt{r})$ over random guessing. To improve the bias to a constant classically requires $\Omega(r)$ repetitions.

To improve the bias more efficiently we apply the second technique : amplitude amplification. Generalizing the ideas involved in Grover's search algorithm [15], the technique of amplitude amplification [30] has been applied to quantum search and several problems in communication complexity [9,16] to achieve polynomial speedups over classical algorithms. We record the technique in the following lemma.

Lemma 7. *Let $\mathcal{A}$ be a quantum algorithm which makes no measurements, and let G be some collection of 'good' basis states. Suppose that on initial state $|0\rangle$ the algorithm produces the state*

$$\mathcal{A} |0\rangle = \sin(\theta) |g\rangle + \cos(\theta) |b\rangle$$

---

[1] Our protocol does not use fingerprinting in the usual sense since we will really be encoding the factorization vectors corresponding to Alice and Bob's inputs rather than their actual inputs.

where $|g\rangle$ is contained in a subspace W_G spanned by states in G, $|b\rangle$ is contained in the orthogonal complement of W_G, and $\theta \in [0, \pi/2]$. Let $\mathcal{B} = -\mathcal{A}S_0\mathcal{A}^{-1}S_G$, where S_G negates the amplitudes of the states in W_G, and S_0 negates the amplitude of $|0\rangle$. Repeatedly applying $\mathcal{B}$ to $\mathcal{A}|0\rangle$ k times produces the state

$$\mathcal{B}^k \mathcal{A} |0\rangle = \sin((2k+1)\theta)\,|g\rangle + \cos((2k+1)\theta)\,|b\rangle$$

We can view the pre-measurement state $\mathcal{A}|\text{init}\rangle$ of the protocol as a unit vector in a two dimensional space spanned by $|g\rangle$ (corresponding to the protocol outputting 1) and $|b\rangle$ (corresponding to the protocol outputting -1). As a two dimensional vector in this space, the vector $\mathcal{A}|\text{init}\rangle$ has angle $\theta = \pi/4 + \delta$ with $|b\rangle$ if $f(x,y) = 1$, and angle $\theta = \pi/4 - \delta$ with $|b\rangle$ if $f(x,y) = -1$, where $\delta > 0$ is some small value. The operator $\mathcal{B}$ above essentially implements a rotation of 2θ towards $|g\rangle$. Thus applying this operator 4 times, results in the vector having angle $\theta = \pi/4 + 8\delta$ with $|b\rangle$ if $f(x,y) = 1$, and angle $\theta = \pi/4 - 8\delta$ with $|b\rangle$ if $f(x,y) = -1$. This will correspond to increasing the bias by a constant factor! As we will show, Alice and Bob can apply $\mathcal{B}$ in blocks of 4 repetitions so that $O(\sqrt{r})$ rounds of application suffice to amplify the bias to a constant.

We begin with the fingerprinting protocol.

Lemma 8. *Fix $\alpha > 1$, and let $\epsilon = 1/2 - 1/(2\alpha\sqrt{r})$. If $rk^\alpha(f) = r$, then $Q_\epsilon(f) = O(\log r)$.*

Proof. Note that the statement follows from Theorem 6 with slightly different error bound. Here we give a direct proof by presenting an explicit protocol, which will be convenient for us to use in the proof of the next theorem.

The protocol we give is simply a distributed version of what is sometimes called the *Hadamard test*. Let x and y be Alice and Bob's respective inputs. By Lemma 1, there are vectors $u_x, v_y \in \mathbb{R}^{r+2}$ satisfying

$$\frac{1}{\alpha\sqrt{r}} \leq f(x,y)\frac{u_x \cdot v_y}{\|u_x\|_2 \|v_y\|_2} \leq \frac{1}{\sqrt{r}} \tag{2}$$

The vectors' normalizations $u_x/\|u_x\|_2$ and $v_y/\|v_y\|_2$ can be encoded as $d = O(\log(r+2))$ qubit quantum states $|\phi_x\rangle, |\psi_y\rangle$ in a straightforward manner, so that the coordinates of the normalized vectors are precisely the amplitudes of the states in the computational basis. Alice and Bob's protocol is as follows:

1. Alice prepares the $d+1$ qubit state $\frac{1}{\sqrt{2}}|0\rangle|\phi_x\rangle + \frac{1}{\sqrt{2}}|0\rangle|0\rangle$ and sends the state to Bob.
2. Bob unitarily transforms the received state to $\frac{1}{\sqrt{2}}|0\rangle|\phi_x\rangle + \frac{1}{\sqrt{2}}|1\rangle|\psi_y\rangle$ and sends the state to Alice.
3. Alice performs a Hadamard transformation on the first qubit, and measures the first qubit. If she measures 0, Alice outputs 1, and otherwise she outputs -1.

We note that the protocol can be implemented so that it acts as the identity on each of Alice's and Bob's private registers. Clearly the total communication is

$O(\log r)$. By a straightforward calculation, the probability that Alice measures 0 in the first qubit is

$$\frac{1}{2}(1 + \langle\phi_x|\psi_y|\phi_x|\psi_y\rangle)$$

The upper bound on the error probability of the protocol follows immediately from (2).

We are now ready to prove our main theorem.

Theorem 8. *Fix $\alpha > 1$. If $rk^\alpha(f) = r$, then $Q_{1/3}(f) = O(\alpha^2 \sqrt{r} \log r)$.*

Proof. Let $\mathcal{A}$ be the pre-measurement steps of the fingerprinting protocol of Lemma 8. Let the set of good states G be the set of basis states for which the first qubit of Alice and Bob's shared qubits is 0, and let W_G be the subspace spanned by these basis states. As in Lemma 7 let S_0 be an operator which negates the amplitude on the initial state[2] $|\mathrm{init}\rangle = |x\rangle\,|0\rangle\,|y\rangle$ of the protocol and let S_G be an operator which negates the amplitude on states in W_G. Finally let j, k be integers whose values we set later. Alice and Bob's protocol is as follows:

1. Alice and Bob apply $\mathcal{A}$ to the initial state $|\mathrm{init}\rangle$.
2. Alice applies S_G to the shared qubits, negating the amplitudes on states in W_G.
3. Alice and Bob apply the inverse $\mathcal{A}^{-1}$ of the basic protocol.
4. Alice applies S_0 to the shared qubits, negating the amplitude on the initial state $|\mathrm{init}\rangle$.
5. Alice and Bob apply $\mathcal{A}$, and then negate all amplitudes.
6. Alice and Bob repeat steps 2–5 an additional $4j - 1$ times so that the final state of the protocol is $\mathcal{B}^{4j}\mathcal{A}\,|\mathrm{init}\rangle$, where $\mathcal{B} = -\mathcal{A}S_0\mathcal{A}S_G$.
7. Alice measures the first qubit. If she measures 0, Alice outputs 1, and otherwise she outputs -1.
8. Alice and Bob repeat steps 1–7 k independent times and output the majority result.

Clearly the communication cost of each of steps 1–5 is $O(\log r)$. Thus the total communication cost of the protocol is $O(jk \log r)$. It remains to set the parameters j, k.

By (2) the probability of measuring a state in W_G after step 1 is $1/2 + f(x, y) \cdot \delta$, where $\delta \in [1/(2\alpha\sqrt{r}), 1/(2\sqrt{r})]$. Thus we can write the state of the protocol after step 1 as

$$\mathcal{A}\,|\mathrm{init}\rangle = \sin(\pi/4 + \theta)\,|\mathrm{g}\rangle + \cos(\pi/4 + \theta)\,|\mathrm{b}\rangle$$

It follows from basic trigonometric identities that $\sin(2\theta) = 2f(x, y) \cdot \delta$. For all $z \in (0, 1]$, $z < \sin^{-1}(z) \leq z\pi/2$. Thus, $1/(2\alpha\sqrt{r}) < f(x, y) \cdot \theta < \pi/(4\sqrt{r})$.

By Lemma 7 the state of the protocol after step 6 can be written as

$$\mathcal{B}^{4j}\mathcal{A}\,|\mathrm{init}\rangle = \sin(\pi/4 + (1 + 4j)\theta)\,|\mathrm{g}\rangle + \cos(\pi/4 + (1 + 4j)\theta)\,|\mathrm{b}\rangle \qquad (3)$$

---

[2] It suffices for S_0 to negate the amplitudes on states where all channel qubits are 0.

Take j to be the largest integer such that $(1 + 4j) < \sqrt{r}$. Given our choice of j, we can re-write 3 as

$$\mathcal{B}^{4j}\mathcal{A}\,|\text{init}\rangle = \sin(\theta')\,|\text{g}\rangle + \cos(\theta')\,|\text{b}\rangle \tag{4}$$

where $\pi/4 + 1/(c\alpha) \leq \theta' \leq \pi/2$ for some constant c. It follows that the probability that Alice outputs $f(x, y)$ is at least $1/2 + 1/(c'\alpha)$ for some constant c'. By a standard argument we can take $k = O(\alpha^2)$, so that the entire protocol has cost $O(\alpha^2\sqrt{r}\log r)$ and the final result is correct with probability at least $2/3$.

When $\alpha = 1$ (that is when r is the rank of f) we can modify the analysis here to get an exact protocol. In particular, if u_x, v_y arise from a rank r factorization, then (2) can be written as the equality $(u_x \cdot v_y)/(\|u_x\|_2\|v_y\|_2) = f(x, y)/\sqrt{r}$. We can adjust the amount by which we increase the norms of u_x, v_y in the Lemma 1, so that in (4) for some integer $j = O(\sqrt{r})$ we have $(1 + 4j)\theta = f(x, y)\pi/4$. This gives an alternative and more direct proof of the following result which is a corollary of Lovett's deterministic upper bound in terms of rank.

Theorem 9. *If $rk(f) = r$, then $Q(f) = O(\sqrt{r}\log r)$.*

Acknowledgements. We thank Shachar Lovett for allowing us to include his argument improving our bound in Lemma 6. We would also like to thank the anonymous referees of a previous version of this paper for helpful comments. The second author thanks Patrick Rall and Daniel Liang for helpful discussions.

References

1. Aaronson, S., Ambainis, A.: Quantum search of spatial regions. Theory Comput. **1**, 47–79 (2005)
2. Alon, N.: Perturbed identity matrices have high rank: proof and applications. Comb. Probab. Comput. **18**, 3–15 (2009)
3. Alon, N., Lee, T., Shraibman, A., Vempala, S.: The approximate rank of a matrix and its algorithmic applications. In: Proceedings of STOC 2013, pp. 675-684 (2013)
4. Anshu, A., Ben-David, S., Garg, A., Jain, R., Kothari, R., Lee, T.: Separating quantum communication and approximate rank. In: Computational Complexity Conference, vol. 24, no. 1–24 p. 33 (2017)
5. Anshu, A., Boddu, D.T.: Quantum log-approximate-rank conjecture is also false. In: Proceedings of FOCS 2019, pp. 982–994 (2019)
6. Bera, D.: Two-sided quantum amplitude amplification and exact-error algorithms. CoRR abs/1605.01828 (2016)
7. Braverman, M., Weinstein, O.: A discrepancy lower bound for information complexity. Algorithmica **76**(3), 846–864 (2015). https://doi.org/10.1007/s00453-015-0093-8
8. Buhrman, H., Cleve, R., Watrous, J., de Wolf, R.: Quantum fingerprinting. Phys. Rev. Lett. **87**(16), 167902 (2001)
9. Buhrman, H., Cleve, R., Wigderson, A.: Quantum vs. classical communication and computation. In: Proceedings of 30th STOC, pp. 63–68 (1998)

10. Buhrman, H., de Wolf, R.: Communication complexity lower bounds by polynomials. In: Proceedings of the 16th IEEE Conference on Computational Complexity, pp. 120–130 (2001)
11. Chattopadhyay, A., Mande, N.S., Sherif, S.: The log-approximate-rank conjecture is false. J. ACM **67**(4), 23:1–23:28 (2020)
12. Forster, J., Krause, M., Lokam, S.V., Mubarakzjanov, R., Schmitt, N., Simon, H.U.: Relations between communication complexity, linear arrangements, and computational complexity. In: FSTTCS: Foundations of Software Technology and Theoretical Computer Science, vol. 21, pp. 171–182 (2001)
13. Gavinsky, D., Lovett, S.: En route to the log-rank conjecture: New reductions and equivalent formulations, In: Proceedings of ICALP 2014, pp. 514–524 (2014)
14. Göös, M., Pitassi, T., Watson, T.: Deterministic Communication vs. Partition Number. In: Proceedings of FOCS 2015, pp. 1077–1088 (2015)
15. Grover, L.K.: A fast quantum mechanical algorithm for database search. In: Proceedings of the Twenty-Eighth Annual ACM Symposium on the Theory of Computing, pp. 212–219 (1996)
16. Høyer, P., de Wolf, R.: Improved quantum communication complexity bounds for disjointness and equality. In: Proceedings of STACS 2002, pp. 299–310 (2002)
17. Kalyanasundaram, B., Schnitger, G.: The probabilistic communication complexity of set intersection. SIAM J. Discrete Math. **5**(4), 545–557 (1992)
18. Klauck, H.: Lower bounds for quantum communication complexity. SIAM J. Comput. **37**(1), 20–46 (2007)
19. Krause, M.: Geometric arguments yield better bounds for threshold circuits and distributed computing. Theor. Comput. Sci. **156**, 99–117 (1996)
20. Kushilevitz, E., Nisan, N.: Communication Complexity. Cambridge University Press, Cambridge (1997)
21. Lee, T., Shraibman, A.: Lower bounds in communication complexity. Found. Trends Theor. Comput. Sci. **3**(4), 263–398 (2009)
22. Lee, T., Shraibman, A.: An approximation algorithm for approximation rank. In: IEEE Conference on Computational Complexity, pp. 351–357 (2009)
23. Linial, N., Mendelson, S., Schechtman, G., Shraibman, A.: Complexity measures of sign matrices. Combinatorica **27**(4), 439–463 (2007)
24. Linial, N., Shraibman, A.: Learning complexity vs communication complexity. Comb. Probab. Comput. **18**(1–2), 227–245 (2009)
25. Linial, N., Shraibman, A.: Lower bounds in communication complexity based on factorization norms. Random Struct. Algorithms **34**, 368–394 (2009)
26. Lovász, L., Saks, M.: Möbius functions and communication complexity. In: Proceedings of the 29th IEEE Symposium on Foundations of Computer Science, pp. 81–90. IEEE (1988)
27. Lovett, S.: Communication is bounded by root of rank. J. ACM **63**(1) pp. 1:1–1:9 (2016)
28. Lovett, S.: Personal communication (2018)
29. Mehlhorn, K., Schmidt, E.: Las Vegas is better than determinism in VLSI and distributed computing. In: Proceedings of the 14th ACM Symposium on the Theory of Computing, pp. 330–337 (1982)
30. Mosca, M., Brassard, G., Høyer, P., Tapp, A.: Quantum amplitude amplification and estimation. In: Quantum Computation and Quantum Information: A Millennium Volume, volume 305 of AMS Contemporary Mathematics Series. American Mathematical Society (2002)
31. Newman, I.: Private versus common random bits in communication complexity. Inf. Process. Lett. **39**, 67–71 (1991)

32. Nielsen, M.A., Chuang, I.L.: Quantum Computation and Quantum Information: 10th Anniversary Edition, 10th edn. Cambridge University Press, New York (2011)
33. Nisan, N., Wigderson, A.: A note on rank vs. communication complexity. Combinatorica **15**(4), 557–566 (1995)
34. Paturi, R., Simon, J.: Probabilistic communication complexity. J. Comput. Syst. Sci. **33**(1), 106–123 (1986)
35. Razborov, A.: On the distributional complexity of disjointness. Theor. Comput. Sci. **106**(2), 385–390 (1992)
36. Razborov, A.: Quantum communication complexity of symmetric predicates. Izvestiya: Math. **67**(1), 145–159 (2003)
37. Rothvoß, T.: direct proof for Lovett's bound on the communication complexity of low rank matrices. CoRR, abs/1409.6366 (2014)
38. Sinha, M., de Wolf, R.: Exponential separation between quantum communication and logarithm of approximate rank. In: Proceedings of FOCS 2019, pp. 966–981 (2019)
39. Zhang, S.: Efficient quantum protocols for XOR functions. In: Proceedings of SODA 2014, pp. 1878–1885 (2014)

Approximation Schemes for Multiperiod Binary Knapsack Problems

Zuguang Gao$^{(\boxtimes)}$ [iD], John R. Birge [iD], and Varun Gupta [iD]

The University of Chicago, Chicago, IL 60637, USA
{zuguang.gao,john.birge,varun.gupta}@chicagobooth.edu

Abstract. An instance of the multiperiod binary knapsack problem (MPBKP) is given by a horizon length T, a non-decreasing vector of knapsack sizes $(c_1, \ldots, c_T)$ where c_t denotes the cumulative size for periods $1, \ldots, t$, and a list of n items. Each item is a triple (r, q, d) where r denotes the reward or value of the item, q its size, and d denotes its time index (or, deadline). The goal is to choose, for each deadline t, which items to include to maximize the total reward, subject to the constraints that for all $t = 1, \ldots, T$, the total size of selected items with deadlines at most t does not exceed the cumulative capacity of the knapsack up to time t. We also consider the multiperiod binary knapsack problem with soft capacity constraints (MPBKP-S) where the capacity constraints are allowed to be violated by paying a penalty that is linear in the violation. The goal of MPBKP-S is to maximize the total profit, which is the total reward of the selected items less the total penalty. Finally, we consider the multiperiod binary knapsack problem with soft stochastic capacity constraints (MPBKP-SS), where the non-decreasing vector of knapsack sizes $(c_1, \ldots, c_T)$ follows an arbitrary joint distribution with the set of possible sample paths (realizations) of knapsack sizes and the probability of each sample path given to the algorithm.

For MPBKP, we exhibit a fully polynomial-time approximation scheme with runtime $\tilde{\mathcal{O}}\left(\min\left\{n + \frac{T^{3.25}}{\epsilon^{2.25}}, n + \frac{T^2}{\epsilon^3}, \frac{nT}{\epsilon^2}, \frac{n^2}{\epsilon}\right\}\right)$ that achieves $(1 + \epsilon)$ approximation; for MPBKP-S, the $(1 + \epsilon)$ approximation can be achieved in $\mathcal{O}\left(\frac{n \log n}{\epsilon} \cdot \min\left\{\frac{T}{\epsilon}, n\right\}\right)$. To the best of our knowledge, our algorithms are the first FPTAS for any multiperiod version of the Knapsack problem since its study began in 1980s. For MPBKP-SS, we prove that a natural greedy algorithm is a 2-approximation when all items have the same size. Our algorithms also provide insights on how other multiperiod versions of the knapsack problem may be approximated.

Keywords: Approximation algorithms · Knapsack problem · Optimization

1 Introduction

Knapsack problems are a classical category of combinatorial optimization problems, and have been studied for more than a century [21]. They have found wide

© Springer Nature Switzerland AG 2021
R. Santhanam and D. Musatov (Eds.): CSR 2021, LNCS 12730, pp. 131–146, 2021.
https://doi.org/10.1007/978-3-030-79416-3_8

applications in various fields [16], such as selection of investments and portfolios, selection of assets, finding the least wasteful way to cut raw materials, etc. One of the most commonly studied problem is the so-called *0–1 knapsack problem*, where a set of n items are given, each with a reward and a size, and the goal is to select a subset of these items to maximize the total reward, subject to the constraint that the total size may not exceed some knapsack capacity. It is well-known that the 0–1 knapsack problem is NP-complete. However, the problem was shown to possess *fully polynomial-time approximation schemes (FPTASs)*, i.e., there are algorithms that achieve $(1 + \epsilon)$ factor of the optimal value for any $\epsilon \in (0, 1)$, and take polynomial time in n and $1/\epsilon$.

The first published FPTAS for the 0–1 knapsack problem was due to [12] where the authors achieve a time complexity $\widetilde{\mathcal{O}}\left(n + (1/\epsilon^4)\right)$ by dividing the items into a class of "large" items and a class of "small" items. The problem is first solved for large items only, using the dynamic program approach, with rewards rounded down using some discretization quantum (chosen in advance), and the small items are added later. [18] proposed a more nuanced discretization method to improve the polylogarithmic factors. Since then, improvements have been made on the dynamic program for large items [15,24]. Most recently, the FPTAS has been improved to $\widetilde{\mathcal{O}}\left(n + (1/\epsilon)^{9/4}\right)$ in [14].

In this paper, we study three extensions of the 0–1 knapsack problem. First, we consider a multiperiod version of the 0–1 knapsack problem, which we call the *multiperiod binary knapsack problem (MPBKP)*. There is a horizon length T and a vector of knapsack sizes $(c_1, \ldots, c_T)$, where c_t is the cumulative size for periods $1, \ldots, t$ and is non-decreasing in t. We are also given a list of n items, each associated with a triple (r, q, d) where r denotes the reward or value of the item, q its size, and d denotes its time index (or, deadline). The goal is to choose a reward maximizing set of items to include such that for any $t = 1, \ldots, T$, the total size of selected items with deadlines at most t does not exceed the cumulative capacity of the knapsack up to time t. The application that motivates this problem is a seller who produces $(c_t - c_{t-1})$ units of a good in time period t, and can store unsold goods for selling later. The seller is offered a set of bids, where each bid includes a price (r), a quantity demanded (q), and a time at which this quantity is needed. The problem of deciding the revenue maximizing subset of bids to accept is exactly MPBKP.

The second extension we consider is the *multiperiod binary knapsack problem with soft capacity constraints (MPBKP-S)* where at each period the capacity constraint is allowed to be violated by paying a penalty that is linear in the violation. The goal of MPBKP-S is to maximize the total profit, which is the total reward of the selected items less the total penalty. In this case, the seller can procure goods from outside at a certain rate if his supply is not enough to fulfill the bids he accepts, and wants to maximize his profit.

The third extension we consider is the *multiperiod binary knapsack problem with soft stochastic capacity constraints (MPBKP-SS)* where the non-decreasing vector of knapsack sizes $(c_1, \ldots, c_T)$ follows some arbitrary joint distribution given as the set of sample paths of the possible realizations and their probabilities. We select the items *before* realizations of any of these random incremental capacities to maximize the total *expected* profit, which is the total reward of selected items less the total expected penalty. In this case, the production of the seller at each time is random, but he has to select a subset of bids before realizing his supply. Again, the seller can procure capacity from outside at a certain rate if his realized supply is not enough to fulfill the bids he accepts, and wants to maximize his expected profit.

We note that MPBKP is also related to a number of other multiperiod versions of the knapsack problem in literature. The multiperiod knapsack problem (MPKP) proposed by [7] has the same structure as MPBKP, except that in [7] each item can be repeated multiple times, i.e., the decision variables for each item is not binary, but any nonnegative integer (in the single-period case, this is called the unbounded knapsack problem [1]). To the best of our knowledge, there has been no further studies on MPKP since [7]. In the multiple knapsack problem (MKP), there are m knapsacks, each with a different capacity, and items can be inserted to any knapsacks (subject to its capacity constraints). It has been shown in [5] that MKP does not admit an FPTAS, but an efficient polynomial time approximation scheme (EPTAS) has been found in [13], with runtime depending polynomially on n but exponentially on $1/\epsilon$. The incremental knapsack problem (IKP) [11] is another multiperiod version of the knapsack problem, where the knapsack capacity increases over time, and each selected item generates a reward on every period after its insertion, but this reward is discounted over time. Unlike MPBKP, items do not have deadlines and can be selected anytime throughout the T periods. A PTAS for the IKP when the discount factor is 1 (time invariant, referred to as IIKP) and $T = \mathcal{O}\left(\sqrt{\log n}\right)$ has been found in [3], and it has been shown that IIKP is strongly NP-hard. Later, [8] proposed the first PTAS for IIKP regardless of T, and [6] proposed an PTAS for IKP when T is a constant. Most recent developments of IKP include [2,9]. Other similar problems and/or further extensions include the multiple-choice multiperiod knapsack problem [19,20,23], the multiperiod multi-dimensional knapsack problem [17], the multiperiod precedence-constrained knapsack problem [22,25], to name a few.

Our main contributions of this paper are two-fold. First, from the perspective of model formulation, we propose the MPBKP and its generalized versions MPBKP-S and MPBKP-SS. Despite the fact that there are a number of multiperiod/multiple versions of knapsack problems, including those mentioned above (many of which are strongly NP-hard), the MPBKP and MPBKP-S we proposed here are the first to admit an FPTAS among any multiperiod versions of the classical knapsack problem since their initiation back in 1980s. With these results, it is thus interesting to see where the boundary lies between these multiperiod problems that admit an FPTAS and those problems that do not admit an FPTAS. Second, the algorithms we propose for both MPBKP and MPBKP-S are generalized from the ideas of solving 0–1 knapsack problems, but with nontrivial modifications as we will address in the following sections. For MPBKP-SS, we propose a greedy algorithm that achieves 2-approximation for the special case when all items have the same size.

The rest of this paper is organized as follows. In Sect. 2, we formally write the three problems in mathematical programming form. The FPTAS for MPBKP is proposed in Sect. 3 and the FPTAS for MPBKP-S is proposed in Sect. 4. Alternative algorithms for both problems are provided in the full version [10]. A greedy algorithm for a special case of MPBKP-SS is proposed in Sect. 5. We provide proof ideas in this paper, while formal proofs can be found in [10].

2   Problem Formulation and Main Results

In this section, we formally introduce the Multiperiod Binary Knapsack Problem (MPBKP), as well as the generalized versions: the Multiperiod Binary Knapsack Problem with Soft capacity constraints (MPBKP-S), and Multiperiod Binary Knapsack Problem with Soft Stochastic Capacity constraints (MPBKP-SS).

2.1   Multiperiod Binary Knapsack Problem (MPBKP)

An instance of MPBKP is given by a set of n items, each associated with a triple (r_i, q_i, d_i), and a sequence of knapsack capacities $\{c_1, \ldots, c_T\}$. For each item i, we get reward r_i if and only if i is included in the knapsack by time d_i. We assume that $r_i \in \mathbb{N}$, $q_i \in \mathbb{N}$ and $d_i \in [T] := \{1, \ldots, T\}$. The knapsack capacity at time t is c_t, and by convention $c_0 = 0$. The MPBKP can be written in the integer program (IP) form:

$$\max_x z = \sum_{i=1}^{n} r_i x_i \tag{1a}$$

$$\text{s.t.} \sum_{j:d_j \leq t} q_j x_j \leq c_t, \quad \forall t = 1, \ldots, T \tag{1b}$$

$$x_i \in \{0, 1\}, \quad \forall i = 1, \ldots, n \tag{1c}$$

where x_i's are binary decision variables, i.e., x_i is 1 if item i is included in the knapsack and is 0 otherwise. In (1), we aim to pick a subset of items to maximize the objective function, which is the total reward of picked items, subject to the constraints that by each time t, the total size of picked items with deadlines up to t does not exceed the knapsack capacity at time t, which is c_t. For each $t \in [T]$, let $\mathcal{I}(t) := \{i \in [n] \mid d_i = t\}$ denote the set of items with deadline t. Note that without loss of generality, we may assume that $\mathcal{I}(t) \neq \emptyset, \forall t$ and $c_t > 0$. We further note that the decision variables x_i's in (1) are binary, but if we relax this to any nonnegative integers, the problem becomes the so-called multiperiod knapsack problem (MPKP) as in [7]. Our first main result is the following theorem on MPBKP.

Theorem 1 (FPTAS for MPBKP). *There exists a deterministic algorithm that achieves $(1 + \epsilon)$-approximation in $\tilde{O}\left(\min\left\{n + \frac{T^{3.25}}{\epsilon^{2.25}}, n + \frac{T^2}{\epsilon^3}, \frac{nT}{\epsilon^2}, \frac{n^2}{\epsilon}\right\}\right)$.*

As we will see shortly, MPBKP can be viewed as a special case of MPBKP-S. In Sect. 3, we will provide an approximation algorithm for MPBKP with runtime $\tilde{O}\left(n + \frac{T^{3.25}}{\epsilon^{2.25}}\right)$. An alternative algorithm with runtime $\tilde{O}\left(n + \frac{T^2}{\epsilon^3}\right)$ is provided in [10]. In Sect. 4, we will provide an approximation algorithm for MPBKP-S with runtime $\tilde{O}\left(\frac{nT}{\epsilon^2}\right)$, which is also applicable to MPBKP.

2.2 Multiperiod Binary Knapsack Problem with Soft Capacity Constraints (MPBKP-S)

In MPBKP-S, the capacity constraints in (1) no longer exist, i.e., the total size of selected items at each time step is allowed to be greater than the total capacity up to that time, however, there is a penalty rate $B_t \in \mathbb{N}$ for each unit of overflow at period t. We assume that $B_t > \max_{i \in [n]: d_i \leq t} \frac{r_i}{q_i}$ to avoid trivial cases (any item with $\frac{r_i}{q_i} \geq B_t$ and $d_i \leq t$ will always be added to generate more profit). In the IP form, MPBKP-S can be written as

$$\max_{x,y} \sum_{i \in [n]} r_i x_i - \sum_{t=1}^{T} B_t y_t \tag{2a}$$

$$\text{s.t.} \quad \sum_{i \in \mathcal{I}(1) \cup \cdots \cup \mathcal{I}(t)} q_i x_i - \sum_{s=1}^{t} y_s \leq c_t, \quad \forall t : 1 \leq t \leq T \tag{2b}$$

$$x_i \in \{0,1\}, \quad y_t \geq 0, \tag{2c}$$

where the decision variables $y_t, t = 1, \ldots, T$ represent the units of overflow at time t, and $c_t - c_{t-1}$ is the incremental capacity at time t. The objective is to choose a subset of the n items to maximize the total profit, which is the sum of the rewards of the selected items minus the sum of penalty paid at each period, and the constraints enforce that the total size of accepted items by the end of each period must not exceed the sum of the cumulative capacity and the units of overflow. Our second main result is the following theorem on MPBKP-S.

Theorem 2 (FPTAS for MPBKP-S). *There exists an algorithm which achieves* $(1 + \epsilon)$-*approximation in* $\mathcal{O}\left(\frac{n \log n}{\epsilon} \cdot \min\left\{\frac{T}{\epsilon}, n\right\}\right)$.

In Sect. 4 we will present an approximation algorithm for solving MPBKP-S with time complexity $\mathcal{O}\left(\frac{nT \log n}{\epsilon^2}\right)$. An alternative FPTAS with runtime $\mathcal{O}\left(\frac{n^2}{\epsilon}\right)$ is provided in [10]. For the ease of presentation, our algorithms and analysis are presented for the case $B_t = B$, but they can be generalized to the heterogeneous $\{B_1, \ldots, B_T\}$ in a straightforward manner. It is worth noting that the algorithm for MPBKP that we introduce in Sect. 3 does not extend to MPBKP-S, and we will make this clear in the beginning of Sect. 4.

2.3 Multiperiod Binary Knapsack Problem with Soft Stochastic Capacity Constraints (MPBKP-SS)

The MPBKP-SS formulation is similar to (2), except that the vector of knapsack sizes $(c_1, \ldots, c_T)$ follows some arbitrary joint distribution given to the algorithm as the set of possible sample path (realization) of knapsack sizes and the probability of each sample path. We use ω to index sample paths which we denote by $\{c_t(\omega)\}$, $p(\omega)$ as the probability of sample path ω, and Ω as the set of possible sample paths. The goal is to pick a subset of items before the realization of ω so as to maximize the expected total profit, which is the sum of the rewards of the selected items deducted by the total (expected) penalty. For a sample $\omega \in \Omega$ let $y_t(\omega)$ be the overflow at time t. Then, we can write the problem in IP form as:

$$\max_{x,y} \sum_{i \in [n]} r_i x_i - \mathbb{E}_\omega \left[B_t \cdot \sum_{t=1}^{T} y_t(\omega) \right] \tag{3a}$$

$$\text{s.t.} \sum_{i \in \mathcal{I}(1) \cup \cdots \cup \mathcal{I}(t)} q_i x_i - \sum_{s=1}^{t} y_s(\omega) \le c_t(\omega), \quad \forall \omega \in \Omega, 1 \le t \le T \tag{3b}$$

$$x_i \in \{0, 1\}, \quad y_t \ge 0 \tag{3c}$$

Our third main result is the following theorem on MPBKP-SS, which asserts that for the special case when all items are of the same size a natural greedy algorithm is 2-competitive. The greedy algorithm is well-defined even in the case where the distribution can only be accessed through an oracle which returns the expected profit for any given set of items. Details will be provided in Sect. 5.

Theorem 3. *If $q_i = q$ for all $i \in [n]$, then there exists a greedy algorithm that achieves 2-approximation for MPBKP-SS in $\mathcal{O}\left(n^2 T |\Omega|\right)$.*

We further note that both MPBKP-S and MPBKP-SS are special cases of non-monotone submodular maximization which is *not* non-negative, for which not many general approximations are known. In that sense, studying these problems would be an interesting direction to develop techniques for it.

3 FPTAS for MPBKP

In this section, we provide an FPTAS for the MPBKP with time complexity $\tilde{\mathcal{O}}\left(n + \frac{T^{3.25}}{\epsilon^{2.25}}\right)$. We will apply the "functional approach" as used in [4]. The main idea is to use the results on function approximations [4,14] as building blocks – for each period we approximate one function that gives, for every choice of available capacity, the maximum reward obtainable by selecting items in that period. We then combine "truncated" version of these functions using (max, +)-convolution. This idea, despite its simplicity, allows us to obtain an FPTAS for MPBKP. Such a result should not be taken as granted – as we will see in the next section, this method does not apply for MPBKP-S, even though it is just a slight generalization of MPBKP.

We begin with some preliminary definitions and notations. For a given set of item rewards and sizes, $\mathcal{I} = \{(r_1, q_1), \ldots, (r_{n'}, q_{n'})\}$, define the function

$$f_{\mathcal{I}}(c) := \max_{x_1, \ldots, x_{n'}} \left\{ \sum_{i \in \mathcal{I}} r_i x_i \; : \; \sum_{i \in \mathcal{I}} q_i x_i \leq c, \; x_1, \ldots, x_{n'} \in \{0, 1\} \right\} \qquad (4)$$

for all $c \geq 0$, and $f_{\mathcal{I}}(c) := -\infty$ for $c < 0$. The function $f_{\mathcal{I}}$ is a nondecreasing step function, and the number of steps is called the *complexity* of that function. Further, for any $\mathcal{I} = \mathcal{I}_1 \sqcup \mathcal{I}_2$, i.e., $\mathcal{I}$ being a disjoint union of $\mathcal{I}_1$ and $\mathcal{I}_2$, we have that $f_{\mathcal{I}} = f_{\mathcal{I}_1} \oplus f_{\mathcal{I}_2}$, where $\oplus$ denotes the $(\max, +)$-*convolution*: $(f \oplus g)(c) = \max_{c' \in \mathbb{R}} (f(c') + g(c - c'))$.

We define the *truncated function* $f_{\mathcal{I}}^{c'}$ as follows:

$$f_{\mathcal{I}}^{c'}(c) = \begin{cases} f_{\mathcal{I}}(c) & c \leq c', \\ -\infty & c > c'. \end{cases} \qquad (5)$$

Recall that we denote the set of items with deadline t by $\mathcal{I}(t)$. We next define the function f_t as follows:

$$f_t := \begin{cases} f_{\mathcal{I}(1)}^{c_1} & t = 1, \\ \left(f_{t-1} \oplus f_{\mathcal{I}(t)} \right)^{c_t} & t \geq 2. \end{cases} \qquad (6)$$

In words, each function value of $f_t(c)$ corresponds to a feasible, in fact an optimal, solution x for items with deadline at most t, as the next proposition shows.

Proposition 1. *Let x^* be the optimal solution for MPBKP (1). We have that the optimal value of (1), $\sum_{i \in [n]} r_i x_i^*$, satisfies $\sum_{i \in [n]} r_i x_i^* = f_T(c_T)$.*

Proposition 1 implies that, to obtain an approximately optimal solution for MPBKP (1), it is sufficient to have a good approximation for the function

$$f_T = \left(\cdots \left(\left(f_{\mathcal{I}(1)}^{c_1} \oplus f_{\mathcal{I}(2)} \right)^{c_2} \oplus f_{\mathcal{I}(3)} \right)^{c_3} \cdots \oplus f_{\mathcal{I}(T)} \right)^{c_T}. \qquad (7)$$

We say that a function $\tilde{f}$ approximates the nonnegative function f with factor $1 + \epsilon$ if $\tilde{f}(c) \leq f(c) \leq (1 + \epsilon) \tilde{f}(c)$ for all $c \in \mathbb{R}$. It should be clear that if $\tilde{f}$ approximates f with factor $1 + \epsilon$ and $\tilde{g}$ approximates g with factor $1 + \epsilon$, then $\tilde{f} \oplus \tilde{g}$ approximates $f \oplus g$ with factor $1 + \epsilon$. We then introduce the following result from [14] for 0–1 Knapsack problem.

Lemma 1 [14]. *Given a set $\mathcal{I} = \{(r_1, q_1), \ldots, (r_n, q_n)\}$, we can obtain $\tilde{f}_{\mathcal{I}}$ that approximates $f_{\mathcal{I}}$ (defined in (4)) with factor $1 + \epsilon$ and complexity $\tilde{O}\left(\frac{1}{\epsilon}\right)$ in $\tilde{O}\left(n + (1/\epsilon)^{2.25}\right)$.*

With the above lemma, we present Algorithm 1 for MPBKP.

We now describe the intuition behind Algorithm 1. We first discard all items with reward $r_i \leq \frac{\epsilon}{n} \max_j r_j$. The maximum we could lose is $n \cdot \frac{\epsilon}{n} \max_j r_j =$

Algorithm 1. FPTAS for MPBKP

    Input: $\epsilon, [n], c_1, \ldots, c_T$ ▷ Set of items to be packed, cumulative capacities
    Output: $\tilde{f}_t$ ▷ Approximation of function f_t
1: Discard all items with $r_i \leq \frac{\epsilon}{n} \max_j r_j$ and relabel the items
2: $r_0 \leftarrow \min_i r_i$ ▷ Lower bound of solution value
3: $m \leftarrow \left\lceil \log_{1+\epsilon} \frac{n^2}{\epsilon} \right\rceil$ ▷ number of distinct rewards to be considered, each in the form $r_0 \cdot (1+\epsilon)^k$
4: Obtain $\tilde{f}_{\mathcal{I}(1)}$ that approximates $f_{\mathcal{I}(1)}$ with factor $(1+\epsilon)$ using Lemma 1
5: $\tilde{f}_1 := \tilde{f}_{\mathcal{I}(1)}^{c_1}$ ▷ $\tilde{f}_1$ has complexity at most $m = \tilde{\mathcal{O}}\left(\frac{1}{\epsilon}\right)$
6: **for** $t = 2, \ldots, T$ **do**
7:     Obtain $\tilde{f}_{\mathcal{I}(t)}$ that approximates $f_{\mathcal{I}(t)}$ with factor $(1+\epsilon)$ using Lemma 1
8:     $l \leftarrow$ complexity of $\tilde{f}_{\mathcal{I}(t)}$ ▷ $l = \tilde{\mathcal{O}}\left(\frac{1}{\epsilon}\right)$
9:     Compute (all breakpoints and their values of) $\hat{f}_t := \left(\tilde{f}_{t-1} \oplus \tilde{f}_{\mathcal{I}(t)}\right)^{c_t}$, taking $m \cdot l$ time

                                      ▷ $\hat{f}_t$ has complexity $\tilde{\mathcal{O}}\left(\frac{1}{\epsilon^2}\right)$

10:    $\tilde{f}_t := r_0 \cdot (1+\epsilon)^{\left\lfloor \log_{1+\epsilon}\left(\frac{\hat{f}_t}{r_0}\right) \right\rfloor}$ ▷ Round $\hat{f}_t$ down to the nearest $r_0 \cdot (1+\epsilon)^k$ for $k = 0, \ldots, m$.

                               ▷ Now $\tilde{f}_t$ has complexity at most $m = \tilde{\mathcal{O}}\left(\frac{1}{\epsilon}\right)$

11: **end for**

$\epsilon \max_j r_j$, which is at most ϵ fraction of the optimal value. We next obtain all $\tilde{f}_{\mathcal{I}(t)}$, for all $t = 1, \ldots, T$, that approximate $f_{\mathcal{I}(t)}$ (as defined in (4)) within a $(1+\epsilon)$ factor. These functions $\tilde{f}_{\mathcal{I}(t)}$ have complexity $\tilde{\mathcal{O}}\left(\frac{1}{\epsilon}\right)$. We start with combining the functions of period 1 and period 2 using $(\max, +)$-convolution. To enforce the constraint that the total size of selected items in period 1 does not exceed the capacity of period 1, we truncate $\tilde{f}_{\mathcal{I}(1)}$ by c_1 (so that any solution using more capacity in period 1 results in $-\infty$ reward) and do the convolution on the truncated function $\tilde{f}_1$. Since both functions are step functions with complexity $\tilde{\mathcal{O}}\left(\frac{1}{\epsilon}\right)$, the $(\max, +)$ convolution can be done in time $\mathcal{O}\left(\frac{1}{\epsilon^2}\right)$. The resulting function $\hat{f}_2$ would have complexity $\mathcal{O}\left(\frac{1}{\epsilon^2}\right)$. To avoid inflating the complexity throughout different periods (which increases computation complexity), the function $\hat{f}_2$ is rounded down to the nearest $r_0 \cdot (1+\epsilon)^k$, where $r_0 := \min_j r_j$ and k is some nonnegative integer. Note that r_0 is a lower bound of any solution value. After discarding small-reward items, we have that $\frac{\max_j r_j}{r_0} \leq \frac{n}{\epsilon}$, which implies that $n \max_j r_j = \frac{n^2}{\epsilon} r_0$ is an upper bound for the optimal solution value. Therefore, after rounding down the function values of $\hat{f}_2$ and obtaining $\tilde{f}_2$, there are at most $\log_{1+\epsilon} \frac{n^2}{\epsilon} \approx \frac{1}{\epsilon} \log \frac{n^2}{\epsilon}$ different values on $\tilde{f}_2$. Now we have brought down the complexity of $\tilde{f}_2$ again to $\tilde{\mathcal{O}}\left(\frac{1}{\epsilon}\right)$, at an additional $(1+\epsilon)$ factor loss in the approximation error. We then move to period 3 and continue this pattern of $(\max, +)$-convolution, truncation, and rounding down. In the end when we reach period T, $\tilde{f}_T$ will only contain feasible solutions to (1), and approximate f_T with total approximation factor of $(1+\epsilon)^T \approx (1+T\epsilon)$. Formally, we have the following lemma which shows the approximation factor of $\tilde{f}_t$ for f_t.

Lemma 2. *Let $\tilde{f}_t$ be the functions obtained from Algorithm 1, and let f_t be defined as in* (6). *Then, $\tilde{f}_t$ approximates f_t with factor $(1+\epsilon)^t$, i.e., $\tilde{f}_t(c) \leq f_t(c) \leq (1+\epsilon)^t \tilde{f}_t(c)$ for all $0 \leq c \leq c_t$.*

Lemma 2 and Proposition 1 together imply that $\tilde{f}_T(c_T)$, obtained from Algorithm 1, approximates the optimal value of MPBKP (1) by a factor of $(1+\epsilon)^T \approx (1+T\epsilon)$. In Algorithm 1, obtaining $\tilde{f}_{\mathcal{I}(t)}$ for all $t = 1, \ldots, T$ takes time $\tilde{O}\left(n + T/\epsilon^{2.25}\right)$; computing the (max, +)-convolution on $\tilde{f}_{t-1} \oplus \tilde{f}_{\mathcal{I}(t)}$ for all t take time $T \cdot m \cdot l = \tilde{O}\left(T/\epsilon^2\right)$. Therefore, Algorithm 1 has runtime $\tilde{O}\left(n + T/\epsilon^{2.25}\right)$. As a result, we have the following proposition.

Proposition 2. *Taking $\epsilon' = T\epsilon$, Algorithm 1 achieves $(1+\epsilon')$-approximation for MPBKP in $\tilde{O}\left(n + \frac{T^{3.25}}{\epsilon'^{2.25}}\right)$.*

4   FPTAS for MPBKP-S

In this section, we provide an FPTAS for the MPBKP-S with time complexity $O\left(\frac{Tn\log n}{\epsilon^2}\right)$. An alternative FPTAS with time complexity $O\left(\frac{n^2 \log n}{\epsilon}\right)$ is provided in [10]. Combining the two, we show that our algorithms achieve $(1+\epsilon)$ approximation ratio in time $O\left(\frac{n\log n}{\epsilon} \cdot \min\left\{\frac{T}{\epsilon}, n\right\}\right)$, which proves Theorem 2. We should note that the algorithm in the previous section does not apply here: we could similarly define a function which gives the maximum profit (=reward−penalty) under a given capacity constraint, but the main obstacle is on the (max, +)-convolution because profits do not "add up". In other words, the total profit we earn by selecting items in the set $\mathcal{S}_1 \cup \mathcal{S}_2$ is not the sum of the profits earned by selecting $\mathcal{S}_1$ and $\mathcal{S}_2$ separately. For this reason, we can no longer rely on the techniques used in function approximation and (max, +)-convolution as in [4,14]. Instead, our main idea is motivated by the techniques that originated from earlier papers [12,18], but adapting their technique to MPBKP-S requires significant modifications as we show in this section. We restrict our presentation to the case $B_t = B$ for readability, but our algorithms and analysis generalize in a straightforward manner when the penalties for buying capacity are heterogeneous $\{B_1, \ldots, B_T\}$ (by replacing B with $\min_{\tau \leq t} B_\tau$ in the calculations of profit/penalty at period t on line 7 of Algorithm 2).

Preliminaries: We first introduce some notation. From now on, let $\mathcal{R}(\mathcal{S}) := \sum_{i \in \mathcal{S}} r_i$. The optimal solution set to (2) is denoted by $\mathcal{S}^*$. The total profit earned can be expressed as a function of the solution set $\mathcal{S}$:

$$\mathcal{P}(\mathcal{S}) = \mathcal{R}(\mathcal{S}) - B \cdot \sum_{t=1}^{T}\left[\sum_{j \in \mathcal{S} \cap \mathcal{I}(t)} q_j - \max\left\{c_t - \sum_{j \in \mathcal{S}, d_j \leq t} q_j, \; c_t - c_{t-1}\right\}\right]^+.$$
$$(8)$$

Let p_i be the profit of item i, which is defined as the profit earned if we select only i, i.e., $p_i = r_i - B \cdot (q_i - c_{d_i})^+$. Without loss of generality, we assume that each item i is by itself profitable, i.e., $p_i \geq 0$, so one profitable solution would be $\{i\}$. Let $P := \max_i p_i$ and $\bar{P} := \sum_{i \in [n]} p_i$. The following bounds on $\mathcal{P}(\mathcal{S}^*)$ follow:

$$P \leq \mathcal{P}(\mathcal{S}^*) \leq \bar{P} \leq nP. \tag{9}$$

Partition of Items: We partition the set of items $[n]$ into two sets: a set of "large" items $\mathcal{I}_L$ and a set of "small" items $\mathcal{I}_S$ such that we can bound the number of large items in any optimal solution. The main idea is to use dynamic programming to pick the large items in the solution, and a greedy heuristic for 'padding' this partial solution with small items. The criterion for small and large items is based on balancing the permissible error $\epsilon \mathcal{P}(\mathcal{S}^*)$ equally in filling large items and filling small items. Instead of first packing all large items and then all small items, we consider items in the order of their deadlines, and for each deadline t, the large items are selected first and then the small items are selected greedily in order of their reward densities. As a result, the approximation error due to large items overall will be $\frac{1}{2}\epsilon \mathcal{P}(\mathcal{S}^*)$, and the error due to the small items with each deadline will be $\frac{1}{2T}\epsilon \mathcal{P}(\mathcal{S}^*)$. This gives a total approximation error of $\frac{1}{2}\epsilon \mathcal{P}(\mathcal{S}^*) + T \cdot \frac{1}{2T}\epsilon \mathcal{P}(\mathcal{S}^*) = \epsilon \mathcal{P}(\mathcal{S}^*)$.

Suppose that we can find some P_0 that satisfies (10).

$$P_0 \leq \mathcal{P}(\mathcal{S}^*) \leq 2P_0. \tag{10}$$

Then, the set of items is partitioned as follows.

$$\mathcal{I}_L := \left\{ i \in [n] \mid p_i \geq \frac{1}{2T}\epsilon P_0 \right\}; \qquad \mathcal{I}_S := \left\{ i \in [n] \mid p_i < \frac{1}{2T}\epsilon P_0 \right\}. \tag{11}$$

This partition is computed in $\mathcal{O}(n)$ time and is not the dominant term in time complexity. Let $n_L = |\mathcal{I}_L|$ and $n_S = |\mathcal{I}_S|$, so that $n_L + n_S = n$. Further, let

$$\mathcal{I}_L(t) := \{i \in \mathcal{S}_L \mid d_i = t\}, \quad \text{and} \quad \mathcal{I}_S(t) := \{i \in \mathcal{S}_S \mid d_i = t\}$$

denote the set of large and small items, respectively, with deadline t. We will assume that the items in $\mathcal{I}_L$ are indexed in non-decreasing order of their deadlines, i.e., $\forall i, j \in \mathcal{I}_L$ such that $j \geq i$, we have that $d_i \leq d_j$. Denote by $I_L(t)$ as the index of the last item with deadline t, i.e., $I_L(t) := \max_{i \in \mathcal{S}_L \cap \mathcal{I}_L(t)} i$. For each time t, we will also sort the small items in $\mathcal{I}_S(t)$ according to their reward densities, i.e., $\forall i < j$ and $i, j \in \mathcal{I}_S(t)$, $\frac{r_i}{q_i} \geq \frac{r_j}{q_j}$. This sorting only takes place once for each guess P_0, and does not affect our overall time complexity result.

Algorithm Overview: Our FPTAS algorithm is given in Algorithm 5 which uses a doubling trick to guess the value of P_0 satisfying (10), and for each guess uses Algorithm 4 as a subroutine. Algorithm 4 is the main algorithm for MPBKP-S, which first selects the items with deadline 1, then the items with deadline 2, and so on. For each deadline t, we maintain two sets of partial solutions: the first, $\tilde{A}_t(p)$, corresponds to an approximately optimal (in terms of leftover capacity

---

Algorithm 2. DP on large items for MPBKP-S

Input: $\mathcal{I}_L, \Delta c$, ▷ Set of (large) items to be packed, additional capacity available for packing
$\widetilde{A}(p)$ for all $p = \{0, 1, \ldots, \lceil \frac{16T}{\epsilon^2} \rceil\} \cdot \kappa$ ▷ A set of partial solutions
Output: $\hat{A}(\mathcal{I}_L, p)$ for all $p = \{0, 1, \ldots, \lceil \frac{16T}{\epsilon^2} \rceil\} \cdot \kappa$ ▷ Set of partial solutions after packing $\mathcal{I}_L$

1: Initialize $\forall p$: $\hat{A}(0, p) := \widetilde{A}(p) + \Delta c$
2: **for** $i = 1, \ldots, I_L$ **do**
3:   **for** $p = \{0, 1, \ldots, \lceil \frac{16T}{\epsilon^2} \rceil\} \cdot \kappa$ **do**
4:     $\hat{A}(i, p) := \hat{A}(i - 1, p)$ ▷ If reject item i
5:   **end for**
6:   **for** $\bar{p} = \{0, 1, \ldots, \lceil \frac{16T}{\epsilon^2} \rceil\} \cdot \kappa$ **do**
7:     $p = \bar{p} + \hat{r}_i - \left\lceil B\left(q_i - \max\left\{0, \hat{A}(i - 1, \bar{p})\right\}\right)^+ \right\rceil_\kappa$
8:     $\hat{A}(i, p) = \max\left\{\hat{A}(i, p), \hat{A}(i - 1, \bar{p}) - q_i\right\}$ ▷ Accept i
9:   **end for**
10:   **for** $p = \{\lceil \frac{16T}{\epsilon^2} \rceil, \lceil \frac{16T}{\epsilon^2} \rceil - 1, \ldots, 1\} \cdot \kappa$ **do**
11:     **if** $\hat{A}(i, p - \kappa) < \hat{A}(i, p)$ **then**
12:       $\hat{A}(i, p - \kappa) = \hat{A}(i, p)$
13:     **end if**
14:   **end for**
15: **end for**

---

carried forward to time $t + 1$) subset of large and small items with deadline at most t and some *rounded profit* p ; and the second $\hat{A}_t(p)$ corresponds to the optimal appending of large items with deadline t to the approximately optimal set of solutions corresponding to $\widetilde{A}_{t-1}$.

Given $\widetilde{A}_{t-1}$, we first select large items from $\mathcal{I}_L(t)$ using dynamic programming to obtain $\hat{A}_t$, which is done in Algorithm 2. In other words, *given* the partial solutions $\widetilde{A}_{t-1}(\bar{p})$ for all $\bar{p} \in \{0, 1, \ldots, \lceil \frac{16T}{\epsilon^2} \rceil\} \cdot \kappa$, $\hat{A}_t(p)$ is the maximum capacity left when earning *rounded profit* (precise definition given in [10]) p by adding items in $\mathcal{I}_L(t)$. We then use a greedy heuristic to pick small items from $\mathcal{I}_S(t)$ to obtain $\widetilde{A}_t$, which is done in Algorithm 3. Specifically, our goal in Algorithm 3 is to obtain the partial solutions $\widetilde{A}_t(\cdot)$ given the partial solutions $\hat{A}_t(\cdot)$ by packing the small items $\mathcal{I}_S(t)$. We initialize $\widetilde{A}_t(\bar{p})$ with $\hat{A}_t(\bar{p})$, and for each $\bar{p}$, we try to augment the solution corresponding to $\hat{A}_t(\bar{p})$ using a subset $\widetilde{\mathcal{I}}_S(t) \subseteq \mathcal{I}_S(t)$ defined as

$$\widetilde{\mathcal{I}}_S(t) := \{i \in \mathcal{I}_S(t) \mid q_i \leq \hat{A}_t(\bar{p})\}.$$

The small items in $\widetilde{\mathcal{I}}_S(t)$ are sorted according to their reward densities, and are added to the solution of $\hat{A}_t(\bar{p})$ one by one. After each addition of a small item, if the new total rounded reward is p, we compare the leftover capacity with current $\widetilde{A}_t(p)$, and update $\widetilde{A}_t(p)$ with the new solution if it has more leftover capacity. We continue this add-and-compare (and possibly update) until we reach the situation where adding the next small item overflows the available capacity.

Algorithm 3. Greedy on small items for MPBKP-S

Input: $\mathcal{I}_S$, $\hat{A}(p)$ for all $p = \left\{0, 1, \ldots, \left\lceil \frac{16T}{\epsilon^2} \right\rceil\right\} \cdot \kappa.$ ▷ Set of (small) items to be packed, a set of partial solutions

Output: $\widetilde{A}(p)$ for all $p = \left\{0, 1, \ldots, \left\lceil \frac{16T}{\epsilon^2} \right\rceil\right\} \cdot \kappa$ ▷ Set of partial solutions after packing $\mathcal{I}_S$

1: Initialize $\forall p \; : \; \widetilde{A}(p) = \hat{A}(p)$
2: **for** $\bar{p} = \left\{0, 1, \ldots, \left\lceil \frac{16T}{\epsilon^2} \right\rceil\right\} \cdot \kappa$ **do**
    // Filter out small items with size larger than $\hat{A}(p)$
3:     $\widetilde{\mathcal{I}}_S \leftarrow \emptyset$
4:     **for** $i \in \mathcal{I}_S$ **do**
5:         **if** $\hat{A}(\bar{p}) \geq q_i$ **then**
6:             $\widetilde{\mathcal{I}}_S \leftarrow \widetilde{\mathcal{I}}_S \cup \{i\}$
7:         **end if**
8:     **end for**
9:     $\tilde{\mathcal{R}}_{0'} = 0, \tilde{q}_{0'} = 0$, and relabel the items in $\widetilde{\mathcal{I}}_S$ as $\left\{1', \ldots, |\widetilde{\mathcal{I}}_S|'\right\}$ (in decreasing order of reward density)
10:     **for** $i' = 1', \ldots, |\widetilde{\mathcal{I}}_S|'$ **do**
11:         $\tilde{\mathcal{R}}_{i'} = \tilde{\mathcal{R}}_{(i-1)'} + r_{i'}$
12:         $\tilde{q}_{i'} = \tilde{q}_{(i-1)'} + q_{i'}$
13:     **end for**
14:     // Add small items using Greedy algorithm
15:     **for** $i' = 1', \ldots, |\widetilde{\mathcal{I}}_S|'$ **do**
16:         **if** $\tilde{q}_{i'} \leq \hat{A}(\bar{p})$ **then**
17:             $p = \left\lfloor \bar{p} + \tilde{\mathcal{R}}_{i'} \right\rfloor_\kappa$
18:             $\widetilde{A}(p) = \max\left\{\widetilde{A}(p), \hat{A}(\bar{p}) - \tilde{q}_{i'}\right\}$
19:         **end if**
20:     **end for**
21: **end for**

Intuitively, for any amount of capacity available to be filled by small items, and a minimum increase in profit, the optimal solution either packs a single item from $\mathcal{I}_S(t) \setminus \widetilde{\mathcal{I}}_S(t)$ in which case the loss by ignoring items in this set is bounded by the maximum reward of any small item, or the optimal solution only contains items from $\widetilde{\mathcal{I}}_S(t)$ in which case the space used by this optimal set of items is lower bounded by the a fractional packing of the highest density items in $\widetilde{\mathcal{I}}_S(t)$. In Algorithm 3, one of the solutions we consider would be the integral items of this fractional solution, where we lose at most $\frac{1}{2T}\epsilon P_0$ in profit, and the solution has used smaller space (has more leftover capacity) than the fractional solution. Accumulation of these errors for t periods will then give us the invariant: the partial solution $\widetilde{A}_t(p)$ obtained as above has more leftover capacity than any solution obtained by selecting items from $\cup_{t'=1}^t \mathcal{I}_L(t')$ with rounded rewards and rounded penalties, and items from $\cup_{t'=1}^t \mathcal{I}_S(t')$ with original (unrounded) rewards such that the rounded total profit is at least $p + \frac{1}{2T}\epsilon P_0 t + \kappa t$.

Our main theorem on the approximation ratio for MPBKP follows.

Algorithm 4. DP on large items and Greedy on small items for MPBKP-S

1: **Define** $\kappa = \frac{\epsilon^2 P_0}{8T}$
2: **Define** $\hat{r}_i = \lfloor r_i \rfloor_\kappa$ ▷ Round down reward
   // $\widetilde{A}_t(p)$ = leftover capacity for the algorithm's partial solution when
   earning (rounded) profit p using items with deadlines at most t (small
   and large)
   // $\hat{A}_t(p)$ = capacity left for the algorithm's partial solution when
   earning (rounded) profit p by selecting large items in $\mathcal{I}_L(t)$ with
   rounded down rewards $\hat{r}$, given the partial solutions $\widetilde{A}_{t-1}(p)$
3: Initialize $\hat{A}(0,p) = \widetilde{A}_0(p) = \begin{cases} 0 & p = 0, \\ -\infty & p > 0. \end{cases}$
4: **for** $t = 1, \ldots, T$ **do**
5:    Run Algorithm 2 with $\mathcal{I}_L = \mathcal{I}_L(t)$, $\Delta c = c_t - c_{t-1}$, and $\widetilde{A}(p) = \widetilde{A}_{t-1}(p)$ for all
      $p = \left\{0, 1, \ldots, \lceil \frac{16T}{\epsilon^2} \rceil \right\} \cdot \kappa$, and obtain $\hat{A}_t(p) := \hat{A}(\mathcal{I}_L, p)$ for all p.
6:    Run Algorithm 3 with $\mathcal{I}_S = \mathcal{I}_S(t)$ and $\hat{A}(p) = \hat{A}(\mathcal{I}_L(t), p)$ for all $p =$
      $\left\{0, 1, \ldots, \lceil \frac{16T}{\epsilon^2} \rceil \right\} \cdot \kappa$, and obtain $\widetilde{A}_t(p) := \widetilde{A}(p)$ for all p.
7: **end for**

Algorithm 5. FPTAS for MPBKP-S in $\mathcal{O}(Tn \log n/\epsilon^2)$

1: $P_0 \leftarrow \bar{P}$
2: $p^* \leftarrow 0$
3: **while** $p^* < (1 - \epsilon)P_0$ **do**
4:    $P_0 \leftarrow \frac{P_0}{2}$
5:    Run Algorithm 4 with the current P_0.
6:    $p^* \leftarrow \max_{\substack{p \in \left\{0, \ldots, \lceil \frac{16T}{\epsilon^2} \rceil \right\} \cdot \kappa \\ \widetilde{A}_T(p) > -\infty}} p$
7: **end while**

Theorem 4 (Partially restating Theorem 2). *Algorithm 5 is a fully poly-nomial approximation scheme for the MPBKP-S, which achieves $(1 + \epsilon)$ approx-imation ratio with running time $\mathcal{O}\left(\frac{Tn \log n}{\epsilon^2}\right)$.*

5   A Greedy Algorithm for a Special Case of MPBKP-SS

In this subsection, we consider the special case of MPBKP-SS when all items have the same size, i.e., $q_i = q, \forall i \in [n]$. We again only present for the case $B_t = B, \forall t \in [T]$. We note that in the deterministic problems (MPBKP or MPBKP-S), when items all have the same size, greedily adding items one by one in decreasing order of their rewards leads to the optimal solution. For MPBKP-SS, as the capacities are now stochastic, we wonder if there is any greedy algorithm performs well. We propose Algorithm 6, where we start with an empty set, and greedily insert the item that brings the maximum increment on expected profit, and we stop if adding any of the remaining items does not increase the expected profit.

Algorithm 6. Greedy algorithm according to profit change

1: $\mathcal{S} \leftarrow \emptyset$
2: $s \leftarrow 1$
3: **while** $s == 1$ **do**
4:      $i^* \leftarrow \text{argmax}_{i \notin \mathcal{S}} \{\mathcal{P}(\mathcal{S} \cup \{i\}) - \mathcal{P}(\mathcal{S})\}$
5:      **if** $\mathcal{P}(\mathcal{S} \cup \{i^*\}) - \mathcal{P}(\mathcal{S}) \geq 0$ **then**
6:          $\mathcal{S} \leftarrow \mathcal{S} \cup \{i^*\}$
7:      **else**
8:          $s \leftarrow 0$
9:      **end if**
10: **end while**
11: $\mathcal{S}_p \leftarrow \mathcal{S}$
12: **Return** $\mathcal{S}_p$

Let $\mathcal{S}^*$ be an optimal solution, i.e., $\mathcal{S}^* \in \arg\max_{\mathcal{S} \subseteq [n]} \mathcal{P}(\mathcal{S}) := \mathcal{R}(\mathcal{S}) - B \cdot \Phi(\mathcal{S})$, where

$$\Phi(\mathcal{S}) := \mathbb{E} \left\{ \sum_{t=1}^{T} \left[\sum_{j \in \mathcal{I}(t) \cap \mathcal{S}} q_j - \max_{0 \leq t' < t} \left\{ c_t - c_{t'} - \sum_{j \in \mathcal{S}: t'+1 \leq d_j \leq t-1} q_j \right\} \right]^+ \right\}$$

is the expected quantity of overflow on set $\mathcal{S}$, and let $\mathcal{S}_p$ be the set output by Algorithm 6. Then, we have the following theorem.

Theorem 5 (Restating Theorem 3). *Algorithm 6 achieves 2-approximation factor for MPBKP-SS when items have the same size, i.e., $\mathcal{P}(\mathcal{S}_p) \geq \frac{1}{2}\mathcal{P}(\mathcal{S}^*)$, in $\mathcal{O}\left(n^2 T |\Omega|\right)$.*

The proof of the 2-approximation could be more nontrivial than one may think. The idea is to look at the greedy solution set $\mathcal{S}_p$ and the optimal solution set $\mathcal{S}^*$, where we will use the dual to characterize the optimal solution on each sample path. By swapping each item in $\mathcal{S}_p$ to $\mathcal{S}^*$ in replacement of the same item or two other items, we construct a sequence of partial solutions of the greedy algorithm as well as modified optimal solution set, while maintaining the invariant that the profit of $\mathcal{S}^*$ is bounded by the sum of two times the profit of items in $\mathcal{S}_p$ swapped into $\mathcal{S}^*$ so far and the additional profit of remaining items in the modified optimal solution set. The formal proof of Theorem 5 is left to [10].

6   Comments and Future Directions

The current work represents to the best of our knowledge the first FPTAS for the two multi-period variants of the classical knapsack problem. For MPBKP, we obtained the runtime $\tilde{\mathcal{O}}\left(n + (T^{3.25}/\epsilon^{2.25})\right)$. This was done via the function approximation approach, where a function is approximated for each period. The runtime increases in T since we conduct T number of rounding downs, one after each $(\max, +)$-convolution. An alternative algorithm with runtime $\tilde{\mathcal{O}}\left(n + \frac{T^2}{\epsilon^3}\right)$ is also provided in [10]. Note that the function we approximated is in the same form as used in the 0-1 knapsack problem [4]. It is thus interesting to ask if we could instead directly approximate the following function:

$$f_{\mathcal{I}}(c) = \max_x \left\{ \sum_{i \in \mathcal{I}} r_i x_i \ : \ \sum_{i \in \cup_{t'=1}^t \mathcal{I}(t')} q_i x_i \le c_t, \forall t \in [T],\ x \in \{0,1\}^n \right\},$$

where $\mathcal{I} = \cup_{t=1}^T \mathcal{I}(t)$ and $c = \{c_1, \ldots, c_T\}$ is a T-dimensional vector. Here we impose all T constraints in the function. The hope is that, if the above function could be approximated, and if we could properly define the $(\max, +)$-convolution on T dimensional vectors (and have a fairly easy computation of it), then we may get an algorithm that depends more mildly on T.

For MPBKP-S and MPBKP-SS, there seems to be less we can do without further assumptions. One direction to explore is parameterized approximation schemes: assuming that in the optimal solution, the total (expected) penalty is at most β fraction of the total reward. Our ongoing work suggests that in this setting an approximation factor of $\left(1 + \frac{\epsilon}{1-\beta}\right)$ may be achieved in $\tilde{\mathcal{O}}\left(n + (T^{3.25}/\epsilon^{2.25})\right)$ time for MPBKP-S.

We further note that the objective function for the three multiperiod variants are in fact submodular (but not non-negative, or monotone). Whether we can get a constant competitive solution in time $\tilde{\mathcal{O}}(n)$, using approaches in submodular function maximization, is also an intriguing open problem.

Finally, motivated by applications, one natural extension that the authors are working on now is when there is a general non-decreasing cost function $\phi_t(\Delta c)$ for procuring capacity Δc at time t, and the goal is to admit a profit maximizing set of items when the unused capacity can be carried forward. Another extension is when there is a bound on the leftover capacity that can be carried forward.

References

1. Andonov, R., Poirriez, V., Rajopadhye, S.: Unbounded knapsack problem: dynamic programming revisited. Eur. J. Oper. Res. **123**(2), 394–407 (2000)
2. Aouad, A., Segev, D.: An approximate dynamic programming approach to the incremental knapsack problem. arXiv preprint arXiv:2010.07633 (2020)
3. Bienstock, D., Sethuraman, J., Ye, C.: Approximation algorithms for the incremental knapsack problem via disjunctive programming. arXiv preprint arXiv:1311.4563 (2013)
4. Chan, T.M.: Approximation Schemes for 0-1 Knapsack. In: Seidel, R. (ed.) 1st Symposium on Simplicity in Algorithms (SOSA 2018). OpenAccess Series in Informatics (OASIcs), vol. 61, pp. 5:1–5:12. Schloss Dagstuhl-Leibniz-Zentrum fuer Informatik, Dagstuhl, Germany (2018)
5. Chekuri, C., Khanna, S.: A polynomial time approximation scheme for the multiple knapsack problem. SIAM J. Comput. **35**(3), 713–728 (2005)

6. Della Croce, F., Pferschy, U., Scatamacchia, R.: On approximating the incremental knapsack problem. Discrete Appl. Math. **264**, 26–42 (2019)
7. Faaland, B.H.: The multiperiod knapsack problem. Oper. Res. **29**(3), 612–616 (1981)
8. Faenza, Y., Malinovic, I.: A PTAS for the time-invariant incremental knapsack problem. In: Lee, J., Rinaldi, G., Mahjoub, A.R. (eds.) ISCO 2018. LNCS, vol. 10856, pp. 157–169. Springer, Cham (2018). https://doi.org/10.1007/978-3-319-96151-4_14
9. Faenza, Y., Segev, D., Zhang, L.: Approximation algorithms for the generalized incremental knapsack problem. arXiv preprint arXiv:2009.07248 (2020)
10. Gao, Z., Birge, J.R., Gupta, V.: Approximation schemes for multiperiod binary knapsack problems. arXiv preprint arXiv:2104.00034 (2021)
11. Hartline, J., Sharp, A.: An incremental model for combinatorial maximization problems. In: Àlvarez, C., Serna, M. (eds.) WEA 2006. LNCS, vol. 4007, pp. 36–48. Springer, Heidelberg (2006). https://doi.org/10.1007/11764298_4
12. Ibarra, O.H., Kim, C.E.: Fast approximation algorithms for the knapsack and sum of subset problems. J. ACM (JACM) **22**(4), 463–468 (1975)
13. Jansen, K.: A fast approximation scheme for the multiple knapsack problem. In: Bieliková, M., Friedrich, G., Gottlob, G., Katzenbeisser, S., Turán, G. (eds.) SOFSEM 2012. LNCS, vol. 7147, pp. 313–324. Springer, Heidelberg (2012). https://doi.org/10.1007/978-3-642-27660-6_26
14. Jin, C.: An improved FPTAS for 0–1 Knapsack. In: Baier, C., Chatzigiannakis, I., Flocchini, P., Leonardi, S. (eds.) 46th International Colloquium on Automata, Languages, and Programming (ICALP 2019). Leibniz International Proceedings in Informatics (LIPIcs), vol. 132, pp. 76:1–76:14. Schloss Dagstuhl-Leibniz-Zentrum fuer Informatik, Dagstuhl, Germany (2019)
15. Kellerer, H., Pferschy, U.: Improved dynamic programming in connection with an FPTAs for the knapsack problem. J. Comb. Optim. **8**(1), 5–11 (2004)
16. Kellerer, H., Pferschy, U., Pisinger, D.: Knapsack Problems. Springer, Heidelberg (2004). https://doi.org/10.1007/978-3-540-24777-7
17. Lau, H.C., Lim, M.K.: Multi-period multi-dimensional knapsack problem and its application to available-to-promise (2004)
18. Lawler, E.L.: Fast approximation algorithms for knapsack problems. Math. Oper. Res. **4**(4), 339–356 (1979)
19. Lin, E.Y., Chen, M.: A dynamic programming approach to the multiple-choice multi-period knapsack problem and the recursive apl2 code. J. Inf. Optim. Sci. **31**(2), 289–303 (2010)
20. Lin, E.Y., Wu, C.M.: The multiple-choice multi-period knapsack problem. J. Oper. Res. Soc. **55**(2), 187–197 (2004)
21. Mathews, G.B.: On the partition of numbers. Proc. Lond. Math. Soc. **1**(1), 486–490 (1896)
22. Moreno, E., Espinoza, D., Goycoolea, M.: Large-scale multi-period precedence constrained knapsack problem: a mining application. Electron. Notes Discrete Math. **36**, 407–414 (2010)
23. Randeniya, R.: Multiple-choice Multi-period Knapsack Problem (MCMKP): Application and Solution Approach
24. Rhee, D.: Faster fully polynomial approximation schemes for knapsack problems. Ph.D. thesis, Massachusetts Institute of Technology (2015)
25. Samavati, M., Essam, D., Nehring, M., Sarker, R.: A methodology for the large-scale multi-period precedence-constrained knapsack problem: an application in the mining industry. Int. J. Prod. Econ. **193**, 12–20 (2017)

Limitations of Sums of Bounded Read Formulas and ABPs

Purnata Ghosal[1] and B. V. Raghavendra Rao[2(✉)]

[1] Department of Computer Science, University of Liverpool, Liverpool, UK
purnata.ghosal@liverpool.ac.uk
[2] Department of Computer Science and Engineering, IIT Madras, Chennai, India
bvrr@cse.iitm.ac.in

Abstract. Proving super-polynomial size lower bounds for various classes of arithmetic circuits computing explicit polynomials is a very important and challenging task in algebraic complexity theory. We study representation of polynomials as sums of weaker models such as read once formulas (ROFs) and read once oblivious algebraic branching programs (ROABPs). We prove:
(1) An exponential separation between sum of ROFs and read-k formulas for some constant k.
(2) A sub-exponential separation between sum of ROABPs and syntactic multilinear ABPs.
Our results are based on analysis of the partial derivative matrix under different distributions. These results highlight richness of bounded read restrictions in arithmetic formulas and ABPs.

Finally, we consider a generalization of multilinear ROABPs known as strict-interval ABPs defined in [Ramya-Rao, MFCS2019]. We show that strict-interval ABPs are equivalent to ROABPs up to a polynomial blow up in size. In contrast, we show that interval formulas are different from ROFs and also admit depth reduction which is not known in the case of strict-interval ABPs.

1 Introduction

Polynomials are one of the fundamental mathematical objects and have wide applications in Computer Science. Algebraic Complexity Theory aims at a classification of polynomials based on their computational complexity. In his seminal work, Valiant [38] laid foundations of Algebraic Complexity Theory and popularized arithmetic circuits as a natural model of computation for polynomials. He proposed the permanent polynomial perm_n:

$$\mathsf{perm}_n = \sum_{\pi \in S_n} \prod_{i=1}^{n} x_{i\pi(i)},$$

as the primary representative of intractability in algebraic computation. In fact, Valiant [38] hypothesized that the complexity of computing perm_n by arithmetic circuits is different from that of the determinant function.

© Springer Nature Switzerland AG 2021
R. Santhanam and D. Musatov (Eds.): CSR 2021, LNCS 12730, pp. 147–169, 2021.
https://doi.org/10.1007/978-3-030-79416-3_9

A primary offshoot of Valiant's hypothesis is the arithmetic circuit lower bound problem: prove a super polynomial lower bound on the size of an arithmetic circuit computing an explicit polynomial of polynomial degree. Here, an explicit polynomial is one whose coefficients are efficiently computable. Baur and Strassen [5] obtained a super linear lower bound on the size of any arithmetic circuit computing the sum of powers of variables. This is the best known size lower bound for general classes of arithmetic circuits.

Lack of progress in the lower bounds for general arithmetic circuits lead the community to investigate restrictions on arithmetic circuits. Restrictions considered in the literature can be broadly classified into two categories: syntactic and semantic. Syntactic restrictions considered in the literature include restriction on fan-out i.e., arithmetic formulas, restriction on depth i.e., bounded depth circuits [14,15,34], and the related model of algebraic branching programs. Semantic restrictions include monotone arithmetic circuits [18,36,40], homogeneous circuits [9], multilinear circuits [30] and non-commutative computation [26].

Grigoriev and Razborov [15] obtained an exponential lower bound for the size of a depth three arithmetic circuit computing the determinant or permanent over finite fields. In contrast, only almost cubic lower bound is known over infinite fields [20]. Explaining the lack of progress on proving lower bounds even in the case of depth four circuits, Agrawal and Vinay [1] showed that an exponential lower bound for the size of depth four circuits implies Valiant's hypothesis over any field. This lead to intense research efforts in proving lower bounds for the size of constant depth circuits. The reader is referred to an excellent survey by Saptharishi et al. [33] for details.

A polynomial $p \in \mathbb{F}[x_1, \ldots, x_n]$ is said to be *multilinear* if every monomial in p with non-zero coefficient is square-free. An arithmetic circuit is said to be multilinear if every gate in the circuit computes a multilinear polynomial. Multilinear circuits are natural models for computing multilinear polynomials. Raz [31] obtained super polynomial lower bounds on the size of multilinear formulas computing the determinant or permanent. Further, he gave a super polynomial separation between multilinear formulas and circuits [30]. In fact, Raz [31] considered a syntactic version of multilinear circuits known as *syntactic multilinear* circuits. An arithmetic circuit C is said to be syntactic multilinear, if for every product gate $g = g_1 \times g_2$ the sub-circuits rooted at g_1 and g_2 are variable disjoint. The syntactic version has an advantage that the restriction can be verified by examining the circuit, whereas the problem of testing if a circuit is multilinear or not is equivalent to the polynomial identity testing problem. Following Raz's work, there has been significant interest in proving lower bounds on the size of syntactic multilinear circuits. Exponential separation of constant depth multilinear circuits is known [11], while the best known lower bound for unbounded depth syntactic multilinear circuits is only almost quadratic [2].

An *Algebraic Branching Program* (ABP) is a model of computation for polynomials that generalizes arithmetic formulas, and were studied by Ben-Or and Cleve [6], who showed that ABPs of constant width are equivalent to arithmetic formulas. Nisan [26] proved exponential size lower bound for the size of an ABP computing the permanent when the variables are non-commutative. It is known that polynomial families computed by ABPs are the same as families of polynomials computed by skew circuits, a restriction of arithmetic circuits where every product gate can have at most one non-input gate as a predecessor [23]. Further, skew arithmetic circuits are known to characterize the complexity of determinant [37]. Despite their simplicity compared to arithmetic circuits, the best known lower bound for size of ABPs is only quadratic [10,21]. Even with the restriction of syntactic multilinearity, the best known size lower bound for ABPs is, again, quadratic [16]. However, a super polynomial separation between syntactically multilinear formulas and ABPs is known [13].

Proving super quadratic size lower bounds for syntactic multilinear ABPs (smABPs for short) remains a challenging task. Given that there is no promising approach yet to prove super quadratic size lower bounds for smABPs, it is imperative to consider further structural restrictions on smABPs and formulas to develop finer insights into the difficulty of the problem. Following the works in [27–29], we study syntactic multilinear formulas and smABPs with restrictions on the number of reads of variables and the order in which variables appear in a smABP.

Models and Results: **(1) Sum of ROFs:** A *read-once formula* (ROF) is a formula where every variable occurs exactly once as a leaf label. ROFs are syntactic multilinear by definition and have received wide attention in the literature. Volkovich [39] gave a complete characterization of polynomials computed by ROFs. Further, Minahan and Volkovich [24] obtained a complete derandomization of the polynomial identity testing problem on ROFs. While most of the multilinear polynomials are not computable by ROFs [39], sum of ROFs, denoted by $\Sigma \cdot \mathsf{ROF}$ is a natural model of computation for multilinear polynomials. Shpilka and Volkovich showed that a restricted form of $\Sigma \cdot \mathsf{ROF}$ requires linear summands to compute the monomial $x_1 x_2 \cdots x_n$. Further, Mahajan and Tawari [22] obtained a tight lower bound on the size of $\Sigma \cdot \mathsf{ROF}$ computing an elementary symmetric polynomial. Ramya and Rao [29] obtained an exponential lower bound on the number of ROFs required to compute a polynomial in VP. In this article, we improve the lower bound in [29] to obtain an exponential separation between read-k formulas and $\Sigma \cdot \mathsf{ROF}$ for a sufficiently large constant k. Formally, we prove:

Theorem 1. *There is a constant $k > 0$ and a family of multilinear polynomials f_{PRY} computable by n-variate read-k formulas such that if $f_{PRY} = f_1 + f_2 + \cdots + f_s$, where $f_1, \ldots, f_s$ are ROFs, then $s = 2^{\Omega(n)}$.*

(2) Sum of ROABPs: A natural generalization of ROFs are read-once oblivious branching programs (ROABPs). In an ROABP, a layer reads at most one variable and every variable occurs in exactly one layer. Arguments in [26] imply that any ROABP computing the permanent and determinant requires exponential size. Kayal et al. [19] obtain an exponential separation between the size of ROABPs and depth three multilinear formulas. In [27], an exponential lower bound for the sum of ROABPs computing a polynomial in VP is given. We improve this bound to obtain a super polynomial separation between sum of ROABPs and smABPs:

Theorem 2. *There is a multilinear polynomial family $\hat{f}$ computable by smABPs of polynomial size such that if $\hat{f} = f_1 + \ldots + f_s$, each $f_i \in \mathbb{F}[x_1, \ldots, x_n]$ being computable by a ROABP of size poly(n), then $s = \exp(\Omega(n^\epsilon))$ for some $\epsilon > 1/500$.*

(3) Strict-interval ABPs and Interval formulas: It may be noted that any sub-program of a ROABP computes a polynomial in an interval $\{x_i, \ldots, x_j\}$ of variables for some $i < j$. A natural generalization of ROABPs would be to consider smABPs where every sub-program computes a polynomial in some interval of variables, while a variable can occur in multiple layers. These are known as *interval* ABPs and were studied by Arvind and Raja [4] who obtained a conditional lower bound for the size of interval ABPs. Ramya and Rao [28] obtained an exponential lower bound for a special case of interval ABPs known as *strict-interval* ABPs. We show that strict-interval ABPs are equivalent to ROABPs upto polynomial size:

Theorem 3. *The class of strict-interval ABPs is equivalent to the class of ROABPs.*

Finally, we examine the restriction of intervals in syntactic multilinear formulas. We show that unlike ROFs, interval formulas can be depth reduced (Theorem 23).

Related Work: To the best of our knowledge, Theorem 1 is the first exponential separation between bounded read formulas and $\Sigma \cdot$ ROF. Prior to this, only a linear separation between bounded read formulas and $\Sigma \cdot$ ROF was known [3].

In [29], Ramya and Rao show an exponential lower bound for the sum of ROFs computing a polynomial in VP. This already establishes that there are explicit families of polynomials computable by quasi-polynomial size multilinear polynomials requiring exponential size $\Sigma \cdot$ ROFs. Our result significantly strengthens this since our hard polynomial is computable by bounded read formulas. Mahajan and Tawari [22] obtain a tight linear lower bound for $\Sigma \cdot$ ROF computing an elementary symmetric polynomial.

Kayal, Nair and Saha [19] obtain a separation between ROABPs and multilinear depth three circuits. The authors define a polynomial, efficiently computed by set multilinear depth three circuits, that has an exponential size ROABP computing it. This polynomial can be expressed as a sum of three ROFs. Later, Ramya and Rao [27] obtain a sub-exponential lower bound against the model of $\Sigma \cdot$ ROABP computing the polynomial defined by Raz and Yehudayoff [32]. Dvir et al. [13] obtain a super-polynomial lower bound on the size of syntactic multilinear formulas computing a polynomial that can be efficiently computed by smABPs. We use the polynomial defined by [13] and adapt their techniques to obtain a separation between smABPs and $\Sigma \cdot$ ROABP.

Organization of the Paper: Sect. 2 contains basic definitions of the models of computations, concepts and explicit polynomials used in the rest of the paper. The rest of the sections each describe results with respect to a particular bounded-read model. Section 3 describes the lower bound on the $\Sigma \cdot$ROF model and Sect. 4 describes the lower bound on the $\Sigma \cdot$ROABP model which follows using the same arguments as in the work of Dvir et al. [13]. Section 5 shows that strict-interval ABP is a fresh way to look at ROABPs since the two models are equivalent. In Sect. 6 we see that Brent's depth reduction result [8] holds for the class of interval formulas.

2   Preliminaries

In this section, we present necessary definitions and notations. For more details, reader is referred to excellent surveys by Shpilka and Yehudayoff [35] and by Saptharishi et al. [33].

Arithmetic Circuits: Let $X = \{x_1, \ldots, x_n\}$ be a set of variables. An arithmetic circuit C over a field $\mathbb{F}$ with input X is a directed acyclic graph (DAG) where the nodes have in-degree zero or two. The nodes of in-degree zero are called input gates and are labeled by elements from $X \cup \mathbb{F}$. Non-input gates of C are called internal gates and are labeled from $\{+, \times\}$. Nodes of out degree zero are called output gates. Typically, a circuit has a single output gate. Every gate v in C naturally computes a polynomial $f_v \in \mathbb{F}[X]$. The polynomial computed by C is the polynomial represented at its output gate. The *size* of a circuit denoted by size(C), is the number of gates in it, and *depth* is the length of the longest root to leaf path in C, denoted by depth(C). An *arithmetic formula* is a circuit where the underlying undirected graph is a tree.

Multilinear polynomials are polynomials such that in every monomial, the degree of a variable is either 0 or 1. Multilinear circuits, where every gate in the circuit computes a multilinear polynomial, are natural models of computation for multilinear polynomials. For a gate v in a circuit C, let var(v) denote the set of all variables that appear in the sub-circuit rooted at v. A circuit C is said to be *syntactic multilinear* if for every product gate $v = v_1 \times v_2$ in C, we have var(v_1) $\cap$ var(v_2) = $\emptyset$. By definition, a syntactic multilinear circuit is also multilinear.

An arithmetic formula F is said to be a *read-once formula* (ROF in short) if every input variable in X labels at most one input gate in F. Polynomials computed by ROFs are known as *read-once polynomials*.

Algebraic branching program (ABP in short) is a model for computation of polynomials defined as analogous to the branching program model of computation for Boolean functions. An ABP P is a layered DAG with layers $L_0, \ldots, L_m$ such $L_0 = \{s\}$ and $L_m = \{t\}$ where s is the start node and t is the terminal node. Each edge is labeled by an element in $X \cup \mathbb{F}$. The output of the ABP P is the polynomial $p = \sum_{\rho \text{ is an } s \text{ to } t \text{ path}} \mathsf{wt}(\rho)$, where $\mathsf{wt}(\rho)$ is the product of edge labels in the path ρ. Further, for any two nodes u and v let $[u, v]_P$ denote the polynomial computed by the sub-program P_{uv} of P with u as the start node and v as the terminal node. Let X_{uv} denote the set of variables that appear as edge labels in the sub-program P_{uv}. The size of an ABP P, denoted by size(P) is the number of nodes in P.

In a *syntactic multilinear* ABP (smABP), every s to t path reads any input variables at most once. An ABP is *oblivious* if every layer reads at most one variable. A *read-once oblivious ABP* (ROABP) is an oblivious smABP where every variable can appear in at most one layer i.e., for every i, there is at most one layer j_i such that x_i occurs as a label on the edges from L_{j_i} to L_{j_i+1}.

An interval on the set $\{1, \dots, n\}$ with end-points $i, j \in [n]$, can be defined as $I = [i, j]$, $i < j$, where $I = \{\ell \mid i, j \in [n], i \leq \ell \leq j\}$. An *interval of variables* X_{ij} is defined such that $X_{ij} \subseteq \{x_\ell \mid \ell \in I, I = [i, j]\}$, where I is an interval on the set $\{1, \dots, n\}$. For an ordering $\pi \in S_n$, we define a π-interval of variables, $X_{ij} \subseteq \{x_{\pi(i)}, x_{\pi(i+1)}, \dots, x_{\pi(j)}\}$. In [4], Arvind and Raja defined a sub-class of syntactically multilinear ABPs known as *interval ABPs* and proved lower bounds against the same. Later, [28] defined a further restricted version of interval ABPs, denoted by strict-interval ABPs, defined as follows.

Definition 4. *[28] A strict interval ABP P is a syntactically multilinear ABP where we have the following:*

1. *For any pair of nodes u and v in P, the indices of variables occurring in the sub-program $[u, v]_P$ is contained in some π-interval I_{uv} called the associated interval of $[u, v]_P$; and*
2. *for any pairs of sub-programs of the form $[u, v]_P$, $[v, w]_P$, the associated π-intervals of variables are disjoint, i.e., $I_{uv} \cap I_{vw} = \emptyset$.*

It may be noted that in a strict interval ABP, intervals associated with each sub-program need not be unique. We assume that the intervals associated are largest intervals with respect to set inclusion such that condition 2 in the definition above is satisfied.

The Partial Derivative Matrix: We need the notion of partial derivative matrices introduced by Raz [31] and Nisan [26] as primary measure of complexity for multilinear polynomials. The partial derivative matrix of a polynomial $f \in \mathbb{X}$ defined based on a partition $\varphi : X \to Y \cup Z$ of the X into two parts. We follow the definition in [31]:

Definition 5. *(Raz [31]) Let $\varphi : X \to Y \cup Z$ be a partition of the input variables in two parts. Let $\mathcal{M}_Y$, $\mathcal{M}_Z$ be the sets of all possible multilinear monomials in the variables in Y and Z respectively. Then we construct the partial derivative matrix M_{f^φ} for a multilinear polynomial f under the partition φ such that the rows of the matrix are indexed by monomials $m_i \in \mathcal{M}_Y$, the columns by monomials $s_j \in \mathcal{M}_Z$ and entry $M_{f^\varphi}(i, j) = c_{ij}$, c_{ij} being the coefficient of the monomial $m_i \cdot s_j$ in f. We denote by $\mathsf{rank}_\varphi(f)$ the rank of the matrix M_{f^φ}.*

We call φ an equi-partition when $|X| = n$, n even and $|Y| = |Z| = n/2$. Raz [31] showed the following fundamental property of rank_φ:

Lemma 6. *Let g and h be multilinear polynomials in $\mathbb{F}[X]$. Then, $\forall \varphi : X \to Y \cup Z$, we have the following.*

Sub-additivity: $\mathrm{rank}_\varphi(g + h) \leq \mathrm{rank}_\varphi(g) + \mathrm{rank}_\varphi(h)$, and
Sub-multiplicativity: $\mathrm{rank}_\varphi(gh) \leq \mathrm{rank}_\varphi(g) \times \mathrm{rank}_\varphi(h)$.

Two Explicit Polynomials: Let f_{RY} denote the family of multilinear polynomials defined by Raz and Yehudayoff in [32]. The family f_{RY} can be computed by polynomial size syntactic multilinear circuits. Dvir et al. [13] defined a polynomial family $\widehat{f}$ that can be computed by polynomial size smABPs and have full-rank under a special type of partitions called arc-partitions. Further details of the families f_{RY} and $\widehat{f}$ are included in the appendix. Polynomials that exhibit maximum rank of the partial derivative matrix under all or a large fraction of equi-partitions can be thought of as *high complexity* or *hard* polynomials. We need two such families found in the literature.

Raz and Yehudayoff [32] defined a multilinear polynomial in VP. To describe this polynomial we denote an interval $\{a \mid i \leq a \leq j, a \in \mathbb{N}\}$, $i, j \in \mathbb{N}$ by $[i, j]$, and consider the sets of variables $X = \{x_1, \ldots, x_{2n}\}$, $W = \{w_{i,\ell,j}\}_{i,\ell,j \in [2n]}$. We denote it as the Raz-Yehudayoff polynomial and define it as follows.

Definition 7 (Raz-Yehudayoff polynomial, [32]). *Let us consider $f_{ij} \in \mathbb{F}[X, W]$ defined over the interval $[i, j]$. For $i \leq j$, the polynomial f_{ij} is defined inductively as follows. If $j - i = 0$, then $f_{ij} = 0$. For $|j - i| > 0$,*

$$f_{ij} = (1 + x_i x_j) f_{i+1,j-1} + \sum_{\ell \in [i+1, j-2]} w_{i,\ell,j} f_{i,\ell} f_{\ell+1,j},$$

where we assume without loss of generality, lengths of $[i, \ell]$, $[\ell + 1, j]$ are even and smaller than $[i, j]$. We define $f_{1,2n}$ as the Raz-Yehudayoff polynomial f_{RY}.

Note that, f_{RY} can be defined over any subset $X' \subseteq X$ such that $|X'|$ is even, by considering the induced ordering of variables in X' and considering intervals accordingly. We denote this polynomial as $f_{\mathsf{RY}}(X')$ for $X' \subseteq X$. Raz and Yehudayoff showed:

Proposition 8. *[32] Let $\mathbb{G} = \mathbb{F}(W)$ be the field of rational functions over the field $\mathbb{F}$ and the set of variables W. Then for every equi-partition $\varphi : X \to Y \cup Z$, $\mathrm{rank}_\varphi(f_{\mathsf{RY}}) = 2^{n/2}$.*

Dvir et al. [13] defined a polynomial that is hard i.e., full rank with respect to a special class of partitions called *arc-partitions*. Suppose $X = \{x_0, \ldots, x_{n-1}\}$ be identified with the set $V = \{0, \ldots, n-1\}$. For $i, j \in V$, the set $[i, j] = \{i, (i+1) \bmod n, (i+2) \bmod n, \ldots, j\}$ is called the *arc* from i to j. An arc pairing is a distribution on the set of all pairings (i.e., perfect matchings) on V obtained in $n/2$ steps as follows. Assuming a pairing $(P_1, \ldots, P_t)$ constructed in $t < n/2$ steps, where $P_1 = (0, 1)$, $[L_t, R_t]$ is the interval spanned by $\cup_{i \in [t]} P_i$ and the random pair P_{t+1} is constructed such that

$$P_{t+1} = \begin{cases} (L_t - 2, L_t - 1) & \text{with probability } 1/3, \\ (L_t - 1, R_t + 1) & \text{with probability } 1/3, \\ (R_t + 1, R_t + 2) & \text{with probability } 1/3, \end{cases}$$

and therefore, $[L_{t+1}, R_{t+1}] = [L_t, R_t] \cup P_{t+1}$.

Given a pairing $\mathcal{P} = \{P_1, \ldots, P_{n/2}\}$ of V, there are exactly $2^{n/2}$ partitions of X, by assigning $\varphi(x_i) \in Y$ and $\varphi(x_j) \in Z$ or $\varphi(x_i) \in Z$ and $\varphi(x_j) \in Y$ independently for each pair $(i, j) \in \mathcal{P}$. An *arc partition* is a distribution on all partitions obtained by sampling an arc pairing as defined above and sampling a partition corresponding to the pairing uniformly at random. We denote this distribution on partitions by $\mathcal{D}$. For a pairing $\mathcal{P} = \{P_1, \ldots, P_{n/2}\}$ let $M_\mathcal{P}$ be the degree $n/2$ polynomial $\prod_{i=1}^{n/2}(x_{\ell_i} + x_{r_i})$ where $P_i = (\ell_i, r_i)$. Dvir et al. [13] defined the arc full rank polynomial $\widehat{f} = \sum_{\mathcal{P} \in \mathcal{D}} \lambda_\mathcal{P} M_\mathcal{P}$, where $\lambda_\mathcal{P}$ is a formal variable. Dvir et al. [13] showed:

Proposition 9. *[13] The polynomial $\widehat{f}$ can be computed by a polynomial size smABP and for every $\varphi \in \mathcal{D}$, $\mathsf{rank}_\varphi(\widehat{f}) = 2^{n/2}$ over a suitable field extension $\mathbb{G}$ of $\mathbb{F}$.*

Now that we are familiar with most of the definitions required for an understanding of the results in this paper, we proceed to discuss our results.

3   Sum of ROFs

In this section, we show that there is an exponential separation between syntactic multilinear read-k formula and $\Sigma \cdot \mathsf{ROF}$. We begin with the construction of a hard polynomial computable be a read-k formula for a large constant k.

A Full Rank Polynomial: Let $X = \{x_1, \ldots, x_n\}$ be the set of input variables of the hard polynomial such that $4 \mid n$. Let $f_{\mathsf{RY}}(X')$ to denote the Raz-Yehudayoff polynomial defined on the variable set X', where X' is an arbitrary subset of X with $|X'|$ even.

Let $r = \Theta(1)$ be a sufficiently large integer factor of n such that r and n/r are both even. For $1 \le i \le n/r$, let $B_i = \{x_{(i-1)r+1}, \ldots, x_{ir}\}$ and $\mathcal{B}$ denote the partition $B_1 \cup B_2 \cup \cdots \cup B_{n/r}$ of X. The polynomial f_{PRY} is defined as follows:

$$f_{\mathsf{PRY}} = f_{\mathsf{RY}}(B_1) \cdot f_{\mathsf{RY}}(B_2) \cdots f_{\mathsf{RY}}(B_{n/r}). \tag{1}$$

By definition of the polynomial f_{PRY}, it can be computed by a constant-width ROABP of polynomial size as well as by a read-k formula where $k = 2^{O(r)}$.

In order to prove a lower bound against a class of circuits computing the polynomial f_{PRY}, we consider the complexity measure of the rank of partial derivative matrix. Like in [30] and many follow-up results, we analyse the rank of the partial derivative matrix of f_{PRY} under a random partition. The reader might have already noticed that there are equi-partitions under which the $\mathsf{rank}_\varphi(f_{\mathsf{PRY}}) = 1$. Thus, we need a different distribution on the equi-partitions under which f_{PRY} has full rank with probability 1. In fact, under any partition φ, which induces an equi-partition on each of the variable blocks B_i, we have $\mathsf{rank}_\varphi(f_{\mathsf{PRY}}) = 2^{n/2}$, i.e., full rank. We define $\mathcal{D}_B$ as the uniform distribution on all such partitions. Formally, we have:

Definition 10. *(Distribution $\mathcal{D}_\mathcal{B}$) The distribution $\mathcal{D}_\mathcal{B}$ is the distribution on the set of all equi-partitions $\hat{\varphi}$ of X obtained by independently sampling an equi-partition φ_i of each variable blocks B_i, for all i such that $1 \leq i \leq n/r$. We express $\hat{\varphi}$ as $\hat{\varphi} = \varphi_1 \circ \ldots \circ \varphi_{n/r}$.*

For any partition φ in the support of $\mathcal{D}_\mathcal{B}$, we argue that the polynomial f_{PRY} has full rank:

Observation 1. *For any $\varphi \sim \mathcal{D}_\mathcal{B}$, $\mathrm{rank}_\varphi(f_{\mathsf{PRY}}) = 2^{n/2}$ with probability 1.*

Proof. Let us fix an equi-partition function $\hat{\varphi} \sim \mathcal{D}_\mathcal{B}$, $\hat{\varphi} : X \rightarrow Y \cup Z$. Let $t = r$. Considering $f_{\mathsf{RY}}(X')$ where $|X'| = t$ and t is even, we can prove the partial derivative matrix of $f_{\mathsf{RY}}(X')$ has rank $2^{n/2}$ under $\hat{\varphi}$ by induction on t. By definition of f_{RY}, for $t = 2$ we have $f_{\mathsf{RY}} = 0$.

So, for the higher values of t, we see the term $(1 + x_1 x_t)$ and $f_{2,t-1}$ are variable disjoint, where $(1 + x_1 x_t)$ has rank ≤ 2, and by the induction hypothesis, $f_{2,t-1}$ has rank $2^{t/2-1}$. Also, by induction hypothesis, for any ℓ, the ranks of partial derivative matrices of $f_{1,\ell}$ and $f_{\ell+1,t}$ are $2^{\ell/2}$ and $2^{(t-\ell)/2}$ respectively.

When $\hat{\varphi}(x_1) \in Y$ and $\hat{\varphi}(x_t) \in Z$, we set $w_{1,\ell,t} = 0$ for all $\ell \in [2, t-1]$ and $\mathrm{rank}_{\hat{\varphi}}(f_{1,t}) = \mathrm{rank}_{\hat{\varphi}}(1 + x_1 x_t) \cdot \mathrm{rank}_{\hat{\varphi}} f_{2,t-1} = 2 \cdot 2^{(t/2-1)} = 2^{t/2}$. When $\hat{\varphi}(x_1) \in Y$ and $\hat{\varphi}(x_t) \in Y$, for an arbitrary $\ell \in [t]$ we set $w_{1,\ell,t} = 1$ and we have $\mathrm{rank}_{\hat{\varphi}}(f_{1,t}) = \mathrm{rank}_{\hat{\varphi}}(f_{1,\ell}) \cdot \mathrm{rank}_{\hat{\varphi}}(f_{\ell+1,t}) = 2^{t/2}$, since $\hat{\varphi}$ is an equi-partition.

By sub-additivity of rank, and since B_i, $i \in [n/r]$ are disjoint sets of variables, we have $\mathrm{rank}_{\hat{\varphi}}(f_{\mathsf{PRY}}) = \prod_{i \in [n/r]} \mathrm{rank}_{\hat{\varphi}}(f_{\mathsf{RY}}(B_i)) = \prod_{i \in [n/r]} 2^{t/2} = 2^{tn/2r} = 2^{n/2}$.

3.1 Rank Upper Bound on ROFs

In the following, we argue that the polynomial f_{PRY} cannot be computed by sum of ROFs of sub-exponential size. More formally,

Theorem 1. *There is a constant $k > 0$ and a family of multilinear polynomials f_{PRY} computable by n-variate read-k formulas such that if $f_{\mathsf{PRY}} = f_1 + f_2 + \cdots + f_s$, where $f_1, \ldots, f_s$ are ROFs, then $s = 2^{\Omega(n)}$.*

We use the method of obtaining an upper bound on the rank of partial derivative matrix for ROFs with respect to a random partition developed by [29]. Though the argument in [29] works for an equi-partition sampled uniformly at random, we show that their structural analysis of ROFs can be extended to the case of our distribution $\mathcal{D}_\mathcal{B}$. We begin with the notations used in [29] for the categorisation of the gates in a read-once formula F. (In this categorisation, the authors have only considered gates with at least one input being a variable.)

- Type- A: These are sum gates in F with both inputs variables in X.
- Type- B: Product gates in F with both inputs variables in X.
- Type- C: Sum gates in F where only one input is a variable in X.
- Type- D: Product gates in F where only one input is a variable in X.

Thus, type-D gates compute polynomials of the form $h \cdot x_i$ where $x_i \in X, h \in \mathbb{F}[X \setminus \{x_i\}]$ are the inputs to the type-D gate. Let a, b, c, d be the number of gates of type-A, B, C and D respectively. Let a'' be the number of Type A gates that compute a polynomial of rank 2 under an equi-partition φ, and a' be the number of Type-A gates that compute a polynomial of rank 1 under φ such that $a = a' + a''$.

The following is an adaption of Lemma 3.3 in [29] for our distribution $\mathcal{D}_\mathcal{B}$.

Lemma 11. *Let $f \in \mathbb{F}[X]$ be an ROP, and φ be an equi-partition sampled uniformly at random from the distribution $\mathcal{D}_\mathcal{B}$. Then with probability at least $1 - 2^{-\Omega(n)}$, $\mathrm{rank}_\varphi(M_f) \leq 2^{n/2 - \Omega(n)}$.*

Proof. We first argue a rank upper bound for an arbitrary f_i. Let Φ_i be the formula computing f_i with gates of the types described as above. Let $\hat{\varphi} = \varphi_1 \circ \ldots \circ \varphi_{n/r}$ sampled from the distribution $\mathcal{D}_B$ uniformly at random.

We use the Lemma 3.1 from [29] which concludes that type-D gates do not contribute to the rank of a ROF.

Lemma 12. *[29, Lemma 3.1] Let F be a ROF computing a read-once polynomial f and $\varphi : X \rightarrow Y \cup Z$ be an partition function on n variables. Then, $\mathrm{rank}_\varphi(f) \leq 2^{a'' + \frac{2a'}{3} + \frac{2b}{3} + \frac{9c}{20}}$.*

Intuitively, Lemma 12 can be applied to a ROF F under a distribution $\hat{\varphi} \sim \mathcal{D}_B$ as follows. If there are a large number of type D gates (say αn, for some $0 \leq \alpha < 1$), then for any such equi-partition $\hat{\varphi}$, $\mathrm{rank}_{\hat{\varphi}}(f) \leq 2^{(1-\alpha)n/2}$. A type C gate, too, contributes a small value (at most 2) to the rank compared to gates of types A and B. Thus, without loss of generality, we assume that the number of type C and D gates is at most αn. Now our analysis proceeds as in [29], only differing in the estimation of a'', a' under an equi-partition $\hat{\varphi} \sim \mathcal{D}_B$.

Let $(P_1, \ldots, P_t)$ be a pairing induced by the gates of types A and B (i.e., the two inputs to a gate of type A or B form a pair). There can be at most $n/2$ pairs, but since we have αn gates of type C and D for some $0 \leq \alpha < 1$, we assume $(1 - \alpha)n$ remaining type A and B gates. Therefore, for $t = (n - \alpha n)/2$, $t \leq n/2$, we have the pairs $P_1, \ldots, P_t$ induced by the type A and B gates in Φ_i.

Now, considering the division of X into $B_1, \ldots, B_{n/r}$, we can divide the pairs into two sets depending on whether a pair lies entirely within a block B_i, $i \in [n/r]$ or the pair has its members in two different blocks B_i and B_j for $i, j \in [n/r]$, $i \neq j$. We define these two sets as $W = \{P_i \mid P_i = (x, y), \exists \ell, x, y \in B_\ell\}$ for pairs lying within blocks and $A = \{P_i \mid P_i = (x, y), \exists j, k, j \neq k, x \in B_j, y \in B_k\}$ for pairs lying across blocks, where x, y are two arbitrary variables in X.

Each pair P_i can be monochromatic or bichromatic under the randomly sampled equi-partition $\hat{\varphi}$ with the probability $\frac{1}{2}$. Presence of monochromatic edges will give us a reduction in the rank of f_i under $\hat{\varphi}$. The analysis on W and A is done separately as follows.

Analysing W, $|W| > t/2$: Let $B_{i_1}, \ldots, B_{i_\ell}$ be the blocks containing at least one pair from W, $\ell \le n/r$. We want to estimate ℓ and count how many of these ℓ blocks have at least one monochromatic pair under $\hat{\varphi}$ from W.

For each B_i, $i \in [t]$, we define the Bernoulli random variable X_i such that,

$$X_i = \begin{cases} 1, & \text{if } \exists P \in W, \ P = (x, y), \ x, y \in B_i, \\ 0, & \text{otherwise.} \end{cases}$$

Let $\Pr[X_i = 1] = \Pr[\exists P \in W, \ P = (x, y), \ x, y \in B_i] = \epsilon$, for some $\epsilon > 0$.

Then we have $\mathsf{E}[X_i] = \epsilon$, and for $\mathcal{X} = X_1 + \ldots + X_{n/r}$, $\mathsf{E}[\mathcal{X}] = \epsilon \cdot n/r$. By the Chernoff's bound defined in [25], we have,

$$\Pr[\mathcal{X} > 2\epsilon n/r] < \exp(\frac{-\epsilon n}{3r}).$$

Now we estimate ϵ as follows:

$$\begin{aligned} \epsilon &= \Pr[X_i = 1] = \Pr[\exists P \in W, \ P = (x, y), \ x, y \in B_i] \\ &= \Pr[x, y \in B_i | \exists P \in W, \ P = (x, y)] \\ &= \frac{\Pr[x, y \in B_i]}{\Pr[\exists P \in W, \ P = (x, y)]} \\ &\ge \Pr[x, y \in B_i] \quad \text{since } \Pr[\exists P \in W, \ P = (x, y)] \le 1 \\ &= \frac{1}{r^2}. \end{aligned}$$

Therefore, $\Pr[\mathcal{X} > 2\epsilon n/r] < \exp(\frac{-\epsilon n}{3r}) \le \exp(-\Omega(n))$, when r is a constant. This implies that at least $2/r^2$ fraction of the blocks have a pair entirely within them with probability $1 - \exp(-\Omega(n))$ and each of these pairs is monochromatic under $\hat{\varphi}$ with the constant probability $1/2$. This gives an upper bound on the rank of f_i,

$$\mathrm{rank}_{\hat{\varphi}}(f_i) \le 2^{n/2 - n/r^3} = 2^{n/2 - \Omega(n)}.$$

Analysing A, $|A| > t/2$: Since each pair of variables in A lies across two blocks, we create a graph $G = (V, E)$ where each $v_i \in V$ represents the block B_i and $E = \{(v_i, v_j) \mid (x, y) \in A, \ x \in B_i, \ y \in B_j, i \ne j\}$.

The graph G has maximum degree r since there can be at most r pairs with one member in a fixed block B_i. If the edges in E form a perfect matching M' in G, then under $\hat{\varphi}$, the edges in E can be either bichromatic or monochromatic. We need to show there will be sufficient number of monochromatic edges to give a tight upper bound for $\mathrm{rank}_{\hat{\varphi}}(f_i)$.

By a result in [7], any graph with maximum degree r has a maximal matching of size $m/(2r - 1)$, where $|E| = m$. Since $|A| \ge t/2$, $m \ge t/2$ and hence the maximal matching is of size $t/2(2r - 1) = \Omega(n)$ when r is a suitable constant. With probability $1/2$, an edge in the maximal matching is bichromatic. Hence, $\le t/2$ number of the edges in the maximal matching are bichromatic with probability $1/2^{t/2} = O(\exp(n^{-1}))$. So, with the high probability of $1 - O(\exp(n^{-1}))$, more than half of the edges in the maximal matching are monochromatic, thus giving us the rank bound,

$$\text{rank}_{\hat{\varphi}}(f_i) \leq 2^{n/2-t/2} = 2^{n/2-\Omega(n)}.$$

Given an upper bound on the rank of ROFs under a random partition from $\mathcal{D}_\mathcal{B}$, we now proceed to prove the Theorem 1 by showing a lower bound on the size of ROFs computing our hard polynomial f_{PRY}.

Proof. (Proof of Theorem 1) By Observation 1, the upper bound on the rank of ROFs given by Lemma 11 and the sub-additivity of rank, we have:

$$s \cdot 2^{n/2-\Omega(n)} \leq 2^{n/2} \implies s = 2^{\Omega(n)}.$$

4   A Separation Between Sum of ROABPs and smABPs

In this section, we prove a sub-exponential lower bound against the size of sum of read-once oblivious ABPs computing the hard polynomial constructed in [13]. This shows a sub-exponential separation between syntactically multilinear ABPs and sum of ROABPs. We prove the following theorem in this section:

Theorem 2. *There is a multilinear polynomial family $\hat{f}$ computable by smABPs of polynomial size such that if $\hat{f} = f_1 + \ldots + f_s$, each $f_i \in \mathbb{F}[x_1, \ldots, x_n]$ being computable by a ROABP of size $poly(n)$, then $s = \exp(\Omega(n^\epsilon))$ for some $\epsilon > 1/500$.*

Our aim is to give an upper bound on the maximum rank of ROABPs under an arc partition. We refer to the rank of the coefficient matrix of the sum of ROABPs against an arc-partition as the *arc-rank*. We analyze the arc-rank of the sum of ROABPs against an arc-partition to give a lower bound on the size of the sum necessary to compute $\hat{f}$.

Let us assume that n is even. In order to prove the lower bound, we need to estimate an upper bound on the arc-rank computed by a ROABP. We define the notion of F-arc-partition, F being a ROABP, as follows:

Definition 13. *Let us consider an arc partition Q constructed from a ROABP F in the following manner: Let the order of variables appearing in the ROABP be $x_{\sigma(1)}, x_{\sigma(2)}, \ldots, x_{\sigma(n)}$, where $\sigma \in S_n$ is a permutation on n indices. Then, $Q = \{(x_{\sigma(i)}, x_{\sigma(i+1)}) \mid i \in [n], i \text{ is odd}\}$ is a F-arc-partition.*

We assume $2K \mid n$. Let $S_1, \ldots, S_K$ be a K-coloring of the variable set X, where $x_1, \ldots, x_n$ are ordered according to the ROABP and for every $i \in [k]$, S_i contains the variables $x_{(i-1)n/K+1}, \ldots, x_{in/K}$ according to that ordering. Then $S_1, \ldots, S_K$ is a K-partitioning of the pairs in the F-arc-partition Q. So pairs in Q are monochromatic, whereas the pairs $(P_1, \ldots, P_{n/2})$ on which a random arc-partition Π sampled from $\mathcal{D}$ is based, might cross between two colors.

Our analysis for the ROABP arc-rank upper bound follows along the lines of the analysis for the arc-rank upper bound given by [13] for syntactic multilinear formulas. For this analysis we define the set of violating pairs for each color c, $V_c(\Pi)$, that is defined as: $V_c(\Pi) = \{\Pi_t \mid |\Pi_t \cup S_c| = 1,\ t \in [n/2]\}$, where $\Pi_1, \ldots, \Pi_{n/2}$ are pairs in Π. The quantity $G(\Pi) = |\{c \mid |V_c(\Pi)| \geq n^{\frac{1}{1000}}\}|$, representing the number of colors with many violations, is similarly defined. We use the following lemma directly from [13]:

Lemma 14. *Let* $K \leq n^{\frac{1}{100}}$, Π *be the sampled arc-partition, and* $G(\Pi)$ *be as defined above. Then, we have,* $\Pr_{\Pi \in \mathcal{D}}[G(\Pi) \leq K/1000] \leq n^{-\Omega(K)}$.

The following measure is used to compute the arc-rank upper bound for ROABPs.

Definition 15. *(Similarity function) Let* φ *be a distribution on functions* $\mathcal{S} \times \mathcal{S} \to \mathbb{N}$, *such that* $\mathcal{S}$ *is the support of the distribution on arc-partitions,* $\mathcal{D}$. *Let* P, Q *be arc-partitions sampled independently and uniformly at random from* $\mathcal{D}$. *Then,* $\varphi(Q, P) : \mathcal{S} \times \mathcal{S} \to \mathbb{N}$ *is the total number of common pairs between two arc-partitions* Q *and* P.

We assume Q to be the F-arc-partition for the ROABP F. For a pair that is not common between Π and Q, we show both the variables in the pair is in the same partition, Y or Z with high probability.

Theorem 16. *Under an arc-partition* Π *sampled from* $\mathcal{D}$ *uniformly at random, if* $p \in \mathbb{F}[X]$ *is the polynomial computed by a ROABP* P, *then, for the similarity function* φ *and* $\delta > 0$,

$$\Pr_{\Pi \sim \mathcal{D}}[\varphi(\Pi, Q) \geq n/2 - n^{\delta}] \leq 2^{-o(n)}.$$

Proof Outline: Our argument is the same as [13]. It is being included here for completeness for the parameters here being somewhat different than [13].

In order to analyze the number of common pairs counted by φ, we consider the K-coloring of F and show that under a random arc-partition Π, the number of crossing pairs are large in number using Lemma 14. Then, we show, this results in large number of pairs having both elements in Y. In order to identify the colors with the high number of crossing pairs, a graphical representation of the color sets is used.

Proof. [13] construct the graph $H(\Pi)$, where each vertex is a color c such that $|V_c(\Pi)| \geq n^{\frac{1}{1000}}$, and vertices c and d have an edge connecting them if and only if $|V_c(\Pi) \cap V_d(\Pi)| \geq n^{\frac{1}{1500}}$. We know for any two colors $c, d \in [K]$, $|V_c(\Pi) \cap V_d(\Pi)| \leq n^{\frac{1}{1000}}$. So, by definition of $H(\Pi)$, the least degree of a vertex in $H(\Pi)$ is 1. Using this, [13] prove the following claim:

Claim. Let the size of the vertex set of $H(\Pi)$, $V(H(\Pi))$, be M. For any subset U of $V(H(\Pi))$ size $N \geq M/2 - 1$, there is some color $h_{j+1}, j \in [N-1]$ such that in the graph induced on all vertices except $\{h_1, \ldots, h_j\}$, the degree of h_{j+1} is at least 1.

By Claim 4, we have $U \subseteq V(H(\Pi))$, $U = \{c_1, \ldots, c_{M/2-1}\}$ such that this is the set of colors having high number of crossing pairs common with colors not in U. Considering the colors sequentially, given Π, we first examine the pairs crossing from color c_1 to other colors, then c_2 and so on. Therefore, to examine the event E_i for color c_i, we have to estimate $\Pr_{\Pi \sim \mathcal{D}}[E_i \mid E_1, \ldots, E_{i-1}, \Pi]$.

Here, E_i is the event $|Y_{c_i} - |S_{c_i}|/2| \le n^{\frac{1}{5000}}$, equivalently expressed as $|S_{c_i}|/2 - n^{\frac{1}{5000}} \le Y_{c_i} \le |S_{c_i}|/2 - n^{\frac{1}{5000}}$. But for an upper bound, it suffices to analyze the $n^{\frac{1}{1500}}$ crossing pairs from S_{c_i} to S_{c_j} instead of considering the entire set. Let the subset of Y_{c_i} constituted by one end of crossing pairs going to color c_j be P_{ij}. Each element x in a crossing pair $P_t = (x, w)$ is a binomial random variable in a universe of size $\ge n^{\frac{1}{1500} = s}$ with probability $1/2$ of being allotted to the subset Y of the universe. This event is independent of how the c_i colored element of other crossing pairs $P_{t'}$ are allotted. So, $|B_{ij}| = b_j$ is a hyper-geometric random variable where B_{ij} contains all such $x \in Y$. By the properties of a hyper-geometric distribution, $\Pr_{b_j}[b_j = a] = O(s^{\frac{-1}{2}}) = O(n^{\frac{-1}{3000}})$, where a is a specific value taken by the size of B_{ij}.

Applying the union bound over all colors c_j for the crossing pairs, and taking $b = \sum_{j \in U \setminus \{i\}} b_j$, we have:

$$\Pr_b[s/2 - n^{\frac{1}{5000}} \le b \le |S_{c_i}|/2 - n^{\frac{1}{5000}}] \le 2n^{\frac{1}{5000}} O(n^{\frac{-1}{3000}}) = n^{-\Omega(1)}.$$

Therefore, $\Pr_{\Pi \sim \mathcal{D}}[E_i \mid E_1, \ldots, E_{i-1}, \Pi] = n^{-\Omega(\delta)}$.

We want an upper bound for $\Pr[|Y_c - |S_c|/2| \le n^{\frac{1}{5000}} \forall c \in [K]]$. We have calculated an upper bound for the colors in $[K]$ that were highly connected to each other in $H(\Pi)$. So, we can now estimate the total probability as follows:

$$\begin{aligned}
&\Pr[|Y_c - |S_c|/2| \le n^{\frac{1}{5000}} \forall c \in [K]] \\
&= \mathsf{E}[n^{-\Omega(G(P))} \mid G(P) > K/1000] + \mathsf{E}[n^{-\Omega(G(P))} \mid G(P) \le K/1000] \\
&= \mathsf{E}[n^{-\Omega(G(P))} \mid G(P) > K/1000] + n^{-\Omega(K)} \text{ by Lemma 14} \\
&\le n^{-\Omega(K)}.
\end{aligned}$$

If we consider $\delta = 1/5000$, then:

$$\Pr_{\Pi \sim \mathcal{D}}[\varphi(\Pi, Q) \ge n/2 - n^{\delta}] \le \Pr[|Y_c - |S_c|/2| \le n^{\frac{1}{5000}} \forall c \in [K]] \le n^{-\Omega(K)}$$

Now, in Lemma 14, $K \le n^{\frac{1}{1000}}$.

Hence, $\Pr_{\Pi \sim \mathcal{D}}[\operatorname{rank}_\varphi(M(p_\Pi)) \ge 2^{n/2 - n^{\delta}}] \le 2^{-cn^{\frac{1}{1000}} \log n} = 2^{-o(n)}$.

Now, using the above Theorem 16, we can prove the lower bound on the size of the sum of ROABP, s.

Proof. Since the polynomial f is such that each multiplicand is of the form $\lambda_e(x_u + x_v)$, if x_u, x_v are both mapped to the same partition Y or Z, it will reduce the rank of the partial derivative matrix by half. Hence, we have the following:

$$\Pr_{\Pi \sim \mathcal{D}}[\operatorname{rank}_\varphi(M(f_\Pi)) \ge 2^{n/2 - n^{\delta}}] = \Pr_{\Pi \sim \mathcal{D}}[\varphi(\Pi, Q) \ge n/2 - n^{\delta}],$$

for some suitable $\delta > 0$.

$$\Pr[\text{rank}(M(f_\Pi)) = 2^{n/2}] \leq \Pr[\exists i \in [s],\ \text{rank}(M((f_i)_\Pi)) \geq 2^{n/2}/s]$$

$$\leq \sum_{i=1}^{s} \Pr[\text{rank}(M((f_i)_\Pi)) \geq 2^{n/2}/s]$$

$$\leq \sum_{i=1}^{s} \Pr[\text{rank}(M((f_i)_\Pi)) \geq 2^{n/2-n^\delta}] \text{ for some } \delta > 0$$

$$\leq s \cdot n^{-\Omega(n^{\frac{1}{1000}})}$$

$$\implies s = 2^{\Omega(n^{\frac{1}{1000}} \log n)} = 2^{\Omega(n^{\frac{1}{500}})}.$$

5   Strict-Interval ABPs

A strict-interval ABP, defined in [28] (See Definition 4), is a restriction of the notion of interval ABPs introduced by [4]. In the original definition given by [28], every sub-program in a strict-interval ABP P is defined on a π-interval of variables for some order π, however, without loss of generality, we assume π to be the identity permutation on n variables. Therefore, an interval of variables $[i, j]$, $i < j$ here is the set $\{x_i, \ldots, x_j\}$. In this section we show that strict-interval ABPs are equivalent to ROABPs upto a polynomial blow-up in size.

Theorem 3. *The class of strict-interval ABPs is equivalent to the class of ROABPs.*

We start with some observations on intervals in $[1, n]$ and the intervals involved in a strict interval ABP. Let P be a strict-interval ABP over the variables $X = \{x_1, \ldots, x_n\}$. For any two nodes u and v in P, let $I_{u,v}$ be the interval of variables associated with the sub-program of P with u as the start node and v as the terminal node. For two intervals $I = [a, b]$, $J = [c, d]$ in $[1, n]$, we say $I \preceq J$, if $b \leq c$. Note that any two intervals I and J in $[1, n]$ are comparable under $\preceq$ if and only if either they are disjoint or the largest element in one of the intervals is the smallest element in the other. This defines a natural transitive relation on the set of all intervals in $[1, n]$. The following is a useful property of $\preceq$:

Observation 2. *Let I, J and J' be intervals over $[1, n]$ such that $I \preceq J$ and $J' \subseteq J$. Then $I \preceq J'$.*

Proof. Let $I = [a, b]$, $J = [c, d]$ and $J' = [c', d']$. As $I \preceq J$, we have $b \leq c$. Further, since $J' \subseteq J$, we have $c \leq c'$ and $d' \leq d$. Therefore, $b \leq c'$ and hence $I \preceq J'$.

We begin with an observation on the structure of intervals of the sub-programs of P. Let v be a node in P. We say v is an *ascending* node, if $I_{s,v} \preceq I_{v,t}$ and a *descending* node if $I_{v,t} \preceq I_{s,v}$.

Observation 3. *Let P be a strict-interval ABP and v any node in P. Then, v is either ascending or descending and not both.*

Proof. Let $I = I_{s,v}$ and $J = I_{v,t}$. Since P is a strict-interval ABP, the intervals I and J are disjoint and hence either $I \preceq J$ or $J \preceq I$ as required.

Consider any s to t path ρ in P. We say that ρ is *ascending* if every node in ρ except s and t is ascending. Similarly, ρ is called descending if every node in ρ except s and t is descending.

Lemma 17. *Let P be a strict interval ABP and let ρ any s to t path in P. Then either ρ is ascending or descending.*

Proof. We prove that no s to t path in P can have both ascending and descending nodes. For the sake of contradiction, suppose that ρ has both ascending and descending nodes. There are two cases. In the first, there is an edge (u, v) in ρ such that u is an ascending node and v is a descending node. Let $I = I_{s,u}, J = I_{u,t}, I' = I_{s,v}$ and $J' = I_{v,t}$. Since $P_{s,u}$ is a sub-program of $P_{s,v}$, we have $I \subseteq I'$, similarly $J' \subseteq J$. By the assumption, we have $I \preceq J$ and $J' \preceq I'$. By Observation 2, we have $I \preceq J'$ and $J' \preceq I'$. By transitivity, we have $I \preceq I'$. However, by the definition of $\preceq$, I and I' are incomparable, which is a contradiction. The second possibility is u being a descending node and v being an ascending node. In this case, $J \preceq I$ and $I' \preceq J'$. Then, by Observation 2, we have $J' \preceq I$ as $J' \subseteq J$. Therefore, $J \preceq J'$ by the transitivity of $\preceq$, a contradiction. This completes the proof.

Lemma 17 implies that the set of all non-terminal nodes of P can be partitioned into two sets such that there is no edges across. Formally:

Lemma 18. *Let P be an interval ABP. There exist two strict-interval ABPs P_1 and P_2 such that*

1. *All non-terminal nodes of P_1 are ascending nodes and all non-terminal nodes of P_2 are descending nodes; and*
2. *$P = P_1 + P_2$.*

Proof. Let P_1 be the sub-program of P obtained by removing all descending nodes from P and P_2 be the sub-program of P obtained by removing all ascending nodes in P. By Lemma 17, the non-terminal nodes in P_1 and P_2 are disjoint and every s to t path ρ in P is either a s to t path in P_1 or a s to t path in P_2 but not both. Thus $P = P_1 + P_2$.

Next we show that any strict-interval ABP consisting only of ascending or only of descending nodes can in fact be converted into an ROABP.

Lemma 19. *Let P be a strict-interval ABP consisting only of ascending nodes or only of descending nodes. Then the polynomial computed by P can also be computed by a ROABP P' of size polynomial in $\mathsf{size}(P)$. The order of variables in P' is $x_1, \ldots, x_n$ if P has only ascending nodes and $x_n, \ldots, x_1$ if P has only descending nodes.*

Proof (Proof of Lemma 19). We consider the case when all non-terminal nodes of P are ascending nodes. Let ρ be any s to t path in P. We claim that the edge labels in ρ are according to the order $x_1, \ldots, x_n$. Suppose that there are edges (u, v) and (u', v') occurring in that order in ρ such that (u, v) is labelled by x_i and (u', v') is labelled by x_j with $j < i$. Let $I' = I_{s,u'}$ and $J' = I_{u',t}$. Since $i \in I'$, $j \in J'$ and $I' \cap J' = \emptyset$, it must be the case that $J' \preceq I'$ and hence u' must be a descending node, a contradiction. This establishes that P is an one ordered ABP. By the equivalence between one ordered ABPs and ROABPs [16,17], we conclude that the polynomial computed by P can also be computed by a ROABP of size polynomial in the size of P.

The argument is similar when all non-terminal nodes of P are descending. In this case, we have $i < j$ in the above argument and hence $I' \preceq J'$, making u' an ascending node leading to a contradiction. This concludes the proof. $\qquad\square$

A permutation π of $[1, n]$ naturally induces the order $x_{\pi(1)}, \ldots, x_{\pi(n)}$. The *reverse* of π is the order $x_{\pi(n)}, x_{\pi(n-1)}, \ldots, x_{\pi(1)}$. Since branching programs are layered, any multilinear polynomial computed by a ROABP where variables occur in the order given by π can also be computed by a ROABP where variables occur in the reverse of π.

Observation 4. *Let P be a ROABP where variables occur in the order of a permutation π. The polynomial computed by P can also be computed by a ROABP of same size as P that reads variables in the reverse order corresponding to π.*

Proof. Let P' be the ROABP obtained by reversing the edges of P and swapping the start and terminal nodes. Since P is a layered DAG, there is a bijection between the set of all s to t paths in P and the set of all s to t paths in P', where the order of occurrence of nodes and hence the edge labels are reversed. This completes the proof. $\qquad\square$

The above observations immediately establish Theorem 3.

Proof (Proof of Theorem 3). Let P be a strict-interval ABP of size S computing a multilinear polynomial f. By Lemma 18 there are strict interval ABPs P_1 and P_2 such that P_1 has only ascending non-terminal nodes and P_2 has only descending non-terminal nodes such that $f = f_1 + f_2$ where f_i is the polynomial computed by P_i, $i \in \{1, 2\}$. By Lemma 19 and Observation 4, f_1 and f_2 can be computed by a ROABPs that read the variables in the order $x_1, \ldots, x_n$. Then $f_1 + f_2$ can also be computed by an ROABP. It remains to bound the size of the resulting ROABP. Note that $\mathsf{size}(P_i) \leq S$. A ROABP for f_i can be obtained by staggering the reads of P_i which blows up the size of the ABP by a factor of n [16,17]. Therefore size of the resulting ROABP is at most $2nS \leq O(S^2)$.

The notion of intervals of variables corresponding to every sub-program can be applied to formulas in the form of Interval Formulas, where every sub-formula corresponds to an interval. In the following section we explore such a model.

6   Interval Formulas

We saw that strict-interval ABPs have the same computational power as ROABPs despite being seemingly a non-trivial generalization. It is naturally tempting to guess that a similar generalization of ROFs might yield a similar result. However, we observe that it is not the case.

We introduce interval formulas as a generalization of read-once formulas. An interval on variable indices, $[i, j]$, $i < j$, is an interval corresponding to the set of variables $X_{ij} \subseteq X = \{x_1, \ldots, x_n\}$, where $X_{ij} = \{x_p \mid x_p \in X, \ i \leq p \leq j\}$. Polynomials are said to be defined on the interval $[i, j]$ when the input variables are from the set X_{ij}. When there is no ambiguity, we refer to X_{ij} as an interval of variables $[i, j]$. Gates in a read-once formula F can also be viewed as reading an interval of variables according to an order π on the variables i.e., there is a permutation $\pi \in S_n$ such that every gate v in F is a sub-formula computing a polynomial on a π-interval of variables. Thus, interval formulas are a different generalization of read-once formulas where every gate v in the formula F reads an interval of variables in a fixed order. We define interval formulas as follows:

Definition 20. *(Interval Formulas) An arithmetic formula F is an interval formula if for every gate g in F, there is an interval $[i, j]$, $i < j$ such that g computes a polynomial in X_{ij} and for every product gate $g = h_1 \times h_2$, the intervals corresponding to h_1 and h_2 must be non-overlapping.*

Thus, if a product gate g in F defined on an interval $I = [i, j]$ takes inputs from gates $g_1, \ldots, g_t$, then the gates $g_1, \ldots, g_t$ compute polynomials on disjoint intervals $[i, j_1], [j_1 + 1, j_2], \ldots, [j_{t-1} + 1, j]$ respectively, where $\forall p$, $j_p < j_{p+1}$ and $i \leq j_p \leq j$. If g_1, g_2, defined on intervals I_1, I_2 are input gates to a sum gate g', then the interval I associated with g' is $I = I_1 \cup I_2$.

A quick observation is that interval formulas are different from ROFs:

Proposition 21. *The set of all polynomials computable by interval formulas is different from that of ROFs*

Proof. By [39], the polynomial $x_1x_2 + x_2x_3 + x_1x_3$ is not an ROF. However, the expression $x_1x_2 + x_2x_3 + x_1x_3$ is itself an interval formula.

In fact, interval formulas are universal, since any sum of monomials can be represented by an interval formula. Our next observation is that the polynomial f_{PRY} defined in Sect. 3 can be computed by an interval formula.

Proposition 22. *The polynomial family f_{PRY} is computable by an interval formula of polynomial size.*

Proof. Recall that $f_{\mathsf{PRY}}(X) = f_{\mathsf{RY}}(B_1) \cdot f_{\mathsf{RY}}(B_2) \cdots f_{\mathsf{RY}}(B_{n/r})$. Since each of the $f_{\mathsf{RY}}(B_i)$ is a constant variate polynomial and the sum of product representation of any multilinear polynomial is an interval formula by definition, we have that $f_{\mathsf{RY}}(B_i)$ is computable by an interval formula of constant size. This $f_{\mathsf{PRY}}(X)$ has a polynomial size interval formula.

It is not known if every ROF can be converted to a ROF of logarithmic depth. However, we argue that interval formulas can be depth-reduced efficiently.

Theorem 23. *Let $f \in \mathbb{F}[X]$ be a polynomial computed by an interval formula F of size s and depth d. Then f can also be computed by an interval formula of size $\mathrm{poly}(s)$ and depth $O(\log s)$.*

We have the following depth reduction result for general arithmetic formulas given by [8]:

Theorem 24. *[8] Any polynomial p computed by an arithmetic formula of size s and depth d, can be computed by a formula of size $\mathrm{poly}(s)$ and depth $O(\log s)$.*

We know that this reduction preserves multilinearity. However, we don't know if Theorem 24 can be modified to preserve the read-k property. We show that the depth reduction algorithm given by Theorem 24 preserves the interval property.

Proof (of Theorem 23). We know that the underlying structure of any arithmetic formula is a tree. The proof by Brent crucially uses the fact that by the *tree-separator lemma* [12], we are guaranteed that there exists a tree-separator node g such that the sub-tree Φ of a formula Φ' of total size s, rooted at the node g, has size $\leq 2s/3$.

The construction proceeds as follows. We replace the gate g by a new formal variable y. Let the resulting polynomial computed by F be $f'(x_1, \ldots, x_n, y)$, where $f(x_1, \ldots, x_n) = f'(x_1, \ldots, x_n, g)$ under the new substitution $y = g$. As f' is linear in y, we have

$$f'(x_1, \ldots, x_n, y) = y f_1(x_1, \ldots, x_n) + f_0(x_1, \ldots, x_n),$$

where $f_0 = f'|_{y=0}$ and $f_1 = f'|_{y=1} - f'|_{y=0}$. Thus, f_0, f_1 can be computed by multilinear formulas of size less than size(F). Now, recursively obtaining small-depth formulas for f_1, f_0, we obtain a $O(\log s)$ depth formula computing f.

However, the above construction does not necessarily preserve the interval property, since the intervals of variables on which f_0, f_1 and g are defined, can be overlapping. We overcome this problem by expressing f_0, f_1 as products of polynomials over disjoint intervals, each of the intervals being disjoint to the interval corresponding to g.

We assume, without loss of generality, that the interval formula F corresponds to the interval $[1, n]$. Let the interval corresponding to g be $I_g = [i, j]$, $i < j$. Now, by definition of f_1 and f_0, they are defined on the same interval of variables. We consider the intervals I_0, I_1 such that $I_0 \cup I_1 = [1, n] \setminus [i, j]$, $I_0 = [j+1, n]$ and $I_1 = [1, i-1]$. We express both f_0, f_1 as products of two polynomials on

I_0 and I_1 respectively. As f_1 and g are multiplicatively related in F, we show that $f_1 = f_{1,1} \times f_{1,0}$ where $f_{1,1}$ is a polynomial on the interval I_1 and $f_{1,0}$ is a polynomial on the interval I_0.

We consider the root to leaf (g) path ρ in the original formula F containing the node g. All the paths meeting ρ at a sum gate represent polynomials additively related to y i.e., contributing towards the computation of f_0 and not f_1. For f_1, we will analyze only the paths meeting ρ at product gates. Let us consider a product gate on ρ computing $h_1 \times h_2$, such that h_2 lies on ρ. Since I is contained in the interval corresponding to h_2, the interval corresponding to h_1, I_{h_1} must be either fully contained in I_1 or I_0.

Constructing an Interval Formula for f_1: We ignore all sum gates on ρ computing $p_1 + p_2$, with p_2 on ρ, by substituting p_1 to zero. The resulting formula is F'. In any product gate computing $h_1 \times h_2$, where h_2 is on ρ, if $I_{h_1} \subset I_0$, we substitute h_1 by 1. We also substitute g by 1. The remaining formula F_1' computes the polynomial $f^{(1)}$.

We repeat this process above, but this time, we substitute h_1 by 1 only when $I_{h_1} \subset I_1$. This remaining formula F_2' computes $f^{(2)}$. By definition of f_1, $f_1 = f^{(1)} \cdot f^{(2)}$. The interval corresponding to F_1' is contained in I_1, the interval corresponding to F_2' is contained in I_0.

Constructing an Interval Formula for f_0: We ignore all product gates on ρ computing $h_1 \times h_2$, with h_2 on ρ, by substituting h_1 by 1. The resulting formula is $\hat{F}$.

In any sum gate computing $p_1 + p_2$, where p_2 is on ρ, if $I_{p_1} \subset I_0$, we substitute p_1 by 0. We also substitute g by 0. The remaining formula $\hat{F}_1$ computes the polynomial $p^{(1)}$.

We repeat this process from the beginning, but substitute p_1 by 0 only when $I_{p_1} \subset I_1$. This remaining formula $\hat{F}_2$ computes $p^{(2)}$. By definition of f_0, $f_0 = p^{(1)} + p^{(2)}$. The interval corresponding to $\hat{F}_1$ is contained in I_1, the interval corresponding to $\hat{F}_2$ is contained in I_0.

Hence, we obtain $f = f^{(1)} f^{(2)} g + p^{(1)} + p^{(2)}$. The recursive relation for calculating depth is as follows: $\text{depth}(F) = \text{depth}(g) + 2 \implies \text{depth}(s) = \text{depth}(2s/3) + 2$, which yields a total depth of $O(\log s)$ for F.

References

1. Agrawal, M., Vinay, V.: Arithmetic circuits: a chasm at depth four. In: 49th Annual IEEE Symposium on Foundations of Computer Science, FOCS 2008, 25–28 October 2008, Philadelphia, PA, USA, pages 67–75. IEEE Computer Society (2008). https://doi.org/10.1109/FOCS.2008.32
2. Alon, N., Kumar, M., Volk, B.L.: Unbalancing sets and an almost quadratic lower bound for syntactically multilinear arithmetic circuits. Comb. **40**(2):149–178 (2020). https://doi.org/10.1007/s00493-019-4009-0

3. Anderson, M., van Melkebeek, D., Volkovich, I.: Derandomizing polynomial identity testing for multilinear constant-read formulae. In: Proceedings of the 26th Annual IEEE Conference on Computational Complexity, CCC 2011, San Jose, California, USA, 8–10 June 2011, pp. 273–282. IEEE Computer Society (2011). https://doi.org/10.1109/CCC.2011.18

4. Arvind, V., Raja, S.: Some lower bound results for set-multilinear arithmetic computations. Chicago J. Theor. Comput. Sci. (2016). http://cjtcs.cs.uchicago.edu/articles/2016/6/contents.html

5. Baur, W., Strassen, V.: The complexity of partial derivatives. Theor. Comput. Sci. **22**, 317–330 (1983). https://doi.org/10.1016/0304-3975(83)90110-X

6. Ben-Or, M., Cleve, R.: Computing algebraic formulas using a constant number of registers. SIAM J. Comput. **21**(1), 54–58 (1992). https://doi.org/10.1137/0221006

7. Biedl, T.C., Demaine, E.D., Duncan, R.C., Fleischer, A., Kobourov, S.: Tight bounds on maximal and maximum matchings. Discret. Math. **285**(1–3), 7–15 (2004). https://doi.org/10.1016/j.disc.2004.05.003

8. Brent, R.P.: The parallel evaluation of general arithmetic expressions. J. ACM **21**(2), 201–206 (1974). https://doi.org/10.1145/321812.321815

9. Bürgisser, P.: Completeness and reduction in algebraic complexity theory. Algorithms and Computation in Mathematics, vol. 1. Springer, Heidelberg (2000). https://doi.org/10.1007/978-3-662-04179-6

10. Chatterjee, P., Kumar, M., She, A., Volk, B.L.: A quadratic lower bound for algebraic branching programs. In: Saraf, S. (eds.) 35th Computational Complexity Conference, CCC 2020, 28–31 July 2020, Saarbrücken, Germany (Virtual Conference). LIPIcs, vol. 169, pp. 2:1–2:21. Schloss Dagstuhl - Leibniz-Zentrum für Informatik (2020). https://doi.org/10.4230/LIPIcs.CCC.2020.2

11. Chillara, S., Engels, C., Limaye, N., Srinivasan, A near-optimal depth-hierarchy theorem for small-depth multilinear circuits. In: Thorup, M., (eds.) 59th IEEE Annual Symposium on Foundations of Computer Science, FOCS 2018, Paris, France, 7–9 October 2018, pp. 934–945. IEEE Computer Society (2018). https://doi.org/10.1109/FOCS.2018.00092

12. Chung, F.R.K.: Separator theorems and their applications. Forschungsinst. für Diskrete Mathematik (1989). http://www.math.ucsd.edu/~fan/mypaps/fanpap/117separatorthms.pdf

13. Dvir, Z., Malod, G., Perifel, S., Yehudayoff, A.: Separating multilinear branching programs and formulas. In: Karloff, H.J., Pitassi, T. (eds.) Proceedings of the 44th Symposium on Theory of Computing Conference, STOC 2012, New York, NY, USA, 19–22 May 2012, pp. 615–624. ACM (2012). https://doi.org/10.1145/2213977.2214034

14. Grigoriev, D., Karpinski, M.: An exponential lower bound for depth 3 arithmetic circuits. In: Vitter, J.S. (eds.) Proceedings of the Thirtieth Annual ACM Symposium on the Theory of Computing, Dallas, Texas, USA, 23–26 May 1998, pp. 577–582. ACM (1998). https://doi.org/10.1145/276698.276872

15. Grigoriev, D., Razborov, A.A.: Exponential lower bounds for depth 3 arithmetic circuits in algebras of functions over finite fields. Appl. Algebra Eng. Commun. Comput. **10**(6), 465–487 (2000). https://doi.org/10.1007/s002009900021

16. Jansen, M.J.: Lower bounds for syntactically multilinear algebraic branching programs. In: Ochmański, E., Tyszkiewicz, J. (eds.) MFCS 2008. LNCS, vol. 5162, pp. 407–418. Springer, Heidelberg (2008). https://doi.org/10.1007/978-3-540-85238-4_33

17. Jansen, M.J., Qiao, Y., Sarma, J.: Deterministic black-box identity testing pi-ordered algebraic branching programs. In: IARCS Annual Conference on Foundations of Software Technology and Theoretical Computer Science, FSTTCS 2010, 15–18 December 2010, Chennai, India, pp. 296–307 (2010). https://doi.org/10.4230/LIPIcs.FSTTCS.2010.296

18. Jerrum, M., Snir, M.: Some exact complexity results for straight-line computations over semirings. J. ACM **29**(3), 874–897 (1982). https://doi.org/10.1145/322326.322341

19. Kayal, N., Nair, V., Saha, C.: Separation between read-once oblivious algebraic branching programs (roabps) and multilinear depth three circuits. In: Ollinger, N., Vollmer, H. (eds.) 33rd Symposium on Theoretical Aspects of Computer Science, STACS 2016, 17–20 February 2016, Orléans, France, volume 47 of LIPIcs, pp. 46:1–46:15. Schloss Dagstuhl - Leibniz-Zentrum für Informatik (2016). https://doi.org/10.4230/LIPIcs.STACS.2016.46

20. Kayal, N., Saha, C., Tavenas, S.: An almost cubic lower bound for depth three arithmetic circuits. In: Electronic Colloquium on Computational Complexity (ECCC), vol. 23, no. 6 (2016). http://eccc.hpi-web.de/report/2016/006

21. Kumar, M.: A quadratic lower bound for homogeneous algebraic branching programs. Comput. Complex. **28**(3), 409–435 (2019). https://doi.org/10.1007/s00037-019-00186-3

22. Mahajan, M., Tawari, A.: Sums of read-once formulas: How many summands are necessary? Theor. Comput. Sci. **708**, 34–45 (2018). https://doi.org/10.1016/j.tcs.2017.10.019

23. Malod, G., Portier, N.: Characterizing valiant's algebraic complexity classes. J. Complex. **24**(1), 16–38 (2008). https://doi.org/10.1016/j.jco.2006.09.006

24. Minahan, D., Volkovich, I.: Complete derandomization of identity testing and reconstruction of read-once formulas. TOCT **10**(3), 10:1–10:11 (2018). https://doi.org/10.1145/3196836

25. Mitzenmacher, M., Upfal, E.: Probability and Computing: Randomized Algorithms and Probabilistic Analysis. Cambridge University Press, Cambridge (2005). https://doi.org/10.1017/CBO9780511813603

26. Nisan, N.: Lower bounds for non-commutative computation (extended abstract). In: Koutsougeras, C., Vitter, J.S. (eds.) Proceedings of the 23rd Annual ACM Symposium on Theory of Computing, New Orleans, Louisiana, USA, 5–8 May 1991, pp. 410–418. ACM (1991). https://doi.org/10.1145/103418.103462

27. Ramya, C., Rao, B.V.R.: Lower bounds for special cases of syntactic multilinear ABPs. In: Wang, L., Zhu, D. (eds.) COCOON 2018. LNCS, vol. 10976, pp. 701–712. Springer, Cham (2018). https://doi.org/10.1007/978-3-319-94776-1_58

28. Ramya, C., Raghavendra Rao, B.V.: Lower bounds for multilinear order-restricted ABPs. In: Rossmanith, P., Heggernes, P., Katoen, J.-P . (eds.) 44th International Symposium on Mathematical Foundations of Computer Science, MFCS 2019, Aachen, Germany, 26–30 August 2019. LIPIcs, vol. 138, pp. 52:1–52:14. Schloss Dagstuhl - Leibniz-Zentrum für Informatik (2019). https://doi.org/10.4230/LIPIcs.MFCS.2019.52

29. Ramya, C., Raghavendra Rao, B.V.: Lower bounds for sum and sum of products of read-once formulas. TOCT **11**(2), 10:1–10:27 (2019.) https://doi.org/10.1145/3313232

30. Raz, R.: Separation of multilinear circuit and formula size. Theory Comput. **2**(6), 121–135 (2006). https://doi.org/10.4086/toc.2006.v002a006

31. Raz, R.: Multi-linear formulas for permanent and determinant are of super-polynomial size. J. ACM **56**(2), 8:1–8:17 (2009). https://doi.org/10.1145/1502793. 1502797
32. Raz, R., Yehudayoff, A.: Balancing syntactically multilinear arithmetic circuits. Comput. Complex. **17**(4), 515–535 (2008). https://doi.org/10.1007/s00037-008-0254-0
33. Saptharishi, R., Chillara, S., Kumar, M.: A survey of lower bounds in arithmetic circuit complexity. Technical report (2016). https://github.com/dasarpmar/lowerbounds-survey/releases
34. Shpilka, A., Wigderson, A.: Depth-3 arithmetic circuits over fields of characteristic zero. Comput. Complex. **10**(1), 1–27 (2001). https://doi.org/10.1007/PL00001609. https://doi.org/10.1007/PL00001609
35. Shpilka, A., Yehudayoff, A.: Arithmetic circuits: a survey of recent results and open questions. Foundations and Trends Theoret. Comput. Sci. **5**(3–4), 207–388 (2010). https://doi.org/10.1561/0400000039. https://doi.org/10.1561/0400000039
36. Srinivasan, S.: Strongly exponential separation between monotone VP and monotone VNP. CoRR, abs/1903.01630 (2019). http://arxiv.org/abs/1903.01630
37. Toda, S.: Classes of arithmetic circuits capturing the complexity of computing the determinant. IEICE Trans. Inf. Syst. **75**(1), 116–124 (1992)
38. Valiant, L.G.: The complexity of computing the permanent. Theor. Comput. Sci. **8**, 189–201 (1979). https://doi.org/10.1016/0304-3975(79)90044-6
39. Volkovich, I.: Characterizing arithmetic read-once formulae. TOCT **8**(1), 2:1–2:19 (2016). https://doi.org/10.1145/2858783
40. Yehudayoff, A.: Separating monotone VP and VNP. In: Charikar, M., Cohen, E. (eds.) Proceedings of the 51st Annual ACM SIGACT Symposium on Theory of Computing, STOC 2019, Phoenix, AZ, USA, 23–26 June 2019, pp. 425–429. ACM (2019). https://doi.org/10.1145/3313276.3316311

On the Computational Complexity
of Reaction Systems, Revisited

Markus Holzer$^{(\boxtimes)}$ and Christian Rauch

Institut für Informatik, Universität Giessen, Arndtstr. 2, 35392 Giessen, Germany
{holzer,christian.rauch}@informatik.uni-giessen.de

Abstract. We study the computational complexity of some important problems on reaction systems (RSs), a biologically motivated model introduced by Ehrenfeucht and Rozenberg in [7], that were overseen in the literature. To this end we focus on the complexity of (i) equivalence and multi-step simulation properties, (ii) special structural and behavioural RS properties such as, e.g., isotonicity, antitonicity, etc., and minimality with respect to reactant and/or inhibitor sets, and (iii) threshold properties. The complexities vary from deterministic polynomial time solvability to coNP- and PSPACE-completeness. Finally, as a side result on the complexity of threshold problems we improve the previously known threshold values for the no-concurrency, the comparability, and the redundancy property studied in [2].

Keywords: Reaction system · Equivalence · (multi-step) simulation · Minimality · Threshold property · Computational complexity

1 Introduction

Reaction systems (RSs) are a biologically inspired novel model of interactive computations, introduced by Ehrenfeucht and Rozenberg in [7]. A RS is based on the idea of interactions on biochemical reactions, whose underlying mechanism is that of facilitation and inhibition. A reaction consists of a set of reactants needed for the reaction to take place, a set of inhibitors which forbids the reaction to take place and a set of products produced by the reaction. Here it is assumed that reactions do not compete with each other, because whenever a resource is available, then it is present in sufficient amounts. This is a significant difference to other concurrent models like, e.g., Petri nets. Then the dynamical behaviour of a RS is given by the product of all reactions that are applicable to a certain state of the system. Since their introduction RSs are studied from at least three perspectives, namely the biological, the computational, and the computational complexity one, see, e.g., [1–3,11] to mention only a few papers that are relevant for this research. Here we focus on the computational complexity view on RSs by studying problems that were either overseen in the literature or left open up to our knowledge.

© Springer Nature Switzerland AG 2021
R. Santhanam and D. Musatov (Eds.): CSR 2021, LNCS 12730, pp. 170–185, 2021.
https://doi.org/10.1007/978-3-030-79416-3_10

There is a vast amount on computational complexity results for RSs. By thematically grouping these results we find reachability problems, preimage and image problems, or more generally ancestor, descendants, and the famous Garden of Eden problem, and fixed points, attractors, and cycle problems. For a comprehensive list of some of these problems we refer to, e.g., [3]. The computational complexity of these problems ranges from tractable to coNP- and PSPACE-complete problems, while only for a handful of problem non-matching upper and lower bounds are known.

During our study of the respective literature on computational complexity results for RSs we realized that at least three different and important problem areas were not covered up to now. These problem areas are equivalence and multi-step simulation problems [8,13], minimality problems w.r.t. reactant and/or inhibitor sets [4,13], and problems on extremal combinatorics [2,5]. A list of the obtained results for problems of the above mentioned areas is given next—for a definition of some of the mentioned properties we refer to the preliminaries section or the subsection that deals with the specific property. For multi-step simulations we prove the following results:

- Deciding whether for two given RSs A and B there exists a k such that each single step of A can be simulated by k-steps of B is PSPACE-hard (Theorem 2) and contained in deterministic exponential space (Theorem 3). If the parameter k is part of the input, then the multi-step simulation problem is is coNP-complete, if k is given in unary, and PSPACE-complete for binary encoded k (Theorems 4 and 5).

The proof techniques developed for the multi-step simulation problems allows us to show that the k-ancestor problem is highly computational intractable, solving a left open problem in [3]:

- Deciding whether for a given RS A, a state T, and a binary integer k to decide whether T has a k-ancestor in A is PSPACE-complete (Theorem 6); this solves a left open problem, because for unary given k it is known that this problem is NP-complete [3].

For certain RS properties we obtain the following results, which also cover minimality problems—roughly speaking a RS is said to be reactant (inhibitor, resource, respectively) minimal if the reactant (inhibitor, union of reactant and inhibitor) sets contain at most one element:

- Deciding the following properties of RS is coNP-complete: isotonicity, antitoninicty, union (sub-)additiveness, and intersection (sub-)additiveness (Theorems 8 and 9). As a direct consequence on the completeness results for union and intersection (sub-)additiveness one obtains that deciding whether for a given RS A there is an equivalent reactant (inhibitor, resource, respectively) minimal system is coNP-complete, too (Corollary 10).

Finally, for the threshold properties the big picture is more diverse. In most cases polynomial solvability is shown, but nevertheless a few properties turn out to be intractable:

– The following threshold properties on the dynamic behaviour of RSs can be solved in deterministic polynomial time: minimal concurrency (Theorem 11), no-concurrency (Corollary 15), inhibitor property (Theorem 17), and decomposability (Theorem 20). In contrast, the always-parallel property and its generalization (Theorem 12 and Corollary 13) turn out to be intractable, that is, coNP-complete. The redundancy property is shown to be contained in P^{NP} and we have to leave open to prove a lower bound.

As a side result, we improve the threshold values for a few properties. To be more specific: the previously claimed threshold in [2, Proposition 11] that was based on a falsely claimed equivalence for the no-concurrency (NC) property is fixed to $(3^n - n) \cdot 2^n + 1$ (Theorem 16), for the comparability (COMP) property the bound is improved from $\frac{n! 3^n}{\lceil \frac{n}{2} \rceil! \lfloor \frac{n}{2} \rfloor!}$ to $(3^n - 2) \cdot 2^n + 1$ (Theorem 18), and for the redundancy (RED) property the first explicit bound of $2^n \cdot n + 1$ (Theorem 19), where n is the size of the background set, is given—compare with [2]. Due to space constraints almost all proofs are omitted; they can be found in the full version of this paper.

2 Preliminaries

We assume the reader to be familiar with the basics in computational complexity theory [10]. In particular we recall the inclusion chain: $P \subseteq NP \subseteq PSPACE$. Here P (NP, respectively) denotes the class of problems solvable by deterministic (nondeterministic, respectively) Turing machines in polytime, and PSPACE refers to the class of languages accepted by deterministic or nondeterministic Turing machines in polynomial space [14]. As usual, the prefix co refers to the complement class. For instance, coNP is the class of problems that are complements of NP problems. Completeness and hardness are always meant with respect to deterministic many-one polytime reducibilities ($\leq_m^{poly}$) unless otherwise stated. Throughout the paper we assume that the input given to the Turing machine is encoded by a function $\langle \cdot \rangle$, such that the necessary parameters can be decoded in reasonable time and space.

Let S be a finite set. A *reaction over* S is a triple $a = (R_a, I_a, P_a)$, where R_a, I_a, and P_a are subsets of S such that $R_a \cap I_a = \emptyset$. We call R_a (I_a, P_a, respectively) the set of *reactants* (*inhibitors*, *products*, respectively). If both R_a and I_a are nonempty we refer to a as a *strict* reaction. In case only R_a (I_a, respectively) is nonempty the reaction is said to be *reactant-strict* (*inhibitor-strict*, respectively). A subset T of S, i.e., $T \subseteq S$, is said to be a *state*. For any state T and any reaction a, we say that reaction a is *enabled* in T, if $R_a \subseteq T$ and $I_a \cap T = \emptyset$. The *T-activity of a set of reactions A*, referred to as

$$en_A(T) = \{\, a \in A \mid a \text{ is enabled in } T \,\},$$

is the set of all reactions of A enabled by T. The *result* $res_a(T)$ *of a reaction a on a set* $T \subseteq S$ is defined as

$$res_a(T) = \begin{cases} P_a, & \text{if } a \text{ is enabled by } T \\ \emptyset, & \text{otherwise.} \end{cases}$$

This notion naturally extends to sets of reactions. The *result of a set of reactions* A *on* $T \subseteq S$ is

$$res_A(T) = \bigcup_{a \in A} res_a(T).$$

Now we are ready to define reactions systems. A *reaction system* (RS) is a pair $\mathcal{A} = (S, A)$, where S is a finite set of symbols, called the *background* set, and A consists of a finite number of reactions over S. Then the *result of* $\mathcal{A}$ *on a state* $T \subseteq S$ is $res_{\mathcal{A}}(T) = res_A(T)$. In case all reactions in A are strict, we call the RS $\mathcal{A}$ a *strict* reaction system.

In order to clarify the notation we give a small example, which we literally take with slight adaptions from [7].

Example 1. Let $\mathcal{A} = (S, A)$ be a RS defined as follows: set $S = \{x_1, x_2, \ldots, x_n\}$ and let A contain the following four different types of reactions:

$$
\begin{aligned}
&(\emptyset, S, \{x_1\}),\\
&(\{x_i\}, \{x_1\}, \{x_1\}) && \text{for } 2 \le i \le n,\\
&(\{x_1, x_2, \ldots, x_{i-1}\}, \{x_i\}, \{x_i\}) && \text{for } 2 \le i \le n,
\end{aligned}
$$

and

$$(\{x_j\}, \{x_i\}, \{x_j\}) \qquad\qquad \text{for } 1 \le i < j \le n.$$

Then $\mathcal{A}$ implements a binary counter, where the subsets T of S define binary numbers: if $x_i \in T$, for $1 \le i \le n$, then the binary number represented by T has a 1 on position $i - 1$; otherwise it has a 0. For instance, in case $n = 4$, the subset $T = \{x_1, x_3\}$ represents the binary number 0101.

The behaviour of the RS is seen as follows: the first reaction starts the counting from number zero, the reactions of the second type (third type, respectively) perform adding 1 to an even (odd, resp) number, while the reactions of the last type sustain the bits x_j that are not affected by carry over results by the addition, and finally the counting process restarts with the smallest number 0 whenever the largest binary number with n times 1 is reached because all reactions are disabled. The reader may easily verify that $res_{\mathcal{A}}(T) = \{x_2, x_3\}$ and $res_{\mathcal{A}}^2(T) = \{x_1, x_2, x_3\}$, where $res_{\mathcal{A}}^2$ refers to the 2-times application of the result function. Hence the k-folded application of the result function runs through all states of the RS in the natural order of binary numbers and restarts the counting with 0 after reaching the largest value.   □

3   Results

We consider the computational complexity of some problems on RSs that can be thematically grouped as follows: (i) equivalence and multi-step simulation problems, (ii) special properties on RSs such as, e.g., isotonicity, antitonicity, etc., and minimality problems, and finally (iii) threshold properties. In most cases we can obtain precise complexity results in terms of completeness statements. The complexity varies form tractable to coNP- and PSPACE-completeness. Moreover, as side results we solve the open problem on the exact complexity of the k-ancestor problem for RSs with binary k [3] and improve and fix the threshold for the no-concurrence property as studied in [2]. We start our investigation with equivalence and multi-step simulation problems.

3.1   The Computational Complexity of Equivalence and Multi-step Simulation

In this subsection we consider the computational complexity of equivalence and simulation of RSs. Two RSs $\mathcal{A}$ and $\mathcal{B}$ are said to be *equivalent* if $res_{\mathcal{A}}(T) = res_{\mathcal{B}}(T)$, for every $T \subseteq S$, where S is the background set of $\mathcal{A}$ and $\mathcal{B}$. The equivalence problem was shown in [6] to be coNP-complete. Up to our knowledge the complexity of multi-step simulation is not known yet. The following definitions were first given in [8]. The basic idea behind multi-step simulation is that the simulating system may use several steps to simulate a single step of the original system. Let k be a natural number and $\mathcal{A} = (S_A, A)$ and $\mathcal{B} = (S_B, B)$ be two RSs with $S_A \subseteq S_B$. Then the RS $\mathcal{A}$ is k-*simulated by* $\mathcal{B}$, if for every $T \subseteq S_A$ we have

$$res_{\mathcal{A}}(T) = res_{\mathcal{B}}^{k}(T) \cap S_A.$$

In this case we simply write $\mathcal{A} \preceq_k \mathcal{B}$. This means that when considering the sequence of states of $\mathcal{A}$ and $\mathcal{B}$ starting from T, then the successor of T in $\mathcal{A}$ coincides with the kth successor of T in $\mathcal{B}$ w.r.t. the elements of S_A, where auxiliary elements of $S_B \setminus S_A$ may also occur. Finally, we say $\mathcal{A}$ is *simulated* by $\mathcal{B}$, for short $\mathcal{A} \preceq \mathcal{B}$, if there is a k such that $\mathcal{A} \preceq_k \mathcal{B}$.

Obviously, 1-simulation and ordinary equivalence coincide, but for k in general this is not the case anymore. What about the computational complexity of k-simulation and simulation? To our knowledge these problems were not studied from a complexity perspective yet. First we consider the simulation problem, where we ask for a k such that the given RSs $\mathcal{A}$ is k-simulated by the given other RS $\mathcal{B}$. The next theorem shows that this is already a highly intractable problem, since it turns out to be PSPACE-hard.

Theorem 2. *The problem for two given RSs $\mathcal{A}$ and $\mathcal{B}$, to decide whether $\mathcal{A} \preceq \mathcal{B}$, asking for the existence of a k with $k \geq 1$ such that $\mathcal{A} \preceq_k \mathcal{B}$, is* PSPACE-*hard.*

Proof. We reduce the PSPACE-complete reachability problem for RSs [3] to the problem in question. Let RS $\mathcal{C} = (S, C)$ and two states T and U with $T, U \subseteq S$ be an instance of the reachability problem, i.e., the question whether T leads to U in $\mathcal{C}$.

Let s be a symbol not contained in S. First we define the RS

$$\mathcal{A} = (\{s\}, \{((\{s\}, \emptyset, \{s\})\})$$

that acts as the identity on all states—in fact there are only two state sets, namely the empty set $\emptyset$ and the background set $\{s\}$. The idea behind the construction of the second RS $\mathcal{B}$ is that if it is started on the state $\{s\}$, it then verifies the reachability question of T to U for the RS $\mathcal{C}$. In case the answer is "yes," it generates the state $\{s\}$; otherwise it produces the empty set $\emptyset$. To this end let $\mathcal{B} = (\{s\} \cup S, B)$, and the set of reactions B contains the following reactions that implement the above described idea: to start the simulation of the reachability the reaction

$$(\{s\}, \emptyset, T)$$

is used, which introduces the elements of T to the state. This reaction is the only one that can be applied to the state set $\{s\}$. The actual simulation of $\mathcal{C}$ is done by the reactions

$$(R_a, I_a \cup \{s\}, P_a), \qquad \text{for every reaction } a = (R_a, I_a, P_a) \text{ of } \mathcal{C}.$$

In case U is reached the reaction

$$(U, (S \setminus U) \cup \{s\}, \{s\})$$

can be applied and the original element $\{s\}$ is produced. Because to a state set containing s only the first reaction is applicable, all other elements from the background set, are not replicated. Thus, by construction, state set T leads to U in $\mathcal{C}$ if and only if state set $\{s\}$ belongs to the result of the function $res_{\mathcal{B}}^k(\{s\})$, for some k. Therefore, $\mathcal{A}$ is k-simulated by $\mathcal{B}$, for some k, if and only if the input $\langle \mathcal{C}, T, U \rangle$ is a positive instance of reachability. Note, that in both RSs $\mathcal{A}$ and $\mathcal{B}$ the empty state set is mapped to itself and has no effect on the k-simulation. This proves PSPACE-hardness, because the RSs $\mathcal{A}$ and $\mathcal{B}$ can be constructed in deterministic logspace from $\mathcal{C}$, T, and U. □

Now the question arises, whether the simulation problem of $\mathcal{A}$ by $\mathcal{B}$ as described in the above theorem belongs to PSPACE. Although one could cycle through all states in polynomial space, it still remains to verify that for every state T of $\mathcal{A}$ the equality $res_{\mathcal{A}}(T) = res_{\mathcal{B}}^k(T)$ holds, and thus depends heavily on k. In case k is exponentially bounded, this check can be performed in polynomial space and thus the simulation is solvable in PSPACE, but as we will see, we can give a double exponential upper and lower bound for k. Thus, the naive algorithm does not suffice to prove containment within PSPACE.

For the lower bound we design the following two RSs. Let $n \geq 1$ and $X = \{x_1, x_2, \ldots, x_n\}$. Then let $\mathcal{A} = (S_A, A)$, where A contains the reactions

$$(\{x_i\}, \emptyset, \{x_i\}) \qquad \text{for } 1 \leq i \leq n.$$

Obviously, $\mathcal{A}$ acts as the identity on its states. The construction of the RS $\mathcal{B}$ that k-simulates $\mathcal{A}$ is more involved. Let $Y = \{y_1, y_2, \ldots, y_n\}$ and $Z = \{z_1, z_2, \ldots, z_n\}$ be such that the set X, Y, and Z are pairwise disjoint. Then define the system $\mathcal{B} = (X \cup Y \cup Z, B)$ and the set of reactions B contains the following actions:

$$(\{x_i\}, \emptyset, \{y_i, z_i\}) \qquad\qquad \text{for } 1 \le i \le n$$

and

$$(\{y_i\}, X \cup \{y_1\}, \{y_1\}) \qquad\qquad \text{for } 2 \le i \le n,$$
$$(\{y_1, y_2, \dots, y_{i-1}\}, X \cup \{y_i, \}, \{y_i\}) \qquad\qquad \text{for } 2 \le i \le n,$$

and

$$(\{y_j\}, X \cup \{y_i\}, \{y_j\}) \qquad\qquad \text{for } 1 \le i < j \le n,$$

together with

$$(\{z_i\}, \{y_j\}, \{z_i\}) \qquad\qquad \text{for } 1 \le i, j \le n,$$

and

$$(\{y_1, y_2, \dots, y_n\} \cup \{z_i\}, \emptyset, \{x_i\}) \qquad\qquad \text{for } 1 \le i \le n.$$

Then $\mathcal{B}$ behaves on states $T \subseteq X$ as follows. We distinguish two cases:

1. If $T = \emptyset$, then no reaction is enabled, since every reaction has a nonempty reactant. So T is mapped onto itself, i.e., $res_{\mathcal{B}}^k(\emptyset) = \emptyset$, for every $k \ge 1$.
2. In case T is non-empty at least one x_j, for $1 \le j \le n$, is present. Thus the reactions of the first type introduce corresponding y_i and z_i, for every x_i in T. The following three types of reactions that implement a counting routine on the y_j's are not enabled since at least one x_j belongs to T. Moreover, all x_i from T disappear. This leads us to the state $\{ y_i, z_i \mid x_i \in T \}$. Then the counting rules on the y_i's are enabled and all the z_j are copied to the next state by the next to the last type of reaction. Whenever the counting procedure on the y_j's reaches a state that contains all y_i, for $1 \le i \le n$, then all the z_j's are replaced by x_j due to the last type of reaction and all y_i disappear since no rule of the counting procedure is applicable anymore. Hence the original state T is reached after $2^n - \ell + 2$ steps from T, where ℓ is the number represented by the set T as described in Example 1. Thus, we have found that T leads to itself in $2^n - \ell + 2$ steps, i.e., $res_{\mathcal{B}}^{2^n - \ell + 2}(T) = T$, if T is non-empty, and $res_{\mathcal{B}}^k(T) \cap X = \emptyset$, for every $1 \le k < 2^n - \ell + 2$.

Thus, in order to k-simulate the RS $\mathcal{A}$ by $\mathcal{B}$ we have to find a common k such that

$$T = res_{\mathcal{B}}^k(T) \cap X, \quad \text{for every } T \subseteq X,$$

because $res_{\mathcal{A}}(T) = T$. Since this is achieved if all of $\mathcal{B}$'s cycles synchronize appropriately. Note that by construction we have the cycle lengths 1 up to 2^n. Since the synchronization happens the first time, if we choose $k = \text{lcm}\{1, 2, \dots, 2^n + 1\}$, the double exponential lower bound follows, because $\text{lcm}\{1, 2, \dots, n\} \ge 2^n$, for $n \ge 7$, see, e.g., [9].

For the double exponential upper bound we argue as follows: observe, that for any state T of the RS $\mathcal{A}$ we find k_T and ℓ_T such that

$$res_{\mathcal{A}}(T) = res_{\mathcal{B}}^{k_T + i_T \cdot \ell_T}(T), \quad \text{for } i_T \ge 1,$$

where both k_T and ℓ_T are bounded exponentially in $\mathcal{B}$'s background set. Thus, in order to fulfill the k-simulation constraint $res_{\mathcal{A}}(T) = res_{\mathcal{B}}^k(T)$, the equations

$$k_T + i_T \cdot \ell_T = k \quad \text{or equivalently} \quad i_T \cdot \ell_T = k - k_T$$

in the unknowns i_T and k, for every $T \subseteq X$, have to be fulfilled, under the additional constraints that $i_T \geq 0$ and $k \geq \max\{\, k_T \mid T \subseteq X \,\}$. Thus, the right hand side of the equations are always non-negative. If we find a solution in i_T's and k satisfying the constraints and for every right hand-side $k - k_T$ is larger than $\ell = \mathrm{lcm}\{\, \ell_T \mid T \subseteq X \}$, then one can find another solution with a smaller k. To see, this, consider one particular equation $i_T \cdot \ell_T = k - k_T$ and subtract ℓ from both sides. Then

$$
\begin{aligned}
i_T \cdot \ell_T - \ell = k - k_T - \ell &\iff i_T \cdot \ell_T - \tfrac{\ell}{\ell_T} \cdot \ell_T = k - k_T - \ell \\
&\iff (i_T - \tfrac{\ell}{\ell_T}) \cdot \ell_T = (k - \ell) - k_T \\
&\iff i_T' \cdot \ell_T = k' - k_T,
\end{aligned}
$$

where $i_T' = (i_T - \tfrac{\ell}{\ell_T})$ and $k' = (k - \ell)$. Thus, the solution for k is at most ℓ. Let $n = |X|$. Since $\mathrm{lcm}\{\, \ell_T \mid T \subseteq X \,\} \leq \mathrm{lcm}\{1, 2, \ldots, 2^n\}$ and $\mathrm{lcm}\{1, 2, \ldots, n\} \leq e^{2n}$, which is a rough estimate and a direct consequence of the prime number theorem, the double exponential upper bound immediately follows, too.

A direct consequence of the double exponential upper bound on the k-simulation parameter k is that the naive algorithm as described above can be implement on a deterministic exponential space bounded Turing machine. The corresponding complexity class is referred to as $\mathsf{EXPSPACE} := \mathsf{DSPACE}(2^{\mathsf{poly}}) = \mathsf{NSPACE}(2^{\mathsf{poly}})$. Thus, we we have shown:

Theorem 3. *The problem for two given RSs $\mathcal{A}$ and $\mathcal{B}$, to decide whether $\mathcal{A} \preceq \mathcal{B}$, that is, asking for the existence of a k with $k \geq 1$ such that $\mathcal{A} \preceq_k \mathcal{B}$, belongs to* $\mathsf{EXPSPACE}$. $\qquad\square$

We have to leave open the exact computational complexity of the simulation problem. If the simulation parameter k is fixed or given as part of the input, we can determine its exact complexity. The problem is contained in coNP since one can universally guess a state T and deterministically verifies that $res_{\mathcal{A}}(T) = res_{\mathcal{B}}^k(T)$ by simply simulating one step of $\mathcal{A}$ and k steps of $\mathcal{B}$. Moreover, since 1-simulation coincides with ordinary equality the coNP lower bound follows from the coNP-completeness of the equivalence problem. Thus, we summarize:

Theorem 4. *The problem for two given RSs $\mathcal{A}$ and $\mathcal{B}$ to decide whether $\mathcal{A} \preceq_k \mathcal{B}$ holds, if k is fixed or even if k is part of the input and encoded in unary, is coNP-complete.* $\qquad\square$

The next theorem shows that if k is given in binary for the k-simulation problem, it becomes PSPACE-complete.

Theorem 5. *The problem for two given RSs $\mathcal{A}$ and $\mathcal{B}$ and a natural number k encoded in binary, to decide whether $\mathcal{A} \preceq_k \mathcal{B}$ holds, is PSPACE-complete.*

As a direct consequence of the previous proof, we can solve the k-ancestor problem for binary given k. The k-ancestor problem is defined as follows: given a RS $\mathcal{A} = (S, A)$, a state $T \subseteq S$, and an integer k as input, is there a state R such that $T = res_{\mathcal{A}}^k(R)$? For k given in unary this problem was shown to be NP-complete [3], but for binary given k the complexity status of this problem was left open.

Theorem 6. *Given a RS $\mathcal{A} = (S, A)$, a state $T \subseteq S$, and a binary integer k as input, it is* PSPACE-*complete to decide if T has a k-ancestor.*

3.2 The Complexity of Special Reaction System Properties

This subsection is devoted to the complexity of some properties that appear in the literature and to some further results on formal language theoretical issues. The first four properties which we are interested in are defined as follows: let $\mathcal{A} = (S, A)$ be a RS. Then $\mathcal{A}$ is said to be *focused*, if there exists a $s \in S$ such that for every $T \subseteq S$, if $res_{\mathcal{A}}(T)$ is nonempty, then $res_{\mathcal{A}}(T) = \{s\}$, *constant*, if there is a $C \subseteq S$ such that $res_{\mathcal{A}}(T) = C$, for every $T \subseteq S$, *isotone* if $T_1 \subseteq T_2$ implies $res_{\mathcal{A}}(T_1) \subseteq res_{\mathcal{A}}(T_2)$, for every subsets T_1 and T_2 of S, and *antitone*, if $T_1 \subseteq T_2$ implies $res_{\mathcal{A}}(T_1) \supseteq res_{\mathcal{A}}(T_2)$, for every subsets T_1 and T_2 of S. If $\mathcal{A}$ is focused (constant, isotone, antitone, respectively), then we simply speak of a focused (constant, isotone, antitone, respectively) RS and the induced mapping by $res_{\mathcal{A}}$ is said to be a focused (constant, isotone, antitone, respectively) RS-function. These properties were used in [4, 15] in the classification of RSs.

First we consider the complexity of the focus property, which turns out be solvable in deterministic polynomial time.

Theorem 7. *The problem for a given RS $\mathcal{A}$ to decide whether the induced function is focused is solvable in deterministic polynomial time.*

The above theorem is in sharp contrast to the coNP-completeness of the constant function problem for RSs, that asks for a given RS $\mathcal{A}$, whether it is constant. Next we turn our attention to the complexity of isotonicity and antitonicity for RSs. Here we find coNP-completeness.

Theorem 8. *The problem for a given RS $\mathcal{A}$ to decide whether the induced function is isotone is* coNP-*complete. The same holds true for checking whether $\mathcal{A}$ induces an antitone function.*

Proof. On input $\mathcal{A}$ the Turing machine universally guesses two states T_1 and T_2 and verifies that whenever $T_1 \subseteq T_2$ holds that $res_{\mathcal{A}}(T_1) \subseteq res_{\mathcal{A}}(T_2)$ is implied. If this is the case the Turing machine halts and accepts; otherwise it halts and rejects. Guessing the states T_1 and T_2, as well as the verification that $T_1 \subseteq T_2$, and $res_{\mathcal{A}}(T_1) \subseteq res_{\mathcal{A}}(T_2)$ hold can be done in polynomial time. Hence the isotonicity problem for RSs belongs to coNP. With a similar argument one can show that also the antitonicity problem can be solved on a polynomially time bounded universal Turing machine, too.

We reduce the UNSAT problem to the isotonicity problem for RSs. Let φ be a 3SAT formula with n variables $x_1, x_2, \ldots, x_n$ and m clauses $C_1, C_2, \ldots, C_m$ with at most three literals per clause. Let $X = \{x_1, x_2, \ldots, x_n\}$. Define the RS $\mathcal{A} = (S, A)$ with $S = X \cup \{x_{n+1}\}$, where x_{n+1} is a new symbol that is not contained in X, and A contains the following reactions:

$$(\emptyset, \{x_{n+1}\}, \{x_1, x_2, \ldots, x_{n+1}\})$$

and

$$(neg(C_j), pos(C_j), \{x_1, x_2, \ldots, x_{n+1}\}), \qquad \text{for } 1 \leq j \leq m,$$

where again $neg(C)$ ($pos(C)$, respectively) denotes the set of variables from X that negatively (positively, respectively) appear in the clause C as literals. Obviously, the RS $\mathcal{A}$ can be constructed from φ within deterministic logspace. Then we find the following situation. Consider two states T_1 and T_2 satisfying $x_{n+1} \notin T_1$ and $T_2 = T_1 \cup \{x_{n+1}\}$. Then $T_1 \subseteq T_2$ and by the construction of $\mathcal{A}$ we have $res_{\mathcal{A}}(T_1) = \{x_1, x_2, \ldots, x_{n+1}\}$, while

$$res_{\mathcal{A}}(T_2) = \begin{cases} \emptyset & \text{if the assignment encoded by } T_2 \text{ satisfies } \varphi \\ \{x_1, x_2, \ldots, x_{n+1}\} & \text{otherwise.} \end{cases}$$

Thus, the isotonicity property on T_1 and T_2 is only fulfilled if the assignment encoded by T_2 restricted to the original variables $x_1, x_2, \ldots, x_n$ does not satisfy φ. Therefore, the RS $\mathcal{A}$ is isotone if this holds for all pairs T_1 and T_2 with the above properties. Hence, we conclude that φ is not satisfiable if and only if $\mathcal{A}$ is isotone.

For the antitonicity problem we slightly modify the above given construction, such that the RS $\mathcal{A}$ now contains the reactions

$$(\{x_{n+1}, \emptyset, \{x_1, x_2, \ldots, x_{n+1}\})$$
and
$$(neg(C_j), pos(C_j), \{x_1, x_2, \ldots, x_{n+1}\}), \qquad \text{for } 1 \le j \le m.$$

Then with a similar argument as in the isotonicity case one can show that the φ is not satisfiable if and only if the new RS $\mathcal{A}$ is antitone. This proves the theorem. □

The study of RSs from a descriptional complexity point of view resulted in a variety of further properties that allowed the authors in [4] to characterize minimal RSs w.r.t. the size of the reactant, inhibitor, and the production sets. In this context the following conditions are of interest: a RS $\mathcal{A} = (S, A)$ is *union additive*, if $res_{\mathcal{A}}(T_1 \cup T_2) = res_{\mathcal{A}}(T_1) \cup res_{\mathcal{A}}(T_2)$, for every subsets T_1 and T_2 of S, *union sub-additivity*, if $res_{\mathcal{A}}(T_1 \cup T_2) \subseteq res_{\mathcal{A}}(T_1) \cup res_{\mathcal{A}}(T_2)$, for every subsets T_1 and T_2 of S, *intersection additivity*, if $res_{\mathcal{A}}(T_1 \cap T_2) = res_{\mathcal{A}}(T_1) \cup res_{\mathcal{A}}(T_2)$, for every subsets T_1 and T_2 of S, and *intersection sub-additivity*, if $res_{\mathcal{A}}(T_1 \cap T_2) \subseteq res_{\mathcal{A}}(T_1) \cup res_{\mathcal{A}}(T_2)$, for every subsets T_1 and T_2 of S.

Theorem 9. *The problem for a given RS $\mathcal{A}$ to determine whether it is union (sub)-additive is coNP-complete. The same is true if the RS is checked for intersection (sub-)additiveness.*

In the remainder of this subsection we consider minimality problems on RSs. A RS $\mathcal{A} = (S, A)$ is *reactant-minimal* (*inhibitor-minimal, resource-minimal*, respectively) if $|R_a| = 1$ ($|I_a| = 1$, $|R_a \cup I_a| = 1$, respectively), for every $a \in A$. A RS $\mathcal{A} = (S, A)$ *describes* a function $f : 2^S \to 2^S$, if $res_{\mathcal{A}}(T) = f(T)$, for every $T \subseteq S$. A function f induced by a RS is *reactant-minimal* (*inhibitor-minimal, resource-minimal*, respectively) if there is a RS that is reactant-minimal (inhibitor-minimal, resource-minimal, respectively) that implements f. Then the following statements hold [4]: (i) Function f is reactant-minimal if and only if f

is union sub-additive. (ii) Function f is inhibitor-minimal if and only if f is intersection sub-additive. (iii) Function f is resource-minimal if and only if f is both union and intersection sub-additive. Then by Theorem 9 we obtain the following corollary:

Corollary 10. *For a given RS $\mathcal{A}$ to decide whether there is a reactant-minimal RS implementing the same function as $\mathcal{A}$ is* coNP-*complete. The completeness remains if one asks for inhibitor- or resource-minimality.* □

3.3   The Complexity of Threshold Problems

Finally we study problems that are induced by properties from extremal combinatorics on RSs [2]. There the main study focused on so called threshold properties. Simply speaking, a threshold property is a predicate over a class of objects which is true for those objects of a size greater or equal than a certain threshold. For instance, a RS $\mathcal{A} = (S, A)$ is said to be *total* (TOT) if for every $T \subseteq S$ with $T \neq \emptyset$ and $T \neq S$, the result function RS $\mathcal{A}$ fulfills $res_{\mathcal{A}}(T) \neq \emptyset$. Whenever a RS with a size n background set has at least $3^n \cdot 2^n - 2^n \cdot (2^n - 1) + 1$ reactions, then this RS is total.[1] Note that this nicely contrasts the coNP-completeness of determining if a given RS $\mathcal{A}$ is total [12]. Up to our knowledge further threshold properties were not investigated w.r.t. their computational complexity. We close this gap in this subsection. Most of the following properties were introduced in [2] if not stated otherwise.

The dynamics of reactions systems lies in the concurrent application of reactions. Thus, one may ask whether a RS is minimal concurrent. This is defined as follows: a RS $\mathcal{A} = (S, A)$ has the *minimal-concurrency (MC)* property, if there exist two reactions $a, b \in A$ and a state $T \subseteq S$ such that $a, b \in en_{\mathcal{A}}(T)$. The following result follows directly from the fact that if two reactions $a = (R_a, I_a, P_a)$ and $b = (R_b, I_b, P_b)$ are enabled on at least one common set the intersection $(R_a \cup R_b) \cap (I_a \cup I_b)$ must be empty, which is testable in polynomial time.

Theorem 11. *The problem for a given $\mathcal{A}$ to decide for minimal-concurrency can be done deterministic polynomial time.* □

When asking if in a RS there are always at least two different reactions applicable in every state the complexity situation completely changes. To be more formal, a RS $\mathcal{A} = (S, A)$ has the *always-parallel* (AP) property if for every state $T \subseteq S$ and every reaction $a \in en_{\mathcal{A}}(T)$ there exists a reaction $b \in A \setminus \{a\}$ such that $b \in en_{\mathcal{A}}(T)$. The next theorem shows that this property is already intractable.

Theorem 12. *The problem for a given $\mathcal{A}$ to decide whether it satisfies the always-parallel property is* coNP-*complete.*

---

[1] If the RS has to be strict the threshold for TOT is $(2^n - 1)(3^n - 3 \cdot 2^n + 2^{\lceil \frac{n}{2} \rceil} + 2^{\lfloor \frac{n}{2} \rfloor}) + 1$, see [2].

Proof. The containment in coNP is straightforward. Universally guess a state T and a reaction a. If a is enabled on T, then search in polynomial time for another reaction that is enabled on T, too. If this is the case, then halt and accept; otherwise halt and reject.

The reduction from UNSAT to the always-parallel problem reads as follows: On input φ with n variables $x_1, x_2, \ldots, x_n$ and m clauses $C_1, C_2, \ldots, C_m$ we construct in deterministic logarithmic space the RS $\mathcal{A} = (X \cup \{\#\}, A)$, where $X = \{x_1, x_2, \ldots, x_n\}$, symbol $\#$ is new and not contained in X, and the set A is made up by the reactions

$$(neg(C_i) \cup \{\#\}, pos(C_i), \{\#\}) \qquad \text{for } 1 \le i \le m$$

and

$$(\{\#\}, \emptyset, \{\#\}).$$

Because $\#$ belongs to every reactant set, all states that do not contain $\#$ never enable any reaction. In this case the always-parallel property is satisfied. For the remaining states $T \subseteq X \cup \{\#\}$ that contain $\#$ the following argumentation applies: If φ is unsatisfiable, then at least one reaction of the first type applies as well as the reaction that copies $\#$. Thus, at least two reactions are enabled. On the other hand, if φ is satisfiable by an assignment, then on the corresponding set $T \cup \{\#\}$ only the $\#$-copy reaction is enabled. This shows that φ is in UNSAT if and only the RS $\mathcal{A}$ has the always-parallel property. Hence, the problem is coNP-hard. □

Slightly adapting the previous proof, this also shows that the generalized always-parallel property, that is, the variant AP_m with $m \ge 2$ such that every subset of its entities enables either 0 or at least m reactions, is coNP-complete, too. Clearly, AP is equivalent to AP_2. To this end it suffices to introduce further $\#$ symbols that have to be present in the reactant sets and that are copied to the next state by an appropriate number of reactions. The details are left to the reader.

Corollary 13. *Let $m \ge 2$ be fixed. Then the problem for a given $\mathcal{A}$ to decide whether it satisfies AP_m is* coNP-*complete.* □

The complexity of the problem does not change even if we ask whether there exists an m such that a given RS has the property AP_m because if a RS has the property AP_m it also has the property AP_{m-1} for every $m \ge 3$.

Yet another threshold property that was studied in [2] is the no-concurrency property that requires that there are at least two reactions that can never be executed at the same time step, or more formally a RS $\mathcal{A} = (S, A)$ has the *no-concurrency* (NC) property, if there exist $a, b \in A$ such that for every $T \subseteq S$, either $a \notin en_{\mathcal{A}}(T)$ or $b \notin en_{\mathcal{A}}(T)$—here we use the literally interpretation of an exclusive or (XOR) in the definition.

In [2] it was mentioned that any RS $\mathcal{A}$ has the NC property if and only if it admits two reactions $a = (R_a, I_a, P_a)$ and $b = (R_b, I_b, P_b)$ such that $R_a \cap I_b \ne \emptyset$. In fact, this equivalence does not hold in general, as we can see from the following counter example. Consider the RS $\mathcal{A} = (S, A)$ with $S = \{1, 2, 3\}$ and the two reactions $a = (\{1, 2\}, \{3\}, P_a)$ and $b = (\{2, 3\}, \{1\}, P_b)$, for some P_a and P_b. Since both reactions are not applicable to the state $\{1, 3\}$ the RS $\mathcal{A}$ does not obey the NC property, but $R_a \cap I_b = \{1, 2\} \cap \{1\} \neq \emptyset$ and $R_b \cap I_a = \{2, 3\} \cap \{3\} \neq \emptyset$. Hence, the above mentioned equivalence of the NC property with the fact that $R_a \cap I_b \neq \emptyset$, for two reactions $a = (R_a, I_a, P_a)$ and $b = (R_b, I_b, P_b)$ does not hold in general. We give an alternative characterization of the NC property that finally allows us to check for no-concurrency in polynomial time. Moreover, we also improve the threshold given in [2] on the NC property. The next theorem states the alternative NC characterization.

Theorem 14. *Let $\mathcal{A} = (S, A)$ be a RS. Then $\mathcal{A}$ has the NC property if and only if it admits two reactions $a = (\emptyset, \{s\}, P_a)$ and $b = (\{s\}, \emptyset, P_b)$, for some $s \in S$ and sets $P_a, P_B \subseteq S$.*

As an immediate consequence of the previous lemma we obtain that deciding the NC property can be done in deterministic polynomial time.

Corollary 15. *It can be verified in deterministic polytime whether a given $\mathcal{A}$ satisfies the no-concurrency property.* ☐

As an additional consequence of Theorem 14 it is clear that the NC property is a threshold property if and only if it is allowed for the reactant and inhibitor set of reactions to be empty. Next we improve the NC threshold, since the threshold proof of the NC property presented in [2, Proposition 11] relies on the falsely claimed equivalence of NC with $R_a \cap I_b \neq \emptyset$ already mentioned above.

Theorem 16. *The threshold for the NC property is $(3^n - n) \cdot 2^n + 1$, where n is the size of the background set.*

Proof. It suffices to count the possible number of reactions on a given background set S with n elements which is $\sum_{i=0}^{n} \binom{n}{i} 2^{n-i} \cdot 2^n$ and subtract for every $s \in S$ the cardinality of one of the sets

$$\{(\emptyset, \{s\}, P) \mid P \subseteq S\} \quad \text{and} \quad \{(\{s\}, \emptyset, P) \mid P \subseteq S\},$$

which is 2^n. Thus, we obtain $\sum_{i=0}^{n} \binom{n}{i} 2^{n-i} \cdot 2^n - 2^n \cdot n + 1$ for the threshold, since for a RS $\mathcal{A}$ that has at least this number of reactions there must be an element $s \in S$ and two sets $P_1, P_2 \subseteq S$ such that $(\emptyset, \{s\}, P_1)$ and $(\{s\}, \emptyset, P_2)$ are both reactions in $\mathcal{A}$. Finally, simplifying the given formula by using elementary properties of the binomial coefficients results in the stated bound. ☐

Let us come back to some other properties studied in [2]. A RS $\mathcal{A} = (S, A)$ is said to have the *inhibitor* (INH) property, if there exist two reactions $a = (P_a, I_a, R_a)$ and $b = (R_b, I_b, R_b)$ in A such that $P_a \cap I_b \neq \emptyset$. Obviously, this property can be verified in deterministic polytime by cycling through all pairs of reactions of the given RS.

Theorem 17. *It can be verified in deterministic polytime whether a given $\mathcal{A}$ satisfies the inhibitor property.*   □

For the next two properties, which are related to each other, we were not able to give precise complexity bounds. These are the properties comparability and redundancy. For the first property we need some notations. Let $\mathcal{A} = (S, A)$ be a RS and let $rac(S)$ refer to the set of all reactions over S. For any pair of reactions $a, b \in rac(S)$, we say that a *covers* b, denoted by $a \geq b$, if and only if $res_a(T) \supseteq res_b(T)$, for every $T \subseteq S$. Then a RS $\mathcal{A} = (S, A)$ has the comparability (COMP) property if for every reaction $a \in rac(S)$, there exists a $b \in A$ such that either $a \geq b$ or $b \geq a$. This property is related to redundancy, which is defined as follows: a RS $\mathcal{A} = (S, A)$ is *redundant* (RED) if there exists $B \subseteq A$ such that (S, A) is equivalent to (S, B). In other words, this means that the set of reactions contains some redundancy, i.e., reactions that do not contribute to the dynamics of the system. If there exists a $B \subseteq A$ such that the RS $\mathcal{B} = (S, B)$ has property COMP, then $\mathcal{A}$ is redundant, as shown in [2]. For both problems the naive algorithm runs on a nondeterministic polynomial time bounded Turing machine with a NP-oracle that checks for coverability or for RSs equivalence. Thus, P^{NP} is an upper bound, but we do not know any non trivial lower bound for these problems. Nevertheless, due to our study of these properties we figured out how to improve the threshold for both properties significantly. The previously known bound [2] for the property COMP was $\frac{n!3^n}{\lceil \frac{n}{2} \rceil! \lfloor \frac{n}{2} \rfloor!}$, where n refers to the size of the background set. To be more precise, instead of an exact threshold for the COMP property an upper and lower bound was given, and the above given value is in fact the upper bound.

Theorem 18. *The threshold value for the property COMP is $(3^n - 2) \cdot 2^n + 1$, where n is the size of the background set.*

For the redundancy property no explicit threshold was given in [2]. Only *via* the COMP property one can deduce an implicit bound for the property RED. We close this gap with the result in the next theorem.

Theorem 19. *The threshold for the RED property is $2^n \cdot n + 1$, where n is the size of the background set.*

Besides the possibility of reducing the number of reactions in a given RS one can also possibly simplify a RS by decomposing it into smaller systems. This notion was introduced in [2] and reads as follows: a RS $\mathcal{A} = (S, A)$ is *non-decomposable*, if there exist *no* RSs $\mathcal{A}_1 = (S_1, A_1)$, $\mathcal{A}_2 = (S_2, A_2)$ such that $S_1 \neq \emptyset \neq S_2$, $S_1 \cup S_2 = S$, $S_1 \cap S_2 = \emptyset$, $A_1 \cup A_2 = A$, and $A_1 \cap A_2 = \emptyset$. In the forthcoming we use the decomposability property instead.

Theorem 20. *It can be verified by a deterministic Turing machine in polytime whether a given $\mathcal{A} = (S, A)$ is decomposable, and if a decomposition of $\mathcal{A}$ exists, then the Turing machine also computes one.*

By repeatedly applying Theorem 20 to a given RS we can compute the decomposition with the maximal number of RSs. Obviously the number of repeats of the decomposition algorithm until no decomposition is possible anymore is bounded by the cardinality of the background set of the input RS. So we obtain the following corollary.

Corollary 21. *For a given RS $\mathcal{A} = (S, A)$ the smallest integer $1 \leq m \leq n$ such that $\mathcal{A}$ cannot be decomposed into m RS can be computed in deterministic polynomial time.* □

A variant of the ND property is the ND_m property [2], for $1 \leq m \leq \lceil \frac{n}{2} \rceil$, which is true for a any RS that its non-decomposable in two RSs both with a background set of at least m elements. For the ND_m, respectively, its negation the D_m property, things get more complicated.

Theorem 22. *For a given RS $\mathcal{A} = (S, A)$ and an integer m, regardless whether the value of m is written in unary or binary, to decide whether the D_m property holds is NP-complete.*

Proof. Let $\mathcal{A}$ and m be given. Then the D_m property is obviously contained in NP, since one guesses a partition of the background set into two sets of size at least m and afterwards verifies with the help of the procedure mentioned in Theorem 20 that is is actually a decomposition.

For the NP-hardness we reduce the NP-complete PARTITION problem to the RS-property D_m, for suitable m. Let $P = \{p_1, p_2, \dots, p_n\}$ with weight function w that maps from P to the positive integers be an instance of the partition problem. Let $w(P) = \sum_{p \in P} w(p)$. If $w(P)$ is odd, then the PARTITION problem has no solution and we produce the RS $\mathcal{A} = (\{1, 2\}, A)$ with

$$A = \{(\{1\}, \emptyset, \{1, 2\}), (\{2\}, \emptyset, \{1, 2\})\}$$

together with $m = 1$ as output. Trivially, this is a "no"-instance. Otherwise, $w(P)$ is even and we proceed as follows: define the RS $\mathcal{A} = (S, A)$, where $S = \cup_{i=1}^{n} S_i$ with

$$S_i = \{\, s_{i,j} \mid 0 \leq j \leq w(p_i) - 1 \,\},$$

for $1 \leq i \leq n$, and the set of reactions contains

$$(\{s_{i,j}\}, \emptyset, \{s_{i,j+1}\}), \qquad \text{for } 1 \leq i \leq n \text{ and } 0 \leq j < w(p_i)$$

and

$$(\{s_{i,j}\}, \emptyset, \{s_{i,0}\}), \qquad \text{for } 1 \leq i \leq n \text{ and } j = w(p_i).$$

It is easy to see that the RS $\mathcal{A}$ can be decomposed into n RSs (S_1, A_1), (S_2, A_2), up to (S_n, A_n) with appropriately chosen reaction sets A_i, for $1 \leq i \leq n$. Additionally, by construction it is not possible to decompose those RSs further. Thus, it follows that P with weight function w belongs to PARTITION if and only if the RS $\mathcal{A}$ has the D_m-property, for $m = \frac{1}{2} \sum_{p \in P} w(p)$. This proves NP-hardness. □

Finally, it is worth mentioning that in [2] further threshold properties were considered. In particular properties that are related to preperiods, periods, attractors, and cycle lengths in RSs. The computational complexity of these problems and variants thereof is already settled in [3].

References

1. Azimi, S., Iancu, B., Petre, I.: Reaction system models for the heat shock response. Fund. Inform. **131**(3–4), 299–312 (2014)
2. Dennunzio, A., Formenti, E., Manzoni, L.: Reaction systems and extremal combinatorics properties. Theoret. Comput. Sci. **598**, 138–149 (2015)
3. Dennunzio, A., Formenti, E., Manzoni, L., Porreca, A.E.: Complexity of the dynamics of reaction systems. Inform. Comput. **267**, 96–109 (2019)
4. Ehrenfeucht, A., Kleijn, J., Koutny, M., Rozenberg, G.: Minimal reaction systems. In: Priami, C., Petre, I., de Vink, E. (eds.) Transactions on Computational Systems Biology XIV. LNCS, vol. 7625, pp. 102–122. Springer, Heidelberg (2012). https://doi.org/10.1007/978-3-642-35524-0_5
5. Ehrenfeucht, A., Main, M., Rozenberg, G.: Combinatorics of life and death for reaction systems. Internat. J. Found. Comput. Sci. **21**(3), 345–356 (2010)
6. Ehrenfeucht, A., Rozenberg, G.: Reaction systems. Fund. Inform. **75**, 263–280 (2007)
7. Ehrenfeucht, A., Rozenberg, G.: Introducing time in reaction systems. Theoret. Comput. Sci. **410**, 310–322 (2009)
8. Manzoni, L., Poças, D., Porreca, A.E.: Simple reaction systems and their classification. Internat. J. Found. Comput. Sci. **25**(4), 441–457 (2014)
9. Nair, M.: On Chebyshev-type inequalities for primes. Amer. Math. Mon. **89**(2), 126–129 (1982)
10. Papadimitriou, C.H.: Computational Complexity. Addison-Wesley, Boston (1994)
11. Salomaa, A.: Functions and sequences generated by reaction systems. Theoret. Comput. Sci. **466**, 87–96 (2012)
12. Salomaa, A.: Functional constructions between reaction systems and propositional logic. Internat. J. Found. Comput. Sci. **24**(1), 147–159 (2013)
13. Salomaa, A.: Two-step simulations of reaction systems by minimal ones. Acta Cybernet. **22**, 247–257 (2015)
14. Savitch, W.J.: Relationships between nondeterministic and deterministic tape complexities. J. Comput. System Sci. **4**(2), 177–192 (1970)
15. Teh, W.C., Atanasiu, A.: Minimal reaction system revisited and reaction system rank. Internat. J. Found. Comput. Sci. **28**(3), 247–261 (2017)

Average-Case Rigidity Lower Bounds

Xuangui Huang$^{(\boxtimes)}$ and Emanuele Viola

Northeastern University, Boston, MA 02115, USA
{stslxg,viola}@ccs.neu.edu

Abstract. It is shown that there exists $f : \{0,1\}^{n/2} \times \{0,1\}^{n/2} \to \{0,1\}$ in $\mathrm{E}^{\mathbf{NP}}$ such that for every $2^{n/2} \times 2^{n/2}$ matrix M of rank $\leq \rho$ we have $\mathbb{P}_{x,y}[f(x,y) \neq M_{x,y}] \geq 1/2 - 2^{-\Omega(k)}$, whenever $\log \rho \leq \delta n / k (\log n + k)$ for a sufficiently small $\delta > 0$, and n is large enough. This generalizes recent results which bound below the probability by $1/2 - \Omega(1)$ or apply to constant-depth circuits.

Keywords: Average-case lower bounds · Matrix rigidity · Correlation bounds

Starting with the seminal paper by Williams [30] a sequence of recent works have proved new lower bounds for functions in various classes which contain super-polynomial non-deterministic time [1,2,4,6,8–11,14,16,22,24,25,27–29], lower bounds that we do not know how to prove by other means. Two sub-sequences of results are relevant to the present work. The first is the sub-sequence establishing *average-case hardness results* for various circuit classes. The concurrent works [9,24] proved incomparable, new average-case lower bounds against AC^0 with parity gates. Both results were improved in [8] to obtain a function that any such circuit of sub-exponential size cannot compute with a sub-exponentially small advantage over random guessing, for a uniform input.

The second is the sub-sequence constructing *rigid matrices* [23], that is, obtaining functions $f(x,y)$, where $|x| = |y| = n/2$ such that the corresponding $2^{n/2} \times 2^{n/2}$ matrix $M_{x,y} = f(x,y)$ is far from low-rank matrices. Using PCPs, [2] gave f such that $\mathbb{P}[M_{x,y} \neq f(x,y)] \geq \Omega(1)$ for any M of rank up to at most $2^{n^{1/4-\epsilon}}$. Low-rank matrices are a generalization of low-degree polynomials [19], but the rank bound in [2] is not strong enough to improve the classic results on polynomials due to Razborov and Smolensky [17,20,21] which hold up to degree $\sqrt{n}$. The subsequent paper [24] achieved nearly-optimal probabilistic degree $n/\mathrm{poly}\log n$ relying on the PCP construction [3]. It also raised the question of constructing PCPs with stronger properties and showed that these would improve the rank bounds in [2] to $2^{n/\Omega(\log^2 n)}$ (under some distribution). Related PCPs were constructed in the subsequent work [4], finally obtaining f such that $\mathbb{P}[M_{x,y} \neq f(x,y)] \geq \Omega(1)$ for any M of rank up to $2^{n/\Omega(\log n)}$.

Supported by NSF CCF award 1813930.

R. Santhanam and D. Musatov (Eds.): CSR 2021, LNCS 12730, pp. 186–205, 2021.
https://doi.org/10.1007/978-3-030-79416-3_11

In this paper we prove a result that generalizes both sub-sequences. We simultaneously achieve the strong average-case hardness parameters of [8] and work in the general model of low-rank matrices.

Theorem 1. *There exists a function* $f\colon \{0,1\}^n \times \{0,1\}^n \to \{-1,1\}$ *in* $\mathbf{E^{NP}}$ *such that for any rank-ρ matrix M, we have*

$$\mathbb{P}_{x,y}[f(x,y) \neq M_{x,y}] \geq 1/2 - 2^{-\Omega(k)}$$

for all large enough n, where $k(\log n + k)\log\rho \leq \delta n$ for a sufficiently small constant $\delta > 0$.

To illustrate the parameters, we can prove lower bounds whenever $k^2 \log\rho \leq \delta n$, for $k \geq \log n$. In particular we can for example bound below the probability by $1/2 - 2^{-n^{\Omega(1)}}$ for log rank $n^{0.99}$. We can also have $\log\rho = n/\Omega(\log n)$ whenever $k = O(1)$, recovering the result from [4].

It seems within reach to improve the tradeoff between k and ρ to obtain lower bounds whenever $k \log\rho \leq \delta n$. Improving the tradeoff even further to obtain lower bounds when $k \log\rho$ is $n^{1+\Omega(1)}$ would give new *data-structure lower bounds*, for functions in $\mathbf{E^{NP}}$, via a connection established in [24].

Independently, Chen and Lyu [7] proved lower bounds whenever $k^{1.5} \log\rho \leq \delta n$. Their proof proceeds in exactly the same way as ours, but in addition they prove a new derandomized XOR lemma where the seed length is just $\sqrt{k}n$ as opposed to kn in our Lemma 8. One can also plug their new XOR lemma in our proof and infer the stronger bound.

Techniques. Our proof builds on the previous work mentioned earlier. We adapt a clever approach in [8] which is based on Levin's proof of Yao's famous XOR lemma, cf. [13]. The approach shows that to prove a strong average-case hardness result it suffices to prove a mild average-case hardness result *for an intermediate model.* The intermediate model in our case consists of *rational sums of low-rank matrices.* We show that a lower bound for this model can be obtained from the rectangular PCP in [4], see Theorem 3.

A little more in detail, we prove a constant-error lower bound for rational sums of low-rank matrices by contradiction using the non-deterministic time-hierarchy theorem following [26]. We fix a unary language in $\mathbf{NTIME}(2^n) \setminus \mathbf{NTIME}(o(2^n))$, and let the lexicographically first rectangular PCP proof for this language be the hard function. Assuming that this hard function has constant correlation with a sum of low-rank matrices, we derive a contradiction by giving a quick non-deterministic algorithm. This algorithm first guesses a sum of low-rank matrices as an approximation of the hard function, i.e. the boolean proof, then performs a series of validity tests that are adapted from [8] to guarantee that this sum is bounded and close to boolean. Then the rectangular property of the PCP is exploited to make sure that when the guessed sum is plugged in as a proof, the bits that the PCP verifier probes can also be written as sums of low-rank matrices, thus the algorithm can quickly evaluate the "acceptance probability" of the guessed sum, based on the fast counting algorithm for low-rank matrices in [2,5]. Now the boundedness and close-to-boolean properties will ensure that this "acceptance probability" is close to that of the boolean proof approximated by the guessed sum, so the algorithm can make a decision for the language based on this value.

We shall first prove our result for infinitely many input lengths n; at the end we shall explain what modifications are sufficient to obtain all sufficiently large n, using results in [8].

1   Preliminaries

For any $n \in \mathbb{N}$, define $[n] = \{1, 2, \ldots, n\}$. For any matrix M, we use $M_{i,j}$ to denote its entry on row i column j. For any $n \times m$ matrix M define the matrix $(-1)^M$ by $\left((-1)^M\right)_{i,j} = (-1)^{M_{i,j}}$ for all $i \in [n]$, $j \in [m]$. For any matrix M, we define its ℓ_p-norm as $\|M\|_p = \left(\mathbb{E}_{i,j}[|M_{i,j}|^p]\right)^{1/p}$, while the ℓ_∞-norm is defined as $\|M\|_\infty = \max_{i,j} |M_{i,j}|$. For any two matrices A and B with the same shape, we define $A \circ B$ to be the Hadamard product (entrywise product) of them over $\mathbb{R}$, which is distributive. We use $\widetilde{O}$ to hide poly(n) terms in runtime.

Definition 1. *For any $\alpha \in \mathbb{Q}$ we define its* bit-complexity *as the maximum of the bit lengths of the denominator and numerator. For a polynomial p with rational coefficients we define its* bit complexity *as the maximum bit complexity among the coefficients.*

Definition 2. *For any given function class $\mathcal{C}$, we call the sum $\widetilde{Q} = C \sum_{i=1}^m b_i \cdot f_i$ an m-sum of $\mathcal{C}$, for $b_i \in \{-1, 1\}$ and $f_i \in \mathcal{C}$ for all $i \in [m]$ and $C \in \mathbb{Q}$. We define the bit-complexity of $\widetilde{Q}$ as the bit-complexity of C.*

In particular, an m-sum of rank-ρ $\mathbb{F}_2$-matrices $\widetilde{Q} \in \mathbb{R}^{n \times n'}$ is given by $\widetilde{Q} = C \sum_{i=1}^m b_i \cdot (-1)^{M^{(i)}}$ where $M^{(i)} \in \mathbb{F}_2^{n \times n'}$ are rank-ρ matrices over $\mathbb{F}_2$.

Definition 3. *Let $f, g \colon \{0, 1\}^n \to [-1, 1]$ be two functions. We define their* correlation *as $\mathrm{corr}(f, g) = \left|\mathbb{E}_{x \sim \{0,1\}^n}[f(x)g(x)]\right|$. We say f ε-correlates with g iff $\mathrm{corr}(f, g) \geq \varepsilon$.*

Definition 4. *We say a matrix M is* bounded *if $M_{i,j} \in [-1, 1]$ for all i, j. Similarly, we say a function f is* bounded *if $f(x) \in [-1, 1]$ for all x.*

Definition 5. *For any boolean function $f \colon \{-1, 1\}^k \to \mathbb{R}$, we identify f with its* multilinear extension *over domain $\mathbb{R}$, defined by its Fourier expansion $f = \sum_{S \subseteq [k]} \beta_S \prod_{i \in S} x_i$, where $\beta_S \in \mathbb{R}$. For any sets X, Y and function $f \colon X^k \to Y$ we define its* extension over matrices $\overline{f} \colon (X^{n \times m})^k \to Y^{n \times m}$ *that maps matrices $M^{(1)}, M^{(2)}, \ldots, M^{(n)} \in X^{n \times m}$ to a matrix $M' \in Y^{n \times m}$ defined by $M'_{i,j} = f(M^{(1)}_{i,j}, M^{(2)}_{i,j}, \ldots, M^{(k)}_{i,j})$ for all $i \in [n], j \in [m]$.*

For example, the Hadamard product $A \circ B$ of matrices A and B is recovered as $\overline{f}(A, B)$ where f is multiplication.

Rectangular PCP. We need the following rectangular PCP to prove this theorem.

Definition 6 (Rectangular PCP, [4]). *For any language L, we say it has an* $(\ell^2, r, q, p, t, s, \tau)$-*rectangular PCP verifier V over alphabet* $\{-1, 1\}$ *if we have the following properties:*

Proof. *the proof* π *of length* ℓ^2 *is viewed as a matrix in* $\{-1, 1\}^{\ell \times \ell}$.
Randomness. *the random string* $R \in \{0, 1\}^r$ *is partitioned into three parts*

$$R = (R_{\text{row}}, R_{\text{col}}, R_{\text{shared}}) \in \{0, 1\}^{r_{\text{rect}}} \times \{0, 1\}^{r_{\text{rect}}} \times \{0, 1\}^{r_{\text{shared}}},$$

where $r_{\text{rect}} = (1 - \tau)r/2$ *and* $r_{\text{shared}} = \tau r$.
Computation. *Given input x and proof oracle* $\pi \in \{-1, 1\}^{\ell \times \ell}$, *with randomness* R, $V^\pi(x; R)$ *runs as follows:*
1. *Use shared randomness* $R_{\text{shared}} \in \{0, 1\}^{r_{\text{shared}}}$ *to:*
    (a) *construct a decision function* $D = D(x; R_{\text{shared}})$: $\{-1, 1\}^q \times \{-1, 1\}^p \to \{0, 1\}$,
    (b) *construct randomness parity check* $(C_1, \ldots, C_p) = (C_1(x; R_{\text{shared}}), \ldots, C_p(x; R_{\text{shared}}))$ *where each* C_i: $\{0, 1\}^{r_{\text{rect}}} \times \{0, 1\}^{r_{\text{rect}}} \to \{-1, 1\}$ *is a parity function, i.e.* $C_i(R_{\text{row}}, R_{\text{col}}) = (-1)^{\langle R_{\text{row}}, u \rangle + \langle R_{\text{col}}, v \rangle + b}$ *for some* $u, v \in \{0, 1\}^{r_{\text{rect}}}$ *and* $b \in \{0, 1\}$, *where* $\langle x, y \rangle$ *is the inner product of x and y.*
2. *Use row randomness* $R_{\text{row}} \in \{0, 1\}^{r_{\text{rect}}}$ *to construct row locations of queries*

$$i^{(1)} = i^{(1)}(x; R_{\text{row}}, R_{\text{shared}}), \ldots, i^{(q)} = i^{(q)}(x; R_{\text{row}}, R_{\text{shared}}).$$

3. *Use column randomness* $R_{\text{col}} \in \{0, 1\}^{r_{\text{rect}}}$ *to construct column locations of queries*

$$j^{(1)} = j^{(1)}(x; R_{\text{col}}, R_{\text{shared}}), \ldots, j^{(q)} = j^{(q)}(x; R_{\text{col}}, R_{\text{shared}}).$$

4. *Output the result*

$$D(\pi_{i^{(1)}, j^{(1)}}, \ldots, \pi_{i^{(q)}, j^{(q)}}, C_1(R_{\text{row}}, R_{\text{col}}), \ldots, C_p(R_{\text{row}}, R_{\text{col}})).$$

Completeness. *If* $x \in L$ *then* $\exists \pi \in \{-1, 1\}^{\ell \times \ell}, \Pr_R[V^\pi(x; R) = 1] = 1.$
Soundness. *If* $x \notin L$ *then* $\forall \pi \in \{-1, 1\}^{\ell \times \ell}, \Pr_R[V^\pi(x; R) = 1] < s.$
Complexity. *The verifier V runs in time* $t \geq r$, *the query complexity is q and parity-check complexity is p.*

Definition 7. *We say an* $(\ell^2, r, q, p, t, s, \tau)$-*rectangular PCP verifier is* smooth *if V queries uniformly on* π *over the choice of randomness* $R \in \{0, 1\}^r$ *and queries* $k \in [q]$.

The above definition means that each location of the proof has equal probability of being queried by a *random* query. A stronger requirement would be that this holds for *each* query. The stronger notion is available in some PCPs (e.g. [15]), but as far as we know not for rectangular PCPs.

Lemma 1 ([4]). *For any constants* $s \in (0, \frac{1}{2})$, $\tau \in (0, 1)$, *and language* $L \in$ **NTIME**(2^n), L *has a smooth* $(\ell^2, r, q, p, t, s, \tau)$-*rectangular PCP verifier* V *over alphabet* $\{-1, 1\}$ *with the following parameters:*

- $r = n + O(\log n)$.
- $q, p = O_s(1)$.
- $\ell^2 = O_s(2^r)$.
- $t = 2^{O(\tau n)}$.

2   Fast Algorithm for "Acceptance Probability"

In this section we prove and collect several facts that allow us to quickly compute the acceptance probability of a rectangular PCP verifier when its proof is a rational sum of low-rank matrices.

First we need the following result to quickly calculate number of 1's in low-rank matrices over $\mathbb{F}_2$ given low-rank decompositions.

Lemma 2 ([2,5]). *Given two matrices* $A \in \mathbb{F}_2^{N \times \rho}$ *and* $B \in \mathbb{F}_2^{\rho \times N}$ *where* $\rho = N^{o(1)}$, *there is a deterministic algorithm that computes the number of 1's in the product matrix* AB *over* $\mathbb{F}_2$ *in time* $T(N, \rho) = N^{2 - \Omega(1/\log \rho)}$.

We prove a general result on evaluating the expectation of a polynomial on sums of low-rank matrices.

Theorem 2. *Let* $\left\{\widetilde{Q}_i\right\}_{i \in [k]}$ *be* k m-*sums of rank-*ρ $\mathbb{F}_2$ *matrices with bit-complexity* c, *and let their low-rank decompositions be* $\widetilde{Q}_i = C_i \sum_{i=1}^{m} b_{i,j} \cdot (-1)^{A^{(i,j)} B^{(i,j)}}$ *where* $C_i \in \mathbb{Q}$ *has bit-complexity* c, $b_{i,j} \in \{-1, 1\}$, $A^{(i,j)} \in \mathbb{F}_2^{N \times \rho}$, *and* $B^{(i,j)} \in \mathbb{F}_2^{\rho \times N}$ *for all* $i \in [k]$ *and* $j \in [m]$. *For any* s-*sparse degree-*d *polynomial on* k *variables* $p \colon \mathbb{R}^k \to \mathbb{R}$ *with bit complexity* c', *given the decompositions we can compute the value of* $\mathbb{E}_{i,j \in [N]}\left[\left(\overline{p}\left(\widetilde{Q}_1, \ldots, \widetilde{Q}_k\right)\right)_{i,j}\right]$ *in time* $O\left(sm^d(T(N, d\rho) + \operatorname{poly}(c, c', d, \log N))\right)$ *if* $d\rho = N^{o(1)}$. *In particular for any boolean function* $f \colon \{-1, 1\}^k \to \{0, 1\}$ *it can be computed in time* $O\left(2^k m^k(T(N, k\rho) + \operatorname{poly}(c, k, \log N))\right)$ *if* $k\rho = N^{o(1)}$.

Proof. To calculate $\mathbb{E}_{i,j}\left[\left(\overline{p}\left(\widetilde{Q}_1, \ldots, \widetilde{Q}_k\right)\right)_{i,j}\right]$, by linearity of expectation it suffices to calculate the expectation for each monomial of p. Wlog, let the monomial $p'(x) = x_1 x_2 \cdots x_d$. Then by the distributive property of Hadamard products we have

$$\overline{p'}\left(\widetilde{Q}_1, \ldots, \widetilde{Q}_k\right) = \widetilde{Q}_1 \circ \widetilde{Q}_2 \circ \cdots \circ \widetilde{Q}_d$$

$$= \circ_{i=1}^{d}\left(C_i \sum_{j=1}^{m} b_{i,j} \cdot (-1)^{A^{(i,j)} B^{(i,j)}}\right)$$

$$= \sum_{(j_1,j_2,\ldots,j_d)\in[m]^d} \left(\prod_{i=1}^{d} C_i b_{i,j_i}\right) \cdot \left(\circ_{i=1}^{d}(-1)^{A^{(i,j_i)}B^{(i,j_i)}}\right)$$

$$= \sum_{(j_1,j_2,\ldots,j_d)\in[m]^d} \left(\prod_{i=1}^{d} C_i b_{i,j_i}\right) \cdot \left((-1)^{\oplus_{i=1}^{d}A^{(i,j_i)}B^{(i,j_i)}}\right),$$

where '$\oplus$' is the addition of $\mathbb{F}_2$-matrices over $\mathbb{F}_2$. Hence by linearity of expectation, it suffices to calculate the expectation of $(-1)^{\oplus_{i=1}^{d}A^{(i,j_i)}B^{(i,j_i)}}$ for each $(j_1,\ldots,j_d) \in [m]^d$. Note that for any $\mathbb{F}_2$-matrix M we have $\mathbb{E}_{\text{row,col}}\left[\left((-1)^M\right)_{\text{row,col}}\right] = 1 - 2\mathbb{E}_{\text{row,col}}[M_{\text{row,col}}]$, thus it suffices to calculate

$$\mathbb{E}_{\text{row,col}}\left[\left(\oplus_{i=1}^{d}A^{(i,j_i)}B^{(i,j_i)}\right)_{\text{row,col}}\right] = \frac{1}{N^2} \cdot \text{ number of 1's in } \oplus_{i=1}^{d} A^{(i,j_i)}B^{(i,j_i)}.$$

Note that $\oplus_{i=1}^{d}A^{(i,j_i)}B^{(i,j_i)}$ is just the product of an $N \times d\rho$ matrix and a $d\rho \times N$ matrix over $\mathbb{F}_2$, where the first matrix is obtained by concatenating the rows of $\left\{A^{(i,j_i)}\right\}_{i\in[d]}$ and the second matrix is obtained by concatenating the columns of $\left\{B^{(i,j_i)}\right\}_{i\in[d]}$. Hence by Lemma 2 the counting can be done in time $T(N, d\rho)$ if $d\rho = N^{o(1)}$. This expectation value has bit-complexity $O(\log N)$, so multiplying it by $\prod_{i=1}^{d} C_i b_{i,j_i}$ and adding to the running sum take time $\text{poly}(c, d, \log N)$. We still need to multiply the result by the coefficients of the monomials in p, thus the runtime becomes $\text{poly}(c, c', d, \log N)$. Therefore the total running time is $O\left(sm^d(T(N, d\rho) + \text{poly}(c, c', d, \log N))\right)$.

For any boolean function f, it can be written as a degree-k multilinear polynomial so there are at most 2^k monomials. Fourier analysis shows that every coefficient of this polynomial is a multiple of 2^{-k}, so its bit complexity is $O(k)$. Therefore the total running time becomes $O(2^k m^k (T(N, k\rho) + \text{poly}(c, k, \log N)))$ if $k\rho = N^{o(1)}$.

The following lemma from [4] shows that randomness parity checks can be written as low-rank matrices. For completeness the proof of this lemma is included in Appendix A.

Lemma 3 ([4, Claim B.1]). *For any parity function $f \colon \{0,1\}^m \times \{0,1\}^m \to \{-1,1\}$ defined by $f(i,j) = (-1)^{\langle i,u \rangle + \langle j,v \rangle + b}$ for some $u, v \in \{0,1\}^m$ and $b \in \{0,1\}$, we can compute in time $O(m2^m)$ two matrices $A \in \mathbb{F}_2^{2^m \times 3}$ and $B \in \mathbb{F}_2^{3 \times 2^m}$ such that $f(i,j) = \left((-1)^{AB}\right)_{i,j}$ for all $i, j \in \{0,1\}^m$.*

We use the following lemma to quickly calculate the "acceptance probability" of a sum of low-rank matrices $\widetilde{\pi}$.

Lemma 4. *Let V be any $(\ell^2, r, q, p, t, s, \tau)$-rectangular PCP verifier over $\{-1,1\}$, and $\widetilde{V}$ be the same as V but with D multilinearly extended over $\mathbb{R}$. Given $\widetilde{\pi} = C\sum_{i=1}^{m} b_i \cdot (-1)^{A^{(i)}B^{(i)}}$ with $C \in \mathbb{Q}$ of bit-complexity $O(n)$, $b_i \in \{-1,1\}$, $A^{(i)} \in \mathbb{F}_2^{\ell \times \rho}$, and $B^{(i)} \in \mathbb{F}_2^{\rho \times \ell}$ for all $i \in [m]$. Assuming that*

$\log((q+p)\rho) = o(r)$, *we can calculate* $\mathbb{E}_R\left[\tilde{V}^{\tilde{\pi}}(1^n; R)\right]$ *in time* $\tilde{O}\left(2^{r_{\mathrm{rect}}+r_{\mathrm{shared}}} \cdot (t + m\rho) + m^{q+p} \cdot 2^{q+p+r-\Omega(r/\log((q+p)\rho))}\right)$.

Using the parameters of the PCP in Lemma 1, the time bound in the above lemma becomes

$$\tilde{O}\left(m^{O(1)}\left(2^{0.51n}\rho + 2^{n-\Omega(n/\log\rho)}\right)\right), \tag{1}$$

which is $O(2^n/n)$ when $n/\log\rho \geq \kappa(\log m + \log n)$ for a constant κ. The proof of Lemma 4 follows closely from the computation process of the PCP in Definition 6, similar to parts of the proof of Lemma 3.1 in [4]. Due to page limits the proof is given in Appendix B.

3 Validity Tests

In this section we discuss two tests on sums of low-rank matrices $\tilde{\pi}$ to ensure that they are close to boolean and somewhat bounded. The following close-to-boolean test simplifies a similar test in [8] due to the smoothness of the PCP verifier. Using the parameters of the PCP in Lemma 1, the time bound in the following lemma becomes $\tilde{O}(m^4 \cdot 2^{n-\Omega(n/\log\rho)})$, which is $O(2^n/n)$ for m and ρ satisfying $n/\log\rho \geq \kappa(\log m + \log n)$ for a constant κ.

Lemma 5 (Close-to-Boolean Test). *Given* $\tilde{\pi} = C\sum_{i=1}^{m} b_i \cdot (-1)^{A^{(i)}B^{(i)}}$ *with* $C \in \mathbb{Q}$ *of bit-complexity* $O(n)$, $b_i \in \{-1, 1\}$, $A^{(i)} \in \mathbb{F}_2^{\ell \times \rho}$, *and* $B^{(i)} \in \mathbb{F}_2^{\rho \times \ell}$. *Assuming that* $\rho = \ell^{o(1)}$, *we can perform a test on* $\tilde{\pi}$ *in time* $\tilde{O}\left(m^4 \cdot \ell^{2-\Omega(1/\log\rho)}\right)$ *such that:*

- *(Completeness) If* $\tilde{\pi}$ *is bounded and there is a proof* $\pi \in \{-1, 1\}^{\ell \times \ell}$ *with* $\|\pi - \tilde{\pi}\|_1 \leq \varepsilon$, *then we have* $\|\pi - \tilde{\pi}\|_2 \leq \sqrt{2\varepsilon}$, *and* $\tilde{\pi}$ *passes the test.*
- *(Soundness) If* $\tilde{\pi}$ *passes the test, there exists a proof* $\pi \in \{-1, 1\}^{\ell \times \ell}$ *with* $\|\pi - \tilde{\pi}\|_2 \leq 2\sqrt{2\varepsilon}$.

Proof. We use Theorem 2 to evaluate the expectation of the degree-4 univariate polynomial $f(x) = (-1 - x)^2(1 - x)^2$ on $\tilde{\pi}$. We accept $\tilde{\pi}$ if $\mathbb{E}_{i,j}[f(\tilde{\pi}_{i,j})] \leq 8\varepsilon$, and reject otherwise. It takes time $O\left(m^4 \cdot (T(\ell, 4\rho) + \mathrm{poly}(n, \log\ell))\right) = \tilde{O}\left(m^4 \cdot \ell^{2-\Omega(1/\log\rho)}\right)$ if $\rho = \ell^{o(1)}$.

For soundness, define $\pi \in \{-1, 1\}^{\ell \times \ell}$ by $\pi_{i,j} = 1$ if $\tilde{\pi}_{i,j} \geq 0$, and -1 otherwise, for all $i, j \in [\ell]$.

Then for all $i, j \in [\ell]$, we have $|(-\pi_{i,j}) - \tilde{\pi}_{i,j}| \geq 1$. As $\{\pi_{i,j}, -\pi_{i,j}\} = \{-1, 1\}$, we have

$$f(\tilde{\pi}_{i,j}) = (-1 - \tilde{\pi}_{i,j})^2(1 - \tilde{\pi}_{i,j})^2 = ((-\pi_{i,j}) - \tilde{\pi}_{i,j})^2(\pi_{i,j} - \tilde{\pi}_{i,j})^2 \geq (\pi_{i,j} - \tilde{\pi}_{i,j})^2.$$

Therefore $\|\pi - \tilde{\pi}\|_2 = \sqrt{\mathbb{E}_{i,j}[(\pi_{i,j} - \tilde{\pi}_{i,j})^2]} \leq \sqrt{\mathbb{E}_{i,j}[f(\tilde{\pi}_{i,j})]} \leq 2\sqrt{2\varepsilon}$.

For completeness, observe that for $\tilde{\pi}_{i,j} \in [-1, 1]$ and $\pi_{i,j} \in \{-1, 1\}$ we have $|(-\pi_{i,j}) - \tilde{\pi}_{i,j}| \leq 2$ and so $(\pi_{i,j} - \tilde{\pi}_{i,j})^2 \leq 2|\pi_{i,j} - \tilde{\pi}_{i,j}|$. Therefore

$f(\widetilde{\pi}_{i,j}) \leq 2^2(\pi_{i,j} - \widetilde{\pi}_{i,j})^2 \leq 8\,|\pi_{i,j} - \widetilde{\pi}_{i,j}|$, thus $\mathbb{E}_{i,j}[f(\widetilde{\pi}_{i,j})] \leq 8\,\|\pi - \widetilde{\pi}\|_1 \leq 8\varepsilon$, so $\widetilde{\pi}$ passes the test. Moreover we have $\|\pi - \widetilde{\pi}\|_2 = \sqrt{\mathbb{E}_{i,j}[(\pi_{i,j} - \widetilde{\pi}_{i,j})^2]} \leq \sqrt{2\mathbb{E}_{i,j}\,|\pi_{i,j} - \widetilde{\pi}_{i,j}|} = \sqrt{2\,\|\pi - \widetilde{\pi}\|_1} \leq \sqrt{2\varepsilon}$.

We also need to test if the sum of low-rank matrices is somewhat bounded. Ideally we would like to ensure that the sum is point-wise bounded. However the quick algorithm in Theorem 2 can only calculate expectation so it is unlikely that we can use it to get a pointwise bound. Fortunately it turns out that for our purpose we don't really need pointwise boundedness. The test we present here generalizes a similar test in [8]. We use the following notion of sampling from the lists I, J.

Definition 8. *Let I, J be any two lists of the same size taking (possibly duplicate) elements from $[\ell]$. We say a real matrix $\widetilde{\pi} \in \mathbb{R}^{\ell \times \ell}$ is power-d bounded for (I, J) if $\mathbb{E}_{i \sim I, j \sim J}[\widetilde{\pi}_{i,j}^d] \leq 1$, where $i \sim I$ means that i is sampled from I uniformly at random.*

Lemma 6 (Boundedness Test). *Let $\widetilde{\pi} = C \sum_{i=1}^{m} b_i \cdot (-1)^{A^{(i)} B^{(i)}}$ be an m-sum with $C \in \mathbb{Q}$ of bit-complexity $O(\log n)$, $b_i \in \{-1, 1\}$, $A^{(i)} \in \mathbb{F}_2^{\ell \times \rho}$, and $B^{(i)} \in \mathbb{F}_2^{\rho \times \ell}$ for all $i \in [m]$. Let I, J be two lists of the same size taking elements from $[\ell]$. Let d be any number. Assuming that $d\rho = |I|^{o(1)}$, we can perform a test on $\widetilde{\pi}$ in time $\widetilde{O}\left(m\rho|I| + m^d|I|^{2-\Omega(1/\log(d\rho))}\right)$ such that:*

- *(Completeness) If $\widetilde{\pi}$ is bounded, it passes the test.*
- *(Soundness) If $\widetilde{\pi}$ passes the test, it is power-d bounded for (I, J).*

Jumping ahead, we will set I (and J) to be the list of the row (column, respectively) indices the verifier probes over row (column, respectively) randomness for each of the q queries and each choice of the shared randomness, so $|I| = |J| = 2^{r_{rect}}$. Using the parameters of the PCP in Lemma 1, each boundedness test runs in time

$$\widetilde{O}(m^{O(1)}(2^{0.49n}\rho + 2^{0.98n - \Omega(n/\log \rho)})).$$

We will use $O(2^{0.02n})$-many boundedness tests so the total runtime is similar to (1), which becomes $O(2^n/n)$ when $n/\log \rho \geq \kappa(\log m + \log n)$ for a constant κ.

Proof. We construct an m-sum $\widetilde{Q} = C \sum_{i=1}^{m} b_i \cdot (-1)^{A'^{(i)} B'^{(i)}}$, where $A'^{(i)} \in \mathbb{F}_2^{|I| \times \rho}$ consists of the rows of $A^{(i)}$ indexed by elements in I and $B'^{(i)} \in \mathbb{F}_2^{\rho \times |I|}$ consists of the columns of $B^{(i)}$ indexed by elements in J. This step takes time $O(m\rho|I|)$.

Note that the uniform distribution over entries of $\widetilde{Q}$ is the same as the distribution over entries of $\widetilde{\pi}$ under I, J, so we have $\mathbb{E}_{i \sim I, j \sim J}[\widetilde{\pi}_{i,j}^d] = \mathbb{E}_{i,j}\left[\left(\widetilde{Q}\right)_{i,j}^d\right]$.

Hence we use Theorem 2 to evaluate the expectation of the polynomial x^d on $\widetilde{Q}$. We accept $\widetilde{\pi}$ if the value is at most 1, and reject otherwise. This step takes time $O(m^d \cdot (T(|I|, d\rho) + \mathrm{poly}(n, d, \log|I|))$ if $d\rho = |I|^{o(1)}$. Therefore the total running time is $O\left(m\rho|I| + m^d|I|^{2-\Omega(1/\log(d\rho))}\right) \mathrm{poly}(n)$.

Completeness and soundness follow from the definition.

We need the following technical lemma for the main theorem. Intuitively it shows that if a real-valued proof is bounded and close to a boolean proof, then its "acceptance probability" is also close to that of the boolean proof. The proof of this lemma is one of the most involved in this paper, and it is given in Appendix C due to page limits.

Definition 9. *Let V be any $(\ell^2, r, q, p, t, s, \tau)$-rectangular PCP verifier. We say a real matrix $\widetilde{\pi} \in \mathbb{R}^{\ell \times \ell}$ is bounded for V if for all $R_{\text{shared}} \in \{0,1\}^{r_{\text{shared}}}$, and all $S \subseteq [q]$, we have*

$$\mathbb{E}_{R_{\text{row}}, R_{\text{col}} \in \{0,1\}^{r_{\text{rect}}}} \left[\prod_{k \in S} \widetilde{\pi}^2_{i^{(k)}, j^{(k)}} \right] \leq 1,$$

where $i^{(k)} = i^{(k)}(1^n; R_{\text{row}}, R_{\text{shared}})$ and $j^{(k)} = j^{(k)}(1^n; R_{\text{row}}, R_{\text{shared}})$ for all $k \in [q]$.

Lemma 7. *Let V be any smooth $(\ell^2, r, q, p, t, s, \tau)$-rectangular PCP verifier over $\{-1, 1\}$, and $\widetilde{V}$ be the same as V but with D multilinearly extended over $\mathbb{R}$. Let π be any matrix in $\{-1, 1\}^{\ell \times \ell}$ and $\widetilde{\pi}$ be any matrix in $\mathbb{R}^{\ell \times \ell}$ that is bounded for V. Then we have*

$$\left| \mathbb{E}_R \left[V^\pi (1^n; R) \right] - \mathbb{E}_R \left[\widetilde{V}^{\widetilde{\pi}} (1^n; R) \right] \right| \leq 2^{O(q+p)} \|\pi - \widetilde{\pi}\|_2 .$$

4   Constant Hardness for Rational Sums of Low-Rank Matrices

In this section we prove our main hardness result against rational sums of low-rank matrices.

Theorem 3. *There is a function $f : \{0,1\}^{n+O(\log n)} \to \{-1, 1\}$ in $\mathbf{E^{NP}}$ that does not $(1 - \Omega(1))$-correlate with any bounded m-sum of rank-ρ matrices with $O(n)$ bit-complexity, for infinitely many n, as long as $n/\log \rho \geq \kappa(\log m + \log n)$ for a constant κ.*

Proof. Fix L to be a unary language in $\mathbf{NTIME}(2^n) \setminus \mathbf{NTIME}(o(2^n))$ [12,18, 31]. Let V be the smooth $(\ell^2, r, q, p, t, s, \tau)$-rectangular PCP verifier over alphabet $\{-1, 1\}$ for L given by Lemma 1, for s and τ to be determined later. Let $\widetilde{V}$ be the same as V but with D multilinearly extended over $\mathbb{R}$.

We use the lexicographically first proof oracle π as our hard function, i.e. our algorithm $f_n : [\ell] \times [\ell] \to \{-1, 1\}$ on input (i, j) searches bit-by-bit for the lexicographically first proof π such that $\forall R, V^\pi(1^n; R) = 1$ if one exists, and outputs $\pi_{i,j}$. Clearly $f_n \in \mathbf{E^{NP}}$. Note that f_n can also be seen as a family of matrices $f_n \in \{-1, 1\}^{\ell \times \ell}$.

Now for the sake of contradiction, we assume that f_n $(1 - \varepsilon)$-correlates with a bounded m-sum of rank-ρ matrices $\widetilde{\pi}$ with bit-complexity $O(n)$, for a constant ε to be determined later. We will show that $L \in \mathbf{NTIME}(2^n/n)$, thus deriving a contradiction.

Algorithm. The nondeterministic algorithm for L goes as follows:

1. Guess $\widetilde{\pi} = C \sum_{i=1}^{m} b_i \cdot (-1)^{A^{(i)} B^{(i)}}$ by guessing $b_i \in \{-1, 1\}$, matrices $A^{(i)} \in \mathbb{F}_2^{\ell \times \rho}$, and $B^{(i)} \in \mathbb{F}_2^{\rho \times \ell}$ for all $i \in [m]$ and $C \in \mathbb{Q}$ with bit-complexity $O(n)$.
2. Perform the close-to-boolean test in Lemma 5 for ε on $\widetilde{\pi}$, reject if it doesn't pass.
3. For each $R_{\text{shared}} \in \{0, 1\}^{r_{\text{shared}}}$, $k \in [q]$:
   (a) Compute the lists

$$I^{(k)} = \left[i^{(k)}(1^n; R_{\text{row}}, R_{\text{shared}}) | R_{\text{row}} \in \{0, 1\}^{r_{\text{rect}}} \right],$$

$$J^{(k)} = \left[j^{(k)}(1^n; R_{\text{col}}, R_{\text{shared}}) | R_{\text{col}} \in \{0, 1\}^{r_{\text{rect}}} \right].$$

   (b) For each $2 \leq d \leq 2q$:
       i. Perform the boundedness test in Lemma 6 for d and $(I^{(k)}, J^{(k)})$ on $\widetilde{\pi}$, reject if it doesn't pass.
4. Use Lemma 4 to calculate $\mathbb{E}_R[\widetilde{V}^{\widetilde{\pi}}(1^n; R)]$, and only accept if $\mathbb{E}_R[\widetilde{V}^{\widetilde{\pi}}(1^n; R)] > \gamma$ where the constant γ is to be determined later.

Runtime. Step 1 takes time $O(m\ell\rho + n)$.

By Lemma 5, Step 2 takes time $\widetilde{O}\left(m^4 \cdot \ell^{2 - \Omega(1/\log \rho)}\right)$ if $\rho = \ell^{o(1)}$.

Step 3(a) takes time $\widetilde{O}(2^{r_{\text{rect}}} \cdot t)$, and we have $|I| = 2^{r_{\text{rect}}}$. Therefore by Lemma 6, Step 3(b) takes time $\widetilde{O}\left(qm\rho 2^{r_{\text{rect}}} + qm^{2q} 2^{2r_{\text{rect}} - \Omega(r/\log(q\rho))}\right)$, if $q\rho = (2^{r_{\text{rect}}})^{o(1)}$, i.e. $\log(q\rho) = o(r)$. Hence the total runtime for Step 3 is

$$\widetilde{O}\left(2^{r_{\text{shared}}} \cdot \left(2^{r_{\text{rect}}} \cdot (t + qm\rho) + q \cdot m^{2q} \cdot 2^{2r_{\text{rect}} - \Omega(r/\log(q\rho))}\right)\right)$$

$$= \widetilde{O}\left(2^{r_{\text{shared}} + r_{\text{rect}}} \cdot (t + qm\rho) + qm^{2q} \cdot 2^{r - \Omega(r/\log(q\rho))}\right).$$

By Lemma 4, Step 4 runs in time $\widetilde{O}(2^{r_{\text{shared}} + r_{\text{rect}}} \cdot (t + m\rho) + m^{q+p} \cdot 2^{q+p+r - \Omega(r/\log((q+p)\rho))})$ if $\log((q + p)\rho) = o(r)$.

For the algorithm to run in time $O(2^n/n)$, it suffices to satisfy all the above requirements and make all the runtime to be $O(2^n/n)$. For convenience we take logarithms on all the time bounds. In summary, it is sufficient to satisfy the following conditions:

1. $\log(m\ell\rho + n) < n - \log n$.
2. $\rho = \ell^{o(1)}$.
3. $\log(m^4\ell^2) - \Omega(\log \ell / \log \rho) + O(\log n) < n - \log n$.
4. $\log(q\rho) = o(r)$.
5. $r_{\text{shared}} + r_{\text{rect}} + \log(t + qm\rho) + O(\log n) < n - \log n$.
6. $\log q + 2q \log m + r - \Omega\left(\frac{r}{\log(q\rho)}\right) + O(\log n) < n - \log n$.
7. $\log((q + p)\rho) = o(r)$.
8. $(q + p)(\log m + 1) + r - \Omega\left(\frac{r}{\log((q+p)\rho)}\right) + O(\log n) < n - \log n$.

We are going to set parameters to meet these conditions at the end.

Correctness. We first prove the following claim.

Claim. If $\widetilde{\pi}$ passes all the tests in the definition of the algorithm then it is bounded for V.

Proof. Fix any $R_{\text{shared}} \in \{0,1\}^{r_{\text{shared}}}$ and $S \subseteq [q]$. Let $d = |S|$. By Hölder's inequality, we get

$$\mathbb{E}_{R_{\text{row}},R_{\text{col}}}\left[\prod_{k \in S} \widetilde{\pi}_{i(k),j(k)}^2\right] \leq \prod_{k \in S}\left(\mathbb{E}_{R_{\text{row}},R_{\text{col}}}\left[\widetilde{\pi}_{i(k),j(k)}^{2d}\right]\right)^{1/d} = \prod_{k \in S}\left(\mathbb{E}_{i \sim I(k),j \sim J(k)}\left[\widetilde{\pi}_{i,j}^{2d}\right]\right)^{1/d}.$$

We have $2d \leq 2q$, therefore the boundedness tests in Step 3 can guarantee that all the terms in the product are bounded by 1, hence $\mathbb{E}_{R_{\text{row}},R_{\text{col}}}\left[\prod_{k \in S} \widetilde{\pi}_{i(k),j(k)}^2\right] \leq 1$, thus by definition $\widetilde{\pi}$ is bounded for V.

If $x = 1^n \in L$, let $\widetilde{\pi} \in \mathbb{R}^{\ell \times \ell}$ be any bounded m-sum of rank-ρ matrices with bit-complexity $O(\log n)$ that $(1-\varepsilon)$-correlates with the hard function f_n, which is the lexicographically first proof π in this case. We can assume wlog $\mathbb{E}_{i,j}[\pi_{i,j}\widetilde{\pi}_{i,j}] \geq 1-\varepsilon$, otherwise we can simply use $-\widetilde{\pi}$. Note that for any $x \in \{-1,1\}$, $y \in [-1,1]$ we have $|x - y| = 1 - xy$. As $\pi_{i,j} \in \{-1,1\}$, $\widetilde{\pi}_{i,j} \in [-1,1]$, we have

$$\|\pi - \widetilde{\pi}\|_1 = \mathbb{E}_{i,j}\left[|\pi_{i,j} - \widetilde{\pi}_{i,j}|\right] = 1 - \mathbb{E}_{i,j}\left[\pi_{i,j}\widetilde{\pi}_{i,j}\right] \leq \varepsilon.$$

Hence $\|\pi - \widetilde{\pi}\|_2 \leq \sqrt{2\varepsilon}$, and moreover $\widetilde{\pi}$ passes the close-to-boolean test. Since $\widetilde{\pi}$ is bounded by assumption it is also bounded for V. Therefore by Lemma 7 we have

$$\mathbb{E}_R\left[\widetilde{V}^{\widetilde{\pi}}(1^n;R)\right] \geq \mathbb{E}_R[V^\pi(1^n;R)] - \left|\mathbb{E}_R\left[V^\pi(1^n;R)\right] - \mathbb{E}_R\left[\widetilde{V}^{\widetilde{\pi}}(1^n;R)\right]\right|$$

$$\geq 1 - 2^{O(p+q)}\|\pi - \widetilde{\pi}\|_2$$

$$\geq 1 - 2^{O(p+q)}\sqrt{\varepsilon}.$$

If $x = 1^n \notin L$, then for any guessed m-sum of matrices $\widetilde{\pi}$ that passes all the tests, by soundness there exists a boolean proof $\pi \in \{-1,1\}^{\ell \times \ell}$ such that $\|\pi - \widetilde{\pi}\|_2 \leq 2\sqrt{2\varepsilon}$, and by the above claim we know that $\widetilde{\pi}$ is bounded for V. Therefore by Lemma 7 we have

$$\mathbb{E}_R[\widetilde{V}^{\widetilde{\pi}}(1^n;R)] \leq \mathbb{E}_R[V^\pi(1^n;R)] + \left|\mathbb{E}_R\left[V^\pi(1^n;R)\right] - \mathbb{E}_R\left[\widetilde{V}^{\widetilde{\pi}}(1^n;R)\right]\right|$$

$$\leq s + 2^{O(p+q)}\|\pi - \widetilde{\pi}\|_2$$

$$\leq s + 2^{O(p+q)}\sqrt{\varepsilon}.$$

We can set γ to be any value between these two. Assuming Condition 1–8 are all met, the above nondeterministic algorithm decides L in time $O(2^n/n)$, a contradiction to our choice of L.

Setting Parameters. We choose an arbitrary small constant s so both p,q are constants. Then we set ε to be any constant smaller than $\left(\frac{1-s}{2^{O(q+p)}}\right)^2$.

Now fix any ρ and m such that $n/\log\rho \geq \kappa(\log m + \log n)$ for a large constant κ to be determined. We are going to verify that Conditions 1–8 are satisfied. Note that by Lemma 1 we have $r = n + O(\log n)$, $\log\ell = n/2 + O(\log n)$, and $\log t = O(\tau n)$. First, $\log\rho \leq n/(\kappa\log n) = o(r)$ and similarly $\rho = \ell^{o(1)}$, so Condition 2, 4, and 7 are all satisfied. As both $\log\rho$ and $\log m$ are at most n/κ, for Condition 1 we have

$$\log(m\ell\rho) + \log n \leq 2n/\kappa + n/2 + O(\log n) < n - \log n,$$

for κ sufficiently large, while for Condition 5 we have

$$(1+\tau)r/2+\log t+\log(qm\rho)+O(\log n) \leq (1/2+O(\tau))n+\log\rho+\log m < n-\log n,$$

for κ sufficiently large and τ sufficiently small. For Condition 8 we have

$$(q+p)(\log m + 1) + r - \Omega\left(\frac{(1-\tau)r}{\log((q+p)\rho)}\right) + O(\log n)$$
$$\leq O(\log m) + n + O(\log n) - \Omega(n/\log\rho)$$
$$\leq n + O(\log m + \log n) - \Omega(\kappa(\log m + \log n))$$
$$< n - \log n,$$

for κ sufficiently large. Similarly Condition 3, 5, and 6 are also satisfied, and we are done.

5   Correlation Bounds via XOR Lemma

In this section we adapt the approach in [8] to our setting, and then prove the main result in this paper, Theorem 1. Due to page limits the proof of Lemma 8 is included in Appendix D.

Definition 10. *For any boolean function $f\colon \{0,1\}^n \to \{-1,1\}$ and number k, we define $f^{\oplus k}\colon \{0,1\}^{nk} \to \{-1,1\}$ by $f^{\oplus k}(x_1,\ldots,x_k) = \prod_{i=1}^{k} f(x_i)$ for all $x_1,\ldots,x_k \in \{0,1\}^n$.*

Lemma 8. *Let $f\colon \{0,1\}^n \to \{-1,1\}$ be any boolean function. Let rational $\varepsilon < 1$ have constant bit-complexity, and for any number $k \geq 1$, let $\varepsilon_k = (\frac{1+\varepsilon}{2})^{k-1}\varepsilon$. Assume that $f^{\oplus k}$ ε_k-correlates with some function $h\colon \{0,1\}^{nk} \to [-1,1]$. Then f ε-correlates with a bounded m-sum of restrictions of h (by fixing some inputs), where $m = O\left(\frac{n}{\varepsilon_k^2}\right)$ and the bit-complexity is $O(k + \log n)$.*

Proof (Proof of Theorem 1 for infinitely many n). Let $f\colon \{0,1\}^{n+O(\log n)} \to \{-1,1\}$ in $\mathrm{E}^{\mathbf{NP}}$ be the function given by Theorem 3. We know that f does not ε-correlate with any bounded m-sum of rank-ρ matrices with $O(n)$ bit-complexity for infinitely many n, for a rational constant ε with constant bit-complexity.

Let $k \leq n$. We set $\varepsilon_k = (\frac{1+\varepsilon}{2})^{k-1}\varepsilon = 2^{-O(k)}$, and $F = f^{\oplus k}$ so $N = k(n + O(\log n))$. For the sake of contradiction we assume that F ε_k-correlates with some rank-ρ matrix h for infinitely many N. We view a matrix as the truth table of a function, so when we take restrictions on the function, we are taking some rows and columns of the matrix but keeping its dimensions, thus the rank doesn't increase. By Lemma 8 we know that f ε-correlates with a bounded m-sum of rank-ρ matrices $\tilde{h}$, where $m = O(n/\varepsilon_k^2) = n2^{O(k)}$ and the bit complexity is $O(k + \log n) = O(n)$.

To get a contradiction for infinitely many n we still need to verify that $n/\log \rho \geq \kappa(\log m + \log n)$ for a sufficiently large constant κ given by Theorem 3. We have $\log n = \log N - \log k - O(\log \log n)$, thus

$$(\log m + \log n)\log \rho = (2\log n + O(k))\log \rho = O((\log N + k)\log \rho).$$

Let $c > 0$ be the constant hidden in the last big-O. We take $\delta = 1/2c\kappa$. Then

$$\kappa(\log m + \log n)\log \rho \leq c\kappa(\log N + k)\log \rho \leq \frac{N}{2k} = \frac{n + O(\log n)}{2} < n.$$

To prove the full version of Theorem 1 that works for all sufficiently large n, we need the following *refuter* from [8].

Theorem 4. *There is a constant $c > 0$ such that the following holds.*

For any non-decreasing time-constructible function $T(n)$ such that $n \leq T(n) \leq 2^{\mathrm{poly}(n)}$, there is an **NTIME**$(T(n))$ *language L and an algorithm R such that:*

> **Input.** *The input for R is a pair $(M, 1^n)$ where M is a nondeterministic algorithm running in time $\leq cT(n)/\log T(n)$.*
> **Output.** *For any fixed M, for all large enough n, $R(M, 1^n)$ outputs a string x such that $|x| \in [n, n + T(n)]$ and $L(x) \neq M(x)$.*
> **Complexity.** *R is a deterministic algorithm running in $O(T(n) \cdot T(T(n)+n))$ time with an* **NP** *oracle.*

We also need a lemma on padding rigid matrices from [2].

Lemma 9. *Let $\mathbf{1}_m$ be the all-ones $m \times m$ matrix. For any square matrix A, A ε-correlates with some rank-ρ matrix if and only if $A \otimes \mathbf{1}_m$ ε-correlates with some rank-ρ matrix, where $\otimes$ is the tensor product of matrices.*

Proof (Proof of Theorem 1). We show how to remove the "infinitely often" part from the previous proof. Fix $T(n) = n^C$ for a large constant C. We are going to use the general version of Lemma 1 in [4] that works for **NTIME**$(T(n))$ languages. Most importantly, we have $r = \log T(n) + O(\log \log T(n)) + O(\log n)$ and $2\log \ell = r + O(1)$. Then the proof of Theorem 3 shows that we have the following results:

1. For any $\mathbf{NTIME}(T(n))$ language L, let V be the smooth $(\ell^2, r, q, p, t, s, \tau)$-rectangular PCP verifier over alphabet $\{-1, 1\}$ for L given by the generalized version of Lemma 1, for small constants s and τ. We define the function $f_{L,x} \colon [\ell] \times [\ell] \to \{-1, 1\}$ such that on input (i, j) it searches bit-by-bit for the lexicographically first proof π such that $\forall R$, $V^\pi(x; R) = 1$ if one exists, and outputs $\pi_{i,j}$. Clearly $f_{L,x} \in \mathbf{E^{NP}}$. Note that $f_{L,x}$ can also be seen as an $\ell \times \ell$ matrix.

2. For any $\mathbf{NTIME}(T(n))$ language L, there exists an explicit nondeterministic algorithm that decides if $x \in L$ in time $O(T(n)/\log T(n))$, for any input x such that $|x| = n$ and $f_{L,x}$ $(1 - \Omega(1))$-correlates with a bounded m-sum of rank-ρ matrices $\tilde{\pi}$ with bit-complexity $O(\log T(n))$, as long as $\log T(n)/\log \rho \geq \kappa(\log m + \log \log T(n))$ for a constant κ.

We consider the language L for $\mathbf{NTIME}(T(n))$ from Theorem 4. Similarly as before, by combining Item 2 with Lemma 8, there exists an explicit nondeterministic $O(T(n)/\log T(n))$-time algorithm M deciding if $x \in L$ for any input x such that $|x| = n$ and $f_{L,x}^{\oplus k}$ ε_k-correlates with some rank-ρ matrix h, for any $k \leq \log T(|x|) = C \log |x|$.

We aim to use the refuter R from Theorem 4 to get a contradiction. For all large enough n, R on $(1^n, M)$ will output an x such that $|x| \in [n, n^C]$ and $L(x) \neq M(x)$. Now the input length of $f_{L,x}^{\oplus k}$ is $k \cdot 2 \log \ell = k(r + O(1)) = k(\log T(|x|) + O(\log \log T(|x|)) + O(\log |x|)) \in [k \cdot C \log n, k \cdot 2C^2 \log n]$. We view $f_{L,x}^{\oplus k}$ as a matrix and use Lemma 9 to get a function F_x with input length $k \cdot 2C^2 \log n$ such that F_x ε_k-correlates with some rank-ρ matrix iff $f_{L,x}^{\oplus k}$ ε_k-correlates with some rank-ρ matrix. Therefore if F_x ε_k-correlates with some rank-ρ matrix then M can decide if $x \in L$, a contradiction.

Our final hard function f works as follows. On input of length $N = k \cdot 2C^2 \log n$ it runs R on $(1^n, M)$ to get an x, then run F_x. Then for all large enough n, f does not ε_k-correlate with any rank-ρ matrices, for any $k \leq C \log n$. R runs in $O(n^{C^2}) = 2^{O(N/k)}$ time with an $\mathbf{NP}$ oracle and $f_{L,x} \in \mathbf{E^{NP}}$, thus $f \in \mathbf{E^{NP}}$. The condition $\log T(n)/\log \rho \geq \kappa(\log m + \log \log T(n))$ in Item 2 can be verified similarly as in the previous proof for a sufficiently small δ.

A   Proof of Lemma 3

Proof. The first column of A is $\langle i, u \rangle$, row-indexed by $i \in \{0, 1\}^m$. The second column of A is all 1, while the third column of A is all b. The second row of B is $\langle j, u \rangle$, column-indexed by $j \in \{0, 1\}^m$, while every other entry in B is 1.

B   Proof of Lemma 4

Proof. The algorithm on input $\tilde{\pi} = \sum_{i=1}^m \alpha_i \cdot (-1)^{A^{(i)} B^{(i)}}$ runs as follows:

1. Initialize the result res to be 0.
2. For each $R_{\mathsf{shared}} \in \{0, 1\}^{r_{\mathsf{shared}}}$:

(a) Compute the decision function $D = D(1^n; R_{shared})$ and randomness parity check
$$(C_1, \ldots, C_p) = (C_1(1^n; R_{shared}), \ldots, C_p(1^n; R_{shared})).$$

(b) For each $k \in [q]$, for each $i \in [m]$,

    i Compute the $2^{r_{rect}} \times \rho$ matrices $A^{(k,i)}$ whose R_{row}-th row is the row of $A^{(i)}$ indexed by $i^{(k)}(1^n; R_{row}, R_{shared})$ for all $R_{row} \in \{0,1\}^{r_{rect}}$.

    ii Compute the $\rho \times 2^{r_{rect}}$ matrices $B^{(k,i)}$ whose R_{col}-th column is the column of $B^{(i)}$ indexed by $j^{(k)}(1^n; R_{col}, R_{shared})$ for all $R_{col} \in \{0,1\}^{r_{rect}}$.

(c) For each $j \in [p]$,

    i Compute the $2^{r_{rect}} \times 3$ matrix $A^{(q+j,1)}$ and the $3 \times 2^{r_{rect}}$ matrix $B^{(q+j,1)}$ with
$$\left((-1)^{A^{(q+j,1)} B^{(q+j,1)}} \right)_{R_{row}, R_{col}} = C_j(R_{row}, R_{col}) \text{ given by Lemma 3.}$$

(d) Now we define q m-sums of rank-ρ matrices, $\widetilde{Q}_k = C \sum_{i=1}^{m} b_i \cdot (-1)^{A^{(k,i)} B^{(k,i)}}$ for each $k \in [q]$, and p 1-sums of rank-3 matrices, $\widetilde{Q}_{q+j} = (-1)^{A^{(q+j,1)} B^{(q+j,1)}}$ for each $j \in [p]$. Apply Theorem 2 to calculate the following value and add it to res:
$$\mathbb{E}_{R_{row}, R_{col}} \left[\left(\overline{D}(\widetilde{Q}_1, \ldots, \widetilde{Q}_q, \widetilde{Q}_{q+1}, \ldots, \widetilde{Q}_{q+p}) \right)_{R_{row}, R_{col}} \right].$$

3. Return res as the value of $\mathbb{E}_R \left[\widetilde{V}^{\widetilde{\pi}}(1^n; R) \right]$.

Correctness of the algorithm follows from Definition 6.

Step 2(b) runs in time $O(2^{r_{rect}} \cdot (t + m\rho))$, while Step 2(c) runs in $O(r 2^{r_{rect}})$ by Lemma 3, which is dominated by the runtime of Step 2(b) since $t \geq r$. By Theorem 2, Step 2(d) takes time $O(2^{q+p} \cdot m^{q+p} \cdot (T(2^{r_{rect}}, (q+p)\rho) + \text{poly}(n, q+p, r))) = O(2^{q+p} \cdot m^{q+p} \cdot T(2^{r_{rect}}, (q+p)\rho)) \text{poly}(n)$, if $(q+p)\rho = (2^{r_{rect}})^{o(1)}$, i.e. $\log((q+p)\rho) = o(r)$. Therefore the running time of the above algorithm is

$$O\left(2^{r_{shared}} \cdot \left(2^{r_{rect}} \cdot (t + m\rho) + 2^{q+p} \cdot m^{q+p} \cdot T(2^{r_{rect}}, (q+p)\rho) \right) \right) \text{poly}(n)$$
$$= O\left(2^{r_{shared}+r_{rect}} \cdot (t + m\rho) + m^{q+p} \cdot 2^{q+p+r-\Omega(r/\log((q+p)\rho))} \right) \text{poly}(n).$$

C   Proof of Lemma 7

Proof. Fix an arbitrary $R_{shared} \in \{0,1\}^{r_{shared}}$. By definition $\mathbb{E}_{R_{row}, R_{col}}$ $\left[\left| V^\pi(1^n; R) - \widetilde{V}^{\widetilde{\pi}}(1^n; R) \right| \right]$ is

$$\mathbb{E}_{R_{row}, R_{col}} \left[\left| D(\pi_{i^{(1)}, j^{(1)}}, \ldots, \pi_{i^{(q)}, j^{(q)}}, C_1(R_{row}, R_{col}), \ldots, C_p(R_{row}, R_{col})) \right. \right.$$
$$\left. \left. - D(\widetilde{\pi}_{i^{(1)}, j^{(1)}}, \ldots, \widetilde{\pi}_{i^{(q)}, j^{(q)}}, C_1(R_{row}, R_{col}), \ldots, C_p(R_{row}, R_{col})) \right| \right],$$
$$(2)$$

where $D = D(1^n; R_{shared})$, $(C_1, \ldots, C_p) = (C_1(1^n; R_{shared}), \ldots, C_p(1^n; R_{shared}))$, $i^{(k)} = i^{(k)}(1^n; R_{row}, R_{shared})$ and $j^{(k)} = j^{(k)}(1^n; R_{row}, R_{shared})$ for all $k \in [q]$.

We write D in its Fourier expansion $D(z_1, \ldots, z_{q+p}) = \sum_{S \subseteq [q+p]} \beta_S \prod_{k \in S} z_k$, where for each $S \subseteq [q+p]$, $\beta_S = \mathbb{E}_{z \in \{-1,1\}^{q+p}} [D(z) \prod_{k \in S} z_k]$. For all $z \in \{-1,1\}^{q+p}$, $D(z) \in \{0,1\}$ and $\prod_{k \in S} z_k \in \{-1,1\}$, thus $|\beta_S| \leq 1$ for any S. Hence by the triangular inequality we can bound (2) by

$$\sum_{S \subseteq [q+p]} \mathbb{E}_{R_{\mathrm{row}}, R_{\mathrm{col}}} \left[\left| \left(\prod_{k \in S \cap [q]} \pi_{i(k), j(k)} - \prod_{k \in S \cap [q]} \widetilde{\pi}_{i(k), j(k)} \right) \prod_{k \in S \setminus [q]} C_k(R_{\mathrm{row}}, R_{\mathrm{col}}) \right| \right]$$

$$= 2^p \sum_{S \subseteq [q]} \mathbb{E}_{R_{\mathrm{row}}, R_{\mathrm{col}}} \left[\left| \prod_{k \in S} \pi_{i(k), j(k)} - \prod_{k \in S} \widetilde{\pi}_{i(k), j(k)} \right| \right], \tag{3}$$

as all the C_k's are $\{-1,1\}$-valued.

Fix any $S \subseteq [q]$. Wlog let $S = \{1, \ldots, d\}$ for some $d \leq q$, then the expectation in (3) can be written as

$$\mathbb{E}_{R_{\mathrm{row}}, R_{\mathrm{col}}} \left[\left| \prod_{u=1}^{d} \pi_{i(u), j(u)} - \prod_{u=1}^{d} \widetilde{\pi}_{i(u), j(u)} \right| \right]$$

$$= \mathbb{E}_{R_{\mathrm{row}}, R_{\mathrm{col}}} \left[\left| \sum_{v=1}^{d} \left(\prod_{u=1}^{v-1} \widetilde{\pi}_{i(u), j(u)} \prod_{u=v}^{d} \pi_{i(u), j(u)} - \prod_{u=1}^{v} \widetilde{\pi}_{i(u), j(u)} \prod_{u=v+1}^{d} \pi_{i(u), j(u)} \right) \right| \right]$$

$$\leq \sum_{v=1}^{d} \mathbb{E}_{R_{\mathrm{row}}, R_{\mathrm{col}}} \left[\left| \prod_{u=1}^{v-1} \widetilde{\pi}_{i(u), j(u)} \prod_{u=v}^{d} \pi_{i(u), j(u)} - \prod_{u=1}^{v} \widetilde{\pi}_{i(u), j(u)} \prod_{u=v+1}^{d} \pi_{i(u), j(u)} \right| \right]$$

$$= \sum_{v=1}^{d} \mathbb{E}_{R_{\mathrm{row}}, R_{\mathrm{col}}} \left[\left| \prod_{u=1}^{v-1} \widetilde{\pi}_{i(u), j(u)} \cdot \left(\pi_{i(v), j(v)} - \widetilde{\pi}_{i(v), j(v)} \right) \cdot \prod_{u=v+1}^{d} \pi_{i(u), j(u)} \right| \right]$$

$$\leq \sum_{v=1}^{d} \left(\mathbb{E}_{R_{\mathrm{row}}, R_{\mathrm{col}}} \left[\left(\pi_{i(v), j(v)} - \widetilde{\pi}_{i(v), j(v)} \right)^2 \right] \right)^{1/2} \left(\mathbb{E}_{R_{\mathrm{row}}, R_{\mathrm{col}}} \left[\prod_{u=1}^{v-1} \widetilde{\pi}_{i(u), j(u)}^2 \prod_{u=v+1}^{d} \pi_{i(u), j(u)}^2 \right] \right)^{1/2}$$

$$\leq \sum_{v=1}^{d} \left(\mathbb{E}_{R_{\mathrm{row}}, R_{\mathrm{col}}} \left[\left(\pi_{i(v), j(v)} - \widetilde{\pi}_{i(v), j(v)} \right)^2 \right] \right)^{1/2}$$

$$= \sum_{k \in S} \left(\mathbb{E}_{R_{\mathrm{row}}, R_{\mathrm{col}}} \left[\left(\pi_{i(k), j(k)} - \widetilde{\pi}_{i(k), j(k)} \right)^2 \right] \right)^{1/2},$$

where the first inequality comes from the triangular inequality, the second inequality follows from the Cauchy-Schwarz inequality, and the last inequality follows from the assumptions that $\pi \in \{-1,1\}^{\ell \times \ell}$ and $\widetilde{\pi}$ is bounded for V.

Summing over S, we can bound (3) by

$$2^p \sum_{S \subseteq [q]} \sum_{k \in S} \left(\mathbb{E}_{R_{\mathrm{row}}, R_{\mathrm{col}}} \left[\left(\pi_{i(k), j(k)} - \widetilde{\pi}_{i(k), j(k)} \right)^2 \right] \right)^{1/2}$$

$$= 2^{p+q-1} \sum_{k \in [q]} \left(\mathbb{E}_{R_{\mathrm{row}}, R_{\mathrm{col}}} \left[\left(\pi_{i(k), j(k)} - \widetilde{\pi}_{i(k), j(k)} \right)^2 \right] \right)^{1/2}$$

$$= 2^{O(p+q)} \mathbb{E}_{k \in [q]} \left[\left(\mathbb{E}_{R_{\mathrm{row}}, R_{\mathrm{col}}} \left[\left(\pi_{i(k), j(k)} - \widetilde{\pi}_{i(k), j(k)} \right)^2 \right] \right)^{1/2} \right]$$

$$\leq 2^{O(p+q)} \left(\mathbb{E}_{R_{\mathrm{row}}, R_{\mathrm{col}}, k} \left[\left(\pi_{i(k), j(k)} - \widetilde{\pi}_{i(k), j(k)} \right)^2 \right] \right)^{1/2},$$

where the first step uses double counting, and the last step follows from Jensen's inequality.

Therefore by averaging over R_{shared}, we have

$$\left| \mathbb{E}_R \left[V^{\pi}(1^n; R) \right] - \mathbb{E}_R \left[\widetilde{V}^{\widetilde{\pi}}(1^n; R) \right] \right|$$

$$\leq \mathbb{E}_{R_{\text{shared}}} \mathbb{E}_{R_{\text{row}}, R_{\text{col}}} \left[\left| V^{\pi}(1^n; R) - \widetilde{V}^{\widetilde{\pi}}(1^n; R) \right| \right]$$

$$\leq 2^{O(p+q)} \mathbb{E}_{R_{\text{shared}}} \left[\left(\mathbb{E}_{R_{\text{row}}, R_{\text{col}}, k} \left[\left(\pi_{i(k), j(k)} - \widetilde{\pi}_{i(k), j(k)} \right)^2 \right] \right)^{1/2} \right]$$

$$\leq 2^{O(p+q)} \left(\mathbb{E}_{R_{\text{shared}}, R_{\text{row}}, R_{\text{col}}, k} \left[\left(\pi_{i(k), j(k)} - \widetilde{\pi}_{i(k), j(k)} \right)^2 \right] \right)^{1/2}$$

$$= 2^{O(p+q)} \left(\mathbb{E}_{i,j} \left[\left(\pi_{i,j} - \widetilde{\pi}_{i,j} \right)^2 \right] \right)^{1/2}$$

$$= 2^{O(p+q)} \| \pi - \widetilde{\pi} \|_2,$$

where the first step uses triangular inequality, the second step uses the above bound for every R_{shared}, the third step comes from Jensen's inequality, and the fourth step follows from the smoothness of V.

D   Proof of Lemma 8

Proof. We prove it by induction on k. For $k = 1$ it is trivial as h is bounded. Now we assume that the hypothesis holds for $k - 1$, and we are proving for k.

For all $x_1 \in \{0,1\}^n$, define $g(x_1) = \mathbb{E}_{y \sim \{0,1\}^{n(k-1)}} \left[f^{\oplus k-1}(y) h(x_1, y) \right]$, where we use y for $(x_2, \ldots, x_k)$ for convenience. If there exists $x_1 \in \{0,1\}^n$ such that $|g(x_1)| \geq \varepsilon_{k-1}$, then we know that $f^{\oplus k-1}$ ε_{k-1}-correlates with h' defined by $h'(y) = h(x_1, y)$, so we can use the induction hypothesis for $k - 1$ to get a bounded m-sum of functions obtained by fixing inputs of h', thus by fixing inputs of h.

Otherwise, for all $x_1 \in \{0,1\}^n$ we have $|g(x_1)| \leq \varepsilon_{k-1} = \frac{2\varepsilon_k}{1+\varepsilon}$. We take m i.i.d. samples $y_1, \ldots, y_m$ uniformly from $\{0,1\}^{n(k-1)}$ for $m = O\left(\frac{n}{(\varepsilon_k)^2}\right)$, then define $\widetilde{g}(x_1) = \mathbb{E}_{i \in [m]} \left[f^{\oplus k-1}(y_i) h(x_1, y_i) \right]$. By Chernoff bound,

$$\Pr_{y_1, \ldots, y_m} \left[|g(x_1) - \widetilde{g}(x_1)| \geq \frac{1-\varepsilon}{(1+\varepsilon)^2} \varepsilon_k \right] \leq 2^{-n-1}.$$

By union bound, there exists a fix assignment to $y_1, \ldots, y_m$ such that for all $x_1 \in \{0,1\}^n$,

$$|g(x_1) - \widetilde{g}(x_1)| \leq \frac{1-\varepsilon}{(1+\varepsilon)^2} \varepsilon_k, \tag{4}$$

thus $|\widetilde{g}(x_1)| \leq |g(x_1)| + |g(x_1) - \widetilde{g}(x_1)| \leq \left(\frac{2}{1+\varepsilon} + \frac{1-\varepsilon}{(1+\varepsilon)^2} \right) \varepsilon_k = \frac{3+\varepsilon}{(1+\varepsilon)^2} \varepsilon_k$.

Let $r = \frac{3+\varepsilon}{(1+\varepsilon)^2} \varepsilon_k$. We define $\widetilde{h}$ by

$$\widetilde{h}(x_1) = \frac{\widetilde{g}(x_1)}{r} = \frac{1}{mr} \sum_{i=1}^{m} f^{\oplus k-1}(y_i) h(x_1, y_i).$$

Now $|\widetilde{h}(x_1)| \leq 1$ for all x_1. We can write $\widetilde{h}$ as $C \sum_{i=1}^{m} b_i h_i$, where

$$C = \frac{1}{mr},$$
$$b_i = f^{\oplus k-1}(y_i), \forall i \in [m],$$
$$h_i \colon x_1 \mapsto h(x_1, y_i), \forall x_1 \in \{0,1\}^n, \forall i \in [m],$$

which is a bounded $O(n/\varepsilon_k^2)$-sum of functions that can be obtained by fixing inputs of h. The bit-complexity of m is $O(k + \log n)$, and $O(k)$ for r, thus the bit-complexity of $\widetilde{h}$ is $O(k + \log n)$.

What remains is to prove that $\mathrm{corr}(f, \widetilde{h}) \geq \varepsilon$. By the definition of g and assumption we have $\mathrm{corr}(f, g) = \mathrm{corr}(f^{\oplus k}, h) \geq \varepsilon_k$. Therefore by the definition of $\widetilde{h}$, the fact that $f(x_1) \in \{-1, 1\}$ for all x_1, and (4), we have

$$
\begin{aligned}
\mathrm{corr}(f, \widetilde{h}) &= \frac{\mathrm{corr}(f, \widetilde{g})}{r} \\
&\geq \frac{1}{r} \left(\mathrm{corr}(f, g) - \mathbb{E}_{x_1} |g(x_1) - \widetilde{g}(x_1)| \right) \\
&\geq \frac{\varepsilon_k - \frac{1-\varepsilon}{(1+\varepsilon)^2} \varepsilon_k}{\frac{3+\varepsilon}{(1+\varepsilon)^2} \varepsilon_k} \\
&= \varepsilon.
\end{aligned}
$$

References

1. Alman, J., Chan, T.M., Williams, R.: Polynomial representations of threshold functions and algorithmic applications. In: IEEE Symposium on Foundations of Computer Science (FOCS) (2016)
2. Alman, J., Chen, L.: Efficient construction of rigid matrices using an NP oracle. In: IEEE Symposium on Foundations of Computer Science (FOCS), pp. 1034–1055 (2019)
3. Ben-Sasson, E., Viola, E.: Short PCPs with projection queries. In: Colloquium on Automata, Languages and Programming (ICALP) (2014)
4. Bhangale, A., Harsha, P., Paradise, O., Tal, A.: Rigid matrices from rectangular PCPs. In: IEEE Symposium on Foundations of Computer Science (FOCS) (2020)
5. Chan, T.M., Williams, R.: Deterministic APSP, orthogonal vectors, and more: quickly derandomizing razborov-smolensky. In: ACM-SIAM Symposium on Discrete Algorithms (SODA), pp. 1246–1255 (2016)
6. Chen, L.: Non-deterministic quasi-polynomial time is average-case hard for ACC circuits. In: IEEE Symposium on Foundations of Computer Science (FOCS) (2019)
7. Chen, L., Lyu, X.: Inverse-exponential correlation bounds and extremely rigid matrices from a new derandomized XOR lemma. In: ACM Symposium on the Theory of Computing (STOC) (2021)
8. Chen, L., Lyu, X., Williams, R.: Almost everywhere circuit lower bounds from non-trivial derandomization. In: IEEE Symposium on Foundations of Computer Science (FOCS) (2020)
9. Chen, L., Ren, H.: Strong average-case circuit lower bounds from non-trivial derandomization. In: ACM Symposium on the Theory of Computing (STOC) (2020)

10. Chen, L., Williams, R.R.: Stronger connections between circuit analysis and circuit lower bounds, via PCPs of proximity. In: IEEE Conference on Computational Complexity (CCC), pp. 19:1–19:43 (2019)
11. Chen, R., Oliveira, I.C., Santhanam, R.: An average-case lower bound against ACC^0. In: LATIN. Lecture Notes in Computer Science, vol. 10807, pp. 317–330. Springer (2018)
12. Cook, S.A.: A hierarchy for nondeterministic time complexity. J. Comput. Syst. Sci. **7**(4), 343–353 (1973)
13. Goldreich, O., Nisan, N., Wigderson, A.: On Yao's XOR-Lemma. In: Goldreich, O. (ed.) Studies in Complexity and Cryptography. Miscellanea on the Interplay between Randomness and Computation. LNCS, vol. 6650, pp. 273–301. Springer, Heidelberg (2011). https://doi.org/10.1007/978-3-642-22670-0_23
14. Murray, C., Williams, R.R.: Circuit lower bounds for nondeterministic quasi-polytime: an easy witness lemma for NP and NQP. In: STOC, pp. 890–901. ACM (2018)
15. Paradise, O.: Smooth and strong PCPss. In: ACM Innovations in Theoretical Computer Science conf. (ITCS), pp. 2:1–2:41 (2020)
16. Rajgopal, N., Santhanam, R., Srinivasan, S.: Deterministically counting satisfying assignments for constant-depth circuits with parity gates, with implications for lower bounds. In: Potapov, I., Spirakis, P.G., Worrell, J. (eds.) Symposium on Mathamatics Foundations of Computer Science (MFCS). LIPIcs, vol. 117, pp. 78:1–78:15. Schloss Dagstuhl - Leibniz-Zentrum für Informatik (2018). https://doi.org/10.4230/LIPIcs.MFCS.2018.78
17. Razborov, A.: Lower bounds on the dimension of schemes of bounded depth in a complete basis containing the logical addition function. Math. Notes Acad. Sci. USSR **41**(4), 598–607 (1987)
18. Seiferas, J.I., Fischer, M.J., Meyer, A.R.: Separating nondeterministic time complexity classes. J. ACM **25**(1), 146–167 (1978)
19. Servedio, R.A., Viola, E.: On a special case of rigidity (2012). http://www.ccs.neu.edu/home/viola/
20. Smolensky, R.: Algebraic methods in the theory of lower bounds for Boolean circuit complexity. In: 19th ACM Symposium on the Theory of Computing (STOC), pp. 77–82. ACM (1987)
21. Smolensky, R.: On representations by low-degree polynomials. In: 34th IEEE Symposium on Foundations of Computer Science (FOCS), pp. 130–138 (1993)
22. Tamaki, S.: A satisfiability algorithm for depth two circuits with a sub-quadratic number of symmetric and threshold gates. Electron. Coll. Comput. Complexity (ECCC) **23**, 100 (2016). http://eccc.hpi-web.de/report/2016/100
23. Valiant, L.G.: Graph-theoretic arguments in low-level complexity. In: Gruska, J. (ed.) MFCS 1977. LNCS, vol. 53, pp. 162–176. Springer, Heidelberg (1977). https://doi.org/10.1007/3-540-08353-7_135
24. Viola, E.: New lower bounds for probabilistic degree and AC0 with parity gates. Electron. Coll. Computat. Complexity (ECCC) **27**, 15 (2020)
25. Vyas, N., Williams, R.: Lower bounds against sparse symmetric functions of ACC circuits: expanding the reach of #SAT algorithms. In: Symposium on Theoretical Aspects of Computer Science (STACS) (2020)
26. Williams, R.: Improving exhaustive search implies superpolynomial lower bounds. In: 42nd ACM Symposium on the Theory of Computing (STOC), pp. 231–240 (2010)
27. Williams, R.: Guest column: a casual tour around a circuit complexity bound. SIGACT News **42**(3), 54–76 (2011)

28. Williams, R.: Natural proofs versus derandomization. In: ACM Symposium on the Theory of Computing (STOC) (2013)
29. Williams, R.: New algorithms and lower bounds for circuits with linear threshold gates. In: ACM Symposium on the Theory of Computing (STOC) (2014)
30. Williams, R.: Nonuniform ACC circuit lower bounds. J. ACM **61**(1), 2:1–2:32 (2014). http://doi.acm.org/10.1145/2559903
31. Zák, S.: A turing machine time hierarchy. Theor. Comput. Sci. **26**, 327–333 (1983)

Analysis of an Efficient Reduction Algorithm for Random Regular Expressions Based on Universality Detection

Florent Koechlin[✉] and Pablo Rotondo

LIGM, Univ Gustave Eiffel, CNRS, ENPC, Champs-sur-Marne, France
{florent.koechlin,pablo.rotondo}@u-pem.fr

Abstract. In this article we study a very simple linear reduction algorithm that is specific to regular expressions. It aims to detect, in a bottom-up fashion, universal subtrees in regular expressions trees, and replace them by the smallest equivalent to Σ^*. Of course, this does not detect every universal subtree, as the universality problem is PSPACE-complete. However, we prove that for the uniform random tree model, this simple algorithm detects a large proportion of universal trees. Furthermore, we prove that on average this algorithm reduces uniform regular expressions to a constant size that is very small, and that can be computed efficiently. For example, for two letters the limit expected size is $\sim$77.8. Our theoretical constants are backed-up by the experimental evidence. This confirms the phenomena reported in [13], and further it completely discards the usefulness of the uniform distribution on regular expressions.

1 Introduction

Regular languages are ubiquitous in computer science. The natural way to specify these languages, when it comes to programming, is through regular expressions. Thus there are many algorithms taking regular expressions as inputs. A notable case is the compilation of regular expressions into automata. There exist several constructions of automata from regular expressions; for instance the Thompson construction, the Glushkov position automaton, the partial derivative automaton and the prefix automaton [1,2,18]. Faced with this choice, it is natural to wonder which one performs better in practice. Average case analysis seeks to give an answer by setting up a probabilistic model, hoping that this reflects real life inputs well. In this context, a natural choice for the model is the uniform distribution on the inputs. This distribution has two advantages: it maximizes the entropy, and it is often susceptible to theoretical analysis.

Regular expressions are represented naturally as expression trees, see Fig. 1. The uniform distribution on the associated trees has been used with success in the literature to study the complexity of automata constructions [3]. However, this model has been recently put into question by the work in [13]. It is shown that a uniform regular expression tree of size n over a two-letter alphabet is

© Springer Nature Switzerland AG 2021
R. Santhanam and D. Musatov (Eds.): CSR 2021, LNCS 12730, pp. 206–222, 2021.
https://doi.org/10.1007/978-3-030-79416-3_12

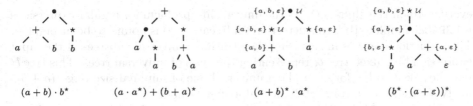

Fig. 1. Four regular expression trees and their associated formulas. The last three are universal, i.e., they recognize every word on the alphabet $\{a, b\}$. Our linear simplification algorithm (see Fig. 2) will detect their universality. For example, the last two expression trees are annotated with the subset (in blue) of symbols a, b, ε recognized and (in green) whether the expression associated to the node is detected to be universal by our algorithm. (Color figure online)

$$\underset{\mathcal{U} \;\; \mathcal{L}}{\overset{+}{\diagup \backslash}} \rightsquigarrow \mathcal{U}, \quad \underset{\mathcal{L} \;\; \mathcal{U}}{\overset{+}{\diagup \backslash}} \rightsquigarrow \mathcal{U}, \quad \underset{\mathcal{U} \;\; \mathcal{T}_\varepsilon}{\overset{\bullet}{\diagup \backslash}} \rightsquigarrow \mathcal{U}, \quad \underset{\mathcal{T}_\varepsilon \;\; \mathcal{U}}{\overset{\bullet}{\diagup \backslash}} \rightsquigarrow \mathcal{U}, \quad \underset{\mathcal{T}_\Sigma}{\overset{\star}{\mid}} \rightsquigarrow \mathcal{U}, \quad \underset{\mathcal{U}}{\overset{\star}{\mid}} \rightsquigarrow \mathcal{U}.$$

Fig. 2. The set of rewriting rules for our bottom-up reduction. Here $\mathcal{U}$ is a special tree representing those identified as universal. Then $\mathcal{L}$ denotes any tree, $\mathcal{T}_\Sigma$ denotes any tree recognizing every letter of Σ, and $\mathcal{T}_\varepsilon$ denotes the class of trees recognizing the empty word ε. Note for example that $\mathcal{U} \in \mathcal{T}_\varepsilon \cap \mathcal{T}_\Sigma$. Hence the last rule is redundant, but written for better understanding.

expected to be equivalent to a tree of constant size $3, 624, 217$, as $n \to \infty$. The equivalent tree can be computed in a bottom-up fashion in linear time. Therefore, the asymptotic average case analysis is doomed to be trivial once the expression has been reduced (in time $O(n)$).

Even if the smaller equivalent expression obtained in [13] is constant size on average, its size ($\approx 3.6 \cdot 10^6$) is not small enough to exclude the usefulness of the uniform model for random generation of regular expressions. But the size of the output expression is likely overestimated due to the fact that the model considered in [13] is very general; it applies to a large range of expression specified by expression trees (e.g., logical formulas, arithmetic expressions, ...). Furthermore, several works on boolean expressions trees [5,10] have shown that considering the finer details of the semantic rules for a concrete case might lead to much stronger results (and a smaller constant).

In this work, we study the application of a simplification algorithm that is specific to regular expression trees and exploits their particular semantic properties[1]. The procedure seeks to recognize when a tree represents a universal expression (i.e., recognizing every possible word). The algorithm must be efficient (linear time) as it is intended to be used in a pre-processing step to simplify the expressions. In Fig. 2 we show the schema of the reduction rules applied by recognizing universality. The reduction works bottom-up. We record for each tree the subset of $\Sigma \cup \{\varepsilon\}$ that is recognized by the expression, depicted in blue in Fig. 1. In addition, the algorithm tries to detect if the expression is universal (i.e. accepts

---

[1] However, this idea may be adaptable to other similar bottom-up procedures.

every word on the alphabet). As the universality problem for regular expressions is PSPACE-complete [16], this detection (in linear time) is bound to be incomplete. The reduction consists in replacing the identified universal subtrees of the input expression by a fixed tree $\mathcal{U}$ representing the class of universal trees. This tree $\mathcal{U}$ may be taken to be, for example, a universal one of minimal size (e.g., $(a + b)^*$ for two letters), or a completely new alphabet symbol.

It could be said that this is the most natural and simplest algorithm to detect in linear time as many universal subtrees as possible. As mentioned previously, this algorithm does not detect all universal trees; consider for instance the tree associated with $L = \Sigma \cdot \Sigma^* + \varepsilon$. The algorithm does not realize that $\Sigma \cdot \Sigma^*$ is only missing ε to be universal. Note that the procedure detects the universality of the three trees on the right of Fig. 1, which in contrast cannot be reduced at all by the absorbing-pattern procedure given in [13].

The goal here is to prove that our reduction is very effective at simplifying random uniform regular expressions. This contrasts with the reductions considered in the literature [4,11,12,15] where the objective is to simplify real life (sensibly chosen and not degenerated) regular expressions, for the purposes of further processing in automata.

Our main contribution is showing, without a shadow of doubt, that the uniform model is not apt for practical use. We already know due to [13] that the size after reduction is bounded by a huge constant[2]. Note that the convergence of the expected size is not immediate in our case because the reduction is significantly different. In this article we prove that our algorithm yields a limit expected size which is significantly smaller. We give a general method to compute this limit to any arbitrary precision, that works for any alphabet size, and is efficient. The limits for $k = |\Sigma|$ from two to five are shown in Table 1, where we have taken $\mathcal{U}$ to be a minimal universal tree of size $2k$. For instance, for an alphabet on two symbols ($k = 2$) this yields a constant ~ 77.797, which is prohibitively small. This is confirmed by our experiments (see Fig. 3) for a wide range of sizes. Further, the experiments suggest that the limit might even be close to being an upper-bound for the sequence of expected values for the size after reduction.

Theorem. *Consider the simple variety of tree expressions encoding regular expressions for an alphabet of fixed size $|\Sigma| = k$, and σ the linear-time simplification induced by the rules in Figure 2. Then the expected size of a uniform random expression of size n after simplification tends to a constant as n tends to infinity.*

Table 1. Limit for the expected size after reduction, for different alphabet sizes.

$|\Sigma|$	2	3	4	5
$\lim \mathbb{E}_n[|\sigma(T)|]$	77.79724 ...	495.59151 ...	2 518.20513 ...	11 694.43727 ...

---

[2] Their algorithm is a sub-reduction of ours in the case of regular expressions.

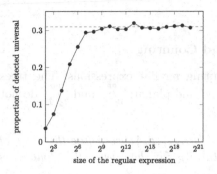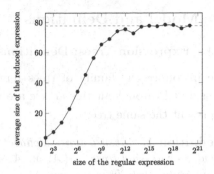

Fig. 3. The proportion of universal expressions detected by the algorithm, and the average size after reduction, observed experimentally[3] on regular expressions on two letters, with 10 , 000 samples for each size. The plots are in log-scale for the input sizes. The theoretical limits are marked by green (dashed) lines. (For the simulations, the uniform trees were sampled using the algorithm in [6]). (Color figure online)

Moreover, Table 1 shows the limits, for alphabets of size k up to 5, computed to five exact digits after the decimal point.

As a second result of interest, we provide bounds for the proportion of universal expression trees. In Proposition 10, we show that there is, asymptotically, a non-negligible proportion of expression trees which represent universal expressions. In particular, the proportion is asymptotically comprised between 0.31 and 0.46 for an alphabet on two letters. See Fig. 3 for the lower bound. Thus the fact that the reduced trees have a limit for their expected size can be thought of, intuitively, as a consequence of the preponderance of universal expressions.

We conclude the introduction by giving a plan of the article. Section 2 introduces the definitions and the basic techniques from Analytic Combinatorics [9] that we will employ. In particular the reduction algorithm, and the main generating functions. Next, in Sect. 3 we study the generating functions of the combinatorial classes associated with the algorithm, and the size after reduction. Theorem 1 gives a recursive system of the combinatorial classes. Using marking techniques on the system, we describe the reduced size in terms of a bivariate generating function. Then Theorem 2 proves that these classes have a limit probability, and we conclude the section with Theorem 3 which describes the expected values. Finally, in Sect. 4 we show how to compute the limits (probability and expectation) efficiently to arbitrary precision[3]. This involves a rewriting of the system in a simpler form (see Sect. 4.1). The procedure[4] works for any value of $k = |\Sigma|$.

The proofs are either sketched or completely omitted in this extended abstract.

---

[3] We could in fact compute them exactly, but their exact expressions are too large and not particularly useful, so we keep the numerical values only.

[4] The code is provided in Sage and Maple at https://igm.univ-mlv.fr/~koechlin/csr_reduction_universality/.

2 Model and Definitions

2.1 Expression Trees: Definitions and Counting

We introduce the family of trees representing regular expressions. The trees considered throughout this article are rooted and planar: $\overset{op}{\underset{T_1\ T_2}{\wedge}}$ and $\overset{op}{\underset{T_2\ T_1}{\wedge}}$ do not represent the same tree.

Definition 1. *Given a finite alphabet $\mathcal{A} = \{a_1,\ldots,a_k\}$, we define the class of regular expression trees $\mathcal{L}_{\mathcal{R}}(\mathcal{A})$ on $\mathcal{A}$, written $\mathcal{L}_{\mathcal{R}}$ for short when the alphabet is clear, inductively from the equation*

$$\mathcal{L}_{\mathcal{R}} = a_1 + \ldots + a_k + \varepsilon + \overset{*}{\underset{\mathcal{L}_{\mathcal{R}}}{|}} + \overset{\cdot}{\underset{\mathcal{L}_{\mathcal{R}}\ \mathcal{L}_{\mathcal{R}}}{\wedge}} + \overset{+}{\underset{\mathcal{L}_{\mathcal{R}}\ \mathcal{L}_{\mathcal{R}}}{\wedge}} . \tag{2.1}$$

The size $|T|$ of an expression tree $T \in \mathcal{L}_{\mathcal{R}}$ is defined to be its number of nodes. In particular the leaves $a_1,\ldots,a_k$ and ε have size 1.

For $n \in \mathbb{N}$, we note $\mathcal{L}_n$ the set of regular expressions of size n. In our model we fix n the size of tree, and draw $T \in \mathcal{L}_n$ uniformly. The probability of picking a particular $T \in \mathcal{L}_n$ is then $1/|\mathcal{L}_n|$. Now we show how to obtain the asymptotics for $|\mathcal{L}_n|$, and how to use this to obtain the probability of recognizing ε.

In order to count the trees, and obtain asymptotics, we make use of the framework of Analytic Combinatorics [9]. In particular, we deal with ordinary generating functions (OGFs for short), which are formal series of the form $C(z) = \sum c_n z^n$, and we use the notation $[z^n]C(z) := c_n$.

The analysis, as is usual in Analytic Combinatorics, follows in two steps: (*i*) a *symbolic step*, where we first characterize (by a formal equation) the OGF of our quantities of interest, and (*ii*) an *analytic step* where we extract the asymptotics of the coefficients by a so-called *Transfer Theorem*.

Analytic Step: Asymptotics for the Coefficients. The Transfer Theorem is the key tool of the analytic step. Given an OGF $C(z) = \sum c_n z^n$ seen as a function on the complex plane $\mathbb{C}$, the behaviour of $C(z)$ at its dominant (closest to the origin) singularities translates into asymptotics for its coefficients $c_n \geq 0$. More precisely, *Pringsheim's Theorem* [9, Theorem IV.6] implies that the radius of convergence $\rho = \rho_C > 0$ of $C(z)$ is a dominant singularity. Then the celebrated *Transfer Theorem* [9, Ch VI.3] states that, under certain analytic conditions, if ρ is the only singularity on the circle $|z| = \rho$ and we have the local estimate $C(z) \sim_{z \to \rho} \lambda(1 - z/\rho)^{-\beta}$, with $\beta \notin \{0, -1, -2, \ldots\}$, around $z = \rho$, then we have the asymptotics $[z^n]C(z) \sim_{n \to \infty} \lambda \rho^{-n} n^{\beta-1}/\Gamma(\beta)$, where Γ is Euler's gamma-function, for the coefficients of $C(z)$.

Formal Generating Series for $\mathcal{L}_{\mathcal{R}}$. The ordinary generating function $L(z)$ associated with $\mathcal{L}_{\mathcal{R}}$ is defined as the formal power series

$$L(z) := \sum_{T \in \mathcal{L}_{\mathcal{R}}} z^{|T|} = \sum_{n \geq 0} \ell_n z^n , \qquad \text{where } \ell_n = |\mathcal{L}_n| .$$

The equation defining the class of expression trees $\mathcal{L}_\mathcal{R}$ translates into a functional equation for its ordinary generating function:

$$L(z) = (k+1)z + zL(z) + 2z(L(z))^2 . \qquad (2.2)$$

This translation from the inductive equation, Eq. (2.1), to the functional equation, Eq. (2.2), comes from general principles of Analytic Combinatorics: the disjoint union of two classes is associated to the sum of their OGFs, and the cartesian product of two classes translates into the product of their OGFs [9].

Asymptotics for the Number of Trees. The Eq. (2.2) is quadratic in $L(z)$. Thus we can solve for the generating function, obtaining $L(z) = \left(1 - z - \sqrt{\Delta(z)}\right)/(4z)$ with $\Delta(z) := -(8k+7)z^2 - 2z + 1$, which is the only combinatorially sound solution. Then $L(z)$ presents a false singularity at $z = 0$, and a unique dominant singularity ρ at the root of $\Delta(z)$ that is closest to the origin. The value of the singularity ρ and the value of $L(\rho)$ are characterized by

$$L(\rho) = \sqrt{\frac{1+k}{2}}, \qquad \rho = \frac{1}{1 + 4L(\rho)} .$$

Since ρ is a simple root of $\Delta(z)$, we derive that $L(z) = h_L - g_L\sqrt{1 - z/\rho} + O\left(\left|1 - \frac{z}{\rho}\right|\right)$ as $z \to \rho$, where $h_L = L(\rho)$, and g_L can be obtained by differentiation, namely $g_L = 2\rho \lim_{z \to \rho} L'(z) \cdot \sqrt{1 - z/\rho}$. Thus the Transfer Theorem[5] yields

$$\ell_n = [z^n]L(z) \sim -g_L\rho^{-n}n^{-3/2}/\Gamma(-1/2) = g_L\rho^{-n}n^{-3/2}/(2\sqrt{\pi}) .$$

Probability of Recognizing ε. We present a basic class of regular expressions that intervene crucially in our work: the tree expressions recognizing the empty word.

The class $\mathcal{T}_\varepsilon$ of tree expressions recognizing ε can be characterized inductively by the following equation where the decomposition into sum of classes is disjoint:

$$\mathcal{T}_\varepsilon = \varepsilon + \overset{\star}{\underset{\mathcal{L}_\mathcal{R}}{|}} + \overset{\bullet}{\underset{\mathcal{T}_\varepsilon \; \mathcal{T}_\varepsilon}{\wedge}} + \overset{+}{\underset{\mathcal{T}_\varepsilon \; \mathcal{L}_\mathcal{R}}{\wedge}} + \overset{+}{\underset{\mathcal{L}_\mathcal{R} \backslash \mathcal{T}_\varepsilon \; \mathcal{T}_\varepsilon}{\wedge}} .$$

Thus we obtain[6] the OGF $T_\varepsilon(z)$ by the principles of Analytic Combinatorics:

$$T_\varepsilon(z) = z + zL(z) + 2zL(z)T_\varepsilon(z) . \qquad (2.3)$$

Solving the linear equation, we can verify that $T_\varepsilon(z) = h_{T_\varepsilon} - g_{T_\varepsilon}\sqrt{1 - z/\rho} + O(1 - z/\rho)$ as $z \to \rho$, with a constant $g_{T_\varepsilon} \neq 0$. Thus the number of tree expressions of size n recognizing ε is asymptotically

$$[z^n]T_\varepsilon(z) \sim g_{T_\varepsilon}\rho^{-n}n^{-3/2}/(2\sqrt{\pi}) .$$

Normalizing by ℓ_n, we obtain the following proposition:

---

[5] Strictly speaking, we should deal with the remainder term.
[6] This is a direct calculation. We remark that the term $z(T_\varepsilon(z))^2$ cancels out.

Proposition 1. *The probability of a random uniform tree of size n recognizing ε converges, as n → ∞, to a positive constant*

$$g_{T_\varepsilon}/g_L = \frac{\sqrt{2k+2}+3/2}{k+\sqrt{2k+2}+3/2}.$$

Numerical values of this constant are given in Table 2.

Table 2. The asymptotic probabilities of recognizing ε for a random uniform tree.

$k = \lvert\Sigma\rvert$	2	3	4	5
g_{T_ε}/g_L	$0.777\ldots$	$0.663\ldots$	$0.590\ldots$	$0.538\ldots$

2.2   The Reduction Process

We consider $\mathcal{R}$ a particular subclass of trees recognizing every word of $\Sigma^\star$:

Definition 2. *The class $\mathcal{R}$ is defined inductively:*

- *if T recognizes every letter of the alphabet, then $\overset{\star}{\underset{T}{\vert}} \in \mathcal{R}$;*
- *if at least one of T_1 or T_2 belongs to $\mathcal{R}$, then $\underset{T_1\ T_2}{\overset{+}{\wedge}} \in \mathcal{R}$;*
- *if $T_1 \in \mathcal{R}$ and $T_2 \in \mathcal{T}_\varepsilon$, then $\underset{T_1\ T_2}{\overset{\bullet}{\wedge}} \in \mathcal{R}$ and $\underset{T_2\ T_1}{\overset{\bullet}{\wedge}} \in \mathcal{R}$.*

In particular if $T \in \mathcal{R}$, then $\overset{\star}{\underset{T}{\vert}} \in \mathcal{R}$.

Note that $\mathcal{R}$ is the class of subtrees reduced to $\mathcal{U}$ by the algorithm. As announced in the introduction, the tree associated with $\Sigma \cdot \Sigma^\star + \varepsilon$ is not in $\mathcal{R}$, in spite of the associated language being universal.

Recognizing Letters. In order to decide whether a tree belongs to $\mathcal{R}$, by using the definition bottom-up, we must be able to decide (also bottom-up) whether a tree T recognizes a given letter a. This is done as follows:

- if $\lvert T\rvert = 1$, then T recognizes a if and only if $T = a$,
- if the root of T is either $\star$ or $+$, then T recognizes a if and only if one of its children recognizes the letter a,
- if the root of T is $\bullet$, then T recognizes a if and only if one of its children recognizes the letter a and the other one recognizes ε.

Example 1. If T_1 recognizes ε and a, while T_2 recognizes ε and b, then $\underset{T_1\ T_2}{\overset{\bullet}{\wedge}}$ recognizes a, b and ε.

Definition 3 (Reduction algorithm[7] σ). *Given a tree T, and a fixed tree $\mathcal{U}$ representing Σ^*, we produce the reduced tree $\sigma_{\mathcal{U}}(T)$ as follows. We begin bottom-up from the leaves, and we keep track of whether the current tree: (1) recognizes each letter of Σ, (2) recognizes ε, (3) is in $\mathcal{R}$. The veracity of all of these predicates is determined bottom-up as described above. Whenever a subtree is in $\mathcal{R}$, we substitute it by $\mathcal{U}$. When $\mathcal{U}$ is clear by context, we simply write $\sigma(T)$.*

2.3   Generating Functions with Additional Parameters

The generating function $L(z)$ only counts the number ℓ_n of trees of a given size n. To keep track of the reduced size of the trees at the same time, we introduce a new "marking" variable u and consider the *bivariate* generating function

$$L(z,u) = \sum_{T \in \mathcal{L}_\mathcal{R}} z^{|T|} u^{|\sigma(T)|}. \tag{2.4}$$

Then the expected size of a tree of size n after reduction can be expressed by:

$$\mathbb{E}_n[|\sigma(T)|] = \frac{[z^n]\partial_u L(z,u)\big|_{u=1}}{[z^n]L(z)}. \tag{2.5}$$

We need information about $L(z,u)$ to evaluate the numerator in Eq. (2.5). In order to find a suitable expression for $L(z,u)$, we will split $\mathcal{L}$ into several subclasses. These subclasses correspond to the different stages in building an element from $\mathcal{R}$, and take the reduction into account (see Sect. 3.1).

3   Analytic Characterization of the Limit

The objective of this section is to show that the expected size of the reduction $\sigma(T)$ of a random tree T converges as the size $n = |T|$ tends to infinity. To do this, we study the analytic properties of $\partial_u L(z,u)|_{u=1}$, the derivative in u of the bivariate generating function defined in Eq. (2.4). The analytic properties presented here will also be needed in Sect. 4, where we show how to exploit them to obtain an efficient and high-precision procedure to compute the limit of the expectation.

3.1   Combinatorial System for the Reduction

Following the construction of the class $\mathcal{R}$, we introduce the following notation: for every subset of letters $X \subseteq \Sigma$, $\mathcal{T}_{X,\varepsilon}$ denotes the set of trees that recognize the empty word, every letter in X, but no letter in $\Sigma \backslash X$. Similarly we denote by $\mathcal{T}_{X,\bar{\varepsilon}}$ the set of tree expression that recognize every letter in X, but no letter in $\Sigma \backslash X$, nor the empty word ε.

---

[7] The full pseudocode of the algorithm is given in Annex A.

Example 2. For instance, $\mathcal{T}_{\{a\},\varepsilon}$ contains the trees for $a^\star$ and $(a \cdot (b^\star) + \varepsilon)$, but does not contain the tree a.

Example 3. The trees in $\mathcal{T}_{X,\varepsilon}$, where $X \subseteq \Sigma$, that have $\bullet$ as a root are of the form $\overset{\bullet}{\underset{T_1\ T_2}{\wedge}}$ where both T_1 and T_2 must recognize ε, and the set of letters recognized by either T_1 or T_2 must be equal to X. Hence it belongs to $\overset{\bullet}{\underset{T_{S,\varepsilon}\ T_{S',\varepsilon}}{\wedge}}$, for some S, S' satisfying $S \cup S' = X$.

Theorem 1. *The combinatorial classes* $(\mathcal{T}_{X,\varepsilon})_{X\subseteq\Sigma}$ *and* $(\mathcal{T}_{X,\bar{\varepsilon}})_{X\subseteq\Sigma}$ *satisfy the inductive definition:*

$$\mathcal{T}_{X,\varepsilon} = \varepsilon \mathbf{1}_{X=\emptyset} + \overset{\star}{\underset{\mathcal{T}_{X,\varepsilon}}{|}} + \overset{\star}{\underset{\mathcal{T}_{X,\bar{\varepsilon}}}{|}} + \sum_{(S,S'):S\cup S'=X} \overset{\bullet}{\underset{\mathcal{T}_{S,\varepsilon}\ \mathcal{T}_{S',\varepsilon}}{\wedge}}$$

$$+ \sum_{(S,S'):S\cup S'=X} \overset{+}{\underset{\mathcal{T}_{S,\varepsilon}\ \mathcal{T}_{S',\varepsilon}}{\wedge}} + \sum_{(S,S'):S\cup S'=X} \overset{+}{\underset{\mathcal{T}_{S,\varepsilon}\ \mathcal{T}_{S',\bar{\varepsilon}}}{\wedge}} + \sum_{(S,S'):S\cup S'=X} \overset{+}{\underset{\mathcal{T}_{S,\bar{\varepsilon}}\ \mathcal{T}_{S',\varepsilon}}{\wedge}},$$

$$\mathcal{T}_{X,\bar{\varepsilon}} = X\mathbf{1}_{|X|=1} + \sum_{S\subseteq\Sigma} \overset{\bullet}{\underset{\mathcal{T}_{X,\bar{\varepsilon}}\ \mathcal{T}_{S,\varepsilon}}{\wedge}} + \sum_{S\subseteq\Sigma} \overset{\bullet}{\underset{\mathcal{T}_{S,\varepsilon}\ \mathcal{T}_{X,\bar{\varepsilon}}}{\wedge}} + \mathbf{1}_{X=\emptyset} \sum_{S,S'\subseteq\Sigma} \overset{\bullet}{\underset{\mathcal{T}_{S,\varepsilon}\ \mathcal{T}_{S',\bar{\varepsilon}}}{\wedge}}$$

$$+ \sum_{(S,S'):S\cup S'=X} \overset{+}{\underset{\mathcal{T}_{S,\bar{\varepsilon}}\ \mathcal{T}_{S',\bar{\varepsilon}}}{\wedge}},$$

where the union $S \cup S'$ *need not be disjoint, but the sums* $+$ *are all disjoint. The notation* $\mathbf{1}_{condition}$ *means that the term appears if and only if the given condition is satisfied.*

Proof (Sketch). This is just an exhaustive enumeration of every possible case for a tree to recognize any set of letters and the empty word, following the principle of Example 3. We have to be careful to produce an unambiguous specification.

Adding $\mathcal{R}$ *to the System.* The class of fully reducible trees $\mathcal{R}$ satisfies the equation:

$$\mathcal{R} = \overset{\star}{\underset{\mathcal{T}_{\Sigma,\bar{\varepsilon}}}{|}} + \overset{\star}{\underset{\mathcal{T}_{\Sigma,\varepsilon}}{|}} + \overset{+}{\underset{\mathcal{R}\ \mathcal{L}}{\wedge}} + \overset{+}{\underset{\mathcal{L}\backslash\mathcal{R}\ \mathcal{R}}{\wedge}} + \overset{\bullet}{\underset{\mathcal{R}\ \mathcal{T}_\varepsilon}{\wedge}} + \overset{\bullet}{\underset{\mathcal{T}_\varepsilon\backslash\mathcal{R}\ \mathcal{R}}{\wedge}}, \qquad (3.1)$$

We want to add this equation to our system. For the terms to remain positive we introduce the class $\mathcal{T}_G := \mathcal{T}_{\Sigma,\varepsilon} \setminus \mathcal{R}$, namely, the class of trees recognizing every letter and the empty word, that are not fully reducible. Then we have the disjoint sum $\mathcal{T}_{\Sigma,\varepsilon} = \mathcal{R} + \mathcal{T}_G$. Hence, in Eq. (3.1), we can expand $\mathcal{L} = \sum_{X\subsetneq\Sigma} \mathcal{T}_{X,\varepsilon} + \mathcal{T}_G + \mathcal{R} + \sum_{X\subseteq\Sigma} \mathcal{T}_{X,\bar{\varepsilon}}$ and $\mathcal{T}_\varepsilon = \sum_{X\subsetneq\Sigma} \mathcal{T}_{X,\varepsilon} + \mathcal{R} + \mathcal{T}_G$. In particular this gives $\mathcal{L} \setminus \mathcal{R} = \mathcal{T}_G + \sum_{X\subsetneq\Sigma} \mathcal{T}_{X,\varepsilon} + \sum_{X\subseteq\Sigma} \mathcal{T}_{X,\bar{\varepsilon}}$, and similarly for $\mathcal{T}_\varepsilon \setminus \mathcal{R}$.

For $\mathcal{T}_G$ we have a similar equation, which we derive from expanding $\mathcal{T}_{\Sigma,\varepsilon} = \mathcal{R} + \mathcal{T}_G$ in the equation for $\mathcal{T}_{\Sigma,\varepsilon}$ and eliminating the terms involving $\mathcal{R}$. In particular there are no trees in $\mathcal{T}_G$ having $\star$ as root, and those having $\bullet$ as root are

$$\sum_{\substack{S\subsetneq\Sigma,S'\subsetneq\Sigma: \\ S\cup S'=\Sigma}} \overset{\bullet}{\underset{\mathcal{T}_{S,\varepsilon}\ \mathcal{T}_{S',\varepsilon}}{\wedge}} + \sum_{S\subseteq\Sigma} \overset{\bullet}{\underset{\mathcal{T}_G\ \mathcal{T}_{S,\varepsilon}}{\wedge}} + \overset{\bullet}{\underset{\mathcal{T}_G\ \mathcal{T}_G}{\wedge}} + \sum_{S\subseteq\Sigma} \overset{\bullet}{\underset{\mathcal{T}_{S,\varepsilon}\ \mathcal{T}_G}{\wedge}},$$

as we must prohibit the subtrees from being in $\mathcal{R}$. For $+$ to be the root, we must prohibit having one subtree that recognizes ε and the other in $\mathcal{R}$. This is easily calculated but a bit long to write out in full.

3.2   Generating Functions and Probability of Full Reduction

From the combinatorial system of Theorem 1, in $(\mathcal{T}_{X,\bar{\varepsilon}}, \mathcal{T}_{X,\varepsilon})_{X \subseteq \Sigma}$, we have introduced a new one in $(\mathcal{R}, \mathcal{T}_G, \mathcal{T}_{\Sigma,\bar{\varepsilon}}, (\mathcal{T}_{X,\bar{\varepsilon}}, \mathcal{T}_{X,\varepsilon})_{X \subsetneq \Sigma})$. In this section we look at the system of their generating functions and prove several basic properties.

We denote by $y_{X,\varepsilon}(z)$ (resp. $y_{X,\bar{\varepsilon}}(z)$) the generating series of $\mathcal{T}_{X,\varepsilon}$ (resp. $\mathcal{T}_{X,\bar{\varepsilon}}$). Similarly, we denote by $R(z)$ and $y_G(z)$ the OGFs of $\mathcal{R}$ and $\mathcal{T}_G$. Note in particular that $y_{\Sigma,\varepsilon}(z) = R(z) + y_G(z)$. Henceforth we will write as a column vector

$$\boldsymbol{y}(z) = [R(z), y_G(z), y_{\Sigma,\bar{\varepsilon}}(z), (y_{X,\bar{\varepsilon}}(z), y_{X,\varepsilon}(z))_{X \subsetneq \Sigma}].$$

Proposition 2. *The vector $\boldsymbol{y}(z)$ satisfies a vectorial system under the form:*

$$\boldsymbol{y}(z) = \boldsymbol{\Phi}(z; \boldsymbol{y}(z)) \tag{3.2}$$

where each component of $\boldsymbol{\Phi}(z; \boldsymbol{y})$ is a polynomial of degree 2 in $\boldsymbol{y}$, of degree 1 in z, such that $\boldsymbol{\Phi}(0; \boldsymbol{y}) = \boldsymbol{0}$.

Remark 1. If $X, Y \subseteq \Sigma$ have the same number of letters $|X| = |Y|$, then $y_{X,\varepsilon}(z) = y_{Y,\varepsilon}(z)$ and $y_{X,\bar{\varepsilon}}(z) = y_{Y,\bar{\varepsilon}}(z)$. This follows from picking any isomorphism permuting the letters of Σ and mapping X to Y. Thus in reality we may rewrite the system in just $1 + 2 \times (k + 1)$ equations where $k = |\Sigma|$, rather than the exponential $1 + 2^{k+1}$ we would have by considering every subset.

Proposition 3. *Every coordinate of the solution $\boldsymbol{y}(z)$ of the system has ρ as a unique dominant singularity, such that near $z = \rho$:*

$$\boldsymbol{y}(z) = \boldsymbol{h}(z) - \boldsymbol{g}(z)\sqrt{1 - z/\rho} \tag{3.3}$$

where $\boldsymbol{h}(z)$ and $\boldsymbol{g}(z)$ are two vectors of analytic functions in a neighbourhood of $z = \rho$. Besides, every coordinate of $\boldsymbol{g}(\rho)$ is strictly positive.

Proof (sketch). We prove that the system (3.2) is strongly connected, so that we can apply Drmota's theorem [7]. The common singularity is already known since it must coincide with the singularity of $L(z)$.

Theorem 2. *The probability that a random tree T of size n belongs to a class C of the extended combinatorial system tends to a positive constant $g_C(\rho)/g_L(\rho)$ as $n \to \infty$. In particular, the limit probability of a full reduction ($C = \mathcal{R}$) is positive.*

Proof (Sketch). Proposition 3 implies that $R(z) = h_R(z) - g_R(z)\sqrt{1 - z/\rho}$ around $z = \rho$, with $g_R(\rho) \neq 0$. Since there is no other singularity on the circle $|z| = \rho$, we obtain the asymptotics from the Transfer Theorem and the proof follows as that of the probability of recognizing ε.

3.3   Extended System for the Expected Reduced Size

To deal with expected values, as explained in Sect. 2.3, we introduce a new variable u which will "mark" the size of the reduced expression. We extend each generating function to two variables $\boldsymbol{y}(z, u)$. It is immediate to see that $R(z, u) = u^{|\mathcal{U}|} R(z)$. Note that for the other classes of trees, the root symbol always remains unchanged after the reduction. Hence we almost have for them the same equations in two variables as in one variable, with an additional factor u to count the root in the size of the reduced tree. We summarize this discussion in the following proposition.

Proposition 4. *Let us write* $\boldsymbol{y} = (R, \tilde{\boldsymbol{y}})$, *and* $\boldsymbol{\Phi} = (\Phi_R, \tilde{\boldsymbol{\Phi}})$. *The vector of bivariate generating functions* $\tilde{\boldsymbol{y}}(z, u)$ *satisfies the vectorial system:*

$$\tilde{\boldsymbol{y}}(z, u) = \tilde{\boldsymbol{\Phi}}(zu; u^p R(z), \tilde{\boldsymbol{y}}(z, u))$$

where $\boldsymbol{\Phi} = (\Phi_R, \tilde{\boldsymbol{\Phi}})$ *is defined in Eq. 3.2, and* $p := |\mathcal{U}|$.

Notice that $L(z, u) = u^p R(z) + (1, \ldots, 1) \cdot \tilde{\boldsymbol{y}}(z, u)$. Following Eq (2.5), we need to differentiate $L(z, u)$ on u then set $u = 1$ to find the expected value. For notation convenience, we will write $Q\tilde{\boldsymbol{y}}(z) := \partial_u \tilde{\boldsymbol{y}}(z, u)\big|_{u=1}$.

Proposition 5. *The vector* $Q\tilde{\boldsymbol{y}}(z) = \partial_u \tilde{\boldsymbol{y}}(z, u)\big|_{u=1}$ *satisfies the linear system:*

$$(\mathrm{Id} - \mathrm{Jac}_{\tilde{\boldsymbol{y}}}[\tilde{\boldsymbol{\Phi}}](z; R(z), \tilde{\boldsymbol{y}}(z)))Q\tilde{\boldsymbol{y}}(z) = \tilde{\boldsymbol{\Phi}}(z; R(z), \tilde{\boldsymbol{y}}(z)) + p\partial_R \tilde{\boldsymbol{\Phi}}(z; R(z), \tilde{\boldsymbol{y}}(z))R(z)$$

Proof. It is straightforward by differentiating Proposition 4. Note that $z\partial_z \tilde{\boldsymbol{\Phi}} = \tilde{\boldsymbol{\Phi}}$.

Hence we can show that $Q\tilde{\boldsymbol{y}}(z)$ has a dominant square-root singularity at $z = \rho$:

Proposition 6. *Every coordinate of the vector* $Q\tilde{\boldsymbol{y}}(z) = \partial_u \tilde{\boldsymbol{y}}(z, u)\big|_{u=1}$ *has a unique dominant singularity at* $z = \rho$. *Further, near* $z = \rho$ *we may write:*

$$Q\tilde{\boldsymbol{y}}(z) = \boldsymbol{h}_{Q\tilde{\boldsymbol{y}}}(z) - \boldsymbol{g}_{Q\tilde{\boldsymbol{y}}}(z)\sqrt{1 - z/\rho}$$

where $\boldsymbol{h}_{Q\tilde{\boldsymbol{y}}}(z)$ *and* $\boldsymbol{g}_{Q\tilde{\boldsymbol{y}}}(z)$ *are two vectors of analytic functions in a neighbourhood of* $z = \rho$, *such that every coordinate of* $\boldsymbol{g}_{Q\tilde{\boldsymbol{y}}}(\rho)$ *is strictly positive.*

Proof (Sketch). We prove that we can inverse the matrix in Proposition 5, and use Proposition 3 to prove that the solution has the right form.

Using Eq. (2.5) and the Transfer Theorem, we can finally conclude:

Theorem 3 (Limit of the expected size). *Consider the simple variety of tree expressions encoding regular expressions for an alphabet of fixed size* $|\Sigma| = k$, *and the linear-time simplification algorithm* σ. *Then the expected size of a uniform random expression of size* n *after simplification by* σ *tends to a constant as* n *tends to infinity:*

$$\lim_{n \to +\infty} \mathbb{E}_n[|\sigma(T)|] = \frac{|\mathcal{U}|g_R(\rho) + \|\boldsymbol{g}_{Q\tilde{\boldsymbol{y}}}(\rho)\|_1}{g_L(\rho)} \tag{3.4}$$

where $\|(v_1, \ldots, v_s)\|_1 = |v_1| + \ldots + |v_s|$.

Remark 2 (size of $\mathcal{U}$. A natural value for the size of $\mathcal{U}$ is $|\mathcal{U}| = 2k$ if we represent universality by any minimal unary-binary tree for Σ^*, or $|\mathcal{U}| = 1$ if we use a special symbol. We remark that one must be careful when changing $\mathcal{U}$ as the vector $(g_R(z), g_{Q\tilde{y}}(z))$, evaluated in Eq. (3.4), also depends on $|\mathcal{U}|$.

4 Practical Computation of the Limit: A Numerical Study

The main goal of this section is to give an effective procedure to compute the constant in Theorem 3, for any size of the alphabet.

According to Eq. (3.4), we need to compute $g_R(\rho)$ and $g_{Q\tilde{y}}(\rho)$. We notice that for any analytic function $w(z)$ under the form $w(z) = h(z) - g(z)\sqrt{1 - z/\rho}$, then $w'(z) = O(1) + \frac{g(\rho)}{2\rho\sqrt{1-z/\rho}}$ for $z \sim \rho$. Hence $g(\rho) = \lim_{z\to\rho} 2\rho\, w'(z)\sqrt{1 - z/\rho}$. Differentiating in z the system in Proposition 5 leads to the following proposition:

Proposition 7. *The vector $g_{Q\tilde{y}}(\rho)$ satisfies the equation:*

$$g_{Q\tilde{y}}(\rho) = \left(\mathrm{Id} - \mathrm{Jac}_{\tilde{y}}[\tilde{\Phi}](\rho; y(\rho))\right)^{-1} \times K_{\Phi}(\rho; y(\rho), g_y(\rho), Q\tilde{y}(\rho))$$

where $K_{\Phi}(z; y, g, h)$ depends on the derivatives of Φ, $p = |\mathcal{U}|$, and it is polynomial in its input vectors.

Hence we need to compute $y(\rho), g_y(\rho)$ and $Q\tilde{y}(\rho)$. This is done in three steps:

1. To compute $y(\rho)$, we rewrite the system in a "triangular form". This is done by exploiting the known functions $L(z)$ and $T_\varepsilon(z)$. Then $y(\rho)$ can be effectively computed by dynamic programming. See Sect. 4.1 for details.
2. Then $g_y(\rho)$ is computed by solving a linear system, as it is an eigenvector of the matrix $\mathrm{Jac}_y[\Phi](\rho, y(\rho))$. This is explained in Sect. 4.2.
3. Finally, setting $z = \rho$ in Proposition 5 yields a simple matrix formula for $Q\tilde{y}(\rho)$. The inverse of the matrix is well-defined, as shown in the proof of Proposition 6.

4.1 Triangular Form of the System

Factorizing terms in the equations for the combinatorial classes for $\mathcal{T}_{X,\bar{\varepsilon}}$ in Theorem 1, the combinatorial classes $\mathcal{T}_\varepsilon$ and $\mathcal{T}_{\bar{\varepsilon}} := \mathcal{L}_\mathcal{R} \backslash \mathcal{T}_\varepsilon$ turn up. This is summarized in the following proposition.

Proposition 8. *The combinatorial class $(\mathcal{T}_{X,\bar{\varepsilon}})_{X\subseteq\Sigma}$ satisfies the inductive definition:*

$$\mathcal{T}_{X,\bar{\varepsilon}} = X\mathbf{1}_{|X|=1} + \overset{\bullet}{\underset{\mathcal{T}_{X,\bar{\varepsilon}} \quad \mathcal{T}_\varepsilon}{\wedge}} + \overset{\bullet}{\underset{\mathcal{T}_\varepsilon \quad \mathcal{T}_{X,\bar{\varepsilon}}}{\wedge}} + \mathbf{1}_{X=\emptyset}\overset{\bullet}{\underset{\mathcal{T}_{\bar{\varepsilon}} \quad \mathcal{T}_{\bar{\varepsilon}}}{\wedge}} + \sum_{(S,S'):S\cup S'=X} \overset{+}{\underset{\mathcal{T}_{S,\bar{\varepsilon}} \quad \mathcal{T}_{S',\bar{\varepsilon}}}{\wedge}} .$$

This yields a new triangular system for the generating functions: we obtain a quadratic equation in $y_{X,\bar{\varepsilon}}(z)$ whose coefficients involve only $y_{S,\bar{\varepsilon}}$ with $S \subsetneq X$, similarly, we have a quadratic equation for $y_{X,\varepsilon}(z)$ whose coefficients involve only $y_{S,\bar{\varepsilon}}$ with $S \subsetneq X$ and also $y_{X,\bar{\varepsilon}}$. Then the numerical solution[8] for $y(\rho)$ can be computed efficiently.

Algorithm to Compute $y(\rho)$. First we compute $L(\rho)$, $T_{\varepsilon}(\rho)$ and $T_{\bar{\varepsilon}}(\rho)$ as explained in Sect. 2.1. Then, given the triangular system, we compute the values of $y_{X,\varepsilon}(\rho)$ and $y_{X,\bar{\varepsilon}}(\rho)$ for each $X \subseteq \Sigma$ by dynamic programming. Each step requires simple operations (sums, products, and a single square-root) on previously computed values. Finally, $R(\rho)$ is computed from Eq. (3.1), while $y_G(\rho) = y_{\Sigma,\varepsilon}(\rho) - R(\rho)$.

4.2   Limit Probabilities as an Eigenvector

The vector $g_y(\rho)$ is characterized in terms of a linear system of equations.

Proposition 9. *The coefficients $g_y(\rho)$ constitute an eigenvector for $\lambda = 1$ for the Jacobian matrix $\mathrm{Jac}_y[\Phi](\rho; y(\rho))$ at $z = \rho$, namely*

$$\mathrm{Jac}_y[\Phi](\rho; y(\rho)) \cdot g_y(\rho) = g_y(\rho).$$

Furthermore, the eigenspace associated to $\lambda = 1$ has dimension 1 and $g_y(\rho)$ is characterized as the only eigenvector satisfying $\|g_y(\rho)\|_1 = g_L(\rho)$.

Probabilities of Each Class. In particular, for two letters we obtain $\lim_n \mathrm{Pr}_n(\mathcal{R}) \doteq 0.310122\ldots$, while $\lim_n \mathrm{Pr}_n(\mathcal{T}_{\Sigma,\varepsilon}) \doteq 0.457051\ldots$. These constitute bounds for the proportion of universal expressions. We summarize in the following:

Proposition 10. *For all n large enough, the proportion $\mathrm{Pr}_n(\mathrm{univ.})$ of trees representing universal expressions over a k-letter alphabet belongs to the intervals:*

k	2	3	4	5
interval	$(0.31, 0.46)$	$(0.13, 0.27)$	$(0.062, 0.15)$	$(0.028, 0.077)$

---

[8] In fact, we can solve the system in $y(z)$ exactly. However the closed-form solutions become huge; for instance, for $\Sigma = \{a, b\}$: $y_{\Sigma,\bar{\varepsilon}}(z) = \frac{1}{4z}\Big(- \sqrt{\Delta(z)} + 2\sqrt{(2z+2)\sqrt{\Delta(z)} - 6z^2 + 2} - \sqrt{(2z+2)\sqrt{\Delta(z)} + 10z^2 + 2 - z - 1}\Big).$

5   Conclusion

We have provided a simple linear algorithm reducing a random regular expression to an equivalent one that on average has a small constant size. This shows that the uniform tree model is most definitely flawed when it comes to producing random regular expressions, as it produces very limited languages. This raises the question of what would be a more suitable model.

A more promising distribution could be the BST-like trees [8], where BST stands for binary search tree. This model is used in practice in the context of model checking (for instance for LTL formula [17]), and it is quite different from the uniform tree model. We considered the effect of simpler reductions in this model in [14], and we found that the situation was more complex than in the uniform model. It would be interesting to study the effects of the simplification procedure of the current article in the context of the BST-like tree distribution for regular expressions.

An interesting aspect to highlight is the combinatorial system characterizing the reduction process (see Sect. 3.1), in particular its simplicity, and the fact that it allows for efficient computation and (big) exact solutions.

The simplification process relies on detecting universality. Our study reveals that universality is abundant in the random uniform model. Moreover, the proportion of universal trees is comprised in a small range (see Proposition 10). We could refine the detection algorithm by considering slight improvements. The goal is to, for instance, recognize the universality of $L = \Sigma \cdot \Sigma^\star + \varepsilon$. An idea is to consider not only whether $\Sigma^\star$ is recognized but also $a \cdot \Sigma^\star$ and $b \cdot \Sigma^\star$. More generally, one can consider for a given depth k, whether the words in $\Sigma^{\leq k}$ are recognized, and also the sets $w \cdot \Sigma^\star$ for a prefix free set of $w \in \Sigma^{\leq k}$. For large k this should lead to better upper and lower bounds for the asymptotic probability of a tree being universal. Hopefully these bounds will coalesce as $k \to \infty$, which would then prove the existence of the limit probability for universality.

A   Detailed Pseudo-code of the Reduction Algorithm σ

In this section we give the full pseudo-code of our reduction algorithm on the alphabet Σ. The reduction corresponds to the first component of the return value of the extended function reduce, which returns the pair

$$\texttt{reduced}(T) = (\sigma(T), S_T),$$

where $S_T \subseteq \{\varepsilon\} \cup \Sigma$ denotes the set of leaves recognized by T.

> **function** reduce(T):
>
> **if** $|T| = 1$ **then**
> > **return** $(T, \{T\})$;
>
> **if** $T = \underset{T_L \ T_R}{\overset{+}{\wedge}}$ **then**
> > $(T'_L, S_L) := \text{reduce}(T_L)$;
> > $(T'_R, S_R) := \text{reduce}(T_R)$;
> > **if** $T'_L = \mathcal{U}$ *or* $T'_R = \mathcal{U}$ **then**
> > > **return** $(\mathcal{U}, S_L \cup S_R)$;
> >
> > **return** $(\underset{T'_L \ T'_R}{\overset{+}{\wedge}}, S_L \cup S_R)$;
>
> **else if** $T = \underset{T_L \ T_R}{\overset{\cdot}{\wedge}}$ **then**
> > $(T'_L, S_L) := \text{reduce}(T_L)$;
> > $(T'_R, S_R) := \text{reduce}(T_R)$;
> > $S := \emptyset$;
> > **if** $\varepsilon \in S_R$ **then**
> > > **if** $T'_L = \mathcal{U}$ **then**
> > > > **return** $(\mathcal{U}, S_L)$;
> > >
> > > $S := S \cup S_L$;
> >
> > **if** $\varepsilon \in S_L$ **then**
> > > **if** $T'_R = \mathcal{U}$ **then**
> > > > **return** $(\mathcal{U}, S_R)$;
> > >
> > > $S := S \cup S_R$;
> >
> > **return** $(\underset{T'_L \ T'_R}{\overset{\cdot}{\wedge}}, S)$;
>
> **else if** $T = \underset{T_0}{\overset{*}{|}}$ **then**
> > $(T', S') := \text{reduce}(T_0)$;
> > **if** $\Sigma \subseteq S'$ **then**
> > > **return** $(\mathcal{U}, \{\varepsilon\} \cup \Sigma)$;
> >
> > **return** $(\underset{T'}{\overset{*}{|}}, \{\varepsilon\} \cup S')$;

References

1. Allauzen, C., Mohri, M.: A unified construction of the glushkov, follow, and antimirov automata. In: Královič, R., Urzyczyn, P. (eds.) MFCS 2006. LNCS, vol. 4162, pp. 110–121. Springer, Heidelberg (2006). https://doi.org/10.1007/11821069_10
2. Antimirov, V.: Partial derivatives of regular expressions and finite automata constructions. In: Mayr, E.W., Puech, C. (eds.) STACS 1995. LNCS, vol. 900, pp. 455–466. Springer, Heidelberg (1995). https://doi.org/10.1007/3-540-59042-0_96

3. Broda, S., Machiavelo, A., Moreira, N., Reis, R.: On the average size of glushkov and partial derivative automata. Int. J. Found. Comput. Sci. **23**(5), 969–984 (2012). https://doi.org/10.1142/S0129054112400400

4. Brüggemann-Klein, A.: Regular expressions into finite automata. Theor. Comput. Sci. **120**(2), 197–213 (1993). https://doi.org/10.1016/0304-3975(93)90287-4, https://www.sciencedirect.com/science/article/pii/0304397593902874

5. Chauvin, B., Gardy, D., Mailler, C.: The growing tree distribution on boolean functions. In: 2011 Proceedings of the Workshop on Analytic Algorithmics and Combinatorics (ANALCO), pp. 45–56 (2011). https://doi.org/10.1137/1.9781611973013.5, https://epubs.siam.org/doi/abs/10.1137/1.9781611973013.5

6. Devroye, L.: Simulating size-constrained galton-watson trees. SIAM J. Comput. **41**(1), 1–11 (2012). https://doi.org/10.1137/090766632

7. Drmota, M.: Systems of functional equations. Random Struct. Algorithms **10**(1–2), 103–124 (1997)

8. Flajolet, P., Gourdon, X., Martinez, C.: Patterns in random binary search trees. Random Struct. Algorithms **11**(3), 223–244 (1997). https://doi.org/10.1002/(SICI)1098-2418(199710)11:3⟨223::AID-RSA2⟩3.0.CO;2-2

9. Flajolet, P., Sedgewick, R.: Analytic Combinatorics. Cambridge University Press, Cambridge (2009)

10. Gardy, D.: Random boolean expressions. In: DMTCS Proceedings, vol. AF, Computational Logic and Applications (CLA '05), pp. 1–36. Discrete Mathematics & Theoretical Computer Science, Episciences. org (2005)

11. Gruber, H., Gulan, S.: Simplifying regular expressions. In: Dediu, A.-H., Fernau, H., Martín-Vide, C. (eds.) LATA 2010. LNCS, vol. 6031, pp. 285–296. Springer, Heidelberg (2010). https://doi.org/10.1007/978-3-642-13089-2_24

12. Ilie, L., Yu, S.: Follow automata. Inform. Comput. **186**(1), 140–162 (2003). https://doi.org/10.1016/S0890-5401(03)00090-7, https://www.sciencedirect.com/science/article/pii/S0890540103000907

13. Koechlin, F., Nicaud, C., Rotondo, P.: Uniform random expressions lack expressivity. In: Rossmanith, P., Heggernes, P., Katoen, J. (eds.) 44th International Symposium on Mathematical Foundations of Computer Science, MFCS 2019, August 26–30, 2019, Aachen, Germany. LIPIcs, vol. 138, pp. 51:1–51:14. Schloss Dagstuhl - Leibniz-Zentrum für Informatik (2019)

14. Koechlin, F., Rotondo, P.: Absorbing patterns in BST-like expression-trees. In: Bläser, M., Monmege, B. (eds.) 38th International Symposium on Theoretical Aspects of Computer Science (STACS 2021). Leibniz International Proceedings in Informatics (LIPIcs), vol. 187, pp. 48:1–48:15. Schloss Dagstuhl - Leibniz-Zentrum für Informatik, Dagstuhl, Germany (2021). https://doi.org/10.4230/LIPIcs.STACS.2021.48, https://drops.dagstuhl.de/opus/volltexte/2021/13693

15. Lee, J., Shallit, J.: Enumerating regular expressions and their languages. In: Domaratzki, M., Okhotin, A., Salomaa, K., Yu, S. (eds.) CIAA 2004. LNCS, vol. 3317, pp. 2–22. Springer, Heidelberg (2005). https://doi.org/10.1007/978-3-540-30500-2_2

16. Meyer, A.R., Stockmeyer, L.J.: The equivalence problem for regular expressions with squaring requires exponential space. In: Proceedings of the 13th Annual Symposium on Switching and Automata Theory (Swat 1972), pp. 125–129. SWAT '72, IEEE Computer Society, USA (1972). https://doi.org/10.1109/SWAT.1972.29

17. Tauriainen, H.: Automated testing of Büchi automata translators for linear temporal logic. Research Report A66, Helsinki University of Technology, Laboratory for Theoretical Computer Science, Espoo, Finland, December 2000
18. Yamamoto, H.: A new finite automaton construction for regular expressions. In: Bensch, S., Freund, R., Otto, F. (eds.) Sixth Workshop on Non-Classical Models for Automata and Applications - NCMA 2014, Kassel, Germany, July 28–29, 2014. Proceedings. books@ocg.at, vol. 304, pp. 249–264. Österreichische Computer Gesellschaft (2014)

Bit-Complexity of Solving Systems of Linear Evolutionary Partial Differential Equations

Ivan Koswara[1] , Gleb Pogudin[2] , Svetlana Selivanova[1(✉)] , and Martin Ziegler[1]

[1] School of Computing, KAIST, 291 Daehak-ro, Yuseong-gu, Daejeon 34141, Republic of Korea
{chaoticiak,sseliv,ziegler}@kaist.ac.kr
[2] LIX, CNRS, École Polytechnique, Institute Polytechnique de Paris, 91128 Palaiseau, France
gleb.pogudin@polytechnique.edu

Abstract. *Finite Elements* are a common method for solving differential equations via discretization. Under suitable hypotheses, the solution $\mathbf{u} = \mathbf{u}(t, \vec{x})$ of a well-posed initial/boundary-value problem for a linear evolutionary system of PDEs is approximated up to absolute error $1/2^n$ by repeatedly (exponentially often in n) multiplying a matrix $\mathbf{A}_n$ to the vector from the previous time step, starting with the initial condition $\mathbf{u}(0)$, approximated by the spatial grid vector $\mathbf{u}(0)_n$. The dimension of the matrix A_n is exponential in n, which is the number of the bits of the output.

We investigate the bit-cost of computing exponential powers and inner products $\mathbf{A}_n^K \cdot \mathbf{u}(0)_n$, $K \sim 2^{\mathcal{O}(n)}$, of matrices and vectors of exponential dimension for various classes of such *difference schemes* $\mathbf{A}_n$. Non-uniformly fixing any polynomial-time computable initial condition and focusing on single but arbitrary entries (instead of the entire vector/matrix) allows to improve naïve exponential sequential runtime EXP: Closer inspection shows that, given any time $0 \leq t \leq 1$ and space $\vec{x} \in [0; 1]^d$, the computational cost of evaluating the solution $\mathbf{u}(t, \vec{x})$ corresponds to the discrete class PSPACE.

Many partial differential equations, including the Heat Equation, admit difference schemes that are (tensor products of constantly many) circulant matrices of constant bandwidth; and for these we show exponential matrix powering, and PDE solution computable in #P. This is achieved by calculating individual coefficients of the matrix' multivariate companion polynomial's powers using Cauchy's Differentiation

Supported by the National Research Foundation of Korea (grant 2017R1E1A1A 03071032) and by the International Research & Development Program of the Korean Ministry of Science and ICT (grant 2016K1A3A7A03950702) and by the NRF Brain Pool program (grant 2019H1D3A2A02102240). GP was supported by NSF grants CCF-1564132, CCF-1563942, DMS-1853482, DMS-1853650, and DMS-1760448, by PSC-CUNY grants #69827-0047 and #60098-0048. We thank Lina Bondar' for a helpful discussion on different versions of the Sobolev Embedding Theorem (Example 2b)).

R. Santhanam and D. Musatov (Eds.): CSR 2021, LNCS 12730, pp. 223–241, 2021.
https://doi.org/10.1007/978-3-030-79416-3_13

Theorem; and shown optimal for the Heat Equation. Exponentially powering twoband circulant matrices is established even feasible in P; and under additional conditions, also the solution to certain linear PDEs becomes computable in P.

Keywords: Reliable computing · Bit-cost · Partial differential equations

1   Introduction and Summary of Contributions

Computable Analysis [31] provides a framework for rigorous computability and complexity investigations of computational problems over real numbers and functions by approximation up to guaranteed absolute error $1/2^n$ [2,10,15,32]. This has been applied to ordinary [1,11,14] and partial [27,28,30] differential equations. It allows to prove asymptotic optimality of numerical algorithms by relating the intrinsic computational bit-cost of a problem to a classical discrete complexity class [8,13].

The present work considers general classes of systems of linear evolutionary partial differential equations (PDEs).

In order to solve some system of ordinary differential equations (ODEs) $\partial_t \mathbf{u} = \vec{f}(\mathbf{u}, t)$, common numerical approaches—such as Euler's Method and its refinements—discretize time $t \in [0; 1]$ into steps $\tau \ll 1$: From the fixed initial value $\mathbf{u}(0) = \mathbf{u}_0$ at $t = 0$ they iteratively proceed to approximations $\mathbf{u}(\tau)$, $\mathbf{u}(2\tau)$, $\ldots \mathbf{u}(M \cdot \tau)$. In order for the last one to approximate $\mathbf{u}(1)$ up to error $1/2^n$, the number $M = 1/\tau \in \mathbb{N}$ of steps is generally exponential in (the number of bits of the output) n; and the problem thus seen to belong to the discrete complexity class EXP^1. Closer inspection improves that to PSPACE [15, §7.2], which has been proven best possible in general [8]. Bit complexity is measured w.r.t. the output precision parameter n.

Evolutionary PDEs generalize ODEs: by replacing the right-hand side function $\vec{f} = \vec{f}(\mathbf{u}, t)$ with an operator $\mathcal{A}$, commonly involving spatial derivatives. Solutions $\mathbf{u}(t)$ accordingly now take values in some function space, rather than in Euclidean. The mathematical theory of PDEs is considerably more involved than that of ODEs [3] regarding existence, uniqueness, and continuous dependence of solutions (=well-posedness in the sense of Hadamard): Recall that one of the Millennium Prize Problem asks such questions for Navier-Stokes' Equation. Computability investigations of PDEs have challenged the Church-Turing Hypothesis [25,30,33]. The present work considers linear evolutionary PDEs with initial and boundary conditions:

$$\begin{cases} \mathbf{u}_t = \mathcal{A}\mathbf{u}, & 0 \le t \le 1, \quad \vec{x} \in \Omega, \\ \mathbf{u}\,|_{t=0} = \varphi(\vec{x}), & \vec{x} \in \Omega, \\ \mathcal{L}\mathbf{u}(t, \vec{x})\,|_{\partial\Omega} = 0, & (t, \vec{x}) \in [0, 1] \times \partial\Omega \end{cases} \tag{1}$$

---

[1] Definitions of the real-valued counterparts of the complexity classes are given in Subsect. 1.3; for simplicity we use same notation as for the "discrete" case for them.

where $\Omega = [0,1]^d$ is the unit cube (for technical simplicity); $\partial\Omega$ is its boundary; the solution $\mathbf{u} = (u_1, \ldots, u_e) = \mathbf{u}(t, \vec{x})$ is an unknown vector function on Ω; $\mathcal{L}$ in the boundary condition is a linear differential operator of order less than the order of the differential operator $\mathcal{A}$. The coefficients of $\mathcal{A} = \sum_{|\vec{j}|} \mathbf{B}_{\vec{j}}(\vec{x}) \cdot \partial^{\vec{j}}$ are $e \times e$ matrices $\mathbf{B}_j$ that may depend on $\vec{x}$, but not on t (autonomous case), $\vec{j} = (j_1, \ldots, j_d)$ denotes a multi-index ($|\vec{j}| = j_1 + j_2 + \ldots + j_d$), $\partial^{\vec{j}} = \partial_1^{j_1} \cdots \partial_d^{j_d}$ the induced differential operator (with $\partial_k^{j_k} = \frac{\partial^{j_k}}{\partial x_k^{j_k}}$), and $\varphi(\vec{x})$ is the initial condition.

Note that the Eq. (1) are linear in the derivatives, but the matrix coefficients $\mathbf{B}_j$ can depend on $\vec{x}$ nonlinearly.

Example 1. An important and rich class of PDEs of form (1) are the (first-order) symmetric hyperbolic systems

$$\mathbf{B}_0(\vec{x}) \cdot \mathbf{u}_t = \sum_{j=1}^{d} \mathbf{B}_j(\vec{x}) \cdot \partial_{x_j} \mathbf{u} \tag{2}$$

with $\mathbf{B}_j(\vec{x}) = \mathbf{B}_j^*(\vec{x})$, $j = 0, 1, \ldots, d$; $\mathbf{B}_0(\vec{x}) > 0$. The corresponding differential operator is $\mathcal{A} = \sum_{j=1}^{d} \mathbf{B}_0^{-1}(\vec{x}) \cdot \mathbf{B}_j(\vec{x}) \cdot \partial_j$. This class includes the linear acoustics, elasticity and Maxwell equations [27]. Also the (second-order) Wave Equation $u_{tt} = \Delta u$ ($= \sum_{j=1}^{n} \frac{\partial^2 u}{\partial x_j^2}$) and many others can be reduced to such a system by introducing extra unknown functions.

The Heat Equation $u_t = \Delta u$ is not of the form (2), but still of the form (1). Periodic boundary conditions on the unit cube are captured by

$$\mathcal{L}\mathbf{u}(t, x_1, \ldots, x_{j-1}, 0, x_{j+1}, \ldots, x_d) := \mathbf{u}(t, x_1, \ldots, x_{j-1}, 0, x_{j+1}, \ldots, x_d) - $$
$$\mathbf{u}(t, x_1, \ldots, x_{j-1}, 1, x_{j+1}, \ldots, x_d) \tag{3}$$

for first-order systems, and include similar conditions on spacial derivatives up to $l-1$ order for l-order systems.

Under suitable hypotheses, Euler's method generalizes from ODEs to evolutionary PDEs (1): by discretizing now both physical time and space, the latter with some grid of sufficiently (=exponentially) small width $h \ll 1$. This turns the initial condition φ into a vector of exponentially large dimension $\mathcal{O}(1/h)$. The right-hand side linear operator $\mathcal{A}$ may be approximated by a matrix $\mathbf{A}$, often referred to as *difference scheme*, see Definition 1. And if the evolution equation is autonomous, said matrix does not depend on time. In this case repeated time-stepping $\mathbf{u}(t) \mapsto \mathbf{u}(t+\tau) = \mathbf{A} \cdot \mathbf{u}(t)$ amounts to repeated (M-fold) multiplication by $\mathbf{A}$, i.e., to (exponential) matrix powering $\mathbf{A}^M$.

Now all three, the discretized initial condition $\mathbf{u}(0) = \varphi$ and the matrix $\mathbf{A}$ and the resulting approximation to $\mathbf{u}(1)$, have dimension $K = \mathcal{O}(1/h)$ exponential in n: leaving no chance for sub-exponential computational cost. More relevant is therefore the following question:

Question 1. Fix polynomial-time computable initial condition, *fix* polynomial-time computable matrix coefficients $\mathbf{B}_j$ in the right-hand side of PDE (1), and similarly for boundary condition $\mathcal{L}$. Now consider only $(t, \vec{x}) \in [0; 1] \times \Omega$ as input: What is the bit-cost (measured w.r.t. the parameter n) of approximating the solution $\mathbf{u}(t, \vec{x})$ at time t and point $\vec{x}$ up to absolute error $1/2^n$?

Thus, non-uniformly fixing all data of exponential 'size' (formally: from spaces of exponential entropy [12,16,32]) and restricting to polynomial 'size' inputs $(t, \vec{x}) \in [0; 1] \times \Omega$ avoids information-theoretic exponential lower complexity bounds.

Compact domains ensure that one can restrict to complexity considerations in terms of one parameter n [31, Theorem 7.2.7] and does not need to resort to second-order complexity [9]. We use n as parameter for producing approximations up to error bound $1/2^n$, not $1/n$, following the conventions of Real Complexity Theory [15, Definition 2.7]; see also Remark 1 below.

Note that PSPACE has been proven best possible for solving a certain *non-linear* ODE [8]. Poisson's elliptic (i.e. non-evolutionary) PDE has been established to similarly characterize $\#P_1$ [13]. Computation on grids of size $\mathcal{O}(N)$ have been shown computable in $\mathcal{O}(\log N)$ parallel runtime [22]. In our terminology of grid width $h \sim 1/2^n$ for guaranteed output approximation error $1/2^n$, this means $N = \mathcal{O}(1/h^d)$ and parallel runtime $\mathcal{O}(nd)$. That would amount to complexity class PAR =PSPACE, were it not for the superpolynomial number $\mathcal{O}(N/\log N) = \mathcal{O}(2^{nd}/nd)$ of processors. Our previous work [17] has rigorously and in the sense of Question 1 established solutions to a large class of linear first-order evolutionary PDEs computable in PSPACE, and under additional hypotheses even in $\#P^{\#P}$.

The main result of the present paper, Theorem 2, improves the latter to $\#P$. It applies, in particular, to the Heat Equation, where we show $\#P_1$ as optimal.

1.1 Main Result and Overview

The complexity considerations in this work refer to (real counterparts, formalized in Subsect. 1.3, of) the classical hierarchy[2] commonly conjectured proper:

$$ NC \subseteq P \subseteq NP \subseteq \#P \subseteq \#P^{\#P} \subseteq \ldots \subseteq PSPACE = PAR \subseteq EXP . \quad (4) $$

The following hypotheses are very natural and hold for many PDEs including many of the ones mentioned in Example 1, see Example 2 for more detail. For the notation of (iii) see Subsect. 1.2; $\|\varphi\|_{C^l(\bar{\Omega})} = \sup_{\vec{x} \in \bar{\Omega}} \sum_{|\vec{j}| \leq l} |\partial^{\vec{j}} \varphi(\vec{x})|$.

Hypotheses 1 *(i) The problem (1) is well-posed (Hadamard) in that the classical solution $\mathbf{u}(t, \vec{x})$ to (1) exists, is unique and depends continuously on the initial data in the following sense:*

$$ \varphi(\vec{x}) \in \mathcal{C}^l(\bar{\Omega}), \quad \mathbf{u}(t, \vec{x}) \in \mathcal{C}^2([0, 1] \times \bar{\Omega}), \quad \|\mathbf{u}\|_{C^2([0,1] \times \bar{\Omega})} \leq C_0 \|\varphi\|_{C^l(\bar{\Omega})}, \quad (5) $$

for some fixed C_0, $l \geq 2$.

---

[2] The reader may forgive us for identifying decision and function complexity classes.

(ii) The initial functions $\varphi(\vec{x})$ and matrix coefficients $\mathbf{B_j}(\vec{x})$ as well as their partial derivatives up to order l are computable in P.

(iii) The system (1) admits a difference scheme $\mathbf{A}_{h(n)}$ (see (6) below) which is computable in P, and its solution $u^{(n)}$ converges to the solution $\mathbf{u}$ of (1) w.r.t. the maximum norm on the uniform grid $G_{h(n)}$ with the step $h = h(n)$:

$$\max_{x \in G_{h(n)}} |\mathbf{u}\,|_{G_{h(n)}} -u^{(n)}| < C \cdot h(n), \quad C \text{ does not depend on } n.$$

Note that technically a difference scheme is a family $\mathbf{A}_{h(n)}$ of matrices of dimension growing exponentially in $n \to \infty$ such as to approximate the operator $\mathcal{A}$ with increasing precision; the approximating solution $u^{(n)}$ is a sequence of vectors of dimension growing exponentially in n. See Definitions 5, 7 of Subsect. 1.3 for adjustment of the complexity classes to this case.

The main results of the present paper are collected in the following

Theorem 2. *a) The solution $\mathbf{u}$ of (1) under Hypotheses 1 is computable in* PSPACE.

b) If additionally the difference scheme $\mathbf{A}_h$ from (iii) is a sum of tensor products of circulant block matrices of constant bandwidth (as formalized in the hypothesis of Theorem 4), then evaluating the solution $(t, \vec{x}) \mapsto \mathbf{u}(t, \vec{x})$ of (1) is computable in #P.

c) Evaluating the solution $\mathbf{u}$ of (2) is computable in P if the matrices $\mathbf{B}_j$ are constant and mutually commute for $j = 0, 1, \ldots d$.

d) For the Heat Equation $u_t = \Delta u$ there exists a polynomial time computable initial condition φ such that the solution u is classical but cannot be computed in polynomial time unless $P_1 = \#P_1$.

Item b) harnesses a particular structure common to difference schemes, formalized in Theorem 4 below. Intuitively, in 1D the locality of the grid discretization of the differential operator yields a difference scheme $\mathbf{A}_{h(n)}$ with constant bandwidth, and periodic boundary conditions yield to a circulant structure. Higher-dimensional Euclidean domains translate to tensor products of such (families of) matrices.

Example 2. Many evolutionary linear PDEs admit difference schemes that satisfy the hypotheses of Theorem 2b), and thus can be computed in #P, including:

a) the Heat Equation with periodic boundary conditions and polynomial time computable initial function: see [19, §2.11] for maximum norm difference scheme convergence, and e.g. [3] for well-posedness.
b) the Wave Equation with periodic boundary conditions and polynomial time computable initial functions. Indeed, the wave equation admits a max-norm convergent difference scheme under additional smoothness assumptions, see Theorem 3.1 of [20] for the two-dimensional case, given that $u(t, x, y) \in C^{(4,5)}([0, T] \times \bar{\Omega})$. The continuous dependence condition (5) can be verified combining the well-known continuous dependence w.r.t. L_2-norms, and Sobolev Embedding Theorem, see e.g. [3, §5.6.3].

c) Note that the two-dimensional acoustics system

$$
\begin{cases}
\rho_0 \dfrac{\partial u}{\partial t} + \dfrac{\partial p}{\partial x} = 0, \\
\rho_0 \dfrac{\partial v}{\partial t} + \dfrac{\partial p}{\partial y} = 0, \\
\dfrac{\partial p}{\partial t} + \rho_0 c_0^2 \left(\dfrac{\partial u}{\partial x} + \dfrac{\partial v}{\partial y} \right) = 0
\end{cases}
$$

can be equivalently reduced to the two-dimensional wave equation (e.g. [4]); see also [5] for other examples of symmetric hyperbolic systems (2) (with constant coefficients $\mathbf{B}_j$), which are equivalent to higher-order wave equations.

Note that without assuming the boundary conditions periodic, the solution u in the examples a), b), c) above can be computed in PSPACE, according to Theorem 2a).

Subsection 1.2 collects notational conventions and recalls some basic definitions. Subsection 1.3 formalizes computational bit-complexity theory of real vectors and matrices of exponential dimension. Section 2 presents the proof (sketches) to our main Theorem 2.

1.2   Notation

We use $n \in \mathbb{N}$ to parametrize the absolute output approximation error bound $1/2^n$; $d \in \mathbb{N}$ is the dimension of the torus $\Omega = [0;1)^d \bmod 1$ as compact spatial domain of the partial differential equation under consideration and $e \in \mathbb{N}$ denotes the dimension of the solution function vector $\mathbf{u}$.

Definition 1 (Difference schemes)

a) *Consider, for any positive integer N, the uniform rectangular grid G_N on Ω defined by the points*

$$
\left(\frac{i_1 - \frac{1}{2}}{2^N}, \frac{i_2 - \frac{1}{2}}{2^N}, \ldots, \frac{i_m - \frac{1}{2}}{2^N} \right)
$$

where $1 \le i_1, i_2, \ldots, i_m \le 2^N$. Let $h = 1/2^N$ be the corresponding spatial grid step and τ be a time step. Denote $G_N^\tau = G_N \times \{l\tau\}_{l=1}^M$, where M is the number of time steps. The choice of steps h and τ, depending on the output precision parameter n, is specified below in Subsect. 2.3. We consider the following grid norm: $|g^{(h)}| = \max_{x \in G_N} |g^{(h)}(x)|$.

b) *For a linear differential operator $\mathcal{A}$, the matrix $\mathbf{A}_{(h)}$ (with the grid step $h = h(n)$) defines the corresponding difference scheme*

$$
u^{(h,(l+1)\tau)} = \mathbf{A_h} u^{(h,l\tau)}, \quad u^{(h,0)} = \varphi^{(h)} \tag{6}
$$

under consideration. Its entries are denoted $(\mathbf{A}_{(h)})_{I,J}$, $1 \le I, J \le K$. Here $K \sim 2^{\mathcal{O}(n)}$ is the dimension of the vectors $\mathbf{u}^{(h,m\tau)}$ approximating the solution

$\mathbf{u}(m\tau,\vec{x},)$ *at time* $m\tau \leq 1$, *i.e., for* $1 \leq m \leq M := 1/\tau \sim 2^n$. $\tau, h \sim 1/2^n$ *denote the temporal and spatial grid widths, respectively. Generally speaking, capital letters denote quantities (ranging up to) exponential in* n.

c) *The solution* $u^{(h)}$ *of the difference scheme* (6) *converges to the solution* u *of* (1) *if there is a constant* C *not depending on* h *and* τ *such that*

$$|u|_{G_h^\tau} - u^{(h)}| \leq Ch^p \ . \tag{7}$$

Due to the Lax Convergence Theorem, a difference scheme (6) converges in the sense of (7) to the solution of the PDEs (1) if and only if it is *approximating* and *stable*; see e.g. [29]. The latter means that the matrix A_h has bounded powers, see Definition 2b) below.

Definition 2 (Matrices)

a) *Equip vectors with the maximum norm, and matrices with the induced operator norm:* $\|\mathbf{u}\| = \max_j |u_j|$, $\|\mathbf{A}\| = \max\{\|\mathbf{A} \cdot \mathbf{u}\|/\|\mathbf{u}\|\}$.

b) *A square matrix* $\mathbf{A}$ *with entries* $\mathbf{A}_{I,J}$ $(0 \leq I, J < K)$ *is said to have* bounded powers *if its powers are uniformly bounded, i.e., iff there exists some* $C \in \mathbb{N}$ *such that* $\|\mathbf{A}^M\| \leq C$ *holds for all* $M \in \mathbb{N}$.
   Similarly for a family $\mathbf{A}_n$ *of square matrices of possibly varying format,* $n \in \mathbb{N}$*: Here* $\|\mathbf{A}_n^M\|$ *must be bounded independently of both* $M \in \mathbb{N}$ *and of* $n \in \mathbb{N}$.

c) *For a (not necessarily commutative) ring* $\mathcal{R}$, *let* $\mathcal{R}_D^{N \times N}$ *denote the vector space of* $N \times N$ *matrices of bandwidth* $< D$:

$$\mathbf{A} \in \mathcal{R}_D^{N \times N} \quad \Leftrightarrow \quad |I - J| \geq D \Rightarrow \mathbf{A}_{I,J} = 0.$$

The (one-)norm of an *integer* multi-index $\vec{\jmath} = (j_1, \ldots, j_L) \in \mathbb{Z}^L$ is $|\vec{\jmath}| = j_1 + \cdots + j_L$. We write "$\vec{\jmath} \geq \vec{0}$" to indicate non-negative (i.e. natural number) multi-indices. $D = \max\{|\vec{\jmath}| : \mathbf{B}_{\vec{\jmath}} \neq 0\}$ denotes the *order* of the PDE (1).

A main tool in our algorithms translates circulant matrix powering to polynomial powering. Indeed, the above notions have immediate counterparts:

Definition 3 (Polynomials). *Fix a (not necessarily commutative) ring* $\mathcal{R}$.

a) *A polynomial in* L *commuting variables of* componentwise degree *less than* $\vec{D} \in \mathbb{N}^L$ *has the form*

$$P(\vec{X}) = P(X_1, \ldots, X_L) = \sum\nolimits_{\vec{0} < \vec{\jmath} \leq \vec{D}} p_{\vec{\jmath}} \vec{X}^{\vec{\jmath}} = \sum\nolimits_{\vec{\jmath} \geq \vec{0}, |\vec{\jmath}| < \vec{D}} p_{\vec{\jmath}} \cdot \prod\nolimits_{\ell=1}^{L} X_\ell^{j_\ell}.$$

Write $\mathcal{R}_{\vec{D}}[X_1, \ldots, X_L] = \mathcal{R}_{\vec{D}}[\vec{X}]$ *for the vector space of such polynomials.*

b) *An* L-*variate* Laurent *polynomial of componentwise degree* $< \vec{D}$ *has the form*

$$P(\vec{X}) = P(X_1, \ldots, X_L) = \sum\nolimits_{-\vec{D} < \vec{\jmath} < \vec{D}} p_{\vec{\jmath}} \vec{X}^{\vec{\jmath}} \ ,$$

that is, including negative powers of the variables. Write $\mathcal{R}_{\vec{D}}[X_1, X_1^{-1}, \ldots, X_L, X_L^{-1}] = \mathcal{R}_{\vec{D}}[\vec{X}, \vec{X}^{-1}]$ *for the vector space of such Laurent polynomials; and* $P[\vec{X}^{\vec{\jmath}}] := p_{\vec{\jmath}}$ *for the coefficient to* $\vec{X}^{\vec{\jmath}}$ *in* P, $\vec{\jmath} \in \mathbb{Z}^L$.

c) *Suppose $\mathcal{R}$ is equipped with a norm $|\cdot|$. Consider the ring $\mathcal{R}[\vec{X}, \vec{X}^{-1}]$ of all (Laurent) polynomials, equipped with the induced norm $\|P\| := \sum_{|\vec{\jmath}|} |p_{\vec{\jmath}}|$. A (Laurent) polynomial P has* bounded powers *iff $\|P^M\|$ is bounded independently of $M \in \mathbb{N}$.*
Similarly for a family P_n of polynomials of possibly varying number of variables, $n \in N$: Here $\|P_n^M\|$ must be bounded independently of both $M \in \mathbb{N}$ and of $n \in N$.

Multivariate degree is understood componentwise and w.r.t. strict inequality $<$ (not $=$ nor $\leq$). For example $X^2 \cdot Y^3 + X^3 \cdot Y^2$ has componentwise degree $< \vec{D} = (4,4)$, but not $< (3,4)$ nor $< (4,3)$. An L-variate Laurent polynomial P of componentwise degree $< \vec{D}$ can be converted to an ordinary polynomial by multiplying P with $X_1^{D_1} \cdots X_L^{D_L}$.

In the sequel, *exponential* growth is to be understood as bounded by $2^{p(k)}$, $k \to \infty$, for some polynomial p.

1.3   Real Complexity Theory

In [15], major classical complexity classes have been adapted from the discrete case to the setting of real numbers and (continuous) real functions. There the integer parameter n governing the output approximation error $1/2^n$ replaces the role of the binary input length. Let us recall the definitions of polynomial/exponential time/space computability of real numbers, (fixed-dimensional) real vectors, sequences of real numbers, and partial real functions [15].

Definition 4. *a) Computing a* **real number** *$r \in \mathbb{R}$ means to output, given $n \in \mathbb{N}$, some numerators $a_n \in \mathbb{Z}$ in binary with $|r - a_n/2^{p(n)}| \leq 1/2^n$ for some polynomial $p \in \mathbb{N}[N]$. Such a computation runs in* polynomial time *(P) if said a_n is output within a number of steps bounded by a polynomial in n. It runs in* exponential time *(EXP) if the number of steps is bounded exponentially in n. The computation runs in* polynomial space *(PSPACE) if the amount of memory is bounded polynomially in n.*
b) Computing a **(finite-dimensional) real vector** *(in P, EXP, PSPACE) means to compute each of its entries separately (in P, EXP, PSPACE).*
c) Computing a **sequence** *$\bar{r} = (r_k) \in \mathbb{R}$ of real numbers means to output, given n and k, some $a_{n,k} \in \mathbb{Z}$ in binary with $|r_k - a_{n,k}/2^{p(n+k)}| \leq 1/2^n$ for some polynomial $p \in \mathbb{N}[N]$. Such a computation runs in* polynomial time *(P) if said $a_{n,k}$ is output within a number of steps bounded by a polynomial in $n+k$. Similarly for* exponential time *(EXP) and* polynomial space *(PSPACE).*
d) Computing a **partial real function** *$f :\subseteq \mathbb{R} \to \mathbb{R}$ (w.r.t. some polynomials $q, p \in \mathbb{N}[N]$) means, given $n \in \mathbb{N}$ and any numerator $a \in \mathbb{Z}$ with $|x - a/2^{q(p(n))}| \leq 1/2^{p(n)}$ in binary for some $x \in \mathrm{dom}(f)$, to output some $b = b_n(a) \in \mathbb{Z}$ in binary with $|f(x) - b_n/2^{p(n)}| \leq 1/2^n$. The computation may behave arbitrarily on inputs a that do not satisfy the hypothesis.*

Remark 1. In addition to following the conventions of Real Complexity Theory [15, Definition 2.7], we prefer error bound $1/2^n$ (as opposed to $1/n$) for four reasons:

a) It corresponds to measuring computational cost of discrete problems, such as of integer factorization, in dependence of the binary (as opposed to unary) length n.
b) It reflects that (for instance Chudnovsky's or Borwein's) algorithms can approximate π in time polynomial in n up to error $1/2^n$ (while error bound $1/n$ is trivial to achieve).
c) It gives rise to the aforementioned and subsequent and many more [15] numerical characterizations of discrete complexity classes.
d) The first-order theory of the two-sorted structure $(\mathbb{Z}, 0, 1, +, >) \cup (\mathbb{R}, 0, 1, +, \times, >)$ with 'error embedding' $\imath : \mathbb{Z} \ni n \mapsto 2^{-n} \in \mathbb{R}$ (capturing *Exact Real Computation*) is decidable, while that with $\mathbb{N}_+ \ni n \mapsto 1/n \in \mathbb{R}$ is not [24, Theorem 4.4].

For real vectors of exponential (in some integer parameter k) dimension D_k, our notions of (time/space) complexity are more subtle and we adapt them to the computation of any desired entry rather than of the entire matrix. The complexity of computing said dimension D_k itself must be taken into account as well (see details in the Appendix). Defining real counterparts to #P is subtle and discussed in detail, since most of our major results refer to it.

When speaking of complexity of computation for **matrices**, we identify a $D \times E$-dimensional matrix $B = (b_{I,J})$ with the $D \times E$-dimensional vector $B_{\langle I,J \rangle}$ for the pairing function $\langle I, J \rangle = J + (I + J) \cdot (I + J + 1)/2$.

Definition 5. *a) Computing a* **sequence** $\vec{r}_k = (r_{k,J})_{J \leq D_k} \in \mathbb{R}^{D_k}$ *of* $\boldsymbol{D_k}$-**dimensional real vectors** *means to output, given* $n, k \in \mathbb{N}$ *(in unary) and* $J \leq \dim(\vec{r}_k) = D_k$ *in binary, some* $a_{n,k,J} \in \mathbb{Z}$ *in binary with* $|r_{k,J} - a_{n,k,J}/2^{p(n+k)}| \leq 1/2^n$ *for some polynomial* $p \in \mathbb{N}[N]$.
Such a computation runs in polynomial time *(P) if said* $a_{n,k,J}$ *is output within a number of steps bounded by a polynomial in* $n + k$ *but independently of* J. *Similarly for* exponential time *(EXP) and* polynomial space *(PSPACE).*

b) More generally, computing a **sequence** $\vec{f}_k :\subseteq \mathbb{R}^d \to \mathbb{R}^{D_k}$ *of* **partial vector functions** *(w.r.t. some polynomials* $q, p \in \mathbb{N}[N]$*) means, given* $n, k, J \in \mathbb{N}$ *and any numerator* $\vec{a} \in \mathbb{Z}^d$ *with* $|\vec{x} - \vec{a}/2^{q(p(n+k)+k)}| \leq 1/2^{p(n+k)}$ *for some* $\vec{x} \in \mathrm{dom}(f_k)$ *and* $J \leq D_k = \dim f_k$, *to output some* $b = b_{n,k,J}(a) \in \mathbb{Z}$ *with* $|f_{k,J}(\vec{x}) - b/2^{p(n+k)}| \leq 1/2^n$. *Here* $\|\vec{y}\| = \max\{|y_1|, \ldots, |y_d|\}$ *denotes the maximum norm.*
Such a computation runs in polynomial time *(P) if said* $b = b_{n,k,J}(a)$ *is output within a number of steps bounded by a polynomial in* $n + k$ *but independently of* J *and* a. *Similarly for* exponential time *(EXP) and* polynomial space *(PSPACE).*

c) A sequence (D_k) *of natural numbers is computable in* unary polynomial time *(Pone) if the mapping* $0^k \mapsto \mathrm{bin}(D_k) \in \{0,1\}^*$ *is computable in time polynomial in the input length.*

Note that a polynomial-time computable real sequence r_k according to (4c) can grow at most exponentially (in k); similarly for both the entries and the dimension D_k of a polynomial-time computable vector sequence according to (4d), and for vector functions according to (5b): in agreement with a sequence D_k computable in unary polynomial time according to (5c) growing at most exponentially in k. Memory-bounded computation here is understood to charge for all, input and working and output tape; hence sequences of reals and real vectors computable in polynomial space also satisfy exponential bounds of growth of value and dimension.

In this way our real counterparts of P and PSPACE above agree with both the conception of real numbers as 'streams' of approximations [31,32] as well as with the oracle-based approach [9,15].

Recall that the discrete complexity class #P consists of all total functions $\psi :$ $\{0,1\}^* \to \mathbb{N}$ such that some non-deterministic polynomial-time Turing machine on input $\vec{x} \in \{0,1\}^*$ has precisely $\psi(\vec{x})$ accepting computations.

Definition 6. *Fix $\psi : \{0,1\}^* \to \mathbb{N}$.*

a) *We say that ψ counts the real number $r \in \mathbb{R}$ if it holds $a_n = \psi(1^n) - \psi(0^n)$ according to Definition 4a. If $\psi \in$ #P, call r computable in #P.*
b) *Say that ψ counts the real sequence $\bar{r} = r_k$ if $a_{n,k} = \psi(1^n 0^k) - \psi(0^n 1^k)$ according to Definition 4c. If $\psi \in$ #P, call $\bar{r}$ computable in #P.*

As in Definitions 4, the case of vectors of exponential dimension is more subtle:

Definition 7. *a) Suppose $\log(D_k)$ grows at most polynomially. We say that ψ counts the sequence $\vec{r}_k$ of D_k-dimensional real vectors if*

$$a_{n,k,J} = \psi(1^n 0^k 1 \operatorname{bin}(J)) - \psi(0^n 1^k 0 \operatorname{bin}(J))$$

according to Definition 5a. If $\psi \in$ #P, call $\vec{r}_k$ computable in #P.
b) *Say that ψ counts the partial function $f :\subseteq \mathbb{R} \to \mathbb{R}$ if*

$$b_n(a) = \psi(1^n 0 \operatorname{bin}(a)) - \psi(0^n 1 \operatorname{bin}(a))$$

according to Definition 4d. If $\psi \in$ #P, call f computable in #P.
c) *Suppose $\log(D_k)$ grows at most polynomially. We say that ψ counts the sequence of partial vector functions $\vec{f}_k :\subseteq \mathbb{R}^d \to \mathbb{R}^{D_k}$ if*

$$b_{n,k,J}(a) = \psi(1^n 0^k \operatorname{bin}(J), \operatorname{bin}(a)) - \psi(0^n 1^k \operatorname{bin}(J), \operatorname{bin}(a))$$

according to Definition 5b. If $\psi \in$ #P, call $\vec{f}_k$ computable in #P.

It is unknown under which arithmetic operations #P is closed; for instance GapP has been introduced as the closure of #P under subtraction [6]. The above real counterparts could thus perhaps more accurately be called "GapP-computability", rather than #P-computability. Since it holds $P^{\#P} = P^{GapP}$, the difference between #P and GapP seems minor, if any, from the perspective of computational cost. For notational convenience, we define *real counting complexity* neglecting such subtleties. Also Definition 4a) actually refers to P_1 rather than P. Similarly, Definition 6a) could perhaps better (but more awkwardly) refer to "#P_1-computability", since it employs restrictions $\psi|_{1^*}$ of $\psi \in$ #P to *unary* arguments. This might cause our notions to slightly deviate from [15, p.184].

Polynomial-time computability implies computability in #P, which in turn implies computability in PSPACE: Note that, since $\psi \in$ #P grows in value at most exponentially in the input length, any real sequence r_k computable in #P according to c) also has $|r_k|$ growing at most exponentially in k.

2   Techniques and Proof Ideas

For the difference scheme approach (6), taking into account Hypotheses 1 (ii), (iii), the computational problem is equivalent to: first raise a matrix $\mathbf{A}$ (or, rather, any desired one from a sequence of matrices $\mathbf{A}_{h(n)}$) of exponential dimension $K \sim 2^{O(n)}$ to an exponential power $M \sim 2^{O(n)}$; then multiply the intermediate result to a K-dimensional sample vector $\varphi^{(h)}$ of the initial condition φ; and finally return an approximation to any desired entry with given index #J of the result vector $\left(\mathbf{A}_{h(n)}^{M(n)} \cdot \varphi^{(h(n))}\right)_J$ up to error $1/2^n$, $0 \le J < K$. Naïvely the intermediate result matrices and vectors have exponential dimension, hence leading to complexity class EXP.

On the other hand, according to Question 1, only one entry #J of the result is required; and both the difference-scheme matrix $\mathbf{A}_{h(n)}$ and the sampled initial vector $\varphi^{(h(n))}$ do not need to be stored, but by hypothesis any desired of its entries can be (re-)computed on-the-fly in P, whenever and however often required: only $J \in \{0, 1, \ldots, K-1\}$ and M are part of the input, given in binary with a linear number $\mathcal{O}(n)$ of bits.

The discrete counterpart to our real problem would ask for any desired entry of a P-computable Boolean matrix of exponential dimension to exponential power; and this can be solved in PSPACE by Savitch's Algorithm. When applying the same approach to the integer or to the present real case, the hypothesis of $\mathbf{A}$ having bounded powers (Definition 2b) becomes crucial: to guarantee that the resulting entries do not blow up, nor do they require excessive initial precision in order to keep the rounding error propagation in check (details omitted). We thus have an improved, namely PSPACE algorithm, and establish Theorem 2a).

Regarding Theorem 2b), Subsect. 2.2 reduces the problem of (recovering any desired entry of a) circulant matrix raised to an exponential power to that of (recovering any desired coefficient of a) polynomials raised to such power; in a way that relates bandwidth to degrees. Subsection 2.1 solves the latter problem in #P by reduction to Riemann integration via Cauchy's Differentiation Theorem. Linear polynomials can even be raised to exponential powers in P according to Proposition 1.

Subsection 2.3 concludes the proof of Theorem 2 a), b) by treating both grid and nongrid points to compute the solution of (1); this proof makes use of Hypotheses 1 (i). Subsections 2.4, 2.5 establish Theorem 2 c), d), respectively.

2.1 Raising Polynomials to Exponential Powers

Consider the problem of raising a fixed P-computable univariate polynomial P of 'small' degree to an exponential M-th power, and read off the J-th coefficient $P[X^J]$ with $J, M \in \mathbb{N}$ in binary as only actual input. Of course having bounded powers (Definition 3c) is again a crucial hypothesis to prevent coefficient explosion and numerical instability. Note that already for the quadratic polynomial $P(X) = (1 + X + X^2)/3$, naïve evaluation of the 'explicit' formula

$$\left(\tfrac{1}{3} + \tfrac{1}{3}X + \tfrac{1}{3}X^2\right)^K [X^J] = 3^{-K} \cdot \sum_{\substack{0 \le \mu, \nu \le K \\ \mu + 2\nu = M}} \frac{K!}{\mu! \cdot \nu! \cdot (K - \mu - \nu)!} \tag{8}$$

involves terms like $K!$ of value, and the sum with a number of terms, doubly exponential in k: not at all obvious to compute in #P.

However, thanks to Cauchy's Integral Theorem, we can express any single desired coefficient of P^M as loop integral over $P^M(z)/z^{M+1}$ for $|z| = 1$ running over the complex unit circle. And due to P having bounded powers, the values of $P^M(z)/z^{M+1}$ are bounded, can be computed with repeated squaring on real *numbers*, and their dependence on z is sufficiently well-behaved: Such Riemann integrals are known computable in #P [15, §5.4]. This generalizes to the multivariate case:

Theorem 3. *Fix a sequence $P_k(\vec{X})$ of L-variate polynomials of componentwise degree $< \vec{D}$ and bounded powers. Let K_k denote a sequence of natural numbers with binary representation computable in time polynomial (and thus of value exponentially bounded) in k.*
Then each coefficient $P_k^{K_k}[\vec{X}^{\vec{J}}]$ of $P_k^{K_k}$, $\vec{0} \le \vec{j} < \vec{D}$, formalized as mapping $\mathrm{bin}(1^k, \mathrm{bin}(\vec{j})) \mapsto P_k^{K_k}[\vec{X}^{\vec{J}}]$, is computable in #P.

Note that $P_k^{K_k}$ has componentwise degree $< \vec{D} \cdot K^k$, increased at most exponentially in k; hence the binary length of $\vec{j}$ is polynomial in k, the binary length of 1^k. Recall Definition 6 for the formal notion of a (sequence/family of) real vectors/matrices/polynomials to be computable in #P. Theorem 3 extends immediately to Laurent polynomials.

Raising linear polynomials is feasible in P: Note that $\binom{N}{K} p^K q^{N-K}$ is the coefficient of X^K of the polynomial $(pX + q)^N$.

Proposition 1. *Fix P-computable $q \in (0; 1)$ and $k \in \mathbb{N}$. Abbreviate $p := 1 - q$. Given $K \le N \in \mathbb{N}$, one can approximate $\binom{N}{K} p^K q^{N-K}$ to absolute error N^{-k} in time polynomial in $\log N$.*

2.2 Circulant Matrices as Polynomials

Consider an $N \times N$ circulant matrix $\mathcal{A}$ with parameters $a_0, \ldots, a_{N-1}$,

$$\mathcal{A} = \left(a_{I-J \bmod N}\right)_{0 \leq I, J < N} = \sum\nolimits_{J=0}^{N-1} a_J \cdot \mathcal{C}_N^J \tag{9}$$

for the 'generator' circulant matrix $\mathcal{C}_N$ with parameters $(0, 1, 0, 0, \ldots, 0)$. Let $P_{\mathcal{A}}(X) = \sum_{j=0}^{N-1} a_j X^j$ denote the *associated polynomial* to circulant matrix $\mathcal{A}$: so $\mathcal{A}^M = P_{\mathcal{A}}^M(\mathcal{C}_N)$, and the entry with indices (I, J) of $\mathcal{A}^M$ is

$$(\mathcal{A}^M)_{I,J} = \sum\nolimits_{K \equiv (I-J) \bmod N} P_{\mathcal{A}}^M[X^K]$$

in the terminology of Definition 3. The associated polynomial is a well-known concept [18]. We will primarily consider circulant matrices A with small bandwidth D. These correspond to the images of *Laurent* polynomials with small degrees D: linear combinations of monomials X^j of possibly negative exponents j ranging from $-D$ to $+D$. More generally, we treat matrices that can be decomposed into components combined by Kronecker product as in Eq. (11).

Lemma 1. *Fix a not necessarily commutative ring $\mathcal{R}$ of characteristic zero. Let* $\mathrm{CIRC}_D^N(\mathcal{R}) \subseteq \mathcal{R}_D^{N \times N}$ *denote the subspace of $N \times N$ circulant matrices of bandwidth $< D$. Recall that $\mathcal{C}_N \in \mathrm{CIRC}_2^N(\mathcal{R})$ denotes the $N \times N$ cyclic permutation matrix from (9) and (Definition 2) we write $P[\vec{X}^{\vec{j}}] \in \mathcal{R}$ for the coefficient to $\vec{X}^{\vec{j}} = X_1^{j_1} \cdots X_L^{j_L}$ in an L-variate Laurent polynomial $P \in \mathcal{R}[\vec{X}, \vec{X}^{-1}]$, $\vec{j} \in \mathbb{Z}^L$.*

a) *For any $P \in \mathcal{R}_D[X, X^{-1}]$ it holds $P(\mathcal{C}_N) \in \mathrm{CIRC}_D^N(\mathcal{R})$. More precisely*

$$P(\mathcal{C}_N)_{I,J} = \sum\nolimits_{n \in \mathbb{Z}} P[X^{J-I+nN}] \in \mathcal{R} ,$$

and the sum is finite. The mapping $\mathcal{R}[X, X^{-1}] \ni P \mapsto P(\mathcal{C}_N) \in \mathrm{CIRC}^N(\mathcal{R})$ is a homomorphism of non-commutative algebras. For $D \leq N/2$ and normed $\mathcal{R}$, the restriction $\mathcal{R}_D[X, X^{-1}] \ni P \mapsto P(\mathcal{C}_N) \in \mathrm{CIRC}_D^N(\mathcal{R})$ is an isometry of vector spaces with respect to the induced norms from Definitions 3 and 2.
b) *Generalizing a), fix $m, L, N_1, \ldots, N_L \in \mathbb{N}$ and consider ring homomorphism*

$$\Phi_{m,\vec{N}} : \mathbb{R}^{m \times m}[X_1, X_1^{-1}, \ldots, X_L, X_L^{-1}] \ni P \mapsto \mathbb{R}^{m \times m} \otimes \mathbb{R}^{N_1 \times N_1} \otimes \cdots \mathbb{R}^{N_L \times N_L}$$

$$\mathbb{R}^{m \times m} \ni \mathbf{B} \mapsto \mathbf{B} \otimes \mathbb{D}_{N_1} \otimes \cdots \otimes \mathbb{D}_{N_L}, \quad X_\ell \mapsto \mathbb{D}_m \otimes \mathbb{D}_{N_1} \otimes \cdots \otimes \mathcal{C}_{N_\ell} \otimes \cdots \otimes \mathbb{D}_{N_L}$$

Then it holds $\left(\Phi_{m,\vec{N}}(P)\right)_{I_1, J_1, \ldots, I_L, J_L} =$

$$\sum\nolimits_{n \in \mathbb{Z}} P[X_1^{J_1 - I_1 + n_1 N_1}, \ldots X_L^{J_L - I_L + n_L N_L}] \in \mathbb{R}^{m \times m}. \tag{10}$$

For $D_\ell \leq N_\ell/2$, $\Phi_{m,\vec{N}}$ restricted to $\mathbb{R}_{\vec{D}}^{m \times m}[\vec{X}, \vec{X}^{-1}]$ maps isometrically and surjectively onto the vector space of matrices of the form (11). □

The following Theorem 4 now is a corollary to Theorem 3: Translate the matrix $\mathbf{A} = \mathbf{A}_k$ into normal form according to Lemma 1b); then further on into a multivariate Laurent polynomial P according to the inverse isometry from Lemma 1c); raise P_k^M to the desired power following Theorem 3; and finally recover the desired coefficient of $\mathbf{A}^M$ according to Eq. (10).

Theorem 4. *Fix $J, L \in \mathbb{N}$ and, for $k = 1, 2, \ldots$, let*

- $K = K(k, j, \ell) \in \mathbb{N}$ *with binary representation computable in time polynomial (and thus of value exponentially bounded) in k, $1 \leq j \leq J$, $1 \leq \ell \leq L$*
- $\mathcal{C}_{k,j,\ell}$ *polynomial-time computable circulant matrices of bandwidth $< D$ and dimensions $K = K(k, \ell)$, $\ell \leq L$*
- Q_j *polynomial time computable real matrices of constant dimension d, $1 \leq j \leq J$*
- $M = M(k)$ *natural numbers with binary representations computable in time polynomial in k.*

Consider the following (sequence of) Kronecker/tensor products:

$$\mathbf{A}_k := \sum\nolimits_{j=1}^{J} Q_j \otimes \mathcal{C}_{k,j,1} \otimes \mathcal{C}_{k,j,2} \otimes \cdots \otimes \mathcal{C}_{k,j,L} \qquad (11)$$

If that matrix sequence has bounded powers, then the sequence $\mathbf{A}_k^{M(k)}$ of matrix powers is computable in #P.   □

Replacing Theorem 3 with Proposition 1, we obtain the following corollary which may be of independent interest:

Corollary 1. *If the circulant matrices $\mathcal{C}_{k,j,\ell}$ in Theorem 4 have bandwidth two, then the sequence $\mathbf{A}_k^{M(k)}$ of matrix powers is computable in polynomial time.*

Difference schemes of bandwidth two correspond to simple transport equations, which admit explicit solutions in Subsect. 2.4. Also, fast exponentiation A^M of a difference scheme A by itself is insufficient: It remains to apply the inner product with the initial condition, which still incurs cost #P [17, Proposition 6e].

2.3   Complexity of Solutions of Evolutionary PDEs

Proof (Theorem 2a,b). To compute the solution $\mathbf{u}$ at a fixed point (t, x) with the prescribed precision 2^{-n} and estimate the bit-cost of the computation, consider the following computation steps.

1. Choose the space and time grid steps, h and τ, in the following way:

- h is any binary-rational number of the form $h = 2^{-N}$, where $N = \mathcal{O}(n)$, satisfying the Inequality (13) below;
- τ is any binary-rational number meeting the Courant inequality $\tau \leq \nu h$, where ν is the Courant number guaranteeing convergence property of the considered difference scheme which can be computed from the coefficients of the system (1).

2. For a grid point $(t, \vec{x})$ put $l = \frac{t}{\tau}$ (note that $l \leq M = \left[\frac{1}{\tau}\right] = \mathcal{O}(2^n)$) and calculate the matrix powers and vector products $\mathbf{A}_{h(n)}^{M(n)} \cdot \varphi^{(h(n))}$. Note that we use matrix powering instead of step-by-step iterations initially suggested by the difference scheme (6).

3. For non-grid points take (e.g.) a multilinear interpolation $\widetilde{u^{(h)}}$ of $u^{(h)}$, which can be computed in polynomial time from the (constant number of) "neighbor" grid points. Due to well-known properties of multilinear interpolations,

$$\sup_{t,\vec{x}} |\widetilde{\mathbf{u}^{(h)}}(t, \vec{x})| \leq \tilde{C} \sup_{G_N^\tau} |u^{(h)}|; \quad \sup_{t,\vec{x}} |\mathbf{u}(t, \vec{x}) - \widetilde{\mathbf{u} \,|_{G_N^\tau}}(t, \vec{x})| \leq \bar{C} \sup_{t,\vec{x}} |\partial^2 \mathbf{u}(t, \vec{x})| \cdot h^2,$$

(12)

where $\tilde{C}$ and $\bar{C}$ are absolute constants.

Based on (12) and on the continuous dependence property, as well as on linearity of the interpolation operator, infer

$$\sup_{t,\vec{x}} |\mathbf{u}(t, \vec{x}) - \widetilde{u^{(h)}}(t, \vec{x})| \leq \sup_{t,\vec{x}} \left(|\mathbf{u}(t, \vec{x}) - \widetilde{\mathbf{u} \,|_{G_N^\tau}}(t, \vec{x})| + |\widetilde{\mathbf{u} \,|_{G_N^\tau}}(t, \vec{x}) - \widetilde{u^{(h)}}(t, \vec{x})| \right)$$

$$\leq \bar{C} C_0 ||\varphi||_{C^l(\bar{\Omega})} h^2 + \tilde{C} C \cdot h \leq 2^{-n}.$$

Thus choosing a grid step $h = 2^{-N}$ such that

$$h \leq C_h \cdot 2^{-n}, \quad C_h = \bar{C} C_0 ||\varphi||_{C^l(\bar{\Omega})} + \tilde{C} C,$$

(13)

will guarantee the computed function $\widetilde{u^{(h)}}$ approximate the solution u with the prescribed precision 2^{-n} (here C_h depends only on the fixed P-computable functions φ, B_i and therefore is a fixed constant. Hence it remains to estimate complexity of matrix powering (which turns out to be in PSPACE or #P in items a) or b) of Theorem 2, respectively) and inner product (which turns out to be in #P), see comments in the beginning of this Section and Theorem 3.

2.4   Symmetric Hyperbolic PDEs with Commuting Coefficients

Towards proving Theorem 2c), first consider the following *scalar* linear PDE with constant coefficients and its explicit solution

$$\lambda_0 u_t = \sum_j \lambda_j \partial_j u, \quad u(\vec{x}, t) = u_0 \left(t \sum_j \lambda_j + \lambda_0 \sum_j x_j \right)$$

(14)

which is obviously computable in P, provided that $\lambda_0, \lambda_1, \ldots, \lambda_d \in \mathbb{R}$ are.

Theorem 2c) considers *vector systems* of such PDEs, with constant symmetric mutually commuting matrix coefficients $\mathbf{B}_{\vec{j}}$. These hypotheses assert a simultaneous diagonalization, that is, a basis of joint eigenvectors: which 'decouples' the system into e independent scalar Eq. (14). It remains to prove that such a joint spectral decomposition can be computed in P: which is wrong in case (even a single symmetric) matrix $\mathbf{B}_{\vec{j}}$ is given as input [34]. Fortunately, in agreement with Question 1, these matrices are not part of the input but fixed computable in P. And for this case we have and apply the following

Theorem 5. *Fix symmetric and mutually commuting matrices $\mathbf{B}_{\vec{\jmath}} = \mathbf{B}_{\vec{\jmath}}^*$ of fixed dimension $e \in \mathbb{N}$ with fixed real polynomial-time computable entries. Then a joint spectral decomposition $\mathbf{B}_{\vec{\jmath}} = \mathbf{T}^\mathsf{T} \cdot \mathbf{D}_{\vec{\jmath}} \cdot \mathbf{T}$ can be computed in time polynomial in the output precision parameter n (and unspecified dependence on the dimension e as well as on the 'input' matrix entries). Here $\mathbf{D}_{\vec{\jmath}}$ is a diagonal matrix with the eigenvalues of $\mathbf{B}$; and $\mathbf{T}$ is an orthonormal matrix consisting of a joint basis of eigenvectors of all $\mathbf{B}_{\vec{\jmath}}$.*

Proof. Let k denote the dimension of B. By [21], the d-tuple of eigenvalues with multiplicity (i.e. the sought diagonal matrix D) can be approximated up to absolute error $1/2^n$ in time polynomial in n. Note that the (integer) multiplicities themselves depend discontinuously on the entries of B and therefore cannot be computed; but, since the matrix is fixed, they can be hardcoded into the thus non-uniform computation. It then remains to compute, for each eigenvalue λ with correct multiplicity $k = k(\lambda)$, an orthonormal basis of the k-dimensional kernel of $B - \lambda \cdot \mathrm{id}$ [23]. Naive Gaussian Elimination does not suffice for this purpose since the tests for in/equality during pivot search are undecidable. Bareiss Algorithm avoids the divisions; but cannot avoid tests either. On the other hand Bareiss performs (more than Gauss, but still) only a number of test polynomially bounded in the dimension d, which is constant in our setting, namely independent of the output precision parameter n. Hence, again, the outcomes of these tests can be hardcoded in this non-uniform computation. The remaining operations of Bareiss are arithmetic, and can performed on approximations. More precisely consider the $d \times d$ matrix $\tilde{B}_{m,\lambda}$ of integer numerators such that dyadic $\tilde{B}_{m,\lambda}/2^m$ approximates $B - \lambda \cdot \mathrm{id}$ 'sufficiently' well: Since the number of arithmetic operations is polynomial in d, it suffices to choose m polynomially larger than n in order to guarantee that the output of Bareiss is still within absolute error $1/2^n$ from the hypothetical exact result. The bit-cost of Bareiss is well-known polynomial in (d and) m.

2.5   #P₁ is Optimal for the Heat Equation

#P is known optimal for *in*definite Riemann Integration [15, Theorem 5.33], while #P₁ is optimal for definite Riemann Integration [15, Theorem 5.32] in the following sense:

Fact 6 *a) There is a P-computable (hence continuous) $h : [0;1] \to [-1;1]$ such that $\int_0^1 h(x)\, dx$ is not computable in P_1 unless $FP_1 = \#P_1$.*
b) For every P-computable analytic function $g : [0;1] \to \mathbb{R}$, $\int_0^1 g(x)\, dx$ is computable in P_1 [15, bottom of page 208].
c) Let $u = u(x,t)$ solve the 1D Heat Equation $u_t = u_{xx}$ on $[0;1]$ with periodic boundary conditions $u(0,t) = u(1,t)$ and $u_x(0,t) = u_x(1,t)$ such that $u_0 = u(x,0)$ is in $\mathcal{C}^2[0;1]$. Then $u(x,t)$ is analytic in x for each fixed $t > 0$, and the 'overall heat' $\int_0^1 u(x,t)\, dx$ does not depend on t; see e.g. [7]. □

In order to establish Theorem 2d), consider $u_0 := h$ from Fact 6a). W.l.o.g. $h(0) = h(1)$: otherwise consider $[0;1/2] \ni x \mapsto h(2x)/2$ and $[1/2;1] \ni x \mapsto h(2-2x)/2$. Then apply Fact 6b) to $g(x) := u(x,t)$ according to Fact 6c).

3   Conclusion and Perspective

We have extended previous rigorous bit-cost investigations of PDEs [13] from elliptic to hyperbolic and parabolic linear PDEs, improving upper complexity bounds PSPACE and $\#P^{\#P}$ [17] to $\#P$. $\#P_1$ (not $\#P$, as claimed in [13, §6]) turned out as necessary for both Poisson and Heat equation. Current (limited) evidence suggests that differential equations might exhibit a dichotomy [26]: either P (Subsection 2.4 of the present paper, Theorem 3 of [17] for the case of linear evolutionary systems of PDEs with *analytic* initial functions and matrix coefficients, and [1] for analytic ODEs) or $\#P/\#P_1$-hard.

References

1. Bournez, O., Graça, D.S., Pouly, A.: Solving analytic differential equations in polynomial time over unbounded domains. In: Murlak, F., Sankowski, P. (eds.) MFCS 2011. LNCS, vol. 6907, pp. 170–181. Springer, Heidelberg (2011). https://doi.org/10.1007/978-3-642-22993-0_18

2. Braverman, M., Cook, S.: Computing over the reals: foundations for scientific computing. Not. AMS **53**(3), 318–329 (2006)

3. Evans, L.: Partial differential equations. Graduate Studies in Mathematics, vol. 19. American Mathematical Society (1998)

4. Godunov, S., Ryaben'kii, V.: Difference schemes: an introduction to the underlying theory. Studies in Mathematics and Its Applications, vol. 19. North-Holland (1987)

5. Gordienko, V.: Hyperbolic systems equivalent to the wave equation. Siberian Math. J. **50**, 14–21 (2009)

6. Hemachandra, L.A., Ogiwara, M.: Is $\#P$ closed under subtraction? Current Trends in Theoretical Computer Science, pp. 523–536 (1993). https://doi.org/10.1142/9789812794499_0039

7. Epstein, M.: Standing waves and separation of variables. In: Partial Differential Equations. ME, pp. 183–208. Springer, Cham (2017). https://doi.org/10.1007/978-3-319-55212-5_9

8. Kawamura, A.: Lipschitz continuous ordinary differential equations are polynomial-space complete. Comput. Complex. **19**(2), 305–332 (2010)

9. Kawamura, A., Cook, S.: Complexity theory for operators in analysis. ACM Trans. Comput. Theor. **4**(2), 5:1–5:24 (2012)

10. Kawamura, A., Ota, H., Rösnick, C., Ziegler, M.: Computational complexity of smooth differential equations. Logical Methods Comput. Sci. **10**(1:6), 15 (2014). https://doi.org/10.2168/LMCS-10(1:6)2014

11. Kawamura, A., Steinberg, F., Thies, H.: Parameterized complexity for uniform operators on multidimensional analytic functions and ODE solving. In: Proceedings 25th International Workshop on Logic, Language, Information, and Computation (WOLLIC), pp. 223–236 (2018). https://doi.org/10.1007/978-3-662-57669-4_13

12. Kawamura, A., Steinberg, F., Ziegler, M.: Complexity theory of (functions on) compact metric spaces. In: Proceedings 31st Annual Symposium on Logic in Computer Science, LICS, pp. 837–846. ACM (2016). https://doi.org/10.1145/2933575.2935311

13. Kawamura, A., Steinberg, F., Ziegler, M.: On the computational complexity of the Dirichlet problem for Poisson's equation. Math. Struct. Comput. Sci. **27**(8), 1437–1465 (2017). https://doi.org/10.1017/S096012951600013X

14. Kawamura, A., Thies, H., Ziegler, M.: Average-case polynomial-time computability of hamiltonian dynamics. In: Potapov, I., Spirakis, P.G., Worrell, J. (eds.) 43rd International Symposium on Mathematical Foundations of Computer Science, MFCS 2018, August 27–31, 2018, Liverpool, UK. LIPIcs, vol. 117, pp. 30:1–30:17. Schloss Dagstuhl - Leibniz-Zentrum für Informatik (2018)
15. Ko, K.I.: Complexity theory of real functions. Progress in Theoretical Computer Science, Birkhäuser, Boston (1991)
16. Kolmogorov, A.N., Tikhomirov, V.M.: E-entropy and E-capacity of sets in functional spaces. In: Shiryayev, A. (ed.) Selected Works of A.N. Kolmogorov, vol. III, pp. 86–170. Springer, Dordrecht (1993). originally, pp. 3–86 in Uspekhi Mat. Nauk, vol. 14:2 (1959)
17. Koswara, I., Selivanova, S., Ziegler, M.: Computational complexity of real powering and improved solving linear differential equations. In: van Bevern, R., Kucherov, G. (eds.) CSR 2019. LNCS, vol. 11532, pp. 215–227. Springer, Cham (2019). https://doi.org/10.1007/978-3-030-19955-5_19
18. Kra, I., Simanca, S.R.: On circulant matrices. Not. AMS **59**(3), 368–377 (2012). https://doi.org/10.1090/noti804
19. LeVeque, R.J. (ed.): Finite Difference Methods for Differential Equations. SIAM, Philadelphia (2007)
20. Liao, H.l., Sun, Z.Z.: Maximum norm error estimates of efficient difference schemes for second-order wave equations. J. Comput. Appl. Math. **235**(8), 2217–2233 (2011). https://doi.org/10.1016/j.cam.2010.10.019
21. Neff, C.A.: Specified precision polynomial root isolation is in NC. J. Comput. Syst. Sci. **48**(3), 429–463 (1994). https://doi.org/10.1016/S0022-0000(05)80061--3
22. Pan, V., Reif, J.: The bit-complexity of discrete solutions of partial differential equations: compact multigrid. Comput. Math. Appl. **20**, 9–16 (1990)
23. Park, S.: Reliable degenerate matrix diagonalization. Technical report CS-TR-2018-415, KAIST (2018)
24. Park, S., et al.: Foundation of computer (algebra) analysis systems: Semantics, logic, programming, verification (2020). https://arxiv.org/abs/1608.05787
25. Pour-El, M.B., Richards, J.I.: The wave equation with computable inital data such that its unique solution is not computable. Adv. Math. **39**, 215–239 (1981)
26. Schaefer, T.J.: The complexity of satisfiability problems. In: Lipton, R.J., Burkhard, W.A., Savitch, W.J., Friedman, E.P., Aho, A.V. (eds.) Proceedings of the 10th Annual ACM Symposium on Theory of Computing, May 1–3, 1978, San Diego, California, USA, pp. 216–226. ACM (1978)
27. Selivanova, S., Selivanov, V.L.: Computing solution operators of boundary-value problems for some linear hyperbolic systems of PDEs. Logical Methods in Computer Science, vol. 13, no. 4 (2017). https://doi.org/10.23638/LMCS-13(4:13)2017
28. Selivanova, S.V., Selivanov, V.L.: Bit complexity of computing solutions for symmetric hyperbolic systems of PDEs (Extended Abstract). In: Manea, F., Miller, R.G., Nowotka, D. (eds.) CiE 2018. LNCS, vol. 10936, pp. 376–385. Springer, Cham (2018). https://doi.org/10.1007/978-3-319-94418-0_38
29. Strikwerda, J.C.: Finite Difference Schemes and Partial Differential Equations. SIAM, Thailand (2004)
30. Sun, S.-M., Zhong, N., Ziegler, M.: Computability of the solutions to navier-stokes equations via effective approximation. In: Du, D.-Z., Wang, J. (eds.) Complexity and Approximation. LNCS, vol. 12000, pp. 80–112. Springer, Cham (2020). https://doi.org/10.1007/978-3-030-41672-0_7

31. Weihrauch, K.: 9. other approaches to computable analysis. In: Computable Analysis. TTCSAES, pp. 249–268. Springer, Heidelberg (2000). https://doi.org/10.1007/978-3-642-56999-9_9

32. Weihrauch, K.: Computational complexity on computable metric spaces. Math. Logic Q. **49**(1), 3–21 (2003)

33. Weihrauch, K., Zhong, N.: Is wave propagation computable or can wave computers beat the Turing machine? Proc. London Math. Soc. **85**(2), 312–332 (2002)

34. Ziegler, M., Brattka, V.: A computable spectral theorem. In: Blanck, J., Brattka, V., Hertling, P. (eds.) CCA 2000. LNCS, vol. 2064, pp. 378–388. Springer, Heidelberg (2001). https://doi.org/10.1007/3-540-45335-0_23

A Secure Three-Input AND Protocol with a Standard Deck of Minimal Cards

Hiroto Koyama[1] , Daiki Miyahara[2,4(✉)] , Takaaki Mizuki[3,4] ,
and Hideaki Sone[3]

[1] School of Engineering, Tohoku University, 6-6 Aramaki-Aza-Aoba, Aoba-ku,
Sendai 980-8579, Japan
`hiroto.koyama.t4@dc.tohoku.ac.jp`
[2] Graduate School of Information Sciences, Tohoku University,
6-3-09 Aramaki-Aza-Aoba, Aoba-ku, Sendai 980–8579, Japan
`daiki.miyahara.q4@dc.tohoku.ac.jp`
[3] Cyberscience Center, Tohoku University, 6-3 Aramaki-Aza-Aoba, Aoba-ku,
Sendai 980-8578, Japan
`mizuki+lncs@tohoku.ac.jp`
[4] National Institute of Advanced Industrial Science and Technology (AIST),
2-4-7 Aomi, Koto-ku, Tokyo 135-0064, Japan

Abstract. Card-based protocols are used to perform cryptographic tasks such as secure multiparty computation using a deck of physical cards. While most of the existing protocols use a two-colored deck consisting of red cards and black cards, Niemi and Renvall in 1999 constructed protocols for securely computing two-input Boolean functions (such as secure logical AND and XOR computations) using a commonly available standard deck of playing cards. Since this initial investigation, two-input protocols with fewer cards and/or shuffles have been designed, and by combining them, one can perform a secure computation of any Boolean circuit. In this paper, we directly construct a simple card-based protocol for the three-input AND computation. Our three-input AND protocol requires fewer cards and shuffles compared to that required when applying any existing two-input AND protocol twice to perform the three-input AND computation. Our protocol is unique in the sense that it is card minimal if we use two cards to encode a single bit.

Keywords: Card-based cryptography · Secure computation · Real-life hands-on cryptography · Logical AND function

1 Introduction

Card-based protocols perform cryptographic tasks such as secure multiparty computation using a deck of physical cards. Most existing studies in this line of research use a two-colored deck consisting of indistinguishable red $\heartsuit$ and black $\clubsuit$ cards, whose backs have the same pattern $\boxed{?}$. The Boolean values are usually encoded as follows:

$$\boxed{\clubsuit}\,\boxed{\heartsuit} = 0, \quad \boxed{\heartsuit}\,\boxed{\clubsuit} = 1.$$

© Springer Nature Switzerland AG 2021
R. Santhanam and D. Musatov (Eds.): CSR 2021, LNCS 12730, pp. 242–256, 2021.
https://doi.org/10.1007/978-3-030-79416-3_14

If two face-down cards represent a bit $x \in \{0,1\}$ according to the above encoding rule, then we call them a *commitment to* x and write it as follows:

Previous research has proposed many card-based protocols that are capable of securely computing Boolean functions such as the logical AND function. As mentioned above, most prior studies have used a two-colored deck of cards (e.g., [1,2,4,5,7,16,24]); however, there are a few protocols that use a *standard deck of playing cards*, as introduced below.[1]

1.1 Card-Based Protocols with a Standard Deck of Cards

In 1999, Niemi and Renvall [18] proposed card-based protocols for secure two-input computations (such as computations of the logical AND and XOR) using a commonly available standard deck of playing cards for the first time. Since then, several protocols using such a standard deck have been proposed [6,9,10].

A typical deck of playing cards consists of 52 distinct cards excluding jokers; hence, we assume the following deck of 52 numbered cards:

$$\boxed{1}\boxed{2}\boxed{3}\boxed{4}\boxed{5}\boxed{6}\cdots\boxed{51}\boxed{52},$$

where all their backs are identical $\boxed{?}$.

Niemi and Renvall [18] considered, based on two cards $\boxed{i}$ and $\boxed{j}$, $1 \le i < j \le 52$, an encoding scheme such that

$$\boxed{i}\,\boxed{j} = 0, \quad \boxed{j}\,\boxed{i} = 1. \tag{1}$$

That is, two cards can be represented as 0 if the left number is smaller than the right one; otherwise, they are represented as 1. In this way, similar to the two-colored case, we can consider a *commitment to* a bit $x \in \{0,1\}$ with two numbered cards $\boxed{i}$ and $\boxed{j}$, denoted by

$$\boxed{?}\boxed{?}.$$
$$[x]^{\{i,j\}}$$

Here, the set $\{i,j\}$ is called a *base* of the commitment. We sometimes write

without the description of a base if it is clear from context or there are multiple possibilities for the base. Note that, swapping the two face-down cards of a given

---

[1] There are other types of cards used for secure computations, such as polarizing cards [29], polygon cards [30], triangle cards [28], and dihedral cards [27].

commitment to $x \in \{0, 1\}$ converts it into a commitment to the negation $\bar{x}$ while keeping the value of x secret; thus, a secure NOT computation is easy.

Based on this encoding, Niemi and Renvall [18] designed card-based protocols. They were followed by a couple of research groups [6,10], whose advancements will be detailed in the next subsection.

1.2 Existing Protocols

Table 1 presents the existing protocols for the secure computation of the two-input AND function with a standard deck of cards. A protocol that terminates in a finite number of steps is said to be *finite*, while a protocol with a runtime that is finite in expectation is said to be *Las Vegas*; the fourth column of Table 1 indicates this.

Niemi and Renvall [18] proposed a Las Vegas two-input AND protocol using five cards; four cards are used to represent input commitments and the remaining card is an auxiliary one.[2] In other words, given two commitments to bits $a, b \in \{0, 1\}$ along with one additional card, their protocol produces a commitment to $a \wedge b$ without leaking (revealing) any information about a and b:

$$\underbrace{\boxed{?}\,\boxed{?}}_{[a]^{\{1,2\}}}\,\underbrace{\boxed{?}\,\boxed{?}}_{[b]^{\{3,4\}}}\,\boxed{5} \quad \rightarrow \quad \cdots \quad \rightarrow \quad \underbrace{\boxed{?}\,\boxed{?}}_{[a \wedge b]}.$$

In the (original) Niemi–Renvall protocol [18], the base of the output commitment is always $\{1, 2\}$ and the expected number of required shuffles is 9.5; however, Koch, Schrempp, and Kirsten [6] demonstrated that the expected number of shuffles can be reduced to 7.5 if we allow the base of the output commitment to be either $\{1, 4\}$, $\{1, 5\}$, or $\{4, 5\}$. In this paper, we refer this modified version as to the *Niemi–Renvall protocol*. As will be seen in Sect. 2.4, this protocol uses only a simple shuffle called a *random cut* (which is described in Sect. 2.2).

Later, Mizuki [10] proposed a finite two-input AND protocol with eight cards.[3] This protocol uses only a shuffle action called a *random bisection cut* (which is described in Sect. 2.3), and the number of required shuffles is four.

In 2021, Koch, Schrempp, and Kirsten [6] proposed a Las Vegas two-input AND protocol with four cards. This protocol uses only a random cut, and its expected number of shuffles is six.

1.3 Contribution

Let us consider how to securely compute the three-input AND function:

$$\underbrace{\boxed{?}\,\boxed{?}}_{[a]^{\{1,2\}}}\,\underbrace{\boxed{?}\,\boxed{?}}_{[b]^{\{3,4\}}}\,\underbrace{\boxed{?}\,\boxed{?}}_{[c]^{\{5,6\}}}\cdots \quad \rightarrow \quad \cdots \quad \rightarrow \quad \underbrace{\boxed{?}\,\boxed{?}}_{[a \wedge b \wedge c]}.$$

---

[2] Niemi and Renvall [18] also provided Las Vegas protocols for the secure computations of XOR and copy.
[3] Mizuki [10] also presented finite XOR and copy protocols.

Table 1. The existing two-input AND protocols with a standard deck of cards

	# of cards	# of shuffles	finite?
Niemi & Renvall [18]	5	7.5 (exp.)	
Mizuki [10]	8	4	✓
Koch & Schrempp & Kirsten [6]	4	6 (exp.)	

Table 2. Three-input AND computations with a standard deck of cards

	# of cards	# of shuffles	finite?
Mizuki [10] (twice)	10	8	✓
Koch & Schrempp & Kirsten [6] (twice)	6	12 (exp.)	
Ours	**6**	**8.5 (exp.)**	

Computation of the three-input AND function can be performed by repeatedly applying one of the three existing protocols [6,10,18] twice: First, one must obtain a commitment to $a \wedge b$ by applying a two-input protocol, and then apply the protocol to the commitments to $a \wedge b$ and c. To perform this, the numbers of required cards and shuffles by Mizuki's protocol [10] or Koch et al.'s protocol [6] are listed in Table 2.

In this study, we solicit a more efficient computation of the three-input AND. The main contribution of this paper is the novelty of our protocol that directly performs a secure three-input AND computation in a single application. Our protocol is partly based on Niemi and Renvall's two-input AND protocol [18]. Our protocol uses only six cards, which are necessary for three input commitments; therefore, it uses the same number of cards as two runs of Koch et al.'s protocol [6]. The expected number of shuffles required for our protocol is 8.5, which is 3.5 fewer than the previous best six-card computation (see Table 2).

Overall, our proposed three-input AND protocol is "card-minimal" in the sense that only six cards are necessary under the encoding rule (1). Additionally, our protocol is simple because it uses only random cuts and random bisection cuts as shuffles.

1.4   Outline

This paper is organized as follows: Sect. 2 describes basic terminology in card-based cryptography and introduces the Niemi–Renvall protocol [18]. In Sect. 3, we describe our three-input AND protocol. We conclude this paper in Sect. 4.

2   Preliminaries

In this section, we introduce the actions involved in card-based protocols as well as practical shuffles, specifically the random cut and the random bisection cut, which will be used in our proposed protocol. Additionally, we explain the two-input AND protocol constructed by Niemi and Renvall [18].

2.1 Actions

In card-based protocols, as they have been formalized in [7,13,15], there are the following three main actions to be applied to a sequence of n cards.

Rearrangement. Apply some permutation $\pi \in S_n$ to a sequence of n cards, where S_n denotes the symmetric group of degree n. We write this action as (perm, π):

$$\underset{?\,?\,\cdots\,?}{\overset{1\;\;2\qquad\quad n}{\boxed{?}\boxed{?}\cdots\boxed{?}}} \xrightarrow{(\mathsf{perm},\pi)} \overset{\pi^{-1}(1)\;\;\pi^{-1}(2)\qquad\quad \pi^{-1}(n)}{\boxed{?}\;\;\boxed{?}\;\cdots\;\boxed{?}}\ .$$

Turn. Turn over the t-th card (from the left) for $t \in \{1, \dots, n\}$ in a sequence to check the number of the card. We write this action as $(\mathsf{turn}, \{t\})$:

$$\overset{1\;\;2\qquad t\qquad n}{\boxed{?}\boxed{?}\cdots\boxed{?}\cdots\boxed{?}} \xrightarrow{(\mathsf{turn},\{t\})} \overset{1\;\;2\qquad t\qquad n}{\boxed{?}\boxed{?}\cdots\boxed{1}\cdots\boxed{?}}.$$

In this example, the numbered card displaying a 1 was revealed.

Shuffle. Apply a permutation $\pi \in \Pi$ chosen uniformly randomly from a permutation set $\Pi \subseteq S_n$. We write this action as (shuf, Π):

$$\overset{1\;\;2\qquad\quad n}{\boxed{?}\boxed{?}\cdots\boxed{?}} \xrightarrow{(\mathsf{shuf},\Pi)} \overset{\pi^{-1}(1)\;\;\pi^{-1}(2)\qquad\quad \pi^{-1}(n)}{\boxed{?}\;\;\boxed{?}\;\cdots\;\boxed{?}}\ .$$

Note that it is not possible for an observer to know which permutation in Π was applied.

2.2 Random Cut

A random cut is a shuffling action (denoted by $\langle\cdot\rangle$) that shifts a sequence of cards cyclically and randomly. If a random cut is applied to a sequence of n cards, then the resulting sequence becomes one of the following n sequences, each of which occurs with a probability of $1/n$:

$$\left\langle \overset{1\;\;2\;\;3\qquad\;\; n-1\;\, n}{\boxed{?}\boxed{?}\boxed{?}\cdots\boxed{?}\boxed{?}} \right\rangle \rightarrow \begin{cases} \overset{1\;\;2\;\;3\qquad\;\; n-1\;\, n}{\boxed{?}\boxed{?}\boxed{?}\cdots\boxed{?}\boxed{?}}, \\[4pt] \overset{2\;\;3\;\;4\qquad\;\;\; n\;\;\, 1}{\boxed{?}\boxed{?}\boxed{?}\cdots\boxed{?}\boxed{?}}, \\ \quad\vdots \\ \overset{n-1\;\, n\;\;\; 1\qquad\; n-3\; n-2}{\boxed{?}\boxed{?}\boxed{?}\cdots\boxed{?}\boxed{?}}, \\[4pt] \overset{n\;\;\; 1\;\;\; 2\qquad n-2\; n-1}{\boxed{?}\boxed{?}\boxed{?}\cdots\boxed{?}\boxed{?}}. \end{cases}$$

This random cut can be written, using a cyclic permutation $\sigma = (1\,2\,3\,\cdots\,n)$, as

$$(\mathsf{shuf}, \{\mathsf{id}, \sigma, \sigma^2, \ldots, \sigma^{n-1}\}),$$

where id denotes the identity permutation. Hereinafter, we use $\mathsf{RC}_{1,2,\ldots,n}$ to represent $\{\mathsf{id}, \sigma, \sigma^2, \ldots, \sigma^{n-1}\}$.

A random cut can be easily performed by human hands, and a secure implementation called the Hindu cut is well known [31,32].

2.3 Random Bisection Cut

A random bisection cut is a shuffling action invented by Mizuki and Sone [16] in 2009. This shuffle (denoted by $[\,\cdot\,|\,\cdot\,]$) bisects a sequence of $2n$ cards and randomly swaps the two halves; the resulting sequence becomes one of the following two sequences:

$$\left[\boxed{?}^{1}\cdots\boxed{?}^{n}\,\middle|\,\boxed{?}^{n+1}\cdots\boxed{?}^{2n}\right] \rightarrow \begin{cases} \overset{1}{\boxed{?}}\cdots\overset{n}{\boxed{?}}\overset{n+1}{\boxed{?}}\cdots\overset{2n}{\boxed{?}}, \\ \underset{n+1}{\boxed{?}}\cdots\underset{2n}{\boxed{?}}\underset{1}{\boxed{?}}\cdots\underset{n}{\boxed{?}}. \end{cases}$$

That is, the resulting sequence either remains the same as the original one or becomes a sequence where the two halves are swapped with a probability of $1/2$. This random bisection cut can be written as follows:

$$(\mathsf{shuf}, \{\mathsf{id}, (1\ n{+}1)(2\ n{+}2)\cdots(n\ 2n)\}).$$

Secure implementations using familiar tools were shown in [31,32]. The random bisection cut has brought many efficient protocols (e.g., [11,12,14,19–21]).

2.4 The Niemi–Renvall Protocol

Given two commitments to $a, b \in \{0,1\}$ along with an additional card, the Niemi–Renvall protocol [18] outputs a commitment to $a \wedge b$. The procedure is as follows.

1. Place the two input commitments and an additional card $\boxed{5}$ and turn it over:

$$\boxed{5}\ \underbrace{\boxed{?}\boxed{?}}_{[a]\{1,2\}}\ \underbrace{\boxed{?}\boxed{?}}_{[b]\{3,4\}}\ \rightarrow\ \boxed{?}\ \underbrace{\boxed{?}\boxed{?}}_{[a]\{1,2\}}\ \underbrace{\boxed{?}\boxed{?}}_{[b]\{3,4\}}.$$

2. Rearrange the third and fourth cards:

$$\overset{1\ \ 2\ \ 3\ \ 4\ \ 5}{\boxed{?}\boxed{?}\boxed{?}\boxed{?}\boxed{?}}\ \rightarrow\ \overset{1\ \ 2\ \ 4\ \ 3\ \ 5}{\boxed{?}\boxed{?}\boxed{?}\boxed{?}\boxed{?}}.$$

Now, let us consider the order of $\boxed{5}$, $\boxed{1}$, and $\boxed{4}$ in the rearranged sequence; the order is $\boxed{5} \rightarrow \boxed{4} \rightarrow \boxed{1}$ (apart from cyclic rotation) if and only if $(a, b) = (1,1)$, i.e., $a \wedge b = 1$ (whereas the order is $\boxed{5} \rightarrow \boxed{1} \rightarrow \boxed{4}$ if and only if $a \wedge b = 0$). Therefore, we try to remove $\boxed{2}$ and $\boxed{3}$ in Steps 3 and 4.

Table 3. The sequence after removing $\boxed{2}$ and $\boxed{3}$

(a,b)	$a \wedge b$	Sequence after Removing $\boxed{2}$ and $\boxed{3}$
$(0,0)$	0	$\boxed{1}\,\boxed{4}\,\boxed{5}$ or $\boxed{4}\,\boxed{5}\,\boxed{1}$ or $\boxed{5}\,\boxed{1}\,\boxed{4}$
$(0,1)$	0	$\boxed{1}\,\boxed{4}\,\boxed{5}$ or $\boxed{4}\,\boxed{5}\,\boxed{1}$ or $\boxed{5}\,\boxed{1}\,\boxed{4}$
$(1,0)$	0	$\boxed{1}\,\boxed{4}\,\boxed{5}$ or $\boxed{4}\,\boxed{5}\,\boxed{1}$ or $\boxed{5}\,\boxed{1}\,\boxed{4}$
$(1,1)$	1	$\boxed{1}\,\boxed{5}\,\boxed{4}$ or $\boxed{4}\,\boxed{1}\,\boxed{5}$ or $\boxed{5}\,\boxed{4}\,\boxed{1}$

3. Apply a random cut to the sequence of all cards:

$$\langle\,\boxed{?}\boxed{?}\boxed{?}\boxed{?}\boxed{?}\,\rangle \;\rightarrow\; \boxed{?}\boxed{?}\boxed{?}\boxed{?}\boxed{?}.$$

4. Turn over the first card; if it is $\boxed{2}$ or $\boxed{3}$, then remove it. Otherwise, turn it over. If there is still a $\boxed{2}$ or $\boxed{3}$ in the sequence, then return to Step 3.
5. The resulting sequence after removing $\boxed{2}$ and $\boxed{3}$ becomes one as presented in Table 3. Apply a random cut to the sequence and then reveal the first card to obtain the output commitment. (This step was developed by Koch et al. [6].)

If the first card is $\boxed{4}$, then we obtain a commitment to the negation of $a \wedge b$. In this case, we apply the NOT computation to obtain a commitment to $a \wedge b$.

Correctness of this protocol is clear from the above description. Regarding security, since a random cut is applied to the sequence in Step 3, the revealed card in Step 4 is chosen randomly from the sequence. Therefore, no information about the input is leaked. For example, if the number of cards in Step 3 is five, the revealed card in Step 4 should be one in $\{\boxed{1},\boxed{2},\boxed{3},\boxed{4},\boxed{5}\}$ with an equal probability regardless of the input values. In the same manner, no information about the input is leaked when the first card is revealed in Step 5. To summarize, this protocol is correct and secure.

3 Our Three-Input AND Protocol

In this section, we present a three-input AND protocol that requires no additional card, i.e., is card-minimal and uses fewer shuffles as compared to the applications of previous protocols:

$$\underbrace{\boxed{?}\boxed{?}}_{[a]\{1,2\}}\underbrace{\boxed{?}\boxed{?}}_{[b]\{3,4\}}\underbrace{\boxed{?}\boxed{?}}_{[c]\{5,6\}} \rightarrow \cdots \rightarrow \underbrace{\boxed{?}\boxed{?}}_{[a\wedge b\wedge c]} .$$

We begin by describing the idea behind our proposed protocol.

3.1 Idea

Observe the following simple fact about $a \wedge b \wedge c$:

$$a \wedge b \wedge c = \begin{cases} 0 & \text{if } c = 0, \\ a \wedge b & \text{if } c = 1. \end{cases}$$

In other words, to construct a three-input AND protocol, it suffices to simulate a two-input AND protocol if $c = 1$ and to always output 0 if $c = 0$. For this, we borrow the idea behind the Niemi–Renvall protocol [18] introduced in Sect. 2.4.

Remember that their protocol swaps the third and fourth cards in Step 2; this action reverses the order of $\boxed{1}$ and $\boxed{4}$ if and only if $(a, b) = (1, 1)$, i.e., the output is 1. That is, if we skip this step and perform the remaining steps, then the output should be always 0. From this observation, it suffices to swap the third and fourth cards if and only if $c = 1$ and then perform the remaining steps of the protocol. Therefore, in the next subsection, we will describe how to swap two cards according to the value of c (without knowing it).

3.2 Swapping by Value of Commitment

For a sequence of two cards along with a commitment to $c \in \{0, 1\}$ whose base is $\{i, j\}$, we want to swap the cards if $c = 1$ and we want to keep them unchanged if $c = 0$, without leaking the value of c:

$$\overset{1\ 2}{\boxed{?}\boxed{?}}\ \underbrace{\boxed{?}\boxed{?}}_{[c]\{i,j\}} \rightarrow \begin{cases} \overset{1\ 2}{\boxed{?}\boxed{?}} & \text{if } c = 0, \\ \overset{2\ 1}{\boxed{?}\boxed{?}} & \text{if } c = 1. \end{cases}$$

We call this the *swap operation* by the commitment to c. The swap operation proceeds as follows (whose procedure is similar to the Mizuki XOR protocol [10]).

1. Assume a target sequence of two cards and a commitment to c (with $i < j$):

$$\boxed{?}\boxed{?}\underbrace{\boxed{?}\boxed{?}}_{[c]\{i,j\}} .$$

2. Rearrange the second and third cards:

$$
\begin{array}{cccc} 1 & 2 & 3 & 4 \\ \boxed{?} & \boxed{?} & \boxed{?} & \boxed{?} \end{array} \quad \rightarrow \quad \begin{array}{cccc} 1 & 3 & 2 & 4 \\ \boxed{?} & \boxed{?} & \boxed{?} & \boxed{?} \end{array}.
$$

3. Apply a random bisection cut to the sequence of all cards:

$$
\left[\boxed{?}\boxed{?} \,\middle|\, \boxed{?}\boxed{?} \right] \quad \rightarrow \quad \boxed{?}\boxed{?}\boxed{?}\boxed{?}.
$$

4. Rearrange the second and third cards:

$$
\begin{array}{cccc} 1 & 2 & 3 & 4 \\ \boxed{?} & \boxed{?} & \boxed{?} & \boxed{?} \end{array} \quad \rightarrow \quad \begin{array}{cccc} 1 & 3 & 2 & 4 \\ \boxed{?} & \boxed{?} & \boxed{?} & \boxed{?} \end{array}.
$$

Observe that, depending on the value of c, the current sequence satisfies the followings:

$$
c = 0 \Rightarrow
\begin{cases}
\begin{array}{cc} 1 & 2 \\ \boxed{?}\boxed{?} \end{array} \boxed{i}\boxed{j} & \text{(Prob. of } 1/2), \\[2ex]
\begin{array}{cc} 2 & 1 \\ \boxed{?}\boxed{?} \end{array} \boxed{j}\boxed{i} & \text{(Prob. of } 1/2).
\end{cases}
$$

$$
c = 1 \Rightarrow
\begin{cases}
\begin{array}{cc} 1 & 2 \\ \boxed{?}\boxed{?} \end{array} \boxed{j}\boxed{i} & \text{(Prob. of } 1/2), \\[2ex]
\begin{array}{cc} 2 & 1 \\ \boxed{?}\boxed{?} \end{array} \boxed{i}\boxed{j} & \text{(Prob. of } 1/2).
\end{cases}
$$

5. Turn over the third and fourth cards; $\boxed{i}\boxed{j}$ or $\boxed{j}\boxed{i}$ should appear with a probability of 1/2.
   (a) If $\boxed{i}\boxed{j}$ appear, then output the first and second cards.
   (b) If $\boxed{j}\boxed{i}$ appear, then swap the first and second cards and then output them.

3.3 Description of Our Protocol

We now present our three-input AND protocol.

1. Assume three input commitments to $a, b, c \in \{0, 1\}$:

$$
\underbrace{\boxed{?}\boxed{?}}_{[a]\{1,2\}} \; \underbrace{\boxed{?}\boxed{?}}_{[b]\{3,4\}} \; \underbrace{\boxed{?}\boxed{?}}_{[c]\{5,6\}}.
$$

2. Apply the swap operation by the commitment to c shown in Sect. 3.2 to the second and third cards, as follows.

(a) Rearrange the sequence:

$$\begin{array}{cccccc} 1 & 2 & 3 & 4 & 5 & 6 \\ \boxed{?} & \boxed{?} & \boxed{?} & \boxed{?} & \boxed{?} & \boxed{?} \end{array} \rightarrow \begin{array}{cccccc} 1 & 4 & 2 & 5 & 3 & 6 \\ \boxed{?} & \boxed{?} & \boxed{?} & \boxed{?} & \boxed{?} & \boxed{?} \end{array}.$$

(b) Apply a random bisection cut to the last four cards:

$$\boxed{?}\boxed{?}\left[\boxed{?}\boxed{?}\middle\|\boxed{?}\boxed{?}\right] \rightarrow \boxed{?}\boxed{?}\boxed{?}\boxed{?}\boxed{?}\boxed{?}.$$

(c) Apply the inverse rearrangement of Step 2(a):

$$\begin{array}{cccccc} 1 & 2 & 3 & 4 & 5 & 6 \\ \boxed{?} & \boxed{?} & \boxed{?} & \boxed{?} & \boxed{?} & \boxed{?} \end{array} \rightarrow \begin{array}{cccccc} 1 & 3 & 5 & 2 & 4 & 6 \\ \boxed{?} & \boxed{?} & \boxed{?} & \boxed{?} & \boxed{?} & \boxed{?} \end{array}.$$

(d) Turn over the fifth and sixth cards: If $\boxed{5}\;\boxed{6}$ appear, then do nothing. If $\boxed{6}\;\boxed{5}$ appear, then swap the second and third cards.

3. Rearrange the sequence so that the first card becomes $\boxed{5}$ appeared in the previous step and then turn it over. Subsequently, apply Steps 3, 4, and 5 of the Niemi–Renvall protocol [18] to the first five cards.

3.4 Correctness and Security

Here, we prove the correctness and security of our three-input AND protocol. Observe that, in the sequence after Step 2 as listed in Table 4, the order of $\boxed{1}\,\boxed{4}$ is $\boxed{4} \rightarrow \boxed{1}$ if and only if $a \wedge b \wedge c = 1$ (whereas it is $\boxed{1} \rightarrow \boxed{4}$ if and only if $a \wedge b \wedge c = 0$). From this, it suffices to remove $\boxed{2}$ and $\boxed{3}$ in the same manner as the Niemi–Renvall protocol [18]. Therefore, our protocol can always output a commitment to $a \wedge b \wedge c$ by applying Steps 3, 4, and 5 of the Niemi–Renvall protocol [18].

More formally, we prove this by using the *KWH-tree* [7] as in Figs. 1 and 2. The KWH-tree is a tree-like diagram that shows the transitions of possible sequences of cards along with their respective polynomials in a box, where actions to be applied to the sequence are appended to an edge. In the figure, the probability of $(a, b, c) = (x, y, z)$ is denoted by X_{xyz}. A polynomial annotating a sequence in a box such as $1/2X_{000}$ represents the conditional probability that the current sequence is the one next to the polynomial, given what can be observed so far on the table.

If the sum of all the polynomials in each box is equal to

$$\sum_{x,y,z \in \{0,1\}} X_{xyz},$$

then it is guaranteed that no information about the input is leaked. The KWH-tree of our protocol for Steps 1 and 2 is shown in Fig. 1. The KWH-tree for Step 3 is shown in Fig. 2. From the figures, it can be easily confirmed that the aforementioned condition is satisfied in each box, i.e., our protocol is secure.

Table 4. The possible sequences after Step 2 in our protocol

(a, b, c)	$a \wedge b \wedge c$	Sequence After Step 2			
$(0,0,0)$	0	1	2	3	4
$(0,0,1)$	0	1	3	2	4
$(0,1,0)$	0	1	2	4	3
$(1,0,0)$	0	2	1	3	4
$(0,1,1)$	0	1	4	2	3
$(1,0,1)$	0	2	3	1	4
$(1,1,0)$	0	2	1	4	3
$(1,1,1)$	1	2	4	1	3

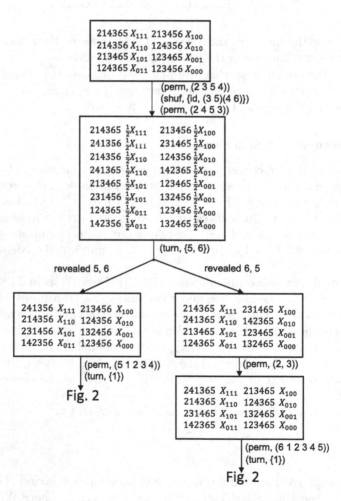

Fig. 1. The KWH-tree for our three-input AND protocol (Steps 1 to 2)

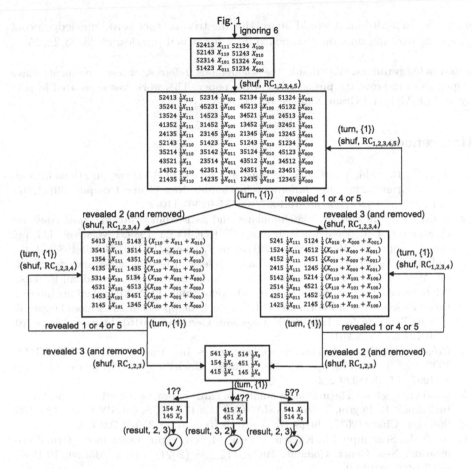

Fig. 2. The KWH-tree for our three-input AND protocol (Step 3), where $X_0 = X_{000} + X_{001} + X_{010} + X_{011} + X_{100} + X_{101} + X_{110}$ and $X_1 = X_{111}$. The result action indicates the positions of the output commitment.

4   Conclusion

In this study, we designed a simple three-input AND protocol using playing cards by only using six cards; in other words, we do not need any additional cards (aside from three input commitments). To the best of our knowledge, this is the first type of protocol that can be used for directly computing a three-input Boolean function with a standard deck of cards.

A natural open problem that presents itself from our research is the construction of efficient AND protocols for more than three inputs. It would also be interesting to investigate whether a finite AND protocol can be constructed using only random cuts even for two inputs. Making use of a standard deck in the "private permutation" setting (e.g., [8,17,22,33]) would be another interesting topic. In addition, it would be worthwhile to construct zero-knowledge proof protocols working only on a standard deck for pencil puzzles, cf. [3,23,25,26].

Acknowledgements. We thank the anonymous referees, whose comments have helped us to improve the presentation of the paper. This work was supported in part by JSPS KAKENHI Grant Number JP19J21153.

References

1. Abe, Y., Hayashi, Y., Mizuki, T., Sone, H.: Five-card AND computations in committed format using only uniform cyclic shuffles. New Gener. Comput. **39**(1), 97–114 (2021). https://doi.org/10.1007/s00354-020-00110-2
2. Boer, B.: More efficient match-making and satisfiability *the five card trick.* In: Quisquater, J.-J., Vandewalle, J. (eds.) EUROCRYPT 1989. LNCS, vol. 434, pp. 208–217. Springer, Heidelberg (1990). https://doi.org/10.1007/3-540-46885-4_23
3. Bultel, X., Dreier, J., Dumas, J.G., Lafourcade, P.: Physical zero-knowledge proofs for Akari, Takuzu, Kakuro and KenKen. In: Demaine, E.D., Grandoni, F. (eds.) 8th International Conference on Fun with Algorithms (FUN 2016). Leibniz International Proceedings in Informatics (LIPIcs), vol. 49, pp. 8:1–8:20. Schloss Dagstuhl-Leibniz-Zentrum fuer Informatik, Dagstuhl, Germany (2016). https://doi.org/10.4230/LIPIcs.FUN.2016.8
4. Crépeau, C., Kilian, J.: Discreet solitary games. In: Stinson, D.R. (ed.) CRYPTO 1993. LNCS, vol. 773, pp. 319–330. Springer, Heidelberg (1994). https://doi.org/10.1007/3-540-48329-2_27
5. Kastner, J., et al.: The minimum number of cards in practical card-based protocols. In: Takagi, T., Peyrin, T. (eds.) ASIACRYPT 2017. LNCS, vol. 10626, pp. 126–155. Springer, Cham (2017). https://doi.org/10.1007/978-3-319-70700-6_5
6. Koch, A., Schrempp, M., Kirsten, M.: Card-based cryptography meets formal verification. New Gener. Comput. **39**(1), 115–158 (2021). https://doi.org/10.1007/s00354-020-00120-0
7. Koch, A., Walzer, S., Härtel, K.: Card-based cryptographic protocols using a minimal number of cards. In: Iwata, T., Cheon, J.H. (eds.) ASIACRYPT 2015. LNCS, vol. 9452, pp. 783–807. Springer, Heidelberg (2015). https://doi.org/10.1007/978-3-662-48797-6_32
8. Manabe, Y., Ono, H.: Secure card-based cryptographic protocols using private operations against malicious players. In: Maimut, D., Oprina, A.-G., Sauveron, D. (eds.) SecITC 2020. LNCS, vol. 12596, pp. 55–70. Springer, Cham (2021). https://doi.org/10.1007/978-3-030-69255-1_5
9. Miyahara, D., Hayashi, Y., Mizuki, T., Sone, H.: Practical card-based implementations of Yao's millionaire protocol. Theor. Comput. Sci. **803**, 207–221 (2020). https://doi.org/10.1016/j.tcs.2019.11.005
10. Mizuki, T.: Efficient and secure multiparty computations using a standard deck of playing cards. In: Foresti, S., Persiano, G. (eds.) CANS 2016. LNCS, vol. 10052, pp. 484–499. Springer, Cham (2016). https://doi.org/10.1007/978-3-319-48965-0_29
11. Mizuki, T., Asiedu, I.K., Sone, H.: Voting with a logarithmic number of cards. In: Mauri, G., Dennunzio, A., Manzoni, L., Porreca, A.E. (eds.) UCNC 2013. LNCS, vol. 7956, pp. 162–173. Springer, Heidelberg (2013). https://doi.org/10.1007/978-3-642-39074-6_16

12. Mizuki, T., Komano, Y.: Analysis of information leakage due to operative errors in card-based protocols. In: Iliopoulos, C., Leong, H.W., Sung, W.-K. (eds.) IWOCA 2018. LNCS, vol. 10979, pp. 250–262. Springer, Cham (2018). https://doi.org/10.1007/978-3-319-94667-2_21
13. Mizuki, T., Shizuya, H.: A formalization of card-based cryptographic protocols via abstract machine. Int. J. Inf. Secur. **13**(1), 15–23 (2014). https://doi.org/10.1007/s10207-013-0219-4
14. Mizuki, T., Shizuya, H.: Practical card-based cryptography. In: Ferro, A., Luccio, F., Widmayer, P. (eds.) Fun with Algorithms. LNCS, vol. 8496, pp. 313–324. Springer, Cham (2014). https://doi.org/10.1007/978-3-319-07890-8_27
15. Mizuki, T., Shizuya, H.: Computational model of card-based cryptographic protocols and its applications. IEICE Trans. Fundam. Electron. Commun. Comput. Sci. **100**(1), 3–11 (2017). https://doi.org/10.1587/transfun.E100.A.3
16. Mizuki, T., Sone, H.: Six-card secure AND and four-card secure XOR. In: Deng, X., Hopcroft, J.E., Xue, J. (eds.) Frontiers in Algorithmics. LNCS, vol. 5598, pp. 358–369. Springer, Berlin, Heidelberg (2009). https://doi.org/10.1007/978-3-642-02270-8_36
17. Nakai, T., Misawa, Y., Tokushige, Y., Iwamoto, M., Ohta, K.: How to solve millionaires' problem with two kinds of cards. New Gener. Comput. **39**(1), 73–96 (2021). https://doi.org/10.1007/s00354-020-00118-8
18. Niemi, V., Renvall, A.: Solitaire zero-knowledge. Fundam. Inf. **38**(1,2), 181–188 (1999). https://doi.org/10.3233/FI-1999-381214
19. Nishida, T., Hayashi, Y., Mizuki, T., Sone, H.: Card-based protocols for any Boolean function. In: Jain, R., Jain, S., Stephan, F. (eds.) TAMC 2015. LNCS, vol. 9076, pp. 110–121. Springer, Cham (2015). https://doi.org/10.1007/978-3-319-17142-5_11
20. Nishida, T., Hayashi, Y., Mizuki, T., Sone, H.: Securely computing three-input functions with eight cards. IEICE Trans. Fundam. Electron. Commun. Comput. Sci. **98**(6), 1145–1152 (2015). https://doi.org/10.1587/transfun.E98.A.1145
21. Nishida, T., Mizuki, T., Sone, H.: Securely computing the three-input majority function with eight cards. In: Dediu, A.-H., Martín-Vide, C., Truthe, B., Vega-Rodríguez, M.A. (eds.) TPNC 2013. LNCS, vol. 8273, pp. 193–204. Springer, Heidelberg (2013). https://doi.org/10.1007/978-3-642-45008-2_16
22. Ono, H., Manabe, Y.: Card-based cryptographic logical computations using private operations. New Gener. Comput. **39**(1), 19–40 (2020). https://doi.org/10.1007/s00354-020-00113-z
23. Robert, L., Miyahara, D., Lafourcade, P., Mizuki, T.: Physical zero-knowledge proof for Suguru puzzle. In: Devismes, S., Mittal, N. (eds.) SSS 2020. LNCS, vol. 12514, pp. 235–247. Springer, Cham (2020). https://doi.org/10.1007/978-3-030-64348-5_19
24. Ruangwises, S., Itoh, T.: AND protocols using only uniform shuffles. In: van Bevern, R., Kucherov, G. (eds.) CSR 2019. LNCS, vol. 11532, pp. 349–358. Springer, Cham (2019). https://doi.org/10.1007/978-3-030-19955-5_30
25. Ruangwises, S., Itoh, T.: Physical zero-knowledge proof for numberlink puzzle and k vertex-disjoint paths problem. New Gener. Comput. **39**(1), 3–17 (2020). https://doi.org/10.1007/s00354-020-00114-y
26. Ruangwises, S., Itoh, T.: Physical zero-knowledge proof for ripple effect. In: Uehara, R., Hong, S.-H., Nandy, S.C. (eds.) WALCOM 2021. LNCS, vol. 12635, pp. 296–307. Springer, Cham (2021). https://doi.org/10.1007/978-3-030-68211-8_24
27. Shinagawa, K.: Card-based cryptography with dihedral symmetry. New Gener. Comput. **39**(1), 41–71 (2021). https://doi.org/10.1007/s00354-020-00117-9

28. Shinagawa, K., Mizuki, T.: Card-based protocols using triangle cards. In: Ito, H., Leonardi, S., Pagli, L., Prencipe, G. (eds.) 9th International Conference on Fun with Algorithms (FUN 2018). Leibniz International Proceedings in Informatics (LIPIcs), vol. 100, pp. 31:1–31:13. Schloss Dagstuhl-Leibniz-Zentrum fuer Informatik, Dagstuhl, Germany (2018). https://doi.org/10.4230/LIPIcs.FUN.2018.31

29. Shinagawa, K., et al.: Secure computation protocols using polarizing cards. IEICE Trans. Fundam. Electron. Commun. Comput. Sci. **99**(6), 1122–1131 (2016)

30. Shinagawa, K., et al.: Card-based protocols using regular polygon cards. IEICE Trans. Fundam. Electron. Commun. Comput. Sci. **100**(9), 1900–1909 (2017)

31. Ueda, I., Miyahara, D., Nishimura, A., Hayashi, Y., Mizuki, T., Sone, H.: Secure implementations of a random bisection cut. Int. J. Inf. Secur. **19**, 445–452 (2020). https://doi.org/10.1007/s10207-019-00463-w

32. Ueda, I., Nishimura, A., Hayashi, Y., Mizuki, T., Sone, H.: How to implement a random bisection cut. In: Martín-Vide, C., Mizuki, T., Vega-Rodríguez, M.A. (eds.) TPNC 2016. LNCS, vol. 10071, pp. 58–69. Springer, Cham (2016). https://doi.org/10.1007/978-3-319-49001-4_5

33. Yasunaga, K.: Practical card-based protocol for three-input majority. IEICE Trans. Fundam. Electron. Commun. Comput. Sci. **103**(11), 1296–1298 (2020). https://doi.org/10.1587/transfun.2020EAL2025

Upper Bound for Torus Polynomials

Vaibhav Krishan[✉]

Indian Institute of Technology Bombay, Powai 400076, Mumbai, India
vkrishan@cse.iitb.ac.in

Abstract. We prove that all functions that have low degree torus polynomials approximating them with small error also have MidBit$^+$ circuits computing them. This serves as a partial converse to the result that all ACC functions have low degree torus polynomials approximating them with small error, by Bhrushundi, Hosseini, Lovett and Rao (ITCS 2019).

Keywords: Torus polynomials · ACC · MidBit

1 Introduction

Proving lower bounds for boolean circuits has been a major quest in complexity theory. Much of the recent work, for example [6–9,13,14,17,18] and reference therein, has focused on proving lower bounds for constant-depth circuits.

The first lower bounds for constant-depth circuits consisting of AND, OR, NOT gates were proven by Furst, Saxe and Sipser [10], and independently by Ajtai [1]. The lower bound was improved by Yao [19], and further improved by Håstad [12]. Razborov [15] and Smolensky [16] proved lower bounds for constant-depth circuits that additionally contain MOD$_p$ gates where p is a prime. Barrington [3] first posed the question of extending lower bounds to the class ACC, where MOD$_m$ gates are allowed for a general m.

Williams [18], in a breakthrough result, proved a lower bound against ACC circuits, where the hard function comes from non-deterministic exponential time NEXP. The lower bound was improved to average case by Chen, Oliveira and Santhanam [8]. The hard function was brought down to non-deterministic quasi-polynomial time NQP by Murray and Williams [14], which was then improved to an average case lower bound by Chen [7].

All the above lower bounds for ACC use a conversion of ACC circuits to SYM$^+$ circuits, a result first proven by Beigel and Tarui [4] and subsequently improved upon by Allender and Gore [2]. SYM$^+$ circuits are depth-two size-$O(2^{(\log n)^{O(1)}})$ circuits where the top gate is a symmetric function and the bottom layer has AND gates of fan-in $(\log n)^{O(1)}$. Green, Köbler and Torán [11] improved the result to show that the symmetric function implemented by the top gate can be the MidBit function. We define the MidBit function below.

Definition 1 (MidBit). *The* MidBit *function over the input* $(x_1, \ldots, x_n)$ *behaves as follows. Consider the sum of inputs* $\sum_{i=1}^{n} x_i$ *and consider it's binary expansion*

© Springer Nature Switzerland AG 2021
R. Santhanam and D. Musatov (Eds.): CSR 2021, LNCS 12730, pp. 257–263, 2021.
https://doi.org/10.1007/978-3-030-79416-3_15

$b_{\ell-1}b_{\ell-2}\ldots b_0$ *with* $\ell = \lfloor \log_2(n) \rfloor + 1$ *many bits,* b_0 *being the least significant bit. Then* $\mathsf{MidBit}(x_1, x_2, \ldots, x_n) = b_{\lfloor \ell/2 \rfloor}$.

SYM^+ circuits where the top gate implements the MidBit function are called MidBit^+ circuits. The formal definition follows.

Definition 2 (MidBit^+). *A* MidBit^+ *circuit is a depth-two circuit, with a* MidBit *gate of fan-in* $2^{(\log n)^{O(1)}}$ *at the top, and* AND *gates of fan-in* $(\log n)^{O(1)}$ *at the bottom.*

Green, Köbler and Torán [11] proved that all ACC circuits can be converted into an equivalent MidBit^+ circuit. MidBit^+ circuits are seemingly simpler in structure than ACC circuits, as MidBit^+ circuits have a fixed depth of two while ACC circuits can be of arbitrary constant depth. Understanding the power of MidBit^+ can be important for obtaining ACC lower bounds, as a lower bound for MidBit^+ circuits will automatically translate to a lower bound against ACC circuits.

1.1 Torus Polynomials

While lower bounds against ACC are known from "high" classes, such as NQP, such lower bounds are not known from "low" circuit classes, such as TC^0, the class of constant-depth polynomial size circuits consisting of Majority gates. In other words, it is not yet known whether the containment $\mathsf{ACC} \subseteq \mathsf{TC}^0$ is strict or not. Also the lower bounds for some classes contained in ACC were obtained by using algebraic methods, for example in [15,16], where it was proved that the concerned class has low degree polynomial approximations of a particular type. Such algebraic methods were not known to carry over to ACC until Bhrushundi et al. [5] introduced torus polynomials, which they proposed as an approach to solving the ACC vs TC^0 problem.

Torus polynomials are polynomials that approximate a function in its fractional part, the integer part is ignored. We first define the torus, based on which we define torus polynomials.

Definition 3 (**Torus**). *For* $\alpha, \beta \in \mathbb{R}$, *define* $\alpha \equiv \beta \mod 1$ *when* $\alpha - \beta \in \mathbb{Z}$ *(here* $\mod 1$ *is an abuse of notation).*
Define $\alpha \mod 1$ *to be the unique* $\beta \in [-1/2, 1/2)$ *such that* $\alpha \equiv \beta \mod 1$.

Definition 4 (**Torus Polynomial**). *Let* $P \in \mathbb{R}[X_1, \ldots, X_n]$ *be a real multilinear polynomial and* $f : \{0, 1\}^n \to \{0, 1\}$ *be a boolean function.* P *is a torus polynomial approximating* f *with error* ε *if*

$$\forall x \in \{0,1\}^n, \left| P(x) - \frac{f(x)}{2} \mod 1 \right| \leq \varepsilon$$

Denote by $\overline{deg}_\varepsilon(f)$ *as the smallest possible degree of a torus polynomial approximating* f *with error* ε.

All boolean functions have degree n torus polynomials approximating them with error 0, by considering their unique multilinear extension. Bhrushundi et al. [5] proved that all functions in ACC have polylogarithmic degree torus polynomials approximating them with small error.

Theorem 1 ([5]). *Let* $f : \{0,1\}^n \to \{0,1\}$ *be a boolean function such that* $f \in \mathsf{ACC}$, *and* $n^{-O(1)} \le \varepsilon < 1/4$. *Then* $\overline{deg_\varepsilon}(f) \le (\log n)^{O(1)}$.

This makes it important to understand the power of torus polynomials, as proving a lower bound on the degree of torus polynomials approximating a certain function with small error will prove a lower bound against ACC. In particular, proving that the Majority function doesn't have low degree torus polynomials approximating it with small error will prove the separation $\mathsf{ACC} \subsetneq \mathsf{TC}^0$. Some evidence towards this was given by Bhrushundi et al. [5] as they proved that *symmetric torus polynomials* need high degree to approximate the Majority function with small error.

1.2 Our Results

Theorem 1 proves that torus polynomials are powerful enough to capture all of ACC. On the other hand, an upper bound on the power of torus polynomials was not yet known. We prove that all functions that have low degree torus polynomials approximating them also have MidBit$^+$ circuits computing them.

Theorem 2. *Let* $n^{-O(1)} \le \varepsilon < 1/8$ *and let* $f : \{0,1\}^n \to \{0,1\}$ *be a boolean function. Then* $\overline{deg_\varepsilon}(f) \le (\log n)^{O(1)} \implies f \in \mathsf{MidBit}^+$.

The proof of Theorem 1 by Bhrushundi et al. [5] has two steps, where they first convert an ACC circuit into an equivalent MidBit$^+$ circuit and then convert the MidBit$^+$ circuit into a torus polynomial. Therefore our result can be thought of as a partial converse to Theorem 1, which we prove in Sect. 2. We also prove that the parity of two functions with low degree torus polynomial approximations, also has low degree torus polynomial approximations. We prove this result in Sect. 3.

2 Torus Polynomials Are Equivalent to MidBit$^+$

We prove an upper bound on the power of torus polynomials, by constructing a MidBit$^+$ circuit for any function that has a low degree torus polynomial approximation. Any MidBit$^+$ circuit can be converted into a low degree torus polynomial, as is evident from Bhrushundi et al. [5], but the error of approximation cannot be bounded. We prove that the MidBit$^+$ circuits we construct satisfy an additional property about the number of AND gates that evaluate to "true" on a particular input. This property, using ideas from Bhrushundi et al. [5], can be used to convert the MidBit$^+$ circuits we construct into low degree torus polynomials with *small error*.

Theorem 3. *Let $\varepsilon < 1/8$. Let $f : \{0,1\}^n \to \{0,1\}$ have a degree d torus polynomial approximating it with error ε. Then there is a MidBit$^+$ circuit computing f of the following form:*

- *Fan-in of each AND gate is bounded by d.*
- *Fan-in of the MidBit gate is $2^{2k-1} - 1$ where $k = O(d\log n + log(1/\varepsilon))$ (the fan-in puts the middle bit at position k).*
- *For $x \in \{0,1\}^n$, let $A(x)$ be the number of AND gates that output 1. Then $A(x) \equiv f(x)2^{k-1} + E(x) \mod 2^k$, where $E(x)$ can be thought of as an error term and is bounded by $0 \leq E(x) \leq 4\varepsilon 2^k$.*

Proof. Let $P = \sum_\alpha c_\alpha X^\alpha$ be a degree d torus polynomial approximating f with error ε. The first step is to perturb the coefficients by a small amount to make them rational with a power of 2 in the denominator. Let k be a natural number, the value of which will be chosen later. Find $q_\alpha \in \{0, 1/2^k, \ldots, (2^k - 1)/2^k\}$ for each c_α such that

$$c_\alpha - q_\alpha \mod 1 \leq \frac{1}{2^k}$$

Consider $P_{disc} = \sum_\alpha q_\alpha X^\alpha$. Let M be the number of monomials in P. Each monomial can contribute at most $1/2^k$ additional error after changing the coefficients, therefore P_{disc} is a torus polynomial that approximates f with error $\varepsilon' = \varepsilon + M/2^k$. Therefore the following holds:

$$\forall x \in \{0,1\}^n, \left| P_{disc}(x) - \frac{f(x)}{2} \mod 1 \right| \leq \varepsilon'$$

The next step is to add a small rational number so that the resulting polynomial is always "more" than the function when considered mod 1. Find the least natural number q such that $\varepsilon' \leq q/2^k$. Construct the polynomial $P_{pos} = P_{disc} + q/2^k$. The following holds now:

$$\forall x \in \{0,1\}^n, 0 \leq P_{pos}(x) - \frac{f(x)}{2} \mod 1 \leq \varepsilon' + \frac{q}{2^k}$$

Now choose a value of k such that

$$\varepsilon' + \frac{q}{2^k} \leq 4\epsilon$$

Note that $q/2^k \leq \varepsilon' + 1/2^k$ as per the choice of q. Also $M \leq (d+1)n^d$. Substitute these as well as the value of ε' to get that the following inequality suffices for the inequality above to hold.

$$\frac{2(d + 1)n^d + 1}{2^k} \leq 2\varepsilon$$

A value of $k = O(d\log n + \log(1/\varepsilon))$ with a large enough constant suffices for this inequality. All coefficients of P_{pos} have 2^k as the common denominator. The next step is to clear out this common denominator to make the coefficients integral. Consider $P_{int} = 2^k \cdot P_{pos}$. It is easy to see that all coefficients of P_{int} are integers. For $x \in \{0,1\}^n$, the following holds:

- $f(x) = 0 \iff P_{int}(x) \equiv E(x) \mod 2^k$,
- $f(x) = 1 \iff P_{int}(x) \equiv 2^{k-1} + E(x) \mod 2^k$,

where $0 \le E(x) \le 4\varepsilon 2^k < 2^{k-1}$. The last inequality holds because $\varepsilon < 1/8$.

Hence the value of $f(x)$ is determined exactly by the kth bit of $P_{int}(x)$. Note that P_{int} can be written as $P_{int} = \sum_\alpha n_\alpha X^\alpha$ where $n_\alpha \in \mathbb{Z}$ and $0 \le n_\alpha < 2^k$.

Use this polynomial to construct the following circuit. For each monomial indexed by α, create n_α many copies of an AND gate, the variables fed to the gate being the variables in the support of α. Note that the fan-in of these AND gates is bounded by d. Feed all these AND gates to a MidBit gate.

There are at most $(d+1)n^d$ many distinct AND gates and each AND gate has at most 2^k many copies. Therefore the fan-in of the MidBit gate is bounded by $2^k(d+1)n^d < 2^{2k-1}$. Add dummy gates which output 0, if needed, to ensure the fan-in of the MidBit gate becomes $2^{2k-1} - 1$.

This circuit will now compute $f(x)$. That the circuit satisfies the third property is easy to see. □

This can now be used to prove Theorem 2 by substituting $d = (\log n)^{O(1)}$ and $n^{-O(1)} \le \varepsilon$, and observing the fan-in of the top MidBit gate and AND gates is as required for the function to be in MidBit^+.

Proof (Theorem 2 of Subsection 1.2). Substitute $d = (\log n)^{O(1)}$ in Theorem 3. Observe that the fan-in of the AND gates is bounded by $d = (\log n)^{O(1)}$. The fan-in of the top MidBit gate is $2^{2k-1} - 1$ for $k = O(d \log n + \log(1/\varepsilon)) = (\log n)^{O(1)}$. This implies that the fan-in of the MidBit is bounded by $2^{(\log n)^{O(1)}}$. This proves $f \in \mathsf{MidBit}^+$. □

3 Closure Under Parity

We prove that if two functions have low degree torus approximations with small error, then the parity of these functions also has low degree torus approximations with slightly larger error. The error grows in an additive fashion.

Theorem 4. *Let $f_1 : \{0,1\}^n \to \{0,1\}$ be a boolean function that has a degree d_1 torus polynomial approximating it with error ε_1. Let $f_2 : \{0,1\}^n \to \{0,1\}$ be another boolean function that has a degree d_2 torus polynomial approximating it with error ε_2. Then $f_1 \oplus f_2$ has a degree $\max(d_1, d_2)$ torus polynomial approximating it with error $\varepsilon_1 + \varepsilon_2$.*

Proof. Let P_1, P_2 be the torus polynomials approximating f_1, f_2 respectively. The polynomial that approximates $f_1 \oplus f_2$ is simply $P_1 + P_2$. Consider these three cases to prove that this approximation is correct:

- Let $f_1(x) = f_2(x) = 0$, hence $f_1(x) \oplus f_2(x) = 0$. In this case

$$|P_1(x) \mod 1| \le \varepsilon_1, |P_2(x) \mod 1| \le \varepsilon_2$$

Therefore

$$|P_1(x) + P_2(x) \mod 1| \le \varepsilon_1 + \varepsilon_2$$

– Let $f_1(x) = 1, f_2(x) = 0$, hence $f_1(x) \oplus f_2(x) = 1$. In this case

$$|P_1(x) - 1/2 \mod 1| \leq \varepsilon_1, |P_2(x) \mod 1| \leq \varepsilon_2$$

Therefore
$$|P_1(x) + P_2(x) - 1/2 \mod 1| \leq \varepsilon_1 + \varepsilon_2$$

Similar analysis works for $f_1(x) = 0, f_2(x) = 1$.
– Let $f_1(x) = f_2(x) = 1$, hence $f_1(x) \oplus f_2(x) = 0$. In this case

$$|P_1(x) + 1/2 \mod 1| \leq \varepsilon_1, |P_2(x) + 1/2 \mod 1| \leq \varepsilon_2$$

Therefore
$$|P_1(x) + 1/2 + P_2(x) + 1/2 \mod 1| \leq \varepsilon_1 + \varepsilon_2$$

Note that $P_1(x) + 1/2 + P_2(x) + 1/2 \equiv P_1(x) + P_2(x) \mod 1$. Hence

$$|P_1(x) + P_2(x) \mod 1| \leq \varepsilon_1 + \varepsilon_2$$

This proves that $P_1 + P_2$ approximates $f_1 \oplus f_2$ within error $\varepsilon_1 + \varepsilon_2$ in all possible cases. Note that the degree of $P_1 + P_2$ is $\max(d_1, d_2)$. Hence this is the desired polynomial to approximate $f_1 \oplus f_2$.   □

4   Future Directions

We have proved that the parity of two functions, that have torus approximations, has a torus approximation as well. It will be interesting to see whether the same can be proven for other ACC functions, such as AND, OR, MOD_m. If all these can be proven, it may provide an alternate proof to the fact that ACC has low degree torus approximations.

Acknowledgements. We would like to thank Nutan Limaye for a key suggestion that made our results cleaner to state as well as for helpful feedback on earlier drafts. We would also like to thank Srikanth Srinivasan and Shachar Lovett for useful discussions. Finally we would like to thank anonymous reviewers from CSR for their helpful comments.

References

1. Ajtai, M.: Σ_1^1- formulae on finite structures. Ann. Pure Appl. Logic **24**(1), 1–48 (1983)
2. Allender, E., Gore, V.: On strong separations from AC^0. In: Budach, L. (ed.) FCT 1991. LNCS, vol. 529, pp. 1–15. Springer, Heidelberg (1991). https://doi.org/10.1007/3-540-54458-5_44
3. Barrington, D.A.: Bounded-width polynomial-size branching programs recognize exactly those languages in NC^1. J. Comput. Syst. Sci. **38**(1), 150–164 (1989)
4. Beigel, R., Tarui, J.: On ACC (circuit complexity). In: [1991] Proceedings 32nd Annual Symposium of Foundations of Computer Science, pp. 783–792. IEEE (1991)

5. Bhrushundi, A., Hosseini, K., Lovett, S., Rao, S.: Torus polynomials: an algebraic approach to ACC lower bounds. In: 10th Innovations in Theoretical Computer Science Conference (ITCS 2019). Schloss Dagstuhl-Leibniz-Zentrum fuer Informatik (2018)
6. Chattopadhyay, A., Mande, N.: A short list of equalities induces large sign rank. In: 2018 IEEE 59th Annual Symposium on Foundations of Computer Science (FOCS), pp. 47–58. IEEE (2018)
7. Chen, L.: Non-deterministic quasi-polynomial time is average-case hard for ACC circuits. In: 2019 IEEE 60th Annual Symposium on Foundations of Computer Science (FOCS), pp. 1281–1304. IEEE (2019)
8. Chen, R., Oliveira, I.C., Santhanam, R.: An average-case lower bound against ACC⁰. In: Bender, M.A., Farach-Colton, M., Mosteiro, M.A. (eds.) LATIN 2018. LNCS, vol. 10807, pp. 317–330. Springer, Cham (2018). https://doi.org/10.1007/978-3-319-77404-6_24
9. Chen, R., Santhanam, R., Srinivasan, S.: Average-case lower bounds and satisfiability algorithms for small threshold circuits. arXiv preprint arXiv:1806.06290 (2018)
10. Furst, M., Saxe, J.B., Sipser, M.: Parity, circuits, and the polynomial-time hierarchy. Math. Syst. Theory 17(1), 13–27 (1984)
11. Green, F., Köbler, J., Torán, J.: The power of the middle bit. In: [1992] Proceedings of the Seventh Annual Structure in Complexity Theory Conference, pp. 111–117. IEEE (1992)
12. Håstad, J.: Computational Limitations of Small-Depth Circuits. MIT press, Cambridge (1987)
13. Kane, D.M., Williams, R.: Super-linear gate and super-quadratic wire lower bounds for depth-two and depth-three threshold circuits. In: Proceedings of the Forty-Eighth Annual ACM symposium on Theory of Computing, pp. 633–643 (2016)
14. Murray, C., Williams, R.: Circuit lower bounds for nondeterministic quasi-polytime: an easy witness lemma for NP and NQP. In: Proceedings of the 50th Annual ACM SIGACT Symposium on Theory of Computing, pp. 890–901 (2018)
15. Razborov, A.A.: Lower bounds on the size of bounded depth circuits over a complete basis with logical addition. Mathe. Notes Acad. Sci. USSR 41(4), 333–338 (1987)
16. Smolensky, R.: Algebraic methods in the theory of lower bounds for Boolean circuit complexity. In: Proceedings of the Nineteenth Annual ACM Symposium on Theory of Computing, pp. 77–82 (1987)
17. Williams, R.: New algorithms and lower bounds for circuits with linear threshold gates. In: Proceedings of the Forty-Sixth Annual ACM Symposium on Theory of Computing, pp. 194–202 (2014)
18. Williams, R.: Nonuniform ACC circuit lower bounds. J. ACM (JACM) 61(1), 1–32 (2014)
19. Yao, A.C.C.: Separating the polynomial-time hierarchy by oracles. In: 26th Annual Symposium on Foundations of Computer Science (SFCS 1985), pp. 1–10. IEEE (1985)

A PCP of Proximity for Real Algebraic Polynomials

Klaus Meer[✉]

BTU Cottbus-Senftenberg, Computer Science Institute,
Platz der Deutschen Einheit 1, 03046 Cottbus, Germany
meer@b-tu.de

Abstract. Let $f : F_q^k \mapsto \mathbb{R}$ be a function given via its table of values, where $F_q := \{0, 1, \ldots, q-1\} \subset \mathbb{R}, k, q \in \mathbb{N}$. We design a randomised verification procedure in the BSS model of computation that verifies if f is close to an algebraic polynomial of maximal degree $d \in \mathbb{N}$ in each of its variables. If f is such an algebraic polynomial there exists a proof certificate that the verifier will accept surely. If f has at least distance $\epsilon > 0$ to the set of max-degree algebraic polynomials on F_q^k, the verifier will reject any proof with probability at least $\frac{1}{2}$ for large enough q. The verification procedure establishes a real number PCP of proximity, i.e., it has access to both the values of f and the additional proof certificate via oracle calls. It uses $O(k \log q)$ random bits and reads $O(1)$ many components of both f and the additional proof string, which is of length $O((kq)^{O(k)})$. The paper is a contribution to the not yet much developed area of designing PCPs of proximity in real number complexity theory.

1 Introduction

Property testing in the last decades has evolved as an important area in theoretical computer science. One of its several versions studies fast randomized verification algorithms that yield an approximate decision making for a decision problem L in the following sense: given an input x and an error bound $\epsilon > 0$, the verification should confirm with probability 1 if $x \in L$ and reject every x that is not ϵ-close[1] to S with large enough constant probability. Most importantly, the verifier may access components of x only by oracle calls and one major goal is to reduce the number of such calls. In addition to this one-sided-error definition other variants have been studied as well. Property testing among other applications has become crucial in relation with designing probabilistically checkable proofs PCPs and proving the famous PCP theorem [1,2]. In the context of PCPs the question is varied; here, the verifier has full access to the input x and oracle access to an additional certificate meant to provide a proof that $x \in L$. In yet another scenario, so called PCPs of proximity (a notion coined in [6]), both x and an additional certificate are accessible via oracle calls and again one major goal is to give an approximate decision using as few as possible oracle calls to

---

[1] Where closeness usually is measured by using the Hamming distance of strings.

© Springer Nature Switzerland AG 2021
R. Santhanam and D. Musatov (Eds.): CSR 2021, LNCS 12730, pp. 264–282, 2021.
https://doi.org/10.1007/978-3-030-79416-3_16

both objects. A thorough presentation of the field can be found, for example, in [12].

Similar questions also played a crucial role in using algebraic methods to derive a real number analogue of the PCP theorem in the computational model by Blum, Shub, and Smale, BSS model henceforth, over the real numbers [5]. Whereas in the first proof of the classical PCP theorem in [1,2] algebraic polynomials defined over finite fields are used crucially, the proof of the real number $PCP_\mathbb{R}$ theorem heavily relies on using trigonometric polynomials on suitable subsets of $\mathbb{R}$ instead. They were in one major step of the proof used as coding objects for real solutions of a quadratic system of polynomial equations; then, a PCP of proximity was designed to show that trigonometric polynomials are useful in this context.

It remained open whether algebraic polynomials can be used as coding objects as well, i.e., whether they can be tested by a real number PCP of proximity using the same (low) amount of resources. The purpose of this paper is to answer this question in the positive. Our construction heavily relies on the use of trigonometric polynomials in [5] for designing segmented verifiers for real number problems and on the structure of algebraic polynomials on arbitrary finite subsets of $\mathbb{R}^k$ as analysed in [11]. The main result complements the one of [11]; therein, a property test for algebraic polynomials on such general domains is designed which uses a non-constant number of oracle queries. The PCP of proximity given in this paper reduces the number of proof components read by the verifier to be constant, having access to an additional proof certificate beside the table of function values. Also from this additional certificate the verifier inspects a constant number of components.

The present paper extends the still small list of objects for which real number PCPs of proximity exist by algebraic polynomials - a central class for many problems studied in BSS computability over $\mathbb{R}$.

1.1 Previous Work and Outline

In the first proof of the PCP theorem in the Turing model [1,2] multivariate algebraic polynomials defined on finite fields were used as coding objects for satisfying assignments of instances of the 3SAT problem. Testing a function given by a table of its values for being close to such a polynomial was one key ingredient in the entire proof. Most importantly, the test had to have a segmented structure, see below, in order to be used in a further proof step called verifier composition.

When trying to do the same in the real number BSS model [7], major difficulties arise from the fact that a function value table can only specify a function on a finite domain which, as subset of the reals, is not any longer a field. Loosing the field structure causes severe problems when using algebraic polynomials in the PCP framework over $\mathbb{R}$.

In [15], a low degree test designed by Friedl et al. in [11] for functions $f : F^k \mapsto \mathbb{R}$ on finite domains $F \subset \mathbb{R}$ was used to prove the existence of short almost transparent proofs for the real number complexity class $NP_\mathbb{R}$. Though the test is able to verify closeness to an algebraic polynomial of maximal degree d in each of its variables, it lacks the main feature of segmentation. This means the following: A verifier is called segmented if it reads $O(1)$ positions in the input table for f as well as $O(1)$ segments of an additional proof certificate π provided by a prover; here, a certificate is a sequence of reals and a segment is a consecutive subsequence of it. The segments are allowed to have non-constant length, but they have to be queried in this structured form. This is a basic requirement in the PCP framework for using such a verifier in the so-called composition step - another important keystone to reduce the query complexity.

In [5] the authors succeeded in proving this theorem using algebraic methods and replacing the role of algebraic polynomials by trigonometric ones.[2] A significant amount of work was devoted to developing a test, more precisely a PCP of proximity, for trigonometric polynomials that is in segmented form. It relies on the fact that at least some features of finite fields can be regained when considering trigonometric polynomials on finite subdomains of $\mathbb{R}$. However, the authors have not been able to design a similar test for algebraic polynomials that can be used as ingredient of a proof of the real PCP theorem.

In the present paper we show that such a test can be designed once we have the machinery from [5] at hand. There are two main previous results needed in our approach. The actual test for closeness to an algebraic polynomial uses (a variant of) a result from [11]. It gives an estimate for the distance of a function $f : F^k \mapsto \mathbb{R}$ to algebraic polynomials by distances of f to the set of functions that are polynomials in at least one variable. The main task is to put this test procedure into segmented form. To do so, we again use trigonometric polynomials for coding certain restrictions of algebraic ones. This gives a segmented test algorithm using $O(k \log |F|)$ random bits and making $O(1)$ inspections of the table of function values for f as well as $O(1)$ inspections into proof segments of length $O(|F|^2)$. In a final standard step, the technique of verifier composition as used in [5] for the real number framework can be applied to reduce the total number of proof components read to be constant. Our test procedure complements the one from [11], which actually is a property testing algorithm.

The paper is organised as follows: Sect. 2 collects some basic definitions and states the main result of this paper. Its proof is described in Sect. 3. In its first subsection, we describe and analyse the closeness test verifying whether a function is close to an algebraic polynomial. Section 3.2 is the main part of the paper. Here, we explain how the closeness test can be put into segmented form necessary for verifier composition. Since the construction heavily relies on the low-degree test for trigonometric polynomials from [5], we first recall those results in the form we need them. Then we show how trigonometric polynomials can be used to code the information needed in the closeness test for algebraic polynomials in such a way that the letter can be performed in segmented form and how this leads to our main result. Section 3.3 briefly explains how the technique of real

---

[2] Note that the real number PCP theorem had been established in the BSS model earlier in [4] by a more combinatorial proof along the lines of [10].

verifier composition leads to our main result. We close with some concluding remarks.

2   Basic Definitions, Main Result

We assume the reader to be familiar with the very basics of the BSS model of computation and its complexity theoretic features over the real numbers, see [7], and [16] for some surveys on more recent developments. Our main objects of interest are real valued functions that are algebraic polynomials on a finite subset of some $\mathbb{R}^k$. More precisely, the following definition is fundamental.

Definition 1. *Let $q, k, d \in \mathbb{N}$ and let $F_q := \{0, 1, \ldots, q-1\} \subset \mathbb{R}$.*

a) *The set $P(k,d)$ denotes all functions $f : F_q^k \mapsto \mathbb{R}$ that have an extension to $\mathbb{R}^k$ being a polynomial with maximal degree d in each of its variables.*

b) *For all $1 \leq i \leq k$ we denote by $P_i(k,d)$ the set of all functions $f : F_q^k \mapsto \mathbb{R}$ that have an extension $\tilde{f} : \mathbb{R}^k \mapsto \mathbb{R}$ being a polynomial of degree at most d in its i-th variable. This means that whenever a point $x \in \mathbb{R}^k$ is fixed, the map $t \mapsto \tilde{f}(x+te_i)$ with $e_i \in \mathbb{R}^k$ being the i-th unit vector is a univariate polynomial in t of degree at most d. It is not hard to see that $P(k,d) = \bigcap\limits_{i=1}^{k} P_i(k,d)$.[3]*

c) *The distance of two functions $f, g : F_q^k \mapsto \mathbb{R}$ is defined as $d(f,g) := \frac{1}{q^k}|\{x \in F_q^k | f(x) \neq g(x)\}|$. Similarly, for the distance between an f and a set A of functions from $F_q^k \mapsto \mathbb{R}$ we define $d(f,A) := min\{d(f,g)|g \in A\}$.*

Below, we shall choose q to be (without loss of generality) a prime and large enough in relation to k, d, and $1/\epsilon$, where $\epsilon > 0$ is a constant error bound, for our results to hold. Our aim is to design a verification procedure that figures out whether a given function value table represents with high probability an algebraic polynomial of certain max-degree. The concept of such a verification procedure in the BSS model and the resources it uses is defined next. We directly focus on the task of designing a PCP of proximity for real algebraic polynomials. The corresponding algorithms use randomisation and are allowed to inspect an additional proof certificate in order to come to a decision.

Definition 2. *a)   Let $r, u : \mathbb{N} \to \mathbb{N}$ be two resource functions. A real proba-bilistic $(r(n), u(n))$-restricted verifier V for testing algebraic polynomials is a randomised real BSS machine; on input $f : F_q^k \mapsto \mathbb{R}$, given via a table of its $n := q^k$ real function values, V first generates uniformly and indepen-dently a string ρ of $O(r(n))$ random bits. Using ρ it makes $O(u(n))$ non-adaptive queries into the table for f and into an additional proof certificate $\pi \in \mathbb{R}^\infty := \bigsqcup_{i \geq 1} \mathbb{R}^i$. A query is made by writing an address on a query tape and then in one step the real number stored at that address in the table or in π, respectively, is returned. The verifier uses the results of the queries and computes in time polynomial in n its decision 'accept' or 'reject'.*

---

[3] This can be shown by easy induction on k, see for example [3], pages 225ff.

b) We call a verifier V segmented if the queries are structured such that V asks $O(1)$ many segments of length at most $O(u(n))$ from the table for f and the certificate π. Note that these objects are given as elements in $\mathbb{R}^\infty$, thus a segment is sequence of contiguous components of real numbers in such an element.

c) Let $\epsilon > 0$. A verifier for testing algebraic polynomials is ϵ-reliable if the following conditions are satisfied. If $f \in P(k,d)$, there is a proof certificate π such that V accepts (f,π) with probability 1. And if $d(f, P(k,d)) > \epsilon$, then for all certificates π the verifier V rejects (f,π) with probability $> \frac{1}{2}$.

Note that the segmentation requirement only is meaningful when the number of queries is not yet constant. Segmentation is crucial for obtaining our main result:

Theorem 1. Let $\epsilon > 0$ be fixed; let q be a prime and let $q, k, d \in \mathbb{N}$ be such that $q \in \Omega(\frac{k^2 d^3}{\epsilon^2})$ and $F_q := \{0, 1, \ldots, q-1\} \subset \mathbb{R}$. There is a $(O(k \log q), O(1))$-restricted verifier that is ϵ-reliable for testing whether a given $f : F_q^k \mapsto \mathbb{R}$ is an algebraic polynomial of maximal degree d in each of its variables. The verifier gets as input a table of the function values of f together with a proof certificate of length $O((kq)^{O(k)})$. Its running time is polynomial in kq.

Using the techniques behind the proof of the real number $\mathrm{PCP}_\mathbb{R}$ theorem in [5], in particular verifier composition, the theorem actually follows easily from the following one, in which the verifier's query complexity is not yet constant and thus segmentation is an important requirement. Proving Theorem 2 therefore is the major task that will be solved in this paper.

Theorem 2. Let $\epsilon > 0$ be fixed, q be a prime and $F_q := \{0, 1, \ldots, q-1\} \subset \mathbb{R}$. Let $q, k, d \in \mathbb{N}$ be s.t. $q \in \Omega(\frac{k^2 d^3}{\epsilon^2})$. There is a segmented and $(O(\frac{1}{\epsilon} k \log q), O(q^2))$-restricted verifier that is ϵ-reliable for testing whether a given $f : F_q^k \mapsto \mathbb{R}$ is an algebraic polynomial of maximal degree d in each of its variables. The verifier gets as input a table of the function values of f together with a proof certificate of length $O((kq)^{O(k)})$, i.e., it is a PCP of proximity. Its running time is polynomial in kq.

An outline for proving Theorems 1 and 2 is as follows. Two steps are needed. The first consists of a test actually verifying closeness of a given f to a max-degree d polynomial. The design and analysis of this test, given in Theorem 3, is based on a variant of a result from [11] relating $d(f, P(k,d))$ to the sum of distances $\sum_{i=1}^{k} d(f, P_i(k,d))$. Its resources are not yet those required. In order to reduce it we want to apply the classical technique of verifier composition. However, to do so the closeness test has to be put into segmented form which it does not have. Whereas a general procedure for putting a verifier into segmented form for the Turing model has been developed in relation with the first proof of the classical PCP theorem, a similar technique in the real number framework needs significant changes. A major part of [5] is devoted to design the relevant steps in relation with a proof of the PCP theorem in the real number BSS model. More precisely, a closeness test in form of a PCP of proximity is given for trigonometric polynomials, i.e., [5] solves for trigonometric problems the question we are now looking for to solve for algebraic polynomials. Our way to succeed with the latter is based on the former. We therefore in the second proof step show that the ideas from [5] as well can be used to put the above closeness test for algebraic polynomials into segmented form. This will prove Theorem 2. Our main Theorem 1 then follows by applying the verifier composition technique for real number verifiers as designed in [5].

3   Segmentation as Main Task

We now explain how the main result is proved. We start with describing and analyzing the test for verifying proximity of a function to an algebraic polynomial of given max-degree. The significantly more intricate part of segmenting this test is done in Sect. 3.2. Therein, we recall the necessary results from [5] for testing trigonometric polynomials and show, how this can be used to put our proximity test into a segmented form. Note that this step is by far not straightforward given the results from [5]. The final step then will apply verifier composition for achieving the claim of Theorem 1.

3.1   Distance to Paraxial Univariate Restrictions

In this subsection we study the distance of an algebraic polynomial to paraxial univariate restrictions. We describe a test for verifying closeness of a function to an algebraic polynomial and give the test analysis.

Proposition 1 below occurs as an intermediate result in a slightly different form within the proof of Theorem 2.2 in [11], page 61. For sake of completeness and since that paper might not be easily available we include a proof here. We also correct some minor typos.

Proposition 1. *Let $q, k, d \in \mathbb{N}$ be such that $q \geq 18kd^3$ and define $F_q := \{0, 1, \ldots, q - 1\}$. Let $f : F_q^k \to \mathbb{R}$ and $f_i \in P_i(k, d)$ for $1 \leq i \leq k$ be given. Then the following holds, where $\mu := \frac{\sqrt{d}}{\sqrt{18q}}$:*

$$\frac{1}{6} \cdot d(f, P(k, d)) \leq 2k\mu + d(f, f_1) + \sum_{j=1}^{k-1} d(f_j, f_{j+1}).$$

Proof. Before we give the proof note that the major change in our assertion with respect to the statement as it occurs on page 61, right column of [11] is the following: Our statement is about an arbitrary set of polynomials $f_i \in P_i(k, d), 1 \leq i \leq k$, whereas in [11] the authors consider $f_i \in P_i(k, d)$ that minimize the distance from f to $P_i(k, d)$. However, in the proof of the intermediate result this optimality of the f_i's is not used.[4]

---

[4] In [11] the optimality is needed because the final statement relates $d(f, P(k, d))$ with the sum of the distances $d(f, P_i(k, d))$. Note that we use the letters k, d for arity and degree instead of n, k in [11].

Now towards the proof of the statement as we need it. The proof in [11] uses a result from [2] and an easy corollary of it, stated as Lemmas 4.4 and 4.5 in [11]. These two results lead in [11], page 61 to the following alternatives, that we take as starting point for proving the proposition.

Lemma 1. *Let F_q, q, k, d be as in the statement of the proposition, define $\mu := \frac{\sqrt{d}}{\sqrt{18q}}$ and let $g : F_q^k \mapsto \mathbb{R}$ be an element of $P_1(k, d)$. Then either*

$$\left| \{c \in F_q \mid d(g|_{x_1=c}, P(k-1, d)) \leq \frac{1}{6}\} \right| \leq 2d - 1 \tag{1}$$

or

$$\left| \{c \in F_q \mid d(g|_{x_1=c}, P(k-1, d)) \leq d(g, P(k, d)) - \mu\} \right| \leq 6\mu q \tag{2}$$

Now let functions $f, f_1, \ldots, f_k$ with $f_i \in P_i(k, d)$ be given. We use the lemma to show by induction on i the following

CLAIM: For all $1 \leq i \leq k$ and any point $(c_1, \ldots, c_i) \in F_q^i$ except for at most a fraction of $6i\mu$, i.e., except for at most $6iq^i\mu$ many points, it is

$$\frac{1}{6}d(f, P(k, d)) \leq i\mu + d(f, f_1) + \sum_{j=1}^{i-1} d(f_j|_{x_1=c_1,\ldots,x_j=c_j}, f_{j+1}|_{x_1=c_1,\ldots,x_j=c_j})$$

$$+ d(f_i|_{x_1=c_1,\ldots,x_i=c_i}, P(k-i, d)).$$

Proof (of the CLAIM). For $f_1 \in P_1(k, d)$ one of the alternatives (1) or (2) holds according to Lemma 1. If (1) is satisfied we have $d(f_1|_{x_1=c_1}, P(k-1, d)) > \frac{1}{6}$ for at least $q-2d+1$ choices of $c_1 \in F_q$. Clearly, $d(f, P(k, d)) \leq 1$, so $\frac{1}{6}d(f, P(k, d)) \leq \frac{1}{6} < d(f_1|_{x_1=c_1}, P(k-1, d))$ for a fraction of at least $1 - \frac{2d-1}{q} \geq 1 - 6\mu$ choices for c_1. The last inequality results from $q > 2d$ and the definition of μ.

If alternative (2) holds, then for all but an exception of at most $6\mu q$ choices for c_1 one has $d(f_1, P(k, d)) < d(f_1|_{x_1=c_1}, P(k-1, d)) + \mu$ and thus $d(f, P(k, d)) \leq d(f, f_1) + d(f_1, P(k, d)) \leq d(f, f_1) + d(f_1|_{x_1=c_1}, P(k-1, d)) + \mu$. In both cases it follows

$$\frac{1}{6}d(f, P(k, d)) \leq 1 \cdot \mu + d(f, f_1) + d(f_1|_{x_1=c_1}, P(k-1, d))$$

as required.[5]

The argument is the same in the induction step: If (1) is true for an $1 \leq i < k$ and the function $f_{i+1}|_{x_1=c_1,\ldots,x_i=c_i}$, which is in $P_1(k-i, d)$ with respect to x_{i+1}, then $\frac{1}{6}d(f, P(k, d)) \leq \frac{1}{6} < d(f_{i+1}|_{x_1=c_1,\ldots,x_{i+1}=c_{i+1}}, P(k-i-1, d))$ except for a fraction of at most $\frac{2d-1}{q} < 6\mu$ choices for $c_{i+1} \in F_q$. In this situation the claim is trivially satisfied.

---

[5] Note that this argument does nowhere rely on whether f_1 is the best approximation from $P_1(k, d)$ to f or not. Similarly below in the induction step.

And if (2) holds, then for all but an exception of at most $6\mu q$ choices for c_{i+1} one has

$$d(f_{i+1}|_{x_1=c_1,\ldots,x_i=c_i}, P(k-i,d)) < d(f_{i+1}|_{x_1=c_1,\ldots,x_{i+1}=c_{i+1}}, P(k-i-1,d)) + \mu$$

and thus by the triangle inequality

$$d(f_i|_{x_1=c_1,\ldots,x_i=c_i}, P(k-i,d)) \le d(f_i|_{x_1=c_1,\ldots,x_i=c_i}, f_{i+1}|_{x_1=c_1,\ldots,x_i=c_i})$$
$$+ \quad d(f_{i+1}|_{x_1=c_1,\ldots,x_i=c_i}), P(k-i,d))$$
$$< d(f_i|_{x_1=c_1,\ldots,x_i=c_i}, f_{i+1}|_{x_1=c_1,\ldots,x_i=c_i})$$
$$+ d(f_{i+1}|_{x_1=c_1,\ldots,x_{i+1}=c_{i+1}}, P(k-i-1,d)) + \mu$$

except for a fraction of at most 6μ choices of $c_{i+1} \in F_q$ (and that for each fixed $(c_1,\ldots,c_i)$ for which the induction hypothesis holds). Together with the induction hypothesis for i this implies

$$\tfrac{1}{6}d(f, P(k,d)) \le (i+1)\mu + d(f, f_1) + \sum_{j=1}^{i} d(f_j|_{x_1=c_1,\ldots,x_j=c_j}, f_{j+1}|_{x_1=c_1,\ldots,x_j=c_j})$$
$$+ \quad d(f_{i+1}|_{x_1=c_1,\ldots,x_{i+1}=c_{i+1}}, P(k-i-1,d)).$$

It remains to upper bound the number of points in F_q^{i+1} for which this is false. The induction hypothesis is false for at most $6iq^i\mu$ points in F_q^i, which contributes at most $6iq^{i+1}\mu$ violations in F_q^{i+1}. Moreover, each of the at least $q^i(1-6\mu)$ points in F_q^i that satisfy the hypothesis contributes at most $6\mu q$ bad choices of $c_{i+1} \in F_q$. Altogether, this yields at most $6iq^{i+1}\mu+6\mu qq^i(1-6i\mu) \le 6(i+1)\mu q^{i+1}$ exceptions. The claim is proved.

To finish the proof of the proposition for $i = k$ we take the average over $\underline{c} \in F_q^k$ of all the inequalities in the claim as follows: for satisfying choices $\underline{c} \in F_q^k$ we take the average of the inequality as it is stated in the claim, for the other at most $6k\mu q^k$ choices we take the inequality with $\frac{1}{6}$ added on the right hand side, which results trivially in a correct inequality. The latter for the average contributes an additional term of $k\mu$ and we finally get

$$\frac{1}{6} \cdot d(f, P(k,d)) \le 2k\mu + d(f, f_1) + \sum_{j=1}^{k-1} d(f_j, f_{j+1}).$$

$\square$

The proposition gives a nearby idea to test closeness of a given $f_0 := f$ to $P(k,d)$. The verifier estimates for suitable f_i the distances $d(f_i, f_{i+1})$ for $0 \le i \le k-1$ by comparing the values in a randomly chosen point $x^{(i+1)} \in F^k$. In one test round a function f_i has to be evaluated in two random points, one for estimating its distance to f_{i-1} and one for the distance to f_{i+1}. These two points have to be independent. Now if f has a large distance to $P(k,d)$ the proposition will result in a sufficient error probability for the test. Before we describe precisely the test and its analysis note the following two aspects: Later on, we shall explain how information about functions f_i is presented to the verifier as part of a proof certificate π. Basically, this will be done through univariate polynomials resulting canonically from the random points in which an f_i is evaluated by the test. However, since this information should be accessed in segmented form, the implementation needs the machinery of testing trigonometric polynomials. We elaborate on this in the next subsection. Secondly, whereas the above outline naively would require to generate k random points in F_q^k we can reduce the amount of randomness by the well known method of two point sampling, see Lemma 2. We only have to guarantee that the two points in which one function f_i is evaluated are independent, so pairwise independence of the k random points suffices.

Lemma 2 *(Two-point sampling, see [9]). Given random elements $x^{(1)}, x^{(2)} \in F_q^k$ there is a deterministic BSS algorithm running in polynomial time in k that computes a sequence of points $tps(x^{(1)}, x^{(2)}) := (x^{(1)}, \ldots, x^{(k)}) \in (F_q^k)^k$ that are pairwise independent random variables being uniformly distributed.*

We are now ready to describe the first test. In the description we assume that the verification algorithm has access to a black box (later: a part of the certificate) for evaluating f_i in the demanded points.

Test: Closeness to $P(k, d)$:

Input: Function value table for an $f : F_q^k \mapsto \mathbb{R}$; black box for evaluating functions $f_i \in P_i(k, d), 1 \le i \le k$ in requested points from F_q^k.

1) Generate two uniformly distributed random points $x^{(1)}, x^{(2)} \in F_q^k$
2) Compute by two-point sampling the sequence $tps(x^{(1)}, x^{(2)}) := (x^{(1)}, \ldots, x^{(k)}) \in (F_q^k)^k$
3) Evaluate and check the following equalities: $f(x^{(1)}) = f_1(x^{(1)}), f_1(x^{(2)}) = f_2(x^{(2)}), f_2(x^{(3)}) = f_3(x^{(3)}), \ldots, f_{k-1}(x^{(k)}) = f_k(x^{(k)})$
4) If one of the equalities is violated reject; otherwise accept. Similarly for several rounds of the test: Reject if at least one equality is violated.

Theorem 3. *Let $f, f_i, 1 \le i \le k$ be as above, and let $\epsilon > 0$ be a constant such that $\mu := \frac{\sqrt{d}}{\sqrt{18q}} \le \frac{\epsilon}{24k}$. This holds, for example, if $q \ge 32k^2 d \frac{1}{\epsilon^2}$. Then the Closeness Test satisfies the following: If $f \in P(k, d)$ and the f_i all equal f, then the test accepts with probability 1. If $d(f, P(k, d)) \ge \epsilon$, then for all choices of $f_1, \ldots, f_k$ performing $m \ge \frac{12}{\epsilon} + 1$ rounds of the test is sufficient to reject with probability $> \frac{1}{2}$.*

One round of the test needs $2k \log q$ random bits and inspects one value of f and f_k, and two values of each $f_i, 1 \le i \le k - 1$. Similarly, in m rounds considering ϵ as a constant the test needs $O(k \log q)$ random bits and inspects $O(1)$ values of each of the functions.

Proof. All statements except the one for the error probability are obvious, so let us assume that $d(f, P(k, d)) \ge \epsilon$. Define $\epsilon_1 := d(f, f_1)$ and $\epsilon_i := d(f_{i-1}, f_i)$

for $2 \leq i \leq k$. Using Proposition 1 and the assumption on μ we have
$\frac{\epsilon}{12} \leq \frac{1}{6}\epsilon - 2k\mu \leq \sum_{i=1}^{k} \epsilon_i$. The test in one round does not realize an error
with probability at most $\prod_{i=1}^{k}(1 - \epsilon_i)$. This holds because the sequence of $x^{(i)}$'s
consists of pairwise independent uniformly distributed random points and for
one such the probability that $f_{i-1}(x^{(i)}) = f_i(x^{(i)})$ is at most $1 - \epsilon_i$. Using the
above and a well known inequality for the Weierstraß product, see [8] for a proof,
we get $\prod_{i=1}^{k}(1 - \epsilon_i) \leq \frac{1}{1+\sum_{i=1}^{k}\epsilon_i} \leq \frac{1}{1+\epsilon/12}$. Thus, in m rounds of the test an error
is not detected with probability at most $\left(\frac{1}{1+\epsilon/12}\right)^m \leq \left(\frac{1}{e}\right)^{\frac{m\epsilon}{12+\epsilon}}$. The latter is
$< \frac{1}{2}$ for $m \geq \frac{12}{\epsilon} + 1$, which proves the theorem. $\square$

Remark 1. In the next subsection below, for a point $x \in F_q^k$ a proof certificate is
expected to represent the univariate restrictions of f_i into the paraxial directions
e_i, i.e., a representation of the univariate polynomial $t \mapsto f_i(x + te_i)$. In this
context, the verifier expects an ideal proof certificate to represent the correct
restriction of f to that paraxial line through x.

3.2 Segmentation of Closeness Test

The goal of this subsection is to transform the above verifier performing the
closeness test into a segmented form. As in the proof of the original PCP theorem [1], segmented verifiers were crucial in [5] to prove the real number PCP$_\mathbb{R}$
theorem when using an algebraic approach. However, the real number setting
causes severe difficulties when working with algebraic polynomials. To circumvent these difficulties, in [5] trigonometric polynomials mapping a vector space
F_q^k over a finite field F_q to $\mathbb{R}$ are considered. A main result in [5] is a segmented
test for verifying, whether a given function f is close to a trigonometric polynomial of certain max-degree. This test is used below to achieve segmentation
of the closeness test as follows: We suppose that all information about the algebraic polynomials $f_i, 1 \leq i \leq k$ which is potentially queried in the closeness test,
is coded by particular values of a trigonometric polynomial s of sufficient max-
degree. This potential information consists of the univariate paraxial restrictions
of an f_i in a point $x \in F_q^k$ having been generated by two-point sampling. The
corresponding trigonometric polynomial s is given as additional information to
the verifier in form of a table of its values, thereby replacing the black box
assumption in the closeness test. To guarantee that this table with high probability represents a trigonometric polynomial the test from [5] is applied first.
Segmentation then is obtained as follows: For each pair $(x^{(1)}, x^{(2)}) \in F_q^{2k}$, the
necessary information about $tps(x^{(1)}, x^{(2)})$ and the corresponding restrictions
can be found when restricting s to a two-dimensional plane. In the ideal case,
this restriction is a bivariate trigonometric polynomial b of low max-degree. The
certificate expected by the verifier in addition contains the coefficients of these

bivariate polynomials as entire segments. The verifier in each round of the closeness test reads this segment, reconstructs b and performs the test. There are several additional technical details such as guaranteeing that b with high probability is the correct restriction of s. This foregoing will result in Theorem 2; it proves that the closeness test can be put into a format ready to apply composition of real number verifiers as developed in [5]. This will suffice to obtain Theorem 1. Note finally that the classical segmentation technique seems not to work directly (at least to the best of our knowledge) in the real number setting because restrictions of trigonometric functions to lines or planes in general do not behave well with respect to the resulting max-degrees of the restrictions. On the other hand, real algebraic polynomials seem not appropriate as coding objects for proving the real number $PCP_{\mathbb{R}}$ theorem along the lines of [1], see [5] for more detailed explanations of the involved problems. This in the end is the reason why we have to take the detour along trigonometric polynomials for proving a result on algebraic ones. Now towards the details.

Trigonometric Polynomials, Previous Results. We define a special version of real-valued trigonometric polynomials with a finite field as domain and recall the results on testing them. Let $F_q = \{0, \ldots, q-1\}$ be a finite field with q being prime. We consider the elements of F_q as subset of $\mathbb{R}$ when appropriate.

Definition 3. *Let F_q be a finite field as above. For $d \in \mathbb{N}$ a trigonometric polynomial $f : F_q^k \mapsto \mathbb{R}$ of max-degree d is given as $f(x_1, \ldots, x_k) = \sum_t c_t \cdot \exp(\frac{2\pi i}{q} \sum_{j=1}^{k} x_j t_j)$, where the sum is taken over all $t := (t_1, \ldots, t_k) \in \mathbb{Z}^k$ with $|t_1| \le d, \ldots, |t_k| \le d$ and $c_t \in \mathbb{C}$ satisfy $c_t = \overline{c_{-t}}$ for all such t.*

Remark 2. Note the following technical detail: if below a verifier is used as BSS algorithm for inputs of varying size, then for different cardinalities q of the finite field F it needs to work with different constants $\cos \frac{2\pi}{q}, \sin \frac{2\pi}{q}$. It is not hard to see that given q one could add two real numbers to the verification proof which in the ideal case represent real and imaginary part of a complex primitive q-th root of unity. The verifier in question deterministically checks in polynomial time whether this is the case and then continues to use these constants for evaluating trigonometric polynomials.

The following theorem is used crucially later on. It deals with the problem to verify whether a function s, given by a table of values, is close to a trigonometric polynomial. The parameters k_s, q_s, d_s used in the statement later on will depend on the parameters k, q, d given with the input f.

Theorem 4 (Testing and correcting trigonometric polynomials; see Theorems 2.4 and 5.2 in [5]). *Let $d_s \in \mathbb{N}$, $h := 10^{15}$, $k_s \ge \frac{3}{2}(2h+1)$, $\tilde{d}_s := 2hk_sd_s$, and let F_{q_s} be a finite field with q_s being a prime number larger than $10^4(2hk_sd_s + 1)^3$. Let $s : F_{q_s}^{k_s} \to \mathbb{R}$ be a function given by a table of its values.*

a) *There exists a probabilistic verification algorithm in the BSS-model of computation over the reals with the following properties:*

  i) *The verifier as input gets the table for s together with a proof string consisting of at most $q_s^{2k_s}$ segments. Each segment has at most $2hk_s d_s + k_s + 1$ many real components. The verifier uniformly generates $O(k_s \log q_s)$ random bits and has a running time that is polynomially bounded in the quantity $k_s \log q_s$, i.e., polylogarithmic in the input size $O(q_s^{k_s})$.*

  ii) *For every table representing a trigonometric max-degree d_s polynomial on $F_{q_s}^{k_s}$ there exists a proof such that the verifier accepts with probability 1.*

  iii) *For any $0 < \epsilon < 10^{-19}$ and for every function value table whose distance to a closest max-degree $\tilde{d}_s := 2hk_s d_s$ trigonometric polynomial is at least 2ϵ, the probability that the verifier rejects is at least ϵ, no matter which proof is given.*

b) *Suppose the verifier under a) has accepted s and the closest trigonometric polynomial of max-degree $\leq \tilde{d}_s$ is $\tilde{s}$ with a distance at most δ for arbitrary fixed and small enough $\delta > 0$. There exists another segmented verifier working as follows:*

  i) *It gets as input the table for s, a point $x \in F_{q_s}^{k_s}$ and an additional proof certificate with at most $O(\sqrt{q_s}^{k_s-1})$ segments of length $2\sqrt{q_s} k_s d_s + 1$ each. The verifier uniformly generates $O(k_s \log q_s)$ random bits and has a running time that is polynomially bounded in the quantity $k_s \sqrt{q_s} d_s$.*

  ii) *If $s \equiv \tilde{s}$ is a trigonometric max-degree d_s polynomial, there is a certificate such that the verifier accepts with probability 1.*

  iii) *If $s(x) \neq \tilde{s}(x)$, the verifier rejects every certificate with probability $\geq \frac{3}{4}$.*

Note that even though the theorem does not give a sharp test in the sense that acceptance of s only implies with high probability closeness to a trigonometric polynomial of larger max-degree than d_s, in the subsequent steps below that rely on this test a verifier still expects from a correct prover to receive data as if s is a max-degree d_s polynomial. This refers, for example, to certain bivariate restrictions of s used further on.

Coding Univariate Algebraic Polynomials. We shall now work out how to code the information a verifier needs in view of Theorem 3 when dealing with a given function $f : F_q^k \to \mathbb{R}$. This will be done using, among other things, a trigonometric polynomial s with suitable values for k_s, q_s, and d_s to which, as part of the verification procedure, Theorem 4 has to be applied.

For each pair $(x^{(1)}, x^{(2)}) \in F_q^{2k}$ which the verifier might randomly generate in the closeness test it needs to access information about the algebraic polynomials f_i when restricted to the paraxial direction e_i. Those restrictions have to be evaluated in some of the points constituting $tps(x^{(1)}, x^{(2)})$. We next describe how to encode this information using a trigonometric polynomial s in such a way that each round of the closeness test needs to inspect a single segment of the function value table for s only. The basic idea is to encode the $d+1$ coefficients of a restriction of the form $t \to f_i(x + te_i)$ as certain values of s in such a way, that the information needed for one round of the test is encoded in one segment representing the bivariate restriction of s to a two-dimensional plane. Consider a fixed pair $(x^{(1)}, x^{(2)}) \in F_q^{2k}$ and let $tps(x^{(1)}, x^{(2)}) =: (x^{(1)}, x^{(2)}, \ldots, x^{(k)}) \in F_q^{(k^2)}$. Suppose the closeness test needs to access the k functions $f_j, 1 \leq j \leq k$ with $f_j \in P_j(k, d)$. More precisely, the following univariate restrictions then are of interest; the index in the table below will later on denote the first (of two) values used to parameterize the plane.

index	univariate restriction	index	univariate restriction
1	$t \to f_1(x^{(1)} + te_1)$	$2i - 1$	$t \to f_i(x^{(i)} + te_i)$
2	$t \to f_1(x^{(2)} + te_1)$	$2i$	$t \to f_i(x^{(i+1)} + te_i)$
3	$t \to f_2(x^{(2)} + te_2)$	$\vdots$	$\vdots$
4	$t \to f_2(x^{(3)} + te_2)$	$2k - 1$	$t \to f_k(x^{(k)} + te_k)$
5	$t \to f_3(x^{(3)} + te_3)$		
$\vdots$	$\vdots$		

Every restriction is a univariate algebraic polynomial of degree d. We encode the coefficients of these polynomials as values of a trigonometric function $s :$ $F_{q_s}^{k_s} \to \mathbb{R}$; here $k_s := 2 + 2k$ and q_s is chosen to be at least as large as q and satisfying the additional requirements of Theorem 4 once d_s has been chosen below. Furthermore, $s(2i - 1, m, x^{(1)}, x^{(2)})$ gives the coefficient of monomial t^m of the polynomial $t \to f_i(x^{(i)} + te_i)$, similarly for $s(2i, m, x^{(1)}, x^{(2)})$. Note that s is considered on $F_{q_s}^{k_s}$ instead of $F_q^{k_s}$ in view of the requirements of the test behind Theorem 4, part a) used later on. It follows that for each pair $(x^{(1)}, x^{(2)}) \in F_q^{2k}$ the information used by the closeness test is coded by the values of s on the plane $E(x^{(1)}, x^{(2)}) := \{(j, m, x^{(1)}, x^{(2)}) \mid j, m \in F_{q_s}\} \subset F_{q_s}^{k_s}$ parameterized by the first two coordinates. On $E(x^{(1)}, x^{(2)})$ we have to specify $(2k - 1)(d + 1)$ many values. The following easy technical result is necessary to bound the size of that part of a certificate which specifies s as well as the size of another part that codes the bivariate trigonometric polynomials obtained as restrictions of s to the planes $E(x^{(1)}, x^{(2)})$.

Lemma 3. *Let $q, k, d \in \mathbb{N}$ be given, $k_s := 2k+2, q_s \geq q$ and let $f_i \in P_i(k, d), 1 \leq i \leq k$.*

a) *There exists a trigonometric polynomial $s : F_{q_s}^{k_s} \to \mathbb{R}$ of max-degree $d_s :=$ $O(q)$ which codes the coefficients of the f_i for all pairs $(x^{(1)}, x^{(2)})$ as described above.*

b) *For all pairs $(x^{(1)}, x^{(2)}) \in F_q^{2k}$, the bivariate trigonometric polynomial defined on $E(x^{(1)}, x^{(2)})$ by $b_{(x^{(1)}, x^{(2)})}(j, m) := s(j, m, x^{(1)}, x^{(2)})$ has max-degree $O(q)$.*

c) *Let $q', k', d' \in \mathbb{N}$ and let $s' : F_{q'}^{k'} \to \mathbb{R}$ be a trigonometric polynomial of max-degree $d' \in \mathbb{N}$ not identically 0. Then the number of zeros of s' is at most $2d'(k' + 1)(q')^{k'-1}$.*

Proof. We just briefly sketch the elementary proofs. The given bounds might not be sharp, but they suffice for later purposes.

a) This can be shown using trigonometric interpolation. Note that the intended coding only prescribes values in points $(j, m, x^{(1)}, x^{(2)})$, where $1 \leq j \leq 2k - 1, 0 \leq m \leq d$ and $x^{(1)}, x^{(2)} \in F_q^k$. Thus, all components range between 0 and $\max\{2k-1, d, q-1\}$. The previous conditions on q (see Proposition 1 and Theorem 3) guarantee that $q - 1$ is the maximal value. In the univariate case q prescribed values can be interpolated by a trigonometric polynomial of degree $\lceil \frac{q-1}{2} \rceil$. Now, a standard tensor-product construction gives a suitable trigonometric Lagrange polynomial in the multivariate case of max-degree $\lceil \frac{q-1}{2} \rceil$ as well, see for example [14].

b) follows immediately from the fact that the plane is parameterised by the two first unit vectors $(1, 0, \dots)^T$ and $(0, 1, 0, \dots)^T \in F_{q_s}^{k_s}$. The definition of the degree of a trigonometric polynomial shows that the max-degree resulting from a restriction to a plane depends on the values of the components of the respective directional vectors. Since here those values only are 0 or 1, the max-degree of the bivariate restrictions is no larger than that of s.

c) This is an easy induction on k', noticing that in the univariate case $2d' + 1$ zeros imply that a degree d' polynomial is identically 0. The induction step can be performed as done, for example, in [3], pp. 222f in the case of algebraic polynomials. $\qquad\square$

Revised Closeness Test. We now describe and analyze an extended form of the closeness test from Sect. 3.1. This will lead to the proof of Theorem 2. Before we do so let us summarize the conditions on the involved parameters resulting from our considerations so far: $q, k, d \in \mathbb{N}$ are given with the function f to be tested, $\epsilon > 0$ is the fixed reliability parameter. Here, $q \geq \max\{18kd^3, 32k^2d/\epsilon\}$. Then, $k_s := 2k + 2, d_s = O(q)$ and $q_s = \Omega(k_s^3 d_s^3)$, and thus $q_s = \Omega(k^3 q^3)$; h is a (huge) constant.

<u>Closeness Test Revised:</u> Let $q, q_s, k, k_s, d, d_s, h$ be as above, $\epsilon > 0$
Input: Function value table for an $f : F_q^k \to \mathbb{R}$; a proof certificate π of length $O((kq)^{O(k)})$. It consists of four parts $\pi_1, \dots, \pi_4$. Here, π_1 is a function value table for an $s : F_{q_s}^{k_s} \to \mathbb{R}$, π_2 is a certificate necessary to perform the tests on s behind Theorem 4, part a), π_3 represents for every pair $(x^{(1)}, x^{(2)}) \in F_q^{2k}$ a bivariate trigonometric polynomial $b_{(x^{(1)}, x^{(2)})} : F_{q_s}^2 \to \mathbb{R}$ of max-degree q, given by its coefficients, and π_4 is the certificate used for the correctness test on s underlying Theorem 4, part b).

Goal: The test should be ϵ-reliable with respect to the question whether $f \in P(k, d)$.

1. Perform the test behind Theorem 4, part a) on s and verify, whether s is sufficiently close to a trigonometric polynomial $\tilde{s} : F_{q_s}^{k_s} \to \mathbb{R}$ of max-degree $\tilde{d}_s := 2hk_s q$ using part π_2 of the certificate. If the test fails reject.

Perform $O(1/\epsilon)$ many rounds of the following steps 2–8:

2. Randomly generate two points $x^{(1)}, x^{(2)} \in F_q^k$.
3. Compute deterministically $(x^{(1)}, x^{(2)}, \ldots, x^{(k)}) := tps(x^{(1)}, x^{(2)}) \in (F_q^k)^k$.
4. Read the coefficients of $b_{(x^{(1)}, x^{(2)})}$ from part π_3 of the certificate.
5. Generate random $j, m \in F_{q_s}$, verify $s(j, m, x^{(1)}, x^{(2)}) = \tilde{s}(j, m, x^{(1)}, x^{(2)})$ using the correctness test for s from Theorem 4, part b).
6. Check whether $s(j, m, x^{(1)}, x^{(2)}) = b_{(x^{(1)}, x^{(2)})}(j, m)$; reject if not.
7. Check consistency: if a single f_i in one or several rounds has to be evaluated in two points lying on the same paraxial line check that the resulting restrictions are the same.
8. Check the equalities from the closeness test. Here, $f(x^{(1)})$ is read from the given table for f, whereas the values of an f_i in $x^{(i)}$ or $x^{(i+1)}$, respectively, are computed deterministically: Compute all coefficients from $b_{(x^{(1)}, x^{(2)})}$, evaluate it for the relevant values of j, m and then compute the value of the coded univariate algebraic polynomial. Reject, if at least one of the equations does not hold.

If at least one round leads to a reject, the verifier rejects, otherwise it accepts.

A verifier performing this revised closeness test fulfills the statement of Theorem 2. More precisely, we get

Theorem 2 (reformulated). *Let $f : F_q^k \to \mathbb{R}$. For $\epsilon > 0$ let a verifier V perform $O(1/\epsilon)$ rounds of the above revised closeness test. Then V is ϵ-reliable concerning the question whether f is an algebraic polynomial of max-degree d. The running time of V is polynomial in kq, it generates $O(k \log q)$ many random bits, uses a certificate of size $O((kq)^{O(k)})$, and reads $O(1)$ segments of maximal length $O((qk)^{O(1)})$.*

Proof. Let us first inspect the resources needed by V. The lengths of certificate π and of segments can be estimated as follows: The table of values for f has size q^k. The test behind Theorem 4, part a) requires the table of values for s, which has size $O(q_s^{k_s})$. Lemma 3 shows that it suffices to require that the trigonometric polynomial represented by the table has max-degree at most q; together with the conditions $q_s \geq 10^4 (2hk_s q + 1)^3$ and $k_s = 2 + 2k$ (and $k_s \geq \frac{3}{2}(2h + 1)$) it follows that the size of the table for s is in $O((kq)^{3k})$. Performing the test for closeness of s to a trigonometric polynomial $\tilde{s}$ uses a certificate π_2 consisting of $q_s^{k_s} = O((kq)^{3k})$ many segments; each segment has length $O(k_s q + k_s) = O(kq)$. Next, for all q^{2k} many pairs $(x^{(1)}, x^{(2)}) \in F_q^{2k}$ part π_3 of the certificate contains the coefficients of the bivariate trigonometric polynomials $b_{(x^{(1)}, x^{(2)})}$. By Lemma 3 their max-degrees are bounded by $O(q)$, so $O(q^2)$ coefficients are sufficient. Thus, π_3 has total length $O(q^{2k+2})$, split into q^{2k} segments of size $O(q^2)$. Finally, in each round the test in Step 5 performs a correction relying on Theorem 4, part b). It requires a certificate π_4 of length $O(\sqrt{q_s}^{k_s} k_s q) = O((kq)^{2k})$, split into segments of length $2\sqrt{q_s} k_s q + 1 = O((kq)^{2.5})$. Altogether, the certificate π has size $O((kq)^{O(k)})$ and the segments have maximal size $O((kq)^{2.5})$. Given the way the restrictions of the f_i's are coded together with Theorem 4 the verifier reads $O(1)$ many segments.

Next, we determine the number of random bits: Step 1 of the test needs $O(k_s \log q_s) = O(k \log q)$ random bits; Step 2 requires as well $O(k \log q)$ random bits in each round. Step 5 generates in each round two elements from F_{q_s} using $O(\log q_s) = O(\log(kq))$ random bits. The subsequent correctness test uses $O(k_s \log q_s) = O(k \log(kq))$ random bits. In Sect. 3.1 we assumed $q \geq 18kd^3$, so the number of random bits can be bounded by $O(k \log q)$.

The running time for Step 1 is $poly(k_s q_s) = poly(k \log q)$, again using that $q_s = \Theta(k^3 q^3)$ and $q \geq 18kd^3$, so $\log q_s = O(\log q)$. The computation of the sequence $tps(x^{(1)}, x^{(2)})$ runs in polynomial time in k. Each $b_{(x^{(1)}, x^{(2)})}$ has $O(q^2)$ many coefficients, so this is also the time to read all of them. The correctness test for s according to Theorem 4, part b) can be performed in $poly(k_s \sqrt{q_s} d) = poly(kq)$ steps. For Step 6, $b_{(x^{(1)}, x^{(2)})}$ has to be evaluated in a (j, m) which needs time $O(q^2)$. Checking validity of the k equations constituting the closeness test then can be done in time $poly(kd)$. The same holds for recording the points in which the restriction of an f_i is evaluated and for determining, whether different such points lie on the same paraxial line. Altogether, verifier V runs in time $poly(kq)$.

Finally, we analyze the failure probability. Clearly, if f is a max-degree d algebraic polynomial and if the information provided by certificate π is correct, then V accepts with probability 1. Suppose then that f is not ϵ-close to an algebraic max-degree d polynomial. As usual in this area, the arguments below suffice to conclude that an independent repetition of some of the tests performed constantly many times yields the required probability bounds. The error sources in the test are the following: either s is not sufficiently close to a trigonometric polynomial of suitable max-degree; or s is close but for the values in which s has to be evaluated the result is not that of the closest trigonometric polynomial $\tilde{s}$; or the restriction of this polynomial to one of the planes E occurring in the revised closeness test does not equal the bivariate polynomial b; or the univariate algebraic polynomials coded by b do not cause f to pass the original closeness test. We now argue that the verifier detects if one of these cases holds with arbitrarily high (constant) probability. Step 1 of the test rejects at least with a positive constant probability $\delta > 0$, if s is not 2δ-close to a trigonometric polynomial $\tilde{s}$ of a corresponding max-degree; thus, $O(1/\delta)$ repetitions suffice to raise this probability to an arbitrary constant close to 1. Next, the correctness test behind Step 5 detects a difference in the values of s and $\tilde{s}$ in the chosen point with probability at least $\frac{3}{4}$. Step 6 verifies whether the restriction s' of $\tilde{s}$ to the plane $E_{(x^{(1)}, x^{(2)})}$ equals $b_{(x^{(1)}, x^{(2)})}$. Here, b has max-degree $O(q)$ and s' has arity $k' := 2$ and max-degree $d' := 2hk_s q$. Note that even though the low-degree test for s gives closeness to a polynomial of higher max-degree $\tilde{s}$, the verifier expects the degree to be the one of a correct s. Therefore, b is taken to have max-degree $O(q)$ only. The potentially larger degree of $\tilde{s}$ enters the error analysis only with respect to the domain $F_{q_s}^2$ from which the test randomly chooses the point (j, m) in which b and s' are evaluated. Using Lemma 3 applied to $s' - b$

(with k', d' as above and $q' := q_s$) a difference between s' and b is realized by the verifier with probability at least $1 - \frac{12hk_s q q_s}{q_s^2}$. Since $q_s \geq 10^4(2hk_s q + 1)^3$ this probability is $\geq 1 - 10^{-3}$. If all tests so far have passed without rejection, then with arbitrarily high constant probability the data encoded by the different b's corresponds to suitable algebraic univariate polynomials of degree d. Now, according to Theorem 3, Step 8 rejects with probability $> \frac{1}{2}$ if f is not ϵ-close to an algebraic polynomial. $\qquad\Box$

3.3   Finishing the Proof

The final step necessary for improving the statement of Theorem 2 to that of Theorem 1 relies on the well known technique of (segmented) verifier composition and the way it has been used in the real number framework in [5].

Proof (of Theorem 1). We apply the well known technique of verifier composition developed in [2] and used also in the BSS framework in [5]. Therefore, we only briefly outline how to apply the technique in the present situation. Consider the verifier V of Theorem 2. Whenever it reads a segment of length $O((kq)^{O(1)})$ it subsequently computes deterministically from the data read in polynomial time in the size of the segment the corresponding answer to one of the questions occuring in the revised closeness test. This question can be expressed as $P_{\mathbb{R}}$-question by formalizing the verifier's computation on the information read in the currently queried segment. For example, this can be done by describing the computation via an instance of the $NP_{\mathbb{R}}$-complete problem of deciding solvability of a system of quadratic polynomial equations over the real numbers; then, the segment read provides a solution assignment to (part of) the variables of that system; see [7] for this basic construction behind the corresponding $NP_{\mathbb{R}}$-completeness proof. Reformulating things that way this part of the verification can be replaced by what is called an inner verifier: Instead of reading the entire segment (this would not reduce the number of proof components required to be seen) the inner verifier performs a verification for the resulting instance of the above mentioned $NP_{\mathbb{R}}$-problem. The advantage is that the input size is reduced, and so is the number of positions of a new corresponding certificate read by the inner verifier. Composition of verifiers means that a new composed verifier uses the old (outer) verifier to determine the segments that should be asked, but replaces the deterministic polynomial time procedure that uses the entire segment to compute the answer to the corresponding query by another inner verification procedure. In order to make this ongoing working the composed verifier in addition has to perform certain consistency checks. These checks guarantee that a prover gives consistent answers to the inner verifier if the outer verifier asks in several queries segments that contain overlapping data. This is only a sketchy outlook. In the above situation, the outer verifier behind Theorem 2 has to be composed several times with inner verifiers. More precisely, one can compose it twice with the verifier designed in [5] that codes satisfying assignments of polynomial systems using low max-degree trigonometric polynomials and after that compose the resulting intermediate verifier with the long transparent verifier for $NP_{\mathbb{R}}$, see also [5]. Since the way how those verifiers work is described in full detail in the cited papers, this should suffice as short outline. That way, finally a verifier is obtained that has $O(k \log q)$ randomness, query complexity $O(1)$ and is ϵ-reliable, i.e., it has the properties stated in Theorem 1.   □

4   Conclusions

In [11] an ϵ-reliable property test for algebraic k-variate polynomials of given max-degree defined on a finite subdomain of suitable size q of the reals was designed. The test uses $O(k \log q)$ random bits and queries $O(k)$ positions in a table for f (considering ϵ as a constant). In this paper we constructed a real number PCP of proximity for this problem using as well $O(k \log q)$ random bits, but only $O(1)$ oracle calls to both the table of f and an additional proof certificate. Here are some subsequent questions, some of which are inspired by the helpful comments of the reviewers: The parameters used in our algorithm likely are far from being optimal. This holds both for the necessary domain size on which f is defined, for the size of the additional certificate, for the number of queries, and for the running time. Can we significantly reduce the constants hidden behind the respective O-statements? Next, it still seems puzzling that the algorithm for algebraic polynomials relies on the use of trigonometric polynomials as coding objects. This clearly makes the approach complicated and technically involved. Are there easier verification algorithms not needing the detour along trigonometric polynomials? In the classical PCP literature, the important initial role played by segmented verifiers has been subsumed by that of so-called robust verifiers, see [6] and also [13]. Since the main technical problem to overcome is segmentation, it is of course interesting to ask whether robust verifiers in the real-number framework would lead to easier proofs as well, including a possibility to avoid trigonometric polynomials. Another impact of classical robust verifiers is to obtain property testers instead of PCPs of proximity. Would this be possible both for algebraic and trigonometric polynomials? We do not have an educated guess at the moment.[6] Next, what about studying algebraic polynomials with a more 'continuous' closeness measure like the L_1-norm, and then allowing small differences between given values and those of a best-approximating polynomial? Finally, there are of course numerous problems in the real number framework where one could ask for the existence of either PCPs of proximity or property testers, for example for algebraic polynomials (as said above), but also for many other real-number problems. Together with the previous question this is also related to the not yet studied question of software testing in the BSS model: Here one could ask for checking in a randomized way whether a given program approximately computes a predetermined function or a function from a given class with not too many errors.

---

[6] Thanks to an anonymous referee for very helpful comments in this respect.

Acknowledgement. Thanks are due to M. Baartse for helpful discussions in the initial phase leading to this work, and to the anonymous referees for their extremely careful reading and the many informative hints to further literature and interesting future questions.

References

1. Arora, S., Lund, C., Motwani, R., Sudan, M., Szegedy, M.: Proof verification and hardness of approximation problems. J. ACM **45**(3), 501–555 (1998)
2. Arora, S., Safra, S.: Probabilistic checking proofs: a new characterization of NP. J. ACM **45**(1), 70–122 (1998)
3. Ausiello, G., Crescenzi, P., Gambosi, G., Kann, V., Marchetti-Spaccamela, A., Protasi, M.: Complexity and Approximation: Combinatorial Optimization Problems and Their Approximability Properties. Springer, Heidelberg (1999)
4. Baartse, M., Meer, K.: The PCP theorem for NP over the reals. Found. Comput. Math. **15**(3), 651–680 (2015)
5. Baartse, M., Meer, K.: An algebraic proof of the real number PCP theorem. J. Complex. **40**, 34–77 (2017)
6. Ben-Sasson, E., Goldreich, O., Harsha, P., Sudan, M., Vadhan, S.P.: Robust PCPs of proximity, shorter PCPs, and applications to coding. SIAM J. Comput. **36**(4), 889–974 (2006)
7. Blum, L., Cucker, F., Shub, M., Smale, S.: Complexity and Real Computation. Springer, Heidelberg (1998)
8. Bromwich, T.J.I.A.: An Introduction to the Theory of Infinite Series. Macmillan, London (1955)
9. Chor, B., Goldreich, O.: On the power of two-point based sampling. J. Complex. **5**, 96–106 (1989)
10. Dinur, I.: The PCP theorem by gap amplification. J. ACM **54** (3) (2007)
11. Friedl, K., Hátsági, Z., Shen, A.: Low-degree tests. In: Proceedings of SODA, pp. 57–64 (1994)
12. Goldreich, O.: Introduction into Property Testing. Cambridge Univ, Press (2017)
13. Harsha, P.: Robust PCPs of Proximity and Shorter PCPs. Massachusetts Institute of Technology, Cambridge, MA, USA (2004). http://www.tcs.tifr.res.in/ prahladh/papers/thesis/
14. Mastroianni, G., Milovanic, G.: Interpolation processes. Springer Monographs in Mathematics (2008)
15. Meer, K.: Almost Transparent Short Proofs for NP. In: Owe, O., Steffen, M., Telle, J.A. (eds.) FCT 2011. LNCS, vol. 6914, pp. 41–52. Springer, Heidelberg (2011). https://doi.org/10.1007/978-3-642-22953-4_4
16. Montaña, J.L., Pardo, L.M. (eds.): Recent Advances in Real Complexity and Computation. AMS Contemporary Mathematics, vol. 604 (2013)

Predictions and Algorithmic Statistics
for Infinite Sequences

Alexey Milovanov[1,2]([⊠]) [iD]

[1] HSE University, Moscow, Russian Federation
[2] Moscow Institute of Physics and Technology, Moscow, Russian Federation
amilovanov@hse.ru
https://solid-lelik.jimdofree.com/

Abstract. We combine Solomonoff's approach to universal prediction with algorithmic statistics and suggest to use the computable measure that provides the best "explanation" for the observed data (in the sense of algorithmic statistics) for prediction. In this way we keep the expected sum of squares of prediction errors bounded (as it was for the Solomonoff's predictor) and, moreover, guarantee that the sum of squares of prediction errors is bounded along any Martin-Löf random sequence.

Keywords: Kolmogorov complexity · Prediciton · Algorithmic statistics

1 Introduction

We consider probability distributions (or measures) on the binary tree, i.e., non-negative functions $P : \{0,1\}^* \to \mathbb{R}$ such that $P(\text{empty word}) = 1$ and $P(x0) + P(x1) = P(x)$ for every string x. We assume that all the values $P(x)$ are rational; P is called *computable* if there exists an algorithm that on input x outputs $P(x)$.

Consider the following prediction problem. Imagine a black box that generates bits according to some unknown computable distribution P on the binary tree. Let $x = x_1 \ldots x_n$ be the current output of the black box. The predictor's goal is to guess the probability that the next bit is 1, i.e., the ratio $P(1|x) = P(x1)/P(x)$.

Ray Solomonoff suggested to use the universal semi-measure M (called also the *a priori probability*) for prediction. Recall that a semi-measure S on the binary tree (a *continuous semi-measure*) is a non-negative function $S : \{0,1\}^* \to \mathbb{R}$ such that $S(\text{empty word}) \leq 1$ and $S(x0) + S(x1) \leq S(x)$ for every string x. Semi-measures correspond to probabilistic processes that output a bit sequence but can hang forever, so an output may be some finite string x; the probability of this event is $S(x) - S(x0) - S(x1)$. A semi-measure S is called *lower semi-computable*, or *enumerable*, if the set $\{(x,r) : r < S(x)\}$ is (computably) enumerable. Here x is a string and r is a rational number. Finally, a lower semi-computable semi-measure M is called *universal* if it is maximal among all semimeasures up to a constant factor, i.e., if for every lower semi-computable

R. Santhanam and D. Musatov (Eds.): CSR 2021, LNCS 12730, pp. 283–295, 2021.
https://doi.org/10.1007/978-3-030-79416-3_17

semi-measure S there exists $c > 0$ such that $\mathrm{M}(x) \geq cS(x)$ for all x. Such a universal semi-measure exists [6,8,9].[1]

Solomonoff suggested to use the ratio $\mathrm{M}(1|x) := \mathrm{M}(x1)/\mathrm{M}(x)$ to predict $P(1|x)$ for an unknown computable measure P. He proved the following bound for the prediction errors.

Theorem 1 ([10]). *For every computable distribution P and for every $b \in \{0,1\}$ the following sum over all binary strings is finite:*

$$\sum_{x} P(x) \cdot (P(b|x) - \mathrm{M}(b|x))^2 < \infty. \tag{1}$$

Moreover, this sum is bounded by $O(\mathrm{K}(P))$, where $\mathrm{K}(P)$ is the prefix complexity of the computable measure P (the minimal length of a prefix-free program corresponding P).

Note that for semimeasure the probabilities to predict 0 and 1 do not sum up to 1, so the statements for $b = 0$ and $b = 1$ are not equivalent (but both are true).

The sum from Theorem 1 can be rewritten as the expected value of the function D on the infinite binary sequences with respect to P, where $D(\omega)$ is defined as

$$D(\omega) = \sum_{x \text{ is a prefix of } \omega} (P(b|x) - \mathrm{M}(b|x))^2.$$

This expectation is finite, therefore for P-almost all ω the value $D(\omega)$ is finite and

$$P(b|x) - \mathrm{M}(b|x) \to 0.$$

when x is an increasing prefix of ω. One would like to have this convergence for all Martin-Löf random sequences ω (with respect to measure P), but this is not guaranteed, since the null set provided by the argument above may not be an *effectively* null set. An example from [5] shows that this is indeed the case.

Theorem 2 ([5]). *There exist a specific universal semi-measure M, computable distribution P and Martin-Löf random (with respect to P) sequence ω such that*

$$P(b|x) - \mathrm{M}(b|x) \not\to 0.$$

for increasing prefixes x of ω.

Lattimore and Hutter generalized Theorem 2 by proving the same statement for a wide class of universal semi-measures [7].

Trying to overcome this problem and get a good prediction for all Martin-Löf random sequences, we suggest the following approach to prediction. For a finite string x we find a distribution Q on the binary tree that is the best (in some sense) explanation for x. The probabilities of the next bits are then predicted as $Q(0|x)$ and $Q(1|x)$.

---

[1] One may even require that the probabilities for finite outputs, i.e., the differences $S(x) - S(x0) - S(x1)$ are maximal, but we do not require this.

This approach combines two advantages. The first is that the series of type (1) also converges, though the upper bound for it (at least the one that we are able to prove) is much greater than $O(\mathrm{K}(P))$. The second property is that the prediction error (defined as in Theorem 2) converges to zero for every Martin-Löf random sequence.

Let us give formal definitions. The quality of the computable distribution Q on the binary tree, considered as an "explanation" for a given string x, is measured by the value $3\mathrm{K}(Q) - \log Q(x)$: the smaller this quantity is, the better is the explanation. One can rewrite this exression as the sum

$$2\mathrm{K}(Q) + [\mathrm{K}(Q) - \log Q(x)].$$

Here the expression in the square brackets can be interpreted as the length of the two-part description of x using Q (first, we specify the hypothesis Q using its shortest prefix-free program, and then, knowing Q, we specify x using arithmetic coding; the second part requires about $-\log Q(x)$ bits). The first term $2\mathrm{K}(Q)$ is added to give extra preference to simple hypotheses; the factor 2 is needed for technical reasons (in fact, any constant greater than 1 will work).

For a given x we select the best explanation (that makes this quality minimal). Then we predict the probability that the next bit after x is b:

$$H(b|x) := \frac{Q(xb)}{Q(x)},$$

where Q is the best explanation for string x (or one of the best explanations if there are several).

In this paper we prove the following results:

Theorem 3. *For every computable distribution P the following sum over all binary strings x is finite:*

$$\sum_x P(x)(P(0|x) - H(0|x))^2 < \infty.$$

Theorem 4. *Let P be a computable measure and let ω be a Martin-Löf random sequence with respect to P. Then*

$$H(0|x) - P(0|x) \to 0$$

for prefixes x of ω as the length of prefix goes to infinity.

We speak about the probabilities of zeros, but both P and Q are measures, so this implies the same results for the probabilities of ones.

We prove that

$$\sum_{x \text{ is a prefix of } \omega} (H(0|x) - P(0|x))^2 < \infty$$

(Theorem 7) that is the strengthening of Theorem 4.

In [3] Hutter suggested a similar approach but without coefficient 3 for $K(Q)$ (see also [2,4]). For this approach he proved an analogue of Theorem 3 with different proof technique.

In [5] the existence of a semi-computable measure satisfying Theorem 4 was proved. However it is unknown (to the best of our knowledge) that this measure also satisfied Theorem 3.

In the next section we prove Theorem 4.

In Sect. 3 we prove Theorem 3.

Finally, in Sect. 4 we consider the case when we know some information about P. More precisely, we know that P belongs to some enumerable set of computable measures. We suggest a similar approach for prediction in this case. We prove analogues of Theorems 4 and 3 (Theorems 8 and 9) for this prediction method. We achieved better (polynomial in complexity of P) error estimations in these theorems.

2   Prediction on Martin-Löf Random Sequences

Recall the Schnorr–Levin theorem [8, ch.5] that says that a sequence ω is random with respect to a computable probability measure P if and only if the ratio $M(x)/P(x)$ is bounded for x that are prefixes of ω.

The same result can be reformulated in the logarithmic scale. Let us denote by $KA(x)$ the *a priori complexity of* x, i.e., $\lceil -\log M(x)\rceil$ (the rounding is chosen in this way to ensure upper semicomputability of KA). We have

$$KA(x) \leq -\log P(x) + O(1)$$

for every computable probability measure P, where $O(1)$ depends on P but not on x. Indeed, since M is maximal, the ratio $P(x)/M(x)$ is bounded. Moreover, since $P(x)$ can be included in the mix for $M(x)$ with coefficient $2^{-K(P)}$, we have

$$KA(x) \leq -\log P(x) + K(P) + O(1)$$

with some constant in $O(1)$ that does not depend on P (and on x). As we have discussed in the previous section, the right-hand side includes the length of the two-part description of x.

Let us call

$$d(x|P) := -\log P(x) - KA(x)$$

the *randomness deficiency* of a string x with respect to a computable measure P. (There are several notions of deficiency, but we need only this one.). Then we get

$$d(x|P) \geq -K(P) - O(1)$$

so the deficiency is almost non-negative. The Schnorr–Levin theorem characterizes Martin-Löf randomness in terms of deficiency:

Theorem 5 (Schnorr–Levin).

(a) *If a sequence ω is Martin-Löf random with respect to a computable disctibution P, then $d(x|P)$ is bounded for all prefixes x of ω.*

(b) *Otherwise (if ω is not random with respect to P), then $d(x|P) \to \infty$ as the length of a prefix x of ω increases.*

Note that there is a dichotomy: the values $d_P(x)$ for prefixes x of ω either are bounded or converge to infinity (as the length of x goes to infinity). We can define randomness deficiency for infinite sequence ω as

$$d(\omega|P) := \sup_{x \text{ is prefix of } \omega} d(x|P);$$

it is finite if and only if ω is random with respect to P.

Let us also recall the following result of Vovk:

Theorem 6 ([12]). *Let P and Q be two computable distributions. Let ω be a Martin-Löf random sequence with respect both to P and Q. Then*

$$P(0|x) - Q(0|x) \to 0$$

for prefixes x of ω as the length of prefix goes to infinity.

We will prove this theorem (and even more exact statement) in the next section.

Proof (Proof of Theorem 4)

Now we have a sequence ω that is Martin-Löf random with respect to some computable measure P, so $D = d(\omega|P)$ is finite. For each prefix x of ω we take the best explanation Q that makes the expression

$$3\mathrm{K}(Q) - \log Q(x)$$

minimal. Note that P is among the candidates for Q, so this expression should not exceed

$$3\mathrm{K}(P) - \log P(x).$$

Since ω is random with respect to P and x is a prefix of ω, Schnorr–Levin theorem guarantees that the latter expression

$$3\mathrm{K}(P) - \log P(x) = \mathrm{KA}(x) + O_P(1)$$

where constant in O_P depends on P but not on x. On the other hand, the inequality $\mathrm{KA}(x) \leq \mathrm{K}(Q) - \log Q(x) + O(1)$ implies that

$$3\mathrm{K}(Q) - \log Q(x) = 2\mathrm{K}(Q) + \mathrm{K}(Q) - \log Q(x) \geq 2\mathrm{K}(Q) + \mathrm{KA}(x) - O(1). \quad (*)$$

So measures Q with large $\mathrm{K}(Q)$ cannot compete with P, and there is only a finite list of candidate measures for the best explanation Q. For some of these Q

the sequence ω is Q-random with respect to Q, so one can use Vovk's theorem to get the convergence of predicted probabilities when these measures are used.

Still we may have some "bad" Q in the list of candidates for which ω is not Q-random. However, the Schnorr–Levin theorem guarantees that for a bad Q we have

$$-\log Q(x) - \mathrm{KA}(x) \to \infty$$

if x is a prefix of ω of increasing length. So the difference between two sides of $(*)$ goes to infinity as the length of x increases, so Q loses to P for large enough x (is worse as an explanation of x). Therefore, only good Q will be used for prediction after sufficiently long prefixes, and this finishes the proof of Theorem 4.

3   On the Expectation of Squares of Errors

In this section we prove Theorem 3. First we will prove some strengthening of Theorem 6

Lemma 1. *Let P and Q be computable distributions. and let M be a universal semi-measure. Assume that for string $x = x_1 \ldots x_n$ and $C > 0$ it holds that $P(x), Q(x) \geq \mathrm{M}(x)/C$. Then:*

$$\sum_{i=1}^{n-1} (P(x_i|x_1 \ldots x_{i-1}) - Q(x_i|x_1 \ldots x_{i-1}))^2 = O(\log C + \mathrm{K}(P,Q)).$$

Proof (Proof of Theorem 6 from Lemma 1). According to one of definitions of Martin-Löf randomness the values $\mathrm{M}(x)/P(x)$ and $\mathrm{M}(x)/Q(x)$ are bounded by a constant. It reminds to use Lemma 1.

Proof (Proof of Lemma 1). Denote

$$p_i = P(x_i|x_1 \ldots x_{i-1}), \quad q_i = Q(x_i|x_1 \ldots x_{i-1}).$$

Note that

$$P(x_1 \ldots x_n) = p_1 p_2 \ldots p_n, \quad Q(x_1 \ldots x_n) = q_1 q_2 \ldots q_n.$$

Now consider the "intermediate" measure R for which the probability of 0 (or 1) after some x is the average of the same conditional probabilites for P and Q:

$$R(0|x_1 \ldots x_{i-1}) = \frac{P(0|x_1 \ldots x_{i-1}) + Q(0|x_1 \ldots x_{i-1})}{2}.$$

The corresponding $r_i = R(x_i|x_1 \ldots x_{i-1})$ are equal to $(p_i + q_i)/2$.

Probability distribution R is computable and $\mathrm{K}(R) \leq \mathrm{K}(P,Q) + O(1)$. Hence, it holds that $R(x) \leq 2^{\mathrm{K}(P,Q)} \mathrm{M}(x) \leq 2^{\mathrm{K}(P,Q)} \cdot C \cdot P(x)$. The similar inequality holds for distribution Q. Therefore:

$$r_1 \cdots r_n \leq C \cdot 2^{\mathrm{K}(P,Q)} \cdot p_1 \cdots p_n$$

and
$$r_1 \cdots r_n \leq C \cdot 2^{K(P,Q)} \cdot q_1 \cdots q_n.$$

Multiplying we obtain:
$$\left(\frac{p_1 + q_1}{2} \cdots \frac{p_n + q_n}{2}\right)^2 \leq C^2 \cdot 2^{2K(P,Q)} \cdot p_1 \cdots p_n \cdot q_1 \cdots q_n.$$

These two inequalities show that the product of arithmetical means of p_i and q_i is not much bigger than the product of their geometrical means, and this is only possible if p_i is close to q_i (logarithm is a strictly convex function).

To make the argument precise, recall the bound for the logarithm function:

Lemma 2. *For $p, q \in (0, 1]$ we have*

$$\log \frac{p+q}{2} - \frac{\log p + \log q}{2} \geq \frac{1}{8 \ln 2}(p - q)^2$$

Proof. Let us replace the binary logarithms by the natural ones; then the factor $\ln 2$ disappears. Note that the left hand side remains the same if p and q are multiplied by some factor $c \geq 1$ while the right side can only increase. So it is enough to prove this for $p = 1 - h$ and $q = 1 + h$ for some $h \in (0, 1)$, and this gives

$$-\frac{\ln(1 - h) + \ln(1 + h)}{2} \geq \frac{1}{2}h^2;$$

and this happens because $\ln(1 - h) + \ln(1 + h) = \ln(1 - h^2) \leq -h^2$.

For the product of n terms we get the following bound:

Lemma 3. *If for $p_1, \ldots, p_n, q_1, \ldots, q_n \in (0, 1]$ we have*

$$\left(\frac{p_1 + q_1}{2} \cdot \ldots \cdot \frac{p_n + q_n}{2}\right)^2 \leq c p_1 \ldots p_n q_1 \ldots q_n,$$

then $\sum_i (p_i - q_i)^2 \leq O(\log c)$, with some absolute constant hidden in $O(\cdot)$-notation.

Proof. Taking logarithms, we get

$$2 \sum_i \log \frac{p_i + q_i}{2} \leq \log c + \sum_i \log p_i + \sum_i \log q_i,$$

and therefore

$$\sum_i \left(\log \frac{p_i + q_i}{2} - \frac{\log p_i + \log q_i}{2}\right) \leq \frac{1}{2} \log c.$$

It remains to use Lemma 2 to get the desired inequality.

To complete the proof of Lemma 1 it remains to take $c := C^2 \cdot 2^{2K(P,Q)}$ in Lemma 3.

Now we prove a strengthening of Theorem 4.

Theorem 7. *Let P be a computable measure, let ω be a Martin-Löf random sequence with respect to P such that $d(\omega|P) = D$.*
   Then

$$\sum_{x \text{ is a prefix of } \omega} (H(0|x) - P(0|x))^2 = O((K(P) + D) \cdot 2^{\frac{3K(P)+D+O(1)}{2}}).$$

Proof. Assume that distribution Q is the best for some $x = x_1 \ldots x_n$. Then

$$3K(Q) - \log Q(x) \le 3K(P) - \log P(x). \tag{2}$$

Since $d(\omega|P) = D$ we obtain that

$$-\log P(x) \le KA(x) + D. \tag{3}$$

Therefore,

$$-\log Q(x) \le 3K(P) - \log P(x) \le 3K(P) + KA(x) + D \text{ , so}$$

$$Q(x) \ge M(x) \cdot 2^{-3K(P)-D} \text{ and}$$

$$P(x) \ge M(x) \cdot 2^{-D}.$$

We want to estimate $\sum_{i=1}^{n} (Q(0|x_1 \ldots x_i) - P(0|x_1 \ldots x_i))^2$ by Lemma 1. We can use this lemma for $C = 2^{3K(P)+D}$.

From (2) and (3) it follows that

$$K(Q) \le \frac{3K(P) + D + O(1)}{2}. \tag{4}$$

Therefore by Lemma 1 we obtain

$$\sum_{i=1}^{n-1} (Q(0|x_1 \ldots x_i) - P(0|x_1 \ldots x_i))^2 = O(K(P) + D).$$

In fact, we can not use this lemma for the last term $(Q(0|x) - P(0|x))^2$. This term we just bound by 1 1.

So, every probability distribution that is the best for some x "contributes" $O(K(P) + D)$ in the sum $\sum_{x \text{ is a prefix of } \omega} (H(0|x) - P(0|x))^2$. There are at most $2^{\frac{3K(P)+D+O(1)}{2}}$ such distribution (by (4)), so we obtain the required estimation.

Recall the following well-known statement

Proposition 1. *Let P be a computable distribution. Then the P-measure of all sequences x such that $d(\omega|P) \ge D$ is not greater than 2^{-D}.*

Proof (Proof of Theorem 3). Denote by Ω the set of all infinite sequences with zeros and ones. Note that

$$\sum_x P(x)(P(0|x) - H(0|x))^2 = \int_{(\Omega,P)} \sum_{x \text{ is a prefix of} \omega} (H(0|x) - P(0|x))^2.$$

By Theorem 7 we can estimate the sum in the integral for sequence ω with $d(x|\omega) = D$ as $O((\mathrm{K}(P) + D) \cdot 2^{\frac{3\mathrm{K}(P)+D+O(1)}{2}})$. By Proposition 1 the measure of sequences with such randomness deficiency is at most 2^{-D}. So we can estimate the integral as

$$\sum_{D=0}^{\infty} O((\mathrm{K}(P) + D) \cdot 2^{\frac{3\mathrm{K}(P)+D+O(1)}{2}})2^{-D} = O(\mathrm{K}(P)2^{\frac{3\mathrm{K}(P)}{2}}).$$

(Recall that the P-measure of sequences that are not Martin-Löf random with respect to P is equal to 0, so they do not affect to the integral.)

4   Prediction for Enumerable Classes of Hypotheses

Assume that we have some information about distribution P. We know that P belongs to some enumerable set $\mathcal{A}$ of computable distributions, (i.e. there is an algorithm that enumerate programs that generate distributions from $\mathcal{A}$). For this case it is natural to consider the following measure of complexity measures in $\mathcal{A}$:

$\mathrm{K}_{\mathcal{A}}(P) := \mathrm{K}(i_P)$, where i_p is the number of P in a computable enumeration of $\mathcal{A}$.

If P has several numbers in an enumeration we choose i_P with the smallest complexity. (This definition does depend on the choice of a computable enumeration but this dependence is bounded by some additive constant.) Clearly, $\mathrm{K}_{\mathcal{A}}(P) \geq \mathrm{K}(P) + O(1)$.

Now we can generalize our prediction method: for prediction of the next bit of x we select $Q \in \mathcal{A}$ with the smallest value of $3\mathrm{K}_{\mathcal{A}}(Q) - \log Q(x)$ and predict the next bit according to Q:

$$H_{\mathcal{A}}(x) := \frac{Q(xb)}{Q(x)}.$$

In this section we show that if set $\mathcal{A}$ has some nice properties than some analogues of previous theorems hold. Even more—we can get a better error estimation. We assume that enumerable set $\mathcal{A}$ has the following property: if $P_1, \ldots, P_k \in \mathcal{A}$ then their mixture $\frac{P_1+\ldots P_k}{k}$ belongs to $\mathcal{A}$. Moreover there exists an algorithm that for given numbers of $P_1, \ldots, P_k$ outputs the number of their mixture.

(Further everywhere $\mathcal{A}$ is an enumerable set of computable distributions with this property)

Remark 1. Consider the following example of set $\mathcal{A}$: the set of all *provable* (in some proof system) computable distributions on the binary tree: so, for every program $p \in \mathcal{A}$ there exists a proof that $p(x)$ halts for every x, $p(x) = p(x0) + p(x1)$ and $p(\text{empty word}) = 1$. We guess that all using in practice computable distributions are provable computable, so, in some sense we get better error estimation "almost free". Our discussion about practice might look unsuitable because our prediction method is not computable. However, it can be considered as limit best prediction based on (really used) MDL-principle.

Theorem 8. *Let $P \in \mathcal{A}$ be a computable measure, let ω be a Martin-Löf random sequence with respect to P such that $d(\omega|P) = D$.*
    Then

$$\sum_{x \text{ is a prefix of } \omega} (H_{\mathcal{A}}(0|x) - P(0|x))^2 = O((\mathrm{K}_{\mathcal{A}}(P) + D) \cdot \mathrm{poly}(\mathrm{K}_{\mathcal{A}}(P) + D).$$

Theorem 9. *For every computable distribution $P \in \mathcal{A}$ the following sum over all binary strings x is finite:*

$$\sum_x P(x)(P(0|x) - H(0|x))^2 < \mathrm{poly}(\mathrm{K}_{\mathcal{A}}(P)).$$

The proofs of these theorems is in general the same as the proofs of Theorems 7 and 3, however some new tools are added. The difference is that we can get better estimation on the number of possible best explanations for prefixes of some sequence.

Lemma 4. *Let x be a finite string. Assume that there are 2^k probabilities $Q_1, \ldots Q_{2^k} \in \mathcal{A}$ such that for every i it holds $\mathrm{K}_{\mathcal{A}}(Q_i) \leq a$ and $Q_i(x) \geq 2^{-b}$. Then there is probability distribution $Q \in \mathcal{A}$ such that*
    $\mathrm{K}_{\mathcal{A}}(Q) \leq a - k + O(\log a + k)$ *and* $Q(x) \geq 2^{-b-k}$.

Note that $3\mathrm{K}_{\mathcal{A}}(Q) - \log Q(x) \leq 3 \cdot (a - k + O(\log a + k) + b + k) \leq 3 \cdot a - b$ for big enough k. This means that string x can not has many "best" explanations.

Proof (of Lemma 4). Let enumerate all distributions of $\mathcal{A}$ with complexity at most a by groups of size 2^{k-1} (the last group can be incomplete). The number of such groups is $O(2^{a-k})$. The complexity of every group is at most $a-k+O(\log a + k)$. Indeed, to describe a group we need its ordinal number in an enumeration and describe this enumeration (we need to know k, a and some enumeration of $\mathcal{A}$).

    One of these complete group contains some Q_i. Define Q as the mixture of the distributions in this group. Since the group has complexity at most $a - k + O(\log a + k)$ the same estimation holds for the complexity of Q. Since some Q_i belongs to the mixture it holds that $Q(x) \geq 2^{-b-k+1}$. Recall that Q belongs to $\mathcal{A}$ because every mixture of distributions from $\mathcal{A}$ belongs to $\mathcal{A}$.

Also we need the following lemma.

Lemma 5. *Let string s be a prefix of string h and let P be a computable distribution such that $d(s|P) = D$. Then $d(h|P) \geq D - 2\log D + O(1)$.*

(So, a prefix of a string that has small deficiency, has (almost as) small deficiency).

In fact the proof of this lemma is the same as the proof of Theorem 124 in [8].

Proof (Proof of Lemma 5). For each k consider the enumerable set of all finite sequences that have deficiency greater than k. All the infinite continuations of these sequences form an open set S_k, and P-measure of this set does not exceed 2^{-k}. Now consider the measure P_k on Ω that is zero outside S_k and is equal to $2^k P$ inside S_k. That means that for every set U the value $P_k(U)$ is defined as $2^k(U \cap P_k)$. Actually, P_k is not a probability distribution according to our definition, since $P_k(\omega)$ is not equal to 1. However, P_k can be considered as a lower semicomputable semimeasure, if we change it a bit and let $P_k(\omega) = 1$ (this means that the difference between 1 and the former value of $P_k(\omega)$ is assigned to the empty string).

Now consider the sum

$$S = \sum_k \frac{1}{2k^2} P_k$$

It is a lower semicomputable semimeasure (the factor 2 in the denominator is used to make the sum $\sum_k \frac{1}{2k^2}$ less than 1); again, we need to increase S so that $S(\Omega) = 1$. Then we have

$$-\log S(x) \leq -\log P(x) - k + 2\log k + O(1)$$

for every string x that has a prefix with deficiency greater than k. Since S does not exceed a priori probability (up to $O(1)$-factor), we get the desired inequality.

Proof (of Theorem 8)

Part 1. We claim that there are only $\text{poly}(D + \mathrm{K}_{\mathcal{A}}(P))$ different distributions that are the best for some prefix of ω.

Let x be a prefix of ω and Q is the best distribution for x. As in the proof of Theorem 7 we obtain

$$\mathrm{K}_{\mathcal{A}}(Q) \leq \frac{3\mathrm{K}_{\mathcal{A}}(P) + D + O(1)}{2}, \tag{5}$$

$$Q(x) \geq M(x) \cdot 2^{-3\mathrm{K}_{\mathcal{A}}(P) - D}$$

and hence

$$d(x|P) \leq 3\mathrm{K}_{\mathcal{A}}(P) + D. \tag{6}$$

Let $Q_1, \ldots, Q_m$ be different and the best distribution for prefixes $x_1, \ldots, x_m$ of ω.

We need to prove that $m = \text{poly}(D + \mathrm{K}_{\mathcal{A}}(P))$.

Fix some natural a and b. We can assume that $K(Q_i) = a$ and

$$b \leq d(z_i|Q_i) < 2b.$$

Indeed, if we prove that there are only $\text{poly}(D + K(P))$ best distributions with fixing complexity and randomness deficiency then the honest estimation of m will be multiplied by $\text{poly}(D + K(P))$ because of (5) and (6).

Let z_i be the shortest prefix among $z_1, \ldots z_m$.

By Lemma 5 every Q_j is "rather good" distribution for z: $d(z|Q_j) \leq b + O(\log b)$ and hence $Q_j(z) \geq Q_i(z) \cdot 2^{-O(\log b)}$. By Lemma 4 there exists a distribution from $R \in \mathcal{A}$ such that

$$K_{\mathcal{A}}(R) \leq a - \log m + O(\log a + \log m) \text{ and}$$

$$R(z) \geq Q_i(z) \cdot 2^{-\log m - O(\log b)}.$$

Since Q_i is not worse distribution then R for z we have:

$$3 \cdot K_{\mathcal{A}}(Q_i) - \log Q_i(z) \leq 3 \cdot K_{\mathcal{A}}(R) - \log R(z).$$

Therefore:

$$3a \leq 3a - 2\log m + O(\log b) \text{ and hence}$$

$$\log m \leq O(\log b) = O(\log(K_{\mathcal{A}}(P) + D).$$

That is proved our claim.

Part 2. To complete the proof we do the same things as in the proof of Theorem 7.

If $x = x_1 \ldots x_n$ is a prefix of ω and Q is the best distribution for x then by Lemma 1

$$\sum_{i=1}^{n-1} (Q(0|x_1 \ldots x_i) - P(0|x_1 \ldots x_i))^2 = O(K_{\mathcal{A}}(P) + D + K(P, Q) = O(K_{\mathcal{A}}(P) + D).$$

(In the last equation we use $K(P, Q) = O(K(P) + K(Q)) = O(K_{\mathcal{A}}(P) + K_{\mathcal{A}}(Q))$.) So, every probability distribution that is the best for some x "contributes" $O(K_{\mathcal{A}}(P) + D)$ in the sum $\sum_{x \text{ is a prefix of } \omega}(H_{\mathcal{A}}(0|x) - P(0|x))^2$. There are $\text{poly}(D + K_{\mathcal{A}}(P))$ such distributions, so we obtain the required estimation.

Proof (Proof of Theorem 9). The proof is the same as the proof of Theorem 3 but with using Theorem 8 instead of Theorem 7.

5   Open Questions

A natural question arises: can we get a better estimation in the last theorem than $O(K(P)2^{\frac{3K(P)}{2}})$? We have exponential (in $K(P)$) estimation because it is our estimation of the number of distributions that are the best for some x. However, the author does not know an example of P-random sequence ω such that there are exponentially many (in terms of $K(P)$ and $d(\omega|P)$) different best distributions for prefixes of ω.

Algorithmic statistics [1, 8, 11] studies good distributions for strings among distributions on finite sets. There exists a family of "standard statistics" that cover all the best distributions for finite strings. It is interesting: are there the same things for distributions on the binary tree?

Acknowledgements. I would like to thank Alexander Shen and Nikolay Vereshchagin and for useful discussions, advice and remarks. This work is supported by 19-01-00563 RFBR grant and by RaCAF ANR-15-CE40-0016-01 grant.

The article was prepared within the framework of the HSE University Basic Research Program.

References

1. Gács, P., Tromp, J., Vitányi, P.M.B.: Algorithmic statistics. IEEE Tran. Inf. Theory **47**(6), 2443–2463 (2001)
2. Hutter, M., Poland, J.: Asymptotics of discrete MDL for online prediction. IEEE Trans. Inf. Theory **51**(11), 3780–3795 (2005)
3. Hutter, M.: Discrete MDL predicts in total variation. Adv. Neural Inf. Process. Syst. **22** (NIPS-2009), 817–825 (2009)
4. Hutter, M.: Sequential predictions based on algorithmic complexity. J. Comput. Syst. Sci. **72**, 95–117 (2006)
5. Hutter, M., Muchnik, A.: Universal convergence of semimeasures on individual random sequences. In: Ben-David, S., Case, J., Maruoka, A. (eds.) ALT 2004. LNCS (LNAI), vol. 3244, pp. 234–248. Springer, Heidelberg (2004)
6. Li M., Vitányi P., An Introduction to Kolmogorov complexity and its applications, 3rd ed., Springer, (1 ed., 1993; 2 ed., 1997), xxiii+790 (2008). ISBN 978-0-387-49820-1
7. Lattimore, T., Hutter, M.: On Martin-Löf convergence of Solomonoff's mixture. In: Chan, T.-H.H., Lau, L.C., Trevisan, L. (eds.) TAMC 2013. LNCS, vol. 7876, pp. 212–223. Springer, Heidelberg (2013). https://doi.org/10.1007/978-3-642-38236-9_20
8. Shen, A., Uspensky, V., Vereshchagin, N.: Kolmogorov Complexity and Algorithmic Randomness. ACM (2017)
9. Solomonoff, R.J.: A formal theory of inductive inference: parts 1 and 2. Inf. Control **7**, 1–22, 224–254 (1964)
10. Solomonoff, R.J.: Complexity-based induction systems: comparisons and convergence theorems. IEEE Trans. Inf. Theory, **IT-24**, 422–432 (1978)
11. Vereshchagin, N., Shen, A.: Algorithmic statistics: forty years later. In: Day, A., Fellows, M., Greenberg, N., Khoussainov, B., Melnikov, A., Rosamond, F. (eds.) Computability and Complexity. LNCS, vol. 10010, pp. 669–737. Springer, Cham (2017). https://doi.org/10.1007/978-3-319-50062-1_41
12. Vovk, V.G.: On a criterion for randomness. Dokl. Akad. Nauk SSSR **294**(6), 1298–1302 (1987)

Lower Bounds and Hardness Magnification for Sublinear-Time Shrinking Cellular Automata

Augusto Modanese[(✉)]

Karlsruhe Institute of Technology (KIT), Karlsruhe, Germany
modanese@kit.edu

Abstract. The minimum circuit size problem (MCSP) is a string compression problem with a parameter s in which, given the truth table of a Boolean function over inputs of length n, one must answer whether it can be computed by a Boolean circuit of size at most $s(n) \geq n$. Recently, McKay, Murray, and Williams (STOC, 2019) proved a hardness magnification result for MCSP involving (one-pass) streaming algorithms: For any reasonable s, if there is no $\mathsf{poly}(s(n))$-space streaming algorithm with $\mathsf{poly}(s(n))$ update time for MCSP$[s]$, then $\mathsf{P} \neq \mathsf{NP}$. We prove an analogous result for the (provably) strictly less capable model of shrinking cellular automata (SCAs), which are cellular automata whose cells can spontaneously delete themselves. We show every language accepted by an SCA can also be accepted by a streaming algorithm of similar complexity, and we identify two different aspects in which SCAs are more restricted than streaming algorithms. We also show there is a language which cannot be accepted by any SCA in $o(n/\log n)$ time, even though it admits an $O(\log n)$-space streaming algorithm with $O(\log n)$ update time.

Keywords: Cellular automata · Hardness magnification · Minimum circuit size problem · Streaming algorithms · Sublinear-time computation

1 Introduction

The ongoing quest for lower bounds in complexity theory has been an arduous but by no means unfruitful one. Recent developments have brought to light a phenomenon dubbed *hardness magnification* [5–7,17,22,23], giving several examples of natural problems for which even slightly non-trivial lower bounds are as hard to prove as major complexity class separations such as $\mathsf{P} \neq \mathsf{NP}$. Among these, the preeminent example appears to be the *minimum circuit size problem*:

Definition 1 (MCSP). *For a Boolean function $f\colon \{0,1\}^n \to \{0,1\}$, let $\mathsf{tt}(f)$ denote the truth table representation of f (as a binary string in $\{0,1\}^+$ of length $|\mathsf{tt}(f)| = 2^n$). For $s\colon \mathbb{N}_+ \to \mathbb{N}_+$, the* minimum circuit size problem MCSP$[s]$ *is*

© Springer Nature Switzerland AG 2021
R. Santhanam and D. Musatov (Eds.): CSR 2021, LNCS 12730, pp. 296–320, 2021.
https://doi.org/10.1007/978-3-030-79416-3_18

the problem where, given such a truth table $\mathrm{tt}(f)$, *one must answer whether there is a Boolean circuit* C *on inputs of length* n *and size at most* $s(n)$ *that computes* f, *that is,* $C(x) = f(x)$ *for every input* $x \in \{0,1\}^n$.

It is a well-known fact that there is a constant $K > 0$ such that, for any function f on n variables as above, there is a circuit of size at most $K \cdot 2^n/n$ that computes f; hence, MCSP$[s]$ is only non-trivial for $s(n) < K \cdot 2^n/n$. Furthermore, MCSP$[s] \in$ NP for any constructible s and, since every circuit of size at most $s(n)$ can be described by a binary string of $O(s(n) \log s(n))$ length, if $2^{O(s(n) \log s(n))} \subseteq \mathsf{poly}(2^n)$ (e.g., $s(n) \in O(n/\log n)$), by enumerating all possibilities we have MCSP$[s] \in$ P. (Of course, such a bound is hardly useful since $s(n) \in O(n/\log n)$ implies the circuit is degenerate and can only read a strict subset of its inputs.) For large enough $s(n) < K \cdot 2^n/n$ (e.g., $s(n) \geq n$), it is unclear whether MCSP$[s]$ is NP-complete (under polynomial-time many-one reductions); see also [13,21]. Still, we remark there has been some recent progress regarding NP-completeness under *randomized* many-one reductions for *certain variants* of MCSP [12].

Oliveira and Santhanam [23] and Oliveira, Pich, and Santhanam [22] recently analyzed hardness magnification in the average-case as well as in the worst-case approximation (i.e., gap) settings of MCSP for various (uniform and non-uniform) computational models. Meanwhile, Meanwhile, McKay, Murray, and Williams [17] showed similar results hold in the standard (i.e., exact or gapless) worst-case setting and proved the following magnification result for (single-pass) *streaming algorithms* (see Definition 6), which is a very restricted uniform model; indeed, as mentioned in [17], even string equality (i.e., the problem of recognizing $\{ww \mid w \in \{0,1\}^+\}$) cannot be solved by streaming algorithms (with limited space).

Theorem 2 ([17]). *Let* $s \colon \mathbb{N}_+ \to \mathbb{N}_+$ *be time constructible and* $s(n) \geq n$. *If there is no* $\mathsf{poly}(s(n))$-*space streaming algorithm with* $\mathsf{poly}(s(n))$ *update time for (the search version of)* MCSP$[s]$, *then* P $\neq$ NP.

In this paper, we present the following hardness magnification result for a (uniform) computational model which is provably *even more restricted* than streaming algorithms: *shrinking cellular automata* (SCAs). Here, Block$_b$ refers to a slightly modified presentation of MCSP$[s]$ that is only needed due to certain limitations of the model (see further discussion as well as Sect. 3.1).

Theorem 3. *For a certain* $m \in \mathsf{poly}(s(n))$, *if* Block$_b$(MCSP$[s]$) $\notin$ SCA$[n \cdot f(m)]$ *for every* $f \in \mathsf{poly}(m)$ *and* $b \in O(f)$, *then* P $\neq$ NP.

Furthermore, we show every language accepted by a sublinear-time SCA can also be accepted by a streaming algorithm of comparable complexity:

Theorem 4. *Let* $t \colon \mathbb{N}_+ \to \mathbb{N}_+$ *be computable by an* $O(t)$-*space random access machine (as in Definition 6) in* $O(t \log t)$ *time. Then, if* $L \in$ SCA$[t]$, *there is an* $O(t)$-*space streaming algorithm for* L *with* $O(t \log t)$ *update and* $O(t^2 \log t)$ *reporting time.*

Finally, we identify and prove *two distinct limitations* of SCAs compared to streaming algorithms (under sublinear-time constraints):

1. They are insensitive to the length of long unary substrings in their input (Lemma 17), which means (standard versions of) fundamental problems such as parity, modulo, majority, and threshold cannot be solved in sublinear time (Proposition 18 and Corollary 20).
2. Only a limited amount of information can be transferred between cells which are far apart (in the sense of one-way communication complexity; see Lemma 23).

Both limitations are inherited from the underlying model of cellular automata. The first can be avoided by presenting the input in a special format (the previously mentioned Block_n) that is efficiently verifiable by SCAs, which we motivate and adopt as part of the model (see the discussion below). The second is more dramatic and results in lower bounds even for languages presented in this format:

Theorem 5. *There is a language L_1 for which $\mathsf{Block}_n(L_1) \notin \mathsf{SCA}[o(N/\log N)]$ (N being the instance length) can be accepted by an $O(\log N)$-space streaming algorithm with $\tilde{O}(\log N)$ update time.*

From the above, it follows that any proof of $\mathsf{P} \neq \mathsf{NP}$ based on a lower bound for solving $\mathsf{MCSP}[s]$ with streaming algorithms and Theorem 2 must implicitly contain a proof of a lower bound for solving $\mathsf{MCSP}[s]$ with SCAs. From a more "optimistic" perspective (with an eventual proof of $\mathsf{P} \neq \mathsf{NP}$ in mind), although not as widely studied as streaming algorithms, SCAs are thus at least as good as a "target" for proving lower bounds against and, in fact, should be an easier one if we are able to exploit their aforementioned limitations. Refer to Sect. 6 for further discussion on this, where we take into account a recently proposed barrier [5] to existing techniques and which also applies to our proof of Theorem 5.

From the perspective of cellular automata theory, our work furthers knowledge in sublinear-time cellular automata models, a topic seemingly neglected by the community at large (as pointed out in, e.g., [19]). Although this is certainly not the first result in which complexity-theoretical results for cellular automata and their variants have consequences for classical models (see, e.g., [15,24] for results in this sense), to the best of our knowledge said results address only *necessary* conditions for separating classical complexity classes. Hence, our result is also novel in providing an implication in the other direction, that is, a *sufficient* condition for said separations based on lower bounds for cellular automata models.

The Model. (One-dimensional) cellular automata (CAs) are a parallel computational model composed of identical *cells* arranged in an array. Each cell operates as a deterministic finite automaton (DFA) that is connected with its left and right neighbors and operates according to the same local rule. In classical CAs, the cell structure is immutable; *shrinking* CAs relax the model in that regard by allowing cells to spontaneously vanish (with their contents being

irrecoverably lost). The array structure is conserved by reconnecting every cell
with deleted neighbors to the nearest non-deleted ones in either direction.

SCAs were introduced by Rosenfeld, Wu, and Dubitzki in 1983 [25], but it
was not until recent years that the model received greater attention by the CA
community [16,20]. SCAs are a natural and robust model of parallel computation
which, unlike classical CAs, admit (non-trivial) sublinear-time computations.

We give a brief intuition as to how shrinking augments the classical CA
model in a significant way. Intuitively speaking, any two cells in a CA can only
communicate by signals, which necessarily requires time proportional to the distance between them. Assuming the entire input is relevant towards acceptance,
this imposes a linear lower bound on the time complexity of the CA. In SCAs,
however, this distance can be shortened as the computation evolves, thus ren-
dering acceptance in sublinear time possible. As a matter of fact, the more cells
are deleted, the faster distant cells can communicate and the computation can
evolve. This results in a trade-off between space (i.e., cells containing information) and time (i.e., amount of cells deleted).

Comparison with Related Models. Unlike other parallel models such as random
access machines, SCAs are *incapable of random access* to their input. In a similar
sense, SCAs are constrained by the *distance* between cells, which is an aspect
usually disregarded in circuits and related models except perhaps for VLSI complexity [4,28], for instance. In contrast to VLSI circuits, however, in SCAs distance is a fluid aspect, changing dynamically as the computation evolves. Also
of note is that SCAs are a *local* computational model in a quite literal sense
of locality that is coupled with the above concept of distance (instead of more
abstract notions such as that from [30], for example).

These limitations hold not only for SCAs but also for standard CAs. Nevertheless, SCAs are more powerful than other CA models capable of sublinear-time
computation such as ACAs [11,19], which are CAs with their acceptance behavior such that the CA accepts if and only if all cells simultaneously accept. This is
because SCAs can *efficiently aggregate results* computed in parallel (by combining them using some efficiently computable function); in ACAs any such form
of aggregation is fairly limited as the underlying cell structure is static.

Block Words. As mentioned above, there is an input format which allows us to
circumvent the first of the limitations of SCAs compared to streaming algorithms
and which is essential in order to obtain a more serious computational model.
In this format, the input is subdivided into *blocks* of the same size and which
are separated by delimiters and numbered in ascending order from left to right.
Words with this structure are dubbed *block words* accordingly, and a set of such
words is a *block language*. There is a natural presentation of any (ordinary) word
as a block word (by mapping every symbol to its own block), which means there
is a block language version to any (ordinary) language. (See Sect. 3.1.)

The concept of block words seems to arise naturally in the context of
sublinear-time (both shrinking and standard) CAs [11,19]. The syntax of block
words is very efficiently verifiable (more precisely, in time linear in the block

length) by a CA (without need of shrinking). In addition, the translation of a language to its block version (and its inverse) is a very simple map; one may frame it, for instance, as an AC^0 reduction. Hence, the difference between a language and its block version is solely in presentation.

Block words coupled with CAs form a computational paradigm that appears to be substantially diverse from linear- and real-time CA computation (see [19] for examples). Often we shall describe operations on a block (rather than on a cell) level and, by making use of block numbering, two blocks with distinct numbers may operate differently even though their contents are the same; this would be impossible at a cell level due to the locality of CA rules. In combination with shrinking, certain block languages admit merging groups of blocks in parallel; this gives rise to a form of reduction we call *blockwise reductions* and which we employ in a manner akin to downward self-reducibility as in [1].

An additional technicality which arises is that the number of cells in a block is fixed at the start of the computation; this means a block cannot "allocate extra space" (beyond a constant multiple of the block length). This is the same limitation as that of linear bounded automata (LBAs) compared to Turing machines with unbounded space, for example. We cope with this limitation by increasing the block length in the problem instances as needed, that is, by padding each block so that enough space is available from the outset.[1] This is still in line with the considerations above; for instance, the resulting language is still AC^0 reducible to the original one (and vice-versa).

Techniques. We give a broad overview of the proof ideas behind our results.

Theorem 3 is a direct corollary of Theorem 24, proven is Sect. 5. The proof closely follows [17] (see the discussion in Sect. 5 for a comparison) and, as mentioned above, bases on a scheme similar to self-reducibility as in [1].

The lower bounds in Sect. 3.2 are established using Lemma 17, which is a generic technical limitation of sublinear-time models based on CAs (the first of the two aforementioned limitations of SCAs with respect to streaming algorithms) and which we also show to hold for SCAs.

One of the main technical highlights is the proof of Theorem 4, where we give a streaming algorithm to simulate an SCA with limited space. Our general approach bases on dynamic programming and is able to cope with the unpredictability of when, which, or even how many cells are deleted during the simulation. The space efficiency is achieved by keeping track of only as much information as needed as to determine the state of the SCA's decision cell step for step.

A second technical contribution is the application of *one-way* communication complexity to obtain lower bounds for SCAs, which yields Theorem 5. Essentially, we split the input in some position i of our choice (which may even be non-uniformly dependent on the input length) and have A be given as input the symbols

---

[1] An alternative solution is allowing the CA to "expand" by dynamically creating new cells between existing ones; however, this may result in a computational model which is dramatically more powerful than standard CAs [18,20].

preceding i while B is given the rest, where A and B are (non-uniform) algorithms with unbounded computational resources. We show that, in this setting, A can determine the state of the SCA's decision cell with only $O(1)$ information from B for every step of the SCA. Thus, an SCA with time complexity t for a language L yields a protocol with $O(t)$ one-way communication complexity for the above problem. Applying this in the contrapositive, Theorem 5 then follows from the existence of a language L_1 (in some contexts referred to as the indexing or memory access problem) that has nearly linear one-way communication complexity despite admitting an efficient streaming algorithm.

Organization. The rest of the paper is organized as follows: Sect. 2 presents the basic definitions. In Sect. 3 we introduce block words and related concepts and discuss the aforementioned limitations of sublinear-time SCAs. Following that, in Sect. 4 we address the proof of Theorem 4 and in Sect. 5 that of Theorem 3. Finally, Sect. 6 concludes the paper.

2   Preliminaries

We denote the set of integers by $\mathbb{Z}$, that of positive integers by $\mathbb{N}_+$, and $\mathbb{N}_+ \cup \{0\}$ by $\mathbb{N}_0$. For $a, b \in \mathbb{N}_0$, $[a, b] = \{x \in \mathbb{N}_0 \mid a \leq x \leq b\}$. For sets A and B, B^A is the set of functions $A \to B$.

We assume the reader is familiar with cellular automata as well as with the fundamentals of computational complexity theory (see, e.g., standard references [2,8,10]). Words are indexed starting with index zero. For a finite, non-empty set Σ, Σ^* denotes the set of words over Σ, and Σ^+ the set $\Sigma^* \setminus \{\varepsilon\}$. For $w \in \Sigma^*$, we write $w(i)$ for the i-th symbol of w (and, in general, w_i stands for another word altogether, *not* the i-th symbol of w). For $a, b \in \mathbb{N}_0$, $w[a, b]$ is the subword $w(a)w(a+1) \cdots w(b-1)w(b)$ of w (where $w[a, b] = \varepsilon$ for $a > b$). $|w|_a$ is the number of occurrences of $a \in \Sigma$ in w. $\mathsf{bin}_n(x)$ stands for the binary representation of $x \in \mathbb{N}_0$, $x < 2^n$, of length $n \in \mathbb{N}_+$ (padded with leading zeros). $\mathsf{poly}(n)$ is the class of functions polynomial in $n \in \mathbb{N}_0$. REG denotes the class of regular languages, and $\mathsf{TISP}[t, s]$ (resp., $\mathsf{TIME}[t]$) that of problems decidable by a Turing machine (with one tape and one read-write head) in $O(t)$ time and $O(s)$ space (resp., unbounded space). Without restriction, we assume the empty word ε is not a member of any of the languages considered.

An ω-*word* is a map $\mathbb{N}_0 \to \Sigma$, and a $\omega\omega$-*word* is a map $\mathbb{Z} \to \Sigma$. We write $\Sigma^\omega = \Sigma^{\mathbb{N}_0}$ for the set of ω-words over Σ. For $x \in \Sigma$, x^ω denotes the (unique) ω-word with $x^\omega(i) = x$ for every $i \in \mathbb{N}_0$. To each $\omega\omega$-word w corresponds a unique pair (w_-, w_+) of ω-words $w_-, w_+ \in \Sigma^\omega$ with $w_+(i) = w(i)$ for $i \geq 0$ and $w_-(i) = w(-i - 1)$ for $i < 0$. (Partial) ω-word homomorphisms are extendable to (partial) $\omega\omega$-*word homomorphisms* as follows: Let $f \colon \Sigma^\omega \to \Sigma^\omega$ be an ω-word homomorphism; then there is a unique $f_{\omega\omega} \colon \Sigma^{\mathbb{Z}} \to \Sigma^{\mathbb{Z}}$ such that, for every $w \in \Sigma^{\mathbb{Z}}$, $w' = f_{\omega\omega}(w)$ is the $\omega\omega$-word with $w'_+ = f(w_+)$ and $w'_- = f(w_-)$.

For a circuit C, $|C|$ denotes the *size* of C, that is, the total number of gates in C. It is well-known that any Boolean circuit C can be described by a binary string of $O(|C| \log|C|)$ length.

Definition 6 (Streaming algorithm). *Let $s, u, r \colon \mathbb{N}_+ \to \mathbb{N}_+$ be functions. An s-space streaming algorithm A is a random access machine which, on input w, works in $O(s(|w|))$ space and, on every step, can either perform an operation on a constant number of bits in memory or read the next symbol of w. A has u update time if, for every w, the number of operations it performs between reading $w(i)$ and $w(i+1)$ is at most $u(|w|)$. A has r reporting time if it performs at most $r(|w|)$ operations after having read $w(|w| - 1)$ (until it terminates).*

Our interest lies in s-space streaming algorithms that, for an input w, have $\mathsf{poly}(s(|w|))$ update and reporting time for sublinear s (i.e., $s(|w|) \in o(|w|)$).

Cellular Automata. We consider only CAs with the standard neighborhood. The symbols of an input w are provided from left to right in the cells 0 to $|w| - 1$ and are surrounded by inactive cells, which conserve their state during the entire computation (i.e., the CA is bounded). Acceptance is signaled by cell zero (i.e., the leftmost input cell).

Definition 7 (Cellular automaton). *A cellular automaton (CA) C is a tuple $(Q, \delta, \Sigma, q, A)$ where: Q is a non-empty and finite set of states; $\delta \colon Q^3 \to Q$ is the local transition function; $\Sigma \subsetneq Q$ is the input alphabet of C; $q \in Q \setminus \Sigma$ is the inactive state, that is, $\delta(q_1, q, q_2) = q$ for every $q_1, q_2 \in Q$; and $A \subseteq Q \setminus \{q\}$ is the set of accepting states of C. A cell which is not in the inactive state is said to be active. The elements of $Q^{\mathbb{Z}}$ are the (global) configurations of C. δ induces the global transition function $\Delta \colon Q^{\mathbb{Z}} \to Q^{\mathbb{Z}}$ of C by $\Delta(c)(i) = \delta(c(i-1), c(i), c(i+1))$ for every cell $i \in \mathbb{Z}$ and configuration $c \in Q^{\mathbb{Z}}$.*

C accepts an input $w \in \Sigma^+$ if cell zero is eventually in an accepting state, that is, there is $t \in \mathbb{N}_0$ such that $(\Delta^t(c_0))(0) \in A$, where $c_0 = c_0(w)$ is the initial configuration (for w): $c_0(i) = w(i)$ for $i \in [0, |w| - 1]$, and $c_0(i) = q$ otherwise. For a minimal such t, we say C accepts w with time complexity t. $L(A) \subseteq \Sigma^+$ denotes the set of words accepted by C. For $t \colon \mathbb{N}_+ \to \mathbb{N}_0$, $\mathsf{CA}[t]$ is the class of languages accepted by CAs with time complexity $O(t(n))$, n being the input length.

For convenience, we extend Δ in the obvious manner (i.e., as a map induced by δ) so it is also defined for every (finite) word $w \in Q^*$. For $|w| \le 2$, we set $\Delta(w) = \varepsilon$; for longer words, $|\Delta(w)| = |w| - 2$ holds.

Some remarks concerning the classes $\mathsf{CA}[t]$: $\mathsf{CA}[\mathsf{poly}] = \mathsf{TISP}[\mathsf{poly}, n]$ (i.e., the class of polynomial-time LBAs), and $\mathsf{CA}[t] = \mathsf{CA}[1] \subsetneq \mathsf{REG}$ for every sublinear t. Furthermore, $\mathsf{CA}[t] \subseteq \mathsf{TISP}[t^2, n]$ (where $t^2(n) = (t(n))^2$) and $\mathsf{TISP}[t, n] \subseteq \mathsf{CA}[t]$.

Definition 8 (Shrinking CA). *A shrinking CA (SCA) S is a CA with a delete state $\otimes \in Q \setminus (\Sigma \cup \{q\})$. The global transition function Δ_S of S is given by applying the standard CA global transition function Δ (as in Definition 7) followed by removing all cells in the state $\otimes$; that is, $\Delta_S = \Phi \circ \Delta$, where $\Phi \colon Q^{\mathbb{Z}} \to Q^{\mathbb{Z}}$ is the (partial) $\omega\omega$-word homomorphism resulting from the extension to $Q^{\mathbb{Z}}$ of the map $\varphi \colon Q \to Q$ with $\varphi(\otimes) = \varepsilon$ and $\varphi(x) = x$ for $x \in Q \setminus \{\otimes\}$. For $t \colon \mathbb{N}_+ \to \mathbb{N}_0$, $\mathsf{SCA}[t]$ is the class of languages accepted by SCAs with time complexity $O(t(n))$, where n denotes the input length.*

Note that Φ is only partial since, for instance, any $\omega\omega$-word in $\otimes^\omega \cdot \Sigma^* \cdot \otimes^\omega$ has no proper image (as it is not mapped to a $\omega\omega$-word). Hence, Δ_S is also only a partial function (on $Q^{\mathbb{Z}}$); nevertheless, Φ is total on the set of $\omega\omega$-words in which $\otimes$ occurs only finitely often and, in particular, Δ_S is total on the set of configurations arising from initial configurations for finite input words (which is the setting we are interested in).

The acceptance condition of SCAs is the same as in Definition 7 (i.e., acceptance is dictated by cell zero). Unlike in standard CAs, the index of one same cell can differ from one configuration to the next; that is, a cell index does not uniquely determine a cell on its own (rather, only in conjunction with a time step). This is a consequence of applying Φ, which contracts the global configuration towards cell zero. More precisely, for a configuration $c \in Q^{\mathbb{Z}}$, the cell with index $i \geq 0$ in $\Delta(c)$ corresponds to that with index $i + d_i$ in c, where d_i is the number of cells with index $\leq i$ in c that were deleted in the transition to $\Delta(c)$. This also implies the cell with index zero in $\Delta(c)$ is the same as that in c with minimal positive index that was not deleted in the transition to $\Delta(c)$; thus, in any time step, cell zero is the leftmost active cell (unless all cells are inactive; in fact, cell zero is inactive if and only if all other cells are inactive). Granted, what indices a cell has is of little importance when one is interested only in the configurations of an SCA and their evolution; nevertheless, they are relevant when simulating an SCA with another machine model (as we do in Sects. 3.3 and 4).

Naturally, $\mathsf{CA}[t] \subseteq \mathsf{SCA}[t]$ for every function t, and $\mathsf{SCA}[\mathrm{poly}] = \mathsf{CA}[\mathrm{poly}]$. For sublinear t, $\mathsf{SCA}[t]$ contains non-regular languages if, for instance, $t \in \Omega(\log n)$ (see below); hence, the inclusion of $\mathsf{CA}[t]$ in $\mathsf{SCA}[t]$ in strict. In fact, this is the case even if we consider only regular languages. One simple example is $L = \{w \in \{0,1\}^+ \mid w(0) = w(|w|-1)\}$, which is in $\mathsf{SCA}[1]$ and regular but not in $\mathsf{CA}[o(n)] = \mathsf{CA}[O(1)]$. One obtains an SCA for L by having all cells whose both neighbors are active delete themselves in the first step; the two remaining cells then compare their states, and cell zero accepts if and only if this comparison succeeds or if the input has length 1 (which it can notice immediately since it is only for such words that it has two inactive neighbors). Formally, the local transition function δ is such that, for $z_1, z_3 \in \{0,1,q\}$ and $z_2 \in \{0,1\}$, $\delta(z_1, z_2, z_3) = \otimes$ if both z_1 and z_3 are in $\{0,1\}$, $\delta(z_1, z_2, z_3) = z_2'$ if $z_1 = q$ or $z_3 = q$, and $\delta(q, z_2', z_2') = \delta(q, z_2', q) = a$; in all other cases, δ simply conserves the cell's state. See Fig. 1 for an example.

Using a textbook technique to simulate a (bounded) CA with an LBA (simply skipping deleted cells), we have:

Proposition 9. *For every function* $t \colon \mathbb{N}_+ \to \mathbb{N}_+$ *computable by an LBA in* $O(n \cdot t(n))$ *time,* $\mathsf{SCA}[t] \subseteq \mathsf{TISP}[n \cdot t(n), n]$.

The inclusion is actually proper (see Corollary 19). Using the well-known result $\mathsf{TIME}[o(n \log n)] = \mathsf{REG}$ [14], it follows that at least a logarithmic time bound is needed for SCAs to recognize languages which are not regular:

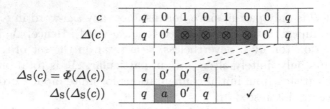

Fig. 1. Computation of an SCA that recognizes $L = \{w \in \{0,1\}^+ \mid w(0) = w(|w|-1)\}$ in $O(1)$ time. Here, the input word is $010100 \in L$.

Corollary 10. $\mathsf{SCA}[o(\log)] \subsetneq \mathsf{REG}$.

This bound is tight: It is relatively easy to show that any language accepted by ACAs in $t(n)$ time can also be accepted by an SCA in $t(n) + O(1)$ time. Since there is a non-regular language recognizable by ACAs [11] in $O(\log n)$ time, the same language is recognizable by an SCA in $O(\log n)$ time.

For any finite, non-empty set Σ, we say a function $f \colon \Sigma^+ \to \Sigma^+$ is *computable in place* by an (S)CA if there is an (S)CA S which, given $x \in \Sigma^+$ as input (surrounded by inactive cells), produces $f(x)$. Additionally, $g \colon \mathbb{N}_+ \to \mathbb{N}_+$ is *constructible in place* by an (S)CA if $g(n) \leq 2^n$ and there is an (S)CA S which, given $n \in \mathbb{N}_0$ in unary, produces $\mathrm{bin}_n(g(n)-1)$ (i.e., $g(n)-1$ in binary). Note the set of functions computable or constructible in place by an (S)CA in at most $t(n)$ time, where n is the input length and $t \colon \mathbb{N}_+ \to \mathbb{N}_+$ is some function, includes (but is not limited to) all functions computable by an LBA in at most $t(n)$ time.

3   Capabilities and Limitations of Sublinear-Time SCAs

3.1   Block Languages

Let Σ be a finite, non-empty set. For $\Sigma_\varepsilon = \Sigma \cup \{\varepsilon\}$ and $x, y \in \Sigma^+$, $\binom{x}{y}$ denotes the (unique) word in $(\Sigma_\varepsilon \times \Sigma_\varepsilon)^+$ of length $\max\{|x|, |y|\}$ for which $\binom{x}{y}(i) = (x(i), y(i))$, where $x(i) = y(j) = \varepsilon$ for $i \geq |x|$ and $j \geq |y|$.

Definition 11 (Block word). *Let $n, m, b \in \mathbb{N}_+$ be such that $b \geq n$ and $m \leq 2^n$. A word w is said to be an (n, m, b)-block word (over Σ) if it is of the form $w = w_0 \# w_1 \# \cdots \# w_{m-1}$ and $w_i = \binom{\mathrm{bin}_n(x_i)}{y_i}$, where $x_0 \geq 0$, $x_{i+1} = x_i + 1$ for every i, $x_{m-1} < 2^n$, and $y_i \in \Sigma^b$. In this context, w_i is the i-th block of w.*

Hence, every (n, m, b)-block word w has m many blocks of length b, and its total length is $|w| = (b+1) \cdot m - 1 \in \Theta(bm)$. For example,

$$w = \binom{01}{0100} \# \binom{10}{1100} \# \binom{11}{1000}$$

is a $(2, 3, 4)$-block word with $x_0 = 1$, $y_0 = 0100$, $y_1 = 1100$, and $y_2 = 1000$. n is implicitly encoded by the entries in the upper track (i.e., the x_i) and we shall see m and b as parameters depending on n (see Definition 12 below), so the structure of each block can be verified locally (i.e., by inspecting the immediate neighborhood of every block). Note the block numbering starts with an arbitrary x_0; this is intended so that, for $m' < m$, an (n, m, b)-block word admits (n, m', b)-block words as infixes (which would not be the case if we required, say, $x_0 = 0$).

When referring to block words, we use N for the block word length $|w|$ and reserve n for indexing block words of different block length, overall length, or total number of blocks (or any combinations thereof). With m and b as parameters depending on n, we obtain sets of block words:

Definition 12 (Block language). *Let $m, b \colon \mathbb{N}_+ \to \mathbb{N}_+$ be non-decreasing and constructible in place by a CA in $O(m(n){+}b(n))$ time. Furthermore, let $b(n) \geq n$ and $m(n) \leq 2^n$. Then, $\mathfrak{B}_b^m$ denotes the set of all $(n, m(n), b(n))$-block words for $n \in \mathbb{N}_+$, and every subset $L \subseteq \mathfrak{B}_b^m$ is an $((n, m, b)$-$)$block language (over Σ).*

An SCA can *verify* its input is a valid block word in $O(b(n))$ time, that is, locally check that the structure and contents of the blocks are consistent (i.e., as in Definition 11). This can be realized using standard CA techniques without need of shrinking (see [11,19] for constructions). Recall Definition 8 does not require an SCA S to explicitly reject inputs not in $L(S)$, that is, the time complexity of S on an input w is only defined for $w \in L(S)$. As a result, when $L(S)$ is a block language, the time spent verifying that w is a block word is only relevant if $w \in L(S)$ and, in particular, if w is a (valid) block word. Provided the state of every cell in S eventually impacts its decision to accept (which is the case for all constructions we describe), it suffices to have a cell mark itself with an error flag whenever a violation in w is detected (even if other cells continue their operation as normal); since every cell is relevant towards acceptance, this eventually prevents S from accepting (and, since $w \notin L(S)$, it is irrelevant how long it takes for this to occur). Thus, for the rest of this paper, when describing an SCA for a block language, we implicitly require that the SCA checks its input is a valid block word beforehand.

As stated in the introduction, our interest in block words is as a special input format. There is a natural bijection between any language and a block version of it, namely by mapping each word z to a block word w in which each block w_i contains a symbol $z(i)$ of z (padded up to the block length b) and the blocks are numbered from 0 to $|z| - 1$:

Definition 13 (Block version of a language). *Let $L \subseteq \Sigma^+$ be a language and b as in Definition 12. The block version $\mathsf{Block}_b(L)$ of L (with blocks of length b) is the block language for which, for every $z \in \Sigma^+$, $z \in L$ holds if and only if we have $w \in \mathsf{Block}_b(L)$ where w is the $(n, m, b(n))$-block word (as in Definition 11) with $m = |z|$, $n = \lceil \log m \rceil$, $x_0 = 0$, and $y_i = z(i)0^{b(n)-1}$ for every $i \in [0, m-1]$.*

Note that, for any such language L, $\mathsf{Block}_b(L) \notin \mathsf{REG}$ for any b (since $b(n) \geq n$ is not constant); hence, $\mathsf{Block}_b(L) \in \mathsf{SCA}[t]$ only for $t \in \Omega(\log n)$ (and constructible b). For $b(n) = n$, $\mathsf{Block}_n(L)$ is the block version with minimal padding.

For any two finite, non-empty sets Σ_1 and Σ_2, say a function $f\colon \Sigma_1^+ \to \Sigma_2^+$ is *non-stretching* if $|f(x)| \leq |x|$ for every $x \in \Sigma_1^+$. We now define k-blockwise maps, which are maps that operate on block words by grouping $k(n)$ many blocks together and mapping each such group (in a non-stretching manner) to a single block of length at most $(b(n) + 1) \cdot k(n) - 1$.

Definition 14 (Blockwise map). *Let $k\colon \mathbb{N}_+ \to \mathbb{N}_+$, $k(n) \geq 2$, be a non-decreasing function and constructible in place by a CA in $O(k(n))$ time. A map $g\colon \mathfrak{B}_b^{km} \to \mathfrak{B}_b^m$ is a k-blockwise map if there is a non-stretching $g'\colon \mathfrak{B}_b^k \to \Sigma^+$ such that, for every $w \in \mathfrak{B}_b^{km}$ (as in Definition 11) and $w_i' = w_{ik}\#\cdots\#w_{(i+1)k-1}$:*

$$g(w) = \binom{\mathrm{bin}_n(x_0)}{g'(w_0')}\#\cdots\#\binom{\mathrm{bin}_n(x_{m-1})}{g'(w_{m-1}')}.$$

Using blockwise maps, we obtain a very natural form of reduction operating on block words and which is highly compatible with sublinear-time SCAs as a computational model. The reduction divides an (n, km, b)-block word in m many groups of k many contiguous blocks and, as a k-blockwise map, maps each such group to a single block (of length b):

Definition 15 (Blockwise reducible). *For block languages L and L', L is (k-)blockwise reducible to L' if there is a computable k-blockwise map $g\colon \mathfrak{B}_b^{km} \to \mathfrak{B}_b^m$ such that, for every $w \in \mathfrak{B}_b^{km}$, we have $w \in L$ if and only if $g(w) \in L'$.*

Since every application of the reduction reduces the instance length by a factor of approximately k, logarithmically many applications suffice to produce a trivial instance (i.e., an instance consisting of a single block). This gives us the following computational paradigm of chaining blockwise reductions together:

Lemma 16. *Let $k, r\colon \mathbb{N}_+ \to \mathbb{N}_0$ be functions, and let $L \subseteq \mathfrak{B}_b^{k^r}$ be such that there is a series $L = L_0, L_1, \ldots, L_{r(n)}$ of languages with $L_i \subseteq \mathfrak{B}_b^{k^{r-i}}$ and such that L_i is $k(n)$-blockwise reducible to L_{i+1} via the (same) blockwise reduction g. Furthermore, let g' be as in Definition 14, and let $t_{g'}\colon \mathbb{N}_+ \to \mathbb{N}_+$ be non-decreasing and such that, for every $w' \in \mathfrak{B}_b^k$, $g'(w')$ is computable in place by an SCA in $O(t_{g'}(|w'|))$ time. Finally, let $L_{r(n)} \in \mathsf{SCA}[t]$ for some function $t\colon \mathbb{N}_+ \to \mathbb{N}_+$. Then, $L \in \mathsf{SCA}[r(n) \cdot t_{g'}(O(k(n) \cdot b(n))) + O(b(n)) + t(b(n))]$.*

Proof. We consider the SCA S which, given $w \in \mathfrak{B}_b^{k^r}$, repeatedly applies the reduction g, where each application of g is computed by applying g' on each group of relevant blocks (i.e., the w_i' from Definition 14) in parallel.

One detail to note is that this results in the same procedure P being applied to different groups of blocks in parallel, but it may be so that P requires more time for one group of blocks than for the other. Thus, we allow the entire process to be carried out asynchronously but require that, for each group of blocks, the respective results be present before each execution of P is started. (One way of realizing this, for instance, is having the first block in the group send a signal across the whole group to ensure all inputs are available and, when it arrives at the last block in the group, another signal is sent to trigger the start of P.)

Using that $t_{g'}$ is non-decreasing and that g' is non-stretching, the time needed for each execution of P is $t_{g'}(|w_i'|) \in t_{g'}(O(k(n) \cdot b(n)))$ (which is not impacted by the considerations above) and, since there are $r(n)$ reductions in total, we have $r(n) \cdot t_{g'}(O(k(n) \cdot b(n)))$ time in total. Once a single block is left, the cells in this block synchronize themselves and then behave as in the SCA S' for $L_{r(n)}$ guaranteed by the assumption; using a standard synchronization algorithm, this requires $O(b(n))$ for the synchronization, plus $t(b(n))$ time for emulating S'. $\square$

3.2   Block Languages and Parallel Computation

In this section, we prove the first limitation of SCAs discussed in the introduction (Lemma 17) and which renders them unable of accepting the languages PAR, MOD_q, MAJ, and THR_k (defined next) in sublinear time. Nevertheless, as is shown in Proposition 21, the *block versions* of these languages can be accepted quite efficiently. This motivates the block word presentation for inputs; that is, this first limitation concerns only the *presentation* of instances (and, hence, is not a *computational* limitation of SCAs).

Let $q > 2$ and let $k \colon \mathbb{N}_+ \to \mathbb{N}_+$ be constructible in place by a CA in at most $t_k(n)$ time for some $t_k \colon \mathbb{N}_+ \to \mathbb{N}_+$. Additionally, let PAR (resp., MOD_q; resp., MAJ; resp., THR_k) be the language consisting of every word $w \in \{0,1\}^+$ for which $|w|_1$ is even (resp., $|w|_1 = 0 \pmod q$; resp., $|w|_1 \geq |w|_0$; resp., $|w|_1 \geq k(|w|)$).

The following is a simple limitation of sublinear-time CA models such as ACAs (see also [26]) which we show also to hold for SCAs.

Lemma 17. *Let S be an SCA with input alphabet Σ, and let $x \in \Sigma$ be such that there is a minimal $t \in \mathbb{N}_+$ for which $\Delta_S^t(y) = \varepsilon$, where $y = x^{2t+1}$ (i.e., the symbol x concatenated $2t + 1$ times with itself). Then, for every $z_1, z_2 \in \Sigma^+$, $w = z_1 y z_2 \in L(S)$ holds if and only if for every $i \in \mathbb{N}_0$ we have $w_i = z_1 y x^i z_2 \in L(S)$.*

Proof. Given w and i as above, we show $w_i \in L(S)$; the converse is trivial. Since w and w_i both have $z_1 y$ as prefix and $\Delta_S^{t'}(y) \neq \varepsilon$ for $t' < t$, if S accepts w in t' steps, then it also accepts w_i in t' steps. Thus, assume S accepts w in $t' \geq t$ steps, in which case it suffices to show $\Delta_S^t(w) = \Delta_S^t(w_i)$. To this end, let α_j for $j \in [0, t]$ be such that $\alpha_0 = x$ and $\alpha_{j+1} = \delta(\alpha_j, \alpha_j, \alpha_j)$. Hence, $\Delta(\alpha_j^{k+2}) = \alpha_{j+1}^k$ holds for every $k \in \mathbb{N}_+$ (and $j < t$) and, by an inductive argument as well as by the assumption on y (i.e., $\alpha_t = \otimes$), $\Delta_S^t(yx^i) = \Delta_S^t(\alpha_0^{2t+i+1}) = \varepsilon$. Using this along with $|y| \geq t$ and $y \in \{x\}^+$, we have $\Delta_S^t(q^t z_1 y x^i) = \Delta_S^t(q^t z_1 y)$ and $\Delta_S^t(y x^i z_2 q^t) = \Delta_S^t(x^i y z_2 q^t) = \Delta_S^t(y z_2 q^t)$; hence, $\Delta_S^t(w) = \Delta_S^t(w_i)$ follows. $\square$

An implication of Lemma 17 is that every unary language $U \in \text{SCA}[o(n)]$ is either finite or cofinite. As $\text{PAR} \cap \{1\}^+$ is neither finite nor cofinite, we can prove:

Proposition 18. PAR $\notin$ SCA[$o(n)$] (where n is the input length).

Proof. Let S be an SCA with $L(S) = $ PAR. We show S must have $\Omega(n)$ time complexity on inputs from the infinite set $U = \{1^{2m} \mid m \in \mathbb{N}_+\} \subset$ PAR. If $\Delta_S^t(1^{2t+1}) = \varepsilon$ for some $t \in \mathbb{N}_0$, then, by Lemma 17, $L(S) \cap \{1\}^+$ is either finite or cofinite, which contradicts $L(S) = $ PAR. Hence, $\Delta_S^t(1^{2t+1}) \neq \varepsilon$ for every $t \in \mathbb{N}_0$. In this case, the trace of cell zero on input $w = 11^{2t+1}1$ in the first t steps is the same as that on input $w' = 11^{2t+1}11$. Since $w \in$ PAR if and only if $w' \notin$ PAR, it follows that S has $\Omega(t) = \Omega(n)$ time complexity on U.   □

Corollary 19. REG $\not\subseteq$ SCA[$o(n)$].

The argument above generalizes to MOD_q, MAJ, and THR_k with $k \in \omega(1)$. For MOD_q, consider $U = \{1^{qm} \mid m \in \mathbb{N}_+\}$. For MAJ and THR_k, set $U = \{0^m1^m \mid m \in \mathbb{N}_+\}$ and $U = \{0^{n-k(n)}1^{k(n)} \mid n \in \mathbb{N}_+\}$, respectively; in this case, U is not unary, but the argument easily extends to the unary suffixes of the words in U.

Corollary 20. MOD_q, MAJ $\notin$ SCA[$o(n)$]. Also, $\mathsf{THR}_k \in$ SCA[$o(n)$] if and only if $k \in O(1)$.

The *block versions* of these languages, however, are not subject to the limitation above:

Proposition 21. For $L \in \{\mathsf{PAR}, \mathsf{MOD}_q, \mathsf{MAJ}\}$, $\mathsf{Block}_n(L) \in$ SCA[$(\log N)^2$], where $N = N(n)$ is the input length. Also, $\mathsf{Block}_n(\mathsf{THR}_k) \in$ SCA[$(\log N)^2 + t_k(n)$].

Proof. Given $L \in \{\mathsf{PAR}, \mathsf{MOD}_q, \mathsf{MAJ}, \mathsf{THR}_k\}$, we construct an SCA S for $L' = \mathsf{Block}_n(L)$ with the purported time complexity. Let $w \in \mathfrak{B}_n^m$ be an input of S. For simplicity, we assume that, for every such w, $m = m(n) = 2^n$ is a power of two; the argument extends to the general case in a simple manner. Hence, we have $N = |w| = n \cdot m$ and $n = \log m \in \Theta(\log N)$.

Let $L_0 \subset \mathfrak{B}_n^m$ be the language containing every such block word $w \in \mathfrak{B}_n^m$ for which, for y_i as in Definition 11 and $y = \sum_{i=0}^{m-1} y_i$, we have $f_L(y) = f_{L,n}(y) = 0$, where $f_{\mathsf{PAR}}(y) = y \bmod 2$, $f_{\mathsf{MOD}_q}(y) = y \bmod q$, $f_{\mathsf{MAJ}}(y) = 0$ if and only if $y \geq 2^{n-1}$, and $f_{\mathsf{THR}_k}(y) = 0$ if and only if $y \geq k(n)$. Thus, (under the previous assumption) we have $L_0 = L'$ (and, in the general case, $L_0 = L' \cap \mathfrak{B}_n^{2^n}$).

Then, L_0 is 2-blockwise reducible to a language $L_1 \subseteq \mathfrak{B}_n^{m/2}$ by mapping every $(n, 2, n)$-block word of the form $\binom{\mathrm{bin}_n(2x)}{y_{2x}}\#\binom{\mathrm{bin}_n(2x+1)}{y_{2x+1}}$ with $x \in [0, 2^{n-1} - 1]$ to $\binom{\mathrm{bin}_n(x)}{y_{2x}+y_{2x+1}}$. To do so, it suffices to compute $\mathrm{bin}_n(x)$ from $\mathrm{bin}_n(2x)$ and add the y_{2x} and y_{2x+1} values in the lower track; using basic CA arithmetic and cell communication techniques, this is realizable in $O(n)$ time. Repeating this procedure, we obtain a chain of languages $L_0, \ldots, L_n$ such that L_i is 2-blockwise reducible to L_{i+1} in $O(n)$ time. By Lemma 16, $L' \in$ SCA[$n^2 + t(n)$] follows, where $t \colon \mathbb{N}_+ \to \mathbb{N}_0$ is such that $L_n \in$ SCA[t]. For $L \in \{\mathsf{PAR}, \mathsf{MOD}_q, \mathsf{MAJ}\}$, checking the above condition on $f_L(y)$ can be done in $t(n) \in O(n)$ time; as for $L = \mathsf{THR}_k$, we must also compute k, so we have $t(n) \in O(n + t_k(n))$.

The general case follows from adapting the above reductions so that words with an odd number of blocks are also accounted for (e.g., by ignoring the last block of w and applying the reduction on the first $m - 1$ blocks).   □

3.3   An Optimal SCA Lower Bound for a Block Language

Corollary 19 already states SCAs are strictly less capable than streaming algorithms. However, the argument bases exclusively on long unary subwords in the input (i.e., Lemma 17) and, therefore, does not apply to block languages. Hence Theorem 5, which shows SCAs are more limited than streaming algorithms *even considering only block languages*:

Theorem 5. *There is a language L_1 for which $\mathsf{Block}_n(L_1) \notin \mathsf{SCA}[o(N/\log N)]$ (N being the instance length) can be accepted by an $O(\log N)$-space streaming algorithm with $\tilde{O}(\log N)$ update time.*

Let L_1 be the language of words $w \in \{0,1\}^+$ such that $|w| = 2^n$ is a power of two and, for $i = w(0)w(1)\cdots w(n-1)$ (seen as an n-bit binary integer), $w(i) = 1$. It is not hard to show that its block version $\mathsf{Block}_n(L_1)$ can be accepted by an $O(\log m)$-space streaming algorithm with $\tilde{O}(\log m)$ update time.

The $O(N/\log N)$ upper bound for $\mathsf{Block}_n(L_1)$ is optimal since there is an $O(N/\log N)$ time SCA for it: Shrink every block to its respective bit (i.e., the y_i from Definition 11), reducing the input to a word w' of $O(N/\log N)$ length; while doing so, mark the bit corresponding to the n-th block. Then shift the contents of the first n bits as a counter that decrements itself every new cell it visits and, when it reaches zero, signals acceptance if the cell it is currently at contains a 1. Using counter techniques as in [27, 29], this requires $O(|w'|)$ time.

The proof of Theorem 5 bases on communication complexity. The basic setting is a game with two players A and B (both with unlimited computational resources) which receive inputs w_A and w_B, respectively, and must produce an answer to the problem at hand while exchanging a limited amount of bits. We are interested in the case where the concatenation $w = w_A w_B$ of the inputs of A and B is an input to an SCA and A must output whether the SCA accepts w. More importantly, we analyze the case where *only B is allowed to send messages*, that is, the case of *one-way* communication.[2]

Definition 22 (One-way communication complexity). *Let $m, f \colon \mathbb{N}_+ \to \mathbb{N}_+$ be functions with $0 < m(N) \le N$. A language $L \subseteq \Sigma^+$ is said to have $(m\text{-})$one-way communication complexity f if there are families of algorithms (with unlimited computational resources) $(A_N)_{N \in \mathbb{N}_+}$ and $(B_N)_{N \in \mathbb{N}_+}$ such that the following holds for every $w \in \Sigma^*$ of length $|w| = N$, where $w_A = w[0, m(N) - 1]$ and $w_B = w[m(N), N - 1]$:*

1. $|B_N(w_B)| \le f(N)$; and
2. $A_N(w_A, B(w_B)) = 1$ (i.e., accept) if and only if $w \in L$.

$\mathfrak{C}^m_{\mathrm{ow}}(L)$ indicates the (pointwise) minimum over all such functions f.

---

[2] One-way communication complexity can also been defined as the maximum over *both* communication directions (i.e., B to A and A to B; see [9] for an example in the setting of CAs). Since our goal is to prove a *lower bound* on communication complexity, it suffices to consider a single (arbitrary) direction (in this case B to A).

Note that A_N and B_N are nonuniform, so the length N of the (complete) input w is known implicitly by both algorithms.

Lemma 23. *For any computable $t\colon \mathbb{N}_+ \to \mathbb{N}_+$ and m as in Definition 22, if $L \in \mathsf{SCA}[t]$, then $\mathfrak{C}^m_{\mathrm{ow}}(L)(N) \in O(t(N))$.*

The proof idea is to have A and B simulate the SCA for L simultaneously, with A maintaining the first half c_A of the SCA configuration and B the second half c_B. (Hence, A is aware of the leftmost active state in the SCA and can detect whether the SCA accepts or not.) The main difficulty is guaranteeing that A and B can determine the states of the cells on the right (resp., left) end of c_A (resp., c_B) despite the respective local configurations "overstepping the boundary" between c_A and c_B. Hence, for each step in the simulation, B communicates the states of the two leftmost cells in c_B; with this, A can compute the states of all cells of c_A in the next configuration as well as that of the leftmost cell α of c_B, which is added to c_A. (See Fig. 2 for an illustration.) This last technicality is needed due to one-way communication, which renders it impossible for B to determine the next state of α (since its left neighbor is in c_A and B cannot receive messages from A). As the simulation requires at most $t(N)$ steps and B sends $O(1)$ information at each step, this yields the purported $O(t(N))$ upper bound.

The attentive reader may have noticed this discussion does not address the fact that the SCA may shrink; indeed, we shall also prove that shrinking does not interfere with this strategy.

Proof. Let S be an SCA for L with time complexity $O(t)$. Furthermore, let Q be the state set of S and $q \in Q$ its inactive state. We construct algorithms A_N and B_N as in Definition 22 and such that $|B_N(w_B)| \le 2\log(|Q|) \cdot t(N)$.

Fix $N \in \mathbb{N}_+$ and an input $w \in \Sigma^N$. For $w_B^0 = w_B q^{2t(N)+2}$ and $w_B^{i+1} = \Delta_{\mathsf{S}}(w_B^i)$ for $i \in \mathbb{N}_0$, B_N computes and outputs the concatenation

$$B_N(w_B) = w_B^0(0)w_B^0(1)w_B^1(0)w_B^1(1)\cdots w_B^{t(N)}(0)w_B^{t(N)}(1).$$

Similarly, let $w_A^0 = q^{2t(N)+2}w_A$ and $w_A^{i+1} = \Delta_{\mathsf{S}}(w_A^i w_B^i(0)w_B^i(1))$ for $i \in \mathbb{N}_0$. A computes $t(N)$ and w_A^i for $i \in [0, t(N)]$ and accepts if there is any j such that $w_A^i(j)$ is an accept state of S and $w_A^i(j') = q$ for all $j' < j$; otherwise, A rejects.

To prove the correctness of A, we show by induction on $i \in \mathbb{N}_0$: $w_A^i w_B^i = \Delta_{\mathsf{S}}^i(q^{2t(n)+2}wq^{2t(n)+2})$. Hence, the $w_A^i(j)$ of above corresponds to the state of cell zero in step i of S, and it follows that A accepts if and only if S does. The induction basis is trivial. For the induction step, let $w' = \Delta_{\mathsf{S}}(w_A^i w_B^i)$. Using the induction hypothesis, it suffices to prove $w_A^{i+1} w_B^{i+1} = w'$. Note first that, due to the definition of w_A^{i+1} and w_B^{i+1}, we have $w' = \Delta_{\mathsf{S}}(w_A^i)\alpha\beta\Delta_{\mathsf{S}}(w_B^i)$, where $\alpha, \beta \in Q \cup \{\varepsilon\}$. Let $\alpha_1 = w_A^i(|w_A^i| - 2)$, $\alpha_2 = w_A^i(|w_A^i| - 1)$, and $\alpha_3 = w_B^i(0)$ and notice $\alpha = \delta(\alpha_1, \alpha_2, \alpha_3)$; the same is true for β and $\beta_1 = \alpha_2$, $\beta_2 = \alpha_3$, and $\beta_3 = w_B^i(1)$. Hence, we have $w_A^{i+1} = \Delta_{\mathsf{S}}(w_A^i)\alpha\beta$, and the claim follows. $\square$

We are now in position to prove Theorem 5.

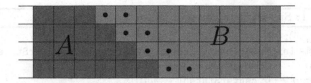

Fig. 2. Simulating an SCA with low one-way communication complexity. (For simplicity, in this example the SCA does not shrink.) B communicates the states of the cells marked with "•". The colors indicate which states are computed by each player. (Color figure online)

Proof (Proof of Theorem 5). We prove that, for our language L_1 of before and $m(n) = n(n + 1)$ (i.e., A_N receives the first n input blocks), $\mathfrak{C}_{\text{ow}}^m(\text{Block}_n(L_1))(N) \geq 2^n - n$. Since the input length is $N \in \Theta(n \cdot 2^n)$, the claim then follows from the contrapositive of Lemma 23.

The proof is by a counting argument. Let A_N and B_N be as in Definition 22, and let $Y = \{0,1\}^{2^n-n}$. The basic idea is that, for the same input w_A, if B_N is given different inputs w_B and w'_B but $B_N(w_B) = B_N(w'_B)$, then $w = w_A w_B$ is accepted if and only if $w' = w_A w'_B$ is accepted. Hence, for any $y, y' \in Y$ with $y \neq y'$, we must have $B_N(w_B) \neq B_N(w'_B)$, where $w_B, w'_B \in \mathfrak{B}_n^{2^n-n}$ are the block word versions of y and y', respectively; this is because, letting $j \in [0, 2^n - n]$ be such that $y(j) \neq y'(j)$ and $z = \text{bin}_n(n + j)$, precisely one of the words zy and zy' is in L_1 (and the other not). Finally, note there is a bijection between Y and the set Y' of block words in $\mathfrak{B}_n^{2^n-n}$ whose block numbering starts with $n + 1$ (i.e., $x_0 = n + 1$, where x_0 is as in Definition 11) and with block entries of the form $a0^{n-1}$ where $a \in \{0,1\}$ (i.e., Y' is essentially the block version of Y as in Definition 13 but where we set $x_0 = n + 1$ instead of $x_0 = 0$). We conclude $\mathfrak{C}_{\text{ow}}^m(\text{Block}_n(L_1))(N) \geq |Y'| = |Y| = 2^n - n$, and the claim follows. □

4   Simulation of an SCA by a Streaming Algorithm

In this section, we recall and prove:

Theorem 4. *Let $t \colon \mathbb{N}_+ \to \mathbb{N}_+$ be computable by an $O(t)$-space random access machine (as in Definition 6) in $O(t \log t)$ time. Then, if $L \in \text{SCA}[t]$, there is an $O(t)$-space streaming algorithm for L with $O(t \log t)$ update and $O(t^2 \log t)$ reporting time.*

Before we state the proof, we first introduce some notation. Having fixed an input w, let $c_t(i)$ denote the state of cell i in step t on input w. Note that here we explicitly allow $c_t(i)$ to be the state $\otimes$ and also disregard any changes in indices caused by cell deletion; that is, $c_t(i)$ refers to the *same cell* i as in the initial configuration c_0 (of Definition 7; see also the discussion following Definition 8). For a finite, non-empty $I = [a, b] \subseteq \mathbb{Z}$ and $t \in \mathbb{N}_0$, let $\text{nndcl}_t(I) = \max\{i \mid i < a, c_t(i) \neq \otimes\}$ denote the nearest non-deleted cell to the left of I; similarly, $\text{nndcr}_t(I) = \min\{i \mid i > b, c_t(i) \neq \otimes\}$ is the nearest such cell to the right of I.

Proof. Let S be an $O(t)$-time SCA for L. Using S, we construct a streaming algorithm A (Algorithm 1) for L and prove it has the purported complexities.

Algorithm 1: Streaming algorithm A

Compute $t(|w|)$;
Initialize lists leftIndex, centerIndex, leftState, and centerState;
leftIndex$[0] \leftarrow -1$; leftState$[0] \leftarrow q$;
centerIndex$[0] \leftarrow 0$; centerState$[0] \leftarrow w(0)$;
next $\leftarrow 1$;
$j_0 \leftarrow 0$;
for $\tau \leftarrow 0, \ldots, t(|w|) - 1$ do

  A $j \leftarrow j_0$;

  B   if next $< |w|$ then

  C   rightIndex $\leftarrow$ next; rightState $\leftarrow w(\text{next})$;
           next $\leftarrow$ next $+ 1$;

       else

  D   rightIndex $\leftarrow |w|$; rightState $\leftarrow q$;
           $j_0 \leftarrow j_0 + 1$;

       end

       while $j \leq \tau$ do

  E   newRightIndex $\leftarrow$ centerIndex$[j]$;
           newRightState $\leftarrow \delta(\text{leftState}[j], \text{centerState}[j], \text{rightState})$;
           leftIndex$[j] \leftarrow$ centerIndex$[j]$; leftState$[j] \leftarrow$ centerState$[j]$;
           centerIndex$[j] \leftarrow$ rightIndex; centerState$[j] \leftarrow$ rightState;
           rightIndex $\leftarrow$ newRightIndex; rightState $\leftarrow$ newRightState;

  F   if rightState $= \otimes$ then goto A $j \leftarrow j + 1$;

       end

       leftIndex$[\tau + 1] \leftarrow -1$; leftState$[\tau + 1] \leftarrow q$;
       centerIndex$[\tau + 1] \leftarrow$ rightIndex; centerState$[\tau + 1] \leftarrow$ rightState;

  G   if centerState$[\tau + 1] = a$ then accept

end
reject;

Construction. Let w be an input to A. To decide L, A computes the states of the cells of S in the time steps up to $t(|w|)$. In particular, A sequentially determines the state of the leftmost active cell in each of these time steps (starting from the initial configuration) and accepts if and only if at least one of these states is accepting. To compute these states efficiently, we use an approach based on dynamic programming, reusing space as the computation evolves.

A maintains lists leftIndex, leftState, centerIndex, and centerState and which are indexed by every step j starting with step zero and up to the current step τ. The lists leftIndex and centerIndex store cell indices while leftState and centerState store the states of the respective cells, that is, leftState$[j] = c_j(\text{leftIndex}[j])$ and centerState$[j] = c_j(\text{centerIndex}[j])$.

Recall the state $c_{j+1}(y)$ of a cell y in step $j + 1$ is determined exclusively by the previous state $c_j(y)$ of y as well as the states $c_j(x)$ and $c_j(z)$ of the left and right neighbors x and z (respectively) of y in the previous step j (i.e., $x = \text{nndcl}_j(y)$ and $z = \text{nndcr}_j(y)$). In the variables maintained by A,

x and $c_j(x)$ correspond to leftIndex$[j]$ and leftState$[j]$, respectively, and y and $c_j(y)$ to centerIndex$[j]$ and centerState$[j]$, respectively. z and $c_j(z)$ are not stored in lists but, rather, in the variables rightIndex and rightState (and are determined dynamically). The cell indices computed (i.e., the contents of the lists leftIndex and centerIndex and the variables rightIndex and newRightIndex) are not actually used by A to compute states and are inessential to the algorithm itself; we use them only to simplify the proof of correctness below (and, hence, do not count them towards the space complexity of A).

In each iteration of the **for** loop, A determines $c_{\tau+1}(z_0^\tau)$, where z_0^τ is the leftmost active cell of S in step τ, and stores it centerState$[\tau + 1]$. next is the index of the next symbol of w to be read (or $|w|$ once every symbol has been read), and j_0 is the minimal time step containing a cell whose state must be known to determine $c_{\tau+1}(z_0^t)$ and remains 0 as long as next $< |w|$. Hence, the termination of A is guaranteed by the finiteness of w, that is, next can only be increased a finite number of times and, once all symbols of w have been read (i.e., the condition in line B no longer holds), by the increment of j_0 in line D.

In each iteration of the **while** loop, the algorithm starts from a local configuration in step j of a cell y = centerIndex$[j]$ with left neighbor x = leftIndex$[j]$ = nndcl$_j(y)$ and right neighbor z = rightIndex$[j]$ = nndcl$_j(y)$. It then computes the next state $c_{j+1}(y)$ of y and sets y as the new left cell and z as the new center cell for step j. As long as it is not deleted (i.e., $c_{j+1}(y) \neq \otimes$), y then becomes the right cell for step $j+1$. In fact, this is the only place (line F) in the algorithm where we need to take into consideration that S is a shrinking (and not just a regular) CA. The strategy we follow here is to continue computing states of cells to the right of the current center cell (i.e., y = centerIndex$[j]$) until the first cell to its right which has not deleted itself (i.e., nndcr$_j(y)$) is found. With this non-deleted cell we can then proceed with the computation of the state of centerIndex$[j + 1]$ in step $j + 1$. Hence, if y has deleted itself, to compute the state of the next cell to its right we must either read the next symbol of w or, if there are no symbols left, use quiescent cell number $|w|$ as right neighbor in step j_0, computing states up until we are at step j again (hence the **goto** instruction).

Correctness. The following invariants hold for both loops in A:

1. centerIndex$[\tau]$ = $\min\{z \in \mathbb{N}_0 \mid c_\tau(z) \neq \otimes\}$, that is, centerIndex$[\tau]$ is the leftmost active cell of S in step j.
2. If $j \leq \tau$, then rightIndex = nndcr$_j$(centerIndex$[j]$) and rightState = c_j(rightIndex).
3. For every $j' \in [j_0, \tau]$:
   - leftIndex$[j']$ = nndcl$_{j'}$(centerIndex$[j']$),
   - leftState$[j']$ = $c_{j'}$(leftIndex$[j']$); and
   - centerState$[j']$ = $c_{j'}$(centerIndex$[j']$).

These can be shown together with the observation that, following the assignment of newRightIndex and newRightState in line E, we have newRightState = c_{j+1}(newRightIndex) and, in case newRightState $\neq \otimes$ and $j < \tau$, then also newRightIndex = nndcr$_j$(centerIndex[$j + 1$]). Using the above, it follows that after the execution of the **while** loop we have $j = \tau + 1$, rightState $\neq \otimes$, and rightState = $c_{\tau+1}$(rightIndex). Since then rightIndex = centerIndex[$j - 1$] = centerIndex[τ], we obtain rightIndex = $\min\{z \in \mathbb{N}_0 \mid c_{\tau+1}(z) \neq \otimes\}$. Hence, as centerState[$\tau + 1$] = rightState = $c_{\tau+1}$(rightIndex) holds in line G, if A then accepts, so does S accept w in step τ. Conversely, if A rejects, then S does not accept w in any step $\tau \leq t(|w|)$.

Complexity. The space complexity of A is dominated by the lists leftState and centerState, which has $O(t(|w|))$ many entries of $O(1)$ size. As mentioned above, we ignore the space used by the lists leftIndex and centerIndex and the variables rightIndex and newRightIndex since they are inessential (i.e., if we remove them as well as all instructions in which they appear, the algorithm obtained is equivalent to A).

As for the update time, note each list access or arithmetic operation costs $O(\log t(|w|))$ time (since $t(|w|)$ upper bounds all numeric variables). Every execution of the **while** loop body requires then $O(\log t(|w|))$ time and, since, there are at most $O(t(|w|))$ executions between any two subsequent reads (i.e., line C), this gives us the purported $O(t(|w|) \log t(|w|))$ update time.

Finally, for the reporting time of A, as soon as $i = |w|$ holds after execution of line C (i.e., A has completed reading its input) we have that the **while** loop body is executed at most $\tau - j + 1$ times before line C is reached again. Every time this occurs (depending on whether line C is reached by the **goto** instruction or not), either j_0 or both j_0 and τ are incremented. Hence, since $\tau \leq t(|w|)$, we have an upper bound of $O(t(|w|)^2)$ executions of the **while** loop body, resulting (as above) in an $O(t(|w|)^2 \log t(|w|))$ reporting time in total.   □

5   Hardness Magnification for Sublinear-Time SCAs

Let $K > 0$ be constant such that, for any function $s\colon \mathbb{N}_+ \to \mathbb{N}_+$, every circuit of size at most $s(n)$ can be described by a binary string of length at most $\ell(n) = Ks(n) \log s(n)$. In addition, let $\bot$ denote a string (of length at most $\ell(n)$) such that no circuit of size at most $s(n)$ has $\bot$ as its description. Furthermore, let Merge[s] denote the following search problem (adapted from [17]):

Given: the binary representation of $n \in \mathbb{N}_+$, the respective descriptions (padded to length $\ell(n)$) of circuits C_0 and C_1 such that $|C_i| \leq s(n)$, and $\alpha, \beta, \gamma \in \{0,1\}^n$ with $\alpha \leq \beta \leq \gamma < 2^n$.
Find: the description of a circuit C with $|C| \leq s(n)$ and such that $\forall x \in [\alpha, \beta-1]$: $C(x) = C_0(x)$ and $\forall x \in [\beta, \gamma-1] : C(x) = C_1(x)$; if no such C exists or $C_i = \bot$ for any i, answer with $\bot$.

Note that the decision version of Merge[s], that is, the problem of determining whether a solution to an instance Merge[s] exists is in Σ_2^p. Moreover, Merge[s] is Turing-reducible (in polynomial time) to a decision problem very similar to Merge[s] and which is also in Σ_2^p, namely the decision version of Merge[s] but with the additional requirement that the description of C admits a given string v of length $|v| \leq s(n)$ as a prefix.[3]

We now formulate our main theorem concerning SCAs and MCSP:

Theorem 24. *Let* $s \colon \mathbb{N}_+ \to \mathbb{N}_+$ *be constructible in place by a CA in* $O(s(n))$ *time. Furthermore, let* $m = m(n)$ *denote the maximum instance length of* Merge$[s]$, *and let* $f, g \colon \mathbb{N}_+ \to \mathbb{N}_+$ *with* $f(m) \geq g(m) \geq m$ *be constructible in place by a CA in* $O(f(m))$ *time and* $O(g(m))$ *space. Then, for* $b(n) = \lfloor g(m)/2 \rfloor$, *if* Merge$[s]$ *is computable in place by a CA in at most* $f(m)$ *time and* $g(m)$ *space, then the search version of* Block$_b$(MCSP$[s]$) *is computable by an SCA in* $O(n \cdot f(m))$ *time, where the instance size of the latter is in* $\Theta(2^n \cdot b(n))$.

We are particularly interested in the repercussions of Theorem 24 *taken in the contrapositive.* Since $\mathsf{P} = \mathsf{NP}$ implies $\mathsf{P} = \Sigma_2^p$, it also implies there is a polytime Turing machine for Merge$[s]$; since a CA can simulate a Turing machine with no time loss, for m as above we obtain:

Theorem 3. *For a certain* $m \in \mathsf{poly}(s(n))$, *if* Block$_b$(MCSP$[s]$) $\notin$ SCA$[n \cdot f(m)]$ *for every* $f \in \mathsf{poly}(m)$ *and* $b \in O(f)$, *then* $\mathsf{P} \neq \mathsf{NP}$.

We now turn to the proof of Theorem 24, which follows [17] closely. First, we generalize blockwise reductions (see Definition 15) to search problems:

Definition 25 (Blockwise reducible (for search problems)). *Let L and L' be block languages that correspond to search problems S and S', respectively. Also, for an instance x, let $S(x)$ (resp., $S'(x)$) denote the set of solutions for x under the problem S (resp., S'). Then L is said to be (k-)blockwise reducible to L' if there is a computable k-blockwise map $g \colon \mathfrak{B}_b^{km} \to \mathfrak{B}_b^m$ such that, for every $w \in \mathfrak{B}_b^{km}$, we have $S(w) = S'(g(w))$.*

Notice Lemma 16 readily generalizes to blockwise reductions in this sense.

Next, we describe the set of problems that we shall reduce Block$_b$(MCSP$[s]$) to. Let $r \colon \mathbb{N}_+ \to \mathbb{N}_+$ be a function. There is a straightforward 1-blockwise reduction from Block$_b$(MCSP$[s]$) to (a suitable block version of) the following search problem Merge$_r[s]$:

Given: the binary representation of $n \in \mathbb{N}_+$ and the respective descriptions (padded to length $\ell(n)$) of circuits $C_1, \ldots, C_r$, where $|C_i| \leq s(n)$ for every i and $r = r(n)$.

Find: (the description of) a circuit C with $|C| \leq s(n)$ and such that, for every i and every $x \in [(i-1) \cdot 2^n/r, i \cdot 2^n/r - 1]$, $C(x) = C_i(x)$; if no such C exists or $C_i = \bot$ for any i, answer with $\bot$.

In particular, for the reduction mentioned above, we shall use $r = 2^n$. Evidently, Merge$_r[s]$ is a generalization of the problem Merge$[s]$ defined previously and, more importantly, every instance of Merge$_r[s]$ is simply a concatenation of $r/2$ many Merge$[s]$ instances where α, β, and γ are given implicitly. Using the

---

[3] This is a fairly common construction in complexity theory for reducing search to decision problems; refer to [10] for the same idea applied in other contexts.

assumption that Merge[s] is computable by a CA in at most $f(m)$ time and $g(m)$ space, we can solve each such instance in parallel, thus producing an instance of Merge$_{r/2}$[s] (i.e., halving r). This yields a 2-blockwise reduction from (the respective block versions of) Merge$_r$[s] to Merge$_{r/2}$[s] (cnf. the proof of Proposition 21). Using Lemma 16 and that Merge$_1$[s] is trivial, we obtain the purported SCA for Block$_b$(MCSP[s]).

Proof. Let n be fixed, and let $r = 2^n$. First, we describe the 1-blockwise reduction from Block$_b$(MCSP[s]) to a block version of Merge$_r$[s] (which we shall describe along with the reduction). Let T_a denote the (description of the) trivial circuit that is constant $a \in \{0,1\}$, that is, $T_a(x) = a$ for every $x \in \{0,1\}^n$. Then we map each block $\binom{\text{bin}_n(x)}{y 0^{b(n)-1}}$ with $y \in \{0,1\}$ to the block $\binom{\text{bin}_n(x)}{T_y \pi}$, where $\pi \in \{0\}^*$ is a padding string so that the block length $b(n)$ is preserved. (This is needed to ensure enough space is available for the construction; see the details further below.) It is evident this can be done in time $O(b(n))$ and (since we just translate the truth-table 0 and 1 entries to the respective trivial circuits) that the reduction is correct, that is, that every solution to the original Block$_b$(MCSP[s]) instance must also be a solution of the produced instance of (the resulting block version of) Merge$_r$[s] and vice-versa.

Next, maintaining the block representation described above, we construct the 2-blockwise reduction from the respective block versions of Merge$_\rho$[s] to Merge$_{\rho/2}$[s], where $\rho = 2^k$ for some $k \in [1, n]$. Let A denote the CA that, by assumption, computes a solution to an instance of Merge[s] in place in at most $f(m)$ time and $g(m)$ space. Then, for $j \in [0, \rho/2 - 1]$, we map each pair $\binom{\text{bin}_n(2j)}{C_0 \pi_0} \# \binom{\text{bin}_n(2j+1)}{C_1 \pi_1}$ of blocks (where $\pi_0, \pi_1 \in \{0\}^*$ again are padding strings) to $\binom{\text{bin}_n(j)}{C \pi}$, where $\pi \in \{0\}^*$ is a padding string (as above) and C is the circuit produced by A for $\alpha = 2j \cdot 2^n/\rho$, $\beta = (2j + 1) \cdot 2^n/\rho$, and $\gamma = (2j + 2) \cdot 2^n/\rho$.

To actually execute A, we need $g(m)$ space (which is guaranteed by the block length $b(n)$) and, in addition, to prepare the input so it is in the format expected by A (i.e., eliminating the padding between the two circuit descriptions and writing the representations of α, β, and γ), which can be performed in $O(b(n)) \subseteq O(g(m)) \subseteq O(f(m))$ time. For the correctness, suppose the above reduces an instance of Merge$_\rho$[s] with circuits $C_1, \ldots, C_\rho$ to an instance of Merge$_{\rho/2}$[s] with circuits $D_1, \ldots, D_{\rho/2}$ (and no $\bot$ was produced). Then, a circuit E is a solution to the latter if and only if $E(x) = D_i(x)$ for every i and $x \in [(i-1) \cdot 2^n/(\rho/2), i \cdot 2^n/(\rho/2) - 1]$. Using the definition of Merge[s], every D_i must satisfy $D_i(x) = C_{2i-1}(x)$ and $D_i(y) = C_{2i}(y)$ for $x \in [(2i - 2) \cdot 2^n/\rho, (2i - 1) \cdot 2^n/\rho - 1]$ and $y \in [(2i - 1) \cdot 2^n/\rho, 2i \cdot 2^n/\rho - 1]$. Hence, E agrees with $C_1, \ldots, C_\rho$ if and only if it agrees with $D_1, \ldots, D_{\rho/2}$ (on the respective intervals).

Since $s(n) \geq n$ and Merge$_1$[s] is trivial (i.e., it can be accepted in $O(b(n))$ time), applying the generalization of Lemma 16 to blockwise reductions for search problems completes the proof.   □

Comparison with [17]. We conclude this section with a comparison of our result and proof with [17]. The most evident difference between the statements of

Theorems 3 and 24 and the related result from [17] (i.e., Theorem 2) is that our results concern CAs (instead of Turing machines) and relate more explicitly to the time and space complexities of Merge[s]; in particular, the choice of the block length is tightly related with the space complexity of computing Merge[s]. As for the proof, notice that we only merge two circuits at a time, which makes for a smaller instance size m (of Merge[s]); this not only simplifies the proof but also minimizes the resulting time complexity of the SCA (as $f(m)$ is then smaller). Also, in our case, we make no additional assumptions regarding the first reduction from $\mathsf{Block}_b(\mathsf{MCSP}[s])$ to $\mathsf{Merge}_r[s]$; in fact, this step can be performed unconditionally. Finally, we note that our proof renders all blockwise reductions explicit and the connection to the self-reductions of [1] more evident. Despite these simplifications, the argument extends to generalizations of MCSP with similar structure and instance size (e.g., MCSP in the setting of Boolean circuits with oracle gates as in [17] or MCSP for multi-output functions as in [12]).

6 Concluding Remarks

Proving SCA Lower Bounds for $\mathsf{MCSP}[s]$. Recalling the language L_1 from the proof of Theorem 5, consider the intersection $L_1[s] = L_1 \cap \mathsf{MCSP}[s]$. Evidently, $L_1[s]$ is comparable in hardness to $\mathsf{MCSP}[s]$ (e.g., it is solvable in polynomial time using a single adaptive query to $\mathsf{MCSP}[s]$). By adapting the construction from the proof of Theorem 24 so the SCA additionally checks the L_1 property at the end in $\mathsf{poly}(s(n))$ time (e.g., using the circuit C produced to check whether $C(x) = 1$ for $x = C(0) \cdots C(n-1)$), we can derive a hardness magnification result for $L_1[s]$ too: If $\mathsf{Block}_b(L_1[s]) \notin \mathsf{SCA}[\mathsf{poly}(s(n))]$ (for every $b \in \mathsf{poly}(s(n))$), then $\mathsf{P} \neq \mathsf{NP}$. Using the methods from Sect. 3.3 and that there are $2^{\Omega(s(n))}$ many (unique) circuits of size $s(n)$ or less,[4] this means that, if $\mathsf{Block}_b(L_1[s]) \in \mathsf{SCA}[t(n)]$ for some $b \in \mathsf{poly}(n)$ and $t: \mathbb{N}_+ \to \mathbb{N}_+$, then $t \in \Omega(s(n))$. Hence, for an eventual proof of $\mathsf{P} \neq \mathsf{NP}$ based on Theorem 3, one would need to develop new techniques (see also the discussion below) to raise this bound at the very least beyond $\mathsf{poly}(s(n))$.

Seen from another angle, this demonstrates that, although we can prove a tight SCA worst-case lower bound for L_1 (Theorem 5), establishing similar lower bounds on instances of L_1 with low circuit complexity (i.e., instances which are also in $\mathsf{MCSP}[s]$) is at least as hard as showing $\mathsf{P} \neq \mathsf{NP}$. In other words, it is straightforward to establish a lower bound for L_1 using arbitrary instances, but it is absolutely non-trivial to establish similar lower bounds for *easy* instances of L_1 where instance hardness is measured in terms of circuit complexity.

---

[4] Let $K > 0$ be constant such that every Boolean function on m variables admits a circuit of size at most $K \cdot 2^m/m$. Setting $m = \lfloor \log s(n) \rfloor$, notice that, for sufficiently large n (and $s(n) \in \omega(1) \cap O(2^n/n)$), this gives us $s(n) \geq K \cdot 2^m/m$, thus implying that every Boolean function on $m \leq n$ variables admits a circuit of size at most $s(n)$. Since there are 2^{2^m} many such (unique) functions, it follows there are $2^{\Omega(s(n))}$ (unique) circuits of size at most $s(n)$.

The Proof of Theorem 24 and the Locality Barrier. In a recent paper [5], Chen et al. propose the concept of a *locality barrier* to explain why current lower bound proof techniques (for a variety of non-uniform computational models) do not suffice to show the lower bounds needed for separating complexity classes in conjunction with hardness magnification (i.e., in our case above a $\mathsf{poly}(s(n))$ lower bound that proves $\mathsf{P} \neq \mathsf{NP}$). In a nutshell, the barrier arises from proof techniques relativizing with respect to *local aspects* of the computational model at hand (in [5], concretely speaking, oracle gates of small fan-in), whereas it is known that a proof of $\mathsf{P} \neq \mathsf{NP}$ must not relativize [3].

The proof of Theorem 24 confirms the presence of such a barrier also in the uniform setting and concerning the separation of P from NP. Indeed, the proof mostly concerns the construction of an SCA where the overall computational paradigm of blockwise reductions (using Lemma 16) is unconditionally compatible with the SCA model (as exemplified in Proposition 21); the $\mathsf{P} = \mathsf{NP}$ assumption is needed exclusively so that the local algorithm for Merge[s] in the statement of the theorem exists. Hence, the result also holds *unconditionally* for SCAs that are, say, augmented with oracle access (in a plausible manner, e.g., by using an additional oracle query track and special oracle query states) to Merge[s]. (Incidentally, the same argument also applies to the proof of the hardness magnification result for streaming algorithms (i.e., Theorem 2) in [17], which also builds on the existence of a similar locally computable function.) In particular, this means the lower bound techniques from the proof of Theorem 5 do not suffice since they extend to SCAs having oracle access to any computable function.

Open Questions. We conclude with a few open questions:

- By weakening SCAs in some aspect, certainly we can establish an unconditional MCSP lower bound for the weakened model which, were it to hold for SCAs, would imply the separation $\mathsf{P} \neq \mathsf{NP}$ (using Theorem 3). *What forms of weakening* (conceptually speaking) are needed for these lower bounds? How are these related to the locality barrier discussed above?
- Secondly, we saw SCAs are strictly more limited than streaming algorithms. Proceeding further in this direction, can we identify *further (natural) models of computation that are more restricted than SCAs* (whether CA-based or not) and for which we can prove results similar to Theorem 24?
- Finally, besides MCSP, what other (natural) problems admit similar SCA hardness magnification results? More importantly, can we identify some *essential property* of these problems that would explain these results? For instance, in the case of MCSP there appears to be some connection to the length of (minimal) witnesses being much smaller than the instance length. Indeed, one sufficient condition in this sense (disregarding SCAs) is sparsity [6]; nevertheless, it seems rather implausible that this would be the sole property responsible for all hardness magnification phenomena.

Acknowledgments. I would like to thank Thomas Worsch for the helpful discussions and feedback.

References

1. Allender, E., Koucký, M.: Amplifying lower bounds by means of self-reducibility. J. ACM **57**(3), 14:1-14:36 (2010)
2. Arora, S., Barak, B.: Computational Complexity: A Modern Approach. Cambridge University Press, Cambridge (2009)
3. Baker, T.P., Gill, J., Solovay, R.: Relativizations of the P =? NP question. SIAM J. Comput. **4**(4), 431–442 (1975)
4. Chazelle, B., Monier, L.: A model of computation for VLSI with related complexity results. J. ACM **32**(3), 573–588 (1985)
5. Chen, L., Hirahara, S., Oliveira, I.C., Pich, J., Rajgopal, N., Santhanam, R.: Beyond natural proofs: hardness magnification and locality. In: Vidick, T. (ed.) 11th Innovations in Theoretical Computer Science Conference, ITCS 2020, Seattle, Washington, USA, 12–14 January 2020. LIPIcs, vol. 151, pp. 70:1–70:48. Schloss Dagstuhl - Leibniz-Zentrum für Informatik (2020)
6. Chen, L., Jin, C., Williams, R.R.: Hardness magnification for all sparse NP languages. In: Zuckerman, D (ed.) 60th IEEE Annual Symposium on Foundations of Computer Science, FOCS 2019, Baltimore, Maryland, USA, 9–12 November 2019, pp. 1240–1255. IEEE Computer Society (2019)
7. Cheraghchi, M., Hirahara, S., Myrisiotis, D., Yoshida, Y.: One-tape turing machine and branching program lower bounds for MCSP. In: Electronic Colloquium on Computational Complexity (ECCC), no. 103 (2020)
8. Delorme, M., Mazoyer, J. (eds.): Cellular Automata: A Parallel Model. Mathematics and Its Applications, vol. 460. Springer, Dordrecht (1999). https://doi.org/10.1007/978-94-015-9153-9
9. Dürr, C., Rapaport, I., Theyssier, G.: Cellular automata and communication complexity. Theor. Comput. Sci. **322**(2), 355–368 (2004)
10. Goldreich, O.: Computational Complexity: A Conceptional Perspective. Cambridge University Press, Cambridge (2008)
11. Ibarra, O.H., Palis, M.A., Kim, S.M.: Fast parallel language recognition by cellular automata. Theor. Comput. Sci. **41**, 231–246 (1985)
12. Ilango, R., Loff, B., Oliveira, I.C.: NP-hardness of circuit minimization for multi-output functions. In: Saraf, S. (ed.) 35th Computational Complexity Conference, CCC 2020, Saarbrücken, Germany, 28–31 July 2020 (Virtual Conference). LIPIcs, vol. 169, pp. 22:1–22:36. Schloss Dagstuhl - Leibniz-Zentrum für Informatik (2020)
13. Kabanets, V., Cai, J.Y.: Circuit minimization problem. In: Yao, F.F., Luks, E.M. (eds.) Proceedings of the Thirty-Second Annual ACM Symposium on Theory of Computing, Portland, OR, USA, 21–23 May 2000, pp. 73–79. ACM (2000)
14. Kobayashi, K.: On the structure of one-tape nondeterministic turing machine time hierarchy. Theor. Comput. Sci. **40**, 175–193 (1985)
15. Kutrib, M.: Complexity of one-way cellular automata. In: Isokawa, T., Imai, K., Matsui, N., Peper, F., Umeo, H. (eds.) AUTOMATA 2014. LNCS, vol. 8996, pp. 3–18. Springer, Cham (2015). https://doi.org/10.1007/978-3-319-18812-6_1
16. Kutrib, M., Malcher, A., Wendlandt, M.: Shrinking one-way cellular automata. Nat. Comput.**16**(3), pp. 383–396 (2017)
17. McKay, D.M., Murray, C.D., Williams, R.R.: Weak lower bounds on resource-bounded compression imply strong separations of complexity classes. In: Charikar, M., Cohen, E. (eds.) Proceedings of the 51st Annual ACM SIGACT Symposium on Theory of Computing, STOC 2019, Phoenix, AZ, USA, 23–26 June 2019, pp. 1215–1225 (2019)

18. Modanese, A.: Complexity-theoretic aspects of expanding cellular automata. In: Castillo-Ramirez, A., de Oliveira, P.P.B. (eds.) AUTOMATA 2019. LNCS, vol. 11525, pp. 20–34. Springer, Cham (2019). https://doi.org/10.1007/978-3-030-20981-0_2

19. Modanese, A.: Sublinear-time language recognition and decision by one-dimensional cellular automata. In: Jonoska, N., Savchuk, D. (eds.) DLT 2020. LNCS, vol. 12086, pp. 251–265. Springer, Cham (2020). https://doi.org/10.1007/978-3-030-48516-0_19

20. Modanese, A., Worsch, T.: Shrinking and expanding cellular automata. In: Cook, M., Neary, T. (eds.) Proceedings of Cellular Automata and Discrete Complex Systems -22nd IFIP WG 1.5 International Workshop, AUTOMATA 2016, Zurich, Switzerland, June 15–17, LNCS, vol. 9664, pp. 159–169. Springer, Cham (2016)

21. Murray, C.D., Williams, R.R.: On the (non) NP-hardness of computing circuit complexity. Theory Comput. 13(1), 1–22 (2017)

22. Oliveira, I.C., Pich, J., Santhanam, R.: Hardness magnification near state-of-the-art lower bounds. In: Shpilka, A. (ed.) 34th Computational Complexity Conference, CCC 2019, New Brunswick, NJ, USA, 18–20 July 2019. LIPIcs, vol. 137, pp. 27:1–27:29. Schloss Dagstuhl - Leibniz- Zentrum für Informatik (2019)

23. Oliveira, I.C., Santhanam, R.: Hardness magnification for natural problems. In: Thorup, M. (ed.) 59th IEEE Annual Symposium on Foundations of Computer Science, FOCS 2018, Paris, France, 7–9 October 2018, IEEE Computer Society, pp. 65–76 (2018)

24. Poupet, V.: A padding technique on cellular automata to transfer inclusions of complexity classes. In: Diekert, V., Volkov, M.V., Voronkov, A. (eds.) Computer Science - Theory and Applications, Second International Symposium on Computer Science in Russia, CSR 2007, Ekaterinburg, Russia, September 3–7, LNCS, vol. 4649, pp. 337–348. Springer, Heidelberg (2007)

25. Rosenfeld, A., Wu, A.Y., Dubitzki, T.: Fast language acceptance by shrinking cellular automata. Inf. Sci. 30(1), 47–53 (1983)

26. Sommerhalder, R., van Westrhenen, S.C.: Parallel language recognition in constant time by cellular automata. Acta Informatica 19, 397–407 (1983)

27. Stratmann, M., Worsch, T.: Leader election in d-dimensional CA in time diam log(diam). Future Gener. Comput. Syst. 18(7), 939–950 (2002)

28. Thompson, C.D.: A complexity theory for VLSI. Ph.D. thesis. Department of Computer Science, Carnegie-Mellon University, August 1980

29. Vollmar, R.: On two modified problems of synchronization in cellular automata. Acta Cybern. 3(4), 293–300 (1977)

30. Yao, A.C.-C.: Circuits and local computation. In: Johnson, D.S. (ed.) Proceedings of the 21st Annual ACM Symposium on Theory of Computing, Seattle, Washington, USA, 14–17 May 1989, pp. 186–196. ACM (1989)

Approximation Algorithms for Connectivity Augmentation Problems

Zeev Nutov[(✉)]

The Open University of Israel, Ra'anana, Israel
nutov@openu.ac.il

Abstract. In CONNECTIVITY AUGMENTATION problems we are given a graph $H = (V, E_H)$ and an edge set E on V, and seek a min-size edge set $J \subseteq E$ such that $H \cup J$ has larger edge/node connectivity than H. In the EDGE-CONNECTIVITY AUGMENTATION problem we need to increase the edge-connectivity by 1. In the BLOCK-TREE AUGMENTATION problem H is connected and $H \cup S$ should be 2-connected. In LEAF-TO-LEAF CONNECTIVITY AUGMENTATION problems every edge in E connects minimal deficient sets. For this version we give a simple combinatorial approximation algorithm with ratio 5/3, improving the 1.91 approximation of [6] (see also [23]), that applies for the general case. We also show by a simple proof that if the STEINER TREE problem admits approximation ratio α then the general version admits approximation ratio $1 + \ln(4 - x) + \epsilon$, where x is the solution to the equation $1 + \ln(4 - x) = \alpha + (\alpha - 1)x$. For the currently best value of $\alpha = \ln 4 + \epsilon$ [7] this gives ratio 1.942. This is slightly worse than the ratio 1.91 of [6], but has the advantage of using STEINER TREE approximation as a "black box". In the ELEMENT CONNECTIVITY AUGMENTATION problem we are given a graph $G = (V, E)$, $S \subseteq V$, and connectivity requirements $r = \{r(u, v) : u, v \in S\}$. The goal is to find a min-size set J of new edges on S such that for all $u, v \in S$ the graph $G \cup J$ contains $r(u, v)$ uv-paths such that no two of them have an edge or a node in $V \setminus S$ in common. The problem is NP-hard even when $r_{\max} = \max_{u,v \in S} r(u, v) = 2$. We obtain ratio 3/2, improving the previous ratio 7/4 of [22]. For the case of degree bounds on S we obtain the same ratio with just +1 degree violation, which is tight, since deciding whether there exists a feasible solution is NP-hard even when $r_{\max} = 2$.

1 Introduction

A graph is **k-connected** if it contains k internally disjoint paths between any two nodes; if the paths are only required to be edge disjoint then the graph is **k-edge-connected**. In CONNECTIVITY AUGMENTATION problems we are given a graph $G_0 = (V, E_0)$ and an edge set E on V, and seek a min-size edge set $J \subseteq E$ such that $G_0 \cup J$ has larger edge/node connectivity than G_0.

- In the EDGE-CONNECTIVITY AUGMENTATION problem G_0 is k-edge-connected and $G_0 \cup J$ should be $(k + 1)$-edge connected.

© Springer Nature Switzerland AG 2021
R. Santhanam and D. Musatov (Eds.): CSR 2021, LNCS 12730, pp. 321–338, 2021.
https://doi.org/10.1007/978-3-030-79416-3_19

– In the 2-CONNECTIVITY AUGMENTATION problem G_0 is a connected graph
and $G_0 \cup J$ should be 2-connected.

A **cactus** is a "tree-of-cycles", namely, a 2-edge-connected graph in which
every block is a cycle (equivalently - every edge belongs to exactly one simple
cycle). By [9], the EDGE-CONNECTIVITY AUGMENTATION problem is equivalent
to the following problem:

CACTUS AUGMENTATION
Input: A cactus $T = (V, E_T)$ and an edge set E on V.
Output: A min-size edge set $J \subseteq E$ such that $T \cup J$ is 3-edge-connected.

It is also known (c.f. [17]) that the 2-CONNECTIVITY AUGMENTATION problem is equivalent to the following problem:

BLOCK-TREE AUGMENTATION
Input: A tree $T = (V, E_T)$ and an edge set E on V.
Output: A min-size edge set $F \subseteq E$ such that $T \cup F$ is 2-connected.

A more general problem than CACTUS AUGMENTATION is as follows. Two
sets A, B **cross** if $A \cap B \neq \emptyset$ and $A \cup B \neq V$. A set family $\mathcal{F}$ on a groundset
V is a **crossing family** if $A \cap B, A \cup B \in \mathcal{F}$ whenever $A, B \in \mathcal{F}$ cross; $\mathcal{F}$ is a
symmetric family if $V \setminus A \in \mathcal{F}$ whenever $A \in \mathcal{F}$. The 2-edge-cuts of a cactus
form a symmetric crossing family, with the additional property that whenever
$A, B \in \mathcal{F}$ cross and $A \setminus B, B \setminus A$ are both non-empty, the set $(A \setminus B) \cup (B \setminus A)$
is not in $\mathcal{F}$; such a symmetric crossing family is called **proper** [10]. Dinitz,
Karzanov, and Lomonosov [9] showed that the family of minimum edge cuts of
a graph G can be represented by 2-edge cuts of a cactus. Furthermore, when
the edge-connectivity of G is odd, the min-cuts form a laminar family and thus
can be represented by a tree. Dinitz and Nutov [10, Theorem 4.2] (see also [21,
Theorem 2.7]) extended this by showing that an arbitrary symmetric crossing
family $\mathcal{F}$ can be represented by 2-edge cuts and specified 1-node cuts of a cactus;
when $\mathcal{F}$ is a proper crossing family this reduces to the cactus representation of
[9]. We say that an edge f **covers** a set A if f has exactly one end in A. The
following problem combines the difficulties of the CACTUS AUGMENTATION and
the BLOCK-TREE AUGMENTATION problems, see [23].

CROSSING FAMILY AUGMENTATION
Input: A graph $G = (V, E)$ and a symmetric crossing family $\mathcal{F}$ on V.
Output: A min-size edge set $J \subseteq E$ that covers $\mathcal{F}$.

In this problem, the family $\mathcal{F}$ may not be given explicitly, but we require
that certain queries related to $\mathcal{F}$ can be answered in polynomial time, see [23].
BLOCK-TREE AUGMENTATION and CROSSING FAMILY AUGMENTATION admit
ratio 2 [12, 26], that applies also for the min-cost versions of the problems.

The inclusion minimal members of a set family $\mathcal{F}$ are called **leaves**. In the
LEAF-TO-LEAF CROSSING FAMILY AUGMENTATION problem, every edge in E

connects two leaves of $\mathcal{F}$. In the LEAF-TO-LEAF BLOCK-TREE AUGMENTATION problem, every edge in E connects two leaves of the input tree T.

Theorem 1. *The leaf-to leaf versions of* CROSSING FAMILY AUGMENTATION *and* BLOCK-TREE AUGMENTATION *admit ratio 5/3.*

Better ratios are known for two special cases. In the TREE AUGMENTATION problem the family $\mathcal{F}$ is laminar, namely, any two sets in $\mathcal{F}$ are disjoint or one contains the other; this problem can be also defined in connectivity terms - make a spanning tree 2-edge-connected by adding a min-size edge set $J \subseteq E$. This problem was vastly studied; see [1,11,16,19,24] and the references therein for additional literature on the TREE AUGMENTATION problem. In the LEAF-TO-LEAF TREE AUGMENTATION problem, every edge in E connects two leaves of the tree; this problem admits ratio 17/12 [20]. The CYCLE AUGMENTATION problem is a particular case of the CACTUS AUGMENTATION problem when the cactus is a cycle; in this case the leaves are the singleton nodes. The CYCLE AUGMENTATION problem admits ratio $\frac{3}{2} + \epsilon$ [15]; our algorithm from Theorem 1 uses some ideas from [15].

Byrka, Grandoni, and Ameli [6] showed that CACTUS AUGMENTATION admits ratio $2 \ln 4 - \frac{967}{1120} + \epsilon < 1.91$, breaching the natural 2 approximation barrier. This was extended to CROSSING FAMILY AUGMENTATION and BLOCK TREE AUGMENTATION in [23]. In the STEINER TREE problem we are given a graph $G = (V, E)$ with edge costs and a set $R \subseteq V$ of terminals, and seek a min-cost subtree of G that spans R. We prove the following.

Theorem 2. *If* STEINER TREE *admits ratio* α *then* CROSSING FAMILY AUGMENTATION *and* BLOCK-TREE AUGMENTATION *admit ratio* $1 + \ln(4 - x) + \epsilon$, *where x is the solution to the equation* $1 + \ln(4 - x) = \alpha + (\alpha - 1)x$.

Currently, $\alpha = \ln 4 + \epsilon$ [7]; in this case we have ratio 1.942 for the problems in the theorem. This is slightly worse than the ratio 1.91 of [6] (see also [23]), but our algorithm is very simple and has the advantage of using STEINER TREE approximation as a "black box". E.g., if ratio $\alpha = 1.35$ can be achieved, then we immediately get ratio $1.895 < 1.9$.

We note that shortly after this paper was written, Cecchetto, Traub & Zenklusen [8] showed that CACTUS AUGMENTATION admits ratio 1.393. This is much better than our ratio 5/3 even for the leaf-to-leaf version of this problem. However, for both the BLOCK-TREE AUGMENTATION and the CROSSING FAMILY AUGMENTATION problems our ratios are still the currently best known ones.

We also consider the following problem:

ELEMENT CONNECTIVITY AUGMENTATION

Input: An undirected graph $G = (V, E)$, a set $S \subseteq V$ of terminals, and connectivity requirements $\{r(u, v) : u, v \in S\}$ on pairs of terminals.

Output: A minimum size set J of new edges on S (any edge is allowed and parallel edges are allowed) such that the graph $G \cup J$ contains $r(u, v)$ uv-paths such that no two of them have an edge or a node in $V \setminus S$ in common.

A particular case when the graph G is bipartite with sides S and $V \setminus S$ is known as the HYPERGRAPH EDGE-CONNECTIVITY AUGMENTATION problem; here S is the set of nodes of the hypergraph and $V \setminus S$ is the set of the hyperedges. This problem is solvable in polynomial time for uniform requirements when $r(u, v) = k$ for all $u, v \in S$ [2] (see also [4] and [5] for a simpler algorithm and proof), and when $r_{max} = 1$, where r_{max} is the maximum requirement. See also [5,13,14] for additional polynomially solvable cases. The non-uniform version of the problem is NP-hard even when the initial graph G is connected and $r_{max} = 2$ [18]. The previous best approximation ratio for the general version was $7/4$, and $3/2$ when $r_{max} = 2$ [22]. In the degree bounded version of the problem we also have degree bounds $\{b(v) : v \in S\}$ and require that $d_J(v) \leq b(v)$ for all $v \in S$, where $d_J(v)$ is the degree of v w.r.t. J. We show that ELEMENT CONNECTIVITY AUGMENTATION admits ratio $3/2$, and that this ratio can be achieved also for the degree bounded version with only additive $+1$ degree violation; a better degree approximation is unlikely, since deciding whether there exists a feasible solution is NP-hard even when $r_{max} = 2$ and $b_{max} = 1$ [18].

Theorem 3. ELEMENT CONNECTIVITY AUGMENTATION *admits approxima-tion ratio* $3/2$. *Moreover, the degree bounded version admits a bicriteria approx-imation algorithm that computes a solution* J *of size at most* $3/2$ *times the optimal such that* $d_J(v) \leq b(v) + 1$ *for all* $v \in S$.

The proof of this theorem is based on a generic algorithm for covering a skew-supermodular set function, as is explained in Sect. 5.

2   The Leaf-to-Leaf Case (Theorem 1)

We prove Theorem 1 for the CROSSING FAMILY AUGMENTATION problem, and later indicate the changes needed to adopt the proof for the BLOCK-TREE AUG-MENTATION problem. We need some definition to describe the algorithm. Let $\mathcal{F}$ be a set family on V. We say that $A \in \mathcal{F}$ **separates** $u, v \in V$ if $|A \cap \{u, v\}| = 1$; u, v are $\mathcal{F}$-**separable** if such A exists and u, v are $\mathcal{F}$-**inseparable** otherwise. Similarly, a set A **separates** edges f, g if one of f, g has both ends in A and the other has no end in A; f, g are $\mathcal{F}$-**separable** if such $A \in \mathcal{F}$ exists, and are $\mathcal{F}$-**inseparable** otherwise. The relation $\{(u, v) \in V \times V : u, v \text{ are } \mathcal{F}\text{-inseparable}\}$ is an equivalence, and we call its equivalence classes $\mathcal{F}$-**classes**. W.l.o.g. we will assume that all $\mathcal{F}$-classes are singletons and that no edge in E has both ends in the same class; in particular, the leaves of $\mathcal{F}$ are singletons, and we denote the leaf set of $\mathcal{F}$ by L. We will also often abbreviate the notation for singleton sets and write v, e instead of $\{v\}, \{e\}$. Given $J \subseteq E$, the **residual instance** $((V^J, E^J), \mathcal{F}^J)$ is defined as follows.

- The **residual family** $\mathcal{F}^J$ of $\mathcal{F}$ w.r.t. J consists of all members of $\mathcal{F}$ that are uncovered by the edges in J. It is known that $\mathcal{F}^J$ is crossing (and symmetric) if $\mathcal{F}$ is.
- V^J is the set of $\mathcal{F}^J$-classes (w.l.o.g, each of them can be shrunk into a single element).

– E^J is obtained from $E \setminus J$ by removing all edges that have both ends in the same $\mathcal{F}^J$-class.

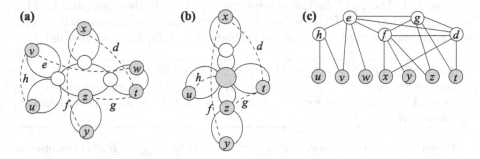

Fig. 1. Illustration of definitions for a CROSSING FAMILY AUGMENTATION instance where $\mathcal{F}$ is represented by a cactus. Here $A \in \mathcal{F}$ if and only if A is a connected component obtained by removing a pair of edges that belong to the same cycle of the cactus. The edges in E are shown by dashed arcs and the terminals in R are gray. The cactus of the residual family w.r.t. to a single edge is obtained by "squeezing" the cycles along the path of cycles between the ends of the edge. (a) The original instance. (b) The residual instance w.r.t. e. (c) The $(R, E, \mathcal{F})$-incidence graph of the instance in (a).

In addition, given a set $R \subseteq V$ of terminals, the **residual set of terminals** R^J is the set of $\mathcal{F}^J$-classes that contain some member of R; see Fig. 1(a,b).

For any edge $e = uv$, there is an $\mathcal{F}^e$-class that contains both u and v; we call this equivalence class the **class of** e and denote it by $C(\mathcal{F}, e)$. Given a set R of terminals (a subset of $\mathcal{F}$-classes), the $(R, E, \mathcal{F})$-**incidence graph** $H = (U, E_H)$ has node set $U = E \cup R$ and edge set

$$E_H = \{ee' : e, e' \in E \text{ are } \mathcal{F}\text{-inseparable}\} \cup \{er : r \in R, e \in E, r \in C(\mathcal{F}, e)\} \ .$$

The idea of the algorithm is to repeatedly add to the partial solution one of the following:

– A single edge that decreases the number of terminals by 2, namely, an edge whose equivalence class contains at least 3 terminals. Equivalently, in the incidence graph, the node that corresponds to such an edge has at least 3 terminal neighbors; see the edge f in Fig. 1(a,c).
– An edge pair that decreases the number of terminals by 3. Assuming the previous case does not apply, this means that the two edges are inseparable and that the sets of terminals in their equavalence classes are disjoint and have size 2 each. Equivalently, in the incidence graph, the two nodes that corresponds to such a pair are adjacent, each has exactly 2 terminal neighbors, but they have no common terminal neighbor; see the edge pair e, d in Fig. 1(a,c), and note that h, e is not such a pair.

Let $R \subseteq V$ and let H be the $(R, E, \mathcal{F})$-incidence graph. Note that R is an independent set in H. It was shown in [23] that for $R = L$ being the set of leaves of $\mathcal{F}$, an edge set $J \subseteq E$ is a feasible solution to CROSSING FAMILY AUGMENTATION if and only if the subgraph $H[J \cup R]$ of H induced by $J \cup R$ is connected. The proof in [23] extends to any $R \subseteq V$ that contains L. This implies that CROSSING FAMILY AUGMENTATION admits an approximation ratio preserving reduction to the following problem (see [3,23] for more details).

SUBSET STEINER CONNECTED DOMINATING SET (SS-CDS)

Input: A graph $H = (U, E_H)$ and a set $R \subseteq U$ of independent terminals.
Output: A min-size node set $S \subseteq U \setminus R$ such that $H[S]$ is connected and S dominates R.

Given a SS-CDS instance and $s \in S = U \setminus R$ let $R(s) = R_H(s)$ denote the set of neighbors of s in H that belong to R. Let opt be the optimal solution value of a problem instance at hand. Before describing the algorithm, we will prove the following lemma.

Lemma 1. *Let* $\mathcal{I} = (H, R)$ *be a SS-CDS instance such that* $|R(s)| = 2$ *for all* $s \in S = U \setminus R$. *Then one of the following holds:*

(i) There are adjacent $a, b \in S$ *with* $R(a) \cap R(b) = \emptyset$.
(ii) opt $\geq |R| - 1$.

Proof. Assume that (i) does not hold for $\mathcal{I}$; we will prove that then (ii) holds. The proof is by induction on $|R|$. In the base case $|R| = 2$ (ii) holds. Assume that the statement is true for $|R| - 1 \geq 2$. Let T be an optimal solution tree and S the set of non-terminals in T. Root T at some node and let $s \in S$ be a non-terminal farthest from the root. The children of s are terminal leaves, and assume w.l.o.g. that $R(s) = \{u, v\}$ is the set of children of s; if s has one child in T, then it has another terminal neighbor in H, that can be attached to s.

Consider the residual instance $\mathcal{I}' = (G' = (V', E'), R')$ and the tree T' obtained by contracting $R(s)$ into the new terminal s', and deleting any $z \in U \setminus (R + s)$ with $R(z) = R(s)$. Then $|R'| = |R| - 1$, $|R'(z)| = 2$ for all $z \in R'$, T' is an optimal solution for $\mathcal{I}'$, and $S' = S - s$ is the set of non-terminals of T'.

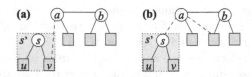

Fig. 2. Illustration to the proof of Lemma 1.

If (i) does not hold for the new instance $\mathcal{I}'$ then (ii) holds for $\mathcal{I}'$, by the induction hypothesis. Then $|S| = |S'| + 1 \geq (|R'| - 1) + 1 = |R| - 1$, and we get that (ii) holds for $\mathcal{I}$. Assume henceforth that (i) holds for $\mathcal{I}'$. We obtain a contradiction by showing that then (i) holds for $\mathcal{I}$. Let $a, b \in V' \setminus R'$ be such that $R'(a) \cap R'(b) = \emptyset$, see Fig. 2. If $s' \notin R'(a) \cup R'(b)$ then clearly (ii) holds for $\mathcal{I}$. Otherwise, if say $s' \in R'(a)$, then we have two cases. If one of u, v, say v, is a neighbor of a in G (see Fig. 2(a)) then $R(a) \cap R(b) = \emptyset$. Otherwise (see Fig. 2(a)), $R(a) \cap R(s) = \emptyset$. In both cases, we obtain a contradiction to the assumption that (i) does not hold for $\mathcal{I}$. □

We also need the following known lemma.

Lemma 2. *Any inclusion minimal cover J of a set family $\mathcal{F}$ is a forest.*

Proof. Suppose to the contrary that J contains a cycle C. Let $e = uv \in C$. Since $P = C \setminus \{e\}$ is a uv-path, then for any $A \in \mathcal{F}$ covered by e, there is $e' \in P$ that covers A. This implies that $J \setminus \{e\}$ also covers $\mathcal{F}$, contradicting the minimality of J. □

Algorithm 1: $(G = (V, E), \mathcal{F}, R)$

---

1 $J \leftarrow \emptyset$
2 **repeat**
3   let H^J be the $(E^J, R^J, \mathcal{F}^J)$-incidence graph
4   **if** H^J has a node $e \in E$ such that $|R^J(e)| \geq 3$ **then do** $J \leftarrow J \cup \{e\}$
5   **else if** H^J has a pair of adjacent nodes $e, f \in E$ with
        $|R^J(e) \cup R^J(f)| = 4$ **then do** $J \leftarrow J \cup \{e, f\}$
6 **until** no edge e or an edge pair e, f as above exists;
7 find an inclusion minimal $\mathcal{F}^J$-cover and add it to J
8 **return** J

The algorithm starts with a partial solution $J = \emptyset$ and has two phases. Phase 1 consists of iterations. At the beginning of each iteration, construct the $(E, R^J, \mathcal{F}^J)$-incidence graph H^J, where initially R is the set of leaves of $\mathcal{F}$. Then, do one of the following:

1. If H^J has a node $e \in E$ with $|R^J(e)| \geq 3$, then add e to J.
2. Else, if there are nonseparable $e, f \in E$ such that $|R^J(e) \cup R^J(f)| = 4$ then add both e, f to J.

Note that if step 1 does not apply then the condition $|R^J(e) \cup R^J(f)| = 4$ is equivalent to $|R^J(e)| = |R^J(f)| = 2$ and $R^J(e) \cap R^J(f) = \emptyset$. If none of the above two cases occurs, then we apply Phase 2, in which we add to J an inclusion minimal cover of $\mathcal{F}^J$; note that all edges in E^J have both endnodes in R^J. A more formal description is given in Algorithm 1.

We show that the algorithm achieves ratio $5/3$. Note that:

- Adding an edge e in step 4 reduces the number of terminals by at least 2, since $|R^J(e)| \geq 3$ and adding e shrinks $R^J(e)$ into one terminal.
- Adding a pair e, f in step 5 reduces the number of terminals by 3, since $|R^J(e) \cup R^J(f)| = 4$ and adding e, f shrinks $R^J(e) \cup R^J(f)$ into one terminal.

Hence the reduction in the number of terminals per added edge is at least $3/2$. Let $\ell = |L|$ be the initial number of terminals. Let $\ell' = |R^J|$ be the number of terminals at the end of Phase 1 (steps 2–6 in Algorithm 1). Let k be the number of edges added during Phase 1. Then $\ell' \leq \ell - \frac{3}{2}k$, hence $k \leq \frac{2}{3}(\ell - \ell')$. The number of edges added at the second phase is at most $\ell' - 1$, by Lemma 2; note that every edge in E^J has both ends in R^J and that $|R^J| = \ell'$. On the other hand, $\mathrm{opt} \geq \frac{\ell}{2}$, and $\mathrm{opt} \geq \ell' - 1$, by part (ii) of Lemma 1. Summarizing, we have the following:

- The solution size is at most $k + \ell' - 1 \leq \frac{2}{3}(\ell - \ell') + \ell' - 1 = (2\ell + \ell' - 3)/3$.
- $\mathrm{opt} \geq \ell/2$ and $\mathrm{opt} \geq \ell' - 1$.

Thus the approximation ratio is bounded by $\frac{(2\ell+\ell'-3)/3}{\max\{\ell/2,\ell'-1\}}$. If $\ell/2 \geq \ell' - 1$ then

$$\frac{(2\ell + \ell' - 3)/3}{\max\{\ell/2, \ell' - 1\}} \leq \frac{(2\ell + (\ell/2 + 1) - 3)/3}{\ell/2} = \frac{(5\ell/2 - 2)/3}{\ell/2} < \frac{5}{3} .$$

Else, $\ell/2 < \ell' - 1$, and then

$$\frac{(2\ell + \ell' - 3)/3}{\max\{\ell/2, \ell' - 1\}} < \frac{(4(\ell' - 1) + \ell' - 3)/3}{\ell' - 1} = \frac{(5\ell' - 7)/3}{\ell' - 1} < \frac{5}{3} .$$

In both cases the ratio is bounded by $5/3$.

We now adjust the proof to the BLOCK-TREE AUGMENTATION problem. Let $G = (V, E)$ be a connected graph. A node v is a **cutnode** of G if $G \setminus \{v\}$ is disconnected; an inclusion maximal node subset whose induced subgraph is connected and has no cutnodes is a **block** of G; equivalently, B is a block if it is the node set of an inclusion maximal 2-connected subgraph or of a bridge. The **block-tree** T of G has node set $C_G \cup B_G$, where C_G is the set of cutnodes of G and B_G is the set of blocks of G; T has an edge between a block and a cutnode that belongs to that block. It is known that every $v \in V \setminus C_G$ belongs to a unique block and that T is a tree. The **block-tree mapping** $\psi : V \to C_G \cup B_G$ is defined by $\psi(v) = v$ is $v \in C_G$ and $\psi(v)$ is the block that contains v if $v \in V \setminus C_G$.

Given a BLOCK-TREE AUGMENTATION instance $(T = (V, E_T), E)$ and a partial solution $J \subseteq E$, the **residual instance** $(T^J = (V^J, E_T^J), E^J)$ is defined as follows.

- T^J is the block tree of $T \cup J$.
- $E^J = \{\psi(u)\psi(v) : uv \in E \setminus J, \psi(u) \neq \psi(v)\}$, where ψ is the block-tree mapping of $T \cup J$.

For a set $R \subseteq V$ of terminals, the **residual set of terminals** is defined by $R^J = \psi(R) = \cup_{r \in R} \psi(r)$. For an edge $e = uv$ let T_e denote the uv-path in T. We say that $e, f \in E$ are T-**inseparable** if the paths T_e, T_f have an edge in common. The (R, E, T)-**incidence graph** $H = (U, E_H)$ has node set $U = E \cup R$ and edge set

$$E_H = \{ef : e, f \in E \text{ are } T\text{-inseparable}\} \cup \{er : r \in R, e \in E, r \in T_e\} .$$

It was shown in [23] that for $R = L$ being the set of leaves of $\mathcal{F}$, an edge set $J \subseteq E$ is a feasible solution to BLOCK-TREE AUGMENTATION if and only if the subgraph $H[J \cup R]$ of H induced by $J \cup R$ is connected. The proof in [23] extends to any $R \subseteq V$ that contains L. This implies that CROSSING FAMILY AUGMENTATION admits an approximation ratio preserving reduction to SS-CDS, see [23] for details. Lemma 2 also extends to this case, as it is known that an if J is an inclusion minimal edge set whose addition makes a connected graph 2-connected, then J is a forest.

With these definitions and facts, the rest of the proof for the BLOCK-TREE AUGMENTATION coincides with the proof given for CROSSING FAMILY AUGMENTATION, concluding the proof of Theorem 1.

3   The General Case (Theorem 2)

Recall that each of the problems CROSSING FAMILY AUGMENTATION and BLOCK-TREE AUGMENTATION admits an approximation ratio preserving reduction to the SS-CDS problem with $R = L$ being the set of terminals. The SS-CDS instances that arise from this reduction have the following property, see [6, 23]:

$(*)$ The neighbors of every $r \in R$ induce a clique.

In fact, SS-CDS with property $(*)$ is equivalent to the NODE WEIGHTED STEINER TREE problem with property $(*)$ with unit node weights for non-terminals (the terminals have weight zero). Clearly, any SS-CDS solution is a feasible NODE WEIGHTED STEINER TREE solution; for the other direction, note that if property $(*)$ holds, then the set of non-terminals in any feasible NODE WEIGHTED STEINER TREE solution is a feasible SS-CDS solution. The relation to the ordinary STEINER TREE problem is given in following lemma.

Lemma 3. ([6]).   *Let S be a* SS-CDS *solution and* $T = (U, J)$ *a* STEINER TREE *solution on instance* (G, R) *with unit edge costs. Then:*

(i) *If* $(*)$ *holds then T can be converted into a* SS-CDS *solution S_J with $|S_J| = |J| - |R| + 1$.*

(ii) *S can be converted into a* STEINER TREE *solution $T_S = (U_S, J_S)$ with $|J_S| = |S| + |R| - 1$.*

Proof. We prove (i). Any STEINER TREE solution $T' = (U', J')$ can be converted into a solution $T = (U, J)$ such that $|J| = |J'|$ and R is the leaf set of T'. For this, for each $r \in R$ that is not a leaf of T', among the edges incident to r in T', choose one and replace the other edges by a tree on the neighbors of r; this is possible by $(*)$. The non-leaf nodes of such T form a SS-CDS as required. For (ii), taking a tree on S and for each $r \in R$ adding an edge from r to S gives a STEINER TREE solution as required. $\square$

Let J^* be an optimal and J an α-approximate STEINER TREE solutions. Let S_J, S^* be SS-CDS solutions, where S_J is derived from J and S^* is an optimal one. Then

$$|S_J| + R - 1 = |J| \leq \alpha|J^*| \leq \alpha|J_{S^*}| = \alpha(|S^*| - 1 + |R|) = \alpha|S^*| + \alpha(|R| - 1) .$$

This implies that if STEINER TREE admits ratio α then SS-CDS with property (∗) admits a polynomial time algorithm that computes a solution S of size $|S| \leq \alpha\text{opt} + (\alpha - 1)|L|$ and achieves ratio $\alpha + (\alpha - 1)\frac{|L|}{\text{opt}} = \alpha + (\alpha - 1)x$, where $x = \frac{|L|}{\text{opt}}$, $0 < x \leq 2$. We will prove the following.

Theorem 4. CROSSING FAMILY AUGMENTATION *and* BLOCK-TREE AUGMENTATION *admit ratio* $1 + \ln\left(4 - \frac{|L|}{\text{opt}}\right) + \epsilon$.

From Lemma 3 and Theorem 4 it follows that we can achieve ratio

$$\max\{\alpha + (\alpha - 1)x, 1 + \ln(4 - x)\} + \epsilon \quad \text{where} \quad x = \frac{|L|}{\text{opt}} .$$

The worse case is when these two ratios are equal, which gives the Theorem 2 ratio. In the case $\alpha = \ln 4 + \epsilon$ [7], we have $x \approx 1.4367$, so $L \approx 1.4367\text{opt}$ and opt $\approx 0.69L$. The ratio in this case is $1 + \ln(4 - x) + \epsilon < 1.942$.

4   Proof of Theorem 4

A set function f is **increasing** if $f(A) \leq f(B)$ whenever $A \subseteq B$; f is **decreasing** if $-f$ is increasing, and f is **sub-additive** if $f(A \cup B) \leq f(A) + f(B)$ for any subsets A, B of the ground-set. Let us consider the following algorithmic problem:

MIN-COVERING

Input: Non-negative set functions ν, τ on subsets of a ground-set U such that ν is decreasing, τ is sub-additive, and $\tau(\emptyset) = 0$.

Output: $A \subseteq U$ such that $\nu(A) + \tau(A)$ is minimal.

We call ν the **potential** and τ the **payment**. The idea behind this interpretation and the subsequent greedy algorithm is as follows. Given an optimization problem, the potential $\nu(A)$ is the (bound on the) value of some "simple" augmenting feasible solution for A. We start with an empty set solution, and iteratively try to decrease the potential by adding a set $B \subseteq U \setminus A$ of minimum "density" – the price paid for a unit of the potential. The algorithm terminates when the price ≥ 1, since then we gain nothing from adding B to A. The ratio of such an algorithm is bounded by $1 + \ln \frac{\nu(\emptyset)}{\text{opt}}$ (assuming that during each iteration a minimum density set can be found in polynomial time). So essentially, the greedy algorithm converts ratio $\alpha = \frac{\nu(\emptyset)}{\text{opt}}$ into ratio $1 + \ln \alpha$.

Fix an optimal solution A^*. Let $\nu^* = \nu(A^*)$, $\tau^* = \tau(A^*)$, so opt $= \tau^* + \nu^*$. The quantity $\frac{\tau(B)}{\nu(A) - \nu(A \cup B)}$ is called the **density** of B (w.r.t. A); this is the price paid by B for a unit of potential. The GREEDY ALGORITHM (a.k.a. RELATIVE GREEDY HEURISTIC) for the problem starts with $A = \emptyset$ and while $\nu(A) > \nu^*$ repeatedly adds to A a non-empty augmenting set $B \subseteq U$ that satisfies the following condition, while such B exists:

Density Condition: $\dfrac{\tau(B)}{\nu(A) - \nu(A \cup B)} \le \min\left\{1, \dfrac{\tau^*}{\nu(A) - \nu^*}\right\}.$

Note that since ν is decreasing, $\nu(A) - \nu(A \cup A^*) \ge \nu(A) - \nu(A^*) = \nu(A) - \nu^*$; hence if $\nu(A) > \nu^*$, then $\frac{\tau(A^*)}{\nu(A) - \nu(A \cup A^*)} \le \frac{\tau^*}{\nu(A) - \nu^*}$ and there exists an augmenting set B that satisfies the condition $\frac{\tau(B)}{\nu(A) - \nu(A \cup B)} \le \frac{\tau^*}{\nu(A) - \nu^*}$, e.g., $B = A^*$. Thus if B^* is a minimum density set and $\frac{\tau(B^*)}{\nu(A) - \nu(A \cup B^*)} \le 1$, then B^* satisfies the Density Condition; otherwise, the density of B^* is larger than 1 so no set can satisfy the Density Condition. The following statement is known, c.f. an explicit proof in [25].

Theorem 5. *The* GREEDY ALGORITHM *achieves approximation ratio*

$$1 + \frac{\tau^*}{\mathrm{opt}} \ln \frac{\nu(\emptyset) - \nu^*}{\tau^*} .$$

This applies also in the case when we can only compute a ρ-approximate minimum density augmenting set, while invoking an additional factor ρ in the ratio.

To use the framework of Theorem 5 we need to define τ and ν. Let $J \subset E$ be an edge set. The payment $\tau(J) = |J|$ is just the size of J. The potential of J is defined by $\nu(J) = |R^J| - 1$, where R is a set of terminals such that $L \subseteq R \subseteq V$, defined in the following lemma. For an edge set F let F_{LL} be the set of edges in F with both ends in L, and F_L the set of edges in F with exactly one end in L.

Lemma 4. *Let F be an optimal solution to* CROSSING FAMILY AUGMENTATION *instance and c a cost function defined by $c(e) = 0$ if $e \in E_{LL}$, $c(e) = 1$ if $e \in E_L$, and $c(e) = 2$ otherwise. Let J be a 2-approximate c-costs solution and let R be the set of ends of the edges in J. Then $|R| \le c(J) + L \le 4|F| - |L| = 4\mathrm{opt} - |L|$.*

Proof. Clearly, $|R| \le c(J) + |L|$. We show that $c(J) \le 4|F| - 2|L|$. Let F' be the set of edges in F that have no end in L. Since $|F'| = |F| - |F_L| - |F_{LL}|$ and $2|F_{LL}| + |F_L| \ge L$

$$c(F) = |F_L| + 2\,F' = |F_L| + 2(|F| - |F_L| - |F_{LL}|) = 2\,F - (|F_L| + 2\,F_{LL}|) \le 2\,F - |L| .$$

Since $c(J) \le 2c(F)$, the lemma follows. $\square$

It is easy to see that ν is decreasing and τ is subadditive. The next lemma shows that the obtained MIN-COVERING instance is equivalent to the CROSSING FAMILY AUGMENTATION instance, and that we may assume that $\tau^* = \mathrm{opt}$ and $\nu^* = 0$.

Lemma 5. *If J is a feasible solution to* CROSSING FAMILY AUGMENTATION *then $\nu(J) = 0$. If J is a feasible* MIN-COVERING *solution then one can construct in polynomial time a feasible* CROSSING FAMILY AUGMENTATION *solution of size $\leq \tau(J) + \nu(J)$. Hence both problems have the same optimal value, and* MIN-COVERING *has an optimal solution J^* such that $\nu(J^*) = 0$ and $\tau(J^*) = $ opt.*

Proof. If J is a feasible CROSSING FAMILY AUGMENTATION solution then $|R^J| = 1$ and thus $\nu(J) = 0$. Let I be a MIN-COVERING solution such that every edge in I has both ends in R; e.g., I can be as in Lemma 4. Then I^J is a feasible solution to the residual problem w.r.t. J and every edge in I^J has both ends in R^J. Let $I' \subseteq I^J$ be an inclusion minimal edge set such that $J \cup I'$ is a feasible solution. By Lemma 2, I' is a forest, hence $|I| \leq |R^J| - 1$. Consequently, $J \cup I'$ is a feasible solution of size at most $|J| + |I'| \leq |J| + |R^J| - 1 = \tau(J) + \nu(J)$. $\square$

Recall also that $\nu(\emptyset) \leq 4\mathsf{opt} - |L|$, by Lemma 4. We will show how to find for any $\epsilon > 0$, a $(1 + \epsilon)$-approximate best density set in polynomial time. It follows therefore that we can apply the greedy algorithm to produce a solution of value $1 + \epsilon$ times of

$$1 + \frac{\tau^*}{\mathsf{opt}} \ln \frac{\nu(\emptyset) - \nu^*}{\tau^*} = 1 + \ln \frac{4\mathsf{opt} - |L|}{\mathsf{opt}} = 1 + \ln \left(4 - \frac{|L|}{\mathsf{opt}} \right) .$$

In what follows note that if $a_1, \ldots, a_q$ and $b_1, \ldots b_q$ are positive reals, then by an averaging argument there exists an index $1 \leq i \leq q$ such that $a_i / b_i \leq \sum_{j=1}^{q} a_j / \sum_{i=1}^{q} b_j$.

Given a CROSSING FAMILY AUGMENTATION instance, a set $R \supseteq L$ of terminals, and $F \subseteq E$, consider the corresponding SS-CDS instance $(H = (U, E_H), R)$ and the set of non-terminals Q that corresponds to F. The density of F is $\frac{|F|}{|R| - |R^F|}$, and in the SS-CDS instance this is computed by taking a maximal forest in the graph induced by Q and the terminals that have a neighbor in Q; then the density is $|Q|$ over the number of trees in this forest. So in what follows we may speak of a density of a subforest of H. Let $T_i = (S_i \cup R_i, E_i)$, $i = 1, \ldots, q$, be the connected components of such a forest, (R_i is the set of terminals in T_i) and let $s_i = |S_i|$ and $r_i = |R_i|$, where $r_i \geq 2$. The density of the forest is $\sum_{i=1}^{q} s_i / \sum_{i=1}^{q} (r_i - 1)$ while the density of each T_i is $s_i / (r_i - 1)$. By an averaging argument, some T_i has density not larger than that of the forest. Consequently, we may assume that the minimum density is attained for a tree, say T.

Let $T = (S \cup R, E)$ be a tree with leaf set R. The density of T is $\frac{s}{r-1}$, where $r = |R|$ is the number of terminals (R-nodes) and s is the number of non-terminals (S-nodes) in T. The usual approach is to show that for any k there exists a subtree T' of T with k terminals (or k non-terminals) such that the density of T' is at most $1 + f(k)$ times the density of T, where $\lim_{k \to \infty} f(k) = 0$. The decomposition lemma that we prove is not a standard one. The difficulty can be demonstrated by the following examples. Consider the case when T is a star with n leaves. Then the density of T is $1/(n-1)$, while a subtree with k leaves has density $1/(k-1)$. If T is a path with n non-terminals, then the density of T is n, while a subtree with $k < n$ non-terminals has density $k/0 = \infty$. In both cases, the density of the subtree may be arbitrarily larger than that of T. To overcome this difficulty, we will decompose T w.r.t. a certain subset P of the non-terminals.

Let $P \subseteq S$. Let $s = |S|$, $r = |R|$, and $p = |P|$. For a subtree T' of T let $S(T')$, $R(T')$, and $P(T')$ denote the set of S-nodes, R-nodes, and P-nodes in T', respectively. We prove the following.

Lemma 6. *Let $k \geq 2$. If $p \geq 3k + 1$ then there exists subtrees $T_1, \ldots, T_q$ of T such that the following holds.*

- $\sum_{i=1}^q |S(T_i)| \leq s + q$.
- *Every R-node belongs to exactly one subtree, hence $\sum_{i=1}^q |R(T_i)| = r$.*
- $|P(T_i)| \in [k, 3k]$ *for all i and $q \leq \frac{p}{k-1}$.*

Proof. Root T at some node in S. For any $v \in S$ chosen as a "local root", the subtree T^v rooted at v is a subtree of T that consist of v and its descendants. Let T^v be an inclusion minimal rooted subtree of T such that $|P(T^v)| \geq k+1$. Note that $v \in P$. Let $B_1, \ldots, B_m$ be the branches hanging on v and let $p_j = |P(B_j)|$. By the definition of T_v, each p_j is in the range $[0, k]$ and $\sum_{j=1}^m p_j \geq k$. We claim that $\{p_1, \ldots, p_m\}$ can be partitioned such that the sum of each part plus 1 is in the range $[k, 3k]$. To see this, apply a greedy algorithm for MULTI-BIN PACKING with bins of capacity $2k$; at the end there is at most one bin with sum $\leq k-1$ (as two such bins can be joined); joining this bin to any other bin gives a partition as required. Now we remove T^v and the S-nodes on the path from v to its closest terminal ancestor, and apply the same procedure on the remaining tree. If the last rooted subtree T^v considered has $|P(T^v)| \leq k - 1$, then this tree can be joined to a subtree T_i with $|P(T_i)| \leq 2k$ derived in previous iteration. Finally, $q \leq \frac{p+q}{k}$ by the construction and since $|P(T_i)| \geq k$ for all i, so $q \leq \frac{p}{k-1}$.   □

Now we let $P = P_1 \cup P_2$, where P_1 is the set of nodes that have degree at least 3 in T and P_2 is the of nodes that have a terminal neighbor in T. Note that $|P_1| \leq r$ and $|P_2| \leq r$. Hence $p \leq 2r$, and clearly $p \leq s$. By an averaging argument and Lemma 6, the density of some T_i is bounded by $s_i/(r_i - 1) \leq \sum_{j=1}^q s_j / \sum_{j=1}^q (r_j - 1) \leq (s+q)/(r-q)$. Thus for $k \geq 3$ we get

$$\frac{s_i}{r_i - 1} \cdot \frac{r - 1}{s} \leq \frac{s+q}{r-q} \cdot \frac{r}{s} \leq \frac{s+p/(k-1)}{r-p/(k-1)} \cdot \frac{r}{s} = \frac{1+1/(k-1)}{1-2/(k-1)} = \frac{k}{k-3} = 1 + \frac{3}{k-3} \ .$$

This implies that we can find a $(1 + \epsilon)$-approximate min-density tree by searching over all trees T' with $|P(T')| \in [k, 3k]$, where given $\epsilon > 0$ we let $k = \lceil 3/\epsilon \rceil + 3$. Specifically, for every $P' \subseteq S$ with $|P'| \in [k, 3k]$, we find an MST T' in the metric completion of the current incidence graph, and then add to T' all the terminals that have a neighbor in P'. Among all subtrees we choose one of minimum density. The time complexity is n^{3k} which is polynomial for any fixed $\epsilon > 0$.

The process of adjusting the proof to the BLOCK-TREE AUGMENTATION is identical to the one in the proof of Theorem 1. This concludes the proof of Theorem 4, and thus also the proof of Theorem 2 is complete.

5   Covering Skew-Supermodular Functions (Theorem 3)

Let $p : 2^S \to \mathbb{Z}$ be a set function and J an edge set on a finite groundset S. We say that J **covers** p if $d_J(A) \geq p(A)$ for all $A \subseteq S$, where $d_J(A)$ denote the set of edges with exactly one end in A. p is **symmetric** if $p(A) = p(S \setminus A)$ for all $A \subseteq S$, and p is **skew-supermodular** (a.k.a. **weakly supermodular**) if for all $A, B \subseteq S$ at least one of the following two inequalities holds:

$$p(A) + p(B) \leq p(A \cap B) + p(A \cup B) \qquad p(A) + p(B) \leq p(A \setminus B) + p(B \setminus A)$$

ELEMENT CONNECTIVITY AUGMENTATION can be reduced to the following problem, with symmetric skew-supermodular set function p, c.f. [5,14,22].

SET FUNCTION EDGE COVER
Input: A set function p on a ground-set S.
Output: A minimum size set J of edges (any edge is allowed) that covers p.

In this problem, the function p may not be given explicitly, and a polynomial time implementation of algorithms requires that some queries related to p can be answered in polynomial time. But the problem is also NP-hard for skew-supermodular p even if p given explicitly, specifically when $p_{\max} = 1$ and $|A| = 3$ for every set A with $p(A) = 1$ [22].

In the degree bounded version of this problem we are also given degree bounds $\{b(v) : v \in S\}$ and require that $d_J(v) \leq b(v)$ for all $v \in S$.

Definition 1. *A function $g : S \to \mathbb{Z}_+$ is a p-transversal if $g(A) \geq p(A)$ for all $A \subseteq S$. Let $T_g = \{v \in S : g(v) \geq 1\}$ denote the support of g. We say that g is a **minimal p-transversal** if for any $v \in T_g$ reducing $g(v)$ by 1 results in a function that is not a p-transversal.*

The following was proved by Benczúr and Frank in [4], see also [22, Lemmas 1.1 and 3.2].

Lemma 7. *Let g be a transversal of a skew-supermodular set function p. Then:*

- *$g(S) = \max\{\sum_{A \in \mathcal{F}} p(A) : \mathcal{F}$ is a subpartition of $S\}$ if g is minimal.*
- *There exists an optimal p-cover J such that every $e \in J$ has both ends in T_g.*

Let $\tau(p)$ denote the size of a minimal p-cover. As $g = \{d_J(v) : v \in S\}$ is a p-transversal for any p-cover J, $\tau(p) \geq g(S)/2$ for any minimal p-transversal g. Thus a natural approach to compute a small p-cover is: repeatedly choose an edge uv with $u, v \in T_g$, such that updating p and reducing $g(u)$ and $g(v)$ by 1, keeps g being a p-transversal. This approach works for many interesting special cases, c.f. [5], but in general such an edge uv may not exist. Formally, given $u, v \in T_g$ define p^{uv} and g^{uv} by:

$$p^{uv}(A) = \max\{p(A) - 1, 0\} \text{ if } |A \cap \{u, v\}| = 1 \text{ and } p^{uv}(A) = p(A) \text{ otherwise;}$$

$$g^{uv}(w) = g(w) - 1 \text{ if } w = u \text{ or if } w = v \text{ and } g^{uv}(w) = g(w) \text{ otherwise.}$$

It is easy to see that if p is (symmetric) skew-supermodular, so is p^{uv}. However, g^{uv} may not be a p^{uv}-transversal if g is. We say that a pair $u, v \in T_g$ is (p, g)-**legal** if g^{uv} is a p^{uv}-transversal; then replacing p, g by p^{uv}, g^{uv} is the **splitting-off** operation at u, v. Intuitively, splitting-off is an attempt to add the edge uv to a partial solution, and to consider the residual problem of covering p^{uv} with the residual lower bound $\lceil g^{uv}(S)/2 \rceil = \lceil g(S)/2 \rceil - 1$. We need the following result due to [22], see also [5] for a short and elegant proof.

Lemma 8. ([22]). *Let p be symmetric skew-supermodular and g a p-transversal. If $p_{\max} \geq 2$ then there exists a (p, g)-legal pair.*

Lemma 8 implies that if no (p, g)-legal pair exists, then any inclusion minimal solution on T_g is a forest, and that any tree on T_g is a feasible solution. In [5, 22] was considered a simple greedy algorithm which repeatedly splits-off legal pairs as long as such exist, and then adds to the partial solution a tree (or any inclusion minimal solution) on T_g.

Algorithm 2: GREEDY(p, g)
(p is symmetric skew-supermodular, g is a minimal p-transversal on S)

1   $M \leftarrow \emptyset$
2   **while** there exists a (p, g)-legal pair u, v **do**
3     $\lfloor$ $g \leftarrow g^{uv}$, $p \leftarrow p^{uv}$, $M \leftarrow M + uv$
4   let F be a tree on $T' = \{v \in S : g(v) = 1\}$
5   **return** $M \cup F$

In the degree bounded version we let $g = \{b(v) : v \in S\}$. If this g is not a p-transversal, then the problem has no feasible solution. To get a degree violation $+1$, at step 4 of the algorithm we choose F to be a path on T'.

In [22] is was shown that for skew-supermodular p this algorithm achieves ratio $7/4$, by characterizing those pairs p, g for which no (p, g)-legal pair exists and deriving a lower bound on $\tau(p)$. We establish a better lower bound than that of [22] and prove the following.

Theorem 6. *Algorithm GREEDY achieves approximation ratio $3/2$. Moreover, if F is chosen to be a path at step 4, then $d_J(v) \leq g(v) + 1$ for all $v \in S$.*

Theorem 6 second statement is obvious, so in the rest of this section we prove the first statement. The following can be deduced from Lemma 8, see [22].

Corollary 1 ([22]). *Let p' be symmetric skew-supermodular and g' a minimal p'-transversal with non-empty support $T' = \{v \in V : g'(v) \geq 1\}$, and suppose that no (p', g')-legal pair exists. Then $p'_{\max} = g'_{\max} = 1$, $|T'| \geq 3$, and for every $A' \subseteq T'$ with $|A'| \in \{1, 2\}$ there is $A \subseteq V$ with $p'(A) = 1$ such that $A \cap T' = A'$. Furthermore, $\tau(p') \geq \frac{2}{3}|T'|$.*

In what follows we will use the following notation to analyze the performance of Algorithm 2:

- k is the number of edges accumulated in M during the while-loop.
- g is the initial minimal p-transversal g and $T = T_g$ is the support of g.
- $t = g(T)$ is the initial value of g.
- $t' = |T'|$ is the transversal value at the beginning of step 4.

We now describe the analysis of the 7/4-approximation of [22]. Note that $t - t' = 2k$ and that $|F| = |T'| - 1 = t - 2k - 1$. Consequently,

$$|M \cup F| = |M| + |F| \leq k + (t - 2k - 1) \leq t - k .$$

We also have the lower bounds $\tau(p) \geq t/2$ and $\tau(p) \geq \frac{2}{3}|T'| = \frac{2}{3}(t - 2k)$. Thus the approximation ratio of Algorithm 2 is bounded by $\frac{t-k}{\max\{t/2, 2(t-2k)/3\}} \leq 7/4$, with $k = t/8$ being the worse case.

One can observe that if $k \geq t/4$ then $t/2 \geq \frac{2}{3}(t - 2k)$, and thus in this case the ratio is bounded by $\frac{t-k}{t/2} \leq \frac{3/4}{1/2} = 3/2$. To get ratio 3/2 for the range $k \leq t/4$ we give a better lower bound on $\tau(p)$. For this, we need the following lemma.

Lemma 9. *Let g be a minimal transversal of a symmetric skew-supermodular set function p. Then there exists an optimal p-cover J such that $d_J(v) \geq g(v)$ if $v \in T_g$ and $d_J(v) = 0$ otherwise.*

Proof. By induction on $\tau(p)$. The base case $\tau(p) = 1$ is trivial. For $\tau(p) \geq 2$, let J be an optimal p-cover such that every $e \in J$ has both ends in T_g; such exists by Lemma 7. Choose some $e = uv \in J$ and let $p' = p^{uv}$. Let g^u be obtained from g by decreasing $g(u)$ by 1, and similarly g^v is defined. Then one of $\{g, g^u, g^v, g^{uv}\}$ is a minimal p'-transversal; denote it by g'. By the induction hypothesis there exists a p'-cover J' such that: $d_{J'}(w) \geq g'(w)$ if $w \in T_{g'}$ and $d_{J'}(w) = 0$ otherwise. It is easy to see that $J = J' \cup \{e\}$ has the required property. $\square$

The following Lemma gives our improved lower bound.

Lemma 10. $\tau(p) \geq \frac{2}{3}(t - k)$.

Proof. Let J be an optimal p-cover as in Lemma 9. Let X be the set of edges in J with both ends in $T \setminus T'$ and Y the set of edges in J with exactly one end in $T \setminus T'$; let $x = |X|$ and $y = |Y|$.

Since $d_J(v) \geq g(v)$ for all $v \in T$, $2x + y \geq t - t' = 2k$, hence $y \geq 2k - 2x$. Let Q be the set of nodes in T' that are uncovered by edges in $X \cup Y$ and let z be the number of edges in J with both ends in Q. Note that $|Q| \geq t' - y = t - 2k - y$. By Corollary 1, for every $A' \subseteq Q$ with $|A'| \in \{1, 2\}$ there is $A \subseteq V$ with $p(A) > 0$ such that $A \cap T' = A'$. This implies that $z \geq \frac{2}{3}|Q| \geq \frac{2}{3}(t - 2k - y)$. Consequently, since $|J| \geq x + y + z$ and $y \geq 2(k - x)$

$$|J| \geq x + y + \frac{2}{3}(t - 2k - y) = x + \frac{1}{3}y + \frac{2}{3}(t - 2k) \geq x + \frac{2}{3}(k - x) + \frac{2}{3}(t - 2k) = \frac{1}{3}x + \frac{2}{3}(t - k) .$$

Since $x \geq 0$ we get $|J| \geq \frac{2}{3}(t - k)$. $\square$

From Lemma 10 it follows that the approximation ratio of the algorithm is bounded by $\frac{t-k}{\max\{t/2, 2(t-k)/3\}} \leq 3/2$, with $k = t/4$ being the worse case.

This concludes the proof of Theorem 6, and thus also of Theorem 3.

References

1. Adjiashvili, D.: Beating approximation factor two for weighted tree augmentation with bounded costs. In: SODA, pp. 2384–2399 (2017)
2. Bang-Jensen, J., Jackson, B.: Augmenting hypergraphs by edges of size two. Math. Program. **84**(3), 467–481 (1999). https://doi.org/10.1007/s101070050033
3. Basavaraju, M., Fomin, F.V., Golovach, P.A., Misra, P., Ramanujan, M.S., Saurabh, S.: Parameterized algorithms to preserve connectivity. In: ICALP, Part I, pp. 800–811 (2014)
4. Benczúr, A.A., Frank, A.: Covering symmetric supermodular functions by graphs. Math. Program. **84**(3), 483–503 (1999). https://doi.org/10.1007/s101070050034
5. Bernáth, A., Király, T.: A unifying approach to splitting-off. Combinatorica **32**(4), 373–401 (2012)
6. Byrka, J., Grandoni, F., Ameli, A.J.: Breaching the 2-approximation barrier for connectivity augmentation: a reduction to steiner tree. In: STOC, pp. 815–825 (2020)
7. Byrka, J., Grandoni, F., Rothvoß, T., Sanitß, L.: Steiner tree approximation via iterative randomized rounding. J. ACM **60**(1), 6:1–6:33 (2013)
8. Cecchetto, F., Traub, V., Zenklusen, R.: Bridging the gap between tree and connectivity augmentation: unified and stronger approaches. CoRR, abs/2012.00086 (2020). To appear in STOC 2021
9. Dinic, E.A., Karzanov, A.V., Lomonosov, M.V.: On the structure of a family of minimal weighted cuts in a graph. In: Studies in Discrete Optimization, pp. 290–306 (1976)
10. Dinitz, Y., Nutov, Z.: A 2-level cactus model for the system of minimum and minimum+1 edge-cuts in a graph and its incremental maintenance. In: STOC, pp. 509–518 (1995)
11. Fiorini, S., Groß, M., Könemann, J., Sanitá, L.: A $\frac{3}{2}$-approximation algorithm for tree augmentation via chvátal-gomory cuts. In: SODA, pp. 817–831 (2018)
12. Fleischer, L., Jain, K., Williamson, D.P.: Iterative rounding 2-approximation algorithms for minimum-cost vertex connectivity problems. J. Comput. Syst. Sci. **72**(5), 838–867 (2006)
13. Frank, A.: Connections in Combinatorial Optimization. Oxford University Press, Oxford (2011)
14. Frank, A., Jordán, T.: Graph connectivity augmentation. In: Thulasiraman, K., Arumugam, S., Brandstadt, A., Nishizeki, T. (eds.) Handbook of Graph Theory, Combinatorial Optimization, and Algorithms, chap. 14, pp. 313–346. CRC Press (2015)
15. Gálvez, W., Grandoni, F., Ameli, A.J., Sornat, K.: On the cycle augmentation problem: hardness and approximation algorithms. In: WAOA, pp. 138–153 (2019)
16. Grandoni, F., Kalaitzis, C., Zenklusen, R.: Improved approximation for tree augmentation: saving by rewiring. In: STOC, pp. 632–645 (2018)
17. Khuller, S.: Approximation algorithms for finding highly connected subgraphs. In: Hochbaum, D. (ed.) Approximation Algorithms for NP-Hard Problems, chap. 6, pp. 236–265. PWS (1995)
18. Király, Z., Cosh, B., Jackson, B.: Local edge-connectivity augmentation in hypergraphs is NP-complete. Discret. Appl. Math. **158**(6), 723–727 (2010)
19. Kortsarz, G., Nutov, Z.: A simplified 1.5-approximation algorithm for augmenting edge-connectivity of a graph from 1 to 2. ACM Trans. Algorithms **12**(2), 23 (2016)

20. Maduel, Y., Nutov, Z.: Covering a laminar family by leaf to leaf links. Discret. Appl. Math. **158**(13), 1424–1432 (2010)
21. Nutov, Z.: Structures of cuts and cycles in graphs; algorithms and applications. Ph.D. thesis, Technion, Israel Institute of Technology (1997)
22. Nutov, Z.: Approximating connectivity augmentation problems. ACM Trans. Algorithms **6**(1), 5:1–5:19 (2009)
23. Nutov, Z.: 2-node-connectivity network design. CoRR, abs/2002.04048 (2020). To appear in WAOA20
24. Nutov, Z.: On the tree augmentation problem. Algorithmica **83**(2), 553–575 (2021)
25. Nutov, Z., Kortsarz, G., Shalom, E.: Approximating activation edge-cover and facility location problems. In: MFCS, pp. 20:1–20:14 (2019)
26. Ravi, R., Williamson, D.P.: An approximation algorithm for minimum-cost vertex-connectivity problems. Algorithmica **18**(1), 21–43 (1997)

On Rooted k-Connectivity Problems in Quasi-bipartite Digraphs

Zeev Nutov[✉]

The Open University of Israel, Ra'anana, Israel
nutov@openu.ac.il

Abstract. We consider the directed ROOTED SUBSET k-EDGE-CONNEC-TIVITY problem: given a digraph $G = (V, E)$ with edge costs, a set $T \subset V$ of terminals, a root node r, and an integer k, find a min-cost subgraph of G that contains k edge disjoint rt-paths for all $t \in T$. The case when every edge of positive cost has head in T admits a polynomial time algorithm due to Frank [9], and the case when all positive cost edges are incident to r is equivalent to the k-MULTICOVER problem. Recently, Chan et al. [2] obtained ratio $O(\ln k \ln |T|)$ for quasi-bipartite instances, when every edge in G has an end (tail and/or head) in $T + r$. We give a simple proof for the same ratio for a more general problem of covering an arbitrary T-intersecting supermodular set function by a minimum cost edge set, and for the case when only every positive cost edge has an end in $T + r$.

1 Introduction

All graphs considered here are directed, unless stated otherwise. We consider the following problem (a.k.a. k-EDGE-CONNECTED DIRECTED STEINER TREE):

ROOTED SUBSET k-EDGE-CONNECTIVITY
Input: A directed (multi-)graph $G = (V, E)$ with edge costs $\{c(e) : e \in E\}$, a set $T \subset V$ of terminals, a root node $r \in V \setminus T$, and an integer k.
Output: A min-cost subgraph that has k-edge disjoint rt-paths for all $t \in T$.

The case when every edge of positive cost has head in T admits a polynomial time algorithm due to Frank [9]. When all positive cost edges are incident to r we get the MIN-COST MULTICOVER problem. The case when all positive cost edges are incident to the same node admits ratio $O(\ln n)$ [24]. More generally, a graph (or an edge set) is called **quasi-bipartite** if every edge in the graph has at least one end (tail or head) in $T + r$.

In the augmentation version of the problem – ROOTED SUBSET (ℓ, k)-EDGE-CONNECTIVITY AUGMENTATION, G contains a subgraph $G_0 = (V, E_0)$ of cost zero that has ℓ-edge disjoint rt-paths for all $t \in T$. Recently, Chan, Laekhanukit, Wei, & Zhang [2] obtained ratio $O(\ln(k - \ell + 1) \ln |T|)$ for the case when G is quasi-bipartite. We provide a simple proof for a slightly more general setting.

An integer valued set function f on a groundset V is **intersecting super-modular** if any $A, B \subseteq V$ that intersect satisfy the **supermodular inequality**

© Springer Nature Switzerland AG 2021
R. Santhanam and D. Musatov (Eds.): CSR 2021, LNCS 12730, pp. 339–348, 2021.
https://doi.org/10.1007/978-3-030-79416-3_20

$f(A) + f(B) \le f(A \cap B) + f(A \cup B)$; if this holds whenever $A \cap B \cap T \ne \emptyset$ for a given set $T \subseteq V$ of terminals, then f is T-**intersecting supermodular**. We say that $A \subseteq V$ is an f-**positive set** if $f(A) > 0$. f is **positively T-intersecting supermodular** if the supermodular inequality holds whenever $A \cap B \cap T \ne \emptyset$ and $f(A), f(B) > 0$. An edge e **covers** a set A if it enters A, namely, if its head is in A and tail is not in A. For an edge set/graph I let $d_I(A)$ denote the number of edges in I that cover A. We say that I **covers** f or that I is an f-**cover** if $d_I(A) \ge f(A)$ for all $A \subseteq V$. We consider the following generic problem.

SET FUNCTION EDGE COVER
Input: A digraph $G = (V, E)$ with edge costs and a set function f on V.
Output: A min-cost f-cover $J \subseteq E$.

Here f may not be given explicitly, but for a polynomial time implementation of algorithms we need that certain queries related to f can be answered in polynomial time. For an edge set I, the **residual function** f^I **of** f is defined by $f^I(A) = \max\{f(A) - d_I(A), 0\}$. It is known that if f is positively T-intersecting supermodular then so is f^I. Let $\mu_f = \max\{f(A) : A \subseteq V\}$ denote the maximum f-value taken over all sets. The **max-family** of a set function g is defined by $\{A \subseteq V : g(A) = \mu_g\}$. An inclusion minimal member of a set-family $\mathcal{F}$ is an $\mathcal{F}$-**core**, or simply a **core**, if $\mathcal{F}$ is clear from the context. Let $\mathcal{C}_{\mathcal{F}}$ denote the family of $\mathcal{F}$-cores. We will assume the following.

Assumption 1. *For any edge set I, the cores of the max-family $\{A : g(A) = \mu_g\}$ of the residual function $g = f^I$ of f, can be found in polynomial time.*

It is known that the ROOTED SUBSET (ℓ, k)-EDGE-CONNECTIVITY AUGMENTATION problem is equivalent to covering the set function f defined by $f(A) = \max\{k - d_{G_0}(A), 0\}$ if $A \cap T \ne \emptyset$ and $f(A) = 0$ otherwise, where $G_0 = (V, E_0)$ is the subgraph of zero cost edges of G. This function is positively T-intersecting supermodular, see [9]. Assumption 1 holds for this function, since the cores as in Assumption 1 can be found by computing for every $t \in T$ the closest to t minimum rt-cut of $G_0 + I$, c.f. [9,28]. Given a set function f on V and a set $T \subseteq V$ of terminals, we say that a graph $G = (V, E)$ (or an edge set) is **quasi-bipartite** if every edge has an end (tail or head) v such that $v \in T$ or such that v does not belong to any f-positive set. Under Assumption 1, we prove the following.

Theorem 1. SET FUNCTION EDGE COVER *with positively T-intersecting supermodular f and quasi-bipartite G admits approximation ratio $2H(\mu_f) \cdot (1 + \ln|T|)$.*

Theorem 1 implies the following extension of the result of Chan et al. [2].

Corollary 1. *The* ROOTED SUBSET (ℓ, k)-EDGE-CONNECTIVITY AUGMENTATION *problem admits approximation ratio $2H(k - \ell) \cdot (1 + \ln|T|)$ if the set of positive cost edges of G is quasi-bipartite.*

As far as we can see, Corollary 1 cannot be deduced from the work of Chan et al. [2]. Our approach is motivated by an earlier result of Frank [9], who showed that ROOTED SUBSET k-EDGE-CONNECTIVITY can be solved in polynomial time provided that every positive cost edge has head in T. For this, he proved that SET FUNCTION EDGE COVER with positively T-intersecting supermodular f can be solved in polynomial time provided that every positive cost edge has head in T. While our ratio is asymptotically similar to the one of [2] – $O(\ln k \cdot \ln |T|)$, but our constant hidden in the $O(\cdot)$ term is better and the proof (of a more general result) is substantially simpler. Moreover, our algorithm is combinatorial while the [2] algorithm is LP based.

We use a method initiated by the author in [28], that extends the Klein-Ravi [21] algorithm for the NODE WEIGHTED STEINER TREE problem, to high connectivity problems. It was applied later in [29, 30] also for node weighted problems, and the same method is used in [2]; a restricted version of this method appeared earlier in [22] and later in [7]. The method was further developed by Fukunaga [11] and Chekuri, Ene, and Vakilian [4] for prize-collecting problems.

In the rest of the introduction we briefly survey some literature on rooted connectivity problems. The DIRECTED STEINER TREE problem admits approximation $O(\ell^3 |T|^{2/\ell})$ in time $O(|T|^{2\ell} n^\ell)$, see [3, 18, 25, 33], and also a tight quasi-polynomial time approximation $O(\log^2 |T| / \log \log |T|)$ [13, 16]; see also a survey in [6]. For similar results for ROOTED SUBSET 2-EDGE-CONNECTIVITY see [15]. DIRECTED STEINER TREE is $\Omega(\log^2 n)$-hard even on very special instances [17] that arise from the GROUP STEINER TREE problem on trees; the latter problem admits a matching ratio $O(\log^2 n)$ [12]. The (undirected) STEINER TREE problem was also studied extensively, c.f. [1, 14] and the references therein. The study of quasi-bipartite instances was initiated for undirected graphs in the 90's [32], and the directed version was shown to admit ratio $O(\ln |T|)$ in [10, 19].

Rooted k-connectivity problems were studied for both directed and undirected graphs, edge-connectivity and node-connectivity, and various types of graphs and costs; c.f. a survey [31]. For undirected graphs the problem admits ratio 2 [20], but for digraphs it has approximation threshold $\max\{k^{1/2-\epsilon}, |T|^{1/4-\epsilon}\}$ [26]. For the undirected node connectivity version, the currently best known ratio is $O(k \ln k)$ [30] and threshold $\max\{k^{0.1-\epsilon}, |T|^{1/4-\epsilon}\}$ [26]. However, the augmentation version when any edge can be added by a cost of 1 is just SET COVER hard and admits ratios $O(\ln |T|)$ for digraphs and $\min\{O(\ln |T|, O(\ln^2 k)\}$ for graphs [23]; a similar result holds when positive cost edges form a star [24].

In digraphs, node connectivity can be reduced to edge-connectivity by a folklore reduction of "splitting" each node v into two nodes v^{in}, v^{out}. However, this reduction does not preserve quasi-bipartiteness. The reduction of [27] that transfers undirected connectivity problems into directed ones, and the reduction of [5] that reduces general connectivity requirements to rooted requirements, also do not preserve quasi-bipartiteness.

2 Covering T-Intersecting Supermodular Functions (Theorem 1)

A set family $\mathcal{F}$ is a T-**intersecting family** if $A \cap B, A \cup B \in \mathcal{F}$ whenever $A \cap B \cap T \neq \emptyset$. It is known that if f is (positively) T-intersecting supermodular then the family $\mathcal{F} = \{A \subseteq V : f(A) = \mu_f\}$ is T-intersecting. An edge set I is a **cover of** $\mathcal{F}$, or an $\mathcal{F}$-**cover**, if $d_I(A) \geq 1$ for all $A \in \mathcal{F}$; I is a p-**cover of** $\mathcal{F}$ if $d_I(A) \geq p$ for all $A \in \mathcal{F}$. The **residual family** $\mathcal{F}^I$ of $\mathcal{F}$ consists of the members of $\mathcal{F}$ that are uncovered by I. It is known that if $\mathcal{F}$ is T-intersecting then so is $\mathcal{F}^I$, c.f. [9]. Let us consider the following problem.

SET FAMILY EDGE COVER

Input: A directed graph $G = (V, E)$ with edge costs and a set family $\mathcal{F}$ on V.
Output: A min-cost $\mathcal{F}$-cover $I \subseteq E$.

In the next section we will prove the following.

Lemma 1. *Consider the* SET FUNCTION EDGE COVER *problem with positively T-intersecting supermodular f, quasi-bipartite G, and optimal solution value τ_f. Let $\mathcal{F} = \{A \subseteq V : f(A) = \mu_f\}$, $\rho_f = \frac{2}{\mu_f}$, and for $I \subseteq E$ let $\nu_f(I)$ denote the number of $\mathcal{F}^I$-cores. There exists a polynomial time algorithm that given $I \subseteq E$ with $\nu_f(I) \geq 1$ finds an edge set $S \subseteq E \setminus I$ such that*

$$\frac{c(S)}{\nu_f(I) - \nu_f(I \cup S)} \leq \rho_f \cdot \frac{\tau_f}{\nu_f(I)} .$$

From Lemma 1 it is a routine to deduce the following corollary, c.f. [21] and [29, Theorem 3.1]; we provide a proof sketch for completeness of exposition.

Corollary 2. SET FUNCTION EDGE COVER *with positively T-intersecting supermodular f and quasi-bipartite G admits a polynomial time algorithm that computes a cover I of $\mathcal{F} = \{A \subseteq V : f(A) = \mu_f\}$ of cost $c(I) \leq \frac{2}{\mu_f} \cdot (1 + \ln \nu_f(\emptyset)) \cdot \tau_f$.*

Proof. Start with $I = \emptyset$ an while $\nu_f(I) \geq 1$ add to I an edge set S as in Lemma 1. Let I_j be the partial solution at the end of iteration j, where $I_0 = \emptyset$, and let S_j be the set added at iteration j; thus $I_j = I_{j-1} \cup S_j$, $j = 1, \ldots, q$. Let $\nu_j = \nu_f(I_j)$, so $\nu_0 = \nu_f(\emptyset)$, $\nu_q = 0$, and $\nu_{q-1} \geq 1$. Then

$$\frac{c_j}{\nu_{j-1} - \nu_j} \leq \rho_f \cdot \frac{\tau_f}{\nu_{j-1}} \qquad j = 1, \ldots, q .$$

This implies $c_q \leq \rho_f \tau_f$ and

$$\nu_j \leq \nu_{j-1}(1 - \frac{c_j}{\rho_f \tau_f}) \qquad j = 1, \ldots, q .$$

Unraveling we get

$$\frac{\nu_{q-1}}{\nu_0} \leq \prod_{j=1}^{q-1} \left(1 - \frac{c_j}{\rho_f \tau_f}\right).$$

Taking natural logarithms and using the inequality $\ln(1+x) \leq x$, we obtain

$$\rho_f \cdot \tau_f \cdot \ln\left(\frac{\nu_0}{\nu_{q-1}}\right) \geq \sum_{j=1}^{q-1} c_j$$

Since $c_q \leq \rho_f \tau_f$ and $\nu_{q-1} \geq 1$, we get $c(I) \leq c_q + \sum_{j=1}^{q-1} c_j \leq \rho_f \tau_f (1 + \ln \nu_0)$. $\square$

To see that Corollary 2 implies Theorem 1, consider the following algorithm.

Algorithm 1: BACKWARD-AUGMENTATION$(f, G = (V, E), c)$

---

1 $I \leftarrow \emptyset$
2 **for** $\ell = \mu_f$ downto 1 **do**
3 $\quad$ Compute a cover I_ℓ of $\mathcal{F}_\ell = \{A \subseteq V : f^I(A) = \ell\}$ as in Corollary 2
4 $\quad$ $I \leftarrow I \cup I_\ell$
5 **return** I

At iteration ℓ we have $c(I_\ell)/\tau_f = 2(1 + \ln|T|)/\ell$, hence the overall ratio is $2(1 + \ln|T|) \cdot \sum_{\ell=\mu_f}^{1} 1/\ell = 2H(\mu_f) \cdot (1 + \ln|T|)$, as required in Theorem 1.

3 Proof of Lemma 1

For $C \in \mathcal{C}_\mathcal{F}$ let $\mathcal{F}(C)$ denote the family of sets in $\mathcal{F}$ that contain no core distinct from C; for $\mathcal{C} \subseteq \mathcal{C}_\mathcal{F}$ let $\mathcal{F}(\mathcal{C}) = \cup_{C \in \mathcal{C}} \mathcal{F}(C)$. An intersecting family $\mathcal{R}$ such that all its members contain the same element is called a **ring**; such $\mathcal{R}$ has a unique core C (the intersection of all sets in $\mathcal{R}$) and a unique inclusion maximal set M (the union of all sets in $\mathcal{R}$). Some additional known properties of rings that we use are summarized in the next lemma; c.f. [28, Lemma 2.6 and Corollary 2.7] for the first property and [8] for the second. Recall that an edge set I is a p-**cover** of a set family $\mathcal{F}$ if $d_I(A) \geq p$ for all $A \in \mathcal{F}$.

Lemma 2. *Let $\mathcal{R}$ be a ring with minimal member C and maximal member M.*

(i) If I is a minimal cover of $\mathcal{R}$ then there is an ordering $e_1, e_2, \ldots, e_q$ of I and a nested family $C = C_1 \subset C_2 \cdots \subset C_q = M$ of sets in $\mathcal{R}$ such that for every $j = 1, \ldots, q$, e_j is the unique edge in I that enters C_j (namely, e_j has had in C_j and tail not in C_j).
(ii) Any p-cover of $\mathcal{R}$ is a union of p edge disjoint covers of $\mathcal{R}$.

The following lemma is a folklore.

Lemma 3. *If $\mathcal{F}$ is a T-intersecting family then $\mathcal{F}(C)$ is a ring for any $C \in \mathcal{C}_\mathcal{F}$; thus $\mathcal{F}(C)$ has a unique maximal member M_C. Furthermore, $M_C \cap M_{C'} \cap T = \emptyset$ for any distinct $C, C' \in \mathcal{C}_\mathcal{F}$.*

Recall that a **spider** is a directed tree such that only its root may have out-degree ≥ 2. We now define an analogue of spiders. While T-intersecting families considered here are more complex than intersecting families considered in [28], the case of quasi-bipartite graphs is simpler than that of general graphs, and this allows to use a simplified and less restrictive form of [28, Definition 2.3].

Definition 1. *An edge set* $S = S(e, \mathcal{C})$ *is a* **spider-cover** *of a subfamily* $\mathcal{C} \subseteq \mathcal{C}_{\mathcal{F}}$ *of cores if for some* $e \in S$ *(the* **head** *of the spider-cover)* $S - e$ *has a partition* $\{S_C : C \in \mathcal{C}\}$ *(the* **legs** *of the spider-cover) such that each* $S_C + e$ *covers* $\mathcal{F}(C)$.

Lemmas 2(i) and 3 imply the following.

Corollary 3. *Let* I *be a quasi-bipartite cover of a* T-intersecting family $\mathcal{F}$. *For* $C \in \mathcal{C}_{\mathcal{F}}$ *let* $I_C \subseteq I$ *be an inclusion minimal cover of* $\mathcal{F}(C)$, *and let* e_C *be the unique (by Lemma 2(i)) edge in* I_C *that covers* M_C. *Then* $I_C \cap I_{C'} = \emptyset$ *or* $I_C \cap I_{C'} = \{e_C\} = \{e_{C'}\}$ *holds for any distinct* $C, C' \in \mathcal{C}_{\mathcal{F}}$.

Consider the MIN-COST p-MULTICOVER problem: Given a collection $\mathcal{S}$ of sets over a set U of elements with costs $\{c(S) : S \in \mathcal{S}\}$, find a min-cost p-**multicover** – a sub-collection $\mathcal{S}' \subseteq \mathcal{S}$ such that every $u \in U$ belongs to at least p sets in $\mathcal{S}'$. The corresponding LP-relaxation is

$$\min\{c \cdot x : \sum_{S \ni u} x_S \geq p \ \forall u \in U, \ 0 \leq x_S \leq 1 \ \forall S \in \mathcal{S}\}.$$

Lemma 4. *Let* x *be a feasible* MIN-COST p-MULTICOVER *LP solution and let* $\tau = c \cdot x$. *Then there is a set* S' *in the support of* x *such that* $\frac{c(S')}{|S'|} \leq \frac{1}{p} \cdot \frac{\tau}{|U|}$.

Proof. Note that $y = x/p$ is a fractional 1-cover of cost τ/p, namely, $c \cdot y = \tau/p$ and $\sum_{S \ni u} y_S \geq 1$ for all $u \in U$. Let $S' = \arg\max_{S \in \mathcal{S}} \frac{|S|}{c(S)}$. Then

$$\frac{|S'|}{c(S')}(c \cdot y) \geq \sum_{S \in \mathcal{S}} \frac{|S|}{c(S)} c(S) y_S = \sum_{S \in \mathcal{S}} |S| y_S = \sum_{u \in U} \sum_{S \ni u} y_S \geq \sum_{u \in U} 1 = |U|$$

Thus $\frac{|S'|}{c(S')} \geq \frac{|U|}{c \cdot y}$, so $\frac{c(S')}{|S'|} \leq \frac{c \cdot y}{|U|} = \frac{1}{p} \cdot \frac{\tau}{|U|}$. $\square$

Lemma 5. *Let* I *be a quasi-bipartite p-cover of cost* τ *of a* T-intersecting family $\mathcal{F}$. *There is a spider-cover* $S \subseteq I$ *of* $\mathcal{C} \subseteq \mathcal{C}_{\mathcal{F}}$ *such that* $\frac{c(S)}{|C|} \leq \frac{1}{p} \cdot \frac{\tau}{|\mathcal{C}_{\mathcal{F}}|}$.

Proof. For every $C \in \mathcal{C}_{\mathcal{F}}$ fix an inclusion minimal p-cover $I_C \subseteq I$ of $\mathcal{F}(C)$. By Lemma 2(ii) I_C has a subpartition of p edge disjoint inclusion minimal covers $I_C^1, \ldots, I_C^p$ of $\mathcal{F}(C)$. Each I_C^j has a unique edge $e_C^j \in I$ that enters M_C. Note that $I_C^j \cap I_{C'}^{j'} = \emptyset$ or $I_C^j \cap I_{C'}^{j'} = \{e_C^j\} = \{e_{C'}^{j'}\}$, by Corollary 3.

Define an auxiliary bipartite graph $(I \cup \mathcal{C}_{\mathcal{F}}, \mathcal{E})$ with edge and node costs as follows.

- Every node $e \in I$ of $\mathcal{H}$ inherits the cost of e in G.
- Each $C \in \mathcal{C}_{\mathcal{F}}$ is connected in $\mathcal{H}$ to each e_C^j by an edge of cost $c(I_C^j) - c(e_C^j)$.

Every $C \in \mathcal{C}_{\mathcal{F}}$ has exactly p neighbors in $\mathcal{H}$, thus the collection of maximal stars in $\mathcal{H}$ defines a (multiset) p-multicover $\mathcal{S}_{\mathcal{H}}$ of $\mathcal{C}_{\mathcal{F}}$ of cost $\leq \tau$; for each star, the corresponding set $S_{\mathcal{H}} \in \mathcal{S}_{\mathcal{H}}$ is the set of leaves of the star, and the cost of $S_{\mathcal{H}}$ is the total cost of the star – of the center and the edges. By Lemma 4, $\frac{c(S_{\mathcal{H}})}{|S_{\mathcal{H}}|} \leq \frac{1}{p} \cdot \frac{\tau}{|\mathcal{C}_{\mathcal{F}}|}$ for some $S_{\mathcal{H}} \in \mathcal{S}_{\mathcal{H}}$. This $S_{\mathcal{H}}$ defines a spider-cover $S \subseteq I$ of C being the set of leaves of $S_{\mathcal{H}}$ with $c(S_{\mathcal{H}}) = c(S)$, and the lemma follows. $\square$

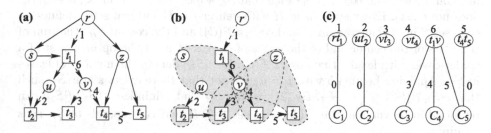

Fig. 1. (a) An instance of ROOTED SUBSET (ℓ, k)-EDGE-CONNECTIVITY AUGMENTATION where $\ell = 1$ and $k = 2$. Terminals are shown by square nodes and zero cost edges by solid lines. (b) The sets M_C, each core C_i consists of a single terminal t_i). (c) The bipartite graph constructed in the proof of Lemma 5.

Let us illustrate the construction from Lemma 5 by an example. Consider the ROOTED SUBSET (ℓ, k)-EDGE-CONNECTIVITY AUGMENTATION instance in Fig. 1 with $\ell = 1$ and $k = 2$. The set function we need to cover is defined by $f(A) = \max\{k - d_{G_0}(A), 0\}$ if $A \cap T \neq \emptyset$ and $f(A) = 0$ otherwise. This function is $0, 1$ valued, so $\mu_f = 1$. The max-family is $\mathcal{F} = \{A \subseteq V : A \cap T \neq \emptyset, d_{G_0}(A) = 1\}$. The cores of $\mathcal{F}$ are the single terminals, so $C_i = \{t_i\}$, $i = 1, \ldots, 5$. The sets $M_i = M_{C_i}$ are shown in (b). The ring family $\mathcal{F}(C_1)$ consists of all subsets of $M_1 = \{t_1, s, u, v\}$ that contain t_1. The single edge rt_1 covers $\mathcal{F}(C_1)$. The ring family $\mathcal{F}(C_5)$ consists of all subsets of $M_5 = \{t_5, v, z\}$ that contain t_5. The single edge t_4t_5 covers $\mathcal{F}(C_5)$. On the other hand, to cover $\mathcal{F}(C_3) = \{\{t_3\}, \{t_3, v\}\}$ we need two edges $\{t_1v, vt_3\}$, and to cover $\mathcal{F}(C_4) = \{\{t_4\}, \{t_4, v\}, \{t_4, z\}, \{t_4, v, z\}\}$ we need two edges $\{t_1v, vt_4\}$.

In the bipartite graph in (c), every star defines a spider-cover of the subfamily of cores of the leaves of the star that has cost equal to the cost of the star. For example, the star with leaf set $\{C_3, C_4, C_5\}$ defines the spider-cover $S = \{t_1v, vt_3, vt_4, t_4t_5\}$, with the partition into legs being the singleton edges of S – t_1v. For this spider-cover we have $\frac{c(S)}{|C|} = 18/3 = 6$. This spider-cover does not satisfy the statement of the lemma, since 6 is larger than $\frac{1}{p} \cdot \frac{\tau}{|\mathcal{C}_{\mathcal{F}}|} = 21/5$ (here we have $\tau = 21$, $\mu_f = p = 1$, and $|\mathcal{C}_{\mathcal{F}}| = 5$). On the other hand, the spider-cover of $C_1 = \{C_1\}$ has the required property, and so is the spider-cover of $C_2 = \{C_2\}$.

Lemma 6. *Consider a* SET FUNCTION EDGE COVER *instance with positively T-intersecting supermodular f and quasi-bipartite $G = (V, E)$. There exists a polynomial time algorithm that for $\mathcal{F} = \{A \subseteq V : f(A) = \mu_f\}$, finds $\mathcal{C} \subseteq \mathcal{C}_\mathcal{F}$ and an $\mathcal{F}(\mathcal{C})$-cover $S \subseteq E$ such that $\frac{c(S)}{|\mathcal{C}|} \leq \frac{1}{\mu_f} \cdot \frac{\tau_f}{|\mathcal{C}_\mathcal{F}|}$.*

Proof. Define an auxiliary complete bipartite graph $\mathcal{H} = (E \cup \mathcal{C}, \mathcal{E})$ with edge and node costs in a similar way as in the proof of Lemma 5. Every node $e \in E$ of $\mathcal{H}$ has cost equal to the cost of e in G, and the cost of an edge $e \in \mathcal{E}$ is the minimum cost of an edge set S_C such that $S_C + e$ covers $\mathcal{F}(\mathcal{C})$ (or ∞, if such S_C does not exist). Every star $S_\mathcal{H}$ in $\mathcal{H}$ with center $e \in E$ and leaf set $\mathcal{C}$ defines an edge set S in G that contains e and covers $\mathcal{F}(\mathcal{C})$, and the cost of $S_\mathcal{H}$ (the sum of the costs of the center and of the edges of $S_\mathcal{H}$) is at most the optimal cost of a spider-cover with head e that covers $\mathcal{F}(\mathcal{C})$. Let $|S_\mathcal{H}|$ denote the number of leaves in $S_\mathcal{H}$. Applying Lemma 5 with $p = \mu_f$ we get that there exists a star $S_\mathcal{H}$ such that $c(S_\mathcal{H})/|S_\mathcal{H}| \leq \frac{1}{\mu_f} \cdot \frac{\tau}{|\mathcal{C}_\mathcal{F}|}$. By [21], a star $S_\mathcal{H}$ that minimizes $c(S_\mathcal{H})/|S_\mathcal{H}|$ can be found in polynomial time. The edge set $S \subseteq E$ that corresponds to $S_\mathcal{H}$ is as required.

It remains to show that under Assumption 1 one can find in polynomial time a minimum cost edge set S_C such that $S_C + e$ covers $\mathcal{F}(\mathcal{C})$. Reset the cost of e to zero and let I be a set of new edges, that contains μ_f edges from r to every terminal in $T \setminus M_C$. Then $\mathcal{F}(\mathcal{C}) = \{A \subseteq V : f^I(A) = k_{f^I}\}$ is a ring. It is known that a min-cost edge-cover of a ring can be found in polynomial time under Assumption 1 (c.f. [9,28]), by a standard primal dual algorithm.   □

An analogue of the following lemma was proved in [28, Lemma 3.3] for intersecting families, and the proof for T-intersecting families is similar.

Lemma 7. *Let $\mathcal{F}$ be a T-intersecting family. If S covers $\mathcal{F}(\mathcal{C})$ for $\mathcal{C} \subseteq \mathcal{C}_\mathcal{F}$ then $\nu(\emptyset) - \nu(S) \geq |\mathcal{C}|/2$, where $\nu(S)$ denotes the number of $\mathcal{F}^S$-cores.*

Proof. The $\mathcal{F}^S$-cores are T-disjoint, and each of them contains some $\mathcal{F}$-core. Every $\mathcal{F}^S$-core that contains a core from $\mathcal{C}$ contains at least two $\mathcal{F}$-cores. Thus the number of $\mathcal{F}^S$-cores that contain exactly one $\mathcal{F}$-core is at most $\nu(\emptyset) - |\mathcal{C}|/2$. Consequently, $\nu(S) \leq \nu(\emptyset) - |\mathcal{C}|/2$.   □

Applying Lemmas 6 and 7 on the residual function $g = f^I$ we get that we can find in polynomial time an edge set $S \subseteq E \setminus I$ such that

$$\frac{c(S)}{\nu_g(\emptyset) - \nu_g(S)} \leq \frac{c(S)}{|\mathcal{C}|/2} \leq \frac{2}{\nu_g} \cdot \frac{\tau_g}{\nu_g(\emptyset)}$$

Lemma 1 follows by observing that $\mu_g = \mu_f$, $\nu_g(\emptyset) = \nu(I)$, $\nu_g(S) = \nu_f(I + S)$, and $\tau_g \leq \tau_f$.

This concludes the proof of Lemma 1, and thus also the proofs Theorem 1 and Corollary 1 are complete.

References

1. Byrka, J., Grandoni, F., Rothvoß, T., Sanitá, L.: Steiner tree approximation via iterative randomized rounding. J. ACM **60**(1), 6:1–6:33 (2013). Preliminary version in STOC 2010
2. Chan, C-H., Laekhanukit, B., Wei, H-T., Zhang, Y.: Polylogarithmic approximation algorithm for k-connected directed Steiner tree on quasi-bipartite graphs. In: APPROX/RANDOM, pp. 63:1–63:20 (2020)
3. Charikar, M., et al.: Approximation algorithms for directed Steiner problems. J. Algorithms **33**(1), 73–91 (1999). Preliminary version in SODA 1998
4. Chekuri, C., Ene, A., Vakilian, A.: Prize-collecting survivable network design in node-weighted graphs. In: Gupta, A., Jansen, K., Rolim, J., Servedio, R. (eds.) APPROX/RANDOM -2012. LNCS, vol. 7408, pp. 98–109. Springer, Heidelberg (2012). https://doi.org/10.1007/978-3-642-32512-0_9
5. Cheriyan, J., Laekhanukit, B., Naves, G., Vetta, A.: Approximating rooted steiner networks. ACM Trans. Algorithms **11**(2):8:1–8:22 (2014). Preliminary version in SODA 2012
6. Even, G.: Recursive greedy methods. In: Gonzalez, T.F. (ed.), Handbook of Approximation Algorithms and Metaheuristics, Second Edition, Volume 1: Methologies and Traditional Applications, pp. 71–84. Chapman & Hall/CRC (2018)
7. Fakcharoenphol, J., Laekhanukit, B.: An $O(\log^2 k)$-approximation algorithm for the k-vertex connected spanning subgraph problem. SIAM J. Comput. **41**(5):1095–1109 (2012). Preliminary version in STOC 2008
8. Frank, A.: Kernel systems of directed graphs. Acta Sci. Math. (Szeged) **41**(1–2), 63–76 (1979)
9. Frank, A.: Rooted k-connections in digraphs. Discret. Appl. Math. **157**(6), 1242–1254 (2009)
10. Friggstad, Z., Könemann, J., Shadravan, M.: A logarithmic integrality gap bound for directed Steiner tree in quasi-bipartite graphs. In: SWAT, pp. 3:1–3:11 (2016)
11. Fukunaga, T.: Spider covers for prize-collecting network activation problem. ACM Trans. Algorithms **13**(4), 49:1–49:31 (2017). Preliminary version in SODA 2015
12. Garg, N., Konjevod, G., Ravi, R.: A polylogarithmic approximation algorithm for the group steiner tree problem. J. Algorithms **37**(1), 66–84 (2000). Preliminary version in SODA 1998
13. Ghuge, R., Nagarajan, V.: A quasi-polynomial algorithm for submodular tree orienteering in directed graphs. In: SODA, pp. 1039–1048 (2020)
14. Goemans, M.X., Olver, N., Rothvoß, T., Zenklusen, R.: Matroids and integrality gaps for hypergraphic Steiner tree relaxations. In: STOC, pp. 1161–1176 (2012)
15. Grandoni, F., Laekhanukit, B.: Surviving in directed graphs: a quasi-polynomial time polylogarithmic approximation for two-connected directed Steiner tree. In: STOC, pp. 420–428 (2017)
16. Grandoni, F., Laekhanukit, B., Li, S.: $O(\log^2 k/\log\log k)$-approximation algorithm for directed Steiner tree: a tight quasi-polynomial-time algorithm. In: STOC, pp. 253–264 (2019)
17. Halperin, E., Krauthgamer, R.: Polylogarithmic inapproximability. In: STOC, pp. 585–594 (2003)
18. Helvig, C.S., Robins, G., Zelikovsky, A.: An improved approximation scheme for the group Steiner problem. Networks **37**(1), 8–20 (2001)

19. Hibi, T., Fujito, T.: Multi-rooted greedy approximation of directed Steiner trees with applications. Algorithmica **74**(2), 778–786 (2016). Preliminary version in WG 2012

20. Jain, K.: A factor 2 approximation algorithm for the generalized Steiner network problem. Combinatorica **21**(1), 39–60 (2001). preliminary version in FOCS 1998

21. Klein, P.N., Ravi, R.: A nearly best-possible approximation algorithm for node-weighted Steiner trees. J. Algorithms **19**(1), 104–115 (1995). Preliminary version in IPCO 1993

22. Kortsarz, G., Nutov, Z.: Approximating k-node connected subgraphs via critical graphs. SIAM J. Comput. **35**(1), 247–257 (2005). Preliminary version in STOC 2004

23. Kortsarz, G., Nutov, Z.: Tight approximation algorithm for connectivity augmentation problems. J. Comput. Syst. Sci. **74**(5), 662–670 (2008). Preliminary version in ICALP 2006

24. Kortsarz, G., Nutov, Z.: Approximating source location and star survivable network problems. Theor. Comput. Sci. **674**, 32–42 (2017). Preliminary version in WG 2015, pp. 203–218

25. Kortsarz, G., Peleg, D.: Approximating the weight of shallow Steiner trees. Discrete Appl. Math. **93**(2–3), 265–285 (1999). Preliminary version in SODA 1997

26. Laekhanukit, B.: Parameters of two-prover-one-round game and the hardness of connectivity problems. In: SODA, pp. 1626–1643 (2014)

27. Lando, Y., Nutov, Z.: Inapproximability of survivable networks. Theor. Comput. Sci. **410**(21–23), 2122–2125 (2009). Preliminary version in APPROX-RANDOM 2008

28. Nutov, Z.: Approximating minimum power covers of intersecting families and directed edge-connectivity problems. Theor. Comput. Sci. **411**(26–28), 2502–2512 (2010). Preliminary version in APPROX-RANDOM 2006, pp. 236–247

29. Nutov, Z.: Approximating Steiner networks with node-weights. SIAM J. Comput. **39**(7), 3001–3022 (2010). Preliminary version in LATIN 2008, pp. 411–422

30. Nutov, Z.: Approximating minimum cost connectivity problems via uncrossable bifamilies and spider-cover decompositions. ACM Trans. Algorithms **9**(1), 1:1–1:16, 2012. Preliminary version in FOCS 2009, pp. 417–426

31. Nutov, Z.: Node-connectivity survivable network problems. In: Gonzalez, T.F. (ed.), Handbook of Approximation Algorithms and Metaheuristics, Second Edition, Volume 2: Contemporary and Emerging Applications, chapter 13. Chapman & Hall/CRC (2018)

32. Rajagopalan, S., Vazirani, V.V.: On the bidirected cut relaxation for the metric Steiner tree problem. In: SODA, pp. 742–751 (1999)

33. Zelikovsky, A.: A series of approximation algorithms for the acyclic directed steiner tree problem. Algorithmica **18**(1), 99–110 (1997)

Input-Driven Pushdown Automata on Well-Nested Infinite Strings

Alexander Okhotin[1]([⊠]) [iD] and Victor L. Selivanov[2] [iD]

[1] Department of Mathematics and Computer Science, St. Petersburg State University, 7/9 Universitetskaya Nab., Saint Petersburg 199034, Russia
alexander.okhotin@spbu.ru
[2] A. P. Ershov Institute of Informatics Systems, Novosibirsk, Russia
vseliv@iis.nsk.su

Abstract. Automata operating on strings of nested brackets, known as input-driven pushdown automata, and also as visibly pushdown automata, have been studied since the 1980s. They were extended to the case of infinite strings by Alur and Madhusudan ("Visibly pushdown languages", STOC 2004). This paper investigates the properties of these automata under the assumption that a given infinite string is always well-nested. This restriction enables a complete characterization of the corresponding ω-languages in terms of classical ω-regular languages and input-driven pushdown automata on finite strings. This characterization leads to a determinization construction for these automata, as well as to the first results on their topological classification.

1 Introduction

Input-driven pushdown automata (IDPDA), also known under the name of *visibly pushdown automata*, are an important special class of pushdown automata, introduced by Mehlhorn [8]. In these automata, the input symbol determines whether the automaton pushes a stack symbol, pops a stack symbol or does not access the stack at all. These symbols are called *left brackets*, *right brackets* and *neutral symbols*, and the symbol pushed at each left bracket is always popped when reading the corresponding right bracket. As shown by von Braunmühl and Verbeek [3], deterministic (DIDPDA) and nondeterministic (NIDPDA) input-driven automata are equivalent in power, and the languages they recognize lie in the deterministic logarithmic space. Input-driven automata enjoy excellent closure properties, which almost rival those of finite automata, and their complexity on terms of the number of states has been extensively studied. For more details, the reader is directed to a survey by Okhotin and Salomaa [10].

A systematic study of input-driven automata was undertaken in a famous paper by Alur and Madhusudan [1,2], who, in particular, generalized these automata to the case of infinite strings (ω-IDPDA), and identified them as a suitable model for verification. By analogy with finite automata, one can consider

This work was supported by the Russian Science Foundation, project 18-11-00100.

R. Santhanam and D. Musatov (Eds.): CSR 2021, LNCS 12730, pp. 349–360, 2021.
https://doi.org/10.1007/978-3-030-79416-3_21

deterministic and nondeterministic ω-IDPDA, with Büchi and Muller acceptance conditions. However, their properties are different: for instance, unlike finite-state Muller ω-automata, Muller ω-IDPDA cannot be determinized [1,2].

This issue with determinization was further studied by Löding et al. [7], who discovered a more sophisticated acceptance condition, under which determinization is possible. The analysis made by Löding et al. [7] clearly indicates that the main difficulty is caused by *ill-nested inputs*, that is, infinite strings that may have *unmatched left brackets*. Recalling an important motivation for ω-IDPDA as a model of infinite computation traces involving a stack [1,2], an ill-nested input often means a particular type of faulty behaviour: a memory leak. A computation trace represented by a *well-nested infinite string* of the form $u_1 u_2 \ldots u_i \ldots$, where each u_i is a well-nested finite string, is the normal behaviour.

This observation motivates the study of a special class of ω-IDPDA operating on well-nested infinite strings. These automata model a normal behaviour of an infinite process without memory leaks, and they avoid the difficulties with determinization discovered by Löding et al. [7].

This paper investigates the properties of well-nested ω-IDPDA, which are defined in Sect. 2 in usual variants: deterministic and nondeterministic, with Büchi and Muller acceptance conditions.

In Sect. 3, a characterization of the corresponding ω-IDPDA languages of well-nested strings in terms of IDPDA on finite strings and the usual ω-regular languages is obtained. This characterization leads to a determinization procedure for well-nested Muller IDPDA, as well as to the proof of their equivalence to nondeterministic Büchi IDPDA.

Besides the determinization question, another property investigated by Löding et al. [7] is the topology of ω-languages recognized by ω-IDPDA. These results motivate the study of Wadge degrees of ω-IDPDA languages. Although this collection is larger than that for the regular ω-languages, there is a hope to obtain an effective extension of Wagner hierarchy [14] to such languages, which would nicely contrast the case of (deterministic) context-free ω-languages [4,6].

The result of this paper, presented in Sect. 4, is that the hierarchy of Wadge degrees of ω-IDPDA languages of well-nested strings essentially coincides with the classical Wagner hierarchy of ω-regular languages.

2   Input-Driven ω-automata

Let Σ be a finite alphabet. The set of finite strings over Σ is Σ^*, the set of one-sided infinite strings is Σ^ω.

In input-driven automata, the alphabet is split into three disjoint sets of *left brackets* Σ_{+1}, *right brackets* Σ_{-1} and *neutral symbols* Σ_0. In this paper, symbols from Σ_{+1} and Σ_{-1} shall be denoted by left and right angled brackets, respectively ($<, >$), whereas lower-case Latin letters from the beginning of the alphabet $(a, b, c, \ldots)$ shall be used for symbols from Σ_0.

In a well-nested string, any left bracket from Σ_{+1} can match any right bracket from Σ_{-1}. Let the set of all well-nested finite strings be denoted by $\mathrm{wn}(\Sigma)$.

An *elementary well-nested string* is a well-nested string that is either a single symbol, or a well-nested string enclosed in a pair of matching brackets: $\mathrm{wn}_0(\Sigma) = \Sigma_0 \cup \Sigma_{+1}\mathrm{wn}(\Sigma)\Sigma_{-1}$.

An infinite string $\alpha \in \Sigma^\omega$ is called *well-nested*, if it is a concatenation of infinitely many elementary well-nested finite strings.

$$\mathrm{wn}^\omega(\Sigma) = \mathrm{wn}_0(\Sigma)^\omega = \{\, x_1 x_2 \ldots x_i \ldots \mid x_i \in \mathrm{wn}_0(\Sigma) \text{for all} i \geqslant 1 \,\}$$

Equivalently, an infinite string α is in $\mathrm{wn}^\omega(\Sigma)$ if it has infinitely many well-nested prefixes. Note that if the above definition had $x_i \in \mathrm{wn}(\Sigma)$, then it would also include all finite strings.

Input-driven automata on finite strings [8] are pushdown automata with the following restriction on their use of the stack. If the input symbol is a left bracket from Σ_{+1}, then the automaton always pushes one symbol onto the stack. For a right bracket from Σ_{-1}, the automaton must pop one symbol. Finally, for a neutral symbol in Σ_0, the automaton may not access the stack at all. The acceptance of a finite string is determined by the state reached in the end of the computation, whether it belongs to a set of accepting states $F \subseteq Q$.

The definition was extended to infinite strings by adopting the notions of acceptance from finite ω-automata.

Definition 1 (Alur and Madhusudan [1]). *A nondeterministic input-driven pushdown automaton on infinite strings (ω-NIDPDA) consists of the following components.*

- *The input alphabet Σ is a finite set split into three disjoint classes: $\Sigma = \Sigma_{+1} \cup \Sigma_{-1} \cup \Sigma_0$.*
- *The set of (internal) states Q is a finite set, with an initial state $q_0 \in Q$.*
- *The stack alphabet Γ is a finite set, and a special symbol $\perp \notin \Gamma$ is used to denote an empty stack.*
- *For each neutral symbol $c \in \Sigma_0$, the transitions by this symbol are described by a function $\delta_c \colon Q \to 2^Q$.*
- *For each left bracket symbol $< \, \in \Sigma_{+1}$, the behaviour of the automaton is described by a function $\delta_< \colon Q \to 2^{Q \times \Gamma}$, which, for a given current state, provides possible actions of the form "push a given stack symbol and enter a given state".*
- *For every right bracket symbol $> \, \in \Sigma_{-1}$, there is a function $\delta_> \colon Q \times \Gamma \to 2^Q$ specifying possible next states, assuming that the given stack symbol is popped from the stack.*
- *Acceptance is determined either by a Büchi condition, with a set of accepting states $F \subseteq Q$, or by a Muller condition, with a set of subsets $\mathcal{F} \subseteq 2^Q$.*

A triple (q, α, x), in which $q \in Q$ is the current state, $\alpha \in \Sigma^\omega$ is the remaining input and $x \in \Gamma^*$ is the stack contents, is called a configuration. *For each configuration, the next configuration is defined as follows.*

$$
\begin{aligned}
(q, c\alpha, x) &\vdash (r, \alpha, x) & (q \in Q,\ c \in \Sigma_0,\ r \in \delta_c(q)) \\
(q, {<}\alpha, x) &\vdash (r, \alpha, sx), & (q \in Q,\ < \, \in \Sigma_{+1},\ (r, s) \in \delta_<(q)) \\
(q, {>}\alpha, sx) &\vdash (r, \alpha, x) & (q \in Q,\ > \, \in \Sigma_{-1},\ s \in \Gamma,\ r \in \delta_>(q, s))
\end{aligned}
$$

A run *on a well-nested string* $\alpha \in \Sigma^\omega$ *is any sequence of configurations* $\rho = C_0, C_1, \ldots, C_i, \ldots$, *with* $C_0 = (q_0, \alpha, \varepsilon)$ *and* $C_{i-1} \vdash C_i$ *for all* $i \geqslant 1$. *The set of states that occur in a run infinitely many times is denoted by* $\inf(\rho)$.

Under a Büchi acceptance condition, *an infinite string* α *is accepted, if there exists a run with at least one of the accepting states repeated infinitely often.*

$$L(\mathcal{B}) = \{ \alpha \mid \exists \rho : \rho \text{ is a run on } \alpha, \, \inf(\rho) \cap F \neq \varnothing \}$$

Under a Muller acceptance condition, *the set of states repeated infinitely often must be among the specified subsets.*

$$L(\mathcal{M}) = \{ \alpha \mid \exists \rho : \rho \text{ is a run on } \alpha, \, \inf(\rho) \in \mathcal{F} \}$$

An automaton is deterministic, if all sets $\delta_c(q)$, $\delta_<(q)$ *and* $\delta_>(q, s)$ *are singletons.*

Actually, the definition by Alur and Madhusudan [1] further allows ill-nested infinite strings: on unmatched right brackets, it uses special transitions by an empty stack, in which the automaton detects the stack emptiness and leaves the stack empty; on unmatched left brackets, an automaton pushes symbols that shall never be popped.

In the case of finite strings, the extension to ill-nested strings does not affect the principal properties, such as determinization and decidability. For infinite strings, as it turns out, this detail makes a difference. In the rest of this paper, all infinite strings are assumed to be well-nested.

3   A Characterization of ω-input-driven Languages of Well-Nested Strings

The results of this paper are based on a characterization of ω-input-driven languages by ω-regular languages, which allows parts of the classical theory of ω-regular languages to be lifted to the case of well-nested infinite strings of brackets.

If an ω-input-driven automaton $\mathcal{M}$ operates on a well-nested string, this means that it reaches the bottom level of brackets infinitely often, reading a well-nested string $x \in \mathrm{wn}_0(\Sigma)$ between every two consecutive visits. The idea is to replace this string by a single symbol of a new alphabet, which represents the essential information on all possible computations of $\mathcal{M}$ on x. Then the language of such encoded strings shall be proved to be ω-regular.

Lemma 1. *Let* $L \subseteq \mathrm{wn}^\omega(\Sigma)$ *be a language of well-nested infinite strings over an alphabet* $\Sigma = \Sigma_{+1} \cup \Sigma_{-1} \cup \Sigma_0$ *recognized by an n-state deterministic input-driven Muller automaton. Then there exists a finite set of pairwise disjoint languages,* $K_1, \ldots, K_m \subseteq \mathrm{wn}_0(\Sigma)$, *with* $m \leqslant 2^{n^2} \cdot n^n$, *each recognized by an input-driven automaton, and a regular ω-language M over the alphabet* $\Omega = \{a_1, \ldots, a_m\}$ *such that L has the following representation.*

$$L = \{ x_1 x_2 \ldots x_j \ldots \mid x_j \in K_{i_j} \text{ for each } j \geqslant 0, \, a_{i_1} a_{i_2} \ldots a_{i_j} \ldots \in M \}$$

The language M is recognized by a deterministic Muller automaton with $n \cdot 2^n$ states.

Proof. Let $\mathcal{M} = (\Sigma, Q, \Gamma, q_0, \langle \delta_\sigma \rangle_{\sigma \in \Sigma}, \mathcal{F})$ be a deterministic input-driven Muller automaton that processes a string $x_1 x_2 \ldots x_j \ldots$, with $x_j \in \mathrm{wn}_0(\Sigma)$ for all $j \geqslant 0$. The goal is to construct a deterministic finite Muller automaton $\mathcal{A}$ operating on a corresponding string $a^{(1)} a^{(2)} \ldots a^{(j)} \ldots$, so that, after reading each prefix $a^{(1)} a^{(2)} \ldots a^{(j)}$, the automaton $\mathcal{A}$ would know the state reached by $\mathcal{M}$ after reading the prefix $x_1 x_2 \ldots x_j$, as well as the set of all states passed through by $\mathcal{M}$ while reading the most recent well-nested substring x_j. With this information, $\mathcal{A}$ can use its Muller acceptance condition to simulate the Muller acceptance condition of $\mathcal{M}$.

The data to be computed for each well-nested substring $x \in \mathrm{wn}_0(\Sigma)$ is represented by a function $f_x \colon Q \to Q \times 2^Q$, which maps a state p, in which $\mathcal{M}$ begins reading x, to a pair (q, S), where q is the state in which $\mathcal{M}$ finishes reading q, and S is the set of all states visited by $\mathcal{M}$ in this computation. In the context of this proof, f represents *the behaviour of $\mathcal{M}$ on x*.

There are $m = 2^{n^2} \cdot n^n$ different behaviour functions. For each behaviour function f, let $K_f \subseteq \mathrm{wn}_0(\Sigma)$ be the set of all strings, on which $\mathcal{M}$ demonstrates the behaviour f. A new Muller finite automaton promised in the statement of the lemma shall be defined over an alphabet $\Omega = \{\, a_f \mid f \colon Q \to Q \times 2^Q \,\}$, so that each symbol a_f stands for any string $x \in K_f$.

Let ρ be the run of $\mathcal{M}$ on $x_1 x_2 \ldots x_j \ldots$, and denote by $q_{j,\ell}$ the state entered after reading ℓ first symbols of x_j.

$$\rho = \underbrace{q_{1,0}, \ldots, q_{1,|x_1|-1}}_{\text{reading } x_1}, \underbrace{q_{2,0}, \ldots, q_{2,|x_2|-1}}_{\text{reading } x_2}, q_{3,1}, \ldots, \underbrace{q_{j,0}, \ldots, q_{q,|x_j|-1}}_{\text{reading } x_j}, \ldots$$

For each substring x_j, let $f_j = f_{x_j}$ be the behaviour function of $\mathcal{M}$ on this substring. Then each f_j, applied to $q_{j,0}$, provides the following data on the computation on x_j.

$$f_j(q_{j,0}) = \left(q_{j+1,0}, \{q_{j,0}, q_{j,1}, \ldots, q_{j,|x_j|-1}, q_{j+1,0}\} \right)$$

Claim 1.1. A state q occurs in the sequence $\{q_{j,\ell}\}$ infinitely many times if and only if there exists a pair (p, S), with $q \in S$, which occurs as $\left(q_{j+1,0}, \{q_{j,0}, q_{j,1}, \ldots, q_{j,|x_j|-1}, q_{j+1,0}\} \right)$ for infinitely many values of j.

Now the task is to simulate a run ρ of $\mathcal{M}$ on $x_1 x_2 \ldots x_j \ldots$ by an automaton $\mathcal{M}'$ operating on the string $a_{f_1} a_{f_2} \ldots a_{f_j} \ldots$, with every substring $x \in \mathrm{wn}_0(\Sigma)$ represented by a symbol a_{f_x}.

The desired Muller finite automaton $\mathcal{M}'$ is defined over the alphabet $\Omega = \{\, a_f \mid f \colon Q \to Q \times 2^Q \,\}$ as follows. Its states are pairs (q, S), where q shall be the state in the corresponding run of $\mathcal{M}$ at this point, whereas S shall be the set of states visited by $\mathcal{M}$ on the longest well-nested suffix of the input.

$$P = \{\, (q, S) \mid q \in Q, \ S \subseteq Q \,\}$$

A transition by a_f applies the behaviour function f to q to determine the next state of $\mathcal{M}$, as well as the set of states visited on the next well-nested substring.

$$\Delta((q, S), a_f) = f(q)$$

Claim 1.2. In the run ρ' of $\mathcal{M}'$ on $a_{f_1} a_{f_2} \ldots a_{f_j} \ldots$, the state reached after reading each a_{f_j} is $\big(q_{j+1,0}, \{ q_{j,0}, q_{j,1}, \ldots, q_{j,|x_j|-1}, q_{j+1,0} \} \big)$.

Putting together this property and Claim 1.1, a state q occurs in ρ infinitely often if and only if there exists a pair (p, S), with $q \in S$, which occurs in ρ' infinitely often.

Then the Muller acceptance condition for $\mathcal{M}'$ is defined by putting together all states of $\mathcal{M}$ visited infinitely often and checking whether this set is in $\mathcal{F}$.

$$\mathcal{G} = \big\{ \, R \mid R \subseteq Q \times 2^Q, \ \bigcup_{(q,S) \in R} S \in \mathcal{F} \, \big\}$$

Claim 1.3. Let $x_1 x_2 \ldots x_j \ldots$, with $x_j \in \mathrm{wn}_0(\Sigma)$, be an infinite string, and let $f_j \colon Q \to Q \times 2^Q$ be the behaviour function on each x_j. Then $\mathcal{M}$ accepts $x_1 x_2 \ldots x_j \ldots$ if and only if $\mathcal{M}'$ accepts $a_{f_1} a_{f_2} \ldots a_{f_j} \ldots$.

In order to complete the proof of the lemma, it remains to prove that each language $K_f \subseteq \mathrm{wn}_0(\Sigma)$ of all finite strings with a given behaviour function f is recognized by an input-driven automaton.

Claim 1.4. There exists a deterministic input-driven automaton $\mathcal{A}$ with the set of states $Q' = \{ q_0' \} \cup \{ (f, b) \mid f \colon Q \to Q \times 2^Q, \ b \in \{0,1\} \}$, which, upon reading an elementary well-nested string $x \in \mathrm{wn}_0(\Sigma)$, enters a state $(f, 0)$, where $f \colon Q \to Q \times 2^Q$ is the behaviour function of $\mathcal{M}$ on x.

Proof. Besides computing the behaviour, the automaton also has to verify that the input string is in $\mathrm{wn}_0(\Sigma)$, and for that reason it has to remember whether it is currently inside any brackets. In a state (f, b), the first component is the desired behaviour function for the longest well-nested suffix $v \in \mathrm{wn}(\Sigma)$, whereas the second component indicates whether the automaton is currently at the outer level of brackets (0) or not (1). No transitions in a state of the form $(f, 0)$ are possible. The stack alphabet is $\Gamma' = \{ (f, <) \mid f \colon Q \to Q \times 2^Q, \ < \in \Sigma_{+1} \} \cup \Sigma_{+1}$, where a symbol of the form $(f, <)$ is pushed while inside the brackets, whereas on the outer level of brackets, the automaton pushes just a left bracket.

Let $f_\varepsilon \colon Q \to Q \times 2^Q$ be the behaviour function on the empty string, defined by $f(q) = (q, \{q\})$ for each $q \in Q$. For each neutral symbol $c \in \Sigma_0$, the behaviour on this symbol, $f_c \colon Q \to Q \times 2^Q$, is defined by $f(q) = (\delta_c(q), \{q, \delta_c(q)\})$. If f is the behaviour function on $u \in \mathrm{wn}(\Sigma)$, and g is the behaviour function on $v \in \mathrm{wn}(\Sigma)$, then the behaviour function on uv is the following kind of function composition that accumulates the states visited.

$$(g \circ f)(p) = (r, S \cup T), \qquad \text{where } f(p) = (q, S), \ g(q) = (r, T)$$

The initial state of $\mathcal{A}$ is q_0'; upon reading a neutral symbol $c \in \Sigma_0$ in the initial state, the automaton enters the state corresponding to the behaviour on c.

$$\delta_c'(q_0') = (f_c, 0)$$

If the automaton reads a left bracket $< \in \Sigma_{+1}$ in the initial state, it begins constructing a new behaviour function on the inner level, remembering the bracket in the stack.

$$\delta'_<(q'_0) = ((f_\varepsilon, 1), <)$$

When a matching right bracket $> \in \Sigma_{-1}$ is read, the automaton pops the left bracket $< \in \Sigma_{+1}$ from the stack, and therefore knows that it is returning to the outer level of brackets. Then it merges the computed behaviour g on the substring inside the brackets with the actions made on the outer pair of brackets.

$$\delta'_>((g, 1), <) = (h, 0), \qquad \text{where } h(p) = (r, S \cup \{p, r\}), \delta_<(p) = (q, s)$$
$$g(q) = (q', S), \ r = \delta_>(q', s)$$

Inside the brackets, upon reading a neutral symbol $c \in \Sigma_0$, behaviour functions are composed as follows.

$$\delta'_c((f, 1)) = f_c \circ f$$

When a left bracket $< \in \Sigma_{+1}$ is encountered inside the brackets, a new simulation is started inside, but the current behaviour function is stored in the stack.

$$\delta'_<((f, 1)) = ((f_\varepsilon, 1), (f, <))$$

Upon reading a matching right bracket $> \in \Sigma_{-1}$, the two simulations are composed.

$$\delta'_>((g, 1), (f, <)) = (h \circ f, 1), \quad \text{where } h(p) = (r, S \cup \{p, r\}), \delta_<(p) = (q, s)$$
$$g(q) = (q', S), \ r = \delta_>(q', s)$$

The transitions of the automaton $\mathcal{A}$ have been defined. For each behaviour function $f \colon Q \to Q \times 2^Q$, setting the set of accepting states to be $F_f = \{(f, 0)\}$ yields the desired IDPDA recognizing the language K_f of all strings in $\mathrm{wn}_0(\Sigma)$, on which $\mathcal{M}$ behaves as specified by f. □

All claims of the lemma have thus been verified. □

The same characterization as in Lemma 1 also extends to nondeterministic input-driven Muller automata, at the expense of using exponentially more types of well-nested strings.

Lemma 2. *Let $L \subseteq \mathrm{wn}^\omega(\Sigma)$ be a language of well-nested infinite strings over an alphabet $\Sigma = \Sigma_{+1} \cup \Sigma_{-1} \cup \Sigma_0$ recognized by an n-state nondeterministic input-driven Muller automaton. Then there exists a finite set of pairwise disjoint languages, $K_1, \ldots, K_m \subseteq \mathrm{wn}_0(\Sigma)$, with $m \leqslant 2^{n^2 \cdot 2^n}$, each recognized by an input-driven automaton, and a regular ω-language M over the alphabet $\Omega = \{a_1, \ldots, a_m\}$ such that L has the following representation.*

$$L = \{ x_1 x_2 \ldots x_j \ldots \mid x_j \in K_{i_j} \text{ for each } j \geqslant 0, \ a_{i_1} a_{i_2} \ldots a_{i_j} \ldots \in M \}$$

The language M is recognized by a nondeterministic Muller automaton with $n \cdot 2^n$ states.

The proof further develops the method in Lemma 1.

The representation of well-nested ω-input-driven languages by ω-regular languages given in Lemmata 1–2 has the following converse representation.

Lemma 3. *Let* $K_1, \ldots, K_m \subseteq \mathrm{wn}_0(\Sigma)$ *be pairwise disjoint languages, each recognized by an input-driven automaton, and let* M *be a regular* ω-language over the alphabet $\Omega = \{a_1, \ldots, a_m\}$. Then, the following language is recognized by a deterministic input-driven Müller automaton.*

$$L = \{\, x_1 x_2 \ldots x_j \ldots \mid x_j \in K_{i_j} \text{ for each } j \geqslant 0,\ a_{i_1} a_{i_2} \ldots a_{i_j} \ldots \in M \,\}$$

Proof (a sketch). Let $\mathcal{M}$ be a deterministic Müller automaton recognizing M, let P be its set of states. For each K_i, let $\mathcal{A}_i$ be a DIDPDA recognizing this language, and let Q_i be its set of states. Then the new automaton uses the set of states $P \cup (Q_1 \times \ldots \times Q_m) \cup \{p_{dead}\}$. For each minimal well-nested substring of the input, $x \in \mathrm{wn}_0(\Sigma)$, the automaton simulates all $\mathcal{A}_i$ in parallel using m-tuples of states from $Q_1 \times \ldots \times Q_m$. Once x is read, at most one automaton $\mathcal{A}_i$ accepts, and then the automaton simulates a single step of $\mathcal{M}$ on a_i. $\square$

A variant of the same construction produces an automaton with a Büchi acceptance condition.

Lemma 4. *Let* $K_1, \ldots, K_m \subseteq \mathrm{wn}_0(\Sigma)$ *be pairwise disjoint languages, each recognized by an input-driven automaton, and let* M *be a regular* ω-language over the alphabet $\Omega = \{a_1, \ldots, a_m\}$. Then, the following language is recognized by a nondeterministic input-driven Büchi automaton.*

$$L = \{\, x_1 x_2 \ldots x_j \ldots \mid x_j \in K_{i_j} \text{ for each } j \geqslant 0,\ a_{i_1} a_{i_2} \ldots a_{i_j} \ldots \in M \,\}$$

Putting the above results together completes the characterization of well-nested ω-input-driven languages by ω-regular languages, and also establishes the equivalence of several kinds of automata on well-nested infinite strings.

Theorem 1. *Let* $\Sigma = \Sigma_{+1} \cup \Sigma_{-1} \cup \Sigma_0$ *be an alphabet. Let* $L \subseteq \Sigma^\omega$ *be a language of infinite strings and assume automata operating only on well-nested infinite strings. Then the following three conditions are equivalent:*

1. L *is recognized by a deterministic input-driven Müller automaton;*
2. L *is recognized by a nondeterministic input-driven Müller automaton;*
3. L *is recognized by a nondeterministic input-driven Büchi automaton;*
4. *there is a finite set of pairwise disjoint languages,* $K_1, \ldots, K_m \subseteq \Sigma_0 \cup \Sigma_{+1} \mathrm{wn}(\Sigma) \Sigma_{-1}$, *each recognized by an input-driven automaton, and a regular* ω-language M over the alphabet $\Omega = \{a_1, \ldots, a_m\}$ such that L has the following representation.

$$L = \{\, x_1 x_2 \ldots x_j \ldots \mid a_{i_1} a_{i_2} \ldots a_{i_j} \ldots \in M,\ x_j \in K_{i_j} \text{ for each } j \geqslant 0 \,\}$$

The exact complexity of the transformation between models is left as a question for future research.

4   Wadge Degrees of ω-Input-Driven Languages of Well-Nested Strings

Here we apply the results of Sect. 3 to relate the Wadge degrees of IDPDA-recognisable languages in $\mathrm{wn}^\omega(\Sigma)$ (w.r.t. the topology induced by the Cantor topology on Σ^ω) to those of regular ω-languages.

Recall that for subsets $L, M \subseteq X$ of a topological space X, the subset L is *Wadge reducible* to M (denoted by $L \leq^X_W M$) if $L = f^{-1}(M)$ for some continuous function f on X. The sets L, M are *Wadge equivalent* ($L \equiv^X_W M$) if $L \leq^X_W M$ and $M \leq^X_W L$. The equivalence classes under the Wadge equivalence are known as *Wadge degrees* in X. Thus, Wadge reducibility is just many-one reducibility by continuous functions, so $L \leq^X_W M$ informally means that L is topologically at most as complicated as M, and the structure of Wadge degrees is a topological classification of sets in X.

Slightly more generally, one can define Wadge reducibility between subsets $L \subseteq X$ and $M \subseteq Y$ of different topological spaces X, Y as follows: $L \leq_W M$, if $L = f^{-1}(M)$ for some continuous function $f\colon X \to Y$. Recall that *Borel sets* in X are generated from the open sets by repeated applications of complement and countable intersection. The Borel sets are organised in the Borel hierarchy; in particular, the $\mathbf{\Pi}_2$-sets in X are the countable intersections of open sets.

The structure of Wadge degrees of Borel sets was first introduced and studied by Wadge [13] for the Baire space $X = N^\omega$ of infinite words over the countable alphabet $N = \{0, 1, \ldots\}$ (open sets in this space are sets of the form $U \cdot N^\omega$, with $U \subseteq N^*$). W. Wadge proved a remarkable fact that the structure of Wadge degrees of Borel sets in the Baire space is very simple (almost well ordered), which leads to an elegant topological classification of the Borel sets. This fact is quite surprising, because most of the degree structures in computability theory and set theory are usually very complicated.

Later, Wadge reducibility in some popular spaces X other that the Baire space, such as Euclidean spaces or the Scott domain, was studied by several authors (see e.g. [5,9] and references therein). Typically, this structure becomes much more complicated than for the Baire space, and does not yield a reasonable topological classification of subsets of X even for the lower levels of the Borel hierarchy.

With minor modifications, the results of Wadge for the Baire space also hold for the Cantor space $X = \Sigma^\omega$ (open sets in this space are sets of the form $U \cdot \Sigma^\omega$, with $U \subseteq \Sigma^*$). This is of interest for theoretical computer science, because it enables to consider Wadge degrees of ω-languages over Σ recognized by different kinds of automata, and to obtain topological classifications of the corresponding ω-languages.

This direction was initiated by Wagner [14], who characterised the structure $(\mathcal{R}_\Sigma; \leq_W)$ of Wadge degrees of regular ω-languages over Σ, now known as the *Wagner hierarchy*. Wagner has shown that this structure is isomorphic to the poset obtained from the poset $\bar{2} \cdot \omega^\omega$ by adding the suprema of incomparable pairs of elements. Here $\bar{2}$ is the poset formed by two incomparable elements; ω^ω is the supremum of the sequence of ordinals $\omega = \{0 < 1 < \cdots\}$, $\omega^2 = \omega \cdot \omega$, $\omega^3 = \omega \cdot \omega^2$, etc.; and $P \cdot Q$ is the poset obtained from given posets P, Q by inserting a copy of P in place of any element of Q. Moreover, all natural decision problems related to the Wagner hierarchy turn out to be decidable, even in polynomial time.

The results on the Wagner hierarchy also hold in the alphabet-independent version, where one considers the structure $(\mathcal{R}; \leq_W)$ of Wadge degrees of regular ω-languages over arbitrary finite alphabets, using the aforementioned version of Wadge reducibility for different spaces. It turns out that, up to isomorphism, their structure is the same as in the previous paragraph, i.e., $(\mathcal{R}; \leq_W) \simeq (\mathcal{R}_\Omega; \leq_W)$, for any finite non-unary alphabet Ω, where the relation $\simeq$ between preorders means isomorphism of the corresponding quotient-posets.

Later, Wadge degrees of several important classes of ω-languages were investigated. It was shown, in particular, that Wadge degrees of pushdown ω-languages are much richer than those of regular ω-languages (see e.g. [4,6] and references therein). Moreover, the related algorithmic problems for the Wadge degrees of pushdown ω-languages become undecidable.

In the context of this paper, a natural question is to characterize the structure $(\omega\text{-IDPDA}(\Sigma); \leq_W)$ of Wadge degrees of languages recognized by input-driven automata. A theorem by Löding et al. [7, Thm. 6] implies that this structure embeds the structure $(\mathcal{R}; \leq_W)$, and even $(\mathcal{R}; \leq_W) \not\simeq (\omega\text{-IDPDA}(\Sigma); \leq_W)$. We postpone the discussion of the structure $(\omega\text{-IDPDA}(\Sigma); \leq_W)$ for arbitrary words (including the ill-nested ones) till a later publication, because it requires techniques different from those used in this paper.

Here, we formulate some immediate implications of the results of Sect. 3 to the relevant structure $(\text{wnIDPDA}(\Sigma); \leq_W^X)$, where $\text{wnIDPDA}(\Sigma)$ is the class of languages of well-nested ω-words recognised by the IDPDA, and $\leq_W^X$ is the Wadge reducibility in the space $X = \text{wn}^\omega(\Sigma)$, considered as a subspace of Σ^ω. Namely, we show that $(\mathcal{R}; \leq_W) \simeq (\text{wnIDPDA}(\Sigma); \leq_W^X)$, for any non-unary alphabet Σ. Note that the structure $(\text{wnIDPDA}(\Sigma); \leq_W^X)$ is different from $(\text{wnIDPDA}(\Sigma); \leq_W)$, where $\leq_W$ is the Wadge reducibility on Σ^ω.

We start with establishing some lemmas. The first one describes the induced topology on $\text{wn}^\omega(\Sigma)$.

Lemma 5. *The set* $\text{wn}^\omega(\Sigma)$ *is a* $\mathbf{\Pi}_2$*-subset of* Σ^ω *homeomorphic to the Baire space.*

Proof. Considering $\text{wn}_0(\Sigma)$ as a countable alphabet, we may think that $\text{wn}_0(\Sigma)^\omega$ is the Baire space. By the remarks in the beginning of Sect. 2, the infinite concatenation $h : \text{wn}_0(\Sigma)^\omega \to \Sigma^\omega$ is a bijection between $\text{wn}_0(\Sigma)^\omega$ and $\text{wn}^\omega(\Sigma)$. Comparing the topologies on $\text{wn}_0(\Sigma)^\omega$ and on Σ^ω, we see that h is a homeomorphism between the Baire space $\text{wn}_0(\Sigma)^\omega$ and the subspace $\text{wn}^\omega(\Sigma)$ of Σ^ω.

It remains to show that $\text{wn}^\omega(\Sigma)$ is a $\mathbf{\Pi}_2$-subset of Σ^ω. This follows from the obvious remark that, for each $\alpha \in \Sigma^\omega$, the condition $\alpha \in \text{wn}^\omega(\Sigma)$ is equivalent to $\forall i \exists u \in \text{wn}(\Sigma)(\alpha[i] \sqsubseteq u \sqsubseteq \alpha)$, where $\alpha[i]$ is the prefix of α of length i, and $\sqsubseteq$ is the prefix relation. $\qquad\square$

The second lemma relates elements of the structure $(\text{wnIDPDA}(\Sigma); \leq_W^X)$ to elements of the structure $(\mathcal{R}; \leq_W)$.

Lemma 6. *For any* $L \in \text{wnIDPDA}(\Sigma)$ *there exists a regular* ω-*language* M *over a finite alphabet* $\Omega = \{a_1, \ldots, a_m\}$ *such that* $L \equiv_W M$.

Proof. Let $K_1, \ldots, K_m$, $\Omega = \{a_1, \ldots, a_m\}$, and M be as in Lemma 1. Without loss of generality, one can assume that every K_i is non-empty (otherwise, it is omitted from the list) and that $K_1 \cup \cdots \cup K_m = \text{wn}_0(\Sigma)$ (otherwise, the language $K_{m+1} = \text{wn}_0(\Sigma) \setminus (K_1 \cup \cdots \cup K_m)$ is added to the list, and the letter a_{m+1} is added to Ω). Here the closure of IDPDA-recognizable languages under Boolean operations is used.

Define $f \colon \text{wn}^\omega(\Sigma) \to \Omega^\omega$ by $f(x_1 x_2 \cdots) = a_{i_1} a_{i_2} \cdots$, where, for each $j \geq 1$, i_j is the unique element of $\{1, \ldots, m\}$ with $x_j \in K_{i_j}$. Then f is continuous and $L = f^{-1}(M)$, so $L \leq_W M$ via f. Define $g \colon \Omega^\omega \to \text{wn}^\omega(\Sigma)$ by $g(a_{i_1} a_{i_2} \cdots) = x_{i_1} x_{i_2} \cdots$, where, for each $j \in \{1, \ldots, m\}$, x_j is a fixed element of K_j. Then g is continuous and $M = f^{-1}(L)$, so $M \leq_W L$ via g. Therefore, $M \equiv_W L$. $\square$

The third lemma is in a sense the converse of the second one.

Lemma 7. *Let* $m \geq 1$, $\Omega = \{a_1, \ldots, a_m\}$, *and let* $\{K_1, \ldots, K_m\}$ *be a partition of* $\text{wn}_0(\Sigma)$ *into non-empty languages recognized by input-driven automata. Then, for every regular* ω-*language* M *over* Ω, *there exists* $L \in \text{wnIDPDA}(\Sigma)$ *such that* $L \equiv_W M$.

Proof. Let $f \colon \text{wn}^\omega(\Sigma) \to \Omega^\omega$ be as in the proof of Lemma 6. By Lemma 3, for the language $L = f^{-1}(M)$ we have $L \equiv_W M$. $\square$

We are ready to characterize the quotient-poset of $(\text{wnIDPDA}(\Sigma) \leq_W^X)$.

Theorem 2. *For any non-unary* $\Sigma = \Sigma_{+1} \cup \Sigma_{-1} \cup \Sigma_0$ *we have:* $(\mathcal{R}; \leq_W) \simeq (\text{wnIDPDA}(\Sigma) \leq_W^X)$.

Proof. Associate with any $L \in \text{wnIDPDA}(\Sigma)$ a language $M \in \mathcal{R}$ as in Lemma 6, so in particular $L \equiv_W M$. Then $L \leq L'$ clearly implies $M \leq_W M'$ for all $L, L' \in \text{wnIDPDA}(\Sigma)$.

Now we show that for any $M \in \mathcal{R}$ there exists $L \in \text{wnIDPDA}(\Sigma)$ such that $L \equiv_W M$. Let $\Omega = \{a_1, \ldots, a_m\}$ be a finite alphabet with $M \in \mathcal{R}_\Omega$. Since Σ is non-unary, there is clearly a partition $\{K_1, \ldots, K_m\}$ of $\text{wn}_0(\Sigma)$ into non-empty languages recognized by input-driven automata. By the proof of Lemma 7, we can take $L = f^{-1}(M)$.

Returning to the map $L \mapsto M$ from the beginning of the proof, it remains to show that $M \leq_W M'$ implies $L \leq_W L'$. Let Ω be any non-unary finite alphabet. By remarks in the beginning of this section, there are $M_1, M_1' \in \mathcal{R}_\Omega$ such that $M \equiv_W M_1$ and $M' \equiv_W M_1'$. By the preceding paragraph, there exist $L_1, L_1' \in \text{wnIDPDA}(\Sigma)$ such that $L_1 \equiv_W M_1$ and $L_1' \equiv_W M_1'$. Since also $L \equiv_W L_1$ and $L' \equiv_W L_1'$, it follows that $L \leq_W L'$. Therefore, $L \mapsto M$ induces a desired isomorphism between the quotient-posets of $(\text{wnIDPDA}(\Sigma) \leq_W^X)$ and $(\mathcal{R}; \leq_W)$. $\square$

This theorem and the remarks above about the quotient-poset of $(\mathcal{R}; \leq_W)$ immediately imply the following:

360   A. Okhotin and V. L. Selivanov

Corollary 1. *The quotient-poset of* (wnIDPDA(Σ) $\leq_W^X$) *is isomorphic to the poset* $\bar{2} \cdot \omega^\omega$, *with adjoined supremums of incomparable pairs of elements.*

This completes the characterization of Wadge degrees of ω-input-driven languages of well-nested strings. Characterizing Wadge degrees for languages of not necessarily well-nested strings is left as a problem for future study.

References

1. Alur, R., Madhusudan, P.: Visibly pushdown languages. In: ACM Symposium on Theory of Computing (STOC 2004), Chicago, USA, 13–16 June 2004, pp. 202–211 (2004). https://doi.org/10.1145/1007352.1007390
2. Alur, R., Madhusudan, P.: Adding nesting structure to words. J. ACM **56**(3), 1–43 (2009). https://doi.org/10.1145/1516512.1516518
3. von Braunmühl, B., Verbeek, R.: Input driven languages are recognized in log n space. In: Annals of Discrete Mathematics, vol. 24, pp. 1–20 (1985). https://doi.org/10.1016/S0304-0208(08)73072-X
4. Duparc, J.: A hierarchy of deterministic context-free ω-languages. Theor. Comput. Sci. **290**(3), 1253–1300 (2003)
5. Duparc, J., Vuilleumier, L.: The Wadge order on the Scott domain is not a well-quasi-order. J. Symb. Log. **85**(1), 300–324 (2020)
6. Finkel, O.: Topological complexity of context-free ω-languages: a survey. In: Dershowitz, N., Nissan, E. (eds.) Language, Culture, Computation. Computing - Theory and Technology. LNCS, vol. 8001, pp. 50–77. Springer, Heidelberg (2014). https://doi.org/10.1007/978-3-642-45321-2_4
7. Löding, C., Madhusudan, P., Serre, O.: Visibly pushdown games. In: Lodaya, K., Mahajan, M. (eds.) FSTTCS 2004. LNCS, vol. 3328, pp. 408–420. Springer, Heidelberg (2004). https://doi.org/10.1007/978-3-540-30538-5_34
8. Mehlhorn, K.: Pebbling mountain ranges and its application to DCFL-recognition. In: de Bakker, J., van Leeuwen, J. (eds.) ICALP 1980. LNCS, vol. 85, pp. 422–435. Springer, Heidelberg (1980). https://doi.org/10.1007/3-540-10003-2_89
9. Motto Ros, L., Schlicht, P., Selivanov, V.: Wadge-like reducibilities on arbitrary quasi-Polish spaces. Math. Struct. Comput. Sci. **25**(8), 1705–1754 (2015)
10. Okhotin, A., Salomaa, K.: Complexity of input-driven pushdown automata. In: SIGACT News, vol. 45, no. 2, pp. 47–67 (2014). https://doi.org/10.1145/2636805.2636821
11. Okhotin, A., Salomaa, K.: Descriptional complexity of unambiguous input-driven pushdown automata. Theor. Comput. Sci. **566**, 1–11 (2015). https://doi.org/10.1016/j.tcs.2014.11.015
12. Okhotin, A., Salomaa, K.: State complexity of operations on input-driven pushdown automata. J. Comput. Syst. Sci. **86**, 207–228 (2017). https://doi.org/10.1016/j.jcss.2017.02.001
13. Wadge, W.: Reducibility and determinateness in the Baire space. PhD thesis, University of California, Berkely (1984)
14. Wagner, K.: On ω-regular sets. Inf. Control **43**, 123–177 (1979)

Large Clique is Hard on Average for Resolution

Shuo Pang$^{(\boxtimes)}$

Mathematics Department, University of Chicago, Chicago, USA
spang@uchicago.edu

Abstract. We prove an $\exp(\Omega(k^{(1-\epsilon)}))$ resolution size lower bound for the k-Clique problem on random graphs, for (roughly speaking) $k < n^{1/3}$. Towards an optimal resolution lower bound of the problem (i.e. of type $n^{\Omega(k)}$), we also extend the $n^{\Omega(k)}$ bound in [2] on regular resolution to a stronger model called *a-irregular resolution*, for $k < n^{1/3}$. This model is interesting in that all known CNF families separating regular resolution from general [1,24] have short proofs in it.

Keywords: Resolution · k-Clique · Random graphs · Regular resolution

1 Introduction

The *k-Clique* problem, given an input graph G and a number k, asks to decide if G contains a k-clique. As one of the fundamental NP-complete problems, its computational hardness has been intensively studied in both algorithmic and lower bound worlds [11,13,16,19,21,23,25]. Proof complexity studies, among many other aspects, the hardness of proving $f(x) = 0$ for a boolean function f and input x, which is a natural and necessary step for understanding the computational hardness of f. The underlying proof system should be sound and efficiently checkable (called the *Cook-Reckhow* systems). Given such a system Λ, the proof-theoretic version of the k-Clique problem is thus, "is there a short Λ-refutation of the CNF encoding of the fact 'G contains a k-clique'?" In this paper, Λ will be resolution or its sub-systems, and we study the average-case problem, i.e. when G is a random graph and we ask if there a short refutation with high probability. The random graph should be k-clique-free w.h.p. (trivially there is no refutation of a correct claim, short or long); the most studied setting is the *Erdős-Rényi* random graph $G(n, p)$, with p below the so-called threshold of containing a k-clique, usually taken as $p = n^{-\frac{2\xi}{k-1}}$ where $\xi > 1$ is a constant.

Previous Work. An $n^{O(k)}$-sized tree-like resolution refutation is not hard to see when G doesn't contain a k-clique. For lower bounds, a $2^{\Omega(k^6/n^5)}$ size lower bound for resolution is known [4], which is meaningful for $k > n^{5/6}$; the optimal

© Springer Nature Switzerland AG 2021
R. Santhanam and D. Musatov (Eds.): CSR 2021, LNCS 12730, pp. 361–380, 2021.
https://doi.org/10.1007/978-3-030-79416-3_22

$n^{\Omega(k)}$ size lower bounds are known for tree-like resolution [5] and regular resolution [2] (for $k < n^{\frac{1}{4}-o(1)}$).

Our Results. We prove an $\exp(k^{1-\epsilon})$ average-case resolution size lower bound of k-Clique problem, when $G \sim G(n,p)$ as above (Corollary 1). This result holds for $k < n^{1/3}$, thus complements the result of [4] for smaller k's (more precisely, it holds for $k = n^{c_0}$ where c_0, ϵ are arbitrary parameters s.t. $\min\{\frac{1}{3} - c_0, \ \epsilon c_0\} > (\log n)^{-1/5}$). Our second result (Theorem 3) extends the $n^{\Omega(k)}$ average-case lower bound to a new model called a-*irregular* resolution, for $k < n^{1/3-\epsilon}$, as a possible step towards the same bound for resolution. Like in almost all previous work, both results are stated and proved for a "strong" encoding of the problem.

Some words on the model. It is a Cook-Reckhow subsystem of resolution with the following motivation: imagine that in general, the "hard part" of a short resolution proof is the derivation of some clauses with nontrivial width (or a variant of width), so that once they are in place the rest is easy. Then how hard is it to derive these clauses? In particular, if in deriving *any* wide clause we can't be too irregular, is there still a short refutation? Formally, it requires

> In deriving any clause of large (block-)width, few (blocks of) variables are irregularly resolved.

Here *large* and *few* will be characterized by the parameter $a \in (0,1)$ ($a = 0/1$ means regular/general resolution), and *block* is used in the main version where a variable partition is part of the input.

Somewhat surprisingly, it turns out (Remark 3) that all known CNF families separating regular resolution from general separate, in fact, regular resolution from this model, with a as small as $n^{-\Omega(1)}$. (On the contrary, the second result holds for constant a.)

Previously, [7] considered another extension of regular resolution called the δ-*regular resolution*, which restricts the number of irregularly-resolved variables on any path. Our restriction seems simpler in the sense that the resulting system is Cook-Reckhow. The two seem incomparable, but it will be interesting to know their exact relation.

Proof Idea. For the first result, we consider a class of clauses (not depending on the refutation) where each one is very small (under certain "measure" on clauses), while we show that in any refutation, the clauses from this class together "have measure 1". Such a clause C has the property that its associated set of *falsifying* assignments, when regarded as a k-product subset in $[n]^k$ in the natural way[1], has many indices $i \in [k]$ s.t. the i-th component is small in a certain sense. The empty clause has full measure 1. We show that, if travel down the proof DAG with some strategy, one always ends in such a clause. By division, there are many such clauses in C (cf. the *bottleneck counting* method [12,17]). It might

---

[1] More precisely, it is a product-subset of $[n/k]^k$; the reason is clear after seeing the strong encoding (Sect. 2) where the vertex set is partitioned into k parts.

be possible to translate this proof into a random restriction-based argument; the current language is chosen since it works consistently for the second result, too.

The second result is built on [2] in a straightforward manner. In a given refutation, we find *one* small clause C (in the above sense) s.t. the sub-proof deriving C is regular after a suitable restriction. The useful graph-theoretic property used in the regular case seems not inheritable to sub-graphs (which occurs from the restriction), but this can be fixed by using a relativized property (Sect. 4.3).

The proof of the first result has the merit of simplicity and the drawback is there, too: the pseudorandom graph-theoretic property used is insufficient for an $n^{\Omega(k)}$ lower bound (Remark 2). On the other hand, we don't know of a similar limitation of the property used in the regular case and the second result.

Paper Structure. We give the necessary preliminaries in Sect. 2. In Sect. 3 we prove the first result. In Sect. 4 we introduce the model and then prove the second result. The paper is concluded in Sect. 5 with some open problems.

2 Preliminaries

Graphs. $G = (V, E)$ denotes a simple, undirected graph. For $v \in V$, $N(v) = \{u \mid (u, v) \in E\}$ is the set of *neighbors of* v. For $A, B \subset V$, $\hat{N}_A(B) = A \cap (\cap_{v \in B} N(v))$ denotes the set of *common neighbors of B in A*. When $A = V$ it simplifies to $\hat{N}(B)$. A *k-clique* in G is a set $C \subset V$ with size k and $\forall u, v \in C, u \neq v \Rightarrow \{u, v\} \in E$. **G** denotes the n-vertex Erdős-Rényi *graph* $G(n, p)$, $0 < p < 1$, which places an edge between any two vertices with probability p independently. For $1 < k < n$, $n^{-\frac{2}{k-1}}$ is the well-known *threshold probability* [6]: $G(n, p)$ contains a k-clique (or not) w.h.p. as $n \to \infty$ if $p > n^{-(1-O(1))\frac{2}{k-1}}$ (or $p < n^{-(1+O(1))\frac{2}{k-1}}$). We take $p = n^{-\frac{2\xi}{k-1}}$, $\xi > 1$ a constant, throughout the paper.

Resolution and the Encoding. A *literal* over a Boolean variable x is either x or its negation $\neg x$, x called the *variable* of the literal. A *clause* $C = l_1 \vee \ldots \vee l_t$ is a disjunction of distinct literals where there is no appearance of $x, \neg x$ together for any variable x (otherwise the clause is 1). t is the *width* of C, denoted as $w(C)$. $\perp$ is the contradiction/empty clause. A *CNF formula* $\tau = C_1 \wedge \ldots \wedge C_m$ is a conjunction of clauses. A *resolution proof from a CNF* τ is an ordered sequences $\Gamma = (D_1, ..., D_L)$ where for all $i \in [L]$, D_i is either a clause in τ (called an *axiom*) or is derived from $D_j, D_k, j, k < i$ by the *resolution rule*: $A \vee x, B \vee \neg x \vdash A \vee B$, where $D_j = A \vee x, D_k = B \vee \neg x, D_i = A \vee B, D_i \neq 1$. x is the *resolved variable*. The *size* of Γ is L, denoted by $|\Gamma|$. Γ is a *resolution refutation* if $D_L = \perp$. A resolution proof is *tree-like* if its underlying top-down proof DAG ($\perp$ at top/root) is a tree, and is *regular* if along any path from the root to axioms, no variable is resolved more than once.

We now introduce two natural k-Clique CNFs from the literature[2]. $Clique(G, k)$ is the encoding of "G contains a k-clique", on variables $x_{i,v}$ ($i \in [k]$, $v \in V$):

---

[2] There is also the so-called *binary* encoding [15], which we will not discuss here.

$$\bigvee_{v \in V} x_{i,v} \qquad \forall i \in [k]; \tag{1}$$

$$\neg x_{i,u} \vee \neg x_{j,v} \qquad \forall i,j \in [k], u,v \in V \ s.t. \ i \neq j, \ \{u,v\} \notin E; \tag{2}$$

$$\neg x_{i,u} \vee \neg x_{i,v} \qquad \forall i \in [k], u,v \in V \ s.t. \ u \neq v. \tag{3}$$

The other one, $Clique_{block}(G,k)$, is the encoding of "G contains a k-transversal clique" w.r.t any fixed balanced vertex-partition:

$$V = V_1 \sqcup ... \sqcup V_k, \quad |V_i| - |V_j| \in \{0, \pm 1\} \text{ for all } i,j \in [k],$$

where a clique C is *transversal* if $\forall l \in [k], |C \cap V_l| \leq 1$.

$$\bigvee_{v \in V_i} x_v \qquad \forall i \in [k]; \tag{4}$$

$$\neg x_u \vee \neg x_v \qquad \forall i \neq j \in [k], u \in V_i, v \in V_j \ s.t. \ \{u,v\} \notin E; \tag{5}$$

$$\neg x_u \vee \neg x_v \qquad \forall i \in [k], \ u,v \in V_i \ s.t. \ u \neq v. \tag{6}$$

In both encodings, the first group of axioms is called *clique axioms*, the second group *edge axioms*, and the third group *functionality axioms*. Clearly, the block encoding claims something stronger (hence is easier to refute) so a lower bound on its refutation length is stronger, too. We have the following observation which seems to be folklore among researchers.[3]

Theorem 1. *For any graph G that contains an $\Omega(k)$-clique, the $\exp(\Omega(k))$ resolution size lower bound holds for $Clique(G,k)$. In particular, the bound holds for the random graph $G(n, \frac{2\xi}{k-1})$ ($\xi > 1$ constant) with high probability.*

Proof. By a reduction to the *functional pigeonhole principle FPHP*. More precisely, if G contains a clique C, take the restriction ρ which sets $x_{i,v}$ to 0 for all $i \in [k], v \notin C$, then the refutation refutes $FPHP^k_{|C|}$. But an $\exp(|C|)$ lower bound for the latter is known (e.g. [20]). Finally, note a random graph from $G(n, \frac{2\xi}{k-1})$ contains $\Omega(k)$-cliques with high probability.

Remark 1. The encoding $Clique(G,k)$ inherits hardness from $FPHP^k_{\Omega(k)}$ which has little to do with the underlying graph. For $Clique_{block}(G,k)$, however, such a reduction on random graphs seems unlikely[4] as it just prohibits permutation on $[k]$. This is one reason $Clique_{block}(G,k)$ is regarded as more technically appropriate (cf. a similar remark in [4]). In the rest of the paper, we concentrate on the CNF $Clique_{block}(G,k)$.

Notation. We view a resolution proof Γ, i.e. a refutation of a CNF τ as a top-down DAG with the $\bot$ on top, and identify a clause C with the partial-assignment that minimally falsifies it. For example, $\{x_1 = 1, x_2 = 0\}$ represents

---

[3] For complete $(k-1)$-partite graphs, a similar reduction is observed earlier by Alexander Razborov (personal communication).

[4] For some specially structured G this is possible; see Remark 2.

clause $C = \neg x_1 \vee x_2$, and the empty assignment represents $\bot$. For clarity, we call such a representation an *object* and use letter P to denote it. Any non-leaf $P \in \Gamma$ is labeled by a query "$x =$?" on a variable x, and an *answer* is $x = 1$ or $x = 0$, leading to one child whose object contains the answer. For the clique problem, more conveniently, we can denote a query by "(l,v)?" intended for "is $x_v = 1$?" where $l \in [k]$, $v \in V_l$, and the answer is $(l,v)^{yes}$ or $(l,v)^{no}$, which chooses a child whose representation includes $x_v = 1$, $x_v = 0$ respectively. For distinction, let us call $l \in [k]$ a *pigeon* and $v \in V_l$ a *vertex*. The semantics of $Clique_{block}(G,k)$ is, therefore, "assign to each pigeon a vertex so that they form a k-transversal-clique."

Definition 1. *Given object P, let $P_1 := \{(l,v) \mid (l,v)^{yes} \in P\}$, $P_0 := \{(l,v) \mid (l,v)^{no} \in P\}$. For a pigeon $l \in [k]$, denote*

$$P_1(l) := \{ v \in V_l \mid (l,v)^{yes} \in P\} \qquad P_0(l) := \{ v \in V_l \mid (l,v)^{no} \in P\}. \quad (7)$$

$$P_{Live}(l) := V_l \backslash P_0(l) \qquad\qquad P_{Live} := \bigcup_{l \in [k]} P_{Live}(l). \quad (8)$$

By definition, $P_1(l) \cap P_0(l) = \emptyset$, $P_1 = \bigcup_{l \in [k]} \{l\} \times P_1(l)$, and $P_0 = \bigcup_{l \in [k]} \{l\} \times P_0(l)$. We use $\mathrm{dom}(P_1), \mathrm{dom}(P_0)$ to denote the projection to $[k]$ from P_1, P_0. A *live-clique in P* is a transversal clique in P_{Live}. A **partial** function $f : [k] \to V$ is a *live-clique assignment in P* if $f(l) \in V_l$ whenever it is defined, and its image is a live-clique in P.

In most situations each $P_1(l)$ has size 0 or 1. Intuitively, such an object P gives a product set $P(1) \times ... \times P(k) \subset V_1 \times ... \times V_k$ where $P(l) = P_1(l)$ if $P_1(l) \neq \emptyset$ and $P(l) = V_l \backslash P_0(l)$ otherwise. For example, if P is the empty assignment (i.e. the $\bot$ clause) then this set is the full $V_1 \times ... \times V_k$; while if $P_1(l)$ is nonempty for many l's, then the corresponding set has many coordinates of size 1. We will think of the "largeness" of P by measuring this set in a certain way (see the discussion under Definition 3).

3 2^k-Type Lower Bound for Resolution

Parameter Regime. Throughout Sect. 3, we use the following parameter regime.

$\xi > 1$ a constant;

$k = n^{c_0}$ where $c_0, \epsilon \in (0, 1/3)$ arbitrary parameters s.t.

$$\min\{\epsilon c_0, 1/3 - c_0\} > (\log n)^{-1/5};$$

$$N = 1 + \max\{\frac{1}{1 - 3c_0}, \frac{1}{\epsilon c_0}\}, \quad t = \frac{18\xi \cdot N}{1/3 - c_0}; \tag{9}$$

$$r = \frac{k}{t}, \quad q = \frac{1}{2}n^{1 - c_0 - 2\delta r} \text{ where } \delta = \frac{2\xi}{k - 1}.$$

W.l.o.g. we can assume k, r, q are integers. The meaning of k, c_0, ϵ is self-evident. r is a sufficiently small portion of k, and q is appropriately below the expected number of common neighbors of an r-subset in a random graph G. N, t are used only for technical reason. Note $(\log n)^{-1/2} < \delta r < \frac{1/3 - c_0}{4N}$.

The reader can assume for simplicity the parameters are in the "typical" case, i.e. ϵ, c_0 and N, t are all constants. We do not try to optimize the parameter range, e.g. the number $(\log n)^{-1/5}$ is just a convenient choice for the estimates in Lemma 2 and (26) to go through.

3.1   Graph Properties

Fix a balanced vertex partition $V = V_1 \sqcup ... \sqcup V_k$.

Definition 2. *A subset $A \subset V$ is called (r, q)-neighbor-dense [2,5] if for any $U \subset V$ with size $\leq r$, it holds that $|\hat{N}_A(U)| \geq q$. G is called $(r, q)^{block}$-neighbor-dense if for every $j \in [k]$, V_j is (r, q)-neighbor-dense.*

Lemma 1. *(Inheritability of neighbor-denseness) For any integers a_1, a_2, b_1, b_2 and fixed G, if $A \subset V$ is $(a_1 + a_2, b_1 + b_2)$-neighbor-dense and $A_1 \subset A$ is not (a_1, b_1)-neighbor-dense, then $A \backslash A_1$ is (a_2, b_2)-neighbor-dense.*

Proof. Take a witness W_1 of size a_1 for A_1 s.t. $|\hat{N}_{A_1}(W_1)| < b_1$. For any $W \subset V$, $|W| \leq a_2$,

$$|\hat{N}_{A \backslash A_1}(W)| \geq |\hat{N}_{A \backslash A_1}(W_1 \cup W)| = |\hat{N}_A(W_1 \cup W)| - |\hat{N}_{A_1}(W_1 \cup W)| \geq (b_1 + b_2) - b_1,$$

where the second inequality used $|W_1 \cup W| \leq a_1 + a_2$ and A is $(a_1 + a_2, b_1 + b_2)$-neighbor-dense.

Lemma 2. *W.p. $> 1 - \exp(-0.5\sqrt{\log n})$, $G \sim G(n, n^{-\frac{2\epsilon}{k-1}})$ is $(2r, q)^{block}$-neighbor-dense with parameters in (9).*

Proof. By standard use Chernoff bound and union bound. For any fixed $j \in [k]$, any $R \subset V$ with $|R| = 2r$, $\mathrm{E}[\,|\hat{N}_{V_j}(R)|\,] \geq (n/k - |R|) \cdot n^{-\delta r} > \frac{2}{3}n^{1-c_0-\delta r} > q$. So

$$\Pr[\,|\hat{N}_{V_j}(R)| < \frac{1}{2}q\,] \leq \exp(-\frac{n^{1-c_0-2\delta r}}{48})$$
$$< \exp(-n^{2c_0+\delta r}) \quad \text{since } \delta r < 1/3 - c_0 \text{ by (9).}$$
$$\tag{10}$$

The first "$\leq$" in above uses Chernoff bound as all different edges are independent. Finally, take a union bound over R's whose total number is at most $n^{2r} < \exp(0.5n^{2c_0}\log n)$, and $\exp(-n^{2c_0+\delta r}) \cdot \exp(0.5n^{2c_0}\log n) < \exp(-0.5\sqrt{\log n})$ since $\delta r > (\log n)^{-1/2}$ in (9).

Remark 2. Some particular graph family is also neighbor-dense, yet being far from pseudo-random. For example, consider a complete $(k-1)$-partite graph G where $2r < k_1 < k$ (r, k as in (9)), with partition $V = W_1 \sqcup ... \sqcup W_{k_1}$ where $|W_i \cap V_j| \approx \frac{n}{k_1 k}$ for all $i \in [k_1], j \in [k]$. Notice, however, for these graphs there is a $2^k n^2 k^2$ refutation (e.g. [5]) which is regular, and thus to obtain strong lower bound $n^{\Omega(k)}$ the property of neighbor-denseness is not enough, even for regular resolution.[5]

---

[5] Although a variant of it seems sufficient for tree-like resolution, cf. [5].

3.2   The Lower Bound Proof

Theorem 2. *For parameters as in* (9) *where* $k = n^{c_0}$, *if* G *is* $(2r, q)^{block}$-*neighbor-dense then any resolution refutation of* $Clique_{block}(G, k)$ *has size* $\geq \exp(\Omega(k^{1-\epsilon})/t^2)$, *where* $\Omega(\cdot)$ *only relies on some absolute constant. In particular, if* $c_0, \epsilon \in (0, 1/3)$ *are constant, then the bound is* $\exp(\Omega(k^{1-\epsilon}))$.

Corollary 1. *(of Theorem 2 and Lemma 2) Within the same parameters as in Theorem 2,* $Clique_{block}(G, k)$ *is sub-exponentially hard for* $G(n, n^{-\delta})$ *on average, where* $\delta = \frac{2\xi}{k-1}$, $\xi > 1$ *constant.*

The rest of this section is devoted to the proof of Theorem 2. To show size lower bound, we design an answering strategy that finds many different objects in Γ. We call this an *adversary strategy* (against the prover Γ; cf. [17]).

Fix any resolution proof Γ of $Clique_{block}(G, k)$. We will first describe an adversary strategy and then analyze the size bound from it.

Adversary Strategy. 1. Random part. Choose a set of $\frac{r}{2}$ pigeons from $[k]$ uniformly randomly, each with probability $\binom{k}{r/2}^{-1}$. Then choose an $\boldsymbol{\alpha}$, a transversal clique assignment to the chosen pigeons, according to the following distribution:

> *(Distribution of $\boldsymbol{\alpha}$) Suppose the chosen pigeons are $l_1, ..., l_{\frac{r}{2}} \in [k]$. Choose $\alpha(l_1)$*
>
> *uniformly from V, then $\alpha(l_2)$ uniformly from $\hat{N}_{V_{l_2}}(\{\alpha(l_1)\})$, $\alpha(l_3)$ uniformly* (11)
>
> *from $\hat{N}_{V_{l_3}}(\{\alpha(l_1), \alpha(l_2)\})$ and so on till $\alpha(l_{\frac{r}{2}})$ is chosen.*

Denote this distribution by $\mathcal{D}$, which is well-defined when G is $(2r, q)^{block}$-neighbor-dense. The strategy is deterministic after α is chosen.

2. Deterministic part. Fix a sample α from above.

Definition 3. *(Narrow pigeons) Given an object P, pigeon $l \in [k]$ is called* **narrow** *in P if:*

$$P_0(l) \text{ is } (r, \frac{1}{2}q)\text{-neighbor-dense.}$$

The set of **useful pigeons** *for P is defined to be $dom(P_1) \cup \{narrow\ pigeons\ in\ P\}$.*

Intuitively, an object is small if it contains $\geq \frac{r}{2}$ many useful pigeons. The **invariance** the strategy keeps is: as long as the number of useful pigeons in the current object is $< r/2$,

$\qquad$ 1. α, P_1 are compatible as functions;

$(*):\qquad$ 2. $\exists$ function β: {narrow pigeons in P} $\to V$ s.t. α, P_1, β are consistent

$\qquad\qquad$ and together is a live-clique assignment for P (Def. 1).

Note at the beginning of any path (top node), $(*)$ trivially holds.

Claim 1. *If for an object P the above (*) holds, then P is not an axiom.*

Proof. Direct check.

The strategy continues as follows. Suppose the invariance (*) holds for current object P where the query is (l, v)?. Answer according to the following:

    (1) If |useful pigeons in $P| \geq r/2$, then *halt*. Otherwise,

    (2) (2a) If $l \in \mathrm{dom}(\alpha \cup P_1 \cup \beta)$, answer according to $\alpha \cup P_1 \cup \beta$;   (12)

        (2b) Otherwise, say "No".

Lemma 3. *Suppose the current object P satisfies (*). Then either we halt, or after extending the path by one more step we still keep (*).*

Proof. For item 1 in (*), it holds for the next object because of (2a) of the strategy. Now we prove item 2. If P has $\geq r/2$ many narrow pigeons then we would halt by (1) of the strategy. Otherwise, by assumption there is β for P_0 as in (*). We prove that the "intermediate" object

$$Q := P \cup \{\text{the new answer}\}$$

satisfies (*), and the lemma follows because (*) is monotone w.r.t. the object.

Assume the new query is (l, v)? In case (2a), the same β for P suffices for Q, trivially from inductive hypothesis. In case (2b), either $P_0(l) \cup \{v\}$ is $(r, \frac{1}{2}q)$-neighbor-dense in G, or it isn't. In the latter case, the pigeon l is still not narrow in Q, and thus (*) holds for Q. In the former case, let $R := \mathrm{Im}(\alpha \cup \beta \cup P_1)$. By assumption,

$$|R| \leq |\alpha| + |\beta \cup P_1| = \frac{r}{2} + |\{\text{useful pegions}\}| < \frac{r}{2} + \frac{r}{2} = r. \qquad (13)$$

Moreover, $P_0(l) \cup \{v\}$ is not $(r, \frac{1}{2}q + 1)$-neighbor-dense by the case assumption. So by Lemma 1, where we take $A := V_l$, $A_1 := P_0(l) \cup \{v\}$, and $a_1 = a_2 = \frac{1}{2}q$, we get that $V_l \backslash (P_0(l) \cup \{v\}) = V_l \backslash Q_0(l) = Q_{\mathrm{Live}}(l)$ is $(r, \frac{1}{2}q - 1)$-neighbor-dense. In particular, as $\frac{1}{q} >> 1$, we can choose a $w \in \hat{N}_{Q_{\mathrm{Live}}(l)}(R)$. Extend β to $\beta \cup \{\beta(l) = w\}$ will keep (*) for Q.

The answering strategy can be now completed: we extend β so that (*) holds until we halt.

The Analysis. Since Γ is a correct proof, the query process must stop. By Claim 1, it could only be halted in Case (1) of (12). Let T be the set of all such halting objects (over all α) in the Γ.

Definition 4. *We say a $\frac{r}{2}$ transversal clique assignment α leads to object P (in T) if when chosen α in the beginning, the adversary strategy halts at P.*

Lemma 4. *Given the distribution* $\boldsymbol{\alpha} \sim \mathcal{D}$ (11), *for any fixed* $P \in T$

$$\Pr[\,\boldsymbol{\alpha} \text{ leads to } P\,] \leq \exp(-\Omega(k^{1-\epsilon})) \qquad (14)$$

where the parameters are as in (9).

Proof. By definition of T and Lemma 3, for P we have $|\{\text{useful pigeons}\}| \geq r/2$. Recall $r = k/t$ in (9). Take another parameter $\epsilon' = \frac{1}{40}\frac{r}{k} = \frac{1}{40t}$ and denote $a_0 := \lceil \epsilon' r \rceil$. By the first part of definition of $\boldsymbol{\alpha}$,

$$\Pr[\,|\text{dom}(\boldsymbol{\alpha}) \cap \{\text{useful pigeons}\}| < \epsilon' r\,] \qquad (15)$$

$$= \sum_{a < a_0} \binom{r/2}{a}\binom{k-r/2}{r/2-a}\Big/\binom{k}{r/2} < a_0 \cdot \binom{r/2}{a_0}\binom{k-r/2}{r/2-a_0}\Big/\binom{k}{r/2}. \qquad (16)$$

Denote $f(a) = \binom{r/2}{a}\binom{k-r/2}{r/2-a}$ then $f(a+1) = f(a) \cdot \frac{(r/2-a)^2}{(a+1)(k-r+a)}$, so $\frac{f(a+1)}{f(a)} = \frac{(r/2-a)^2}{(a+1)(k-r+a)} > \frac{(1/2-2\epsilon')^2 r^2}{2\epsilon' r k} > 2$ when $a < 2a_0$. Also note $\binom{k}{r/2} = \sum_{a=0}^{r/2} f(a)$. Thus $(16) < a_0 \cdot \frac{f(a_0)}{f(2a_0)} < \epsilon' r \cdot 2^{-\epsilon' r} < \exp(-\Omega(k/t^2))$. Therefore,

$$\Pr[\,\boldsymbol{\alpha} \text{ leads to } P\,] \leq$$
$$\exp(-\Omega(k/t^2)) + \Pr[\,\boldsymbol{\alpha} \text{ leads to } P, |\text{dom}(\boldsymbol{\alpha}) \cap \{\text{useful pigeons}\}| \geq \epsilon' r\,]$$

We bound the second term below. There are two cases:

$$|\text{dom}(\boldsymbol{\alpha}) \cap \text{dom}(P_1)| \geq \frac{\epsilon' r}{2}, \quad \text{Or} \qquad (17)$$

$$|\text{dom}(\boldsymbol{\alpha}) \cap (\{\text{narrow pigeons}\}\backslash\text{dom}(P_1))| \geq \frac{\epsilon' r}{2}. \qquad (18)$$

Here as usual, $\text{dom}(\boldsymbol{\alpha})$ denotes the domain of $\boldsymbol{\alpha}$ (a subset of $[k]$). In the following, α' will denote an arbitrary choice of $\boldsymbol{\alpha}$ that satisfies the item's condition.

1. In the first case, (17), α' has to assign exactly the same vertices as P_1 to pigeons in $\text{dom}(P_1) \cap \text{dom}(\alpha')$. Since G is $(2r,q)^{block}$-neighbor-dense where $q = \frac{1}{2}n^{1-2\delta r}$, so in particular, there are $\geq \frac{1}{2}n^{1-c_0-2\delta r}$ many choices of vertices for *each* such pigeon. By definition (11), $\boldsymbol{\alpha}$ chooses among them uniformly. Thus

$$\Pr[\,\boldsymbol{\alpha} \text{ leads to } P \text{ and } |\text{dom}(\boldsymbol{\alpha}) \cap \text{dom}(P_1)\}| \geq \epsilon' r/2\,] \qquad (19)$$

$$\leq \sum_{S \subset [k],\, |S| \geq \epsilon' r/2} \Pr[\text{dom}(\boldsymbol{\alpha}) \cap \text{dom}(P_1) = S \wedge \text{ for all } i \in S, \boldsymbol{\alpha}(i) = P_1(i)] \qquad (20)$$

$$= \sum_{S \subset [k],\, |S| \geq \epsilon' r/2} \Pr[\text{dom}(\boldsymbol{\alpha}) \cap \text{dom}(P_1) = S\,] \cdot \Pr[\text{ for all } i \in S, \boldsymbol{\alpha}(i) = P_1(i)]$$

$$\qquad (21)$$

$$\leq \sum_{S \subset [k],\, |S| \geq \epsilon' r/2} \Pr[\,\text{dom}(\boldsymbol{\alpha}) \cap \text{dom}(P_1) = S\,] \cdot (\frac{1}{2}n^{1-c_0-2\delta r})^{\epsilon' r/2}$$

$$\leq 1 \cdot n^{-c_0 \epsilon' r} = n^{-\Omega(c_0 k/t^2)} \qquad (22)$$

where (21) is from the independence of the two parts in the definition of $\boldsymbol{\alpha} \sim \mathcal{D}$.

2. In the latter case, (18), let B denote {narrow pigeons (in P)}\dom$(P_1)$. In the process of choosing vertices to a pigeon $i \in \text{dom}(\alpha') \cap B$, vertices in $P_0(i)$ must not be chosen (by (2a) in the strategy). On the other hand, for any such pigeon i, it is narrow in P so $P_0(i)$ is $(r, \frac{1}{2}q)$-neighbor-dense. Therefore,

$$\hat{N}_{P_0(i)}(\text{Im}(\alpha'|_{\text{dom}(\alpha'\backslash\{i\})})) \geq \frac{1}{2}q = \frac{1}{4}n^{1-c_0-2\delta r}. \tag{23}$$

So for such i, as $|V_i| = n^{1-c_0}$,

$$\Pr[\alpha(i) \notin P_0(i) \mid i \in \text{dom}(\alpha)] \leq 1 - \frac{n^{1-c_0-2\delta r}}{4n^{1-c_0}} = 1 - \frac{1}{4}n^{-2\delta r}. \tag{24}$$

Now we can bound the overall probability of this case by

$$\sum_{S \subset B, \, |S| \geq \epsilon' r/2} \Pr[\text{ dom}(\alpha) \cap B = S \text{ and } \alpha(i) \notin P_0(i) \text{ for all } i \in S] \tag{25}$$

Similar to estimation (19), from (24) we have

$$(25) \leq (1 - \frac{1}{4}n^{-2\delta r})^{\epsilon' r/2} < \exp(-\Omega(k^{1-\epsilon}/t^2)), \tag{26}$$

where the last inequality uses $k = n^{c_0}$, $2\delta r < \epsilon c_0$ in (9).

Finally, note $c_0 < \log n$ so the sum of probability is $\exp(-\Omega(k^{1-\epsilon}/t^2))$.

Since any choice of α results in halting at some object in T, Lemma 4 implies $|T| \geq \exp(\Omega(k^{1-\epsilon})) = \exp(\Omega(n^{(1-\epsilon)c_0}))$. In particular, there are at least this many different objects in Γ. Theorem 2 is proved.

4 n^k-Type Lower Bounds for a-Irregular Resolution

4.1   The Model

Like before, let us view a resolution proof Γ as a top-down DAG ($\perp$ on top). A variable x is called *irregular* on a path $\mathfrak{p}$ in Γ if it is queried more than once on $\mathfrak{p}$. x is *irregular under clause* C if there is *some* path from C on which x is irregular.

We are going to introduce the model of a-irregular resolution. Its main version assumes a variable partition in input. Let's start with a simpler one.

Definition 5. *For $a \leq 1$, a resolution proof Γ on m variables is **unblocked** a-**irregular** if for any clause $C \in \Gamma$,*

$$w(C) \geq am \Rightarrow |\{variables\ irregular\ under\ C\}| \leq am \tag{27}$$

So regular resolution is 0-irregular, and general resolution is 1-irregular.

We continue to the main version. Given m variables and $\kappa : \text{Var} \rightarrow [k]$ a partition of variables $(1 \leq k \leq m)$, we say x *belongs to block* $\kappa(x)$. Define the *block-size* of a variable set to be $|X|^b := |\kappa(X)|$, and the *block-width* to be

$$w^b(C) = |\text{Var}(C)|^b. \tag{28}$$

Definition 6. *(Main model)* For $a \leq 1$, κ as above, a resolution proof Γ is a-**irregular** for κ if for any clause $C \in \Gamma$,

$$w^b(C) \geq ak \Rightarrow |\{\text{variables irregular under } C\})|^b \leq ak. \tag{29}$$

It is easy to see that this model is at least as strong as "resolution refutations with small block-width (for the same variable partition)"; and it always subsumes the unblocked $\frac{ak}{m}$-irregular model, regardless of the partition.

The unblocked a-irregular resolution (Definition 5) is already exponentially stronger than regular even for $a = k^{-\Omega(1)}$, and the situation is clearer for the main model. It turns out that the known exponential separations between the regular and general resolution [1,24] are, actually, separations between regular and the a-irregular resolution with a natural partition κ and $a = k^{-\Omega(1)}$.

Remark 3. We next give the details of how the a-irregular resolution can handle the hard instances from the known general-regular separations. The instances are *Stone formulas* [1], *Lifted pebbling formulas* [24], and a variation of the *Ordering principle* [1].

1. *Stone formulas.* Under the notation of [1], the $m = \Omega(n^2)$ many variables arc $\{P_{i,s} \mid t \in S\}$ for each $i \in V(G)$ and $\{R_t \mid t \in S\}$, where $|S| = \Omega(|V(G)|) = \Omega(n)$. The variables are naturally partitioned into $k = n+1$ blocks according to the vertex index, plus a block of all stones. Axioms have block-width ≤ 4, and the short resolution in [1] (their Lemma 4.1) is $5/k$-irregular for κ, since every clause in that resolution has block-width ≤ 4.

This short resolution proof is also unblocked $m^{-1/2}$-irregular; actually, only the stone variables $\{R_t\}$ are irregularly resolved since a path in the proof naturally corresponds to a path in G and G is acyclic. There are $O(n) = O(m^{1/2})$ many stone variables.

2. *Lifted pebbling formulas.* It is realized in [24] that the Stone formulas can be regarded as a "lifted" version of the so-called *Pebbling formulas*, Pebb_G, on the same graph G. They give a similar but different family of CNFs: in short, given boolean variables $x_1, ..., x_n$, consider a variable change by encoding every literal x_i^ϵ by $\wedge_{j \in N(i)}(\neg s_{i,j} \vee r_j^\epsilon)$ $(\epsilon \in \{0,1\}$ and $x^0 := \neg x)$, where $\{s_{i,j}, r_j^b\}$ are fresh variables corresponding to a bipartite graph H on components $[n]$, J $([n] \cap J = \emptyset)$, and we add the default axioms $\vee_{j \in N(i)} s_{i,j}$ for all i. If the left degree of H is d, then there are $nd + |J|$ many variables. The Stone Formulas are the resulting CNF expression from this variable change on Pebb_G, when H is complete; [24] showed the same separation holds if take H to be a more economic sparse bipartite expanders[6], with $d = \Theta(n/|J|)$ (their Theorem 12). For us, the

---

[6] The actual construction has one more twist called *mirroring*, which we ignore here.

short resolution refutation in example 1 now only simplifies, so the block width is still constant where blocks are the same $\{s_{i,j}\}$ (for each $i \in V(G)$ and $\{r_j\}$.

Similarly to example 1, the short proof is also unblocked $1/d$-irregular, and in applications $d \geq \Theta(\log \log n)$ (their Theorem 13).

3. GT_n', a variation of the so-called *Ordering principle*. It has $m = n(n-1)$ variables $x_{i,j}$, $i \neq j \in [n]$, with the intended meaning $x_{i,j} \Leftrightarrow$ element i (in some n-element set) is greater than element j. We refer the reader to [1] to this CNF family; what's important for us is that if partition the variables according to the second subscript j, into $k = n$ blocks, then the axioms have constant block-width, and the short refutation (Corollary 3.4 in [1]) is $4/k$-irregular. Namely, that refutation first resolves $x_{i_1,i_2} \lor x_{i_2,i_3} \lor x_{i_3,i_1} \lor \rho(i_1,i_2,i_3)$ with $x_{i_1,i_2} \lor x_{i_2,i_3} \lor x_{i_3,i_1} \lor \neg\rho(i_1,i_2,i_3)$ for all i_1, i_2, i_3, where $\rho(\cdot)$ refers to some literal and all clauses have block-width ≤ 4, then it uses the short refutation from [22] to finish, in which all clauses are either the so-called $C_m(j)$'s (in notation of [22]) or axioms, all with block-width ≤ 4. So, the refutation is $4/k$-irregular for this partition.

Remark 4. In all the examples above, the variable partition we work with not only has a natural semantic meaning but also makes axioms have constant block-width.[7] This might be considered together with the technique of variable substitutions in form $x_i = f(y_{i,1}, ..., y_{i,t})$ with $y_{i,j}$'s being distinct new variables (a.k.a. lifting; see [8,10,14,18] and the references therein), where "blocks of variables" appear naturally. For the lifted CNF, it is reasonable to expect that the block-width measure on proofs w.r.t. this variable partition (i.e. according to the i-index) reflects the hardness of the easier, unlifted CNF which itself is often narrow. In our context, this explains the power of the model in examples 1, 2: with the correct variable partition, it has sufficient power to recover and handle the unlifted CNF, which is just easy for regular resolution. This perspective seems to say nothing about example 3, though.

The main theorem of this section is the following.

Theorem 3. *Fix the natural partition $\kappa_0 : x_v \mapsto i$ if $v \in V_i$ in $V_1 \sqcup ... \sqcup V_k$. Let $\xi > 1$ be constant and $\epsilon > 0$ be any parameter s.t. $(\log n)^{-1/2} < \epsilon < 1/200$. Then for any k s.t. $\xi^2(100/\epsilon)^3 < k < n^{1/3-40\epsilon}$, w.h.p. over $\boldsymbol{G} \sim G(n, n^{-2\xi/(k-1)})$, any $\frac{\epsilon}{\xi}$-irregular resolution proof for $Clique_{block}(G,k)$ has size $n^{k\epsilon^3/(200\xi)^2}$.*

4.2   More Graph Properties

Definition 7. *(relativized neighbor-denseness, Definition 2) Given G and $a, b \in \mathbb{N}_+$, for $A, B \subset V$, B is called $(a,b)^A$-neighbor-dense if $\forall U \subset A$, $|U| \leq a \Rightarrow |\hat{N}_B(U)| \geq b$. When $A = V$, we simply say B is (a,b)-neighbor-dense.*

Note $A' \subset A$, then $(a,b)^A$-neighbor-denseness implies $(a,b)^{A'}$-neighbor-denseness.

---

[7] Other partitions might also seem natural but fail the second property, and we do not know the power of the model with them.

Another pseudorandom property which played an important role in the proof for regular case says: for any (r, q)-neighbor-dense sets in G, all witness sets of its non-(tr, q')-neighbor-denseness are non-trivially concentrated (for suitable t, q').

Definition 8. *([2]) $W \subset V$ is called (tr, r, q', s)-mostly-dense in G, if $\exists S \subset V$, $|S| = s$ such that: $\forall U \subset V$ of size $\leq tr$, $|\hat{N}_W(U)| < q' \Rightarrow |U \cap S| \geq r$. For convenience, we say G itself is (tr, r, q', s)-mostly-dense if every (r, q)-neighbor-dense set is (tr, r, q', s)-mostly-dense (when q is clear from the context).*

Proposition 1. *(mostly-denseness is inheritable w.r.t witness S) Suppose $A \subset V$, and W is (tr, r, q', s)-mostly-dense. Then $\exists S_1 \subset A$ of size $\leq s$ such that, for any $U \subset A$, $|U| \leq tr$, if $|\hat{N}_W(U)| < q'$ then $|U \cap S_1| \geq r$.*

Proof. Take S_1 to be $S \cap A$, where S is as in Definition 8.

As usual, denote $\frac{2\xi}{k-1}$ by δ. For simplicity, we always take $\xi > 1$ to be constant. The main result of [2] is the following.

Theorem 4. *For any parameter $\epsilon \in (0, 1/2)$ and constant $\xi > 1$, if $k < n^{1/4-\epsilon}$ and $k\sqrt{\xi} < n^{1/2-\epsilon}$, then:*
(1) (their Theorem 6.1) W.h.p., $\mathbf{G} \sim G(n, n^{-\frac{2\xi}{k-1}})$ is (tr, tq)-neighbor-dense and (tr, r, q', s)-mostly-dense, with $t = \frac{64\xi}{\epsilon}$, $r = \frac{4k}{t^2}$, $q = \frac{n^{1-\delta tr}}{4t}$, $s = (\frac{n}{\xi})^{1/2}$ and $q' = 3\epsilon s^{1+\epsilon} \log s$.
(2) (their theorem 5.4) Let $t: 4 \leq t \leq k$ be any parameter and $r = 4k/t^2$. If G is (tr, tq)-neighbor-dense and (tr, r, q', s)-mostly-dense, then any regular refutation of $Clique(G, k)$ requires size $\frac{1}{2} \min\{s^{\epsilon r/2}, (1 - rs^{-(1+\epsilon)}/2ek)^{-q'}\}$.

We will need Theorem 4(1) for the following parameters. Theorem 4(2) will be actually re-proved and refined following the original method (Lemma 8 and 9).

Parameter Regime. In the rest of Sect. 4, we use a parameter regime that is similar to that of [2]. As before, let $\xi > 1$ be a constant and $\delta = \frac{2\xi}{k-1}$.

$$\epsilon = \text{any parameter in } \left((\log n)^{-1/2}, \ 1/200\right);$$

$$t = \frac{64\xi}{\epsilon}, \quad k \in \left(\frac{3t^2}{\epsilon}, \ n^{1/3-40\epsilon}\right); \tag{30}$$

$$r = \frac{4k}{t^2}, \quad q = \frac{1}{32t} n^{1-8\delta tr}/k, \quad q' = \frac{1}{4} qn^{-\delta tr};$$

$$s = k^2 n^{9\delta tr+\epsilon}, \quad p = n^{-(9\delta tr+2\epsilon)}/k.$$

Again, we can assume k, r, q, q', s are integers, and their meaning is clear from Definition 8. p is used for a biased-coin in the argument. As before, the "typical" case is when ϵ is a small constant, and the bound is $n^{\Omega(k/\xi^2)}$. Our choice of $p = n^{-(9\delta tr+2\epsilon)}/k$ is larger compared to the original choice for Theorem 4(2) (which is about $n^{-(1+\epsilon)/2}$); this makes the two bounds in proof of Lemma 9 more balanced thus allows to slightly improve the range of k from $n^{1/4}$ to $n^{1/3}$.

Theorem 5. *With parameter regime* (30), *w.h.p.* $\boldsymbol{G} \sim G(n, n^{-\delta})$ *is*

$$(i).(8tr, 4tq)^{block}\text{-neighbor-dense; and}$$
$$(ii).(tr, r, q', s)\text{-mostly-dense.} \tag{31}$$

Proof. (i) is identically proved as Lemma 2. (ii) is Theorem 4(1) except for a difference in parameters; we only have to point out that parameters (30) satisfy $n^{\epsilon/2+1} < qn^{-\delta tr}s/tr$ so can be safely replaced to their proof.

Theorem 3 thus reduces to the following.

Theorem 6. *Recall* $V(G) = V_1 \sqcup ... \sqcup V_k$, *and* κ_0 *is the "canonical" partition that maps* v *to* i *if* $v \in V_i$. *If* G *satisfies* (31) *with parameters* (30), *then any* $\frac{1}{t}$-*irregular resolution for* $(Clique_{block}(G, k), \kappa_0)$ *requires size* $n^{\epsilon k/6t^2}$.

4.3   The Lower Bound Proof (Theorem 6)

Proof Overview. As briefed in the introduction the idea is simple—use a suitable restriction to simplify the refutation to be regular. This would finish the proof (by a self-reduction) if the induced sub-graph were also pseudo-random, which is not the case. It is not far, though: the only additional observation is to use a weaker, relative pseudo-randomness (Lemma 6 and 9).

As before, we give an adversary strategy followed by its analysis, where we will need to open up the argument in the regular case.

Definition 9. *(Narrow pigeons, with new parameters (cf. Def. 3) Suppose* Γ *is a resolution proof,* $P \in \Gamma$ *an object. A pigeon* $l \in [k]$ *is* **narrow** *in* P *if*

$$P_0(l) \text{ is } (4tr, 2tq)\text{-neighbor-dense, where recall } 2tq = \frac{1}{4}n^{1-8t\delta r}/k.$$

Let narrow$_P$ denote the set of narrow pigeons in P.

Adversary Strategy. Stage I (to find a restriction). Travel down the proof, and keep a live-clique assignment β_P (Definition 1) s.t.

$$\beta_P \supset P_1, \qquad \text{dom}(\beta_P) = \text{dom}(P_1) \cup \text{narrow}_P. \tag{32}$$

where P is the current object. Suppose the query at P is "(l_1, v_1)?". If

$$|\text{narrow}_P \cup \text{dom}(P_1)| \geq tr \tag{33}$$

then go to Stage II; otherwise if $l_1 \in \text{dom}(P_1) \cup \text{narrow}_P$, answer according to β_P; otherwise, answer No.

When not transit to Stage II, we show (32) holds for the next node.

Claim 2. *If* G *is* $(8\delta tr, 4tq)^{block}$-*neighbor-dense,* $l \notin \text{narrow}_P$, *then* $P_{Live}(l)$ *is* $(4\delta tr, 2tq)$-*neighbor-dense.*

Proof. Apply Lemma 1 to $A \leftarrow V_l$, $A_1 \leftarrow P_0(l)$, $a_1 = a_2 = 4\delta tr$, $b_1 = b_2 = 2tq$.

Denote the next node by P^+. There is at most one new narrow pigeon l_1, so by Claim 2, $|\hat{N}_{P_{Live}(l_1)}(\text{Im } \beta_P)| \geq 2tq > 1$. Take a $v \in \hat{N}_{P_{Live}(l_1)}(\text{Im } \beta_P) \setminus \{v_1\}$, extend β_P by $l \to v$ then restrict it to narrow$_{P^+} \cup \text{dom}(P_1^+)$ as β_{P^+}. (32) holds for P^+.

This completes Stage I.

Claim 3. *The query-answer process must transit to Stage II at some node P.*

Proof. Similar to Claim 1: if (33) fails then P falsifies no axiom.

Stage II. Suppose we transit to this stage at node P^*.

(i). The restriction. Note $|P_1 \cup \text{narrow}_P|$ increases by at most 1 per step in Stage I (it might decrease), so it must be the case that $|\text{narrow}_{P^*} \cup P_1^*| = tr = k/t$. Now $|P^*|^b \geq k/t$ and since Γ is $\frac{1}{t}$-irregular, all irregular variables below P^* belong to some fixed block set I_{P^*} of size $\leq tr$.

Claim 4. *There exists a live-clique assignment $\tilde{\beta}$ for P^* s.t.*

$$\tilde{\beta} \text{ extends } \beta_{P^*} \quad and \quad \text{dom}(\tilde{\beta}) = \text{dom}(\beta_{P^*}) \cup I_{P^*}. \quad (34)$$

Proof. Extend the function β_{P^*} on $I_{P^*} \setminus \text{dom}(\beta_{P^*}) \subset I_{P^*} \setminus \text{narrow}_{P^*}$ one by one. In each step, the function to be extended has image size $\leq (|\text{dom}((P^*)_1)| + |\text{narrow}_{P^*}|) + |I_{P^*}| \leq 2tr$, so it is possible to find a common neighbor in $P_{\text{Live}}(l)$ for any $l \notin \text{narrow}_{P^*}$ by Claim 2.

(ii). Self-reduction to $\tilde{G}$. Fix a $\tilde{\beta}$ in Claim 4. Let

$$\tilde{G} := G\left[\bigcup_{l \in [k] \setminus \text{dom}(\tilde{\beta})} \tilde{V}_l \right], \quad \text{where } \tilde{V}_l = \hat{N}_{P_{\text{Live}}^*(l)}(\text{Im } \tilde{\beta}), l \in [k] \setminus \text{dom}(\tilde{\beta}). \quad (35)$$

Restrict more appropriate variables to 0 so that axioms become $Clique_{block}(\tilde{G}, k - |\text{dom}(\tilde{\beta})|)$. The restricted proof under P^* is regular. Denote it by Γ^*.

(iii). Strategy on Γ^* (the regular case; cf. [2]). Suppose we travel down Γ^* from the root P^* along a path $\mathfrak{p}$ to node Q, and is faced by a query "(l_1, v_1)?".

1. If $\exists v \in \tilde{V}_{l_1}$ s.t. (l_1, v) was answered Yes along $\mathfrak{p}$, answer No (*forgotten-forced answer*);
2. Otherwise, if $v_1 \notin \hat{N}(\text{Im } Q_1)$, answer No (*edge-forced answer*);
3. Otherwise, answer Yes w.p. p, No w.p. $1 - p$ independently (*random answer*).

This completes the adversary strategy.

Pseudorandomness of $\tilde{G}$. Recall $\tilde{G}$ is the induced subgraph (35). Assume w.l.o.g.

$$\text{dom}(\tilde{\beta}) = [\tilde{k} + 1, k].$$

The vertex-set size $|\tilde{V}|$ will not be important, as the lower bound depends only on the pseudorandomness from Lemma 5 and 6.

Lemma 5. *Assume* G *is* $(8tr, 4tq)^{block}$*-neighbor-dense* $(t, q$ *as in* $(30))$*. Then* $\forall l \in [\widetilde{k}]$, $\widetilde{V}_l$ *is* $(2tr, 2tq)^V$*-neighbor-dense in* G *(the upper "V" stressed here).*

In particular, $\widetilde{G}$ itself is $(2tr, 2tq)^{block}$-neighbor-dense.

Proof. Fix such an l. As in the proof of Claim 2, we apply Lemma 1 to $A \leftarrow V_l$ and $A_1 \leftarrow (P^*)_0(l)$ with $a_1 = a_2 = 4tr$, $b_1 = b_2 = 2tq$, where $l \notin \text{dom}(\widetilde{\beta}) \supset \text{dom}(\text{narrow}_{P^*})$. As a result we have

$$P^*_{\text{Live}}(l) \text{ is } (4tr, 2tq)\text{-neighbor-dense in } V. \tag{36}$$

Now for any $R \subset V$ of size $\leq 2tr$, $|\text{Im}(\widetilde{\beta}) \cup R| \leq 2tr + 2tr = 4tr$, so

$$|\hat{N}_{\widetilde{V}_l}(R)| \stackrel{\text{by def.}}{=} |\hat{N}_{P^*_{\text{Live}}(l)(\text{Im}(\widetilde{\beta}))}(R)| = |\hat{N}_{P^*_{\text{Live}}(l)}(\text{Im }\widetilde{\beta} \cup R)| \stackrel{\text{by (36)}}{\geq} 2tq. \tag{37}$$

The Lemma is proved.

Lemma 6. *Assume* G *is* (tr, r, q', s)*-mostly-dense. The relativized mostly-denseness holds for* $(G, \widetilde{G})$*: for all* $(r, q)^V$*-neighbor-dense set* $W \subset \widetilde{V}$, $\exists S \subset \widetilde{V}$ *of size* $\leq s$ *s.t.* $\forall U \subset \widetilde{V}$, *if* $|U| \leq tr$ *and* $|\hat{N}_W(U)| < q'$ *then* $|S \cap U| \geq r$.

Proof. Since G is (tr, r, q', s)-mostly-dense and W is $(r, q)^V$-neighbor-dense, W is (tr, r, q', s)-mostly-dense. In Proposition 1 take $A \leftarrow \widetilde{V}$, as a result there exists $S_1 \subset A = \widetilde{V}$ that satisfies the condition in the lemma.

Remark 5. This relative property is weaker than (tr, r, q', s)-mostly-denseness of $\widetilde{G}$: $\{(r, q)^V$-neighbor-dense sets in $\widetilde{V}\} \subset \{(r, q)^{\widetilde{V}}$-neighbor-dense sets in $\widetilde{V}\})$.

The Analysis. Now we use the method in [2] to show regular resolution lower bound on $\widetilde{G}$. The key part is Lemmas 8 and 9 in below.

Notation. Let $\mathfrak{p}$ denote the random path from P^* to axioms in *strategy on* Γ^*. A path (not necessarily from P^* to axioms) is *eligible* if it can be traveled through with nonzero probability. If Z is a node on a path $\mathfrak{p}$, $\mathfrak{p}(Z)$ denotes the sub-path from Z. For an eligible $\mathfrak{p}$, similar to Definition 1, let

$$\mathfrak{p}_1 = \{ (l, v) \mid (l, v)^{yes} \text{ is answered along } \mathfrak{p} \}, \text{ and similarly } \mathfrak{p}_0; \tag{38}$$
$$\text{rand}(\mathfrak{p}) = \{ (l, v) \mid (l, v)? \text{ is answered randomly along } \mathfrak{p}\}. \tag{39}$$

$\mathfrak{p}_0(l) := \{v \mid (l, v) \in \mathfrak{p}_0\}$. A subset of $\{(l, v) \mid v \in \widetilde{V}_l, l \in [\widetilde{k}]\}$ is called a *query set.*

Definition 10. *Let* X *be a query set. A path* $\mathfrak{p}$ *is* X^{yes}*-compatible if* $X \cap \mathfrak{p}_0 = \emptyset$, *and is* X^{no}*-compatible if* $X \cap \mathfrak{p}_1 = \emptyset$.

So, if Γ^* is regular then $\mathfrak{p}_1 \cap \mathfrak{p}_0 = \emptyset$, meaning $\mathfrak{p}$ is $\mathfrak{p}_1^{yes}$- and $\mathfrak{p}_0^{no}$-compatible.

It is easy to verify: any eligible path $\mathfrak{p}$ to axioms must end in a *clique axiom*

$$C_l := \bigvee_{v \in V_l} x_v \qquad l \in [\,\widetilde{k}\,]. \tag{40}$$

Lemma 7. *If $\mathfrak{p}$ is an eligible path to axiom C_l in (40), then along $\mathfrak{p}$ there is no forgotten-forced answer to l. In particular, $\mathfrak{p}$ is X^{no}-compatible for $X = \{l\} \times \widetilde{V}_l$.*

Proof. By regularity.

So it suffices to upper bound the probability $\Pr[\,\mathfrak{p}$ ends in $C_l]$, $\forall l \in [\widetilde{k}]$, which is done by the following two lemmas. Note Lemma 8 actually holds without assuming regularity.

Lemma 8. *For any query set X and eligible path $\mathfrak{q}$ from P^* to Z,*

$$\Pr\Big[\mathfrak{p}(Z) \text{ is } X^\theta\text{-compatible}, \ |\mathrm{rand}(\mathfrak{p}(Z)) \cap X| \geq a \ \big| \ \mathfrak{p} \supset \mathfrak{q}\Big] \leq \begin{cases} p^a, & \text{if } \theta = yes, \\ (1-p)^a, & \text{if } \theta = no. \end{cases}$$

Proof. We prove for $\theta = no$; the other is the same. Suppose $\mathfrak{p}$ is in the support of the event in the Lemma. On $\mathfrak{p}(Z)$, any query (l,v)? with $(l,v) \in X$ must be answered No by compatibility. Let $\Pr_{\mathfrak{q},Z,a}$ denote the probability in the lemma (X fixed). When Z is an axiom then $a = 0$ so the conclusion is obvious.

We pass the probability $\Pr_{\mathfrak{q},Z,a}$ to the one or two possible successor(s) of Z, hence use reverse-induction on the length of $\mathfrak{q}$. Suppose the query at Z is (l_1, v_1)?. If $(l_1, v_1) \notin X$ or the answer is a forced-No (which can be decided given $\mathfrak{q}, Z$), then the probability passes to the successor(s) with a unchanged. Otherwise, the answer is a random-No, and $\Pr_{\mathfrak{q},Z,a} = (1-p) \cdot \Pr_{\mathfrak{q}',Z',a-1}$, where $\mathfrak{q}'$ extends $\mathfrak{q}$ by $Z \to Z'$, and Z' is the unique possible successor. The inductive hypothesis on $\mathfrak{q}'$ completes the proof.

Lemma 9. $\forall l \in [\,\widetilde{k}\,]$,

$$\Pr[\,\mathfrak{p} \text{ ends in axiom } C_l, \ (\forall Z \text{ on } \mathfrak{p}) \ |Z_1| < r/2\,] < |\Gamma^*|^2 \cdot n^{-\epsilon k/3t^2 - 1}. \tag{41}$$

Proof. (cf. [2]) Due to item (1) in Stage II's strategy, there are at most $\widetilde{k}$ Yes-answers along any support of $\mathfrak{p}$. Given such a $\mathfrak{p}$, divide it into consecutive segments $\mathfrak{p}^1 \cup ...\mathfrak{p}^{2t}$, such that $|(\mathfrak{p}^i)_1| \leq \lceil \frac{\widetilde{k}}{2t} \rceil \leq tr/2$, $\forall i \in [2t]$. Here recall $(\mathfrak{p}^i)_1$ is defined by (38). Below we consider $(\mathfrak{p}^i)_0(l)$; note by choice of l, $\bigcup_{i \in [2t]} (\mathfrak{p}^i)_0(l) = \widetilde{V}_l$.

We claim that one of $(P^i)_0(l)$, say $(\mathfrak{p}^{i^*})_0(l)$, is $(r,q)^V$-neighbor-dense (similar to lemma 1). This can be seen by contradiction: otherwise, they give a union of $2t$ many sets of size r, together having $< q \cdot 2t$ many common neighbors in $\widetilde{V}_l$ - contradicting Lemma 5. Fix such an i^* for $\mathfrak{p}$.

Let Z, Z' be the start and end nodes of $\mathfrak{p}^{i^*}$ (decided by $\mathfrak{p}$). For simplicity, denote (Z, Z') by pair($\mathfrak{p}$), and let $A = \mathrm{Im}(Z_1) \cup \mathrm{Im}((\mathfrak{p}^{i^*})_1)$. Abbreviate the event

"$\mathfrak{p}$ ends in C_l, and $(\forall P \text{ on } \mathfrak{p}) \ |P_1| < r/2$" (i.e. the event in the lemma)

as $\mathfrak{p}^<$. Since $\mathfrak{p}$ ends in C_l, by regularity of Γ^*, $(\mathfrak{p}^{i*})_0(l) = Z'_0(l) \backslash Z_0(l)$. So,

$$\text{LHS of (41)} = \Pr[\,\mathfrak{p}^<, \, |\hat{N}_{\boldsymbol{Z'}_0 \backslash \boldsymbol{Z}_0}(\boldsymbol{A})| \geq q'\,] + \Pr[\,\mathfrak{p}^<, \, |\hat{N}_{\boldsymbol{Z'}_0 \backslash \boldsymbol{Z}_0}(\boldsymbol{A})| < q'\,] \quad (42)$$

$$= \sum_{Z,Z' \in \Gamma} \left(\Pr[\,\mathfrak{p}^<, \, \text{pair}(\mathfrak{p}) = (Z, Z'), |\hat{N}_{Z'_0 \backslash Z_0}(\boldsymbol{A})| \geq q'\,]\right.$$

$$\left. + \Pr[\,\mathfrak{p}^<, \text{pair}(\mathfrak{p}) = (Z, Z'), |\hat{N}_{Z'_0 \backslash Z_0}(\boldsymbol{A})| < q'\,]\right).$$

For fixed $(Z, Z') \in \Gamma$, we bound the above two terms separately.

First Term. By Lemma 7, any No-answer in $(\mathfrak{p}^i)_0(l)$ is random or edge-forced. By definition of A, the $\geq q'$ many No-answers to $\hat{N}_{Z'_0 \backslash Z_0}(\boldsymbol{A})$ along $\mathfrak{p}^{0,i*}(l)$ are all random. Also, by Lemma 7, any path to C_l is X^{no}-compatible, with $X := \{l\} \times \tilde{V}$. So the event of this term implies event

$$E := \text{``}\mathfrak{p} \text{ is } X^{no}\text{-compatible, } |\text{rand}(\mathfrak{p}) \cap X| \geq q'.\text{''}$$

By Lemma 8 (with $Z \leftarrow P^*$),

$$\Pr[E] \leq (1-p)^{q'} < \exp(-pq') < \exp(-n^{1-2\epsilon-20\delta tr}/(64tk^2)) < n^{-\epsilon k} \quad (43)$$

by (30) since $\delta tr < \epsilon$, $k < n^{1/3-40\epsilon}$ and that $\epsilon > (\log n)^{-1/3}$.

Second Term. By choice of i^*, $Z'_0 \backslash Z_0$ is $(r, q)^V$-neighbor-dense. Now $|A| \leq r/2 + tr/2 < tr$. By (tr, r, q', s)-mostly-denseness of G and Lemma 6, $\exists S \subset \tilde{V}$ of size $\leq s$ s.t. $|A \cap S| \geq r$. As $|\text{Im}(Z_1)| \leq r/2$ in the event $\mathfrak{p}^<$, if let $\boldsymbol{S}_1 = \text{Im}((\mathfrak{p}^{i*})_1) \cap S$ then $\mathfrak{p}^< \Rightarrow |\boldsymbol{S}_1| \geq r/2$. Therefore, as every Yes-answer is random, this term is bounded by

$$\sum_{S_1 \subset S, \, |S_1| = r/2} \Pr[\,\{l_1\} \times S_1 \subset \mathfrak{p}(Z)_1 \cap \text{rand}(\mathfrak{p}(Z))]. \quad (44)$$

For any fixed S_1, this is $< p^{r/2}$ by Lemma 8 (where the compatibility condition is from the fact after Definition 10). Now $\binom{s}{\frac{r}{2}} p^{r/2} < (2et^2 n^{-\epsilon})^{k/t^2} < n^{-\epsilon k/3t^2 - 10}$, by the choice of s, p in (30).

The lemma follows by a union bound over $Z, Z' \in \Gamma^*$ in (42).

Theorem 6 is now a straightforward corollary.

Proof. (*of Theorem* 6) Recall G is $(8tr, 4tq)^{block}$-neighbor-dense and (tr, r, q', s)-mostly-dense, and Γ is $\frac{1}{t}$-irregular resolution w.r.t. the canonical partition. By Claim 3, we only need to bound $|\Gamma^*| (\leq |\Gamma|)$. Consider an eligible path $\mathfrak{p}$ down from P^*. If for some Q on $\mathfrak{p}$, $|Q_1| \geq r/2$, we call $\mathfrak{p}$ *type-1*; otherwise it is *type-2*.

For a type-1 $\mathfrak{p}$, fix such a node Q. $\mathfrak{p}$ is Q_1^{yes}-compatible (by $Q_1 \subset \mathfrak{p}_1$ and the fact after Def. 10), $|Q_1| \geq \frac{r}{2}$. Yes-answers are random in Stage II so Lemma 8 applies to the sub-path from P^* to Q (with $X \leftarrow Q_1$). By a union bound over all possible Q's, this implies a type-1 path appears w.p. $\leq |\Gamma^*| \cdot p^{\frac{r}{2}} < |\Gamma^*| \cdot n^{-\epsilon k/t^2}$.

For a type-2 $\mathfrak{p}$, by Lemma 9 it appears w.p. $< k|\Gamma^*|^2 n^{-\epsilon k/3t^2 - 1}$ (unioned over $l \in [k]$). Together, type 1,2 appear with probability 1, so $|\Gamma^*| \geq n^{\epsilon k/6t^2}$.

5   Conclusion and Open Problems

We proved the $\exp(\Omega(k^{1-\epsilon}))$ resolution lower bound for $Clique_{block}(G,k)$ on random graphs, for $k < n^{1/3}$. We also defined the model of a-irregular resolution, discussed its relative power to regular and general resolution and extended the $n^{\Omega(k)}$ lower bound to this model. Some open problems are in order.

1. Prove the $n^{\Omega(k)}$ lower bound for general resolution. This improvement (from 2^k to n^k) is especially meaningful for small values of k, say $O(\log n)$.

2. Are there candidate families separating $\Omega(1)$-irregular model from general resolution? A possible starting point is to note that the concept of *block-width* has appeared in special forms in the study of many interesting CNFs (see e.g. [3,9]), either with or without lifting (although it seems unclear, in this context, how useful the lifting technique is; see Remark 4). Regarding the relation with the model in [7], does their SETH result hold for our unblocked model?

3. Extend the 2^k-type result to stronger systems, for example, *Res(k)* (where k has a completely different meaning) and algebraic systems like *Polynomial Calculus* and *Cutting Planes*. Does it hold for resolution but with other pseudo-random graphs (e.g. *Ramsey graphs* [15])?

Acknowledgment. I am indebted to Alexander Razborov for many helpful communications and feedback on the early draft. My thanks also go to Aaron Potechin, Jakob Nordström, and Ilario Bonacina, for various comments and references, and to the anonymous referees for their extensive feedback and suggestions that undoubtedly help improve readability.

References

1. Alekhnovich, M., Johannsen, J., Pitassi, T., Urquhart, A.: An exponential separation between regular and general resolution. In: Proceedings of the Thiry-Fourth Annual ACM Symposium on Theory of Computing, pp. 448–456. ACM (2002)
2. Atserias, A., Bonacina, I., de Rezende, S.F., Lauria, M., Nordström, J., Razborov, A.: Clique is hard on average for regular resolution. In: Proceedings of the 50th Annual ACM SIGACT Symposium on Theory of Computing, pp. 866–877 (2018)
3. Atserias, A., Müller, M.: Automating resolution is NP-hard. J. ACM (JACM) **67**(5), 1–17 (2020)
4. Beame, P., Impagliazzo, R., Sabharwal, A.: The resolution complexity of independent sets and vertex covers in random graphs. Comput. Complex. **16**(3), 245–297 (2007). https://doi.org/10.1007/s00037-007-0230-0
5. Beyersdorff, O., Galesi, N., Lauria, M.: Parameterized complexity of DPLL search procedures. ACM Trans. Comput. Log. (TOCL) **14**(3), 20 (2013)
6. Bollobás, B., Erdös, P.: Cliques in random graphs. In: Mathematical Proceedings of the Cambridge Philosophical Society, vol. 80, pp. 419–427. Cambridge University Press (1976)
7. Bonacina, I., Talebanfard, N.: Strong ETH and resolution via games and the multiplicity of strategies. Algorithmica **79**(1), 29–41 (2016). https://doi.org/10.1007/s00453-016-0228-6

8. Garg, A., Göös, M., Kamath, P., Sokolov, D.: Monotone circuit lower bounds from resolution. In: Proceedings of the 50th Annual ACM SIGACT Symposium on Theory of Computing, pp. 902–911. ACM (2018)
9. Göös, M., Koroth, S., Mertz, I., Pitassi, T.: Automating cutting planes is NP-hard. In: Proceedings of the 52nd Annual ACM SIGACT Symposium on Theory of Computing, pp. 68–77 (2020)
10. Göös, M., Pitassi, T.: Communication lower bounds via critical block sensitivity. SIAM J. Comput. **47**(5), 1778–1806 (2018)
11. Hajiaghayi, M.T., Khandekar, R., Kortsarz, G.: Fixed parameter inapproximability for clique and setcover in time super-exponential in opt. arXiv preprint arXiv:1310.2711 (2013)
12. Haken, A.: The intractability of resolution. Theor. Comput. Sci. **39**, 297–308 (1985)
13. Hastad, J.: Clique is hard to approximate within $n^{1-\epsilon}$. In: Proceedings of 37th Conference on Foundations of Computer Science, pp. 627–636. IEEE (1996)
14. Huynh, T., Nordstrom, J.: On the virtue of succinct proofs: amplifying communication complexity hardness to time-space trade-offs in proof complexity. In: Proceedings of the Forty-Fourth Annual ACM Symposium on Theory of Computing, pp. 233–248 (2012)
15. Lauria, M., Pudlák, P., Rödl, V., Thapen, N.: The complexity of proving that a graph is Ramsey. Combinatorica **37**(2), 253–268 (2017). https://doi.org/10.1007/s00493-015-3193-9
16. Nešetřil, J., Poljak, S.: On the complexity of the subgraph problem. Commentationes Mathematicae Universitatis Carolinae **026**, 2 (1985)
17. Pudlák, P.: Proofs as games. Am. Math. Mon. **107**(6), 541–550 (2000)
18. Raz, R., McKenzie, P.: Separation of the monotone NC hierarchy. In: Proceedings 38th Annual Symposium on Foundations of Computer Science, pp. 234–243. IEEE (1997)
19. Razborov, A.: Lower bounds on the monotone complexity of some Boolean functions. In: Soviet Math. Doklady, vol. 31, pp. 354–357 (1985). English translation
20. Razborov, A.A.: Proof complexity of pigeonhole principles. In: Kuich, W., Rozenberg, G., Salomaa, A. (eds.) DLT 2001. LNCS, vol. 2295, pp. 100–116. Springer, Heidelberg (2002). https://doi.org/10.1007/3-540-46011-X_8
21. Rossman, B.: On the constant-depth complexity of k-clique. In: Proceedings of the Fortieth Annual ACM Symposium on Theory of Computing, pp. 721–730. ACM (2008)
22. Stålmarck, G.: Short resolution proofs for a sequence of tricky formulas. Acta Informatica **33**(3), 277–280 (1996)
23. Vassilevska, V.: Efficient algorithms for clique problems. Inf. Process. Lett. **109**(4), 254–257 (2009)
24. Vinyals, M., Elffers, J., Johannsen, J., Nordström, J.: Simplified and improved separations between regular and general resolution by lifting. In: Pulina, L., Seidl, M. (eds.) SAT 2020. LNCS, vol. 12178, pp. 182–200. Springer, Cham (2020). https://doi.org/10.1007/978-3-030-51825-7_14
25. Zuckerman, D.: Linear degree extractors and the inapproximability of max clique and chromatic number. In: Proceedings of the Thirty-Eighth Annual ACM Symposium on Theory of Computing, pp. 681–690. ACM (2006)

On Closed-Rich Words

Olga Parshina[1]([✉]) and Svetlana Puzynina[1,2]

[1] Saint Petersburg State University, Saint Petersburg, Russia
parolja@gmail.com, s.puzynina@gmail.com
[2] Sobolev Institute of Mathematics, Novosibirsk, Russia

Abstract. A word is called closed if it has a prefix which is also its suffix and there is no internal occurrences of this prefix in the word. In this paper we study the maximal number of closed factors in a word of length n. We show that it is quadratic and give lower and upper bounds for a constant.

Keywords: Closed word · Return word · Rich word

1 Introduction

Various questions that concern counting factors of a specific form in a word of length n have been studied in combinatorics on words. Several studies have been devoted to the words that are extremal with respect to the proportion of factors with a given property. For example, an extensive study has been performed on the problem of counting the maximal repetitions (runs) in a word of length n. It has been shown in [17] that the maximal number of runs in a word is linear, and it was conjectured to be n. Subsequently, there was much research performed to find the bound [8]. Recently, the conjecture has been proved with a remarkably simple argument, considering numerous attempts to solve it [4]. We remark that questions about counting regular factors in a word are often non-trivial. For example, the problem of bounding the number of distinct squares in a string: A.S. Fraenkel and J. Simpson showed in 1998 [15] that a string of length n contains at most $2n$ distinct squares, and conjectured that the bound is actually n. After several improvements, the bound of $\frac{11}{6}n$ has been proved in [9], but the conjecture remains unsolved.

A related problem concerns counting palindromic factors. It is easy to see that a word of length n can contain at most $n + 1$ distinct palindromes (see e.g. [10]). Such words are called *rich in palindromes*, and there also exist infinite words such that all their factors are rich. Words rich in palindromes have been characterized in [16]. Words containing few palindromes were studied in [6,14]. Recently some related questions about counting generalizations of palindromes have been studied, e.g. privileged factors [22] and k-abelian palindromes [7].

The first author is supported by Ministry of Science and Higher Education of the Russian Federation, agreement 075–15–2019–1619. The second author is supported by Foundation for the Advancement of Theoretical Physics and Mathematics "BASIS".

© Springer Nature Switzerland AG 2021
R. Santhanam and D. Musatov (Eds.): CSR 2021, LNCS 12730, pp. 381–394, 2021.
https://doi.org/10.1007/978-3-030-79416-3_23

We are interested in counting the factors that are called *closed*. A finite word is called *closed* if it has length ≤ 1 or it is a complete first return to some proper factor, i.e. it starts and ends with the same word that has no other occurrences but these two. Otherwise the word is called *open*. The terminology closed and open was introduced by G. Fici in [12]; for more information on closed words see [13]. The notion of closed word is actually the same as the notion of complete return word. The name return word is usually referred to factors of an infinite word and is used to study its properties. It can be regarded as a discrete analogue of the first return map in dynamical systems. For example, F. Durand characterized primitive substitutive words using the notion of a return word [11]. Return words also provide a nice characterization of the family of Sturmian words [24]. The explicit formulae for the functions of closed and open complexities for the family of Arnoux–Rauzy words, which encompass Sturmian words, were obtained in [21]. In [20], the authors prove a refinement of the Morse–Hedlund theorem (see [19]) providing a criterion of periodicity of an infinite word in terms of closed and open complexities.

The concept of closed factor has recently found applications in string algorithms. The *longest closed factor array* (LCF array) of a string x stores for every suffix of x the length of its longest closed prefix. It was introduced in [3] in connection with closed factorizations of a string. Among other things, the authors presented algorithms for the factorization of a given string into a sequence of longest closed factors and for computing the longest closed factor starting at every position in the string. In [5], the authors present the algorithm of reconstructing a string from its LCF array. See also [1] for some generalizations.

It is easy to show that each word of length n contains at least $n + 1$ distinct closed factors [2]. In this paper, we study closed-rich words, i.e., words containing the maximal number of distinct closed factors among words of the same length. We prove an upper bound of $\sim \frac{3-\sqrt{5}}{4}n^2$ on this number (see Theorem 1'), and we show that a word can contain $\sim \frac{n^2}{6}$ distinct closed factors (see Proposition 3). We also extend the notion of closed-rich words to infinite words, requiring that each factor contains a quadratic number of distinct closed factors. We find a sufficient condition on an infinite word to be closed-rich (see Proposition 5), and provide some families of infinite closed-rich words.

2   Preliminaries

Let $\mathbb{A}$ be a finite set called an alphabet. A finite or an infinite word $w = w_0 w_1 \cdots$ on $\mathbb{A}$ is a finite or infinite sequence of symbols from $\mathbb{A}$. For a finite word $w = w_0 \cdots w_{n-1}$, its *length* is $|w| = n$. We let ε denote the empty word, and we set $|\varepsilon| = 0$. A word v is a *factor* of a finite or an infinite word w if there exist words u and y such that w can be represented as their concatenation $w = uvy$. If $u = \varepsilon$, then v is a prefix, and if $y = \varepsilon$, then v is a suffix of w. If a finite word w has a proper prefix v which is also its suffix, then v is called a *border* of w. If the longest border of a word w occurs in w only twice (as a prefix and as a suffix), then w is *closed*. By convention, if w is the empty word or a letter, then it is closed.

It is not hard to see that a word of length n contains at least $n + 1$ distinct closed factors; G. Fici and Z. Lipták characterized words having exactly $n + 1$ closed factors [2] . In the same paper they showed that there are words containing $\Theta(n^2)$ many distinct closed factors. The example they provided is a binary word with $\sim \frac{n^2}{32}$ closed factors. We say that a finite word w is *closed-rich* if it contains at least as many distinct closed factors as any other word of the same length and on the alphabet of the same cardinality.

If there exists an integer t such that for each i ($i < |w| - t$ in the case w is finite) we have $w_{i+t} = w_i$, then t is called a *period* of w. Let $s = \frac{|w|}{t}$ and let u be the prefix of w of length t. We say that w has *exponent* s and write $w = u^s$. The notation $w = u^{k+}$ means that w has exponent $s > k$ for an integer k. The word u is called the *fractional root* of w. The word w is *primitive* if its only integer exponent is 1. Hereinafter we always assume t to be the shortest period of w, and thus, s to be the largest exponent of w.

The following properties follow directly from the definitions.

Proposition 1. *Any word with exponent at least two is closed.*

Proposition 2. *Let w be a word of exponent 3 and of length n. Then all its factors of length at least $\frac{2n}{3}$ are closed, and moreover, all of them except for one of length $\frac{2n}{3}$ are unioccurrent.*

De Bruijn graph of order n on an alphabet $\mathbb{A}$ is the directed graph whose set of vertices (resp. edges) consists of all words over $\mathbb{A}$ of length n (resp. $n + 1$). There is a directed edge from u to v labeled w if u is a prefix of w and v a suffix of w. We call a Hamiltonian path in this graph a *de Bruijn word*.

Proposition 3. *Let $n = 3 \cdot |\mathbb{A}|^k$ for an integer k, and v be a de Bruijn word of length $\frac{n}{3}$. Then $w = v^3$ has $\sim \frac{n^2}{6}$ distinct closed factors.*

Proof. Due to Proposition 2, all factors that are longer than $\frac{2n}{3}$ are closed and distinct (there are $\frac{n^2}{18}$ of those). All words of length $\frac{n}{3} + \log(\frac{n}{3}) \le l \le \frac{2n}{3}$ are also closed with corresponding border of length $l - n/3$ (there are $\sim (\frac{n}{3})^2$ of distinct factors of these lengths).

If a factor of length less than $\frac{n}{3} + \log(\frac{n}{3})$ is closed, then its border is shorter than $\log(\frac{n}{3})$, because all factors of de Bruijn word of length at least $\log(\frac{n}{3})$ are unioccurrent. Thus, there are not more than $\frac{n}{3} \cdot \log(\frac{n}{3})$ closed factors that are shorter than $\frac{n}{3} + \log(\frac{n}{3})$.

The construction from the previous proposition gives only words of length $n = 3 \cdot |\mathbb{A}|^k, k \ge 0$, but it could be easily modified to other lengths. For lengths n divisible by 3 we can e.g. take cubes of prefixes of de Bruijn words (we omit technical details here). Words of lengths not divisible by 3 can be obtained by shortening a word of next length divisible by 3—clearly, a prefix of length 1 or 2 can add at most linear number of closed factors. However, if we change one letter in the middle of a word, the total number of closed factors can change dramatically:

Example 1. Let us show that the number of closed factors can change from linear to quadratic when changing only one letter in a word. It is easy to see that the word $a^n b a^n b a^n$ has quadratic number of closed factors. After replacing the leftmost occurrence of b with a, we obtain the word $a^n a a^n b a^n$ with linear number of closed factors.

For a finite word w, we let $\mathrm{Cl}(w)$ denote the number of distinct closed factors of w. We now provide a trivial upper bound on $\mathrm{Cl}(w)$.

Proposition 4. *For each word w of length $n \geq 7$, one has $\mathrm{Cl}(w) \leq \frac{n^2}{4}$.*

Proof. For a word u and a letter a, let us denote by t the longest repeated suffix of ua, and z denote the longest repeated suffix of t. Clearly, the number of new closed factors ending in the last letter of ua is at most $|t| - |z| \leq \left\lfloor \frac{|ua|}{2} \right\rfloor$ when u is non-empty. For $n = 0$ we have one closed factor (the empty word), for $n = 1$ we add one closed factor (a letter). So, building w letter by letter, we get at most $2 + \sum_{i=2}^{n} \lfloor \frac{i}{2} \rfloor = 2 + \frac{n(n+1)}{4} - \lceil \frac{n}{2} \rceil$ closed factors in w. The claim follows.

In the next section we prove a tighter upper bound with the leading coefficient $\frac{3-\sqrt{5}}{4} \approx 0.19$. We believe it can be improved to $\frac{1}{6} = 0.1\overline{6}$.

3 Finite Closed-Rich Words

The main goal of this section is to prove Theorem 1 providing an upper bound on the number of closed factors that a finite word can contain. We start with some auxiliary lemmas. Sometimes it will be convenient for us to consider cyclic words. For a normal word w, we can consider a corresponding cyclic word as the class of all its cyclic shifts. Then by a closed factor of a cyclic word we mean a closed factor of some its shift.

Lemma 1. *Let u be a primitive finite word of length k, then its cyclic square has at most k^2 distinct closed factors.*

Proof. Let $\hat{u}$ be the cyclic square of u with the first letter $\hat{u}_0 = u_0$. The following observations constitute the proof of the lemma. Basically, we count (left) borders giving rise to distinct closed words.

   Each occurrence of a factor of $\hat{u}$ is a (left) border of at most one closed factor. Since $\hat{u}$ is a cyclic square, in order to count borders giving rise to distinct closed words, we can only consider the factors of $\hat{u}$ starting in its first half $\hat{u}_0 \cdots \hat{u}_{k-1} = u_0 \cdots u_{k-1}$ (borders starting in the second half of $\hat{u}$ give rise to the same closed words). The border of a closed factor of $\hat{u}$ cannot be longer than k, otherwise u is not a primitive word. The number of factors of $\hat{u}$ that start in its first half and are not longer than k is k^2. The statement follows.

Lemma 2. *Let w be a word with exponent at least 3, i.e., if we denote its period by k and its length by n, then $n \geq 3k$. Then the number of closed factors in the word is at most $k^2 + (n - 3k)k + \frac{1}{2}(k+1)k$.*

Proof. We count closed factors by their lengths: k^2 is the upper bound for the number of closed factors of length at most $2k$ (given by the bound for a cyclic word of length $2k$ from Lemma 1); $(n-3k)k$ counts k closed factors of each length $2k+1, \ldots, n-k$; and $\frac{1}{2}(k+1)k$ counts long ones as the sum of an arithmetic progression (k factors for length $n - k + 1$, $k - 1$ for $n - k + 1$, $\ldots$, and 1 for length n).

Corollary 1. *For words of length n and of exponent greater than 3, we have less than $\frac{n^2}{6}$ factors asymptotically.*

Proof. We estimate the bound from Lemma 2: for words with exponent $t \geq 3$ we count closed factors and get the function $c(t) = n^2(\frac{1}{t} - \frac{3}{2t^2})$. The maximum value of $c(t)$ is $\frac{n^2}{6}$ and it is achieved for $t = 3$.

We say that a finite word w' is a *cyclic shift* (or a *conjugate*) of a finite word w if there exist words u and v such that $w = uv$ and $w' = vu$.

Lemma 3. *Let w be a word of exponent $\alpha \geq 3$ and v its primitive root, so that $w = v^\alpha$. Then for any cyclic shift v' of v, the word v'^α contains the same number of closed factors as w.*

Proof. The proof follows from the counting in the proof of Lemma 2: the set of closed factors consists of all long factors (their numbers are clearly the same since we only take lengths into account) and short closed factors which are also closed factors of the cyclic square, so their sets are the same.

The following example shows that the statement of Lemma 3 does not hold for squares. Moreover, it is possible that taking a cyclic shift of the root changes the number of closed factors in a word from linear to quadratic.

Example 2. The number of closed factors in $a^{n/2}ba^nba^{n/2}$ is quadratic, while in its cyclic shift a^nba^nb it is linear.

The set of distinct closed factors of a word w can naturally be split into two sets, the set of closed words of length at least 2, which have border, and the set of closed words of length at most 1, i.e., letters and the empty word. We let $\mathrm{Cl}'(w)$ denote the number of closed words of length at least 2 ("long" closed factors), and $\mathrm{Cl}^0(w)$ denote the number of "short" closed factors, so that $\mathrm{Cl}(w) = \mathrm{Cl}'(w) + \mathrm{Cl}^0(w)$.

Now we can state the main result of this section.

Theorem 1. *For a finite word w of length n, the following holds:*

$$\mathrm{Cl}'(w) < \frac{3 - \sqrt{5}}{4}n^2 + \frac{\sqrt{5} - 1}{4}n.$$

Clearly, since $\mathrm{Cl}(w) = \mathrm{Cl}'(w) + |\mathbb{A}| + 1 \leq \mathrm{Cl}'(w) + n + 1$, we can rewrite the statement of Theorem 1 as follows.

Theorem 1'. *For a finite word w of length n, the following holds:*

$$\mathrm{Cl}(w) < \frac{3 - \sqrt{5}}{4} n^2 + \frac{3 + \sqrt{5}}{4} n + 1.$$

For a finite word w of length n, we denote its prefix of length $n - 1$ by w^-. The following lemma constitutes the key part of the proof of Theorem 1.

Lemma 4. *Let w be a word of length n. If $\mathrm{Cl}'(w) - \mathrm{Cl}'(w^-) \geq Cn$ for some $\frac{1}{3} < C \leq \frac{1}{2}$, then $\mathrm{Cl}'(w) \leq \frac{(1-C)^2}{2} n^2 + \frac{1-C}{2} n$.*

Proof. In the proofs of the lemma and of Theorem 1, we only talk about long closed factors (of length at least 2). Let t denote the longest repeated suffix of w, z denote the longest repeated suffix of t, and c be the letter preceding the last occurrence of z in w, so that cz is the shortest unrepeated suffix of t. Clearly,

$$\mathrm{Cl}'(w) - \mathrm{Cl}'(w^-) \leq |t| - |z|. \tag{1}$$

Two cases are possible: the last and the penultimate occurrences of t in w might intersect, or not. If they intersect, we have a power as a suffix of w; let l be the period of this power. In this case we denote the suffix of t of length l by x. If they do not intersect, we set $x = t$.

There are several possibilities of how the word w can look like depending on whether the occurrences of t intersect or not and whether the penultimate occurrence of t starts from the beginning of the word or not. The case $w = x^{3+}$ is treated in Lemma 2 and is not considered here. Other possibilities follow.

1. $w = x^s$, $2 \leq s < 3$ where x is the shortest period of w;
2. $w = xvx$ for a non-empty word v;
3. $w = uxvx$ for non-empty words u, v;
4. $w = ux^s$ for $2 \leq s < 3$ and a non-empty word u.

Let us treat each case separately.

1. Let $w = x^s$, $2 \leq s < 3$. We use the following notation: $|x| = l$, $r = n - 2l$. In this case the longest border t ending in the last position of w is of length $r + l$. We should consider two subcases depending on the length of cz.

(a) The shortest unrepeated suffix cz of t occurs in w three times (see Fig. 1). Due to inequality (1), in this case $l \geq Cn$.

Let us count closed factors of w; it is easier to do by counting their borders. Let us count borders that are suffixes of the rightmost occurrences of closed factors of w. It is easy to see that every factor $w_i \cdots w_j$ for $i \geq l$ and $j \geq l + r$ is a border of the factor $w_{i-l} \ldots w_j$. So, such border (suffix of a rightmost occurrence of a closed factor) can start at the earliest at position l. There are $\sum_{i=l}^{n-1} (n-i) - \sum_{i=l}^{l+r-1} (l+r-i)$ such borders in w. The second sum stands for borders

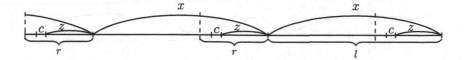

Fig. 1. The case $w = x^s, 2 \le s < 3, |w| = n = 2l + r, |cz| \le r$.

occurring inside $w_l \dots w_{l+r-1}$; each one of them defines the same closed factor of w as the borders occurring in $w_{n-r} \dots w_{n-1}$. Since the word w is a power, borders appearing in $w_0 \dots w_{l-1}$ do not define any closed factors different from the ones already counted in the sum above. Hence,

$$\mathrm{Cl}'(w) \le \sum_{i=l}^{n-1}(n-i) - \sum_{i=l}^{l+r-1}(l+r-i) = \frac{1}{2}((n-l)(n-l+1) - (r+1)r)$$

$$= \frac{1}{2}((n-l)^2 + (n-l) - r^2 - r) = -\frac{3}{2}l^2 + \left(n + \frac{1}{2}\right)l.$$

The last equality is due to the equality $n - l = l + r$. This expression reaches its maximum when $l = \frac{1}{3}n + \frac{1}{6}$ and is equal to $\frac{1}{6}n^2 + \frac{1}{6}n + \frac{1}{24}$. Thus, in this case $\mathrm{Cl}'(w) \le \frac{1}{6}n^2 + \frac{1}{6}n + \frac{1}{24}$.

(b) The shortest unrepeated suffix cz of t occurs in w twice (Fig. 2).

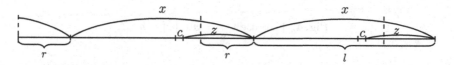

Fig. 2. The case $w = x^s, 2 \le s < 3, |w| = n = 2l + r, |cz| > r$.

Due to inequality (1), in this case $l + r - |z| \ge Cn$.

Let us count the borders that are suffixes of the rightmost occurrences of closed factors of w. All of them are located in the suffix of length $n - l - |z|$ of w. In a similar to 1.(a) way, using the relations $r = n - 2l$ and $|z| \le l + r - Cn$ we obtain the following.

$$\mathrm{Cl}'(w) \le \sum_{i=l+r-|z|}^{n-1}(n-i) - \sum_{i=l+r-|z|}^{n-l-1}(n-l-i) = \frac{l^2}{2} + l|z| + \frac{l}{2} \le -\frac{l^2}{2} + (1-C)nl + \frac{l}{2}.$$

This function reaches its maximum when $l = (1 - C)n + \frac{1}{2}$. Since we deal with integers in all inequalities, we obtain $\mathrm{Cl}'(w) \le \frac{(1-C)^2}{2}n^2 + \frac{(1-C)}{2}n$.

2. If $w = xvx$ for some word v (Fig. 3), inequality (1) gives $l - |z| \ge Cn$.

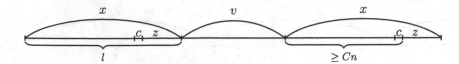

Fig. 3. The case $w = xvx, |w| = n$.

Let us count borders that are suffixes of the rightmost occurrences of closed factors of w. In fact, any such border cannot contain the first occurrence of cz in w (the one starting with $w_{l-|z|}$). Thus, we have the following inequality.

$$\mathrm{Cl}'(w) \le \sum_{j=l-|z|}^{n-1} (n-j) = \frac{1}{2}(n-l+|z|)(n-l+|z|+1)$$

$$\le \frac{1}{2}(n-Cn)(n-Cn+1) \le \frac{(1-C)^2}{2}n^2 + \frac{(1-C)}{2}n.$$

3. Let $w = uxvx$ for non-empty words u, v (Fig. 4).

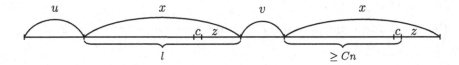

Fig. 4. The case $w = uxvx, |w| = n$.

Due to inequality (1), in this case $l \ge Cn + |z|$. We will count borders of closed factors that start in u, and the borders that are factors of xvx separately.

Let us show that the border of a closed factor starting in u cannot contain x as a factor. Suppose it is the case and consider the next occurrence of x in this closed word. It must end before the index $n - Cn$, otherwise its suffix cz is not unioccurrent in x. By the same reasoning it cannot start before $|u| + Cn$. The distance between these two points is $n - Cn - |u| - Cn = (1 - 2C)n - |u|$. Provided with $1/3 < C \le 1/2$ we have $1 - 2C < C$. Thus, $(1 - 2C)n - |u| < Cn - |u| < Cn < l$, and x cannot be placed in the indicated gap.

We will make a more generous rounding up saying that the number of borders beginning in u is not greater than the number of factors in the prefix of w of length $n - Cn$. This number is $\sum_{j=0}^{|u|-1} (n - Cn - j) = n|u| - Cn|u| - \frac{|u|^2}{2} + \frac{|u|}{2}$.

The number of closed factors in xvx is not greater than $\sum_{i=|u|+l-|z|}^{n-1} (n-i) = \frac{1}{2}(n - |u| - l + |z|)(n - |u| - l + |z| + 1) \le \frac{1}{2}(n - |u| - Cn)(n - |u| - Cn + 1)$.

Summing the two expressions, we obtain $\mathrm{Cl}'(w) \leq \frac{(1-C)^2}{2}n^2 + \frac{(1-C)}{2}n$.

4. Let $w = ux^s$, $2 < s < 3$. Analogously to the first case we should consider two situations, when the shortest unrepeated suffix cz is shorter than r, and when it is longer than r.

(a) The shortest unrepeated suffix cz of t occurs in x^s three times.

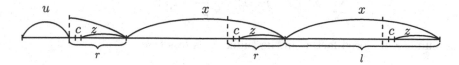

Fig. 5. The case $w = ux^s, 2 \leq s < 3, |cz| \leq r, |w| = n$.

Here $n = |u| + 2l + r$, and $l \geq Cn$.

The border of every closed factor that starts in u must end before the index $|u| + l - 1$, otherwise x would occur in x^2 three times, what contradicts the assumption on x to be the smallest period of x^s (see e.g. Problem 8.1.6. in [18]). Thus, the number of closed factors starting in u is not greater than the sum

$$\sum_{j=0}^{|u|-1} (|u| + l - 1 - j) = \frac{|u|^2}{2} - \frac{|u|}{2} + l|u|.$$

We will count the rightmost occurrences of borders of closed words that are factors of x^s. It is enough to count the borders that end in the last occurrence of x in w. Again, the borders cannot start before the index $n - l - r$, otherwise x would occur in x^2 three times. Thus, the number of closed factors of x^s is not greater than

$$\sum_{j=n-l-r}^{n-1} (n - j) - \frac{r(r+1)}{2} = \frac{l^2}{2} + \frac{l}{2} + lr.$$

Summing the two expressions, we get $\mathrm{Cl}'(w) \leq \frac{|u|^2}{2} - \frac{|u|}{2} + l|u| + \frac{l^2}{2} + \frac{l}{2} + lr = l(|u| + 2l + r) - \frac{3}{2}l^2 + \frac{l}{2} - \frac{|u|}{2} + \frac{|u|^2}{2}$. Using the inequality $|u| \leq n - 2l$, we obtain

$$\mathrm{Cl}'(w) < ln - \frac{3}{2}l^2 + \frac{l}{2} - \frac{n}{2} + l + \frac{n^2}{2} + 2l^2 - 2ln = \frac{(n-l)^2}{2} - \frac{n}{2} + \frac{3l}{2} < \frac{(1-C)^2}{2}n^2 + \frac{n}{4}.$$

(b) The shortest unrepeated suffix cz of t occurs in x^s twice (Fig. 6).

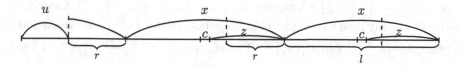

Fig. 6. The case $w = ux^s, 2 \leq s < 3, |w| = n$.

In this case $l+r-|z| \geq Cn$. Since $n = |u|+2l+r$, we have $n-l-|u|-|z| \geq Cn$, and thus,

$$l + |u| + |z| \leq (1 - C)n. \tag{$*$}$$

As in the case 4.(a), the number of closed factors starting in u is

$$\sum_{j=0}^{|u|-1} (|u| + l - j - 1) = \frac{|u|^2}{2} + l|u| - \frac{|u|}{2}.$$

The number of closed factors in the suffix x^s is less than

$$\sum_{j=n-l-|z|}^{n-1} (n - j) - \frac{|z|(|z| + 1)}{2} = \frac{l^2}{2} + l|z| + \frac{l}{2}.$$

Thus, the number of closed factors in this case is

$$\mathrm{Cl}'(w) \leq \frac{|u|^2}{2} + l|u| - \frac{|u|}{2} + \frac{l^2}{2} + l|z| + \frac{l}{2} = l(l + |u| + |z|) - \frac{l^2}{2} + \frac{|u|^2}{2} + \frac{l}{2} - \frac{|u|}{2}.$$

Using $(*)$ and $|u| \leq n - 2l$ we obtain

$$\mathrm{Cl}'(w) \leq (1 - C)nl + \frac{3l^2}{2} + \frac{n^2}{2} - 2ln + \frac{l}{2} - \frac{n}{2} + l$$

$$= \frac{3l^2}{2} + \left(-Cn - n + \frac{3}{2}\right)l + \frac{n^2}{2} - \frac{n}{2} < \frac{3l^2}{2} - (C + 1)nl + \frac{n^2}{2} + \frac{n}{4}.$$

This expression reaches its minimum when $l = \frac{1+C}{3}n$.

Let us compare the values at the endpoints of its domain $\left(Cn, \frac{n}{2}\right)$.

When $l = Cn$, the expression is $\frac{(1-C)^2}{2}n^2 + \frac{n}{4}$. When $l = \frac{n}{2}$, the expression is $\frac{3-4C}{8}n^2 + \frac{n}{4}$. Let us note that the latter value is smaller than the former for all possible values of C. Thus, in this case $\mathrm{Cl}'(w) \leq \frac{(1-C)^2}{2}n^2 + \frac{n}{4}$.

The maximal bound among the obtained ones is $\frac{(1-C)^2}{2}n^2 + \frac{(1-C)}{2}n$.

We let $\mathrm{pref}_i(w)$ and $\mathrm{suff}_i(w)$ denote the prefix and the suffix of w of length i, respectively.

Proof (of Theorem 1). Let us suppose that w is a word of length n with more than $\frac{Cn(n+1)}{2} + \left(\frac{\sqrt{5}}{2} - 1\right)n$ long closed factors, for some $C \in (\frac{1}{3}, \frac{1}{2})$. Clearly, for the proof we only need to consider C in these bounds due to Propositions 3 and 4. In other words, $\sum_{j=1}^{n}(\mathrm{Cl}'(\mathrm{pref}_j(w)) - \mathrm{Cl}'(\mathrm{pref}_{j-1}(w)) \geq \frac{Cn(n+1)}{2} + \left(\frac{\sqrt{5}}{2} - 1\right)n$. It would mean that one of the terms in the sum, let us say the i-th one, is at least Ci.

Let us consider the largest index i satisfying $\mathrm{Cl}'(\mathrm{pref}_i(w)) - \mathrm{Cl}'(\mathrm{pref}_{i-1}(w)) \geq Ci$, i.e., for all $j > i$, we have $\mathrm{Cl}'(\mathrm{pref}_j(w)) - \mathrm{Cl}'(\mathrm{pref}_{j-1}(w)) < Cj$. Using

Lemma 4, we can bound the number of distinct long closed factors of w the following way.

$$\mathrm{Cl}'(w) < \frac{(1-C)^2}{2}i^2 + \frac{(1-C)}{2}i + \sum_{j=i+1}^{n} Cj = \frac{(1-C)^2}{2}i^2 + \frac{(1-C)}{2}i$$

$$+ C\frac{(n+i+1)(n-i)}{2} = \left(\frac{C^2}{2} - \frac{3C}{2} + \frac{1}{2}\right)i^2 + \left(\frac{1}{2} - C\right)i + \frac{C}{2}n^2 + \frac{C}{2}n.$$

The last expression in the formula is smaller than $\frac{Cn(n+1)}{2} + \left(\frac{\sqrt{5}}{2} - 1\right)n$ when $C \geq \frac{3-\sqrt{5}}{2}$. Thus, the word w has less than $\frac{Cn(n+1)}{2} + \left(\frac{\sqrt{5}}{2} - 1\right)n$ long closed factors. Moreover, this expression reaches its maximum when $C = \frac{3-\sqrt{5}}{2}$.

Thus, the number of long closed factors in a word of length n is bounded by $\frac{3-\sqrt{5}}{4}n^2 + \frac{\sqrt{5}-1}{4}n$.

4   Infinite Rich Words

We say that an infinite word w is *closed-rich* if there is a constant C such that for each $n \in \mathbb{N}$ each factor of w of length n contains at least Cn^2 distinct closed factors. We remark that in this definition we do not require the constant to be optimal; however, a natural question is optimizing this constant (see Question 2). In this section, we show that infinite closed-rich words exist, and provide some families of examples.

Proposition 5. *Let w be an infinite word, and let $C > 2$, $\alpha < 1$ be two constants. If for each n each factor of w of length n contains a factor of exponent at least C and of length of period at least αn, then w is infinite closed-rich.*

Proof. Let v be a factor of w of length n. By the condition of the lemma, it contains a factor u of period $k \geq \alpha n$ and of exponent $C' \geq C > 2$, hence its length $l = C'k \geq C\alpha n$.

To count closed factors of u we use the following two observations. All factors of u of length greater than $l - k$ are distinct. Each factor of u of length at least $2k$ has exponent at least 2 and hence is closed by Proposition 1.

If $C' \leq 3$, there are at least $\sum_{j=2k}^{l}(l - j) \geq \frac{(C'-2)^2}{2}k^2 \geq \frac{(C-2)^2(\alpha n)^2}{2}$.

If $C' > 3$, then, in addition to the closed factors longer than $l - k$, the word u has k distinct closed factors of each length between $2k$ and $l - k$. Thus, there are at least $(l - 3k)k + \sum_{j=l-k}^{l}(l - j) \geq (C' - 3)k^2 + \frac{k^2}{2} = \frac{C'-5}{2}k^2 > \frac{(\alpha n)^2}{2} = \frac{\alpha^2 n^2}{2}$.

Therefore, the constant in the definition of infinite rich words is given by $\min(\frac{\alpha^2}{2}, \frac{(C-2)^2\alpha^2}{2})$.

A *morphism* φ is a map on the set of all finite words on the alphabet $\mathbb{A}$ such that $\varphi(uv) = \varphi(u)\varphi(v)$ for all finite words u, v on $\mathbb{A}$. The domain of the morphism φ can be naturally extended to infinite words by $\varphi(w_0 w_1 w_2 \cdots) = \varphi(w_0)\varphi(w_1)\varphi(w_2) \cdots$. A morphism φ is primitive if there exists a positive integer l such that the letter a occurs in the word $\varphi^l(b)$ for each pair of letters $a, b \in \mathbb{A}$. A fixed point of a morphism φ is an infinite word w such that $\varphi(w) = w$.

Example 3. Let w be a fixed point of the morphism $\varphi : a \to abbba, b \to abbbb$. We show that it is infinite closed-rich. Indeed, each block $\varphi^k(c)$ for $c \in \{a, b\}$ has length 5^k and contains a cube with the period 5^{k-1}. Clearly, each factor of length at least $2 \cdot 5^k - 1$ contains a block $\varphi^k(c)$. The maximal length, where we cannot guarantee the next block $\varphi^{k+1}(c)$, is $2 \cdot 5^{k+1} - 2$. Thus, we can apply Proposition 5 with $C = 3$ and $\alpha = \frac{1}{(2 \cdot 5^2)} = 0.02$. We remark that this word can also be seen as a Toeplitz word [23] with pattern $baaa?$, and that this construction can be easily generalized to other morphic and Toeplitz words.

A large subclass of the family of Sturmian words also turns out to be infinite closed-rich. *Sturmian words* are usually defined as infinite words with the smallest possible number of distinct factors of each length among aperiodic words ($n + 1$ factor of each length $n \geq 1$). Sturmian words are known to be rich in palindromes [16]. Sturmian words admit various characterizations. The one we use here is via standard words. Let $(d_1, d_2, \ldots, d_n, \ldots)$ be a sequence of integers, with $d_1 \geq 0$ and $d_n > 0$ for $n > 1$. To such a sequence, we associate a sequence $(s_n)_{n \geq 1}$ of words by

$$s_{-1} = 1, \qquad s_0 = 0, \qquad s_n = s_{n-1}^{d_n} s_{n-2} \quad (n \geq 1).$$

The sequence $(s_n)_{n \geq -1}$ is a *standard sequence*, and the sequence $(d_1, d_2, \ldots)$ is its *directive sequence*. This sequence defines a limit: $s = \lim_{n \to \infty} s_n$.

It is well known that a word is Sturmian if and only if it has the same set of factors as the limit of some standard sequence. For more information on Sturmian words we refer to Chap. 2 of [18].

Proposition 6. *Let D be an integer and s be a Sturmian word with a directive sequence $(d_i)_{i \geq 1}$ such that $d_i \leq D$ for every $i \geq 1$. Then s is infinite closed-rich.*

Proof. For a sketch of proof, consider a factorization of s to standard words s_k and s_{k-1}. If $d_{k+1} \geq 2$, then between each two consecutive s_{k-1} in the factorization there are at least two s_k (in fact, d_k or d_{k+1}); that gives a power s_k^α with $2 < \alpha \leq D + 1$. In the case when $d_{k+1} = 1$, we can factorize s_k to $s_{k-1}^{d_k} s_{k-2}$ and get a power s_{k-1}^γ with $2 < \gamma \leq D + 1$. By Proposition 5, one can see that the quadratic number of closed factors is achieved inside these powers. Here the constant C from the definition of infinite closed-rich words depends on D. Note that if the sequence of d_i's is unbounded, then the Sturmian word can be made not rich due to presence of powers with big exponents and relatively short periods.

5   Concluding Remarks

In this paper, we showed an upper bound $\sim \frac{3 - \sqrt{5}}{4} n^2$ for the number of closed factors in a finite word of length n, and we constructed examples with $\sim \frac{n^2}{6}$. We conjecture that this gives an asymptotic bound:

Conjecture 1. Finite closed-rich words contain $\sim \frac{n^2}{6}$ closed factors.

Based on numerical experiments, we also conjecture that they are cubes or words of exponent close to 3. Table 1 shows the maximal number of closed factors that a binary word of given length can contain.

Table 1. The maximal number of closed factors for binary words of length n.

n	1	2	3	4	5	6	7	8	9	10	11	12
$\max_{|w|=n} \mathrm{Cl}(w)$	2	3	4	6	8	10	12	15	18	21	25	29

n	13	14	15	16	17	18	19	20	21	22	23	24
$\max_{|w|=n} \mathrm{Cl}(w)$	33	37	42	48	54	60	66	72	79	86	93	101

Similar calculations have been made in [2], but there were some errors. We made corrections for values $n = 16, 17, \ldots, 20$. For the lengths we computed, closed-rich words are cubes or close to cubes by their structure. For example, the word $u = (100101)^3$ of length 18 has 60 closed factors (one can easily verify it). The word u^- has 54 closed factors, and the word $u1$ has 66 closed factors. The following question remains open even for binary alphabet:

Question 1. What is the exact formula for the maximal number of distinct closed factors in a finite rich word?

In the last section we defined infinite closed-rich words as words for which there exists a constant C such that each factor of length n contains Cn^2 distinct closed factors. A question that naturally arises is that of optimizing the constant:

Question 2. What is the supremum of the constant for infinite rich words?

References

1. Alamro, H., Alzamel, M., Iliopoulos, C.S., Pissis, S.P., Sung, W.K., Watts, S.: Efficient identification of k-closed strings. Int. J. Found. Comput. Sci. **31**(05), 595–610 (2020)
2. Badkobeh, G., Fici, G., Lipták, Z.: On the number of closed factors in a word. In: Dediu, A.-H., Formenti, E., Martín-Vide, C., Truthe, B. (eds.) LATA 2015. LNCS, vol. 8977, pp. 381–390. Springer, Cham (2015). https://doi.org/10.1007/978-3-319-15579-1_29
3. Badkobeh, G., et al.: Closed factorization. Discret. Appl. Math. **212**, 23–29 (2016)
4. Bannai, H., Tomohiro, I., Inenaga, S., Nakashima, Y., Takeda, M., Tsuruta, K.: The "runs" theorem. SIAM J. Comput. **46**(5), 1501–1514 (2017)
5. Bannai, H., et al.: Efficient algorithms for longest closed factor array. In: Iliopoulos, C., Puglisi, S., Yilmaz, E. (eds.) SPIRE 2015. LNCS, vol. 9309, pp. 95–102. Springer, Cham (2015). https://doi.org/10.1007/978-3-319-23826-5_10
6. Brlek, S., Hamel, S., Nivat, M., Reutenauer, C.: On the palindromic complexity of infinite words. Int. J. Found. Comput. Sci. **15**, 293–306 (2004)

7. Cassaigne, J., Karhumäki, J., Puzynina, S.: On k-abelian palindromes. Inf. Comput. **260**, 89–98 (2018)
8. Crochemore, M., Ilie, L., Tinta, L.: The "runs" conjecture. Theor. Comput. Sci. **412**(27), 2931–2941 (2011)
9. Deza, A., Franek, F., Thierry, A.: How many double squares can a string contain? Discret. Appl. Math. **180**, 52–69 (2015)
10. Droubay, X., Justin, J., Pirillo, G.: Episturmian words and some constructions of de Luca and Rauzy. Theor. Comput. Sci. **255**(1), 539–553 (2001)
11. Durand, F.: A characterization of substitutive sequences using return words. Discret. Math. **179**(1–3), 89–101 (1998)
12. Fici, G.: A classification of trapezoidal words. In: Ambroz, P., Holub, S., Masáková, Z. (eds.) Words 2011. EPTCS, vol. 63, pp. 129–137 (2011)
13. Fici, G.: Open and closed words. Bull. Eur. Assoc. Theor. Comput. Sci. **123**, 140–149 (2017)
14. Fici, G., Zamboni, L.Q.: On the least number of palindromes contained in an infinite word. Theor. Comput. Sci. **481**, 1–8 (2013)
15. Fraenkel, A.S., Simpson, J.: How many squares can a string contain? J. Comb. Theor. Ser. A **82**(1), 112–120 (1998)
16. Glen, A., Justin, J., Widmer, S., Zamboni, L.Q.: Palindromic richness. Eur. J. Comb. **30**(2), 510–531 (2009)
17. Kolpakov, R.M., Kucherov, G.: Finding maximal repetitions in a word in linear time. In: FOCS 1999, pp. 596–604. IEEE Computer Society (1999)
18. Lothaire, M.: Algebraic Combinatorics on Words. Encyclopedia of Mathematics and its Applications, vol. 90. Cambridge University Press (2002)
19. Morse, M., Hedlund, G.A.: Symbolic dynamics. Am. J. Math. **60**(4), 815–866 (1938)
20. Parshina, O., Postic, M.: Open and closed complexity of infinite words. arXiv:2005.06254 (2020)
21. Parshina, O., Zamboni, L.Q.: Open and closed factors in Arnoux-Rauzy words. Adv. Appl. Math. **107**, 22–31 (2019)
22. Peltomäki, J.: Introducing privileged words: privileged complexity of Sturmian words. Theor. Comput. Sci. **500**, 57–67 (2013)
23. Toeplitz, O.: Ein beispiel zur theorie der fastperiodischen funktionen. Math. Ann. **98**, 281–295 (1928)
24. Vuillon, L.: A characterization of Sturmian words by return words. Eur. J. Comb. **22**(2), 263–275 (2001)

Shelah-Stupp's and Muchnik's Iterations Revisited

Paweł Parys[✉] [iD]

Institute of Informatics, University of Warsaw, Warsaw, Poland
parys@mimuw.edu.pl

Abstract. Iteration is a model-theoretic construction that replicates a given structure in an infinite, tree-like way. There are two variants of iteration: basic iteration (a.k.a. Shelah-Stupp's iteration), and Muchnik's iteration. The latter has an additional unary predicate (not present in basic iteration), which makes the structure richer. These two variants lead to two hierarchies of relational structures, generated from finite structures using MSO-interpretations and either basic iteration or Muchnik's iteration. Caucal and Knapik (2018) have shown that the two hierarchies coincide at level 1, and that every level of the latter hierarchy is closed under basic iteration (which in particular implies that the former hierarchy collapses at level 1). We prove the same results using a different, significantly simpler method.

Keywords: Shelah-Stupp's iteration · Muchnik's iteration · Hierarchy · MSO decidability · Infinite relational structures

1 Introduction

The story about iterations starts with the monadic second-order (MSO) logic. While defining sets of words or trees, this logic is equiexpressive with finite-state automata, thus defines regular languages. The MSO logic is of course decidable over finite structures. Moreover, it have been shown decidable over natural numbers with successor [2,12,24], and over the infinite complete binary tree [18].

After these fundamental results, a long series of other examples of infinite structures with decidable MSO theory has emerged. They include natural numbers with successor and an additional unary predicate [13,21,23], transition graphs of pushdown automata [16] or higher-order pushdown automata [3,4], HR-equational hypergraphs [7] and VR-equational hyperhraphs [9], prefix-recognizable graphs [5], and trees generated by higher-order recursion schemes [15,17].

Besides the particular classes of structures with decidable MSO theory, some operations that preserve MSO-decidability, creating a more complex structure

Work supported by the National Science Centre, Poland (grant no. 2016/E/ST6/00041).

R. Santhanam and D. Musatov (Eds.): CSR 2021, LNCS 12730, pp. 395–405, 2021.
https://doi.org/10.1007/978-3-030-79416-3_24

from a simpler one, were proposed; such operations are called MSO-*compatible*. Among those, we have generalised unions of Shelah [20], MSO-interpretations (or, more generally, MSO-transductions) [1,8,14], unfolding of directed graphs into trees [11], and iteration. We concentrate here on the last operation on this list, namely on iteration.

While iterating a structure, we create infinitely many copies of it, and we organize them in a shape of a tree. Children of every node of the tree are indexed by elements of the structure itself. Thus, elements of the iterated structure can be seen as nonempty (finite) words whose letters are elements of the original structure: the last letter is an element in one of the copies, and the prefix without the last letter is an index of this copy. Relations are preserved within each copy, and a new binary "son" relation is added, connecting an element in one copy with all elements belonging to a child of this copy indexed by the former element (i.e., every word w with words of the form wa). This construction is first mentioned by Shelah [20], who refers to an unpublished paper of Stupp [22], which contains the proof of the fact that this operation is indeed MSO-compatible; thus the above operation is called *Shelah-Stupp's iteration*, or *basic iteration*. The resulting structure may be extended by a unary "clone" predicate, which holds in the unique element of every copy that is an index of this copy among its siblings (i.e., in words of the form waa); this way we obtain Muchnik's iteration, which is also MSO-compatible. This result is attributed to Muchnik, but was presented by Semenov [19] and Walukiewicz [25].

The MSO-compatible operations allow to create hierarchies of classes of structures with decidable MSO theories, containing most of the examples mentioned so far. In the most known Caucal's hierarchy of directed graphs [4], one starts from finite graphs, and repeatedly applies unfolding and MSO-interpretation. In an equivalent definition [3], unfolding is replaced by Muchnik's iteration. We consider here a generalization of the latter hierarchy from directed graphs to arbitrary relational structures. Thus, starting from finite structures, we construct structures on the next level of the hierarchy by applying Muchnik's iteration to structures on the previous level, followed by arbitrary MSO-interpretations. Another hierarchy can be constructed using basic iteration instead of Muchnik's iteration.

The latter two hierarchies were considered by Caucal and Knapik [6], who prove that

- the two hierarchies coincide at level 1, and
- every level of the hierarchy involving Muchnik's iteration is closed under basic iteration (which in particular implies that the hierarchy involving basic iteration collapses at level one).

We prove the same results using a different, significantly simpler method.

The proof of Caucal and Knapik is quite indirect: it utilizes prefix-recognizable structures, as well as higher-order pushdown automata. Moreover, their constructions are rather involved. We, instead, work directly with the definition of iterations. Recalling that elements of the iterated structure can be seen as words, our approach is based on a very simple idea saying that a word of words can be encoded in a word (we only need some separator to mark the glue points). Based on this idea, we prove that (modulo existence of the aforementioned separator) a composition of a Muchnik's iteration (applied first) with a basic iteration (applied later) can be encoded in a single Muchnik's iteration. From this statement, the aforementioned second main result of Caucal and Knapik (i.e., closure under basic iteration) easily follows.

2   Preliminaries

The MSO Logic and MSO-Interpretations. A *signature* Σ (of a relational structure) is a finite set of relation names, $R_1, \ldots, R_r$, together with a natural number called an *arity* assigned to each of the names. A *(relational) structure* $\mathcal{S} = (U^{\mathcal{S}}, R_1^{\mathcal{S}}, \ldots, R_r^{\mathcal{S}})$ over such a signature Σ is a set $U^{\mathcal{S}}$, called the *universe*, together with *relations* $R_i^{\mathcal{S}}$ over $U^{\mathcal{S}}$, for all relation names in the signature; the arity of the relations is as specified in the signature.

We assume two countable sets of variables: $\mathcal{V}^{FO}$ of first-order variables (denoted using lowercase letters $x, y, \ldots$) and $\mathcal{V}^{MSO}$ of set variables (denoted using capital letters $X, Y, \ldots$). Atomic formulae are

- $R(x_1, \ldots, x_n)$, where R is a relation name of arity n (coming from a fixed signature Σ), and $x_1, \ldots, x_n$ are first-order variables, and
- $x \in X$, where x is a first-order variable, and X a set variable.

Formulae of the *monadic second-order logic*, MSO, are built out of atomic formulae using the Boolean connectives $\vee, \wedge, \neg$, first-order quantifiers $\exists x$ and $\forall x$ for $x \in \mathcal{V}^{FO}$, and set quantifiers $\exists X$ and $\forall X$ for $X \in \mathcal{V}^{MSO}$. We use the standard notion of *free variables*.

In order to evaluate an MSO formula φ over a signature Σ in a relational structure $\mathcal{S}$ over the same signature, we also need a *valuation* ν, which is a partial function that maps

- variables $x \in \mathcal{V}^{FO}$ to elements of the universe of $\mathcal{S}$, and
- variables $X \in \mathcal{V}^{MSO}$ to subsets of the universe of $\mathcal{S}$.

The valuation should be defined at least for all free variables of φ. We write $\mathcal{S}, \nu \models \varphi$ when φ is *satisfied* in $\mathcal{S}$ with respect to the valuation ν; this is defined by induction on the structure of φ, in the expected way.

We write $\varphi(x_1, \ldots, x_n)$ to denote that the free variables of φ are among $x_1, \ldots, x_n$. Then, given elements $u_1, \ldots, u_n$ in the universe of a structure $\mathcal{S}$, we say that $\varphi(u_1, \ldots, u_n)$ is satisfied in $\mathcal{S}$ if φ is satisfied in $\mathcal{S}$ under the valuation mapping x_i to u_i for all $i \in \{1, \ldots, n\}$.

An MSO-*interpretation* $I = (\delta, (\varphi_R)_{R \in \Sigma_2})$ from Σ_1 to Σ_2 consists of an MSO-formula $\delta(x)$ over Σ_1, and of MSO-formulae $\varphi_R(x_1, \ldots, x_n)$ over Σ_1, for every relation name $R \in \Sigma_2$, where n is the arity of R. Having such an MSO-interpretation, we can apply it to a structure $\mathcal{S}$ over Σ_1; we obtain a structure $I(\mathcal{S})$ over Σ_2, where the universe $U^{I(\mathcal{S})}$ consists of those elements v of the universe of $\mathcal{S}$ for which $\delta(v)$ is satisfied in $\mathcal{S}$, and where every relation $R^{I(\mathcal{S})}$ consists of the tuples $(v_1, \ldots, v_n) \in (U^{I(\mathcal{S})})^n$ for which $\varphi_R(v_1, \ldots, v_n)$ is satisfied in $\mathcal{S}$.

Iterations and Hierarchies. For a set A, by A^* (or A^+) we denote the set of all finite words (or all nonempty finite words, respectively) over alphabet A. In the sequel, we write $[a_1 a_2 \ldots a_k]$ for a word consisting of letters $a_1, a_2, \ldots, a_k$, and we use $\circ$ to denote concatenation of words (this notation allows us to unambiguously describe words of words).

Let $\mathcal{S} = (U^{\mathcal{S}}, R_1^{\mathcal{S}}, \ldots, R_r^{\mathcal{S}})$ be a relational structure over a signature Σ, and let $\sharp, \& \notin \Sigma$ be new relation names, where $\sharp$ is binary and $\&$ unary. The *basic (a.k.a. Shelah-Stupp's) iteration of $\mathcal{S}$*, denoted $\mathcal{S}^\sharp$, is a relational structure over $\Sigma \cup \{\sharp\}$, where

- the universe $U^{\mathcal{S}^\sharp}$ is $(U^{\mathcal{S}})^+$ (i.e., the set of nonempty words whose letters are elements of $U^{\mathcal{S}}$),
- the relation $R_i^{\mathcal{S}^\sharp}$ contains all tuples of the form $(w \circ [a_1], \ldots w \circ [a_n])$ such that $(a_1, \ldots, a_n) \in R_i^{\mathcal{S}}$ (where $w \in (U^{\mathcal{S}})^*$ and $a_1, \ldots, a_n \in U^{\mathcal{S}}$), for every $i \in \{1, \ldots, r\}$, and
- the relation $\sharp^{\mathcal{S}^\sharp}$ contains all pairs of the form $(w, w \circ [a])$ (where $w \in (U^{\mathcal{S}})^+$ and $a \in U^{\mathcal{S}}$).

The *Muchnik's iteration of $\mathcal{S}$*, denoted $\mathcal{S}^{\sharp,\&}$, is a relational structure over $\Sigma \cup \{\sharp, \&\}$, where

- the universe $U^{\mathcal{S}^{\sharp,\&}}$ and the relations $R_i^{\mathcal{S}^{\sharp,\&}}$ and $\sharp^{\mathcal{S}^{\sharp,\&}}$ are defined as in $\mathcal{S}^\sharp$,
- the relation $\&^{\mathcal{S}^{\sharp,\&}}$ contains all elements of the form $w \circ [aa]$ (where $w \in (U^{\mathcal{S}})^*$ and $a \in U^{\mathcal{S}}$).

In the sequel, we also use \$ as an alternative for the $\sharp$ symbol (and then we write $\mathcal{S}^\$$ instead of $\mathcal{S}^\sharp$).

Example 2.1. This example is borrowed from Caucal and Knapik [6]. Consider a structure $\mathcal{S}$ with universe $\{1, 2, 3\}$, and with the following binary relations α, β, depicted by arrows:

$$1 \xleftarrow{\beta} 3$$
$$\alpha \searrow 2 \nearrow \alpha$$

A fragment of Muchnik's iteration $\mathcal{S}^{\sharp,\&}$ has the following shape (the basic iteration $\mathcal{S}^\sharp$ looks similarly, except that the $\&$ predicate should be removed):

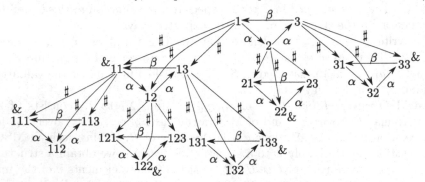

$\square$

For every $n \in \mathbb{N}$ we define two families of relational structures, $\mathfrak{hgt}_n$ and $\mathfrak{hgt}_n^{\text{basic}}$, as follows:

$$\mathfrak{hgt}_0 = \mathfrak{hgt}_0^{\text{basic}} = \text{finite relational structures,}$$

$$\mathfrak{hgt}_{n+1} = \{I(\mathcal{S}^{\sharp, \&}) \mid \mathcal{S} \in \mathfrak{hgt}_n \wedge I \text{ is an MSO-interpretation}\},$$

$$\mathfrak{hgt}_{n+1}^{\text{basic}} = \{I(\mathcal{S}^{\sharp}) \mid \mathcal{S} \in \mathfrak{hgt}_n^{\text{basic}} \wedge I \text{ is an MSO-interpretation}\}.$$

More formally, the classes contain also all structures isomorphic to the structures present in the above definition.

Recall that the composition of two MSO-interpretations is again an MSO-interpretation [10], that there exists an identity MSO-interpretation, and that a structure MSO-interpreted in a finite structure is again finite. It follows that the above definition (where iterations and MSO-interpretations appear alternatingly, starting from an iteration) covers any sequence of iterations and MSO-interpretations applied to a finite structure; then the level n counts the number of iterations.

3   Equality on Level 1

In this section we concentrate on the first main result of the paper:

Theorem 3.1 (cf. Caucal and Knapik [6, Corollary 14]**).** *The classes* $\mathfrak{hgt}_1$ *and* $\mathfrak{hgt}_1^{\text{basic}}$ *coincide.*

In order to prove this result, Caucal and Knapik use a passage through prefix-recognizable structures. Below, we give a straightforward, direct proof.

Proof (Theorem 3.1). Clearly $\mathfrak{hgt}_1^{\text{basic}}$ is contained in $\mathfrak{hgt}_1$. In order to prove the other inclusion, consider a structure in $\mathfrak{hgt}_1$. It is of the form $I(\mathcal{S}^{\sharp, \&})$ for some finite structure $\mathcal{S}$ and MSO-interpretation I. We are going to prove that there is a finite structure $\mathcal{T}$ and an MSO-interpretation J such that $\mathcal{S}^{\sharp, \&} = J(\mathcal{T}^{\sharp})$. This shows that $I(\mathcal{S}^{\sharp, \&})$, which equals $I(J(\mathcal{T}^{\sharp}))$, belongs to $\mathfrak{hgt}_1^{\text{basic}}$, because a composition of two MSO-interpretations is again an MSO-interpretation.

As $\mathcal{T}$ we take $\mathcal{S}$ enriched with additional predicates (i.e., unary relations). Namely, for each element a of the universe, we have a predicate P_a that holds only in this element (it is important that the universe is finite, so we need only finitely many new predicates). The interpretation J leaves unchanged the universe, all the relations from the signature of $\mathcal{S}$, and the relation $\sharp$; they are simply inherited from $\mathcal{T}^{\sharp}$. We only need to define in $\mathcal{T}^{\sharp}$ the "clone" predicate $\&$. It should hold in elements of the form $w \circ [aa]$. But such equality of the last two letters can be easily expressed in an MSO formula, by taking a disjunction over all possible elements a of the universe. Indeed, recall that the new P_a predicates check in $\mathcal{T}^{\sharp}$ whether the last letter is a, and that the $\sharp$ relation allows us to cut off the last letter. $\qquad\square$

4 Closure Under Basic Iteration

We now come to the second main result of the paper:

Theorem 4.1 (cf. Caucal and Knapik [6, Theorem 15]). *For every $n \geq 1$, the class $\mathfrak{hgr}_n$ is closed under basic iteration.*

In particular, because $\mathfrak{hgr}_1^{\text{basic}} = \mathfrak{hgr}_1$ (cf. Theorem 3.1), and because a composition of two MSO-interpretations is an MSO-interpretation, we obtain that $\mathfrak{hgr}_2^{\text{basic}} = \mathfrak{hgr}_1$, and likewise $\mathfrak{hgr}_n^{\text{basic}} = \mathfrak{hgr}_1$ for all $n \geq 1$.

The key point in our proof of Theorem 4.1 is that a composition of a Muchnik's iteration with a basic iteration can be encoded in a single Muchnik's iteration. As already said in the introduction, this amounts to encoding a word of words in a single word. In this encoding, we need a separator to be inserted between concatenated words. We thus consider an operation of adding a distinguished element (which will become the separator) to an arbitrary structure.

Let $\mathcal{S} = (U^{\mathcal{S}}, R_1^{\mathcal{S}}, \dots, R_r^{\mathcal{S}})$ be a relational structure over a signature Σ, and let $\dagger \notin \Sigma$ be a new unary relation name. The *single-element extension of* $\mathcal{S}$, denoted $\mathcal{S}_\dagger$, is a relational structure over $\Sigma \cup \{\dagger\}$, where

- the universe $U^{\mathcal{S}_\dagger}$ is $U^{\mathcal{S}} \uplus \{a_\dagger\}$, for some fresh element $a_\dagger$,
- $R_i^{\mathcal{S}_\dagger} = R_i^{\mathcal{S}}$ for all $i \in \{1, \dots, r\}$, and
- $\dagger^{\mathcal{S}_\dagger} = \{a_\dagger\}$.

In other words, we add a new element to the universe of $\mathcal{S}$, but not to any of the relations; the new unary predicate $\dagger$ holds only in the new element.

Below, the $\equiv$ symbol stands for isomorphism of structures.

Lemma 4.2. *Fix a signature Σ. There exists an MSO-interpretations I such that for every relational structure $\mathcal{S}$ over Σ,*

$$(\mathcal{S}^{\sharp,\&})^{\$} \equiv I((\mathcal{S}_\dagger)^{\sharp,\&}).$$

Proof. Let $\mathcal{S} = (U^{\mathcal{S}}, R_1^{\mathcal{S}}, \dots, R_r^{\mathcal{S}})$. Recall that the universe of the double iteration $(\mathcal{S}^{\sharp,\&})^{\$}$ is $((U^{\mathcal{S}})^+)^+$, and the universe of $(\mathcal{S}_\dagger)^{\sharp,\&}$ is $(U^{\mathcal{S}} \uplus \{a_\dagger\})^+$. We define an injective mapping $\mathsf{flat} \colon ((U^{\mathcal{S}})^+)^+ \to (U^{\mathcal{S}} \uplus \{a_\dagger\})^+$ by

$$\mathsf{flat}([w_1 w_2 \dots w_k]) = w_1 \circ [a_\dagger] \circ w_2 \circ [a_\dagger] \circ \cdots \circ [a_\dagger] \circ w_k.$$

It thus concatenates the words $w_1, \dots, w_k$ being letters of $[w_1 w_2 \dots w_k]$, inserting the separator $a_\dagger$ between them.

The universe-restricting formula δ of the interpretation I should select elements of the image of flat. These are words such that there are no two $a_\dagger$ letters in a row, and the first and the last letter are not $a_\dagger$. This property can be easily expressed in MSO (recall that the $\dagger$ predicate in $(\mathcal{S}_\dagger)^{\sharp,\&}$ checks whether the last letter is $a_\dagger$, and the $\sharp$ relation allows to cut off the last letter of a word).

The interpretation I should not change relations R_i in any way, because the last letter of the last letter of an element of $((U^{\mathcal{S}})^+)^+$ (taken into account by the relations R_i in $(\mathcal{S}^{\sharp,\&})^{\$})$ is mapped by flat to the last letter of an element of $(U^{\mathcal{S}} \uplus \{a_\dagger\})^+$ (taken into account by the relations R_i in $(\mathcal{S}_\dagger)^{\sharp,\&}$).

The relations $\sharp$ and $\&$ should remain unchanged as well. Indeed, the $\sharp$ relation in $(\mathcal{S}^{\sharp,\&})^{\$}$ contains pairs of the form $(v \circ [w], v \circ [w \circ [a]])$. They are mapped by flat to $(\mathsf{flat}(v) \circ [a_\dagger] \circ w, \mathsf{flat}(v) \circ [a_\dagger] \circ w \circ [a])$ (or just $(w, w \circ [a])$, if v is empty), which are exactly pairs contained in the relation $\sharp$ in $(\mathcal{S}_\dagger)^{\sharp,\&}$, while restricted to the image of flat. Likewise, the $\&$ predicate in $(\mathcal{S}^{\sharp,\&})^{\$}$ holds in elements of the form $(v \circ [w \circ [aa]])$. They are mapped by flat to $(\mathsf{flat}(v) \circ [a_\dagger] \circ w \circ [aa])$ (or just $(w \circ [aa])$, if v is empty), which are exactly the elements of the image of flat for which the $\&$ predicate holds in $(\mathcal{S}_\dagger)^{\sharp,\&}$.

Finally, the relation $\$$ in $(\mathcal{S}^{\sharp,\&})^{\$}$ contains pairs of the form $(v, v \circ [w])$, which are mapped by flat to $(\mathsf{flat}(v), \mathsf{flat}(v) \circ [a_\dagger] \circ w)$. Thus, the formula $\varphi_\$(x, y)$ in I should say that x is obtained from y by cutting off the suffix starting from the last $a_\dagger$. As for δ, this can be easily expressed in MSO. $\square$

Remark 4.3. In Lemma 4.2 it is important that Muchnik's iteration is applied first, that is, that we take $(\mathcal{S}^{\sharp,\&})^{\$}$ and not $(\mathcal{S}^{\sharp})^{\$,\&}$. Indeed, the "clone" predicate of the second iteration would say that the last two words w_{k-1}, w_k in an encoding $w_1 \circ [a_\dagger] \circ \cdots \circ [a_\dagger] \circ w_{k-1} \circ [a_\dagger] \circ w_k$ are equal. We are unable to say this in MSO.

Lemma 4.2 eliminates a composition of iterations, at the cost of using the single-element extension. We thus need to prove that the classes $\mathfrak{hgr}_n$ are closed under the latter operation:

Lemma 4.4. *For every $n \geq 1$, the class $\mathfrak{hgr}_n$ is closed under taking single-element extensions.*

To this end, we need the following lemma:

Lemma 4.5. *Fix a signature Σ. There exists an MSO-interpretation I such that for every relational structure $\mathcal{S}$ over Σ,*

$$(\mathcal{S}^{\sharp,\&})_\dagger \equiv I((\mathcal{S}_\dagger)^{\sharp,\&}).$$

Proof. Let $\mathcal{S} = (U^{\mathcal{S}}, R_1^{\mathcal{S}}, \ldots, R_r^{\mathcal{S}})$. Recall that $(\mathcal{S}^{\sharp,\&})_\dagger$ extends $\mathcal{S}^{\sharp,\&}$ by a single fresh element, while in $(\mathcal{S}_\dagger)^{\sharp,\&}$ the fresh element can be used anywhere as a letter of a word. We define an injective mapping $\mathsf{inj} \colon (U^{\mathcal{S}})^+ \uplus \{a_\dagger\} \to (U^{\mathcal{S}} \uplus \{a_\dagger\})^+$ (i.e., from the universe of $(\mathcal{S}^{\sharp,\&})_\dagger$ to the universe of $(\mathcal{S}_\dagger)^{\sharp,\&}$) by

$$\mathsf{inj}(w) = \begin{cases} [a_\dagger] & \text{if } w = a_\dagger, \\ w & \text{otherwise.} \end{cases}$$

The universe-restricting formula δ of the interpretation I should select elements of the image of inj. These are words not using $a_\dagger$ as a letter, plus the length-1 word $[a_\dagger]$. Of course we can select such words in MSO (where we can use the $\dagger$ predicate to check whether the last letter is $a_\dagger$, and the $\sharp$ relation to cut off the last letter of a word). All the relations R_i, as well as $\sharp$ and $\&$, should remain unchanged by the interpretation I. $\square$

The next two lemmata allow us to swap an MSO-interpretation with the operations of iteration or single-element extension.

Lemma 4.6. *Let I be an* MSO*-interpretation from a signature Σ to a signature Π. There exists an* MSO*-interpretation J such that for every relational structure $\mathcal{S}$ over Σ,*

$$(I(\mathcal{S}))^{\sharp} = J(\mathcal{S}^{\sharp}).$$

Proof. First, observe that we can write an MSO formula $\psi(X)$ saying that X is one of the copies in the iteration, (i.e., for some word w, the set X contains all words of the form $w \circ [a]$).

Let $I = (\delta, (\varphi_R)_{R \in \Sigma})$. Let $\delta'(x, X)$ be a formula obtained from $\delta(x)$ by relativizing to the set X (by saying that, we mean that all quantified objects should come from the set X, as well as x should belong to X). Then, let $\delta''(x) \equiv \exists X.(\psi(X) \wedge \delta'(x, X))$; this formula says that the last letter of x satisfies δ. Using δ'' we can easily write the universe-restricting formula of J, which should say that all letters of the considered word satisfy δ.

Likewise, we relativize every formula $\varphi_R(x_1, \ldots, x_n)$ to a set X (saying in particular that all $x_1, \ldots, x_n$ belong to X), obtaining a formula $\varphi'_R(x_1, \ldots, x_n, X)$. Then, we take $\varphi''_R(x_1, \ldots, x_n) \equiv \exists X.(\psi(X) \wedge \varphi'_R(x_1, \ldots, x_n, X))$. This formula says that all $x_1, \ldots, x_n$ belong to the same copy in the iteration, and that their last letters satisfy the formula φ_R; this is exactly the definition of R in $(I(\mathcal{S}))^{\sharp}$, so we can take φ''_R to the interpretation J.

Finally, J should leave the relation $\sharp$ unchanged. $\square$

The above lemma was also given by Caucal and Knapik [6, Lemma 16], but their proof contains a flaw. Namely, they propose to just make a conjunction of $\varphi_R(x_1, \ldots, x_n)$ with a formula saying that all $x_1, \ldots, x_n$ belong to the same copy in the iteration. This is not enough (for example, φ_R may contain a subformula, unrelated to the arguments $x_1, \ldots, x_n$, saying that all elements of the structure are connected by some relation α, which is true in some $\mathcal{S}$, but not in $\mathcal{S}^{\sharp}$); all quantifiers in a formula have to be relativized to the same copy, as in our proof.

Lemma 4.7. *Let I be an* MSO*-interpretation from a signature Σ to a signature Π. There exists an* MSO*-interpretation J such that for every relational structure $\mathcal{S}$ over Σ,*

$$(I(\mathcal{S}))_{\dagger} = J(\mathcal{S}_{\dagger}).$$

Proof. Let $I = (\delta, (\varphi_R)_{R \in \Sigma})$. We relativize the formulae δ and φ_R to elements not satisfying $\dagger$ (i.e., elements of the original structure $\mathcal{S}$), obtaining δ' and φ'_R; in particular, these formulae say that their arguments do not satisfy $\dagger$. As the universe-restricting formula of J we take $\delta'(x) \vee \dagger(x)$, and we use formulae φ'_R to define relations. Additionally, we leave the $\dagger$ predicate unchanged. $\square$

We can now finish the proofs of Lemma 4.4 and Theorem 4.1.

Proof (Lemma 4.4). Induction on n. The base case of $n = 0$ is trivial: the single-element extension of a finite structure is again finite. For the induction step, assume that $\mathfrak{hgr}_n$ is closed under taking single-element extensions, and consider a structure from $\mathfrak{hgr}_{n+1}$; we have to prove that the single-element extension of this structure also belongs to $\mathfrak{hgr}_{n+1}$. The structure is of the form $I(\mathcal{S}^{\sharp,\&})$ for some $\mathcal{S} \in \mathfrak{hgr}_n$, and for some MSO-interpretation I. First, by Lemma 4.7, we can write $(I(\mathcal{S}^{\sharp,\&}))_\dagger = J((\mathcal{S}^{\sharp,\&})_\dagger)$, for some MSO-interpretation J. Then, by Lemma 4.5, we have $(\mathcal{S}^{\sharp,\&})_\dagger \equiv K((\mathcal{S}_\dagger)^{\sharp,\&})$, for some MSO-interpretation K. By the induction hypothesis, $\mathcal{S}_\dagger \in \mathfrak{hgr}_n$, so $J(K((\mathcal{S}_\dagger)^{\sharp,\&})) \in \mathfrak{hgr}_{n+1}$. $\square$

Proof (Theorem 4.1). Consider a structure from $\mathfrak{hgr}_n$, where $n \geq 1$; we have to prove that the basic iteration of this structure also belongs to $\mathfrak{hgr}_n$. The structure is of the form $I(\mathcal{S}^{\sharp,\&})$ for some $\mathcal{S} \in \mathfrak{hgr}_{n-1}$, and for some MSO-interpretation I. First, by Lemma 4.6, we can write $(I(\mathcal{S}^{\sharp,\&}))^{\$} = J((\mathcal{S}^{\sharp,\&})^{\$})$, for some MSO-interpretation J. Then, by Lemma 4.2, we have $(\mathcal{S}^{\sharp,\&})^{\$} \equiv K((\mathcal{S}_\dagger)^{\sharp,\&})$, for some MSO-interpretation K. By Lemma 4.4 we know that $\mathcal{S}_\dagger \in \mathfrak{hgr}_{n-1}$, so $J(K((\mathcal{S}_\dagger)^{\sharp,\&})) \in \mathfrak{hgr}_n$. $\square$

Remark 4.8. In Lemma 4.2 we have shown that a composition of Muchnik's iteration with basic iteration can be encoded in the Muchnik's iteration of the single-element extension. Notice that by removing the "clone" predicate &, the same proof gives us another statement: a composition of two basic iterations can be encoded in a single basic iteration of the single-element extension.

Caucal and Knapik [6, Section 5] ask whether inside every $\mathfrak{hgr}_n$ class there exists a finer hierarchy, where one climbs up from one layer to the next layer via basic iteration. By the above, we know that two basic iterations can be rewritten using a single one; thus the answer to this question is negative, assuming that the classes would be closed under taking single-element extensions (which is a very natural assumption).

5 Conclusions

Caucal and Knapik [6] have proved that the $\mathfrak{hgr}_n^{\text{basic}}$ hierarchy, involving basic iterations, collapses at level 1, where it coincided with $\mathfrak{hgr}_1$, the first level of the hierarchy involving Muchnik's iteration. We have done the same, using much simpler methods. Moreover, we have given an additional insight on the nature of the two kinds of iterations: a composition of two basic iterations boils down to a single basic iteration, and a composition of Muchnik's iteration with basic iteration boils down to just a Muchnik's iteration. Simultaneously, Caucal's hierarchy (being a graph version of $\mathfrak{hgr}_n$) is strict, which implies that a composition of two Muchnik's iteration cannot be reduced to a single Muchnik's iteration, even if we only consider directed graphs instead of arbitrary relational structures.

References

1. Arnborg, S., Lagergren, J., Seese, D.: Easy problems for tree-decomposable graphs. J. Algorithms **12**(2), 308–340 (1991). https://doi.org/10.1016/0196-6774(91)90006-K

2. Büchi, J.R.: On a decision method in restricted second order arithmetic. In: Proceedings of the 1960 International Congress on Logic, Methodology and Philosophy of Science, pp. 1–11. Stanford University Press (1962)
3. Carayol, A., Wöhrle, S.: The caucal hierarchy of infinite graphs in terms of logic and higher-order pushdown automata. In: Pandya, P.K., Radhakrishnan, J. (eds.) FSTTCS 2003. LNCS, vol. 2914, pp. 112–123. Springer, Heidelberg (2003). https://doi.org/10.1007/978-3-540-24597-1_10
4. Caucal, D.: On infinite terms having a decidable monadic theory. In: Diks, K., Rytter, W. (eds.) MFCS 2002. LNCS, vol. 2420, pp. 165–176. Springer, Heidelberg (2002). https://doi.org/10.1007/3-540-45687-2_13
5. Caucal, D.: On infinite transition graphs having a decidable monadic theory. Theor. Comput. Sci. **290**(1), 79–115 (2003). https://doi.org/10.1016/S0304-3975(01)00089-5
6. Caucal, D., Knapik, T.: Shelah-Stupp's iteration and Muchnik's iteration. Fundam. Informaticae **159**(4), 327–359 (2018). https://doi.org/10.3233/FI-2018-1667
7. Courcelle, B.: The monadic second-order logic of graphs, II: infinite graphs of bounded width. Math. Syst. Theory **21**(4), 187–221 (1989). https://doi.org/10.1007/BF02088013
8. Courcelle, B.: The monadic second-order logic of graphs V: on closing the gap between definability and recognizability. Theor. Comput. Sci. **80**(2), 153–202 (1991). https://doi.org/10.1016/0304-3975(91)90387-H
9. Courcelle, B.: The monadic second-order logic of graphs VII: graphs as relational structures. Theor. Comput. Sci. **101**(1), 3–33 (1992). https://doi.org/10.1016/0304-3975(92)90148-9
10. Courcelle, B., Engelfriet, J.: Graph Structure and Monadic Second-Order Logic - A Language-Theoretic Approach, Encyclopedia of Mathematics and its Applications, vol. 138. Cambridge University Press (2012). http://www.cambridge.org/knowledge/isbn/item5758776/
11. Courcelle, B., Walukiewicz, I.: Monadic second-order logic, graph coverings and unfoldings of transition systems. Ann. Pure Appl. Log. **92**(1), 35–62 (1998). https://doi.org/10.1016/S0168-0072(97)00048-1
12. Elgot, C.C.: Decision problems of finite automata design and related arithmetics. Trans. Am. Math. Soc. **98**(1), 21–51 (1961). https://doi.org/10.2307/1993511
13. Elgot, C.C., Rabin, M.O.: Decidability and undecidability of extensions of second (first) order theory of (generalized) successor. J. Symb. Log. **31**(2), 169–181 (1966). https://doi.org/10.2307/2269808
14. Engelfriet, J.: A characterization of context-free NCE graph languages by monadic second-order logic on trees. In: Ehrig, H., Kreowski, H.-J., Rozenberg, G. (eds.) Graph Grammars 1990. LNCS, vol. 532, pp. 311–327. Springer, Heidelberg (1991). https://doi.org/10.1007/BFb0017397
15. Knapik, T., Niwiński, D., Urzyczyn, P.: Higher-order pushdown trees are easy. In: Nielsen, M., Engberg, U. (eds.) FoSSaCS 2002. LNCS, vol. 2303, pp. 205–222. Springer, Heidelberg (2002). https://doi.org/10.1007/3-540-45931-6_15
16. Muller, D.E., Schupp, P.E.: The theory of ends, pushdown automata, and second-order logic. Theor. Comput. Sci. **37**, 51–75 (1985). https://doi.org/10.1016/0304-3975(85)90087-8
17. Ong, C.L.: On model-checking trees generated by higher-order recursion schemes. In: Proceedings of 21th IEEE Symposium on Logic in Computer Science (LICS 2006), 12–15 August 2006, Seattle, WA, USA, pp. 81–90. IEEE Computer Society (2006). https://doi.org/10.1109/LICS.2006.38

18. Rabin, M.O.: Decidability of second-order theories and automata on infinite trees. Trans. Am. Math. Soc. **141**, 1–35 (1969). https://doi.org/10.1090/S0002-9947-1969-0246760-1

19. Semenov, A.L.: Decidability of monadic theories. In: Chytil, M.P., Koubek, V. (eds.) MFCS 1984. LNCS, vol. 176, pp. 162–175. Springer, Heidelberg (1984). https://doi.org/10.1007/BFb0030296

20. Shelah, S.: The monadic theory of order. Ann. Math. **102**(3), 379–419 (1975). https://doi.org/10.2307/1971037

21. Siefkes, D.: Decidable extensions of monadic second order successor arithmetic. In: Automatentheorie und Formale Sprachen, Tagung, Math. Forschungsinst, Oberwolfach, 1969, pp. 441–472. Bibliograph. Inst., Mannheim (1970)

22. Stupp, J.: The lattice-model is recursive in the original model. The Hebrew University, Jerusalem, Institute of Mathematics (1975)

23. Thomas, W.: Das Entscheidungsproblem für einige Erweiterungen der Nachfolger-Arithmetik. Ph.D. thesis, Universität Freiburg (1975)

24. Trakhtenbrot, B.: Finite automata and the logic of monadic predicates. Doklady Akademii Nauk SSSR **140**, 326–329 (1961)

25. Walukiewicz, I.: Monadic second-order logic on tree-like structures. Theor. Comput. Sci. **275**(1–2), 311–346 (2002). https://doi.org/10.1016/S0304-3975(01)00185-2

On Separation Between the Degree of a Boolean Function and the Block Sensitivity

Nikolay V. Proskurin$^{(\boxtimes)}$

HSE University, Moscow, Russian Federation
nproskurin@hse.ru

Abstract. In this paper we study the separation between two complexity measures: the degree of a Boolean function as a polynomial over the reals and the block sensitivity. We show that the upper bound on the largest possible separation between these two measures can be improved from $d^2(f) \geq bs(f)$, established by Tal [19], to $d^2(f) \geq (\sqrt{10} - 2)bs(f)$. As a corollary, we show that the similar upper bounds between some other complexity measures are not tight as well, for instance, we can improve the recent sensitivity conjecture result by Huang [10] $s^4(f) \geq bs(f)$ to $s^4(f) \geq (\sqrt{10} - 2)bs(f)$. Our techniques are based on the paper by Nisan and Szegedy [14] and include more detailed analysis of a symmetrization polynomial.

In our next result we show the same type of improvement for the separation between the approximate degree of a Boolean function and the block sensitivity: we show that $\deg_{1/3}^2(f) \geq \sqrt{6/101}bs(f)$ and improve the previous result by Nisan and Szegedy [14] $\deg_{1/3}(f) \geq \sqrt{bs(f)/6}$. In addition, we construct an example showing that the gap between the constants in the lower bound and in the known upper bound is less than 0.2.

In our last result we study the properties of a conjectured fully sensitive function on 10 variables of degree 4, existence of which would lead to improvement of the biggest known gap between these two measures. We prove that there is the only univariate polynomial that can be achieved by symmetrization of such a function by using the combination of interpolation and linear programming techniques.

Keywords: Degree of a boolean function · Approximate degree · Block sensitivity

1 Introduction

Let $f\colon \{0,1\}^n \to \{0,1\}$ be a Boolean function. We can represent f in many ways, for example, as a polynomial over the reals. It is easy to show that every Boolean

The article was prepared within the framework of the HSE University Basic Research Program.

R. Santhanam and D. Musatov (Eds.): CSR 2021, LNCS 12730, pp. 406–421, 2021.
https://doi.org/10.1007/978-3-030-79416-3_25

function can be uniquely represented by such a polynomial (see [11, exercise 2.23]), so we can introduce a complexity measure that is the degree of the polynomial that represents f, denoted by $d(f)$. Another representation of f related to polynomials is the approximating one: a polynomial is called a ε-approximation of f if for any $x \in \{0,1\}^n$ we have $|f(x) - p(x)| \le \varepsilon$. Such polynomials make sense for any $0 < \varepsilon < \frac{1}{2}$, and it is often assumed that $\varepsilon = \frac{1}{3}$. By $\deg_\varepsilon(f)$ we denote the minimum degree among the polynomials that ε-approximates f.

Exact and approximation degrees are closely related to the model called decision trees. The main measure in this model is a decision tree complexity $D(f)$, which is equal to the amount of bits in input we need to ask in order to give the value of f on such input. Other complexity measures include a sensitivity $s(f)$ and a block sensitivity $bs(f)$. If we denote

$$x^{(R)} = \begin{cases} 1 - x_i & i \in R \\ x_i & i \notin R \end{cases}$$

then a local block sensitivity $bs(f,x)$ is the largest amount of disjoint blocks $R_1, \ldots, R_t$ such that $f(x) \ne f(x^{(R_i)})$ for every $i = 1, \ldots, t$. A block sensitivity in general is the maximum over the local block sensitivities for $x \in \{0,1\}^n$. A local sensitivity and sensitivity defined similarly with a restriction that all the blocks must be of size 1. See the [4] for an overview of these and other complexity measures in the decision tree model.

One of the questions involving various complexity measures is determining the relations between them. For example, recently Huang resolved [10] the well-known sensitivity conjecture and established that $s^4(f) \ge bs(f)$. As for polynomials, the first result of this kind was made by Nisan and Szegedy: they analyzed symmetrizations of Boolean functions and showed that $2d^2(f) \ge bs(f)$ [14]. Later, Tal improved this bound by a constant factor by studying a function composition, proving that $d^2(f) \ge bs(f)$ [19]. However, the best known example with low degree and high block sensitivity is due to Kushilevitz [9, Example 6.3.2], in which $bs(f) = n = 6^k$ while $d(f) = 3^k = n^{\log_6 3} \simeq n^{0.61}$. That means there is still a large gap between the upper and lower bounds in this separation. Our result is the next constant factor improving:

Theorem 1. *For all Boolean functions $\{0,1\}^n \to \{0,1\}$, we have*

$$d^2(f) \ge (\sqrt{10} - 2)bs(f) \simeq 1.16bs(f) \tag{1}$$

As a corollary of this result, we also improve some other relations between complexity measures, including the Huang's result: we prove that $s^4(f) \ge (\sqrt{10} - 2)bs(f)$.

As for approximating polynomials, Nisan and Szegedy proved that $\deg_{1/3}(f) \ge \sqrt{bs(f)/6}$ and provided an example (namely, the OR_n function), for which the constrain is tight up to a constant factor. Later, similar results were archived for this and other Boolean functions [2,5,18]. In presented papers, authors were not interested in a constant factor in bounds. In our result, we improve the constant in the lower bound and prove that

Theorem 2. *For all Boolean functions* $\{0,1\}^n \to \{0,1\}$, *we have*

$$\deg^2_{1/3}(f) \geq \sqrt{\frac{6}{101}} bs(f) \simeq 0.24 bs(f) \tag{2}$$

We also provide an example of a Boolean function (namely, the NAE_n) that can be approximated with a polynomial of degree asymptotically tight to our bound and with a low constant factor in it; in fact, the difference is less than 0.2, which shows that the lower bound is not far from optimal.

Another way to approach the problem of the separation between $d(f)$ and $bs(f)$ is to provide examples of functions of low degree and high sensitivity. The first known example was given by Nisan and Szegedy: like in the Kushilevitz's function, f is fully sensitive, depends on $n = 3^k$ variables and $d(f) = 2^k = n^{\log_3 2} \simeq n^{0.63}$. Both examples achieved by composing the base function with itself arbitrary amount of times, and one can show that in the fully sensitive case in such a composition both $d(f)$ and $bs(f)$ remain the same in terms of n. This technique was later studied by Tal in [19]. In [14] the base polynomial consists of 3 variables and has the degree of 2, while in the [9, Example 6.3.2] it has 6 variables and the degree of 3. In both examples $2n = d(d+1)$, so the next natural step is the fully sensitive $\tilde{f}$ on 10 variables with $d(\tilde{f}) = 4$. Existence of such a function would lead to the new best example of the separation with $bs(\tilde{f}) = n$ and $d(\tilde{f}) = n^{\log_{10} 4} \simeq n^{0.60}$. While we do not provide an example of $\tilde{f}$, we prove that the only polynomial that can be achieved by symmetrization of it is:

$$\tilde{p}(x) = -\frac{x^4}{144} + \frac{5x^3}{36} - \frac{125x^2}{144} + \frac{125x}{72} \tag{3}$$

Our techniques for the lower bounds are based on Nisan and Szegedys' paper. We use the same symmetrization approach but with the more detailed analysis of a symmetrization polynomial: we apply better bounds and study higher order derivatives. As for the upper bounds, we analyze the Chebyshev polynomials of the first kind for approximating polynomials and use the combination of interpolation and linear programming for exact polynomials.

In Sect. 2 we provide necessary definitions and theorems. In Sects. 3, 4 and 5 we prove the lower bound for exact polynomials, the result for approximating polynomials and the property of the low degree function $\tilde{f}$ respectively.

2 Preliminaries

In this paper we assume that in input of a Boolean function 1 corresponds to the logical true while 0 corresponds to the logical false. The weight of an input is the amount of the positive bits in it. We use the notation $||P|| = \sup_{x \in [-1;1]} |P(x)|$ and denote the set of polynomials of degree at most d by $\mathcal{P}_d$.

A symmetrization of a polynomial $p \colon \mathbb{R}^n \to \mathbb{R}$ defined as follows:

$$p^{sym}(x) = \frac{1}{n!} \sum_{\pi \in S_n} p(\pi(x)) \tag{4}$$

S_n denotes the group of permutations of size n and $\pi(x)$ denotes the new input, with bits from x moved according to the permutation π.

The following lemma allows us to represent p^{sym} as a univariate polynomial of small degree:

Lemma 1 (Symmetrization lemma [13]). *If $p \colon \mathbb{R}^n \to \mathbb{R}$ is a multilinear polynomial, then there exists a univariate polynomial $\tilde{p} \colon \mathbb{R} \to \mathbb{R}$ of degree at most the degree of p such that:*

$$p^{sym}(x_1, \ldots, x_n) = \tilde{p}(x_1 + \ldots + x_n), \quad \forall x \in \{0, 1\}^n$$

Note that the value of $\tilde{p}(k)$ for $k = 0, 1, \ldots, n$ is equal to the fraction of inputs such that the weight of x is equal to k and $p(x) = 1$.

From the proof of this lemma we can also get the explicit formula for the $\tilde{p}$:

$$\tilde{p}(x) = c_0 + c_1 \binom{x}{1} + c_2 \binom{x}{2} + \ldots + c_d \binom{x}{d}, \quad d \leq \deg p \tag{5}$$

By definition for the binomial coefficients. put:

$$\binom{x}{k} = \frac{x \cdot (x-1) \cdot \ldots \cdot (x-k+1)}{k!}$$

The original work of Nisan and Szegedy used the following theorem to bound the degree of a polynomial:

Theorem 3 ([7,16]). *Let $p \colon \mathbb{R} \to \mathbb{R}$ be a polynomial such that $b_1 \leq p(k) \leq b_2$ for every integer $0 \leq k \leq n$ and a derivative satisfies $|p'(\eta)| \geq c$ for some real $0 \leq \eta \leq n$; then*

$$\deg(p) \geq \sqrt{\frac{nc}{c + b_2 - b_1}}.$$

However, it is obvious that $\sqrt{\dfrac{c}{c + b_2 - b_1}} < 1$, so any bound achieved using this theorem would be weaker than Tal's $d^2(f) \geq bs(f)$. In order to make progress, we are going to use the following theorem by Ehlich and Zeller, as well as the Markov brothers' inequality:

Theorem 4 ([7]). *Let $p \colon \mathbb{R} \to \mathbb{R}$ be a polynomial of degree d. Suppose $n \in \mathbb{N}$ satisfies:*

1. $\rho = \dfrac{d^2(d^2 - 1)}{6n^2} < 1$, *and*

2. $\forall k = 0, 1, \ldots, n \colon x_k = -1 + \dfrac{2k}{n}, \; |p(x_k)| < 1$

then $\|p\| \leq \dfrac{1}{1 - \rho}$.

Theorem 5 (Markov brothers' inequality). *For any $p \in \mathcal{P}_d$ and $k < d$:*

$$||p^{(k)}|| \leq \frac{d^2 \cdot (d^2 - 1) \cdot \ldots \cdot (d^2 - k + 1)}{1 \cdot 3 \cdot \ldots \cdot (2k - 1)} ||p|| \tag{6}$$

A Boolean function f is called fully sensitive at 0 iff $f(0) = 0$ and $s(f, 0) = n$. The next theorem by Nisan and Szegedy explains why it is enough for us to focus only on fully sensitive functions.

Theorem 6 ([14]). *For every Boolean function f there exists the fully sensitive at 0 function $\tilde{f}$ that depends on $bs(f)$ variables and $d(\tilde{f}) \leq d(f)$.*

The simple proofs of Theorems 1, 4 and 6 are given in Appendix A. While the proof of the Markov brothers' inequality is significantly harder, two "book-proof"s of it are given in [17].

3   Exact Polynomials

In this section we study the separation between the degree of a Boolean function as an exact polynomial and the block sensitivity. The result organized as follows. Firstly, we will prove the warm-up result, in which we introduce a new approach for bounding degree of a polynomial. Secondly, we will prove a series of lemmas for the main result and prove Theorem 1.

3.1   Warm-Up

In order to show new techniques, first we are going to prove a simpler result, which, however, is still better than the previously known upper bound for this separation:

Theorem 7. *For all Boolean functions $\{0, 1\}^n \to \{0, 1\}$, we have*

$$d^2(f) \geq \sqrt{6/5} bs(f) \simeq 1.09 bs(f) \tag{7}$$

First of all, we need to derive a new approach to bound the degree of a polynomial that would be stronger than Theorem 3.

Theorem 8. *Let $p \colon \mathbb{R} \to \mathbb{R}$ be a polynomial of degree d such that:*

1. *$\forall i = 0, 1, \ldots, n \colon 0 \leq p(i) \leq 1$, and*
2. *$\sup_{x \in [0;n]} |p^{(k)}(x)| \geq c$.*

Then either $d^2 \geq \sqrt{6}n$ or the ratio $x = \dfrac{d^2}{n}$ satisfies the following inequality:

$$\left(1 - \frac{x^2}{6}\right) \frac{(2k - 1)!!}{2^{k-1}} c < x^k \tag{8}$$

$(2k-1)!!$ denotes the double factorial: $(2k-1)!! = (2k-1) \cdot (2k-3) \cdot \ldots \cdot 3 \cdot 1$.

Proof. Suppose $d^2 < \sqrt{6n}$, otherwise the first statement holds. In terms of Theorem 4:

$$\rho = \frac{d^2(d^2-1)}{6n^2} < \frac{d^4}{6n^2} < 1 \qquad (9)$$

Let $P(x)$ be defined as follows:

$$P(x) = p\left(\frac{n}{2}(x+1)\right) - \frac{1}{2}$$

By Theorem 4, $\|P\| \leq \frac{1}{2} \cdot \frac{1}{1-\rho}$. On the other hand, $\|P^{(k)}\| \geq \frac{n^k}{2^k} \cdot c$. Combined with Inequality 5, we get

$$c\left(\frac{n}{2}\right)^k \leq \|P^{(k)}\| \leq \frac{d^2 \cdot (d^2-1) \cdot \ldots \cdot (d^2-k+1)}{1 \cdot 3 \cdot \ldots \cdot (2k-1)}\|P\| < \frac{d^{2k}}{(2k-1)!!} \cdot \frac{1}{2(1-\rho)}$$

$$(1-\rho)\frac{(2k-1)!!}{2^{k-1}}c \leq \left(\frac{d^2}{n}\right)^k \qquad (10)$$

By substituting (9) and $x = \frac{d^2}{n}$ into (10), we obtain exactly the inequality (8).

The same approach was used by Beigel [1, lemma 3.2], however, he didn't parameterized his result and proved it only for the first derivative. If we apply his result with a trivial bound $\sup_{x\in[0;n]}|p'(x)| \geq 1$, it follows that $1 - \frac{x^2}{6} < x$ for $x = \frac{d^2}{n}$. As $x > 0$, the solution is $x \geq \sqrt{15} - 3 \simeq 0.87$, which is stronger than the original bound, but weaker than the Tal's result.

Now we are ready to prove the warm-up result.

Proof (Theorem 7). Because of reduction 6, we can assume without loss of generality that f is fully sensitive at 0, and so a polynomial p derived from Lemma 1 satisfies $p(0) = 0$ and $p(1) = 1$. Also, it is obvious that if ρ from Theorem 4 is not less than 1, then $d^2 \geq \sqrt{6n} > \sqrt{6/5n}$, so we assume that $\rho < 1$.

Suppose $p \in \mathcal{P}_2$, i.e. $p(x) = ax^2 + bx + c$. From the values of $p(0)$, $p(1)$ and $p(2)$ we obtain the following constrains:

1. $p(0) = c = 0 \Rightarrow c = 0$.
2. $p(1) = a + b + c = 1 \Rightarrow a + b = 1$.
3. $p(2) = 4a + 2b \Rightarrow -2 \leq 2a \leq -1$.

As a result, $|p''(x)| = |2a| \geq 1$.

In general case, let $q(x)$ be the quadratic polynomial that equals to $p(x)$ at $x \in \{0, 1, 2\}$, and $\tilde{p}(x) = p(x) - q(x)$. From our definition it follows that $\tilde{p}(0) = \tilde{p}(1) = \tilde{p}(2) = 0$, so there exists $\xi \in [0; 2]$: $\tilde{p}''(\xi) = 0$. Then $|p''(\xi)| = |q''(\xi) + \tilde{p}''(\xi)| \geq 1$, and as a direct consequence $\sup_{x\in[0;n]}|p''(\xi)| \geq 1$. Substituting $c = 1$ and $k = 2$ in (8), we obtain the following inequality for $x = \frac{d^2}{n}$:

$$\left(1 - \frac{x^2}{6}\right)\frac{3}{2} < x^2 \tag{11}$$

Since $x \geq 0$, we get $x \geq \sqrt{6/5}$ and $d^2 \geq \sqrt{6/5}n$.

3.2   Proof of Theorem 1

To increase the constant factor from $\sqrt{6/5}$ to $\sqrt{10} - 2$, we should analyze higher order derivatives. In order to do so, we are going to use representation (5). We also need to establish a series of lemmas.

Lemma 2. *Suppose $f: \{0,1\}^n \to \{0,1\}$ is fully sensitive at 0 and $n \geq 4$; then for symmetrization polynomial p we have $\sup_{x \in [0;n]} |p'''(x)| \geq 1 - p(3)$.*

Lemma 3.

$$\sum_{k=4}^{\infty} \frac{1}{k2^{k-2}} < \frac{1}{8} \tag{12}$$

The proofs are omitted to Appendix A. With this lemmas we are ready for the main proof.

Proof (Theorem 1). If $n < 4$, then the theorem follows from Tal's bound $d^2(f) \geq bs(f)$. As in Theorem 7, we assume that f is fully sensitive at 0 and $\rho < 1$.
  It is easy to show that

$$\left.\binom{x}{k}'\right|_{x=0} = \frac{(-1)^{k+1}}{k}, \quad k \in \mathbb{N}$$

If we combine this with representation (5), we get the formula for the first derivative of the symmetrization polynomial at $x = 0$:

$$p'(0) = \sum_{k=1}^{d}(-1)^{k+1}\frac{c_k}{k} \tag{13}$$

We can bound the first three coefficients:

1. $p(1) = c_1 = 1 \Rightarrow c_1 = 1$.
2. $p(2) = 2c_1 + c_2 \leq 1 \Rightarrow c_2 \leq -1$.
3. $p(3) = 3c_1 + 3c_2 + c_3 \geq 0 \Rightarrow c_3 \geq p(3)$.

If $p(3) < \frac{3}{8}$, then by Lemma 2 $\sup_{x \in [0;n]} |p'''(x)| > \frac{5}{8}$. Substituting $c = \frac{5}{8}$ and $k = 3$ in (8), we get the inequality for $x = \frac{d^2}{n}$:

$$\left(1 - \frac{x^2}{6}\right)\frac{75}{32} < x^3$$

Inequality implies that $x > 1.2$, which satisfies the statement of the theorem. In remaining case, we have $c_3 \geq \dfrac{3}{8}$. Substituting all the constrains in (13), we obtain:

$$p'(0) \geq \frac{3}{2} + \frac{1}{8} + \sum_{k=4}^{d}(-1)^{k+1}\frac{c_k}{k} \tag{14}$$

Suppose $|c_k| < \dfrac{1}{2^{k-2}}$ for $k > 3$, then

$$p'(0) \geq \frac{3}{2} + \frac{1}{8} - \sum_{k=4}^{d}\frac{1}{k2^{k-2}} > \frac{3}{2} + \frac{1}{8} - \sum_{k=4}^{\infty}\frac{1}{k2^{k-2}} > \frac{3}{2} + \frac{1}{8} - \frac{1}{8} = \frac{3}{2}$$

Inequality (8) with $c = \dfrac{3}{2}$ and $k = 1$ implies $\left(1 - \dfrac{x^2}{6}\right)\dfrac{3}{2} \leq x$ for $x = \dfrac{d^2}{n}$. The solution is $x \geq (\sqrt{10} - 2)$, which means that $d^2 \geq (\sqrt{10} - 2)n$.

The only case left to consider is if there exists such $k > 3$ that $|c_k| \geq \dfrac{1}{2^{k-2}}$. To deal with it, we first need to show that $\sup_{x \in [0;n]} |p^{(k)}(x)| \geq c_k$. If $d = k$, then the derivative is a constant and equals to c_k because $\binom{x}{k}^{(k)} = 1$. In other case, let $q(x)$ be the polynomial that consists of all the terms from (5) up to one with the c_k. Then $\tilde{p}(x) = p(x) - q(x)$ equals to 0 for $x = 0, 1, \ldots, k$, so $\exists \xi \in [0; k]$: $\tilde{p}^{(k)}(\xi) = 0$ and $|p^{(k)}(\xi)| = |c_k|$.

Now, for $k = 4, 5$ we obtain the following inequalities from (8):

$$\left(1 - \frac{x^2}{6}\right)\frac{105}{32} \leq x^4 \quad \Rightarrow \quad x > 1.24 > (\sqrt{10} - 2)$$

$$\left(1 - \frac{x^2}{6}\right)\frac{945}{128} \leq x^5 \quad \Rightarrow \quad x > 1.3 > (\sqrt{10} - 2)$$

For $k > 5$, we need to show that the solution from the inequality would not be worse than for $k = 5$. Notice that every time we increase k in inequality (8) we multiply the left hand side by $\dfrac{2k+1}{4} > \sqrt{6}$ and the right hand side by x. But if we recall that $\rho < 1$, we get $x < \sqrt{6}$, and thus the inequality becomes tighter. As a result, the statement holds for $k > 5$ as well.

3.3   Corollaries

While our improvement may seem insignificant, it shows that the currently known bound between $d(f)$ and $bs(f)$ is not tight. Next two corollaries shows that the same holds for some other pairs of complexity measures.

Corollary 1. *For all Boolean functions* $\{0,1\}^n \to \{0,1\}$, *we have*

$$s^4(f) \geq (\sqrt{10} - 2)bs(f) \tag{15}$$

Proof. In the proof of the sensitivity conjecture [10], Huang established the following bound for $d(f)$:

$$s^2(f) \geq d(f) \tag{16}$$

Combined with Theorem 1, it follows that

$$s^4(f) \geq d^2(f) \geq (\sqrt{10} - 2)bs(f)$$

Corollary 2. *For all Boolean functions* $\{0,1\}^n \to \{0,1\}$, *we have*

$$d^3(f) \geq (\sqrt{10} - 2)D(f) \tag{17}$$

Proof. Combining Theorem 1 with the bound $D(f) \leq bs(f) \cdot d(f)$ from the paper [12], we get

$$d^3(f) \geq d(f) \cdot (\sqrt{10} - 2)bs(f) \geq (\sqrt{10} - 2)D(f)$$

4   Approximating Polynomials

In this section, we improve the constant factor in the separation between the degree of an approximating polynomial and the block sensitivity and provide an example of a polynomial for NAE_n function that shows that not only our bound is asymptotically tight, but the difference between the best known constant in the lower bound and the constant in our example is relatively small as well.

4.1   Lower Bound

Before the proof we need to derive a similar to Theorem 8 lemma, but this time for approximating polynomials.

Lemma 4. *Let* $p: \mathbb{R} \to \mathbb{R}$ *be a polynomial of degree* d *such that:*

1. $\forall i = 0, 1, \ldots, n: -\dfrac{1}{3} \leq p(i) \leq \dfrac{4}{3}$, *and*
2. $\sup_{x \in [0;n]} |p^{(k)}(x)| \geq c$.

Then either $d^2 \geq \sqrt{6}n$ *or the ratio* $x = \dfrac{d^2}{n}$ *satisfies the following inequality:*

$$\left(1 - \frac{x^2}{6}\right) \frac{(2k-1)!!}{2^k} \cdot \frac{6c}{5} < x^k \tag{18}$$

Proof. The only difference between this lemma and Theorem 8 is the bounds for $p(k)$, therefore if we define $P(x)$ the same as earlier, we get the weaker upper bound: $\|P\| \leq \dfrac{5}{6} \cdot \dfrac{1}{1-\rho}$. The remaining part of the proof is the same as in Theorem 8.

Proof (Theorem 2). Using reduction 6, we can assume that the symmetrization polynomial of f satisfies $-\frac{1}{3} \leq p(0) \leq \frac{1}{3}$ and $\frac{2}{3} \leq p(1) \leq \frac{4}{3}$. As always, we can only consider the case $\rho < 1$. Also, if $n < 5$, then the theorem follows from the original bound $6\deg_{1/3}^2(f) \geq bs(f)$.

Suppose that $p \in \mathcal{P}_3$. By Lagrange's interpolation formula for $x \in \{0, 1, 2\}$ and $x \in \{0, 2, 5\}$:

$$p(x) = \frac{(x-1)(x-2)}{2}p(0) + x(2-x)p(1) + \frac{x(x-1)}{2}p(2) \tag{19}$$

$$p(x) = \frac{(x-2)(x-5)}{10}p(0) - \frac{x(x-5)}{6}p(2) + \frac{x(x-2)}{15}p(5) \tag{20}$$

From (19) we get $\forall x \quad p''(x) = p(0) - 2p(1) + p(2) \leq -1 + p(2)$, and if $p(2) \leq \frac{14}{15}$, then $p''(x) \leq -\frac{1}{15}$. Otherwise, from (20) we get $\forall x \quad p(x) = \frac{1}{5}p(0) - \frac{1}{3}p(2) + \frac{2}{15}p(5) \leq \frac{11}{45} - \frac{1}{3}p(2) \leq -\frac{1}{15}$. If $\deg p > 3$, we can use the same reduction as in the proof of Theorem 7.

Applying (18) with $c = \frac{1}{15}$ and $k = 2$, we obtain the following inequality:

$$\left(1 - \frac{x^2}{6}\right)\frac{3}{50} < x^2 \tag{21}$$

It now follows that $x \geq \sqrt{\frac{6}{101}}$ and $\deg_{1/3}^2(f) \geq \sqrt{\frac{6}{101}}bs(f)$.

4.2   Upper Bound

A function $NAE_n: \{0,1\}^n \rightarrow \{0,1\}$ equals to 1 iff $x \in \{0^n, 1^n\}$, i.e. all the bits in the input are the same. The next theorem provides a polynomial that approximates NAE_n and gives the upper bound for $\deg_{1/3}(f)$ in terms of the block sensitivity.

Theorem 9. *Define $d = \lceil\sqrt{c(n-2)}\rceil$ with a constant c satisfying the following inequality:*

$$2c + \frac{2}{3}c^2 - \frac{2c}{3(n-2)} > 1 \tag{22}$$

Then there exists a polynomial of degree d if d is even and $d+1$ otherwise that is a $\frac{1}{3}$-approximation of NAE_n.

Proof. In our construction, we use the Chebyshev polynomials of the first kind, defined as $T_k(x) = \cos(k\arccos x)$. We need the following properties of them; proof of property 3 is omitted to Appendix A, and property 4 is [17, lemma 5.17] for $k = 1$ and $k = 2$.

1. If k is even, then $T_k(x) = T_k(-x)$.
2. $\forall x \in [-1; 1] \quad |T_k(x)| \leq 1$.

3. $T_k''(\theta) \geq T_k''(1)$ for $\theta \geq 1$.

4. $T_k'(1) = k^2$ and $T_k''(1) = \dfrac{k^4 - k^2}{3}$.

By definition, put

$$p(x) = 1 - \frac{2T_k\left(\frac{2x-n}{n-2}\right)}{3T_k\left(\frac{n}{n-2}\right)} \tag{23}$$

It is clear from property 1 that $p(0) = p(n) = \dfrac{1}{3}$. If we show that $T_k\left(\dfrac{n}{n-2}\right) \geq 2$, then by property 2 for all $1 \leq k \leq n-1$ we have

$$\left| \frac{2T_k\left(\frac{2x-n}{n-2}\right)}{3T_k\left(\frac{n}{n-2}\right)} \right| \leq \frac{2 \cdot 1}{3 \cdot 2} = \frac{1}{3} \quad \Rightarrow \quad \frac{2}{3} \leq p(k) \leq \frac{4}{3}$$

and $q(x_1, \ldots, x_n) = p(n - x_1 - \ldots - x_n)$ is indeed the $\frac{1}{3}$-approximation of NAE_n.

Substituting $x = 1$ in the Taylor series for $T_k\left(\dfrac{n}{n-2}\right)$, we obtain

$$T_k\left(\frac{n}{n-2}\right) = T_k(1) + \frac{2}{n-2}T_k'(1) + \frac{2}{(n-2)^2}T_k''(\theta)$$

$$\geq T_k(1) + \frac{2}{n-2}T_k'(1) + \frac{2}{(n-2)^2}T_k''(1) \tag{24}$$

The last inequality holds because of property 3. Combining property 4 and (22), we get:

$$T_k\left(\frac{n}{n-2}\right) \geq 1 + 2c + \frac{2}{3(n-2)^2}(c^2(n-2)^2 - c(n-2))$$

$$= 1 + 2c + \frac{2}{3}c^2 - \frac{2c}{3(n-2)} \geq 2 \tag{25}$$

Because the last term in the left hand side of (22) tends to zero as n tends to infinity, the optimal c tends to the solution of the following inequality: $2x + \dfrac{2}{3}x^2 > 1$. The solution is $x > \dfrac{1}{2}(\sqrt{15} - 3) \simeq 0.43$, so the difference between c and the best known lower bound is less than 0.2, which shows that the bound Theorem 2 is close to be tight.

5   Fully Sensitive Function of Small Degree

The last result of this paper is about an example of a function with low degree and high block sensitivity. We study properties of conjectured function $\tilde{f}$ on 10 variables with $d(\tilde{f}) = 4$. By applying the same composition scheme as in the previous examples, we can generalize $\tilde{f}$ for the arbitrary large n. While we do not provide an example of $\tilde{f}$, we prove that if $\tilde{f}$ is fully sensitive at 0, then by applying Lemma 1 to it, the only univariate polynomial we can get is (3).

We prove this statement in two steps. Firstly, we achieve such a polynomial by interpolation. Secondly, we prove the uniqueness using the linear programming.

5.1 Interpolation

The first part of the proof is to construct polynomial (3). We do this by establishing the extremal property of all the symmetrizations of degree 4 for $n \geq 8$.

Theorem 10. *Let $f: \{0,1\}^n \rightarrow \{0,1\}$ be fully sensitive at 0 and $n \geq 8$; then for symmetrization polynomial p we have $\sup_{x \in [0;n]} |p^{(4)}(x)| \geq \frac{1}{6}$. Moreover, the only polynomial for which inequality is tight is (3).*

Proof. Using the Lagrange's interpolation formula for $x \in \{0,1,2,7,8\}$ and $x \in \{0,1,2,5,7\}$, we get the following representations:

1.

$$\forall x \quad p^{(4)}(x) = -\frac{24}{1 \cdot 1 \cdot 6 \cdot 7} + \frac{24}{2 \cdot 1 \cdot 5 \cdot 6}p(2) - \frac{24}{7 \cdot 6 \cdot 5 \cdot 1}p(7)$$
$$+ \frac{24}{8 \cdot 7 \cdot 6 \cdot 1}p(8) = -\frac{4}{7} + \frac{2}{5}p(2) - \frac{4}{35}p(7) + \frac{1}{14}p(8) \leq -\frac{1}{10} - \frac{4}{35}p(7) \tag{26}$$

2.

$$\forall x \quad p^{(4)}(x) = -\frac{24}{1 \cdot 1 \cdot 4 \cdot 6} + \frac{24}{2 \cdot 1 \cdot 3 \cdot 5}p(2) - \frac{24}{5 \cdot 4 \cdot 3 \cdot 2}p(5)$$
$$+ \frac{24}{7 \cdot 6 \cdot 5 \cdot 2}p(7) = -1 + \frac{4}{5}p(2) - \frac{1}{5}p(5) + \frac{2}{35}p(7) \leq -\frac{1}{5} + \frac{2}{35}p(7) \tag{27}$$

If $p(7) \leq \frac{7}{12}$, then from (26) we get $p^{(4)}(x) \leq -\frac{1}{6}$, otherwise we get the same result from (27). Inequality is tight iff $p(7) = \frac{7}{12}$, $p(2) = 1$, $p(5) = 0$ and $p(8) = 1$. Combined with $p(0) = 0$ and $p(1) = 1$, we obtain that there is the only polynomial of degree at most 5 that satisfies all the constrains. By applying the Lagrange's interpolation formula, we get polynomial (3).

Note that (3) is indeed a symmetrization polynomial for some function, because for $k = 0, \ldots, 10$ the values $p(k)$ represent the fraction of inputs for corresponding weights (i.e. $\binom{n}{k} \cdot p(k) \in \mathbb{N}_0$).

5.2 Linear Programming

The second part of the proof is to show that (3) is the only symmetrization polynomial for $n = 10$ and $d(f) \leq 4$. This part is done with a linear programming solver. In our case we are going to use the scipy.optimize.linprog, the full code for the problem is available at Google Colab [15].

Theorem 11. *The only symmetrization polynomial for the fully sensitive at 0 function of 10 variables with degree at most 4 is* (3).

Proof. Suppose $p(x) = c_1 x + c_2 x^2 + c_3 x^3 + c_4 x^4$ is the needed polynomial. We use the necessary (but not sufficient) conditions for $p(x)$ to create a linear programming task. Namely, we require $p(1) = 1$, $\forall k = 2, \dots, 10$: $0 \leq p(k) \leq 1$ and some additional constrain for c_4. We require a solver to minimize c_4.

If we add a constrain $c_4 > 0$, solver proves that problem is infeasible. Without any constrains for c_4, solver states that the solution is $-\dfrac{1}{144}$, i.e. the minimum value for $p^{(4)}(x)$ for which solver can find the solution is $-\dfrac{1}{144} \cdot 24 = -\dfrac{1}{6}$. But we know that the only polynomial for this value is (3), thus we obtain it's uniqueness.

6 Conclusion

Although we made improvements in relations between some complexity measures, we strongly suspect that the current results are still not tight. For example, the choice of points for interpolation in many theorems was not really justified, so we think that finding a pattern for the choice of interpolation set is one of the keys for the further improvements. Also, we suspect that by using Bernstein's inequality (see [3, theorem 5.1.7]) in Theorem 1 with or instead of the Markov brothers' inequality for the first derivative one might improve the result as well.

Another open question occurs if we add further restrictions for the function f, for example, if we want f to be symmetrical. It was proved by von zur Gathen and Roche that if $n = p - 1$ for prime p, then $d(f) = n$, and as a corollary in general $d(f) = n - \mathcal{O}(n^{0.525})$ [8]. It is conjectured that $d(f) \geq n - 3$, but very little progress was made since. For instance, the result by Cohen and Shpilka [6] states that if $n = p^2 - 1$, then $d(f) \geq n - \sqrt{n}$.

Acknowledgments. Author would like to thank Vladimir V. Podolskii for the proof idea for Theorem 7.

A Omitted Proofs

Proof (Lemma 1). Define $d = \deg p^{sym}$ and $P_k = \sum_{|S|=k} \prod_{i \in S} x_i$ where S are chosen from the subsets of $[n] = \{1, 2, \dots, n\}$. Suppose that S is a monomial in p; then by definition symmetrization adds up all the monomials of size $|S|$ to p^{sym}

equal amount of times: in order to get a specific monomial S' one should fix the permutation of variables in S' and in $[n] \setminus S'$, thus amount of every monomial is equal. Therefore, we can rewrite p^{sym} as

$$p^{sym}(x) = c_0 + c_1 P_1(x) + \ldots + c_d P_d(x)$$

Note that we only interested in $x \in \{0,1\}^n$; therefore, every term in P_k is equal to 1 iff every variable in it is equal to 1. Thus it is obvious that if $z = x_1 + x_2 + \ldots + x_n$, then

$$p^{sym}(x) = c_0 + c_1 \binom{z}{1} + \ldots + c_d \binom{z}{d} = \tilde{p}(z)$$

$\deg \tilde{p} \leq \deg p$ because $\deg p^{sym} \leq \deg p$.

Proof (Theorem 4). Define $K = \inf\{k : ||p|| \leq 1 + k\}$. From Inequality 5 we get

$$||p''(x)|| \leq \frac{d^2(d^2 - 1)}{3}(1 + K) \qquad (28)$$

Let ξ be the point of maximum on $[-1; 1]$, i.e. $||p|| = |p(\xi)|$. The cases $\xi = \pm 1$ are trivial because $|p(\xi)| \leq 1$, so we can assume that ξ is an inner point and $p'(\xi) = 0$. Also, because $\forall k = 0, 1, \ldots, n - 1$ $x_{k+1} - x_k = \frac{2}{n}$ there exists such k that $|x_k - \xi| \leq \frac{1}{n}$. Applying Taylor series for $p(x)$, we obtain

$$p(x_k) = p(\xi) + (x_k - \xi)p'(\xi) + \frac{(x_k - \xi)^2}{2}p''(\theta) = p(\xi) + \frac{(x_k - \xi)^2}{2}p''(\theta), \quad \theta \in [-1; 1]$$

Substituting (28) in the last equality, we get another bound for $||p||$:

$$||p|| = |p(\xi)| = \left| p(x_k) - \frac{(x_k - \xi)^2}{2}p''(\theta) \right| \leq |p(x_k)| + \left| \frac{(x_k - \xi)^2}{2}p''(\theta) \right| \qquad (29)$$

$$\leq 1 + \frac{1}{2n^2} \cdot \frac{d^2(d^2 - 1)}{3}(1 + K) = 1 + \rho(1 + K)$$

By definition $K \leq \rho(1 + K)$ and as a corollary $K \leq \frac{\rho}{1 - \rho}$ and $||p|| \leq 1 + K$

$$\leq \frac{1}{1 - \rho}.$$

Proof (Theorem 6). Let x be the input such that $bs(f) = bs(f, x)$, and let $S_1, S_2, \ldots, S_t$ be the blocks on which we achieve such block sensitivity. Without loss of generality we can assume that $f(0) = 0$, otherwise we can introduce the new function $g(x) = 1 - f(x)$.

Let $\tilde{f}$ be defined as follows

$$\tilde{f}(y_1, \ldots, y_t) = f(x \oplus y_1 S_1 \oplus \ldots \oplus y_t S_t) \qquad (30)$$

i.e. we create a new input for f such that every bit x_j is equal to $x_j \oplus y_i$ if $x_j \in S_i$ or x_j is left unchanged otherwise.

$d(\tilde{f}) \leq d(f)$ because $\tilde{f}$ is a linear substitution in f. On the other hand, $\tilde{f}$ is fully sensitive at 0 as $\tilde{f}(0) = f(0) = 0$ and $f(e_j) = f(x^{(S_i)}) = 1$. Thus $\tilde{f}$ satisfies the statement as $t = bs(f)$.

Proof (Lemma 2). Suppose that $p \in \mathcal{P}_3$. Using the Lagrange's interpolation formula for $x \in \{0, 1, 3, 4\}$, we get the following representation:

$$p(x) = \frac{x(x-3)(x-4)}{6} - \frac{x(x-1)(x-4)}{6}p(3) + \frac{x(x-1)(x-3)}{12}p(4)$$

$$\forall x \quad p'''(x) = 1 - p(3) + \frac{p(4)}{2} \geq 1 - p(3)$$

In general case, let $q \in \mathcal{P}_3$ be equal to p on the same set of points. Similarly to Theorem 7, if $\tilde{p}(x) = p(x) - q(x)$, then $\exists \xi \in [0;4]$: $\tilde{p}'''(\xi) = 0$ and $|p'''(\xi)| \geq 1 - p(3)$.

Proof (Lemma 3). The Maclaurin series for the natural logarithm converges for $-1 \leq x < 1$. Substituting $x = -\frac{1}{2}$, we can calculate and bound our sum:

$$\ln(1+x) = \sum_{k=1}^{\infty} \frac{(-1)^{k+1}}{k} x^k$$

$$\ln \frac{1}{2} = \sum_{k=1}^{\infty} \frac{(-1)^{k+1}}{k} \left(-\frac{1}{2}\right)^k = -\sum_{k=1}^{\infty} \frac{1}{k 2^k} \quad \Rightarrow \quad \sum_{k=1}^{\infty} \frac{1}{k 2^k} = \ln 2$$

$$\sum_{k=4}^{\infty} \frac{1}{k 2^{k-2}} = 4 \sum_{k=4}^{\infty} \frac{1}{k 2^k} = 4 \left(\sum_{k=1}^{\infty} \frac{1}{k 2^k} - \frac{2}{3}\right) = 4 \left(\ln 2 - \frac{2}{3}\right) < \frac{1}{8}$$

Proof (Theorem 9, property 3). As $T_k(\cos x) = \cos kx$, we can see that all the roots of the Chebyshev polynomial lie on $[-1; 1]$. By the Rolle's theorem, all the roots of any derivative of the Chebyshev polynomial also lie on $[-1; 1]$. From [17, lemma 5.17] we get that $T_k'''(1) = \frac{k^2(k^2-1)(k^2-2)}{15} > 0$, so $T_k'''(x) > 0$ for $x \geq 1$ and $T_k''(\theta) \geq T_k''(1)$ for $\theta \geq 0$.

References

1. Beigel, R.: Perceptrons, pp, and the polynomial hierarchy. Comput. Complex. **4**, 339–349 (1994). https://doi.org/10.1007/BF01263422
2. Bogdanov, A., Mande, N.S., Thaler, J., Williamson, C.: Approximate degree, secret sharing, and concentration phenomena. In: Achlioptas, D., Végh, L.A. (eds.) Approximation, Randomization, and Combinatorial Optimization. Algorithms and Techniques, APPROX/RANDOM 2019, September 20–22, 2019, Massachusetts Institute of Technology, Cambridge, MA, USA. LIPIcs, vol. 145, pp. 71:1–71:21. Schloss Dagstuhl - Leibniz-Zentrum für Informatik (2019). https://doi.org/10.4230/LIPIcs.APPROX-RANDOM.2019.71,https://doi.org/10.4230/LIPIcs.APPROX-RANDOM.2019.71

3. Borwein, P., Erdelyi, T.: Polynomials and Polynomial Inequalities. Graduate Texts in Mathematics, Springer, New York (1995). https://doi.org/10.1007/978-1-4612-0793-1. https://books.google.ru/books?id=386CC7JnuuwC

4. Buhrman, H., de Wolf, R.: Complexity measures and decision tree complexity: a survey. Theor. Comput. Sci. **288**(1), 21–43 (2002). https://doi.org/10.1016/S0304-3975(01)00144-X

5. Bun, M., Thaler, J.: Dual lower bounds for approximate degree and Markov-Bernstein inequalities. In: Fomin, F.V., Freivalds, R., Kwiatkowska, M., Peleg, D. (eds.) ICALP 2013. LNCS, vol. 7965, pp. 303–314. Springer, Heidelberg (2013). https://doi.org/10.1007/978-3-642-39206-1_26

6. Cohen, G., Shpilka, A.: On the degree of symmetric functions on the boolean cube. Electron. Colloquium Comput. Complex. **17**, 39 (2010). http://eccc.hpi-web.de/report/2010/039

7. Ehlich, H., Zeller, K.: Schwankung von polynomen zwischen gitterpunkten. Mathematische Zeitschrift, pp. 41–44 (1964)

8. von zur Gathen, J., Roche, J.R.: Polynomials with two values. Comb. **17**(3), 345–362 (1997). https://doi.org/10.1007/BF01215917

9. Hatami, P., Kulkarni, R., Pankratov, D.: Variations on the sensitivity conjecture. Theory Comput. **4**, 1–27 (2011). https://doi.org/10.4086/toc.gs.2011.004

10. Huang, H.: Induced subgraphs of hypercubes and a proof of the sensitivity conjecture. Ann. Math. **190**(3), 949–955 (2019). https://www.jstor.org/stable/10.4007/annals.2019.190.3.6

11. Jukna, S.: Boolean Function Complexity - Advances and Frontiers, Algorithms and Combinatorics, vol. 27. Springer, Heidelberg (2012). https://doi.org/10.1007/978-3-642-24508-4

12. Midrijanis, G.: Exact quantum query complexity for total Boolean functions. arXiv preprint quant-ph/0403168 (2004)

13. Minsky, M., Papert, S.: Perceptrons - An Introduction to Computational Geometry. MIT Press (1987)

14. Nisan, N., Szegedy, M.: On the degree of Boolean functions as real polynomials. Comput. Complex. **4**, 301–313 (1994). https://doi.org/10.1007/BF01263419

15. Proskurin, N.: Symmetrization linprog (2020). https://colab.research.google.com/drive/1XKJSYLElVxGgZuwHaFy4BdoTN4JIgKXJ?usp=sharing

16. Rivlin, T.J., Cheney, E.W.: A comparison of uniform approximations on an interval and a finite subset thereof. SIAM J. Numer. Anal. **3**(2), 311–320 (1966). http://www.jstor.org/stable/2949624

17. Shadrin, A.: Twelve proofs of the Markov inequality. Approximation theory: a volume dedicated to Borislav Bojanov, pp. 233–298 (2004)

18. Spalek, R.: A dual polynomial for OR. CoRR abs/0803.4516 (2008). http://arxiv.org/abs/0803.4516

19. Tal, A.: Properties and applications of Boolean function composition. In: Kleinberg, R.D. (ed.) Innovations in Theoretical Computer Science, ITCS 20113, Berkeley, CA, USA, January 9–12, 2013, pp. 441–454. ACM (2013). https://doi.org/10.1145/2422436.2422485

Approximation and Complexity of the Capacitated Geometric Median Problem

Vladimir Shenmaier[⊠][iD]

Sobolev Institute of Mathematics, Novosibirsk, Russia

Abstract. In the Capacitated Geometric Median problem, we are given n points in d-dimensional real space and an integer m, the goal is to locate a new point in space (center) and choose m of the input points to minimize the sum of Euclidean distances from the center to the chosen points. We show that this problem admits an "almost exact" polynomial-time algorithm in the case of fixed d and an approximation scheme PTAS in high dimensions. On the other hand, we prove that, if the dimension of space is not fixed, Capacitated Geometric Median is strongly NP-hard and does not admit a scheme FPTAS unless P = NP.

Keywords: Facility location · Geometric median · Outliers · Euclidean distances · Approximation scheme · NP-hardness

1 Introduction

We study the question of the polynomial-time solvability and approximimability of the Capacitated Geometric Median problem, which is formulated as follows.

Capacitated Geometric Median. Let X be a set of n points in space $\mathbb{R}^d$, m be a positive integer, and $\|.\|$ denote the Euclidean norm. Find a center $c \in \mathbb{R}^d$ and an m-element subset $S \subseteq X$ to minimize the value of

$$cost(S, c) = \sum_{x \in S} \|x - c\|.$$

This problem may also be referred to as *Geometric Median with outliers*. In fact, it consists of finding a center $c \in \mathbb{R}^d$ minimizing the total distance from c to m nearest input points.

In an equivalent version, we need to find a center $c \in \mathbb{R}^d$ and a subset $S \subseteq X$ of the maximun cardinality for which the value of $cost(S, c)$ does not exceed a given upper bound. It is easy to see that this "inverse" version is reduced to a series of instances of the original Capacitated Geometric Median problem, with

The study was carried out within the framework of the state contract of the Sobolev Institute of Mathematics (project 0314-2019-0014).

R. Santhanam and D. Musatov (Eds.): CSR 2021, LNCS 12730, pp. 422–434, 2021.
https://doi.org/10.1007/978-3-030-79416-3_26

different values of m. On the other hand, the original problem is reduced to a series of instances of the inverse version, with different cost bounds.

As in the usual Geometric Median problem, where $m = n$, the input points represent clients or demand points, whereas the desired center represents a location for placing a facility to serve the clients. The $n - m$ clients which are removed from this service in the solution are called *outliers*. The problem with outliers arises naturally in the following situations.

- The facility to be placed at the center we are looking for has a limited capacity and may not serve all the demand points.
- There is an upper limit for the transportation cost, so we need to remove a minimum possible number of clients from the service to satisfy this limit.
- The data contains noise and errors. In this case, a few most distant clients may exert a disproportionately strong influence over the final solution and correspond to the least robust input points.
- The discovered outliers do not fit the rest of the data and they are worthy of further investigation. In particular, once identified, they can be used to discover anomalies in the data.

Related Work. Besides the practical considerations, the problem we study is theoretically interesting. Strictly speaking, no polynomial-time algorithms are known even for the usual Geometric Median problem, where we find the best center for the whole set X, i.e., the *geometric median* of X. Finding this center is complicated by the fact that, even for 5-element sets, the geometric median is not expressible by radicals over the rationals [4]. However, one can say that the usual Geometric Median problem is polynomially solvable "almost exactly" since, e.g., the randomized algorithm from [10] computes a $(1 + \varepsilon)$-approximate solution of this problem in time $\mathcal{O}(dn \log^3(n/\varepsilon))$. Moreover, by using constructions based on random sampling, a $(1 + \varepsilon)$-approximate geometric median can be found in time almost or completely independent of n [5,10,17].

In the case of arbitrary $m \leq n$, the Capacitated Geometric Median problem becomes much harder due to the exponential number of m-subsets $S \subseteq X$. Applying random sampling to this problem seems to be effective only if m is sufficiently large. In the general case, when m may be arbitrarily small, any bounded number of random samples may "miss" good m-subsets.

ElGindy and Keil [12] consider the mentioned above version of the problem where it is required to find a maximum-cardinality subset satisfying a given upper bound for the cost value. They suggest an $\mathcal{O}(n^{2.5} \log^4 n)$-time exact algorithm for the 2-dimensional case with rectilinear distances.

Two well-known single location problems closest to Capacitated Geometric Median are Smallest m-Enclosing Ball and m-Variance. The first consists of finding m input points minimizing the radius of the Euclidean ball enclosing these points. In the second, we find m input points minimizing the sum of squared distances from these points to their mean. In high dimensions, both problems are strongly NP-hard [15,21,22] but admit approximation schemes PTAS with running time $\mathcal{O}(dn^{\lceil 1/\varepsilon \rceil})$ [1] and $\mathcal{O}(dn^{\lceil 2/\varepsilon \rceil + 1})$ [20], respectively.

Another close problem is Geometric 2-Median, which consists of finding two centers in space $\mathbb{R}^d$ minimizing the total Euclidean distance from the input points to nearest centers. In high dimensions, this problem admits fast approximation schemes based on random sampling and coresets [5,6,9,17]. However, the computational complexity of Geometric 2-Median is an open question.

A natural generalization of Capacitated Geometric Median is the problem of finding k disjoint clusters of total cardinality m in an n-element input set and selecting centers of these clusters to minimize the total distance from the cluster centers to cluster elements. The discrete version of this problem, in which all the centers must be selected from a given finite set, and also other similar problems are considered in [7,8,11,16].

Our Contributions. Both algorithmic and complexity results for Capacitated Geometric Median are given. First, we describe a randomized algorithm which, with constant probability, computes a $(1+\varepsilon)$-approximate solution of the problem in time $\mathcal{O}\big(dn^{\lfloor(d+1)/2\rfloor}m^{\lceil(d+3)/2\rceil}\log^3(m/\varepsilon)\big)$. Thus, in the case of fixed d, the problem is solvable "almost exactly" in polynomial time. Next, we show that, in high dimensions, Capacitated Geometric Median admits an approximation scheme PTAS with running time $\mathcal{O}\big(dn^{\lceil\log(2/\varepsilon)/\varepsilon\rceil+1}\big)$.

On the other hand, we prove that, if the dimension of space is not fixed, Capacitated Geometric Median is strongly NP-hard and does not admit an approximation scheme FPTAS unless P = NP. In fact, it is proved for the special case when $m = n/2$. The proof is done by a reduction from the Maximum Bisection problem in a 3-regular graph.

2   Algorithms

In this section, we present two polynomial-time algorithms for finding close-to-optimal solutions of the Capacitated Geometric Median problem, in fixed and in high dimensions.

2.1   Algorithm for Fixed Dimensions

The first algorithm is based on the property that an optimal subset consists of m input points nearest to some point in space. It allows to solve the problem by enumerating the cells of the m-order Voronoi diagram of the input set and finding geometric medians of the subsets defining these cells. A similar idea is used in [2,23] for solving a number of related vector-subset problems.

Definition 1. *Given an n-element set $X \subset \mathbb{R}^d$ and a non-empty subset $S \subset X$, the* Voronoi cell *of S is the set*

$$V(S,X) = \big\{z \in \mathbb{R}^d \,\big|\, \|z-x\| < \|z-y\| \ for\ all\ x \in S,\ y \in X \setminus S\big\}.$$

Given an integer $m \in \{1, \ldots, n-1\}$, the m-order Voronoi diagram of X is the collection $V_m(X)$ of all the non-empty Voronoi cells $V(S,X)$, where $S \subset X$ and $|S| = m$, labeled by S.

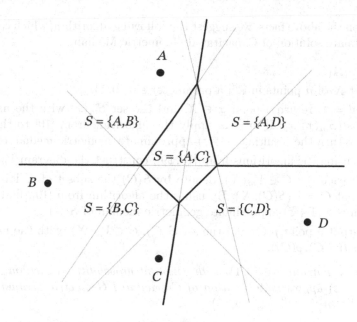

Fig. 1. The m-order Voronoi diagram of $X = \{A, B, C, D\}$ for $m = 2$

Every Voronoi cell $V(S, X)$ consists of the points of $\mathbb{R}^d$ for which the distances to the elements of S are less than those to the other elements of X. This means that $V(S, X)$ is the polytope formed by the intersection of all the open half-spaces $\{z \in \mathbb{R}^d \mid \|z - x\| < \|z - y\|\}$, $x \in S$, $y \in X \setminus S$ (see Fig. 1).

It is easy to see that the closure $\overline{V(S, X)}$ of any non-empty cell $V(S, X)$ consists of the points $z \in \mathbb{R}^d$ satisfying the inequalities $\|z - x\| \leq \|z - y\|$ for all $x \in S$, $y \in X \setminus S$. It immediately implies the following observation.

Fact 1. *If* $p \in \overline{V(S, X)}$, *where* $S \subset X$ *and* $|S| = m$, *then the set* S *consists of* m *points of* X *closest to* p.

Given a point $p \in \mathbb{R}^d$, let $S_m(p)$ be a set of m points of X closest to p (in the case of ambiguity, we choose any of such sets). Obviously, if the distances from p to different points of X are not equal, then $S_m(p)$ is uniquely defined and $p \in V(S_m(p), X)$. It follows that the cells of $V_m(X)$ cover at least all the points of $\mathbb{R}^d$ lying outside the hyperplanes $\{z \in \mathbb{R}^d \mid \|z - x\| = \|z - y\|\}$, $x, y \in X$, $x \neq y$. Hence, the closures of these cells cover the whole space $\mathbb{R}^d$.

Fact 2. [19] *For any n-element set* $X \subset \mathbb{R}^d$, $d \geq 3$, *and* $m \in \{1, \ldots, n - 1\}$, *the diagram* $V_m(X)$ *consists of* $s = \mathcal{O}\left(n^{\lfloor (d+1)/2 \rfloor} m^{\lceil (d+1)/2 \rceil}\right)$ *cells and can be constructed in time* $\mathcal{O}(s \log n + n^2 m^d)$.

Fact 3. [18] *For any n-element set* $X \subset \mathbb{R}^2$ *and* $m \in \{1, \ldots, n-1\}$, *the diagram* $V_m(X)$ *consists of* $\mathcal{O}(mn)$ *cells and can be constructed in time* $\mathcal{O}(m^2 n \log n)$.

Based on the above facts, we suggest the following algorithm, which computes an approximate solution of Capacitated Geometric Median.

Algorithm $\mathcal{A}_1$.

Input: a set X of n points in $\mathbb{R}^d$; a parameter $\varepsilon \in (0,1)$.

Step 0. If $d = 1$, return a point $x \in X$ and the set $S_m(x)$ with the minimum value of $cost(S_m(x), x)$. If $m = n$, apply the algorithm from [10] to the whole set X and return the resulting $(1 + \varepsilon)$-approximate geometric median of X.

Step 1. By using the algorithms from [18,19], construct the diagram $V_m(X)$.

Step 2. For each cell $C \in V_m(X)$, denote by $S(C)$ the subset of X labeling C, i.e., such that $C = V(S(C), X)$. By using the algorithm from [10], find a point $p(C)$ which is a $(1 + \varepsilon)$-approximate geometric median of $S(C)$.

Step 3. Output a point $p(C)$ and the set $S(C)$, $C \in V_m(X)$, with the minimum value of $cost(S(C), p(C))$.

Theorem 1. *For any $\varepsilon \in (0,1)$, with constant probability, Algorithm $\mathcal{A}_1$ computes a $(1 + \varepsilon)$-approximate solution of Capacitated Geometric Median in time $\mathcal{O}\big(dn^{\lfloor (d+1)/2 \rfloor} m^{\lceil (d+3)/2 \rceil} \log^3(m/\varepsilon)\big)$.*

Proof. If $d = 1$, the statement easily follows from the obvious fact that, for any set of points on the real line, one of the points of this set is its geometric median. If $m = n$, the statement is a direct corollary of the result of [10]. Next, let a point $c^* \in \mathbb{R}^d$ and a subset $S^* \subset X$ be an optimal solution of the problem in the case when $d \geq 2$ and $m \leq n - 1$.

Since the closures of cells of $V_m(X)$ cover the whole space $\mathbb{R}^d$, there exists a cell $C \in V_m(X)$ whose closure contains c^*. Then, by Fact 1, the set $S(C)$ consists of m points of X closest to c^*. So

$$cost(S^*, c^*) \geq cost(S(C), c^*) \geq cost(S(C), \mu(S(C))),$$

where $\mu(S(C))$ is the geometric median of $S(C)$. Therefore, the point $\mu(S(C))$ and the set $S(C)$ are also an optimum solution of the problem. But the set $S(C)$ is computed at Step 2 of Algorithm $\mathcal{A}_1$. Hence, the objective function value on the output of this algorithm is at most

$$cost(S(C), p(C)) \leq (1 + \varepsilon)\, cost(S(C), \mu(S(C))) = (1 + \varepsilon)\, cost(S^*, c^*).$$

The time complexity of Algorithm $\mathcal{A}_1$ follows from Facts 2, 3, and the result of [10]. The probability of success is defined by that of the algorithm from [10]. The theorem is proved.   □

2.2   Algorithm for High Dimensions

If the dimension of space is not fixed, a more productive idea for finding approximate solutions of Capacitated Geometric Median is based on using the framework from [24,25], which allows to compute a polynomial-cardinality set of points containing approximations of every point of space with respect to the distances to all n input points.

Definition 2. *Given a finite set $X \subset \mathbb{R}^d$ and $\varepsilon > 0$, a $(1 + \varepsilon)$-approximate centers collection or, shortly, a $(1 + \varepsilon)$-collection for X is a set $K \subseteq \mathbb{R}^d$ such that, for every point $p \in \mathbb{R}^d$, there is a point $p' \in K$ for which the distances from p' to all the elements of X are at most $1 + \varepsilon$ of those from p.*

Fact 4. *[25] For any n-element set $X \subset \mathbb{R}^d$ and each fixed $\varepsilon \in (0, 1]$, there exists a $(1 + \varepsilon)$-collection for X which consists of $N(n, \varepsilon) = \mathcal{O}\left(n^{\lceil \log(2/\varepsilon)/\varepsilon \rceil}\right)$ elements and can be constructed in time $\mathcal{O}(dN(n, \varepsilon))$.*

Note that the cardinality of the $(1 + \varepsilon)$-collection mentioned in Fact 4 does not depend on d, which is useful when we consider the case of high dimensions. This result gives a universal approximation-preserving reduction of geometric center-based problems with continuity-type objective functions to their discrete versions, where the desired centers are selected from a polynomial-cardinality set of points (see [24, 25] for details). In the case of Capacitated Geometric Median, this reduction leads to the following approximation algorithm.

Algorithm $\mathcal{A}_2$.

Input: a set X of n points in $\mathbb{R}^d$; a parameter $\varepsilon \in (0, 1]$.

Step 1. By using the algorithm from [25], construct a $(1 + \varepsilon)$-collection K for X.

Step 2. Output a point $c \in K$ and the set $S_m(c)$ with the minimum value of $cost(S_m(c), c)$.

Theorem 2. *For any fixed $\varepsilon \in (0, 1]$, Algorithm $\mathcal{A}_2$ finds a $(1 + \varepsilon)$-approximate solution of Capacitated Geometric Median in time $\mathcal{O}\left(dn^{\lceil \log(2/\varepsilon)/\varepsilon \rceil + 1}\right)$.*

Proof. Let a point $c^* \in \mathbb{R}^d$ and a subset $S^* \subseteq X$ be an optimal solution of the problem. By the definition of a $(1 + \varepsilon)$-collection, the set K contains a point c such that $\|c - x\| \leq (1 + \varepsilon)\|c^* - x\|$ for all $x \in X$. It follows the inequality $cost(S^*, c) \leq (1 + \varepsilon) \, cost(S^*, c^*)$. On the other hand, by the construction of the set $S_m(c)$, we have $cost(S_m(c), c) \leq cost(S^*, c)$. Hence, the objective function value on the output of Algorithm $\mathcal{A}_2$ is at most $(1 + \varepsilon) \, cost(S^*, c^*)$.

It remains to estimate the time complexity of this algorithm. By Fact 4, the set K consists of $N(n, \varepsilon) = \mathcal{O}\left(n^{\lceil \log(2/\varepsilon)/\varepsilon \rceil}\right)$ elements and can be constructed in time $\mathcal{O}(dN(n, \varepsilon))$. Each set $S_m(.)$ can be computed in linear time by using the known algorithm for finding an mth smallest value in an array [3]. It follows that Algorithm $\mathcal{A}_2$ runs in time $\mathcal{O}(dnN(n, \varepsilon))$. The theorem is proved. $\qquad\square$

Remark 1. The above technique gives also an approximation scheme PTAS for the more general problem in which we need to find a center $c \in \mathbb{R}^d$ and an m-element subset $S \subseteq X$ minimizing the value of

$$cost_{w,\alpha}(S, c) = \sum_{x \in S} w(x)\|x - c\|^{\alpha(x)},$$

where $w(.)$ are any non-negative weights and $\alpha(.)$ are any non-negative degrees bounded by arbitrary positive constant α. Indeed, it is easy to show that a $(1 + \varepsilon)^\alpha$-approximate solution of this problem can be computed in time $\mathcal{O}\left(dn^{\lceil \log(2/\varepsilon)/\varepsilon \rceil + 1}\right)$ by the version of Algorithm $\mathcal{A}_2$ which outputs a point $c \in K$ and the set $S_m(c)$ with the minimum value of $cost_{w,\alpha}(S_m(c), c)$.

3   Complexity

In this section, we prove that the Capacitated Geometric Median problem is strongly NP-hard and does not admit an approximation scheme FPTAS unless P = NP. The proof is done by a reduction from the well-known APX-hard problem of finding a maximum bisection in a 3-regular graph [13].

Max-Bisection|3. Given a 3-regular n-vertex undirected graph $G = (V, E)$, where n is even. Find a partition of the set of its vertices into two equal-size subsets S and $V \setminus S$ to maximize the number $cut(S, V \setminus S)$ of the edges with one endpoint in S and the other in $V \setminus S$.

Let us define the instance of Capacitated Geometric Median corresponding to an instance of Max-Bisection|3. Suppose that G is any 3-regular undirected graph with a set of vertices V and a set of edges E, an input of Max-Bisection|3. Fix arbitrary orientation on the edges of this graph, i.e., for every edge $e \in E$, choose an endpoint of this edge which it is "outgoing from" and one which it is "incoming to". Next, we map each vertex $v \in V$ to the point $x_v \in \mathbb{R}^{E \cup V}$ with the following coordinates: $x_v(e) = 1$ for every edge $e \in E$ outgoing from v and $x_v(e) = -1$ for incoming ones; $x_v(v) = M$, where M is some large integer which will be specified later; all the other coordinates are zero (see Fig. 2). Define the instance of Capacitated Geometric Median corresponding to the graph G as the set $X = \{x_v | v \in V\}$ and the value $m = n/2$.

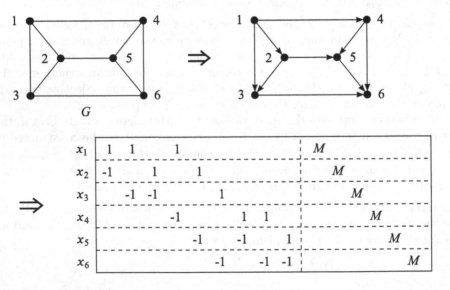

Fig. 2. Reduction scheme: constructing the vectors x_v, $v \in V$

Idea of the Reduction. Setting the coordinates $x_v(v)$, $v \in V$, to a large value M ensures that the geometric median of any subset $Y \subseteq X$ becomes very close to its mean (Lemma 1) and the total distance from the mean to the elements of Y has an almost affine dependence on the sum of pairwise squared distances between these elements (see the proof of Lemma 2). At the same time, each pairwise squared distance $\|x_v - x_u\|^2$ equals $2M^2 + 4 + (1 + 1)^2 = 2M^2 + 8$ for adjacent vertices v, u and $2M^2 + 6$ for non-adjacent ones. Based on these observations, we will prove that, given an m-element subset of vertices $S \subseteq V$, the value of $cost(X_S, \mu(X_S))$, where $X_S = \{x_v | v \in S\}$ and $\mu(X_S)$ is the geometric median of X_S, monotonously depends on the number of inner edges in S and, therefore, on the value of $cut(S, V \setminus S)$ (Lemma 3). So, if the set X_S is an optimal solution of the Capacitated Geometric Median problem on the set X, then the set S is an optimal bisection in the graph G.

For any finite set $Y \subset \mathbb{R}^{E \cup V}$, denote by $c(Y)$ and $\mu(Y)$ the mean and the geometric median of this set, respectively:

$$c(Y) = \frac{1}{|Y|} \sum_{x \in Y} x \quad \text{and} \quad \mu(Y) = \underset{c \in \mathbb{R}^{E \cup V}}{\arg\min} \ cost(Y, c).$$

Consider any subset of vertices $S \subseteq V$ with arbitrary cardinality $m \geq 3$. Let ℓ_S be the number of inner edges in S. For every vertex $v \in S$, define the vector $z_v = x_v - c(X_S)$ and the value $\zeta_v = \sum_{e \in E} z_v^2(e)$. Then $\sum_{v \in S} z_v$ is the zero vector and the following estimates hold.

Property 1. *For every vertex $v \in S$, we have (a) $\zeta_v < 3.38$;*
(b) $\|z_v\| = \sqrt{A^2 + \zeta_v} < A + \dfrac{1.69}{A}$, where $A = M\sqrt{1 - \dfrac{1}{m}}$.

Proof. (a) Given a vector $x \in \mathbb{R}^{E \cup V}$, let $E(x) = \{e \in E \mid x(e) \neq 0\}$. Then it is easy to see that the set $E(c(X_S))$ consists exactly of the edges connecting S with $V \setminus S$, while the set $E(z_v)$ consists of all the elements of $E(c(X_S))$ and also the edges connecting v with the other vertices of S. Hence, by the 3-regularity of the graph G, we have $|E(c(X_S))| = 3m - 2\ell_S$ and $|E(z_v)| = 3m - 2\ell_S + \Delta_S^v$, where Δ_S^v is the degree of v in S. For Δ_S^v coordinates $e \in E(z_v)$, the values of $z_v(e)$ are ± 1; for $3 - \Delta_S^v$ coordinates, these values are $\pm(1 - 1/m)$; for the other coordinates from $E(z_v)$, these values are $\pm 1/m$. Then

$$\zeta_v = \Delta_S^v + (3 - \Delta_S^v)\left(1 - \frac{1}{m}\right)^2 + \frac{3(m-1) - 2\ell_S + \Delta_S^v}{m^2}$$
$$= 3\left(1 - \frac{1}{m}\right)^2 + \frac{2\Delta_S^v}{m} - \frac{\Delta_S^v}{m^2} + \frac{3}{m} - \frac{3}{m^2} - \frac{2\ell_S}{m^2} + \frac{\Delta_S^v}{m^2} = 3 - \frac{3}{m} - \frac{2\ell_S}{m^2} + \frac{2\Delta_S^v}{m}.$$

But $\ell_S \geq \Delta_S^v$ and $\Delta_S^v \leq 3$, so $\zeta_v \leq 3 - \dfrac{3}{m} - \dfrac{2\Delta_S^v}{m^2} + \dfrac{2\Delta_S^v}{m} \leq 3 + \dfrac{3}{m} - \dfrac{6}{m^2}$. The latter is maximized when $m = 4$, therefore, we have $\zeta_v \leq 27/8 < 3.38$.

(b) It is easy to see that $\|z_v\|^2 = M^2\left(1 - \dfrac{1}{m}\right)^2 + (m - 1)\left(\dfrac{M}{m}\right)^2 + \zeta_v = A^2 + \zeta_v$, so $\|z_v\| = \sqrt{A^2 + \zeta_v} < A + \dfrac{\zeta_v}{2A}$. By (a), the latter is less than $A + \dfrac{1.69}{A}$. The property is proved. $\qquad\square$

Next, we formulate the main geometric statement underlying the proposed reduction. It claims that the distance between $c(X_S)$ and $\mu(X_S)$ is close to zero for large M.

Lemma 1. *Suppose that $A \geq 200m$. Then $\|\mu(X_S) - c(X_S)\| < \dfrac{40}{mA}$.*

The proof of Lemma 1 is omitted in this preliminary version. Based on this lemma, we estimate the value of $cost(X_S, \mu(X_S))$.

Lemma 2. *Suppose that $A \geq 200m$. Then, for some $\gamma \in [-1, 1]$, we have*

$$cost(X_S, \mu(X_S)) = mA + \frac{3(m-1)}{2A} + \frac{\ell_S}{mA} + \gamma \frac{121m}{A^3}.$$

Proof. Let $y = \mu(X_S) - c(X_S)$ and $\delta = \|y\|$. Then, by the cosine theorem and Property 1, the value of $cost(X_S, \mu(X_S))$ equals

$$\sum_{v \in S} \sqrt{\|z_v\|^2 + \|y\|^2 - 2\langle y, z_v \rangle} = \sum_{v \in S} \sqrt{A^2 + \zeta_v + \delta^2 - 2\langle y, z_v \rangle},$$

where $\langle ., . \rangle$ is the dot product. Next, Property 1, Lemma 1, and the condition for A imply that

$$|\zeta_v + \delta^2 - 2\langle y, z_v \rangle| < 3.38 + \left(\frac{40}{mA}\right)^2 + 2\frac{40}{mA}\left(A + \frac{1.69}{A}\right) < 31$$

and $\left|\dfrac{\zeta_v + \delta^2 - 2\langle y, z_v \rangle}{A^2}\right| < \dfrac{31}{A^2} < 0.001$. On the other hand, by using Taylor's theorem (in the Lagrange remainder form), we obtain the equation

$$\sqrt{1 + \varepsilon} = 1 + \frac{\varepsilon}{2} - \theta \frac{\varepsilon^2}{7.99} \text{ for some } \theta \in [0, 1] \text{ if } |\varepsilon| \leq 0.001.$$

Therefore, we have

$$cost(X_S, \mu(X_S)) = \sum_{v \in S} A\left(1 + \frac{\zeta_v + \delta^2 - 2\langle y, z_v \rangle}{2A^2} - \theta_v \frac{(\zeta_v + \delta^2 - 2\langle y, z_v \rangle)^2}{7.99 A^4}\right),$$

where $\theta_v \in [0, 1]$. But $\sum_{v \in S} z_v$ is the zero vector, so the sum of terms $\langle y, z_v \rangle$ is zero. Taking into account the inequalities $\delta < \dfrac{40}{mA}$ and $|\zeta_v + \delta^2 - 2\langle y, z_v \rangle| < 31$, it follows that $cost(X_S, \mu(X_S)) =$

$$\sum_{v \in S} \left(A + \frac{\zeta_v}{2A}\right) + \theta_1 \frac{40^2}{2mA^3} - \theta_2 \frac{31^2 m}{7.99 A^3} = mA + \sum_{v \in S} \frac{\zeta_v}{2A} + \gamma \frac{121m}{A^3},$$

where $\theta_1, \theta_2 \in [0, 1]$ and $\gamma \in [-1, 1]$.

Next, we estimate the value of $\sum_{v \in S} \zeta_v$. Given a vertex $v \in S$, define the vector $\tilde{x}_v \in \mathbb{R}^E$ such that $\tilde{x}_v(e) = x_v(e)$ for every $e \in E$. Then each term ζ_v equals the squared distance from the vector $\tilde{x}_v$ to the mean of these vectors. But it is well known (e.g., see [14, 20]) that the sum of such squared distances equals the sum of all the pairwise squared distances divided by $2m$:

$$\sum_{v \in S} \zeta_v = \frac{1}{2m} \sum_{v \in S} \sum_{u \in S} \|\tilde{x}_v - \tilde{x}_u\|^2.$$

At the same time, by the 3-regularity of the graph G and by the construction of vectors $\tilde{x}_v$, each pairwise squared distance $\|\tilde{x}_v - \tilde{x}_u\|^2$ equals

$$\begin{cases} 8 \text{ if the vertices } v, u \text{ are adjacent,} \\ 6 \text{ otherwise.} \end{cases}$$

So we have $\sum_{v \in S} \zeta_v = \dfrac{6(m^2 - m) + 2\ell_S \cdot 2}{2m} = 3(m - 1) + \dfrac{2\ell_S}{m}$. It follows the required equation. The lemma is proved. $\qquad\square$

Lemma 3. *Let $A \geq 200m$. Then $cost(X_S, \mu(X_S)) = f(M, m, cut(S, V \setminus S), \gamma)$, where $f(M, m, t, \gamma) = mA + \dfrac{3m}{2A} - \dfrac{t}{2mA} + \gamma \dfrac{121m}{A^3}$ and $\gamma \in [-1, 1]$.*

Proof. By the 3-regularity of the graph G, we have $cut(S, V \setminus S) = 3m - 2\ell_S$. Then $\ell_S = (3m - cut(S, V \setminus S))/2$ and, by Lemma 2, we obtain the required equation. The lemma is proved. $\qquad\square$

Theorem 3. *Capacitated Geometric Median is strongly NP-hard and does not admit an approximation scheme FPTAS unless $P = NP$.*

Proof. Suppose that the graph G consists of $n \geq 6$ vertices and set $M = 124n$. Then $m = n/2 \geq 3$ and $A \geq M\sqrt{2/3} > 200m$. By Lemma 3, it follows that $cost(X_S, \mu(X_S)) = f(M, m, cut(S, V \setminus S), \gamma)$, where $\gamma \in [-1, 1]$. On the other hand, since $A > 200m$, the absolute value of the term $\gamma \dfrac{121m}{A^3}$ in the expression for f is at most $\dfrac{2 \cdot 121}{200^2} < 0.01$ times of the value $\dfrac{1}{2mA}$, the minimum possible non-zero change of the term $\dfrac{cut(S, V \setminus S)}{2mA}$. Therefore, if the minimum median cost over all the m-element subsets of X is attained on the set X_S, then S is a maximum bisection in the graph G.

Thus, Max-Bisection|3 is reduced to Capacitated Geometric Median. Taking into account that M is an integer bounded by a polynomial in the length of the input, it gives the strong NP-hardness of our problem.

Moreover, by the above, if $cut(S, V \setminus S) \leq cut(T, V \setminus T) - 1$, where S and T are any m-element subsets of vertices, then

$$cost(X_S, \mu(X_S)) - cost(X_T, \mu(X_T)) > \frac{1 - 2 \cdot 0.01}{2mA} > \frac{0.98}{nM} > \frac{0.007}{n^2}.$$

At the same time, Lemma 3 and the inequality $A > 200m$ imply that, both cost values, for X_S and X_T, are less than $m\left(A + \dfrac{3}{400m} + \dfrac{121}{(200m)^3}\right) < mM = 62n^2$. It follows that Capacitated Geometric Median is NP-hard to approximate within a factor of $1 + \dfrac{0.007}{62n^4}$. But, for arbitrary polynomial $poly(n)$, any approximation scheme FPTAS allows to get a $\left(1 + \dfrac{1}{poly(n)}\right)$-approximation in polynomial time. Hence, the existence of such schemes is impossible unless P $=$ NP. The theorem is proved.   □

Remark 2. A slightly more complicated reduction to Capacitated Geometric Median can be constructed from the problem of finding an m-element independent set in a general graph (with arbitrary vertex degrees). Since the latter problem is W[1]-hard with respect to the parameter m, this reduction additionally gives the W[1]-hardness of our problem.

4 Conclusion

The question of the polynomial-time solvability and approximability of the Capacitated Geometric Median problem is studied. We give a simple "almost exact" polynomial-time algorithm for this problem in fixed dimensions and also an approximation scheme PTAS for the general case. On the other hand, we prove that the problem is strongly NP-hard and does not admit a scheme FPTAS unless P $=$ NP. A possible direction for future work is constructing an efficient polynomial-time approximation scheme (EPTAS). Another interesting question is the complexity of the closely related Geometric 2-Median problem.

References

1. Agarwal, P.K., Har-Peled, S., Varadarajan, K.R.: Geometric approximation via coresets. Combinatorial and Computational Geometry, MSRI Publications **52**, 1–30 (2005). http://library.msri.org/books/Book52/files/01agar.pdf
2. Aggarwal, A., Imai, H., Katoh, N., Suri, S.: Finding k points with minimum diameter and related problems. J. Algorithms **12**(1), 38–56 (1991). https://doi.org/10.1016/0196-6774(91)90022-Q
3. Aho, A.V., Hopcroft, J.E., Ullman, J.D.: The Design and Analysis of Computer Algorithms. Addison-Wesley, Boston (1974). https://doi.org/10.1002/zamm.19790590233
4. Bajaj, C.: The algebraic degree of geometric optimization problems. Discrete Comput. Geom. **3**(2), 177–191 (1988). https://doi.org/10.1007/BF02187906
5. Bădoiu, M., Har-Peled, S., Indyk, P.: Approximate clustering via core-sets. In: Proceedings of the 34th ACM Symposium on Theory of Computing (STOC 2002), pp. 250–257 (2002). https://doi.org/10.1145/509907.509947
6. Bhattacharya, A., Jaiswal, R., Kumar, A.: Faster algorithms for the constrained k-means problem. Theory Comput. Syst. **62**(1), 93–115 (2018). https://doi.org/10.1007/s00224-017-9820-7

7. Charikar, M., Khuller, S., Mount, D.M., Narasimhan, G.: Algorithms for facility location problems with outliers. In: Proceedings of the 12th ACM-SIAM Symposium on Discrete Algorithms (SODA 2001), pp. 642–651 (2001). https://dl.acm.org/doi/10.5555/365411.365555

8. Chen, K.: A constant factor approximation algorithm for k-median clustering with outliers. In: Proceedings of the 19th ACM-SIAM Symposium on Discrete Algorithms (SODA 2008), pp. 826–835 (2008). https://dl.acm.org/doi/10.5555/1347082.1347173

9. Chen, K.: On coresets for k-median and k-means clustering in metric and Euclidean spaces and their applications. SIAM J. Comput. **39**(3), 923–947 (2009). https://doi.org/10.1137/070699007

10. Cohen, M.B., Lee, Y.T., Miller, G., Pachocki, J., Sidford, A.: Geometric median in nearly linear time. arXiv:1606.05225 [cs.DS] (2016). https://arxiv.org/abs/1606.05225

11. Cohen-Addad, V., Feldmann, A.E., Saulpic, D.: Near-linear time approximations schemes for clustering in doubling metrics. In: Proceedings of the 60th Symposium on Foundations of Computer Science (FOCS 2019), pp. 540–559 (2019). https://doi.org/10.1109/FOCS.2019.00041

12. ElGindy, H., Keil, J.M.: Efficient algorithms for the capacitated 1-median problem. ORSA J. Comput. **4**(4), 418–425 (1992). https://doi.org/10.1287/ijoc.4.4.418

13. Feige, U., Karpinski, M., Langberg, M.: A note on approximating max-bisection on regular graphs. Inf. Proc. Letters **79**(4), 181–188 (2001). https://doi.org/10.1016/S0020-0190(00)00189-7

14. Inaba, M., Katoh, N., Imai, H.: Applications of weighted Voronoi diagrams and randomization to variance-based k-clustering. In: Proceedings of the 10th ACM Symposium on Computational Geometry, pp. 332–339 (1994). https://doi.org/10.1145/177424.178042

15. Kel'manov, A.V., Pyatkin, A.V.: NP-completeness of some problems of choosing a vector subset. J. Appl. Industr. Math. **5**(3), 352–357 (2011). https://doi.org/10.1134/S1990478911030069

16. Krishnaswamy, R., Li, S., Sandeep, S.: Constant approximation for k-median and k-means with outliers via iterative rounding. In: Proceedings of the 50th ACM Symposium on Theory of Computing (STOC 2018), pp. 646–659 (2018). https://doi.org/10.1145/3188745.3188882

17. Kumar, A., Sabharwal, Y., Sen, S.: Linear-time approximation schemes for clustering problems in any dimensions. J. ACM. **57**(2), 1–32 (2010). https://doi.org/10.1145/1667053.1667054

18. Lee, D.T.: On k-nearest neighbor Voronoi diagrams in the plane. IEEE Trans. Comput. **31**(6), 478–487 (1982). https://doi.org/10.1109/TC.1982.1676031

19. Mulmuley, K.: Output sensitive and dynamic constructions of higher order Voronoi diagrams and levels in arrangements. J. Comp. Syst. Sci. **47**(3), 437–458 (1993). https://doi.org/10.1016/0022-0000(93)90041-T

20. Shenmaier, V.V.: An approximation scheme for a problem of search for a vector subset. J. Appl. Industr. Math. **6**(3), 381–386 (2012). https://doi.org/10.1134/S1990478912030131

21. Shenmaier, V.V.: The problem of a minimal ball enclosing k points. J. Appl. Industr. Math. **7**(3), 444–448 (2013). https://doi.org/10.1134/S1990478913030186

22. Shenmaier, V.V.: Complexity and approximation of the smallest k-enclosing ball problem. European J. Comb. **48**, 81–87 (2015). https://doi.org/10.1016/j.ejc.2015.02.011

23. Shenmaier, V.V.: Solving some vector subset problems by Voronoi diagrams. J. Appl. Industr. Math. **10**(4), 560–566 (2016). https://doi.org/10.1134/S199047891604013X
24. Shenmaier, V.V.: A structural theorem for center-based clustering in high-dimensional Euclidean space. In: Nicosia, G., Pardalos, P., Umeton, R., Giuffrida, G., Sciacca, V. (eds.) LOD 2019. LNCS, vol. 11943, pp. 284–295. Springer, Cham (2019). https://doi.org/10.1007/978-3-030-37599-7_24
25. Shenmaier, V.V.: Some estimates on the discretization of geometric center-based problems in high dimensions. In: Kochetov, Y., Bykadorov, I., Gruzdeva, T. (eds.) MOTOR 2020. CCIS, vol. 1275, pp. 88–101. Springer, Cham (2020). https://doi.org/10.1007/978-3-030-58657-7_10

A Generic Convolution Algorithm for Join Operations on Tree Decompositions

Johan M. M. van Rooij[✉]

Department of Information and Computing Sciences, Utrecht University,
PO Box 80.089, 3508 TB Utrecht, The Netherlands
J.M.M.vanRooij@uu.nl

Abstract. The fast subset convolution algorithm by Björklund et al. (STOC 2007) has made quite an impact on parameterised complexity. Amongst the many applications are dynamic programming algorithms on tree decompositions, where the computations in so-called join nodes are a recurring example of where convolution-like operations are used. As such, several generalisations of the original fast subset convolution algorithm have been proposed, all based on concepts that strongly relate to either Möbius transforms or to Fourier transforms.

We present a new convolution generalisation that uses both Möbius transforms and Fourier transforms on the same transformation domain. This results in new faster algorithms on tree decompositions for a broad class of vertex subset problems known as the $[\sigma, \rho]$-domination problems. We solve them in $\mathcal{O}(s^{t+2}tn^2(t\log(s) + \log(n)))$ arithmetic operations, where t is the treewidth, s is the (fixed) number of states required to represent partial solutions of the specific problem, and n is the number of vertices in the graph. This improves the previous best bound of $\mathcal{O}(s^{t+2}(st)^{2(s-2)}n^3)$ arithmetic operations (van Rooij, Bodlaender, Rossmanith, ESA 2009). Specifically, this removes the dependence of the degree of the polynomial on s.

1 Introduction

We consider the complexity of computing a generalised convolution operation. Let C be a finite set and let C^k be the set of k-tuples from C. Given two functions $f, g : C^k \to \mathbb{N}$ (or tables), we consider computing $h : C^k \to \mathbb{N}$ defined through:

$$h(\vec{c}) = \sum_{\vec{c}_f \oplus \vec{c}_g = \vec{c}} f(\vec{c}_f)g(\vec{c}_g) \tag{1}$$

where $\oplus$ is the coordinate-wise application of a binary operator defined on a subset of $C \times C$.

The fast subset convolution algorithm by Björklund et al. [2] is probably the most well-known algorithm that applies to a problem in this setting. If we set $C = \{0, 1\}$ and let $\oplus$ equal standard addition (with $1 \oplus 1$ undefined, and as such ignored in the sum), then $(C^k, \oplus)$ is isomorphic to the subset lattice over a k-element finite set and Eq. 1 defines subset convolution.

© Springer Nature Switzerland AG 2021
R. Santhanam and D. Musatov (Eds.): CSR 2021, LNCS 12730, pp. 435–459, 2021.
https://doi.org/10.1007/978-3-030-79416-3_27

We consider similar algorithms in a more general setting for sets C that are the disjoint union of one or more sets C_i, where each C_i can be thought of as an independent copy of $[r] = \{0, 1, \ldots, r-1\}$. On each such subset C_i the $\oplus$-operator is defined by one of the following four operation options:

1. standard addition (with $x \oplus y$ undefined if $x + y \geq r$);
2. addition with a maximum of $r - 1$: $x \oplus y = \min(x + y, r - 1)$;
3. addition modulo r;
4. taking the maximum: $x \oplus y = \max(x, y)$.

For $x \in C_i$ and $y \in C_j$ ($i \neq j$), $x \oplus y$ is always undefined.

1.1   Motivation: Faster Algorithms on Tree Decompositions

The success of the fast subset convolution algorithm is due to its many applications in exact and parameterised algorithms (e.g., see [2]). One of these applications where convolution-like operations arise naturally are dynamic programming algorithms on tree decompositions. These algorithms construct memoisation tables in a bottom-up fashion over the tree decomposition. For many vertex or edge subset problems, these tables have entries for each 'state' vertices can have in a relevant partial solution (e.g., in the solution or not, or in some partition) often refined by the number of neighbouring vertices or incident edges in the partial solution (see Sect. 4.1). In this setting, combining partial solutions from two subtrees of the tree decomposition comes down to computing a generalised convolution operation, often involving options 1, 2 or 4 above.

The main focus of this paper are the $[\sigma, \rho]$-domination problems [23]. Let $\sigma, \rho \subseteq \mathbb{N}$, both either finite or cofinite, then a $[\sigma, \rho]$-dominating set in a graph $G = (V, E)$ is a subset $D \subseteq V$ such that: for every $v \in D$, $|N(v) \cap D| \in \sigma$, and for every $v \in V \setminus D$, $|N(v) \cap D| \in \rho$. Given σ, ρ and a graph G several problem variants are defined: whether a $[\sigma, \rho]$-dominating set exists (existence), what the largest or smallest such set is (optimisation), how many $[\sigma, \rho]$-dominating set there are (counting), or how many minimum or maximum size such sets exist (counting optimisation). Details of standard dynamic programming algorithms for these problem variants, and the definition of the amount of *states* that these standard algorithms use can be found in Sect. 4.1.

Theorem 1. *Given a graph G with a tree decomposition T of G of width t, the following problem variants of a $[\sigma, \rho]$-domination problem involving s states can be solved in:*

- *Existence: $\mathcal{O}(s^{t+2}t^2 n \log(s))$ operations on $\mathcal{O}(t \log(s))$-bit numbers.*
- *Optimisation: $\mathcal{O}(s^{t+2}t n^2(t \log(s) + \log(n)))$ operations on $\mathcal{O}(t \log(s) + \log(n))$-bit numbers.*
- *Counting: $\mathcal{O}(s^{t+2}t^2 n \log(s))$ operations on $\mathcal{O}(n)$-bit numbers.*
- *Counting optimisation: $\mathcal{O}(s^{t+2}t n^2(t \log(s) + \log(n)))$ operations on $\mathcal{O}(n)$-bit numbers.*

Using convolution-like operations to speed-up algorithms for dynamic programming on tree decompositions was first done by van Rooij et al. [22]. They showed how to solve DOMINATING SET in $\mathcal{O}^*(3^t)$ time and the $[\sigma, \rho]$-domination problems in general in $\mathcal{O}(s^{t+2}(st)^{2(s-2)}n^3)$ time. The first result is optimal in some sense, as Lokshtanov et al. [19] showed that any $\mathcal{O}^*((3-\epsilon)^t)$-time algorithm would violate the *Strong Exponential-Time Hypothesis* (SETH). It is sometimes conjectured that this also holds for the other $[\sigma, \rho]$-domination problems, which motivates looking for algorithms for the $[\sigma, \rho]$-domination problem with improved polynomial factors in the running time.

Since then several results have appeared that improve running times of dynamic programming algorithms on tree decompositions. For vertex-subset problems, e.g., see Borradaile and Le [10] for DISTANCE-r DOMINATING SET and Katsikarelis et al. [16] for DISTANCE-r INDEPENDENT SET (d-scattered set), both with matching SETH-based lower bounds. Maybe the most notable results are the *Cut and Count* technique [14] giving randomised $\mathcal{O}^*(c^t)$-time algorithms for many graph connectivity problems, mostly also supported by matching SETH-based lower bounds. All these results use convolution-like operations to speed up computations in the so-called join nodes of a nice tree decomposition (see Definition 5 in Sect. 4.1), which can be computed along the lines presented in this paper. We should mention that the *Cut and Count* results have been derandomised at the cost of a greater base of the exponent c using the *rank-based approach* [8, 12] and *determinant-based approach* [8, 24].

1.2   Generalising Subset Convolution

Several generalisations of the fast subset convolution algorithm have been published since Björklund et al.'s orginial work [2]. Cygan and Pilipczuk showed that fast subset convolution can also be obtained by a series of fast multiplications of large polynomials that encode the functions f and g [15]. This result shows that fast subset convolution can be based on Fourier transforms (used for the polynomial multiplication) in contrast to the original algorithm that uses Möbius transforms. They also generalised it to what they call *disjoint set sum*, which equals our problem with $C = [r]$ and using standard addition (option 1) for the $\oplus$-operator. Their approach does not extend to the case where C consists of multiple disjoint sets possibly with a different $\oplus$-operator on each.

At the same time, van Rooij et al. [22] generalised fast subset convolution using counting arguments that follow the same principles as Möbius transforms. Similar to our work, they allow the base set C to consist of multiple independent subsets C_i, and convolution can be defined by options 1 and 2 defined above. However, their approach results in large polynomial factors in the running times that have the size of the ground set C in the exponent of the polynomial.

Another variant of convolution is due to Cygan et al. (in the appendix of [13]), also applied to tree decompositions. Here, addition with modulus is used. Their proof only considers moduli 2 and 4, however, with a little modification it is easily generalised. This approach is fully built on Fourier transforms. We use this approach as a basic ingredient for our results.

From here on, we consider a slightly more general version of the convolution problem described above. Let n, M be positive integers. For $f, g : C^k \times [n] \to [M]$, we will compute:

$$h(\vec{c}, \kappa) = \sum_{\vec{c}_f \oplus \vec{c}_g = \vec{c}} \sum_{\kappa_f + \kappa_g = \kappa} f(\vec{c}_f, \kappa_f) g(\vec{c}_g, \kappa_g) \tag{2}$$

where M is assumed large enough so that $[M]$ can replace $\mathbb{N}$. The addition of $[n]$ to the domain allows us to embed the min-plus or max-plus semiring into the problem in a standard way. That is, if $f', g' : C^k \to [n]$ and we want to compute $h(\vec{c}) = \min_{\vec{c}_f \oplus \vec{c}_g = \vec{c}} \{f'(\vec{c}_f) + g'(\vec{c}_g)\}$, then we can set $f(\vec{c}, \kappa) = 1$ if $f'(\vec{c}) = \kappa$ and $f(\vec{c}, \kappa) = 0$ otherwise, then use Eq. 2 and extract the result.

Theorem 2. *Let C and $\oplus$ as above, $s = |C|$. Given $f, g : C^k \times [n] \to [M]$, the function $h : C^k \times [n] \to [M]$ defined in Eq. 2 can be computed in $\mathcal{O}(s^{k+1} k n (k \log(s) + \log(n)))$ arithmetic operations.*

We want to emphasise that extending earlier work using Möbius transform on C by using Fourier-transform-based polynomial multiplication to speed up the second sum in Eq. 2 is not the goal of this paper. Our use of both transforms is in a more subtle, integrated manner, where both transforms are applied to the C^k part of the domains of f and g.

2   Preliminaries on Möbius and Fourier Transforms

We start with a general introduction on Möbius and Fourier transforms. Most of this section can be found in several standard works, but the details are required to prove Theorem 2 in Sect. 3.

2.1   Möbius Inversion Using Fast Zeta and Fast Möbius Transforms

The zeta and Möbius transforms are transformations that apply to functions on partially ordered sets P. We consider such functions $f : P \to \mathbb{F}_p$ where $\mathbb{F}_p$ is the field of integers modulo p.

Definition 1 (Zeta and Möbius transform). *Let P be a partially ordered set. Given a function $f : P \to \mathbb{F}_p$, the zeta transform $\zeta(f)$ and the Möbius transform $\mu(f)$ are defined as:*

$$\zeta(f)(x) = \sum_{y \leq x} f(y) \qquad\qquad \mu(f)(x) = \sum_{y \leq x} \mu(y, x) f(y)$$

where $\mu(x, y)$ is the Möbius function of P defined on pairs $x, y \in P$ with $x \leq y$ by letting $\mu(x, x) = 1$ and for $x < y$ by letting $\mu(x, y) = -\sum_{x < z \leq y} \mu(z, y)$ recursively.

The *zeta transform* $\zeta(f)$ and the *Möbius transform* are inverses.

Lemma 1 (Möbius inversion). *Let $f : P \to \mathbb{F}_p$, then $\mu(\zeta(f))(x) = f(x)$.*

Proof. Let $x, y \in P$ and consider the sum $\sum_{x \leq z \leq y} \mu(z, y)$. If $x = y$, then this sum equals $\mu(x, x) = 1$. If $x < y$, then this sum equals $\mu(x, y) + \sum_{x < z \leq y} \mu(z, y) = 0$ by definition of $\mu(x, y)$. As such, if we expanding the definitions we obtain:

$$\mu(\zeta(f))(x) = \sum_{y \leq x} \mu(y, x) \sum_{z \leq y} f(z) = \sum_{z \leq x} f(z) \sum_{z \leq y \leq x} \mu(y, x) = \sum_{z \leq x} f(z)[z = x] = f(x)$$

Here, the third equality follows from the above reasoning, where $[z = x]$ is Iverson notation that is 1 if $z = x$ and 0 otherwise. ∎

We will not give Möbius transforms explicitly: for partial orders P, we compute zeta transforms $\zeta(f)$ of functions $f : P \to \mathbb{F}_p$, and given $\zeta(f)$, we compute f. By Lemma 1, this reconstruction is an algorithm for the Möbius transform.

Möbius inversion is often used on lattices. A *meet-semilattice* is a partial order P with, for any $x, y \in P$, the *meet* $x \wedge y$ (greatest lower bound) properly defined. Similarly, a *join-semilattice* is a partial order P with, for any $x, y \in P$, the *join* $x \vee y$ (smallest upper bound) properly defined. A lattice is both a meet and a join semi-lattice, e.g., consider $[r]^k$ with the coordinate-wise natural order and where the meet and join are the coordinate-wise minimum and maximum.

We consider partial orders that are a Cartesian product P^k of a smaller partial orders P. For $\vec{x}, \vec{y} \in P^k$, $\vec{x} = [x_1, x_2, \ldots, x_k]$, $\vec{y} = [y_1, y_2, \ldots, y_k]$, we write $\vec{x} \leq \vec{y}$ if and only if $x_i \leq y_i$ for all i. Our partial orders P have the property that for every $x \in P$, the downward closed set $\{y \in P | y \leq x\}$ forms a join-semilattice. If this property holds for every $x \in P$, then every $\vec{x} \in P^k$, $\{\vec{y} \in P^k | \vec{y} \leq \vec{x}\}$ where the join operation is defined coordinate-wise, is a join-semilattice as well.

For some very simple partial orders P, zeta and Möbius transforms using $\mathcal{O}(|P|)$ arithmetic operations exist. E.g., for the partial order $[r]$, for any integer $r \geq 1$, the zeta transform of $f : [r] \to \mathbb{F}_p$ can be computed recursively in $\mathcal{O}(r)$ arithmetic operations by $\zeta(f)(0) = f(0)$ and $\zeta(f)(x) = \zeta(f)(x - 1) + f(x)$. It is not hard to verify that μ defined by $\mu(f)(0) = f(0)$ and $\mu(f)(x) = f(x) - f(x-1)$ is its inverse: the Möbius transform of f. From such partial orders P, Möbius and zeta transforms on the lattices P^k can be easily constructed.

Lemma 2 (Fast zeta and Möbius transforms). *Given algorithms for the zeta and Möbius transform for functions $g : P \to \mathbb{F}_p$ that use $\mathcal{O}(|P|)$ arithmetic operations, the zeta and Möbius transform for $f : P^k \to \mathbb{F}_p$ can be computed in $\mathcal{O}(|P^k|k)$ arithmetic operations.*

Proof. Let $f_0 = f$ and $\vec{x} = [x_1, \ldots, x_k]$. Denote by $\vec{x}[x_i \leftarrow y]$ the tuple $\vec{x}$ with the value on the i-th coordinate replaced by y. We compute $\zeta(f)$ by an algorithm in k steps: at the i-th step, we compute f_i using the formula below. This formula essentially is $|P|^{k-1}$ times independent applications of the zeta transform to functions $g : P \to \mathbb{F}_p$, when considering all but the i-th coordinate of $\vec{x}$ as fixed.

$$f_i(\vec{x}) = \sum_{j \leq x_i} f_{i-1}(\vec{x}[x_i \leftarrow j])$$

By induction on the step number i, one easily sees that f_i satisfies the equation below, from which we can obtain $\zeta(f)$ since $f_k = \zeta(f)$. The result for $\zeta(f)$ follows because each of the k each steps computes $|P|^{k-1}$ zeta transforms that each require $\mathcal{O}(|P|)$ arithmetic operations.

$$f_i(\vec{x}) = \sum_{y_1 \leq x_1} \sum_{y_2 \leq x_2} \cdots \sum_{y_i \leq x_i} f([y_1, y_2, \ldots, y_i, x_{i+1}, x_{i+2}, \ldots, x_k])$$

For $\mu(f)$, we use that $\mu(f)$ is the inverse of $\zeta(f)$: the sequence $f_0, f_1, \ldots, f_k$ used to compute $\zeta(f)$ from f can be reversed to compute f from $\zeta(f)$. Let $f_k = \zeta(f)$. Assuming that f_i can be computed from f_{i-1} using $|P|^{k-1}$ zeta transforms, f_{i-1} can be computed from f_i using $|P|^{k-1}$ Möbius transforms that consider all but the i-th coordinate fixed. Hence, we can reconstruct $f_0 = f$ in the same way. $\square$

Zeta and Möbius transforms are important to us due to the following lemma.

Lemma 3 (Generalised covering product). *Let P be a finite partial order such that, for every $\vec{x} \in P^k$, the set $\{\vec{y} \in P^k | \vec{y} \leq \vec{x}\}$ forms a join-semilattice, and let $f, g : P^k \to \mathbb{F}_p$.*

Define the generalised covering product *$h : P^k \to \mathbb{F}_p$ of f and g through:*

$$h(\vec{x}) = \sum_{\vec{y_1} \vee \vec{y_2} = \vec{x}} f(\vec{y_1}) \, g(\vec{y_2})$$

Then $\mu(\zeta(f) \cdot \zeta(g))(\vec{x}) = h(\vec{x})$, where the product $\zeta(f) \cdot \zeta(g)$ is defined by pointwise multiplication.

Proof. We will prove that $(\zeta(f) \cdot \zeta(g))(\vec{x}) = \zeta(h)(\vec{x})$, then the result follows from Lemma 1.

$$(\zeta(f) \cdot \zeta(g))(\vec{x}) = \left(\sum_{\vec{y} \leq \vec{x}} f(\vec{y}) \right) \left(\sum_{\vec{y} \leq \vec{x}} g(\vec{y}) \right) = \sum_{\vec{y_1}, \vec{y_2} \leq \vec{x}} f(\vec{y_1}) \, g(\vec{y_2}) \qquad (3)$$

Here, we first use the definition of the ζ-transform and then work out all the product terms. The result equals $\zeta(h)(\vec{x})$ as we now show by working out the definition of $\zeta(h)(\vec{x})$.

$$\zeta(h)(\vec{x}) = \sum_{\vec{z} \leq \vec{x}} \sum_{\vec{y_1} \vee \vec{y_2} = \vec{z}} f(\vec{y_1}) \, g(\vec{y_2}) = \sum_{\vec{y_1}, \vec{y_2} \leq \vec{x}} f(\vec{y_1}) \, g(\vec{y_2}) \qquad (4)$$

For the last equality, we reorder terms using that for any two $\vec{y_1}, \vec{y_2} \leq \vec{x}$ there is a unique $\vec{z}$ such that $\vec{y_1} \vee \vec{y_2} = \vec{z}$; this is well-defined as the set $\{\vec{y} \in P^k | \vec{y} \leq \vec{x}\}$ forms a join-semilattice. $\square$

As a direct result, we generalise the covering product of Björklund et al. [2].

Corollary 1. *The generalised covering product for $f, g : [r]^k \to \mathbb{F}_p$ defined in the statement of Lemma 3 can be computed in $\mathcal{O}(r^k k)$ arithmetic operations.*

Proof. By Lemma 2 and the discussion above it, there are algorithms for the zeta and Möbius transform for functions $f : [r]^k \to \mathbb{F}_p$ that require $\mathcal{O}(r^k k)$ arithmetic operations. Now apply Lemma 3. $\square$

2.2   The Discrete Fourier Transforms Using Modular Arithmetic

We use the discrete Fourier transform on $\mathbb{F}_p$, the field of integers modulo a prime number p, in contrast to most literature that use complex numbers $\mathbb{C}$. This allows us to avoid any analysis of rounding errors, which is convenient when combining Fourier transforms with Möbius transforms.

An r-th root of unity is an element $\omega_r \in \mathbb{F}_p$ such that $\omega_r^r = 1$ while $\omega_r^l \neq 1$ for all $l < r$.

Definition 2 (Discrete Fourier transform). *Let $\vec{a} = (a_i)_{i=0}^{r-1}$ be a sequence of numbers in $\mathbb{F}_p$, and let ω_r be an r-th root of unity in $\mathbb{F}_p$. The discrete Fourier transform and inverse discrete Fourier transform are transformations between sequences of length r in $\mathbb{F}_p$ defined as follows:*

$$DFT(\vec{a})_i = \sum_{j=0}^{r-1} \omega_r^{ij} a_j \qquad\qquad DFT^{-1}(\vec{a})_i = \frac{1}{r} \sum_{j=0}^{r-1} \omega_r^{-ij} a_j$$

Note that the modulus p needs to be chosen such that $\mathbb{F}_p$ contains the required root of unity: we always assume that this is the case; see Sect. 5 for a short discussion on how to achieve this.

The discrete Fourier and inverse discrete Fourier transforms are inverses as their names suggest.

Proposition 1. $DFT^{-1}(DFT(\vec{a}))_i = a_i$

Proof. In the derivation below, we first fill in the definitions and rearrange the terms before splitting the sum based on $k = i$ and $k \neq i$.

$$DFT^{-1}(DFT(\vec{a}))_i = \frac{1}{r} \sum_{j=0}^{r-1} \omega_r^{-ij} \sum_{k=0}^{r-1} \omega_r^{jk} a_k = \frac{1}{r} \sum_{k=0}^{r-1} a_k \sum_{j=0}^{r-1} (\omega_r^{k-i})^j$$

$$= \frac{1}{r} a_i \sum_{j=0}^{r-1} (\omega_r^{i-i})^j + \frac{1}{r} \sum_{\substack{k=0 \\ k \neq i}}^{r-1} a_k \sum_{j=0}^{r-1} (\omega_r^{k-i})^j = a_i \frac{1}{r} \sum_{j=0}^{r-1} 1 + \frac{1}{r} \sum_{\substack{k=0 \\ k \neq i}}^{r-1} a_k \cdot 0 = a_i$$

Then, the first part of the sum is trivial as $\omega_r^{i-i} = \omega_r^0 = 1$, while the second part cancels as $\sum_{j=0}^{r-1}(\omega_r^{k-i})^j = \frac{1-\omega_r^{(k-i)r}}{1-\omega_r^{k-i}}$ is a geometric series with $\omega_r^{(k-i)r} = (\omega_r^r)^{k-i} = 1^{k-i} = 1$.   □

Fast algorithms for the (inverse) discrete Fourier transform are called *fast Fourier transforms* (FFT's); see the Cooley-Tukey FFT algorithm [11] and Rader's FFT algorithm [20].

Proposition 2 (Fast Fourier transform). *The discrete Fourier transform and its inverse for sequences of length r can be computed in $\mathcal{O}(r \log r)$ arithmetic operations.*

The discrete Fourier transform can be naturally extended from sequences to higher dimensional structures. Let $\mathbb{Z}_r$ be the commutative ring of integers modulo r (the modulus r can be non-prime), and let $\mathbb{Z}_r^k$ be the $\mathbb{Z}_r$-module of k-tuples with elements from $\mathbb{Z}_r$.

Definition 3 (Multidimensional discrete Fourier transform). *Let* $Z = \mathbb{Z}_{r_1} \times \mathbb{Z}_{r_2} \times \cdots \times \mathbb{Z}_{r_k}$, *and let* $R = \prod_{i=1}^{k} r_i$. *Also, let* $\vec{A} = (a_{\vec{x}})_{\vec{x} \in Z}$ *be a tensor of rank* k *with elements in* $\mathbb{F}_p$ *indexed by the* k-*tuple* $\vec{x} = [x_1, x_2, \ldots, x_k]$. *The multidimensional discrete Fourier transform and inverse multidimensional discrete Fourier transform are defined as follows:*

$$DFT_k(\vec{A})_{\vec{x}} = \sum_{y_1=0}^{r_1-1} \omega_{r_1}^{x_1 y_1} \sum_{y_2=0}^{r_2-1} \omega_{r_2}^{x_2 y_2} \cdots \sum_{y_k=0}^{r_k-1} \omega_{r_k}^{x_k y_k} a_{\vec{y}} \qquad (5)$$

$$DFT_k^{-1}(\vec{A})_{\vec{x}} = \frac{1}{R} \sum_{y_1=0}^{r_1-1} \omega_{r_1}^{-x_1 y_1} \sum_{y_2=0}^{r_2-1} \omega_{r_2}^{-x_2 y_2} \cdots \sum_{y_k=0}^{r_k-1} \omega_{r_k}^{-x_k y_k} a_{\vec{y}} \qquad (6)$$

where $\vec{y} = [y_1, y_2, \ldots, y_k]$. *When* $r = r_1 = r_2 = \ldots = r_k$, *this simplifies to:*

$$DFT_k(\vec{A})_{\vec{x}} = \sum_{\vec{y} \in \mathbb{Z}_r^k} \omega_r^{\vec{x} \cdot \vec{y}} a_{\vec{y}} \qquad\qquad DFT_k^{-1}(\vec{A})_{\vec{x}} = \frac{1}{r^k} \sum_{\vec{y} \in \mathbb{Z}_r^k} \omega_r^{-\vec{x} \cdot \vec{y}} a_{\vec{y}}$$

where the expressions in the exponents are the dot products on the tuples $\vec{x}$ *and* $\vec{y}$ *in* $\mathbb{Z}_r^k$.

Note that the dot products are in exponents of which the base is an r-th root of unity, hence they are computed modulo r: this agrees with the notation where $\vec{x}$ and $\vec{y}$ are taken from $\mathbb{Z}_r^k$.

Proposition 3 (Fast multidimensional discrete Fourier transform). *Let* $Z = \mathbb{Z}_{r_1} \times \mathbb{Z}_{r_2} \times \cdots \times \mathbb{Z}_{r_k}$, *and let* $R = \prod_{i=1}^{k} r_i$. *Also, let* $\vec{A}$ *be a tensor of rank* k *with elements in* $\mathbb{F}_p$, $\vec{A} = (a_{\vec{x}})_{\vec{x} \in Z}$. *The multidimensional discrete Fourier transform and inverse multidimensional discrete Fourier transform of* $\vec{A}$ *can be computed in* $\mathcal{O}(R \log(R))$ *time.*

Proof. Let $\vec{x}[x_i \leftarrow y]$ be the tuple $\vec{x}$ with the i-th coordinate of $\vec{x}$ replaced by y. We compute $DFT_k(\vec{A})$ with an algorithm in k steps. Let $\vec{A}_0 = \vec{A}$, and at the i-th step of the algorithm, compute $\vec{A}_i$ using the left equation below:

$$(\vec{A}_i)_{\vec{x}} = \sum_{j=0}^{r_i-1} \omega_{r_i}^{x_i j} a_{\vec{x}[x_i \leftarrow j]} \qquad\qquad (\vec{A}_i)_{\vec{x}} = \frac{1}{r_i} \sum_{j=0}^{r_i-1} \omega_{r_i}^{-x_i j} a_{\vec{x}[x_i \leftarrow j]}$$

If $k = 1$, this equals the one dimensional discrete Fourier transform. If one repeatedly substitutes the formula for $\vec{A}_{i-1}$ in the formula for $\vec{A}_i$ starting at $i = k$, one obtains the right-most part of Eq. 5. Therefore, $\vec{A}_k$ is the k-dimensional Fourier transform of $\vec{A}$.

For the inverse, almost the same procedure can be followed. Let $\vec{A}_0 = \vec{A}$, and at the i-th step, compute $\vec{A}_i$ use the left equation above, finally obtaining the result $\vec{A}_k$. Here, one again obtains the right-most part of Eq. 6 by repeated substitution.

Notice that both algorithms preform, at each step, $\frac{R}{r_i}$ standard 1-dimensional (inverse) discrete Fourier transforms on a sequence of length r_i. By Proposition 2 this can be done in $\mathcal{O}(\frac{R}{r_i}r_i \log(r_i)) = \mathcal{O}(R \log(r_i))$ time. This leads to a total running time of $\mathcal{O}(R \sum_{i=1}^{k} \log(r_i)) = \mathcal{O}(R \log(R))$.   □

In the above proof, the sequences $\vec{A}_0, \vec{A}_1, \dots, \vec{A}_k$ are created using 1-dimensional (inverse) discrete Fourier transforms. Because the 1-dimensional discrete Fourier transform and 1-dimensional inverse discrete Fourier transform are inverses, it directly follows that the sequence $\vec{A}_0, \vec{A}_1, \dots, \vec{A}_k$ used in the k-dimensional discrete Fourier transform algorithm equals the sequence $\vec{A}_k, \vec{A}_{k-1}, \dots, \vec{A}_0$ used in the inverse k-dimensional discrete Fourier transform. I.e., as their names suggests, the k-dimensional transforms are inverses.

We use the multidimensional fast discrete Fourier transform because of the well-known convolution theorem stated below for the multidimensional case.

Lemma 4 (Multidimensional convolution theorem). *Let* $Z = \mathbb{Z}_{r_1} \times \mathbb{Z}_{r_2} \times \cdots \times \mathbb{Z}_{r_k}$, *and let* $\vec{A} = (a_{\vec{x}})_{\vec{x} \in Z}$, $\vec{B} = (b_{\vec{x}})_{\vec{x} \in Z}$ *be tensors of rank* k *with elements in* $\mathbb{F}_p$. *Let the tensor multiplication* $\vec{A} \cdot \vec{B}$ *be defined point wise, and let for* $a_{\vec{x}}$ *and* $b_{\vec{y}}$ *the sum* $\vec{x} + \vec{y}$ *be defined as the sum in* Z *(coordinate-wise with the* i-th *coordinate modulo* r_i). *Then:*

$$DFT_k^{-1} \left(DFT_k(\vec{A}) \cdot DFT_k(\vec{B}) \right)_{\vec{x}} = \sum_{\vec{z_1} + \vec{z_2} \equiv \vec{x}} a_{\vec{z_1}} b_{\vec{z_2}}$$

Proof. We prove the lemma for the simplified case where $r = r_1 = r_2 = \dots = r_k$ and hence $Z = \mathbb{Z}_r^k$, the more general case goes analogously but is notation-wise much more tedious as one needs to differentiate between multiple moduli and their corresponding roots of unity.

The proof follows the same pattern as in Proposition 1. That is, we first fill in the definitions (7) and rearrange the terms (8). Next (9), we observe that the sum over all $\vec{j} \in \mathbb{Z}_r^k$ can be written as a product of k smaller sums, each involving just one coordinate of $\vec{j}$.

$$DFT_k^{-1}\left(DFT_k(\vec{A}) \cdot DFT_k(\vec{B})\right)_{\vec{x}} = \frac{1}{r^k} \sum_{\vec{y} \in \mathbb{Z}_r^k} \omega_r^{-\vec{x} \cdot \vec{y}} \left(\sum_{\vec{z_1} \in \mathbb{Z}_r^k} \omega_r^{\vec{y} \cdot \vec{z_1}} a_{\vec{z_1}} \right) \left(\sum_{\vec{z_2} \in \mathbb{Z}_r^k} \omega_r^{\vec{y} \cdot \vec{z_2}} b_{\vec{z_2}} \right) \tag{7}$$

$$= \frac{1}{r^k} \sum_{\vec{z_1}, \vec{z_2} \in \mathbb{Z}_r^k} a_{\vec{z_1}} b_{\vec{z_2}} \sum_{\vec{y} \in \mathbb{Z}_r^k} \omega_r^{\vec{y} \cdot (\vec{z_1} + \vec{z_2} - \vec{x})} \tag{8}$$

$$= \frac{1}{r^k} \sum_{\vec{z_1}, \vec{z_2} \in \mathbb{Z}_r^k} a_{\vec{z_1}} b_{\vec{z_2}} \prod_{i=1}^{k} \left(\sum_{j=0}^{r-1} \omega_r^{j((\vec{z_1})_i + (\vec{z_2})_i - x_i)} \right) \tag{9}$$

Here, x_i, $(\vec{z_1})_i$ and $(\vec{z_2})_i$ are the i-th components of $\vec{x}$, $\vec{z_1}$ and $\vec{z_2}$ respectively.

When $x_i \equiv (\vec{z_1})_i + (\vec{z_2})_i$ modulo r in the parenthesised sum of Eq. 9, this sum becomes $\sum_{j=0}^{r-1} \omega_r^0$ and thus equals r. Otherwise, when $x_i \not\equiv (\vec{z_1})_i + (\vec{z_2})_i$ the parenthesised sum is again a geometric series: $\sum_{j=0}^{r-1} (\omega_r^{(\vec{z_1})_i + (\vec{z_2})_i - x_i})^j$ that solves to $\frac{1-(\omega_r^{(\vec{z_1})_i+(\vec{z_2})_i-x_i})^r}{1-\omega_r^{(\vec{z_1})_i+(\vec{z_2})_i-x_i}} = 0$ as $(\omega_r^{(\vec{z_1})_i+(\vec{z_2})_i-x_i})^r = (\omega_r^r)^{(\vec{z_1})_i+(\vec{z_2})_i-x_i} = 1$ in the numerator. Continuing from (9), we get:

$$\mathrm{DFT}_k^{-1}\Big(\mathrm{DFT}_k(\vec{A})\cdot\mathrm{DFT}_k(\vec{B})\Big)_{\vec{x}} = \frac{1}{r^k}\sum_{\vec{z_1},\vec{z_2}\in\mathbb{Z}_r^k} a_{\vec{z_1}} b_{\vec{z_2}} \prod_{i=1}^{k} r[x_i = (\vec{z_1})_i + (\vec{z_2})_i] = \sum_{\vec{z_1}+\vec{z_2}\equiv\vec{x}} a_{\vec{z_1}} b_{\vec{z_2}}$$

where the Iverson notation $[z = x]$ is 1 if $z = x$ and 0 otherwise. The last step follows since the only non-zero terms of the sum are where $\vec{z_1} + \vec{z_2} \equiv \vec{x}$. □

Using all the above, we obtain the following lemma that needed in Sect. 3.

Lemma 5 (Part cyclic, part non-cyclic convolution). *Let $N = [q_1] \times [q_2] \times \cdots \times [q_l]$, and let $Q = \prod_{i=1}^{l} q_i$. Let $Z = \mathbb{Z}_{r_1} \times \mathbb{Z}_{r_2} \times \cdots \times \mathbb{Z}_{r_k}$, and let $R = \prod_{i=1}^{k} r_i$. Let $f, g : Z \times N \to \mathbb{F}_p$, and let $h : Z \times N \to \mathbb{F}_p$ be the part cyclic, part non-cyclic convolution of f and g defined as:*

$$h(\vec{x}, \vec{i}) = \sum_{\vec{y_1}+\vec{y_2}\equiv\vec{x}} \sum_{\vec{j_1}+\vec{j_2}=\vec{i}} f(\vec{y_1}, \vec{j_1}) g(\vec{y_2}, \vec{j_2})$$

where the sum $\vec{y_1} + \vec{y_2} \equiv \vec{x}$ is component-wise modulo r_i at coordinate i (sum in Z), and the sum $\vec{j_1} + \vec{j_2} = \vec{i}$ is component-wise without modulus (sum in N). Then, this convolution h can be computed in $\mathcal{O}(R\,Q\,2^l(\log(R) + \log(Q) + l))$ arithmetic operations.

Proof. We reduce the problem to a standard multidimensional convolution (with modulus) by padding the input with zeroes. To be precise, let $Z' = \mathbb{Z}_{2q_1} \times \mathbb{Z}_{2q_2} \times \cdots \times \mathbb{Z}_{2q_l}$ (N with for each coordinate twice as many entries and with modulo additions), and let $f', g' : Z \times Z' \to \mathbb{F}_p$ be equal to f and g on the intersection of their domains (where N is interpreted as subset of Z' by interpreting each $[q_i]$ as subset of $\mathbb{Z}_{2q_i}$) and zero otherwise. Use Proposition 3 and Lemma 4 to compute the standard multidimensional convolution of f' and g'. Because $Z \times Z'$ has $RQ2^l$ elements, this requires $\mathcal{O}(RQ2^l(\log(R) + \log(Q) + l))$ arithmetic operations. Because the padded zeroes prevent the circular convolution effect, we can extract h by taking the restriction of the result to $Z \times N$. □

3   Fourier Meets Möbius: Proving Theorem 2

Given $f, g : C^k \times [n] \to [M]$, we need to compute the following function h to prove Theorem 2:

$$h(\vec{c}, \kappa) = \sum_{\vec{c_f} \oplus \vec{c_g} = \vec{c}} \sum_{\kappa_f + \kappa_g = \kappa} f(\vec{c_f}, \kappa_f) g(\vec{c_g}, \kappa_g) \tag{10}$$

To limit the (already heavy) notational burden, we focus on a concrete set $C = C_\sigma \uplus C_\rho$ with the $\oplus$-operator defined on C_σ as addition with a maximum (option 2 in the introduction) and on C_ρ as standard addition (option 1). This is the situation required to prove Theorem 1 for a $[\sigma, \rho]$-domination problem with *cofinite* σ and *finite* ρ. The result with σ finite and/or ρ cofinite can be proved analogously. We will discuss the general case after first proving Lemma 6 below.

In terminology, we go a little bit to the setting of Theorem 1 by using some terminology common to dynamic programming on tree decompositions (see Sect. 4.1). That is, we refer to C as the set of *labels* (used to identify the state a vertex has in a partial solution), and we refer to an element $\vec{c} \in C^k$ as a *state colouring* (assignment of labels to the vertices in a bag of the decomposition).

To distinguish between the two parts of the label set $C = C_\sigma \uplus C_\rho$, we write $C_\sigma = \{|0|_\sigma, |1|_\sigma, \ldots, |\ell-1|_\sigma, |\geq\ell|_\sigma\}$ and $C_\rho = \{|0|_\rho, |1|_\rho, \ldots, |\ell-1|_\rho, |\ell|_\rho\}$. When we write $|l|_\rho$ or $|l|_\sigma$, with a variable l, we mean the labels that are not equal to $|\geq\ell|_\sigma$; this allows us to refer to other labels by expressions such as $|l-1|_\rho$. The symbol ℓ is reserved to indicate the last labels $|\geq\ell|_\sigma$ and $|\ell|_\rho$ and is used similarly to form labels such as $|\ell-1|_\rho$. Now, the $\oplus$-operator equals addition in each half of the label set, where the label $|\geq\ell|_\sigma$ emphasises that it absorbs any result greater or equal to ℓ (addition with maximum ℓ). We use Möbius transforms relative to C and C^k using the following partial order P on C: all labels are incomparable except that, for all $|l|_\sigma \in C_\sigma$, we impose $|l|_\sigma \leq |\geq\ell|_\sigma$. That is, $|\geq\ell|_\sigma$ is the maximum label in C_σ, while any pair of labels that does not involve $|\geq\ell|_\sigma$ is incomparable.

Besides the $\oplus$-operator, we will also use the standard $+$-operator on state colourings: $\vec{c}_f + \vec{c}_g = \vec{c}$. Here, the underlying operation is the standard addition operator within each half of the label set, which is undefined if the $|\geq\ell|_\sigma$-label is involved. That is, if $(\vec{c}_f)_j = |x_f|_\rho$, $(\vec{c}_g)_j = |x_g|_\rho$, then $(\vec{c})_j = |x_f + x_g|_\rho$, which is defined only if $|x_f + x_g|_\rho \in C_\rho$. Addition with the $+$-operator is similar for the σ-labels, but restricted to $C_\sigma \setminus \{|\geq\ell|_\sigma\}$.

Given a state colouring $\vec{c} \in C^k$, we write $\vec{c} = [\vec{c}^\sigma, \vec{c}^\rho] = [\vec{c}^{\geq\ell_\sigma}, \vec{c}^{\ell_\sigma}, \vec{c}^\rho]$ to differentiate between the coordinates with label from C_σ and C_ρ, and also to further differentiate between coordinates with the label $|\geq\ell|_\sigma$ and coordinates with a label from $\{|0|_\sigma, |1|_\sigma, \ldots, |\ell-1|_\sigma\}$. By splitting $\vec{c}$ in this way we notationally split the different coordinates. Be aware that this is just notation: we do not intent to reorder the coordinates of $\vec{c} \in C^k$ (e.g., $\vec{c}^\rho$ can contain the first coordinate of $\vec{c}$ and the last, while $\vec{c}^\sigma$ contains the ones in between).

Using this notation splitting the state colouring $\vec{c} = [\vec{c}^{\geq\ell_\sigma}, \vec{c}^{\ell_\sigma}, \vec{c}^\rho]$, the zeta transform of a function $f : C^k \times X \to [M]$, where X can be any set, becomes:

$$\zeta(f)(\vec{c}, \vec{x}) = \sum_{\vec{d} \leq \vec{c}} f(\vec{d}, \vec{x}) = \sum_{\vec{d}_1 \leq \vec{c}^{\geq\ell_\sigma}} f([\vec{d}_1, \vec{c}^{\ell_\sigma}, \vec{c}^\rho], \vec{x}) \tag{11}$$

Proposition 4. *Let* $s = |C|$. *Given a function* $f : C^k \times X \to [M]$, *the zeta transform* $\zeta(f)$ *of* f *based on the partial order* P *defined above can be computed in* $\mathcal{O}(s^k k |X|)$ *arithmetic operations. Also, given* $\zeta(f)$, f *can be reconstructed in* $\mathcal{O}(s^k k |X|)$ *arithmetic operations.*

Proof. We will show that for $k = 1$ and hence $\vec{c} = [c] \in C^1$, we have zeta and Möbius transforms on $f(\vec{c}, \vec{x})$ that require $\mathcal{O}(s|X|)$ arithmetic operations. The result then follows from Lemma 2 and the fact that the transforms operate independent of the parameter $\vec{x}$.

By definition of P, $\zeta(f)$ can be computed from f and vice versa through:

$$\zeta(f)(c, \vec{x}) = \begin{cases} f(c, \vec{x}) & \text{if } c \neq |\geq\ell|_\sigma \\ \sum_{z \in C_\sigma} f(z, \vec{x}) & \text{if } c = |\geq\ell|_\sigma \end{cases}$$

$$f(c, \vec{x}) = \begin{cases} \zeta(f)(c, \vec{x}) & \text{if } c \neq |\geq\ell|_\sigma \\ \zeta(f)(c, \vec{x}) - \sum_{z \in C_\sigma \setminus \{|\geq\ell|_\sigma\}} \zeta(f)(z, \vec{x}) & \text{if } c = |\geq\ell|_\sigma \end{cases}$$

Each transform uses $\mathcal{O}(s|X|)$ arithmetic operations as the sums are computed only when $c = |\geq\ell|_\sigma$. $\qquad\square$

Lemma 6. *Let $C = C_\sigma \cup C_\rho$ and $\oplus$ as defined above, and let $s = |C|$. Given $f, g : C^k \times [n] \to [M]$, the function $h : C^k \times [n] \to [M]$ defined in Eq. 10 can be computed in $\mathcal{O}(s^{k+1} kn(k \log(s) + \log(n)))$ arithmetic operations.*

Proof. Consider what happens to Eq. 10 when we apply the zeta transform based on the partial order P (Eq. 11) to it:

$$\zeta(h)(\vec{c}, \kappa) = \sum_{\vec{d_1} \leq \vec{c}^{\geq\ell}_\sigma} h([\vec{d_1}, \vec{c}^{<\ell}_\sigma, \vec{c}^p], \kappa) \tag{12}$$

$$= \sum_{\vec{d_1} \leq \vec{c}^{\geq\ell}_\sigma} \sum_{\vec{d_f} \oplus \vec{d_g} = [\vec{d_1}, \vec{c}^{<\ell}_\sigma, \vec{c}^p]} \sum_{\kappa_f + \kappa_g = \kappa} f(\vec{d_f}, \kappa_f) g(\vec{d_g}, \kappa_g) \tag{13}$$

Continuing from here, we decompose $\vec{d_f}$ and $\vec{d_g}$ coordinate-wise in the same way as we have decomposed $\vec{c}$ as $[\vec{c}^{\geq\ell}, \vec{c}^{<\ell}, \vec{c}^p]$ (to be clear: we split the coordinates of $\vec{d_f}$ and $\vec{d_g}$ based on the labels in $\vec{c}$, not based on the actual labels in $\vec{d_f}$ and $\vec{d_g}$). Let $\vec{d_f} = [\vec{d_{1f}}, \vec{d_{2f}}, \vec{d_{3f}}]$, $\vec{d_g} = [\vec{d_{1g}}, \vec{d_{2g}}, \vec{d_{3g}}]$. Now, observe that in $\vec{d_f} \oplus \vec{d_g}$, any pair $\vec{d_{1f}}$ and $\vec{d_{1g}}$ on the coordinates of $\vec{c}^{\geq\ell}_\sigma$ is summed over exactly once because for any pair there is exactly one $\vec{d_1}$ such that $\vec{d_{1f}} \oplus \vec{d_{1g}} = \vec{d_1}$. Also observe that because the other coordinates of $\vec{d_f}$ and $\vec{d_g}$ correspond to the coordinates from the $\vec{c}^{<\ell}_\sigma$ and $\vec{c}^p$ parts of $\vec{c}$ their $\oplus$-sum is the standard (non-cyclic) +-addition on labels. As such, we obtain:

$$\zeta(h)(\vec{c}, \kappa) = \sum_{\vec{d_{1f}}, \vec{d_{1g}} \leq \vec{c}^{\geq\ell}_\sigma} \sum_{\vec{d_{2f}} + \vec{d_{2g}} = \vec{c}^{<\ell}_\sigma} \sum_{\vec{d_{3f}} + \vec{d_{3g}} = \vec{c}^p} \sum_{\kappa_f + \kappa_g = \kappa} f(\vec{d_f}, \kappa_f) g(\vec{d_g}, \kappa_g)$$

$$\tag{14}$$

We note that, in the first sum, we sum over all $\vec{d_{1f}}, \vec{d_{1g}} \leq \vec{c}^{\geq\ell}_\sigma$ which is consistent with earlier notation, but by definition of P equals all $\vec{d_{1f}}, \vec{d_{1g}} \in C_\sigma$.

For our fast join operation, we need the sums $\vec{d_{2f}} + \vec{d_{2g}}$ and $\vec{d_{3f}} + \vec{d_{3g}}$ to be cyclic in order to use cyclic convolution. Therefore, we replace the functions f

and g by functions f' and g' which domains will include an additional parameter for the sums of the labels in a state colouring: this will be to separate non-cyclicly convolved results from the cyclic ones afterwards. This technique has many names: using ranking in ranked-Möbius transforms in [2], or using index vectors [22], or using accumulators [13]. Here, we do so by defining the sum of the labels of a state colouring as the sum of the number in the labels, ignoring whether they are from C_σ or C_ρ and excluding the $|\geq \ell|_\sigma$ label. That is, let the projection function π on labels to be $\pi(|l|_\sigma) = \pi(|l|_\rho) = l$. Then, for a state colouring $\vec{c} \in C^k$, define $\Sigma(\vec{c}) = \Sigma([\vec{c}^{\geq \ell_\sigma}, \vec{c}^{<\ell_\sigma}, \vec{c}^\rho]) = \sum_{j=1}^{|\vec{c}^{<\ell_\sigma}|} \pi((\vec{c}^{\sigma<\ell})_j) + \sum_{j=1}^{|\vec{c}^\rho|} \pi((\vec{c}^\rho)_j)$. For example, $\Sigma([|0|_\sigma, |0|_\rho, |\geq\ell|_\sigma]) = 0$ and $\Sigma([|2|_\sigma, |1|_\rho, |0|_\sigma]) = 3$. As such, we define f' and g' as:

$$f'(\vec{c}_f, \kappa_f, \iota_f) = \begin{cases} f(\vec{c}_f, \kappa_f) & \text{if } \Sigma(\vec{c}_f) = \iota_f \\ 0 & \text{otherwise} \end{cases} \tag{15}$$

Now, we can continue from (14) replacing the sums with sums coordinate-wise modulo ℓ_σ and $\ell_\rho + 1$, using the additional parameter to prevent the modular-cycling in the result: if cycling occurs at some coordinate, the sums do not add up any more. That is, we now compute $\zeta(h)$ using $\zeta(h)(\vec{c}, \kappa) = \zeta(h')(\vec{c}, \kappa, \Sigma(\vec{c}))$ where $\zeta(h')$ is a placeholder defined as:

$$\zeta(h')(\vec{c}, \kappa, \iota) =$$
$$\sum_{\vec{d}_{1f}, \vec{d}_{1g} \leq \vec{c}^{\geq \ell_\sigma}} \sum_{\vec{d}_{2f} + \vec{d}_{2g} \equiv \vec{c}^{<\ell_\sigma}} \sum_{\vec{d}_{3f} + \vec{d}_{3g} \equiv \vec{c}^\rho} \sum_{\kappa_f + \kappa_g = \kappa} \sum_{\iota_f + \iota_g = \iota} f'(\vec{d}_f, \kappa_f, \iota_f) g'(\vec{d}_g, \kappa_g, \iota_g)$$
$$\tag{16}$$

Here $\vec{d}_{2l} + \vec{d}_{2r} \equiv \vec{c}^{<\ell_\sigma}$ is modulo ℓ_σ and $\vec{d}_{3l} + \vec{d}_{3r} \equiv \vec{c}^\rho$ is modulo $\ell_\rho + 1$ (the difference is due to the existence of the $|\geq\ell|_\sigma$-label).

Next, we continue from Eq. 16 by changing the order of summation, taking the outermost sum inwards, and by reordering the resulting inner terms:

$$= \sum_{\vec{d}_{2f} + \vec{d}_{2g} \equiv \vec{c}^{<\ell_\sigma}} \sum_{\vec{d}_{3f} + \vec{d}_{3g} \equiv \vec{c}^\rho} \sum_{\kappa_f + \kappa_g = \kappa} \sum_{\iota_f + \iota_g = \iota}$$
$$\left(\sum_{\vec{d}_{1f} \leq \vec{c}^{\geq \ell_\sigma}} f'(\vec{d}_f, \kappa_f, \iota_f) \right) \left(\sum_{\vec{d}_{1g} \leq \vec{c}^{\geq \ell_\sigma}} g'(\vec{d}_g, \kappa_g, \iota_g) \right) \tag{17}$$

$$= \sum_{\vec{d}_{2f} + \vec{d}_{2g} \equiv \vec{c}^{<\ell_\sigma}} \sum_{\vec{d}_{3f} + \vec{d}_{3g} \equiv \vec{c}^\rho} \sum_{\kappa_f + \kappa_g = \kappa} \sum_{\iota_f + \iota_g = \iota}$$
$$\zeta(f')([\vec{c}^{\geq \ell_\sigma}, \vec{d}_{2f}, \vec{d}_{3f}], \kappa_f, \iota_f) \, \zeta(g')([\vec{c}^{\geq \ell_\sigma}, \vec{d}_{2g}, \vec{d}_{3g}], \kappa_g, \iota_g) \tag{18}$$

Where in the last step, we apply the definition of the ζ-transform for partial order P (Eq. 11). As a result, we obtain Eq. 18 which is a standard convolution sum that can be evaluated using Lemma 5 given that the coordinates with label $|\geq\ell|_\sigma$ in $\vec{c}$ are fixed.

To evaluate Eq. 18 obtaining all function values of $\zeta(h')(\vec{c},\kappa,\iota)$, let us partition the k coordinates of C^k into three parts, $[k] = X_{\geq \ell_\sigma} \cup X_{\lessdot_\sigma} \cup X_\rho$, and say that a state colouring $\vec{c}$ is compatible with this partition if: all coordinates in $X_{\geq \ell_\sigma}$ have the $| \geq \ell|_\sigma$-label in $\vec{c}$; all coordinates in $X_{\lessdot_\sigma}$ have a label from $C_\sigma \setminus \{ | \geq \ell|_\sigma \}$ in $\vec{c}$; and all coordinates in X_ρ have a label from C_ρ in $\vec{c}$. Then, given such a partition of $[k]$, Lemma 5 evaluates Eq. 18 for all $\vec{c}$ compatible with $(X_{\geq \ell_\sigma}, X_{\lessdot_\sigma}, X_\rho)$. Consequently, we can compute $\zeta(h')(\vec{c},\kappa,\iota)$ for all values of $\vec{c}$, κ, and ι by enumerating all partitions of $[k]$ into $(X_{\geq \ell_\sigma}, X_{\lessdot_\sigma}, X_\rho)$ and evaluating Eq. 18 using Lemma 5 for each subset of compatible $\vec{c}$ values, and then taking the results together.

As a result, we can evaluate Eq. 10 using a fast transform that takes the following steps in the following amount of operations:

- Use Eq. 15 to replace $f, g : C^k \times [n] \to [M]$ by the functions $f', g' : C^k \times [n] \times [sk] \to [M]$ taking the sums of the labels as a parameter in their domains. This takes $\mathcal{O}(s^{k+1}kn)$ time, as $\vec{c}_f$ takes $\mathcal{O}(s^k)$ values, κ_f takes $\mathcal{O}(n)$ values and ι_f takes $\mathcal{O}(sk)$ values.
- Compute $\zeta(f')$ and $\zeta(g')$ in $\mathcal{O}(s^{k+1}k^2 n)$ arithmetic operations using Proposition 4.
- Enumerate all partitions of $[k]$ into $(X_{\geq \ell_\sigma}, X_{\lessdot_\sigma}, X_\rho)$. For each such partition, compute the entries of $\zeta(h')(\vec{c},\kappa,\iota)$ that are compatible with this partition using for all $\vec{c}$ at once using Eq. 18 by applying Lemma 5. Then, combine the results to obtain $\zeta(h')(\vec{c},\kappa,\iota)$ for all $\vec{c}$, κ and ι.

For each partition $(X_{\geq \ell_\sigma}, X_{\lessdot_\sigma}, X_\rho)$ of $[k]$, this takes $\mathcal{O}((|C_\sigma| - 1)^{|X_{\lessdot_\sigma}|} |C_\rho|^{|X_\rho|} (nsk)2^2(k\log(s) + \log(n)))$ arithmetic operations by Lemma 5. By summing over all partitions and using the multinomial theorem, we find at total of $\mathcal{O}(s^{k+1}kn(k\log(s) + \log(n)))$ arithmetic operation for this whole step, as:

$$
\sum_{\substack{(X_{\geq \ell_\sigma}, X_{\lessdot_\sigma}, X_\rho) \\ \text{partition of } [k]}} (|C_\sigma| - 1)^{|X_{\lessdot_\sigma}|} |C_\rho|^{|X_\rho|}
$$

$$
= \sum_{\substack{x_1 + x_2 + x_3 \\ = k}} \binom{k}{x_1, x_2, x_3} 1^{x_1} (|C_\sigma| - 1)^{x_2} |C_\rho|^{x_3} = (|C_\sigma| + |C_\rho|)^k = s^k
$$

$$(19)$$

- Extract non-cycling values using $\zeta(h)(\vec{c},\kappa) = \zeta(h')(\vec{c},\kappa,\Sigma(\vec{c}))$.
- Compute the Möbius transform of the result obtaining h through $\mu(\zeta(h)) = h$. This is done $\mathcal{O}(s^k kn)$ arithmetic operations using Proposition 4.

By summing over these steps we conclude that the algorithm requires a total of $\mathcal{O}(s^{k+1}kn(k\log(s) + \log(n)))$ arithmetic operations. $\square$

Now let us return to proving Theorem 2. First notice that it is not hard to modify the above proof to the case where C consists of a different number of independent subsets C_i. This changes in how many sets we need to partition the coordinates of C^k: for each subset C_i to which we apply standard addition (option 1 in the introduction) we need one partition set, while for each subset C_i to which we apply addition with a maximum (option 2) we need two partition sets (one extra to separate the equivalents of the $|{\geq}\ell|_\sigma$ label). While this would make the proof notationally much more complicated, it does note fundamentally alter the proof: it mainly adds to the mentioned partition of the coordinates of C^k and adds more terms to the multinomial used to prove the running time.

Also, in the proof above, we demonstrated options 1 and 2 for the $\oplus$-operator as defined in the introduction, focussing on our main contribution. However, option 3 would follow from the same reasoning as the proof explicitly prevents cyclic (modular) convolution using the functions f' and g', not preventing this only simplifies it. Finally, option 4 would follow by using the construction of Lemma 3 directly and using the standard ordering on $[r]$. □

4 Faster Algorithms for $[\sigma, \rho]$-domination Problems

We start with a introduction on dynamic programming on tree decompositions for $[\sigma, \rho]$-domination problems before using Theorem 2 to prove Theorem 1.

4.1 Graphs and Tree Decompositions

A *terminal graph*[1] $G_X = (V, E, X)$ is a graph $G = (V, E)$ with an ordered sequence of distinct vertices that we call its terminals: $X = \{x_1, x_2, \ldots, x_k\}$ with each $x_j \in V$. Two terminal graphs $G_X = (V_1, E_1, X_1)$ and $H_X = (V_2, E_2, X_2)$ with the same number of terminals k, but disjoint vertex and edge sets, can be *glued* together to form the terminal graph $G_X \otimes H_X$ by identifying, for all $1 \leq i \leq k$, terminal x_i from X_1 with x_i from X_2,. I.e., if $X = X_1 = X_2$ by identification, then $G_X \otimes H_X = (V_1 \cup V_2, E_1 \cup E_2, X)$. A *completion* of a terminal graph G_X is a non-terminal graph G that can be obtained from G_X by gluing a terminal graph H_X on G_X and ignoring which vertices are terminals in the result.

The treewidth of a (non-terminal) graph is a measure of how-tree like the graph is. From an algorithmic viewpoint this is a very useful concept because, where many $\mathcal{NP}$-hard problems on general graphs are linear time solvable on trees by dynamic programming, often similar style dynamic programming algorithms exist for graphs whose treewidth is bounded by a constant. We outline the basics on treewidth and specifically on dynamic programming on tree decompositions below. For more details, see e.g. work by Bodlaender [4–7,9].

Definition 4 (tree decomposition and treewidth). *A tree decomposition of an undirected graph $G = (V, E)$ is a tree T in which each node $i \in T$ has an associated set of vertices $X_i \subseteq V$ (called a bag), with $\bigcup_{i \in T} X_i = V$, such that the following properties hold:*

- *for every edge $\{u, v\} \in E$, there exist a bag X_i such that $\{u, v\} \subseteq X_i$;*

---

[1] This is also known as a k-boundary graph.

– *for every vertex* v *in* G, *the bags containing* v *form a connected subtree: if* $v \in X_i$ *and* $v \in X_j$, *then* $v \in X_k$ *for all nodes* k *on the path from* i *to* j *in* T.

The width *of a tree decomposition* T *is defined as* $\max_{i \in T}\{|X_i|\} - 1$: *the size of the largest bag minus one. The* treewidth *of a graph* G *is the minimum width over all tree decomposition of* G.

For a tree decomposition T with assigned root node $r \in T$, we define the terminal graph $G_i = (V_i, E_i, X_i)$ for each node $i \in T$: let V_i be the union of X_i with all bags X_j where j is a descendant of i in T, and let $E_i \subseteq E$ be the set of edges with at least one endpoint in $V_i \setminus X_i$ (and as a result of Definition 4 with both endpoints in V_i). Now, G_i contains all edges between vertices in $V_i \setminus X_i$, and all edges between $V_i \setminus X_i$ and X_i, but no edges between two vertices in X_i.[2] Observe that, G is the completion of G_i formed through $G_i \otimes ((V \setminus V_i) \cup X_i, E \setminus E_i, X_i)$, and X_i can be seen as the *separator* separating $V_i \setminus X_i$ from $V \setminus V_i$ in G (where either side of the separator can be empty).

We now describe dynamic programming on a tree decomposition T. Given a graph problem that we are trying to solve $\mathcal{P}$, define a *partial solution* of $\mathcal{P}$ on G_i to be the *restriction* to the subgraph G_i of a solution of $\mathcal{P}$ on a completion of G_i (any completion of G_i, not only G itself). We say that the partial solution S' on G_i can be *extended* to a full solution S on a completion of G_i, where $S \setminus S'$ is the *extension* of S'. As an example, consider the MINIMUM DOMINATING SET problem: a solution for this problem is a vertex subset D in G such that for all $v \in V$ there is a $d \in D$ with $v \in N[d]$. A partial solution is a subset $D \subseteq V_i$ such that for all vertices in $v \in V_i \setminus X_i$ there is a $d \in D$ with $v \in N[d]$: for vertices in X_i there does not need to be a dominating neighbour in $d \in D$ as d can also be in an extension of D. A dynamic programming algorithm on a tree decomposition computes, for each node $i \in T$ in a bottom-up fashion, a *memoisation table* A_i containing all *relevant* (described in the next paragraph) partial solutions on G_i obtaining a solution to $\mathcal{P}$ in the root of T.

To restrict the number of partial (relevant) solutions stored, an equivalence relation is defined on them: two partial solutions S'_1 and S'_2 on G_i are *equivalent* with respect to $\mathcal{P}$ if any extension of S_1 also is an extension of S'_2 and vice versa. When given two equivalent partial solutions S'_1 and S'_2 for an optimisation problem (minimisation or maximisation), we say that S'_1 *dominates* S'_2 if for any extension S_E of S'_1 and S'_2, the solution value of $S'_1 \cup S_E$ is equal or better than the solution value of $S'_2 \cup S_E$. Clearly, a dynamic programming algorithm on a tree decomposition needs to store only one partial solution per equivalence class, and if we consider an optimisation problem it can store a partial solution that dominates all other partial solutions within its equivalence class.

---

[2] Often G_i is defined *including* all edges between vertices in X_i. This alternative setup allows us to not do any bookkeeping of the number of neighbours between vertices in X_i: they become neighbours higher up in the tree. In the standard setup, we need to correct for double counting of neighbours in X_i inside the convolution mechanism used for the join operation. In our setup, we do not, and the join operation because a convolution-like operation as defined in the main body of the paper.

Table 1. Examples of $[\sigma, \rho]$-domination problems (based on [23]).

σ	ρ	Standard description
$\{0\}$	$\{0, 1, \ldots\}$	Independent Set/Stable Set
$\{0, 1, \ldots\}$	$\{1, 2, \ldots\}$	Dominating Set
$\{0\}$	$\{0, 1\}$	Strong Stable Set/2-Packing/Distance-2 Independent Set
$\{0\}$	$\{1\}$	Perfect Code/Efficient Dominating Set
$\{0\}$	$\{1, 2, \ldots\}$	Independent Dominating Set
$\{0, 1, \ldots\}$	$\{1\}$	Perfect Dominating Set
$\{1, 2, \ldots\}$	$\{1, 2, \ldots\}$	Total Dominating Set
$\{1\}$	$\{1\}$	Total Perfect Dominating Set
$\{0, 1, \ldots\}$	$\{0, 1\}$	Nearly Perfect Set
$\{0, 1\}$	$\{0, 1\}$	Total Nearly Perfect Set
$\{0, 1\}$	$\{1\}$	Weakly Perfect Dominating Set
$\{0, 1, \ldots, p\}$	$\{0, 1, \ldots\}$	Induced Bounded Degree Subgraph
$\{0, 1, \ldots\}$	$\{p, p + 1, \ldots\}$	p-Dominating Set
$\{p\}$	$\{0, 1, \ldots\}$	Induced p-Regular Subgraph

Mostly, it is convenient to formulate a dynamic programming algorithm on a special kind of tree decomposition called a *nice tree decomposition* [18].[3]

Definition 5 (nice tree decomposition). *A nice tree decomposition is a tree decomposition T with assigned root node $r \in T$ with $X_r = \varnothing$, in which each node is of one of the following types:*

- Leaf node: *a leaf i of T with $X_i = \varnothing$.*
- Introduce node: *an internal node i of T with one child node j and $X_i = X_j \cup \{v\}$ for some $v \in V \setminus V_j$.*
- Forget node: *an internal node i of T with one child node j and $X_i = X_j \setminus \{v\}$ for some $v \in X_j$.*
- Join node: *an internal node i of T with two child nodes l and r with $X_i = X_l = X_r$.*

Given a tree decomposition consisting of $\mathcal{O}(n)$ nodes, a nice tree decomposition of $\mathcal{O}(n)$ nodes of the same width can be found in $\mathcal{O}(n)$ time [18]. Using this transformation, a dynamic programming algorithm on a nice tree decomposition can be used on general tree decompositions. After computing A_i for all nodes $i \in T$, the solution to $\mathcal{P}$ can be found as the unique value in A_r, where r is the root of T: here $G_i = G$ and there is only a single equivalence class as $X_i = \varnothing$.

We need Theorem 2 to speed up the computations for the *join nodes* of a nice tree decomposition. The computations for these nodes often dominates the running time of the entire dynamic programming algorithm. E.g., consider [1]

---

[3] Different version of the original definition [18] exists in literature (e.g., [14,21]): the restrictions on the vertices in a bag of a leaf node and the root node often vary, and sometimes an additional type of node called an *edge introduce node* is used.

where an $\mathcal{O}^*(4^t)$ algorithm for MINIMUM DOMINATING SET for graphs with a tree decomposition of width t is given, while all computations except the computation for the join nodes can be performed in $\mathcal{O}^*(3^t)$ time.

4.2   Dynamic Programming for $[\sigma, \rho]$-Domination Problems

Recall from the introduction that, for $\sigma, \rho \subseteq \mathbb{N}$ both either finite or cofinite, a $[\sigma, \rho]$-dominating set in a graph $G = (V, E)$ is a subset $D \subseteq V$ such that: for every $v \in D$, $|N(v) \cap D| \in \sigma$, and for every $v \in V \setminus D$, $|N(v) \cap D| \in \rho$. These $[\sigma, \rho]$-domination problems were introduced by Telle [23] and generalise many well-known graph problems such as MAXIMUM INDEPENDENT SET, MINIMUM DOMINATING SET, and INDUCED BOUNDED DEGREE SUBGRAPH; see Table 1.

When solving a $[\sigma, \rho]$-domination problem by dynamic programming on a tree decomposition, the equivalence classes for partial solutions stored in the memoisation table A_i (see Sect. 4.1) can be uniquely identified though:

– the vertices in X_i that are in the partial solution D;
– for every vertex in X_i (both in D and not in D), the number of neighbours in D.

This corresponds exactly to the bookkeeping required to verify whether a partial solution locally satisfies the requirements imposed by the specific $[\sigma, \rho]$-domination problem. As such, we can identify every equivalence class using an assignment of *labels* (sometimes also called *states*) that capture the above properties to the vertices in X_i: such an assignment is called a *state colouring*. Given $\sigma, \rho \subseteq \mathbb{N}$, define the set of labels $C = C_\sigma \cup C_\rho$ as given below. Note that these labels and the notation is identical what we used in Sect. 3): this is on purpose and allows the direct application of Theorem 2 later on.

$$
C_\sigma = \begin{cases}
\{|0|_\sigma, |1|_\sigma, |2|_\sigma, \ldots, |\ell-1|_\sigma, |\ell|_\sigma\} & \text{if } \sigma \text{ finite} & \text{where } \ell = \max\{\sigma\} \\
\{|\geq 0|_\sigma\} & \text{if } \sigma = \mathbb{N} \\
\{|0|_\sigma, |1|_\sigma, |2|_\sigma, \ldots, |\ell-1|_\sigma, |\geq \ell|_\sigma\} & \text{if } \sigma \neq \mathbb{N} \text{ cofinite} & \text{where } \ell = \max\{\mathbb{N} \setminus \sigma\} + 1
\end{cases}
$$

$$
C_\rho = \begin{cases}
\{|0|_\rho, |1|_\rho, |2|_\rho, \ldots, |\ell-1|_\rho, |\ell|_\rho\} & \text{if } \rho \text{ finite} & \text{where } \ell = \max\{\rho\} \\
\{|\geq 0|_\rho\} & \text{if } \rho = \mathbb{N} \\
\{|0|_\rho, |1|_\rho, |2|_\rho, \ldots, |\ell-1|_\rho, |\geq \ell|_\rho\} & \text{if } \rho \neq \mathbb{N} \text{ cofinite} & \text{where } \ell = \max\{\mathbb{N} \setminus \rho\} + 1
\end{cases}
$$

Recall from Sect. 3 that when we write $|l|_\rho$ or $|l|_\sigma$, with a variable l, we mean the labels that are not equal to $|\geq \ell|_\rho$ or $|\geq \ell|_\sigma$. We refer to other labels by expressions such as $|l-1|_\rho$ and reserve the symbol ℓ to indicate the last labels $|\ell|_\sigma, |\geq \ell|_\sigma, |\ell|_\rho$, or $|\geq \ell|_\rho$.

Let C^{X_i} be the set of assignments of labels from C to the vertices in X_i. A label from C_σ for a vertex $v \in X_i$ indicates that v is in the solution set D in the partial solution, a label from C_ρ indicates that v is not. Furthermore, the numbers in the labels indicate the number of neighbours that v has in D; the $\geq$ symbol in the label $|\geq 1|_\rho$ indicates that v has this number of neighbours (one in this case) in D or more. E.g., for MINIMUM DOMINATING SET we have $\sigma = \mathbb{N}$ and $\rho = \mathbb{N} \setminus \{0\}$ and thus $C = \{|\geq 0|_\sigma, |0|_\rho, |\geq 1|_\rho\}$.

Now, the elements from C^{X_i} bijectively correspond to the above defined equivalence classes of partial solutions on G_i. Consequently, we can index the memoisation table A_i by C^{X_i}. To keep the dynamic programming recurrences in this paper simple, we will not store partial solutions in A_i, only the required partial solution values or counts. That is, from here on, let the table A_i be a function $A_i : C^{X_i} \to \{0, 1, .., M\} \cup \{\infty\}$ that assigns a number to each equivalence class of partial solutions. In an existence variant of a problem, we let $A_i(\vec{c})$, for $\vec{c} \in C^{X_i}$, be 0 or 1 indicating whether a partial solution of this equivalence class exists. In an optimisation variant, $A_i(\vec{c})$ indicates the size of a dominating partial solution in this equivalence class, or ∞ if no such partial solution exists. For convenience reasons[4], we let $A_i(\vec{c})$, for $\vec{c} \in C^{X_i}$, contain the size of the partial solution D' restricted to $V' \setminus X'$, i.e., the size of a corresponding partial solution equals $A_i(\vec{c})$ plus the number of σ labels in $\vec{c}$. In a counting variant, $A_i(\vec{c})$ indicates the number of partial solutions in the equivalence class of $\vec{c}$. Notice that for an existence variant, we can bound M by 1; for an optimisation variant, we can bound M by n; and for a counting variant, we can bound M by 2^n.

Below, we give explicit recurrences for A_i when solving a minimisation variant of a $[\sigma, \rho]$-domination problem by dynamic programming on a nice tree decomposition T. Modifying the recurrences to the existence or counting variant of the problem is an easy exercise. Extensions to the recurrences in which partial solutions are stored (for existence and optimisation variants) are easy to make, but tedious to write down formally. This is also to true for the extension to the optimisation counting variant where one needs to keep track of both the size and the number of such partial solutions.

Leaf Node. Let i be a leaf node of T. Since $X_i = \varnothing$, the only partial solution is $\varnothing$ with size zero: this size is stored for the empty vector $[]$.

$$A_i([]) = 0$$

Introduce Node. Let i be an introduce node of T with child node j. Let $X_i = X_j \cup \{v\}$ for some $v \in V \setminus V_j$. For $\vec{c} \in C^{X_j}$ and $c_v \in C$ the label for vertex v denote by $[\vec{c}, c_v]$ the vector $\vec{c}$ with the element c_v appended to it such that $[\vec{c}, c_v] \in C^{X_i}$. Now:

$$A_i([\vec{c}, c_v]) = \begin{cases} A_j([\vec{c}]) & \text{if } c_v \in \{|0|_\sigma, |\geq 0|_\sigma\} \text{ or } c_v \in \{|0|_\rho, |\geq 0|_\rho\} \\ \infty & \text{otherwise} \end{cases}$$

Here, G_i equals G_j with one added isolated vertex v. Hence, v can be in the partial solution or not, and both choices do not influence the partial solution size on $V_i \setminus X_i$ (which equals $V_j \setminus X_j$). Note that only one of the labels from $\{|0|_\sigma, |\geq 0|_\sigma\}$ and one from $\{|0|_\rho, |\geq 0|_\rho\}$ is used, and which depends on the specific $[\sigma, \rho]$-domination problem that we are solving.

Forget Node. Let i be a forget node of T with child node j. Let $X_i = X_j \setminus \{v\}$ for some $v \in X_j$. By definition of G_i, G_i contains edges between v and vertices

---

[4] In this way, we do not have to correct for double counting in join nodes in the rest of this paper.

in X_i while G_j does not. To account for these edges, we start by updating the given table A_j such that it accounts for the additional edges: that is, for an edge $\{u, v\}$ with $u \in X_i$, we adjust the counts of the number of neighbours expressed in the state colourings for u and v. We do so before we construct table A_i.

Let $[\vec{c}, c_u, c_v] \in C^{X_j}$ be such that c_u and c_v are labels for u and v respectively. For every edge $\{u, v\}$ with $u \in X_i$, we update A_j twice, once for u and once for v. We update A_j for u as follows:

$$A_j([\vec{c}, c_u, c_v]) := \begin{cases} A_j([\vec{c}, c_u, c_v]) & \text{if } c_v \in C_\rho \\ \infty & \text{if } c_v \in C_\sigma, c_u \in \{|0|_\rho, |0|_\sigma\} \\ A_j([\vec{c}, |l-1|_\rho, c_v]) & \text{if } c_v \in C_\sigma, c_u = |l|_\rho, l > 0 \\ A_j([\vec{c}, |l-1|_\sigma, c_v]) & \text{if } c_v \in C_\sigma, c_u = |l|_\sigma, l > 0 \\ \min\{A_j([\vec{c}, |\ell-1|_\rho, c_v]), A_j([\vec{c}, |{\geq}\ell|_\rho, c_v])\} & \text{if } c_v \in C_\sigma, c_u = |{\geq}\ell|_\rho \\ \min\{A_j([\vec{c}, |\ell-1|_\sigma, c_v]), A_j([\vec{c}, |{\geq}\ell|_\sigma, c_v])\} & \text{if } c_v \in C_\sigma, c_u = |{\geq}\ell|_\sigma \end{cases}$$

No update needs to be done if v is not in the partial solution D (first line). If c_u indicates that u has no neighbours in D while $v \in D$, then no such partial solution exists (second line). Otherwise, the counts in the label of u need to account for the extra neighbour. In the last four lines, we perform the required label update for all other labels giving special attention to the case where a $|{\geq}\ell|_\sigma$ or $|{\geq}\ell|_\rho$ label is used. Here, the minimum needs to be taken over two equivalence classes that through the added edge become equivalent: we take the minimum because we are solving the minimisation variant. Updating A_j for v goes identically with the roles of u and v switched, and as stated above, we perform this update for all edges incident to v in G_j.

Next, we compute A_i and start keeping track of equivalence classes based on X_i instead of based on X_j. To do so, we select a dominating solution from the partial solution equivalence classes for which v has a number of neighbours in D that corresponds to the specific $[\sigma, \rho]$-domination problem:

$$A_i(\vec{c}) = \min_{c_v \text{ a valid label}} A_j([\vec{c}, c_v])$$

Here, a valid label c_v is any label that corresponds to having the correct number of neighbours in D as defined by the specific $[\sigma, \rho]$-domination problem: c_v is a label $|l|_\sigma$ or $|l|_\rho$ for which $l \in \sigma$ or $l \in \rho$, respectively, or c_v is a label $|{\geq}\ell|_\rho$ or $|{\geq}\ell|_\sigma$ in case of cofinite σ or ρ.

Join Node. We give a simple procedure for the join node. We reconsider this step in the next section where we apply Theorem 2 to the join operation to prove Theorem 1.

Let A_i be the memoisation table for a join node i of T with child nodes l and r. A trivial algorithm to compute A_i would loop over all pairs of state colourings $\vec{c}_l, \vec{c}_r$ of X_i that agree on which vertices are in the solution set D, and then consider two corresponding partial solutions D_l on G_l and D_r on G_r and infer the state colouring $\vec{c}_i$ of the partial solution $D_l \cup D_r$ on G_i. It then stores in A_i the minimum size of a solution for each equivalence class for G_i. Note that the agreement on which vertices are in D is necessary for $D_l \cup D_r$ to be a valid partial solution: otherwise vertices that are no longer in X_i can obtain additional neighbours in D. At the same time the agreement is not a too tight restriction as any partial solution D on G_i can trivially be decomposed into partial solutions on G_l and G_r that agree on which vertices on X_i are in D.

Root Node. In the root node r of T (which is a forget node), $X_r = \varnothing$, $G_r = G$ and consequently $A_r([])$ is the minimum size of a $[\sigma, \rho]$-dominating set on G. This is the result we set out to compute.

Lemma 7. *Let $\mathcal{P}$ be the minimisation variant of a $[\sigma, \rho]$-domination problem with label set C using $s = |C|$ labels. Let $\mathcal{A}$ be an algorithm for the computations in a join node for problem $\mathcal{P}$ that, given a join node i with $|X_i| = k$ and the memoisation tables A_l and A_r for its child nodes, computes the memoisation table A_i in $\mathcal{O}(f(n, k))$ arithmetic operations. Then, given a graph G with a tree decomposition T of width t, $\mathcal{P}$ can be solved on G in $\mathcal{O}((s^{t+1}t + f(n, t+1))n)$ arithmetic operations.*

Proof. First transform T into a nice tree decomposition T' with $\mathcal{O}(n)$ nodes. If we show that the table A_j associated to any node j of T' can be computed in $\mathcal{O}(s^k k + f(n, k))$ arithmetic operations, then the result follows as $k \leq t + 1$. Consider the recurrences in the dynamic programming algorithm exposed above. The result trivially holds for leaf and root nodes, and also for the join nodes by definition of $\mathcal{A}$. It is easy to see that in the recurrences for the introduce and forget nodes, every value is computed using a constant amount of work. Since the tables are of size s^k, and for a forget node we need to do at most k update steps as we can add at most k edges, the result follows.   □

It is not difficult to modify the above algorithm to obtain:

Proposition 5. *Lemma 7 holds irrespective of $\mathcal{P}$ being a existence, maximisation, minimisation, counting, counting minimisation or counting maximisation variant of a $[\sigma, \rho]$-domination problem.*

4.3   Proof of Theorem 1

Reconsider the procedure for the join node in the dynamic programming algorithm in the previous section. It is not hard to show that one does not need to consider the partial solutions representing an equivalence class: given the state colourings $\vec{c}_l$ and $\vec{c}_r$, the state colouring of $D_l \cup D_r$ can be inferred directly. This is done as follows. Since G_l, G_r and $G_i = G_l \otimes G_r$ do not contain edges between vertices in X_i, for any vertex $v \in X_i$, the number of neighbours in D_l and D_r add up to the resulting number in $D_r \cup D_l$. As such, for any vertex v, if v has label $|l|_\sigma$ in $\vec{c}_r$ and $|l'|_\sigma$ in $\vec{c}_l$, then any combined partial solution has label $|l+l'|_\sigma$ for v in the state colouring that identifies the equivalence class (or label $|\geq \ell|_\sigma$ when $l + l' \geq \ell$ and the $|\geq \ell|_\sigma$ label is in C_σ). The same holds for labels $|l|_\rho$ from C_ρ.

By this reasoning, we have shown that the join operation comes down to computing $A_i(\vec{c}) = \min_{\vec{c}_l \oplus \vec{c}_r = \vec{c}} \{A_l(\vec{c}_f) + A_r(\vec{c}_g)\}$: when σ or ρ is finite, then combining labels behaves as standard addition within C_σ or C_ρ (option 1 in Sect. 1); while when σ or ρ is cofinite, then combining labels behaves as addition with maximum ℓ within C_σ or C_ρ (option 2 in Sect. 1). As such, we can compute this minimum by as described just above the statement of Theorem 2 in Sect. 1.2 by embedding the computations of the minimum in a larger structure and applying the convolution from Theorem 2.

Corollary 2. *Given a graph G with a tree decomposition T of G of width t, the optimisation variant of a $[\sigma, \rho]$-domination problem involving $s = |C|$ labels can be solved in $\mathcal{O}(s^{t+2}tn^2(t\log(s) + \log(n)))$ arithmetic operations on $\mathcal{O}(t\log(s) + \log(n))$-bit numbers.*

Proof. Following the above reasoning, we can perform a join operation by transforming the memoisation tables A_l and A_r into functions $f, g : C^{t+1} \times [n] \to [M]$ and then using Theorem 2. Note that the $t + 1$ is because $t \geq k - 1$ (the minus one in Definition 4).

The result follows by plugging this into Lemma 7 and observing that all arithmetic operations can be done using $\mathcal{O}(t\log(s) + \log(n))$-bit numbers: the sum of all the entries in A_l and A_r is at most s^k, hence $t\log(s)$ bits, while we need the additional $\log(n)$ bits to store partial solution sizes. □

Theorem 1. *Given a graph G with a tree decomposition T of G of width t, the following problem variants of a $[\sigma, \rho]$-domination problem involving s states (or labels) can be solved in:*

- *Existence: $\mathcal{O}(s^{t+2}t^2n\log(s))$ operations on $\mathcal{O}(t\log(s))$-bit numbers.*
- *Optimisation: $\mathcal{O}(s^{t+2}tn^2(t\log(s)+\log(n)))$ operations on $\mathcal{O}(t\log(s)+\log(n))$-bit numbers.*
- *Counting: $\mathcal{O}(s^{t+2}t^2n\log(s))$ operations on $\mathcal{O}(n)$-bit numbers.*
- *Counting optimisation: $\mathcal{O}(s^{t+2}tn^2(t\log(s) + \log(n)))$ operations on $\mathcal{O}(n)$-bit numbers.*

Proof. The result for the optimisation problem follows from Corollary 2.

For the counting optimisation problem, we use the same construction, only without needing the embed the min-plus semiring in a larger structure: at every step of the algorithm, we let $A_i(\vec{c}, \kappa)$ be the number of partial solutions of size κ that correspond to the equivalence class identified by $\vec{c}$. The join then also comes down to evaluating Eq. 2 directly, resulting in the same amount of arithmetic operations. For the existence and counting problems, we can also use Eq. 2 directly while not needing to add the size-parameter κ to $A_i(\vec{c})$ as solution sizes do not matter. Redoing the analysis of the resulting algorithm gives in the claimed amount of arithmetic operations.

For both counting problems variants, we need $\mathcal{O}(n)$-bit numbers as there can be $\mathcal{O}(2^n)$ solutions to count. For the existence problem, we need $\mathcal{O}(t\log(s))$ as the sum of all entries in a $\mathcal{O}(s^{t+1})$ table with zero-one entries can be at most $\mathcal{O}(s^{t+1})$. □

5   A Note on Modular Arithmetic

Recall from the introduction, that we consider operations on functions/tables $f, g : C^k \times [n] \to [M]$ for some (possibly large) integer M. To use the presented algebraic transforms, we embed M in $\mathbb{F}_p$: we can do the same computations in $\mathbb{F}_p$ as long as $p > M$. However, we need to choose p appropriately such that the resulting field $\mathbb{F}_p$ has the root(s) of unity required for the Fourier transforms. Below we give a short description of how this can be done.

Let $r_1, r_2, \ldots, r_k$ be distinct integers. We look for a prime number p such that $\mathbb{F}_p$ contains, for all i, an r_i-th root of unity. To find the prime p, we consider candidates $m_j = 1 + jR$, where $R = \prod_{i=1}^{k} r_i$, for j large enough such that $m_j > M$. By the prime number theorem for arithmetic progressions, the sequence $(m_j)_{j=1}^{\infty}$ contains $\mathcal{O}(\frac{1}{\phi(R)} \frac{x}{\ln(x)})$ prime numbers less than x, where ϕ is Euler's totient function. Since prime testing can be done in polynomial time, we can look for the first candidate $m_j > M$ that is prime and choose p as such.

By Euler's theorem, for any $x \in \mathbb{F}_p$, with p chosen as in the previous paragraph: $1 = x^{\phi(p)} = x^{p-1} = x^{jR}$. As such, for any $x \in \mathbb{F}_p$, x^l with $l = \frac{jR}{r_i}$ is an r_i-th root of unity if $(x^l)^i \neq 1$ for all $i < r_i$. Finding an appropriate x is not difficult for small r_i as an $\frac{1}{r_i}$-th fraction of all elements $x \in \mathbb{F}_p$ results in x^l being an r_i-th root of unity. To see this, consider a generator g of the multiplicative subgroup of $\mathbb{F}_p$. The sequence $g^1, g^2, \ldots, g^{p-1}$ equals all elements in $\mathbb{F}_p \setminus \{0\}$. Putting this sequence to the power l gives $g^l, g^{2l}, \ldots, g^{(p-1)l}$ which, by choice of l, equals $\omega_{r_i}^1, \omega_{r_i}^2, \ldots, \omega_{r_i}^{(p-1)}$, where ω_{r_i} is an r_i-th root of unity in $\mathbb{F}_p$. Clearly, this forms l times the sequence $\omega_{r_i}^1, \omega_{r_i}^2, \ldots, \omega_{r_i}^{r_i}$, as $\omega_{r_i}^{r_i} = 1$.

Notice that for cyclic convolution, we need an r-th root of unity when a cycle length of r is needed. For non-cyclic convolution, we have more freedom in choosing the prime p as using any r-th root of unity with r larger (but in the same order of magnitude) than the domain size suffices: this results in additional padding with zeroes in the proof of Lemma 5.

6   Conclusion

In this paper, we have shown how Möbius and Fourier transforms can be combined on the same domain the obtain a fast algorithm very general form of convolution. Our application was speeding-up computations for dynamic programming algorithms on tree decompositions. This led us to the currently fastest algorithm for the general case of the $[\sigma, \rho]$-domination problems on tree decompositions. Additionally, we generalised the covering product from [2] from being defined on the subset lattice to more general partial orders (Lemma 3 and Corollary 1).

Acknowledgements. The author would like to thank Gerard Tel and Rolf Plagmeijer for several useful discussions.

References

1. Alber, J., Bodlaender, H.L., Fernau, H., Kloks, T., Niedermeier, R.: Fixed parameter algorithms for dominating set and related problems on planar graphs. Algorithmica **33**(4), 461–493 (2002)
2. Björklund, A., Husfeldt, T., Kaski, P., Koivisto, M.: Fourier meets Möbius: fast subset convolution. In: Johnson, D.S., Feige, U. (eds.) 39th Annual ACM Symposium on Theory of Computing, STOC 2007, pp. 67–74. ACM Press (2007)
3. Björklund, A., Husfeldt, T., Kaski, P., Koivisto, M., Nederlof, J., Parviainen, P.: Fast zeta transforms for lattices with few irreducibles. ACM Trans. Algorithms **12**(1), 41–419 (2015)
4. Bodlaender, H.L.: Dynamic programming on graphs with bounded treewidth. In: Lepistö, T., Salomaa, A. (eds.) ICALP 1988. LNCS, vol. 317, pp. 105–118. Springer, Heidelberg (1988). https://doi.org/10.1007/3-540-19488-6_110
5. Bodlaender, H.L.: A tourist guide through treewidth. Acta Cybernet. **11**(1–2), 1–22 (1993)
6. Bodlaender, H.L.: Treewidth: algorithmic techniques and results. In: Prívara, I., Ružička, P. (eds.) MFCS 1997. LNCS, vol. 1295, pp. 19–36. Springer, Heidelberg (1997). https://doi.org/10.1007/BFb0029946
7. Bodlaender, H.L.: A partial k-arboretum of graphs with bounded treewidth. Theoret. Comput. Sci. **209**(1–2), 1–45 (1998)
8. Bodlaender, H.L., Cygan, M., Kratsch, S., Nederlof, J.: Deterministic single exponential time algorithms for connectivity problems parameterized by treewidth. Inf. Comput. **243**, 86–111 (2015)
9. Bodlaender, H.L., Koster, A.M.C.A.: Combinatorial optimization on graphs of bounded treewidth. Comput. J. **51**(3), 255–269 (2008)
10. Borradaile, G., Le, H.: Optimal dynamic program for r-domination problems over tree decompositions. In: Guo, J., Hermelin, D. (eds.) 11th International Symposium on Parameterized and Exact Computation, IPEC 2016. Leibniz International Proceedings in Informatics, vol. 63, pp. 8:1–8:23. Schloss Dagstuhl - Leibniz-Zentrum fuer Informatik (2017)
11. Cooley, J.W., Tukey, J.W.: An algorithm for the machine calculation of complex fourier series. Math. Comput. **19**(90), 297–301 (1965)
12. Cygan, M., Kratsch, S., Nederlof, J.: Fast hamiltonicity checking via bases of perfect matchings. In: Boneh, D., Roughgarden, T., Feigenbaum, J. (eds.) 42nd Annual ACM Symposium on Theory of Computing, STOC 2010, pp. 301–310. ACM Press (2013)
13. Cygan, M., Nederlof, J., Pilipczuk, M., Pilipczuk, M., van Rooij, J.M.M., Wojtaszczyk, J.O.: Solving connectivity problems parameterized by treewidth in single exponential time. arXiv.org. The Computing Research Repository abs/1103.0534 (2011)
14. Cygan, M., Nederlof, J., Pilipczuk, M., Pilipczuk, M., van Rooij, J.M.M., Wojtaszczyk, J.O.: Solving connectivity problems parameterized by treewidth in single exponential time. In: Ostrovsky, R. (ed.) 52nd Annual IEEE Symposium on Foundations of Computer Science, FOCS 2011, pp. 150–159. IEEE Computer Society (2011)
15. Cygan, M., Pilipczuk, M.: Exact and approximate bandwidth. Theoret. Comput. Sci. **411**(40–42), 3701–3713 (2010)
16. Katsikarelis, I., Lampis, M., Paschos, V.T.: Structurally parameterized d-scattered set. In: Brandstädt, A., Köhler, E., Meer, K. (eds.) WG 2018. LNCS, vol. 11159, pp. 292–305. Springer, Cham (2018). https://doi.org/10.1007/978-3-030-00256-5_24

17. Kennes, R.: Computational aspects of the moebius transform of a graph. IEEE Trans. Syst. Man Cybern. **22**, 201–223 (1991)
18. Kloks, T.: Treewidth, Computations and Approximations. LNCS, vol. 842. Springer, Heidelberg (1994). https://doi.org/10.1007/BFb0045375
19. Lokshtanov, D., Marx, D., Saurabh, S.: Known algorithms on graphs of bounded treewidth are probably optimal. In: Randall, D. (ed.) 22st Annual ACM-SIAM Symposium on Discrete Algorithms, SODA 2011, pp. 777–789. Society for Industrial and Applied Mathematics (2011)
20. Rader, C.M.: Discrete fourier transforms when the number of data samples is prime. Proc. IEEE **56**(6), 1107–1108 (1968)
21. van Rooij, J.M.M.: Exact Exponential-Time Algorithms for Domination Problems in Graphs. Ph.D. thesis, Department of Information and Computing Sciences, Utrecht University, Utrecht, The Netherlands (2011)
22. van Rooij, J.M.M., Bodlaender, H.L., Rossmanith, P.: Dynamic programming on tree decompositions using generalised fast subset convolution. In: Fiat, A., Sanders, P. (eds.) ESA 2009. LNCS, vol. 5757, pp. 566–577. Springer, Heidelberg (2009). https://doi.org/10.1007/978-3-642-04128-0_51
23. Telle, J.A.: Complexity of domination-type problems in graphs. Nordic J. Comput **1**(1), 157–171 (1994)
24. Włodarczyk, M.: Clifford algebras meet tree decompositions. Algorithmica **81**(2), 497–518 (2019)
25. Yates, F.: The design and analysis of factorial experiments. Technical Communication No. 35, Commonwealth Bureau of Soil Science (1937)

A Generalization of a Theorem
of Rothschild and van Lint

Ning Xie[1], Shuai Xu[2(✉)], and Yekun Xu[1]

[1] Florida International University, 11200 SW 8th Street, Miami, USA
{nxie,yxu040}@cis.fiu.edu
[2] Case Western Reserve University, Cleveland, OH 44106, USA
sxx214@case.edu

Abstract. A classical result of Rothschild and van Lint asserts that if every non-zero Fourier coefficient of a Boolean function f over $\mathbb{F}_2^n$ has the same absolute value, namely $|\hat{f}(\alpha)| = 1/2^k$ for every α in the Fourier support of f, then f must be the indicator function of some affine subspace of dimension $n - k$. In this paper we slightly generalize their result. Our main result shows that, roughly speaking, Boolean functions whose Fourier coefficients take values in the set $\{-2/2^k, -1/2^k, 0, 1/2^k, 2/2^k\}$ are indicator functions of two disjoint affine subspaces of dimension $n - k$ or four disjoint affine subspace of dimension $n - k - 1$. Our main technical tools are results from additive combinatorics which offer tight bounds on the affine span size of a subset of $\mathbb{F}_2^n$ when the doubling constant of the subset is small.

Keywords: Fourier analysis · Boolean functions · Additive combinatorics · Affine subspaces

1 Introduction

One of the most fruitful approaches in functional analysis is to represent functions as sums of simple and well-structured objects, such as sine wave functions and polynomials. Such representations often provide additional insights on the combinatorial structures of or complexity measures associated with the subjects under consideration. This paradigm in theoretical computer science has witnessed harmonic analysis on the cube, or the discrete Fourier transform of Boolean functions, emerged in the past three decades as a powerful and versatile tool that finds numerous applications in complexity theory (such as PCP and circuit complexity), property testing, learning, cryptography, coding theory, social choice theory and others; see [25] for a comprehensive survey.

Fourier coefficients and function values are two equivalent ways to represent a function. That is, the Fourier spectrum of a function completely determines the function-value at any point on the cube. However, knowing only the *values* of the Fourier spectrum but without the information of the locations of these values in the Fourier space in general leaves the function undetermined to a large extent,

© Springer Nature Switzerland AG 2021
R. Santhanam and D. Musatov (Eds.): CSR 2021, LNCS 12730, pp. 460–483, 2021.
https://doi.org/10.1007/978-3-030-79416-3_28

even restricted to Boolean functions. To see this, consider the following examples. Generally speaking, we view two Boolean functions as the same function if they are *isomorphic*. More formally, we say that two Boolean functions $f, g : \mathbb{F}_2^n \to \{0, 1\}$ are *isomorphic* to each other if there is an invertible linear transformation $L : \mathbb{F}_2^n \to \mathbb{F}_2^n$ such that $g(x) = Lf(x)$ for every $x \in \mathbb{F}_2^n$, where $Lf(x) := f(Lx)$. Now consider the following two families of Boolean functions $\{f_k : \mathbb{F}_2^k \to \{0, 1\} \mid k \in \mathbb{N}, k \geq 3\}$ and $\{g_k : \mathbb{F}_2^k \to \{0, 1\} \mid k \in \mathbb{N}, k \geq 3\}$, with the Fourier expansions of $f_k(x) = \frac{3}{4} - \frac{1}{4}\chi_{\{1\}}(x) - \frac{1}{4}\chi_{\{2\}}(x) - \frac{1}{4}\chi_{\{1,2\}}(x)$ and $g_k(x) = \frac{3}{4} - \frac{1}{4}\chi_{\{1,2\}}(x) - \frac{1}{4}\chi_{\{1,3\}}(x) - \frac{1}{4}\chi_{\{2,3\}}(x)$. One can check easily that both f_k and g_k are indeed Boolean functions and the multisets of non-zero Fourier coefficients are both $\{\frac{3}{4}, -\frac{1}{4}, -\frac{1}{4}, -\frac{1}{4}\}$. On the other hand, the Fourier dimension—dimension of the subspace spanned by vectors at which the function's Fourier coefficients are non-zero—of f_k is 2 while the Fourier dimension of g_k is 3. Since the Fourier spectrum transforms according to $(L^T)^{-1}$ when the function undergoes the linear transformation L, it follows that there is no invertible linear transformation L that maps f_k to g_k, i.e. they are not isomorphic to each other. Another such example is the class of *address functions* $f_n : \mathbb{F}_2^n \to \{-1, 1\}$, where $n = k + 2^k$ for some positive integer k, together with the class of functions $g_n : \mathbb{F}_2^n \to \{-1, 1\}$ formed by tensoring some *bent function* on $2k$-bits with a δ-function on $n - 2k$ bits. Then both f_n and g_n have 2^{2k} non-zero Fourier coefficients, with $2^{2k-1} + 2^{k-1}$ of them taking value $1/2^k$ and $2^{2k-1} - 2^{k-1}$ of them taking value $-1/2^k$; moreover, since the Fourier dimension of f_n is n and the Fourier dimension of g_n is $2k < n$, these two functions are not isomorphic to each other.

Nevertheless, there are a few exceptions to the general phenomenon in the sense that knowing only the values of the Fourier spectrum completely determine the Boolean function, up to an isomorphism. One such example is the indicator function of an affine subspace, which enjoys a very simple Fourier spectrum. Specifically, if f is the indicator function of an affine subspace in $\mathbb{F}_2^n$ of dimension $n - k$, then it is straightforward to check that every non-zero Fourier coefficient of f is either $1/2^k$ or $-1/2^k$. What about the converse? Namely, if we know that the non-zero Fourier coefficients of a Boolean function all have magnitude $1/2^k$, then what can be said about the function?

1.1   Rothschild and van Lint Theorem

Rothschild and van Lint [28] (see also Chap. 13, Lemma 6 in [24]) proved the following theorem:

Theorem 1. *Let $n \geq 1$ and $0 \leq k \leq n$. Let $f = \mathbb{1}_S$ be the indicator function of a set $S \subseteq \mathbb{F}_2^n$ of size $|S| = 2^{n-k}$. If for every $\alpha \in \mathbb{F}_2^n$, $|\hat{f}(\alpha)|$ is equal to either zero or $1/2^k$, then S is an affine subspace of dimension $n - k$.*

In other words, Rothschild and van Lint Theorem shows that, up to an invertible linear transform, we have a complete characterization when the Fourier coefficients of a *Boolean* function are all from the set $\{-1/2^k, 0, 1/2^k\}$: the Boolean function must be the indicator of some affine subspace of co-dimension k.

A natural question is: how far can we extend such a nice characterization in terms of the values of Fourier coefficients only? Following [14], for a rational number x, the *granularity* $\text{gran}(x)$ of x is defined to be the least nonnegative integer k such that $x = m/2^k$, where m is an (odd) integer. A function $\mathbb{F}_2^n \to \mathbb{R}$ is said to be *k-granular* if the maximum granularity of its Fourier coefficients is k— that is, $k = \max_\alpha \{\text{gran}(\hat{f}(\alpha))\}$. For a Boolean function, its granularity is known to be intimately correlated with its *Fourier sparsity* [14]—the number of non-zero Fourier coefficients; see discussion in Sect. 1.4 for more details. Therefore, one can view Rothschild and van Lint Theorem as a characterization of k-granular Boolean functions with minimum support size (that is, $\hat{f}(\mathbf{0}) = |\{x : f(x) = 1\}|/2^n = 1/2^k$).

1.2   Our Results

In this work, we slightly generalize Rothschild and van Lint Theorem to give a complete characterization of k-granular Boolean functions of support size $2^n \cdot 2/2^k = 2^{n-k+1}$. Roughly speaking, our main theorem is the following:

Theorem 2 (Informal statement). *For large enough integers $n \geq k$, if a Boolean function $f : \mathbb{F}_2^n \to \{0,1\}$ has all its Fourier coefficients in the set $\{0, \frac{\pm 1}{2^k}, \frac{\pm 2}{2^k}\}$, then f is the indicator function of disjoint union of two affine subspaces of dimension $n - k$.*

Our Main Theorem is based on the following Main Lemma, which deals with the general case of $k \geq 5$, together with case analysis[1] for small values of k.

Lemma 1 (Main). *Let $k \geq 5$ and $n \geq k$ be integers. Let $f : \mathbb{F}_2^n \to \{0,1\}$ be a Boolean function such that $\hat{f}(\mathbf{0}) = 1/2^{k-1}$ and any other Fourier coefficients are either zero or equal to $\pm \frac{1}{2^k}$, then f is the indicator function of a disjoint union of two dimension $n - k$ affine subspaces.*

1.3   Proof Overview and Our Techniques

The original form of Rothschild and van Lint Theorem was stated to characterize subspaces in affine geometry and projective geometry. For completeness and more importantly, because the first step in our proof of the main theorem follows a similar strategy, we present a slightly different proof using the notation of Fourier analysis.

A proof of Rothschild and van Lint Theorem. We prove the theorem by induction on n. It is trivial to see that the theorem holds for $n = 1$ (for both $k = 0$ and $k = 1$). Let $n \geq 2$. Clearly there is nothing to prove for $k = 0$ and $k = n$, so we assume $0 < k < n$. Note that $\hat{f}(\mathbf{0}) = |S|/2^n = 1/2^k$, then by Parseval's identity, there exists a non-zero α such that $\hat{f}(\alpha) = 1/2^k$ or $-1/2^k$. Assume that

---

[1] The need for a nasty case analysis stems from a key lemma in the proof, namely Lemma 4, which holds only when $k \geq 5$.

$\hat{f}(\alpha) = 1/2^k$ and the case of $\hat{f}(\alpha) = -1/2^k$ is similar. Applying an invertible linear transform L that maps α to e_1, where e_1 stands for the standard basis vector $(1, 0, \ldots, 0)$. Note that both the Fourier spectrum of f and any affine subspace are invariant under invertible linear transformations, hence it suffices to argue about $g := Lf$. Now we have $\hat{g}(0) = \hat{g}(e_1) = 1/2^k$. Applying a linear restriction over the first bit of the input to get sub-functions g_0 and g_1 (see Proposition 2 in Appendix A for details). By (2), $\hat{g}_1(0) = \hat{g}(0) - \hat{g}(e_1) = 0$, which implies that g_1 is the zero-function. This implies that S is completely contained in the support of g_0 and moreover, by (3), $\hat{g}_0(\beta) = 2\hat{f}(0, \beta)$ for every $\beta \in \mathbb{F}_2^{n-1}$. In other words, g_0 is a Boolean function over $\mathbb{F}_2^{n-1}$ and $|\hat{g}(\beta)|$ is equal to either zero or $1/2^{k-1}$, therefore the induction hypothesis applies to g_0. It follows that S is an affine subspace of dimension $n - 1 - (k - 1) = n - k$. This completes the proof of Theorem 1.

Reducing the Dimension of the Function Domain. The proof of the Main Theorem is much more involved than that of Rothschild and van Lint Theorem. In fact, the proof we described above of Theorem 1 is the first step toward proving the main theorem. The reduction step in the proof of Theorem 1 can be regarded as reducing the dimension of function domain while keeping all the support of the function. Equivalently, one may view the reduction step as decomposing the original function f as a *tensor product* between a "core-function" g and a "δ-function" h (see Sect. 2 for definition of tensor product of Boolean functions). Namely, $f(x, y) = g(x) \otimes h(y)$, where $h : \mathbb{F}_2^m \to \{0, 1\}$ is the δ-function: $h(y) = 1$ if $y = 0^m$ and $h(y) = 0$ for all other vectors. That is, f is "reduced" to a core-function g with dimension $n - m$. To this end, we say a function $f : \mathbb{F}_2^n \to \{0, 1\}$ is *reducible* if there exists an invertible linear transformation L such that Lf can be decomposed as the tensor product of a function $g : \mathbb{F}_2^{n-m} \to \{0, 1\}$ and a δ-function h over $\mathbb{F}_2^m$ with $m \geq 1$. f is said to be *irreducible* if f is not reducible.[2] Now we are ready to present our Main theorem more precisely.

Theorem 3 (Main). *Let $k \geq 1$, $n > k$ be two integers, and let $f : \mathbb{F}_2^n \to \{0, 1\}$ be a non-trivial[3] Boolean function with all its Fourier coefficients taking values in $\{0, \frac{\pm 1}{2^k}, \frac{\pm 2}{2^k}\}$. Then we have the following complete characterization*

- *If $\hat{f}(0) = \frac{1}{2^k}$, then f is the indicator function of an affine subspace of dimension $n - k$ (Rothschild and van Lint Theorem);*
- *If $\hat{f}(0) = \frac{1}{2^{k-1}}$ and f is irreducible, then f is either the indicator function of disjoint union of two affine subspaces of dimension $n - k$, or the indicator function of disjoint union of four affine subspaces of dimension $n - k - 1$. Moreover, the latter case is only possible when $k = 4$.*

Back to our problem, since $\hat{f}(0) = 1/2^{k-1}$, it is easy to see that whenever there is a non-zero α such that $|\hat{f}(\alpha)| = 1/2^{k-1}$, we can restrict f either to the

---

[2] To put it differently, a function f defined on $\mathbb{F}_2^n$ is irreducible if and only if the minimum dimension of the affine subspace containing the support of f is n.

[3] A Boolean function is *trivial* if $f \equiv 0$ or $f \equiv 1$.

subspace $\langle \alpha, x \rangle = 0$ or to the affine subspace $\langle \alpha, x \rangle = 1$ while keeping the entire support of f. We repeat this process until we reach a Boolean function f with $\hat{f}(\mathbf{0}) = 1/2^{k-1}$ and all other non-zero Fourier coefficients have magnitude $1/2^k$.

Additive Structures of the Fourier Spectrum. The starting point of our main argument is the following well-known *characterization* of Boolean functions in terms of their Fourier spectra: a function $f : \mathbb{F}_2^n \to \mathbb{R}$ on the cube is Boolean if and only if

$$\hat{f}(\alpha) = \sum_{\beta \in \mathbb{F}_2^n} \hat{f}(\beta)\hat{f}(\alpha + \beta)$$

holds for every $\alpha \in \mathbb{F}_2^n$. Our main observation is that, since the non-zero Fourier coefficients f can take only two values when f is irreducible, denoting $A := \{\alpha \mid \hat{f}(\alpha) = 1/2^k\}$ and $B := \{\beta \mid \hat{f}(\beta) = -1/2^k\}$, then these two sets—viewed as subsets of abelian group $\mathbb{F}_2^n$—must exhibit strong *additive structures*. Indeed, one can show that $B + B \subseteq A \cup \{\mathbf{0}\}$ and consequently $|B + B|/|B| \le (1 + |A|)/|B|$.

What can be said about a set B if its *doubling constant* $K := |B + B|/B$ is small? This is a classical problem extensively studied in additive combinatorics. Additive combinatorics is a burgeoning mathematics sub-area which finds exciting applications in theoretical computer science in recent years [1–3,5,30]. Green and Tao [16] proved that, when the underlying ambient group is $\mathbb{F}_2^n$, then B is contained in a subspace of size $2^{2K+O(K \log K)}|B|$, which is asymptotically optimal. Unfortunately, such *asymptotic* "high end" bounds are not accurate enough to be useful for our problem. In fact, we make crucial use of a "low end" additive combinatorics result of Even-Zohar [10], which provides tight bounds on the size of affine span of B in terms of its doubling constant. It is worth noting that all aforementioned applications of additive combinatorics in theoretical computer science employ theorems regarding *asymptotic* behaviors of certain combinatorial objects. We hope researchers may find further applications of such "low end" additive combinatorics results in other places.

1.4   Motivations and Related Work

To the best of our knowledge, besides the work of Rothschild and van Lint, there is no previous structural result on Boolean functions in terms the *magnitudes* of their Fourier coefficients only. Friedgut [12] showed that if the total influence of a Boolean function is small, then it is close to some junta—a function that depends only on a bounded number of variables. Friedgut *et al.* [13] studied Boolean functions whose Fourier mass are concentrated on the lowest two levels and proved that such functions are close to parity functions or negations of parity functions. For a special class of Boolean functions, the so-called *linear threshold functions*, a celebrated result of Chow [8] states that these functions are completely determined by their lowest two level Fourier coefficients; see [9,26] for recent robust versions as well as algorithmic versions of Chow's theorem. Note that all previous structural theorems mentioned above, except Chow's, are "robust" in the following sense: the structural results are robust against small perturbations in the Boolean function's Fourier spectrum. Our main result is automatically robust: by Parseval's identity, small distance in Fourier spectrum implies small distance in function space; consequently, any Boolean function whose Fourier coefficients are close to being in the form stated in our Main Theorem must also be close to having the affine subspace structures asserted in the theorem.

Apart from studying to what extent can the values of Fourier coefficients themselves determine a Boolean function, an important motivation of this research is to study the behaviors of *Fourier sparse* Boolean functions [14]. Gopalan *et al.* [14] proved that, if a Boolean function f has only s non-zero Fourier coefficients, then every Fourier coefficient of f is of the form $m/2^k$, where m is an integer and $k/2 \leq \log s \leq k$. That is, the granularity and Fourier sparsity of a Boolean function are, up to a constant factor, identical. Our result may be regarded as characterizing Boolean functions of Fourier granularity k when all Fourier coefficients of f are between $-2/2^k$ and $2/2^k$.

Probably the most prominent open problem in communication complexity is the so-called *Log-rank Conjecture* proposed by Lovász and Saks [21], which asserts that the deterministic communication complexity of any $F : \mathbb{F}_2^n \times \mathbb{F}_2^n \to \{0,1\}$, $\mathsf{D}^{\mathsf{CC}}(F)$, is upper bounded by a polynomial of the logarithm of the rank of the communication matrix $M_F = [F(x,y)]_{x,y}$, where the rank is taken over the reals. Even after more than 30 years of extensive study, we are still very far from resolving it; the current best bound is Lovett's $\mathsf{D}^{\mathsf{CC}}(F) = O(\sqrt{r} \log r)$ [22], where r is the rank of M_F. Recently, studying the Log-rank conjecture for a special class of two-party functions, the so-called *XOR functions*, has attracted much attention [7,17,20,32,34–36]. The corresponding conjecture for this special class of functions is sometimes called *Log-rank XOR conjecture*. Specifically, F is an XOR function if there exists an $f : \mathbb{F}_2^n \to \{0,1\}$ such that for all x and y, $F(x,y) = f(x+y)$. The beautiful connection between the Log-rank XOR conjecture and Fourier analysis of Boolean functions is that, if F is an XOR function, then the rank of M_F is just the Fourier sparsity of f [4]. Moreover, it is now known that resolving the Log-rank XOR conjecture is equivalent to finding a *parity decision tree* of depth polylog(s), or poly(k) for any Boolean function f [17,34,36], where s is the Fourier sparsity and k is the granularity of f.

The parity kill number of a Boolean function f is defined as

$$C_{\oplus,\min}(f) := \min\{\text{co-dim}(S) \mid S \text{ is an affine subspace on which } f \text{ is constant}\}$$

Tsang *et al.* [34] demonstrated that, to resolve the Log-rank XOR conjecture, it is sufficient to prove that the kill number of any Boolean function f is upper bounded by polylog(s) or poly(k). See [6,27] for recent developments on constructing Boolean functions with large kill numbers. Our main result can be regarded as showing that any Boolean function with granularity k and $\hat{f}(\mathbf{0}) \leq 2/2^k$ has kill number at most $k + 1$. In fact, by induction on m and folding $\hat{f}(\mathbf{0})$ with any other non-zero Fourier coefficient, we immediately have the following corollary.

Corollary 1. *Let* $f : \mathbb{F}_2^n \to \{0,1\}$ *be a Boolean function with granularity* k *and* $\hat{f}(0) = m/2^k$. *Then the kill number of* f *is at most* $k + m - 1$.

Of course, Corollary 1 is still very far from showing the desired kill number bound $\mathrm{poly}(k)$ as m can be as large as 2^{k-1}, but it is hoped that further investigations along this approach may lead to more interesting results.

1.5 Organization

The rest of the paper is organized as follows. Preliminaries and notations that we use throughout the paper are summarized in Sect. 2. We prove our Main Lemma, which deals with the cases when k is at least 5 in Sect. 3, while the small value cases are discussed in Sect. 4. Then, by combining these two ingredients, we prove our Main Theorem in Sect. 5. Finally we end with a brief section of conclusions and open questions.

2 Preliminaries

All logarithms in this paper are to the base 2. Let $n \geq 1$ be a natural number, then $[n]$ denotes the set $\{1, \ldots, n\}$. We use $\mathbb{F}_2$ for the field with 2 elements $\{0,1\}$, where addition and multiplication are performed modulo 2. We view elements in $\mathbb{F}_2^n$ as n-bit binary strings, i.e. elements in $\{0,1\}^n$, interchangeably. If x and y are two n-bit strings, then $x + y$ (or $x - y$) denotes bitwise addition (i.e. XOR) of x and y. For positive integers m and n, if $y \in \mathbb{F}_2^m$ and $z \in \mathbb{F}_2^n$, then we write $x = (y, z)$ to denote the binary string $x \in \mathbb{F}_2^{m+n}$ obtained from concatenating y and z together. We view $\mathbb{F}_2^n$ as a vector space equipped with an inner product $\langle x, y \rangle$, which we take to be the standard dot product: $\langle x, y \rangle = \sum_{i=1}^{n} x_i y_i$, where all operations are performed in $\mathbb{F}_2$.

2.1 Boolean Functions and Fourier Analysis

We often use f to denote a real function defined on $\mathbb{F}_2^n$ and write $\mathrm{supp}(f) = \{x \in \mathbb{F}_2^n \mid f(x) \neq 0\}$ for the *support* of f. Sometimes we view f as a 2^n-dimensional vector, e.g. write $f = \mathbf{0}$ and $f = \mathbf{1}$ to denote the trivial all-zero function and all-one function, respectively. In this paper, a function f is *Boolean* if its range is $\{0,1\}$.

For every $\alpha \in \mathbb{F}_2^n$, one can define a *linear function* (or *parity function*) mapping $\mathbb{F}_2^n$ to $\{0,1\}$ as $\ell_\alpha(x) = \langle \alpha, x \rangle$. Let $\chi_\alpha = (-1)^{\ell_\alpha}$, which are commonly known as *characters*. For functions $f, g: \mathbb{F}_2^n \to \mathbb{R}$ the inner product is defined as $\langle f, g \rangle := \mathbb{E}_{x \in \mathbb{F}_2^n}(f(x)g(x))$. For $\alpha = (\alpha_1, \ldots, \alpha_n) \in \mathbb{F}_2^n$, the corresponding character function χ_α is defined as $\chi_\alpha(x_1, \ldots, x_n) = \prod_{i:\, \alpha_i = 1}(-1)^{x_i} = (-1)^{\langle \alpha, x \rangle}$. For $\alpha, \beta \in \mathbb{F}_2^n$, the inner product between χ_α and χ_β is 1 if $\alpha = \beta$, and 0 otherwise. Therefore the characters form an orthonormal basis for real-valued functions over $\mathbb{F}_2^n$, and we can expand any f defined on $\mathbb{F}_2^n$ using $\{\chi_\alpha\}_{\alpha \in \mathbb{F}_2^n}$ as a basis.

Definition 1 (Fourier Transform). *Let* $f \colon \mathbb{F}_2^n \to \mathbb{R}$. *The* Fourier transform $\hat{f} \colon \mathbb{F}_2^n \to \mathbb{C}$ *of* f *is defined to be* $\hat{f}(\alpha) = \mathbb{E}_x(f(x)\chi_\alpha(x))$. *The quantity* $\hat{f}(\alpha)$ *is called the* Fourier coefficient *of* f *at* α.

The Fourier inversion formula is given by $f(x) = \sum_{\alpha \in \mathbb{F}_2^n} \hat{f}(\alpha)\chi_\alpha(x)$, and the Parseval's identity is $\sum_{\alpha \in \mathbb{F}_2^n} \hat{f}(\alpha)^2 = \mathbb{E}_x(f(x)^2)$. The Fourier sparsity of f, denoted by $\|\hat{f}\|_0$, is the number of nonzero Fourier coefficients of f.

Fourier Characterization of Boolean Functions. Our proof crucially relies on the following characterization of Boolean functions in terms of their Fourier spectra. We give a proof for completeness.

Proposition 1 (Folklore). *A function* $f \colon \mathbb{F}_2^n \to \mathbb{R}$ *defined on the hypercube is Boolean if and only if for every* $\alpha \in \mathbb{F}_2^n$,

$$\hat{f}(\alpha) = \sum_{\beta \in \mathbb{F}_2^n} \hat{f}(\beta)\hat{f}(\alpha + \beta). \tag{1}$$

Proof. This follows from the fact that f is Boolean if and only if $f^2(x) - f(x) = 0$ for every x. Now expand the left-hand side in terms of Fourier coefficients and notice that, since the right-hand side is the **0**-function, all of its Fourier coefficients all zero. Comparing each pair of the corresponding Fourier coefficients on both sides gives the desired equality. $\qquad\square$

Linear Restrictions. The following is a folklore theorem regarding the effect of linear restrictions on the Fourier spectrum of a function defined over the Boolean hypercube. We include a proof in Appendix A for completeness.

Proposition 2. *Let* $f \colon \mathbb{F}_2^n \to \mathbb{R}$ *be a function defined on the Boolean hypercube. Let* $f_0, f_1 \colon \mathbb{F}_2^{n-1} \to \mathbb{R}$ *be the "sub-functions" obtained from restricting the first bit of the input to 0 and 1, respectively; that is,* $f_0(y) := f(0, y)$ *and* $f_1(y) := f(1, y)$ *for all* $y \in \mathbb{F}_2^{n-1}$. *Then the Fourier spectra of* f_0 *and* f_1 *satisfy that, for all* $\beta \in \mathbb{F}_2^{n-1}$,

$$\hat{f_0}(\beta) = \hat{f}(0, \beta) + \hat{f}(1, \beta), \qquad \hat{f_1}(\beta) = \hat{f}(0, \beta) - \hat{f}(1, \beta). \tag{2}$$

Conversely, the Fourier spectrum of f *satisfies*

$$\hat{f}(0, \beta) = \frac{1}{2}(\hat{f_0}(\beta) + \hat{f_1}(\beta)), \qquad \hat{f}(1, \beta) = \frac{1}{2}(\hat{f_0}(\beta) - \hat{f_1}(\beta)). \tag{3}$$

Tensor Product. The statement as well as the proof of Main Theorem requires the standard notion of tensor products between functions.

Definition 2 (Tensor Product of Boolean Functions). *Let* $f \colon \mathbb{F}_2^{n_1} \to \{0, 1\}$ *and* $g \colon \mathbb{F}_2^{n_2} \to \{0, 1\}$ *be two Boolean functions on* n_1 *and* n_2 *variables respectively. Then the tensor product of* f *and* g, *denoted by* $f \otimes g$, *is a Boolean function over* $\mathbb{F}_2^{n_1+n_2}$ *such that* $f \otimes g(x, y) = f(x) \cdot g(y)$ *for all* $x \in \mathbb{F}_2^{n_1}$ *and* $y \in \mathbb{F}_2^{n_2}$.

It is easy to verify the following fact.

Fact 4. *If $h = f \otimes g$ is the tensor product of two Boolean function defined above, then the Fourier spectrum h satisfies that $\hat{h}(\alpha, \beta) = \hat{f}(\alpha) \cdot \hat{g}(\beta)$, for every $\alpha \in \mathbb{F}_2^{n_1}$ and $\beta \in \mathbb{F}_2^{n_2}$.*

Given a Boolean function $f : \mathbb{F}_2^{n_1} \to \{0, 1\}$, two commonly used functions to tensor with f are the all-one function $g_1 = 1$ whose Fourier spectrum is $\hat{g}_1(\mathbf{0}) = 1$ and $\hat{g}_1(\alpha) = 0$ for any $\alpha \neq \mathbf{0}$; and the "δ-function" g_2 defined by $g_2(x) = 1$ if and only if $x = 0^{n_2}$, whose Fourier spectrum is $\hat{g}_2(\alpha) = 1/2^{n_2}$ for every α. Note that tensoring f with g_1 is equivalent to setting each to the 2^{n_2} sub-functions, defined by restricting y to different values in $\mathbb{F}_2^{n_2}$, to f; and tensoring f with g_1 is to set the sub-function with $y = \mathbf{0}$ to f and set all other sub-functions to the all-zero function.

Invertible Linear Transformations and Linear Shifts. Let $L : \mathbb{F}_2^n \to \mathbb{F}_2^n$ be an invertible linear transformation. If $f : \mathbb{F}_2^n \to \{0, 1\}$ is a Boolean function, then define $g := Lf$, the function obtained from applying the linear transformation L to f, as $g(x) = f(Lx)$ for all $x \in \mathbb{F}_2^n$. The Fourier spectrum of g is given by $\hat{g}(\alpha) = \hat{f}((L^T)^{-1}\alpha)$, where L^T stands for the transpose of L viewed as an $n \times n$ matrix. One can check that the set of Fourier coefficients as well as the property of being the indicator function of an (affine) linear subspace are invariant under invertible linear transformations. If $a \in \mathbb{F}_2^n$ is a non-zero vector, and let $h(x) := f(x + a)$ be the linear shift of f, then the Fourier spectrum of h is given by $\hat{h}(\alpha) = \chi_a(\alpha)\hat{f}(\alpha)$ for every $\alpha \in \mathbb{F}_2^n$.

2.2 Additive Combinatorics

Additive combinatorics is the sub-field of mathematics concerned with subsets of integers or more generally abelian groups, and studies the interplay between the structural properties of a subset and its combinatorial estimates associated with arithmetic operations. Recently additive combinatorics has found many applications in computer science, see the excellent exposition [23] and the textbook [33] for comprehensive treatments.

Throughout this paper, G is the abelian group $\mathbb{F}_2^n$ for some positive integer n and the underlying field is $\mathbb{F}_2$. If $A = \{a_1, \ldots, a_m\} \subset G$, then span($A$) stands for the *linear span* of A: span(A) = $\{\sum_{i \in S} a_i \mid S \subseteq [m]\}$, where summation over the empty set is understood to be the 0 element by convention. For any $x \in G$ and $A \subset G$, we write $x + A$ to denote the set $\{x + a \mid a \in A\}$. If A and B are two subsets of G, then $A + B$ denotes the *sumset* $\{a + b \mid a \in A \text{ and } b \in B\}$. Similarly, $A - B := \{a - b \mid a \in A \text{ and } b \in B\}$, although $A - B$ is always the same as $A + B$ in this paper as the underlying ambient group is $\mathbb{F}_2^n$. If $A = B$ then we write $2A := A + A$ and in general write $kA := \underbrace{A + \cdots + A}_{k \text{ times}}$ for integer $k \geq 1$.

The following Lemma of Łaba is useful for our proofs.

Lemma 2 ([19], Theorem 2.5). *Let G be an abelian group and $A \subset G$ be a subset of G such that $|A - A| < \frac{3}{2}|A|$. Then $A - A$ is a subgroup of G.*

3 Proof of the Main Lemma

First recall our Main Lemma states the following.

Lemma 1. *Let $k \geq 5$ and $n \geq k$ be integers. Let $f : \mathbb{F}_2^n \to \{0, 1\}$ be a Boolean function such that $\hat{f}(0) = 1/2^{k-1}$ and any other Fourier coefficients are either zero or equal to $\pm\frac{1}{2^k}$, then f is the indicator function of a disjoint union of two dimension $n - k$ affine subspaces.*

In Appendix B, we compute the Fourier spectrum of a Boolean function that is supported on two disjoint affine subspaces such that the two affine subspaces are of the same dimension and their Fourier spectra have minimum intersection. Our strategy for the proof of the Main Lemma is to show that if the Fourier coefficients of a Boolean function satisfy the condition prescribed in the Main Lemma, then its Fourier spectrum matches the one we show in Appendix B.

Let us define

$$A = \{\alpha \in \mathbb{F}_2^n \mid \hat{f}(\alpha) = \frac{1}{2^k}\}$$

and

$$B = \{\beta \in \mathbb{F}_2^n \mid \hat{f}(\beta) = -\frac{1}{2^k}\}.$$

Without loss of generality[4], from now on, we may assume $f(0) = 1$. We begin with calculating the cardinalities of sets A and B.

Claim. For any $k \geq 1$ and $n \geq k$, we have $|A| = 3t$ and $|B| = t$, where $t = 2^{k-1} - 1$.

Proof. Since $\hat{f}(0) = 1/2^{k-1}$, by Parseval's identity $\hat{f}(0) = 1/2^{k-1} = \sum_{\alpha \in \mathbb{F}_2^n} \hat{f}^2(\alpha)$, we have $|A| + |B| = 2^{k+1} - 4$.

On the other hand,

$$1 = f(0) = \sum_{\alpha \in \mathbb{F}_2^n} \hat{f}(\alpha)\chi_\alpha(0) = \frac{1}{2^{k-1}} + \sum_{\alpha \in A} \frac{1}{2^k} + \sum_{\beta \in B} (-\frac{1}{2^k}),$$

which gives $|A| - |B| = 2^k - 2$. Therefore we have $|A| = 3(2^{k-1} - 1)$ and $|B| = 2^{k-1} - 1$.   □

For convenience, we let $A = \{\alpha_1, \ldots, \alpha_{3t}\}$ and $B = \{\beta_1, \ldots, \beta_t\}$ in the following.

---

[4] This is because if $f(0) = 0$, then let $a \in \mathbb{F}_2^n$ be any vector such that $f(a) = 1$. We can apply a linear shift a to f to get a new Boolean function, $h(x) = f(x + a)$ for every x, so that $h(0) = 1$. Note that the conclusions in our Main Theorem are invariant under linear shifts. Moreover, since $\hat{h}(\alpha) = \chi_a(\alpha)\hat{f}(\alpha)$ for every $\alpha \in \mathbb{F}_2^n$, we have $\hat{h}(0) = 1/2^{k-1}$ and $|\hat{h}(\alpha)| = |\hat{f}(\alpha)|$ for any other nonzero α. Therefore, the assumptions apply to h as well.

3.1   Some Additive Properties of Sets A and B

We now study the additive properties of sets A and B. Note that the Fourier coefficients of f are non-zero only at $\mathbf{0}$ and in sets A and B; moreover, the Fourier coefficients are uniform for points in A or B. Therefore, by Proposition 1, we expect that there are nice additive structures within A and B.

Definition 3. *We call $(\alpha, \beta, \alpha + \beta)$ a triangle if α, β and $\alpha + \beta$ are all in the support of $\hat{f}$; that is $\alpha, \beta, \alpha + \beta \in A \cup B \cup \{\mathbf{0}\}$.*

Lemma 3. *For any $\beta_i \in B$, there are exactly t triangles passing through β_i; namely, the t triangles are $(\beta_i, \beta_i, \mathbf{0})$ and $\{(\beta_i, \beta_j, \beta_i + \beta_j)\}_{j=1, j \neq i}^{t}$. In the language of set addition, we have $2B \subseteq A \cup \{\mathbf{0}\}$.*

Proof. For any $\beta_i \in B$, by Proposition 1,

$$\hat{f}(\beta_i) = -\frac{1}{2^k} = \sum_{\gamma \in \mathbb{F}_2^n} \hat{f}(\gamma)\hat{f}(\beta_i + \gamma)$$

$$= 2\hat{f}(\mathbf{0})\hat{f}(\beta_i) + \sum_{\substack{j=1 \\ j \neq i}}^{t} \hat{f}(\beta_j)\hat{f}(\beta_i + \beta_j) + \sum_{\ell=1}^{3t} \hat{f}(\alpha_\ell)\hat{f}(\beta_i + \alpha_\ell)$$

$$\geq 2 \cdot \frac{1}{2^{k-1}} \cdot (-\frac{1}{2^k}) + 2(t-1)(-\frac{1}{2^k})(\frac{1}{2^k})$$

$$= -\frac{1}{2^k},$$

where the inequality in the second last line becomes equality if and only if the following two conditions hold: 1) for every $1 \leq j \leq t$, $j \neq i$, $\beta_i + \beta_j \in A$; and 2) there is no triangle of the form $(\beta_i, \alpha_j, \alpha_\ell)$. Hence the lemma follows.[5]   □

Corollary 2. *The set B is a sum-free set; namely, for any three elements $\beta_1, \beta_2, \beta_3 \in B$, $\beta_1 + \beta_2 \neq \beta_3$. Equivalently, $2B \cap B = \emptyset$.*

Proof. This follows directly from Lemma 3 and the fact sets A and B are disjoint.   □

Corollary 3. *We have $2B \cap 3B = \emptyset$.*

Proof. Suppose not, then there exist $\beta_1, \beta_2, \beta_3, \beta_4, \beta_5$ in B such that $\beta_1 + \beta_2 = \beta_3 + \beta_4 + \beta_5$. These five elements must be distinct as otherwise they would give rise to a triangle in B. But then we have a $(\alpha_1, \alpha_2, \beta_5)$ triangle, where $\alpha_1 := \beta_1 + \beta_2$ and $\alpha_2 := \beta_3 + \beta_4$, contradicting to Lemma 3.   □

---

[5] There is a factor 2 in the second summation because if $\beta_i + \beta_j \in A$, then the triangle $(\beta_i, \beta_j, \beta_i + \beta_j)$ appears twice in the summation $\sum_{\gamma \in \mathbb{F}_2^n} \hat{f}(\gamma)\hat{f}(\beta_i + \gamma)$: once with $\gamma = \beta_j$ and the other with $\gamma = \beta_i + \beta_j$.

Let us define
$$R = 2B \cap A = 2B \setminus \{0\}$$
and
$$L = A \setminus R.$$
Note that L and R are disjoint and $A = L \cup R$. For any $\rho \in R$, let
$$N(\rho) = \{\beta_i \in B \mid \exists \beta_j \in B \text{ s.t. } \rho = \beta_i + \beta_j\}$$
be the set of points in B which has a triangle passing through ρ. Define a set $\Gamma \subset \mathbb{F}_2^n$ as
$$\Gamma = \{\gamma = \rho + \beta \mid \rho \in R, \beta \in B \text{ and } \beta \notin N(\rho)\}.$$
Observe that Γ is nonempty: since for every $\rho \in R$, all its β-neighbors can be paired together, so $|N(\rho)|$ is an even number, but $|B| = 2^{k-1} - 1$ is odd.

Claim. We have $\Gamma = 3B \setminus B$.

Proof. On one hand, by the definition of set Γ, $\Gamma \subseteq 3B$; since R and B are disjoint and $\mathbf{0} \notin R$, we have $\Gamma \cap B = \emptyset$, and hence $\Gamma \subseteq 3B \setminus B$. On the other hand, let γ be any element in $3B$; that is $\gamma = \beta_1 + \beta_2 + \beta_3$, where $\beta_1, \beta_2, \beta_3 \in B$. When will γ actually be in B? This happens only if any two of these three elements are identical, then $\gamma = \beta_i$ for some $i \in \{1, 2, 3\}$, thus $\gamma \in B$. Moreover, assume that these three elements are distinct and suppose $\gamma \in B$, i.e. $\gamma = \beta_j$ for some $j > 3$. Let $\rho := \beta_1 + \beta_2$, then $\rho = \beta_3 + \gamma = \beta_3 + \beta_j$; that is $\gamma = \rho + \beta_3$ and $\beta_3 \in N(\rho)$. Therefore, if $\gamma \in 3B \setminus B$, then we must have $\beta_3 \notin N(\rho)$ and consequently $\gamma \in \Gamma$. It follows that $3B \setminus B \subseteq \Gamma$. This completes the proof of the claim. $\qquad\square$

It is easy to see that Γ is disjoint from the Fourier support of f.

Claim. For every element $\gamma \in \Gamma$, we have $\hat{f}(\gamma) = 0$.

Proof. Recall that, the support of $\hat{f}$ is $A \cup B \cup \{0\}$. Suppose $\hat{f}(\gamma) \neq 0$, that is $\gamma \in \text{supp}(\hat{f})$. Since $A = L \cup R$, from Claim 3.1, we know that $\Gamma \cap B = \emptyset$; from Claim 3.1 and Corollary 3, we know that $\Gamma \cap 2B = \Gamma \cap (R \cup \{0\}) = \emptyset$. So there is only one possibility left, which is $\gamma \in L$. However, if this were the case, because $\gamma = \rho + \beta$ with $\rho \in R$, it would give rise to a (γ, ρ, β)-triangle with $\gamma, \rho \in A$, contradicting Lemma 3, so γ is not in L, hence $\hat{f}(\gamma) = 0$. $\qquad\square$

3.2   Even-Zohar's Tight Bound on $F(K)$

Let G be an abelian group and $A \subset G$ be a subset. The fundamental Freiman theorem [11] in additive combinatorics states that if G is $\mathbb{Z}$ and $|A + A| \leq K|A|$ for some constant K, then there exist functions $d(K)$ and $\ell(K)$ such that A is contained in a $d(K)$-dimensional arithmetic progression of length at most $\ell(K)|A|$. The ratio $\sigma[A] := |A + A|/|A|$ is commonly known as the *doubling constant* of set A. Hence Freiman theorem asserts that if a set of integers has small doubling constant, then the set is well-structured. Ruzsa [29] established an analog of Freiman's theorem for finite abelian groups with torsion r. Specifically, he proved that any subset A with doubling constant K is contained in a subgroup of G of size at most $K^2 r^{K^4} |A|$. The question for groups $\mathbb{F}_2^n$ was first studied by Green and Ruzsa [15] and the bound was later improved by Sanders [31]. An asymptotically tight bound was first proved in [16] and [18].

For a subset $A \subset \mathbb{F}_2^n$, let $\langle A \rangle$ denote the *affine span* of A; namely, the smallest affine subspace that contains A. If $\sigma[A] = K$, then let $F(K) := \max_{A:\sigma[A]=K} |\langle A \rangle|/|A|$ denote the maximum relative size of the affine span of A. Even-Zohar [10] gave the tight bound of $F(K)$ for all values of doubling constant K.

Theorem 5 ([10], **Theorem 2**). *Let A be a subset of $\mathbb{F}_2^n$ with doubling constant K, i.e. $|2A|/|A| \leq K$. If s is the unique positive integer satisfying the inequalities*

$$\frac{\binom{s}{2} + s + 1}{s + 1} \leq K < \frac{\binom{s+1}{2} + s + 2}{s + 2}, \tag{4}$$

then $|\langle A \rangle|/|A| \leq F(K)$, where $F(K)$ is given by

$$F(K) = \begin{cases} \frac{2^s}{\binom{s}{2} + s + 1} \cdot K & \text{if } \frac{\binom{s}{2}+s+1}{s+1} \leq K < \frac{s^2+s+1}{2s}, \\ \frac{2^{s+1}}{s^2+s+1} \cdot K & \text{if } \frac{s^2+s+1}{2s} \leq K < \frac{\binom{s+1}{2}+s+2}{s+2}. \end{cases} \tag{5}$$

3.3 Characterizing $2B$ and span(B)

Note that the doubling constant of set B satisfies that

$$\sigma[B] = \frac{|R| + 1}{|B|} \leq \frac{|A| + 1}{|B|} = 3 + \frac{1}{t}, \tag{6}$$

and recall that $t = 2^{k-1} - 1$. Therefore, when $k \geq 5$, $K = \sigma[B] \leq \frac{46}{15}$. Plugging this K into (4) gives that $s \leq 5$ and consequently $F(K) \leq 2K < 7$. That is, we have $|\langle B \rangle| < 7|B|$.

The most important step in our proof is establishing the following lemma, which almost completely characterizes the structure of set B.

Lemma 4. *If $k \geq 5$, then $|\operatorname{span}(B)| = 2^k = 2(|B|+1)$ and $2B$ is a subspace of dimension $k - 1$.*

We prove Lemma 4 in the following two subsections, distinguishing between the case when $\langle B \rangle$ is an affine subspace and the case when $\langle B \rangle$ is a subspace.

If $\langle B \rangle$ is an Affine Subspace. In the case that $\langle B \rangle$ is an affine subspace, let $\langle B \rangle = a + H$ be the affine subspace, where H is a subspace of $\mathbb{F}_2^n$, $a \in H^\perp$ and $a \neq \mathbf{0}$. Therefore $\operatorname{span}(B) = H \cup (a + H)$. Note that we now have $2\ell B \subseteq H$ and $(2\ell - 1)B \subseteq a + H$ for every integer $\ell \geq 1$. Moreover, $|\operatorname{span}(B)| = 2|\langle B \rangle| < 14|B|$. Since $\operatorname{span}(B)$ is a subspace and $|B| = 2^{k-1} - 1$, so there are only three possibilities: $|\operatorname{span}(B)| = 8(|B|+1)$, $|\operatorname{span}(B)| = 4(|B|+1)$ and $|\operatorname{span}(B)| = 2(|B|+1)$. In the following, we are going to eliminate the first two possibilities.

Claim. Set L is nonempty.

Proof. Suppose not, then $2B = A \cup \{0\} \subset H$. Recall that by Claim 3.1, $\Gamma = 3B \setminus B$, so $\Gamma \subseteq a + H$ and is disjoint from set A. It follows that for any $\gamma \in \Gamma$, $\hat{f}(\gamma) = 0$ (or directly from Claim 3.1). However, applying Proposition 1 to $\hat{f}(\gamma)$, we see that by the definition of set Γ, $\gamma = \rho + \beta$ with $\rho \in A$, $\beta \in B$ and $\beta \notin N(\rho)$. Hence there is at least one negative term contribution on the right-hand side in (1) for $\hat{f}(\gamma)$, but since both $2B$ and $2A$ are disjoint from Γ, there is no positive term on the right-hand side in (1), a contradiction. □

We discuss the following two possibilities separately.

The case when $|H| = 4(|B| + 1)$. First note that if this were the case, then $F(K) = |\langle B \rangle|/|B| = 4(1 + \frac{1}{|B|})$. By Theorem 5, the doubling constant of B is at least $K = |2B|/|B| > 2.5$, or $|2B| > 2.5|B|$. Therefore $|L| \leq 0.5|B|$. On the other hand, $4B = 2B + 2B$ and $4B \subseteq H$ so $\sigma[2B] = |4B|/|2B| < \frac{4 + \frac{1}{16}}{2.5} < 7/4$. Then by Theorem 5 again, $|4B| = |\langle 2B \rangle|$, that is $4B = H$.

We next claim that $L \subseteq H$. To see this, let λ be an arbitrary element in L; applying Proposition 1 to $\hat{f}(\lambda)$ gives

$$\frac{1}{2^k} = \hat{f}(\lambda) = 2\hat{f}(\lambda)\hat{f}(0) + \sum_{\lambda' \in L} \hat{f}(\lambda')\hat{f}(\lambda + \lambda') + \text{other terms}.$$

The first term and the second summation can contribute at most $\frac{1}{2^{2k}}(2|L| + 2) \leq \frac{|B| + 2}{2^{2k}} < \frac{1}{2^k}$. Therefore, the "other terms" on the right-hand side must contain terms of the form $\hat{f}(\alpha_1)\hat{f}(\alpha_2)$, where α_1 and α_2 are two distinct points in A and $\lambda = \alpha_1 + \alpha_2$. That is $\lambda \in 2B + 2B$, hence it follows that $L \subseteq 4B = H$.

Let $D := H \setminus (2B \cup L)$. We have $|D| = 4(|B| + 1) - 3|B| - 1 = |B| + 3 > 0$. Let δ be any point in D. First, since $\delta \notin 2B \cup B$, $\hat{f}(\delta) = 0$. Second, since $\delta \in H$, there is no negative term in the right-hand side of $0 = \hat{f}(\delta) = \sum_{\gamma \in \mathbb{F}_2^n} \hat{f}(\gamma)\hat{f}(\delta + \gamma)$, because if $\gamma \in B$, then $\delta + \gamma \in a + H$ but there is no positive Fourier coefficient in $a + H$ (since $L \subset H$). On the other hand, consider the set $\{\delta + \alpha \mid \alpha \in 2B \cup L\}$. Since $|D| < |H|/2$, this set has non-empty intersection with $2B \cup L$. Therefore, there are positive terms in $\sum_{\gamma \in \mathbb{F}_2^n} \hat{f}(\gamma)\hat{f}(\delta + \gamma)$, this contradicts the fact that $\hat{f}(\delta) = 0$.

The case when $|H| = 2(|B| + 1)$. This case is similar to the previous one. First, if this were the case, then $F(K) = |\langle B \rangle|/|B| = 2(1 + \frac{1}{|B|})$. It follows that, by Theorem 5, the doubling constant of B is at least $K = |2B|/|B| > 7/4$, and hence $|4B|/|2B| \leq |H|/|2B| < 3/2$, and by Theorem 5 again $4B = H$. The rest is identical to the case when $|H| = 4(|B| + 1)$.

Proof of Lemma 4 when $\langle B \rangle$ *is an Affine Subspace.* Now that the only possibility left is $|\operatorname{span}(B)| = 2 \cdot (|B| + 1)$, and because $\langle B \rangle$ is an affine subspace, it follows that $2B \subseteq H$ and hence $|2B| \leq |B| + 1$. Applying Łaba's lemma, Lemma 2, to set B gives that $2B$ is a subspace. Since $|2B| \geq |B|$, it follows that $2B = H$, a dimension $k-1$ subspace.   □

If $\langle B \rangle$ is a Subspace. If the affine span $\langle B \rangle$ is a subspace, and since $|\langle B \rangle| < 7|B|$, then we either have $|\langle B \rangle| = 4(|B| + 1)$ or $|\langle B \rangle| = 2(|B| + 1)$ (because $B \cap 2B = \emptyset$ and $|2B| \geq |B|$, $|\langle B \rangle| \geq 2|B|$). In the following we exclude the first case.

Recall that $R = 2B \setminus \{0\}$ is the set of non-zero points in the Fourier support of f that can be written as a sum of two β-points in B. Let $R = \{\lambda_1, \ldots, \lambda_m\}$, where m is the cardinality of R.

Claim. If $\langle B \rangle$ is a subspace, then $m \leq 2.5t$.

Proof. For the sake of contradiction, suppose that $m > 2.5t$. For every $\lambda_i \in R$, let d_i be the number of β_j's that form a triangle with λ_i. Then we have $\sum_{i=1}^{m} d_i = t(t-1)$ and $d_i \geq 2$ for every $1 \leq i \leq m$. By a standard averaging argument, there is some λ_i with $d_i \leq 0.4t$. By the definition of set Γ, it follows that $|\Gamma| \geq t - d_i = 0.6t$. Recall that $\Gamma = 3B \setminus B$ so $\Gamma \subset \langle B \rangle = \text{span}(B)$, and Γ is disjoint from either $2B$ or B, thus $|\langle B \rangle| \geq |2B| + |B| + |\Gamma| > 4.1t$, contradicting our assumption that $|\langle B \rangle| = 4(|B| + 1)$.   □

Proof of Lemma 4 when $\langle B \rangle$ is a Subspace. Now since $m \leq 2.5t$, the doubling constant of B is at most $|2B|/|B| \leq 2.5 + 1/|B| < 21/8$, then by Theorem 5, $|\langle B \rangle|/|B| < 42/11 < 4$, therefore we must have $|\langle B \rangle| = 2(|B| + 1) = 2^k$. Once again, applying Laba's lemma to set B shows that $2B$ is a subspace of dimension $k-1$.   □

3.4   Completing the Proof of the Main Lemma

By Lemma 4, $2B$ is a dimension $k-1$ subspace; without loss of generality, we may assume that

$$H = 2B = \text{span}(e_1, \ldots, e_{k-1}). \tag{7}$$

Since $|\text{span}(B)| = 2^k = 2|2B|$, and $B \cap 2B = \emptyset$, B is an affine shift of H with one point δ missing. Since $\delta \notin H$, so without loss of generality, we may assume e_k is the missing point. That is

$$B = (e_k + \text{span}(e_1, \ldots, e_{k-1})) \setminus \{e_k\} \quad \text{and} \tag{8}$$
$$R = 2B \setminus \{0\} = \text{span}(e_1, \ldots, e_{k-1}) \setminus \{0\} = e_k + B. \tag{9}$$

Now by Claim 3.1, we have $\Gamma = \{e_k\}$ and consequently $\hat{f}(e_k) = 0$. Our last task is to determine the structure of set L. Recall that $A = R \cup L$ and $|A| = 3t$, and because we now have $R = 2B \setminus \{0\}$, therefore $|L| = 2t = 2^k - 2$.

Claim. For any $\lambda \in L$, $e_k + \lambda \in L$.

Proof. Applying Proposition 1 to the Fourier coefficient of f at e_k and noting that $R = e_k + B$, we have

$$\hat{f}(e_k) = 0 = \sum_{\gamma \in \mathbb{F}_2^n} \hat{f}(\gamma)\hat{f}(e_k + \gamma)$$

$$= 2\sum_{\rho \in R} \hat{f}(\rho)\hat{f}(e_k + \rho) + \sum_{\lambda \in L} \hat{f}(\lambda)\hat{f}(e_k + \lambda)$$

$$\leq 2t \cdot (-\frac{1}{2^{2k}}) + 2t \cdot \frac{1}{2^{2k}}$$

$$= 0,$$

where equality holds in the second last line only if for every $\lambda \in L$, $\hat{f}(e_k+\lambda) = \frac{1}{2^k}$. That is, $e_k + \lambda \in A(= L \cup R)$. As each element in R has already been taken into account in the first summation in the second line, therefore we necessarily have $e_k + \lambda \in L$. $\square$

Claim. For any $\lambda \in L$ and $\rho \in R$, $\hat{f}(\lambda + \rho) = 0$.

Proof. Applying Proposition 1 to $\hat{f}(\rho)$, where ρ is an arbitrary element in R, we have

$$\hat{f}(\rho) = \frac{1}{2^k} = 2 \cdot \hat{f}(0)\hat{f}(\rho) + \sum_{\beta \in B} \hat{f}(\beta)\hat{f}(\rho + \beta) + \sum_{\rho' \in R, \rho' \neq \rho} \hat{f}(\rho')\hat{f}(\rho + \rho') + \sum_{\lambda \in L} \hat{f}(\lambda)\hat{f}(\lambda + \rho)$$

$$= 2 \cdot \frac{2}{2^k} \cdot \frac{1}{2^k} + (t-1) \cdot (-\frac{1}{2^k}) \cdot (-\frac{1}{2^k}) + (t-1) \cdot (\frac{1}{2^k}) \cdot (\frac{1}{2^k}) + \sum_{\lambda \in L} \frac{1}{2^k} \cdot \hat{f}(\lambda + \rho)$$

$$\geq \frac{1}{2^k}, \qquad \text{(as } \lambda + \rho \notin B, \text{ therefore } \hat{f}(\lambda + \rho) \geq 0)$$

where we have a factor of $(t - 1)$ in the second line because $\rho + e_k \in B$ and equality holds in the last line only if $\hat{f}(\lambda + \rho) = 0$ for every $\lambda \in L$ and every $\rho \in R$. $\square$

Claim. For any $\lambda, \lambda' \in L$, $\lambda + \lambda' \in L$ except that $\lambda + \lambda' = 0$ or e_k.

Proof. Applying Proposition 1 to $\hat{f}(\lambda)$, where λ is an arbitrary element in L, we have

$$\hat{f}(\lambda) = \frac{1}{2^k} = 2 \cdot \hat{f}(0)\hat{f}(\lambda) + \sum_{\beta \in B} \hat{f}(\beta)\hat{f}(\lambda + \beta) + \sum_{\rho \in R} \hat{f}(\rho)\hat{f}(\lambda + \rho) + \sum_{\lambda' \in L, \lambda' + \lambda \notin \{0, e_k\}} \hat{f}(\lambda')\hat{f}(\lambda + \lambda')$$

$$= 2 \cdot \frac{2}{2^k} \cdot \frac{1}{2^k} + 0 + 0 + \sum_{\lambda' \in L, \lambda' + \lambda \notin \{0, e_k\}} \frac{1}{2^k} \cdot \hat{f}(\lambda + \lambda')$$

$$\leq \frac{4}{2^{2k}} + (2t - 2) \cdot (\frac{1}{2^k}) \cdot (\frac{1}{2^k})$$

$$= \frac{1}{2^k},$$

where equality holds in the second last line only if $\lambda + \lambda' \in L$ for every $\lambda' \in L$, except when λ' is equal to λ or $\lambda + e_k$.[6] $\square$

---

[6] The second term vanishes because the only triangles passing through a point $\beta_i \in B$ are of the type $(\beta_i, \beta_j, \rho_\ell)$ where $\rho_\ell \in R$; the third term vanishes because of Claim 3.4.

Put Claim 3.4, Claim 3.4 and Claim 3.4 together, and since $|L| = 2^k - 2$ we conclude that $H' := L \cup \{\mathbf{0}, e_k\}$ is a subspace of dimension k. Moreover, as $\text{span}(B) = \text{span}(e_1, \ldots, e_k)$ is a subspace of dimension k, and $L \cap \text{span}(B) = \emptyset$, we thus have $H' \cap \text{span}(B) = \{\mathbf{0}, e_k\}$. Therefore, without loss of generality, we may take $H' = \text{span}(e_k, \ldots, e_{2k-1})$ and consequently finally have

$$L = \text{span}(e_k, \ldots, e_{2k-1}) \setminus \{\mathbf{0}, e_k\}. \tag{10}$$

It is straightforward to check[7] that the Fourier spectrum calculated in Sect. B for a disjoint union of two dimension $n - k$ affine subspaces is identical to the Fourier spectrum of f, which is completely specified by sets in (8), (9) and (10). Therefore the proof of the Main Lemma is complete.

4   Dealing with Small Values of k

When $k = 2$ or $k = 3$, note that since Claim 3 holds for every $k \geq 2$, this will enable us to prove the same results as Main Lemma by slightly different arguments. That is, when $k = 2$ or $k = 3$, support of f is also a disjoint union of two dimension $n - k$ affine subspaces. However, when $k = 4$ one can not prove the same characterization as Main Lemma. In fact, there are two possibilities: one is that f is still the indicator function of two disjoint dimension $n - 4$ affine subspaces; the other is that support of f are *four* disjoint $n - 5$ affine subspaces. Furthermore, we show that this is the only counterexample to Main Lemma for all k. Now we give the precise statements for small values of k and their proofs.

Lemma 5. *Let $2 \leq k \leq 4$ and $n \geq k$ be integers. Let $f : \mathbb{F}_2^n \to \{0,1\}$ be a Boolean function such that $\hat{f}(\mathbf{0}) = 1/2^{k-1}$ and any other Fourier coefficients are either zero or equal to $\pm\frac{1}{2^k}$. If $k = 2$ or $k = 3$, then f is the indicator function of a disjoint union of two dimension $n - k$ affine subspaces; If $k = 4$, then f is either the indicator function of a disjoint union of two dimension $n-k$ affine subspaces, or the indicator function of a disjoint union of four dimension $n - k - 1$ affine subspaces.*

4.1   Proof of the Case $k = 2$

In this case, $|A| = 3$ and $|B| = 1$. For convenience, suppose that $\hat{f}(\mathbf{0}) = \frac{1}{2}$, $\hat{f}(\beta) = -\frac{1}{4}$ and $\hat{f}(\alpha_1) = \hat{f}(\alpha_2) = \hat{f}(\alpha_3) = \frac{1}{4}$, where $\beta, \alpha_1, \alpha_2, \alpha_3$ are four distinct non-zero vectors.

We claim that there exists an α_i, $1 \leq i \leq 3$, such that $\hat{f}(\beta + \alpha_i) = 0$. To see this, suppose $\hat{f}(\beta + \alpha_i) \neq 0$ for every $1 \leq i \leq 3$. Because the four vectors are distinct, $\beta + \alpha_i \neq \mathbf{0}$; furthermore, since $\alpha_i \neq \mathbf{0}$, so $\beta + \alpha_i \neq \beta$. It follows that $\beta + \{\alpha_1, \alpha_2, \alpha_3\} = \{\alpha_1, \alpha_2, \alpha_3\}$; that is, adding β to A permutes the

---

[7] The second line in (11) corresponds to set B, third line in (11) corresponds to set R, and the fourth and fifth lines of (11) correspond to set L.

three elements in the set. But now adding these three elements together gives $3\beta_1 + \sum \alpha_i = \sum \alpha_i$, a contradiction since $\beta_1 \neq \mathbf{0}$.

Without loss of generality, assume $\hat{f}(\beta + \alpha_1) = 0$ and denote $\beta + \alpha_1$ by γ. Now applying Proposition 1 to γ gives:

$$\hat{f}(\gamma) = 0 = \sum_\alpha \hat{f}(\alpha)\hat{f}(\alpha + \gamma)$$

$$= 2 \cdot \hat{f}(\beta_1)\hat{f}(\alpha_1) + \hat{f}(\alpha_2)\hat{f}(\alpha_2 + \gamma) + \hat{f}(\alpha_3)\hat{f}(\alpha_3 + \gamma)$$

$$= 2 \cdot (-\frac{1}{4}) \cdot \frac{1}{4} + \hat{f}(\alpha_2)\hat{f}(\alpha_2 + \gamma) + \hat{f}(\alpha_3)\hat{f}(\alpha_3 + \gamma)$$

$$\leq 0,$$

where equality holds in the last line only if $\gamma = \alpha_2 + \alpha_3$ so that

$$\hat{f}(\alpha_2)\hat{f}(\alpha_2 + \gamma) = \hat{f}(\alpha_3)\hat{f}(\alpha_3 + \gamma) = \hat{f}(\alpha_2)\hat{f}(\alpha_3) = \frac{1}{4} \cdot \frac{1}{4}.$$

After taking an invertible linear transformation if necessary, we may take $\alpha_1 = e_1, \beta = e_1 + e_2, \alpha_2 = e_3$ and $\alpha_3 = e_2 + e_3$, then it is easy to verify that this is identical to the Fourier spectrum in (11) for the case of $k = 2$.

4.2 Proof of the Case $k = 3$

In this case, $|A| = 9$ and $|B| = 3$. Denote set B by $\{\beta_1, \beta_2, \beta_3\}$. Then by Corollary 2, $\beta_1 + \beta_2 + \beta_3 \neq 0$, therefore $R = \{\beta_1 + \beta_2, \beta_1 + \beta_3, \beta_2 + \beta_3\}$, and $\Gamma = \{\beta_1 + \beta_2 + \beta_3\}$. Hence Lemma 4 is established and the rest of the proof is identical to that of the Main Lemma in Sect. 3.4 for the general $k \geq 5$ case.

4.3 Proof of the Case $k = 4$

First of all, it is easy to see that when $k = 4$, the indicator function of a disjoint union of 2 affine subspaces of dimension $n - k = n - 4$ is still a Boolean function with desired Fourier spectrum, for every $n \geq 4$. Next we construct another Boolean function, which demonstrates that Main Lemma is no longer valid for $k = 4$.

Construction 6. Let $G = \mathbb{F}_2^6$ with $e_1, \cdots, e_6$ as the standard basis and let $A, B \subset G$ be two disjoint subsets given as follows:

– $B = \{e_i \mid 1 \leq i \leq 6\} \cup \{\sum_{i=1}^6 e_i\}$;
– $A = \{e_i + e_j \mid 1 \leq i < j \leq 6\} \cup \{\sum_{i \in S} e_i \mid S \subset [6], |S| = 5\}$.

Clearly $A = 2B \setminus \{0\}$, $|B| = 2^{4-1} - 1 = 7$ and $|A| = \binom{7}{2} = 3|B|$, which satisfy the size requirements for A and B for $k = 4$. To see that sets A and B in Construction 6 satisfy all the additive properties imposed by Proposition 1, one can explicitly compute a "core" function $f_{CE} : \mathbb{F}_2^6 \to \mathbb{R}$ with $A \cup B \cup \{0\}$ being its Fourier support to verify that f is indeed a *Boolean* function and

$\mathrm{supp}(f_{CE}) = \{\mathbf{0}\} \cup \{\sum_{i \in S} e_i \mid S \subset [6], |S| = 5\} \cup \{\sum_{i=1}^{6} e_i\}$. That is, f is equal to 1 on vectors of weights 0, 5 and 6, and is equal to 0 on all other vectors. Note that $\mathrm{supp}(f_{CE})$ consists of 8 distinct vectors and is a disjoint union of four affine subspaces of dimension $n - 4 - 1 = 1$ each. Moreover, it can be checked that $\mathrm{supp}(f_{CE})$ is not the union of any two disjoint affine subspaces of dimension 2.

Our next claim shows that, up to an invertible linear transformation, Construction 6 is essentially the only counterexample to the Main Lemma.

Claim. When $k = 4$, either f is the indicator function of a disjoint union of two affine subspaces of dimension $n - k$, or the Fourier spectrum of f is given by Construction 6 under some invertible linear transformation, and consequently f is the indicator function of a disjoint union of four affine subspaces of dimension $n - k - 1$.

Proof. When $k = 4$, we have $|B| = 2^{4-1} - 1 = 7$. By inequality (6), $\sigma[B] = |2B|/|B| \leq 22/7$. But if $|2B| \leq 21$, then plugging $K = \sigma[B] \leq 3$ into (4) gives that $s \leq 5$ and consequently $F(K) \leq 2K < 7$. That is, we would have $|\langle B \rangle| < 7|B| = 49$. Then following the same argument, we would be able to establish Lemma 4 for the case $k = 4$ as well, i.e. to have $|\mathrm{span}(B)| = 2^k = 2(|B|+1)$ and $2B$ is a subspace of dimension $k - 1$, thereby recovering the regular configuration of f being the indicator function of two disjoint affine subspaces of dimension $n - k$.

Therefore, from now on, we assume that $|2B| = 22$. On the other hand, $|A| = 3|B| = 21$; combining this with Lemma 3 (i.e. $2B \subseteq A \cup \{\mathbf{0}\}$), we must have $A = 2B \setminus \{\mathbf{0}\}$. By the upper bound on $|\langle B \rangle|$ given in Theorem 5, we have $|\langle B \rangle| \leq 2^6 = 64$. But if $|\langle B \rangle| < 64$ (hence $|\langle B \rangle| = 32$ or $|\langle B \rangle| = 16$), then the proof of Lemma 4 would follow again.

Hence, the counter-example is possible only when the dimension of $\mathrm{span}(B)$ is at least 6. Without loss of generality, we may assume $B = \{e_i \mid 1 \leq i \leq 6\} \cup \{\beta\}$. We will determine vector β next.

If $\beta \notin \mathrm{span}(e_1, \cdots, e_6)$, then without loss of generality, let $\beta = e_7$. Now $A = 2B \setminus \{\mathbf{0}\} = \{e_i + e_j \mid 1 \leq i < j \leq 7\}$. But applying Proposition 1 to the vector $e_1 + e_2 + e_3$ gives that $\hat{f}(e_1 + e_2 + e_3) = -6/2^{2k}$, contradiction to the fact that $\hat{f}(e_1 + e_2 + e_3) = 0$ because $e_1 + e_2 + e_3 \notin A \cup B$. It follows that $\beta \in \mathrm{span}(e_1, \cdots, e_6)$.

Note that every weight-2 vector $e_i + e_j$, $1 \leq i < j \leq 6$, is in A. On the other hand, since $|A| = \binom{|B|}{2}$, it follows that for every $\alpha_k \in A$, there exist a unique pair $\beta_i, \beta_j \in B$ such that $\beta_i + \beta_j = \alpha_k$. Combining these two facts, we conclude that none of the weight-3 vector of the form $e_i + e_j + e_k$ is in B, for every $1 \leq i < j < k \leq 6$, as it would gives two ways to obtain vectors such as $e_i + e_j$ by adding two vectors from B, thus making $|A| < \binom{|B|}{2}$. By Claim 3, none of the weight-4 vectors can be in B either, which leaves only the possibilities of weight-5 or weight-6 vector for β.

If β is a weight-5 vector, without loss of generality, we may assume $\beta = \sum_{i=1}^{5} e_i$. Then B would contain vectors of weight-1 and weight-5 only, consequently A would contain vectors of weight-2, weight-4 and weight-6 only. Now applying Proposition 1 to the vector $e_1 + e_2 + e_3$ yields $\hat{f}(e_1 + e_2 + e_3) < 0$, contradicting to the fact that $\hat{f}(e_1 + e_2 + e_3) = 0$ as $e_1 + e_2 + e_3 \notin A \cup B$. Therefore, we have $\beta = \sum_{i=1}^{6} e_i$, completing the proof of the claim.   □

5 Proof of the Main Theorem

Clearly, if $\hat{f}(\mathbf{0}) = \frac{1}{2^k}$, then, because $|\hat{f}(\alpha)| \leq \hat{f}(\mathbf{0})$ for every α, all non-zero Fourier coefficients of f have absolute value $\frac{1}{2^k}$. Therefore, Rothschild and van Lint Theorem applies and f is the indicator function of an affine subspace of dimension $n - k$. Therefore, from now on, we assume $\hat{f}(\mathbf{0}) = \frac{1}{2^{k-1}}$.

The first step in our proof of the Main Theorem is to follow a similar procedure employed in the proof of Theorem 1. That is, whenever possible, we reduce the values of n and k simultaneously. This proceeds as follows. Suppose there exists a non-zero α with $\hat{f}(\alpha) = \frac{1}{2^{k-1}}$ or $-\frac{1}{2^{k-1}}$. Without loss of generality, assume that $\hat{f}(\alpha) = \frac{1}{2^{k-1}}$. Apply an invertible linear transform L that maps α to e_1 and let $g := Lf$. Now we have $\hat{g}(\mathbf{0}) = \hat{g}(e_1) = \frac{1}{2^{k-1}}$. Apply the restriction on the first bit of the input to get sub-functions g_0 and g_1. Then by (2), $\hat{g}_1(\mathbf{0}) = \hat{g}(\mathbf{0}) - \hat{g}(e_1) = 0$, which implies that $g_1 \equiv 0$. This implies that supp(f) is completely contained in the support of g_0 and moreover, by (3), $\hat{g}_0(\beta) = 2\hat{f}(0, \beta)$ for every $\beta \in \mathbb{F}_2^{n-1}$. In other words, g_0 is a Boolean function over $\mathbb{F}_2^{n-1}$ and $|\hat{g}(\beta)|$ is equal to either zero, or $\frac{1}{2^{k-1}}$, or $\frac{1}{2^{k-2}}$. That is, by performing a linear restriction, we reduce both the dimension n and the parameter k by one, so that the Main Theorem holds for Boolean functions over $\mathbb{F}_2^n$ as long as it holds for Boolean functions over $\mathbb{F}_2^{n-1}$.

When we arrive at a point that such a linear restriction is no longer possible; equivalently, f is irreducible, then $\hat{f}(\mathbf{0})$ is the only Fourier coefficient whose absolute value is $\frac{1}{2^{k-1}}$. Therefore, the Main Lemma for $k \geq 5$ or Lemma 5 for $2 \leq k \leq 4$ applies.

6 Concluding Remarks and Open Problems

In this work, we extend a classical result of Rothschild and van Lint to give a complete characterization of Boolean functions whose Fourier coefficients take values only in the set $\{-2/2^k, -1/2^k, 0, 1/2^k, 2/2^k\}$. Our work may be regarded as a first step toward understanding the structures of Boolean functions of granularity k. A major motivation for such studies is to prove a polynomial upper bound on the kill number for any k-granular Boolean function, thus resolving the Log-rank XOR conjecture. Another interesting question is to find other sets of Fourier coefficients which uniquely or almost uniquely determine the structures of their corresponding Boolean functions.

Acknowledgment. We would like to thank anonymous referees for their valuable comments and suggestions which help us correcting errors, simplifying proofs and improving presentations. Ning Xie's research was partially supported by grant ARO W911NF1910362.

A   A Proof of Proposition 2

Recall that Proposition 2 on the Fourier spectra of sub-functions obtained from linear restrictions is the following:

Proposition 2. *Let* $f : \mathbb{F}_2^n \to \mathbb{R}$ *be a function defined on the Boolean hypercube. Let* $f_0, f_1 : \mathbb{F}_2^{n-1} \to \mathbb{R}$ *be the "sub-functions" obtained from restricting the first bit of the input to 0 and 1, respectively; that is,* $f_0(y) := f(0, y)$ *and* $f_1(y) := f(1, y)$ *for all* $y \in \mathbb{F}_2^{n-1}$. *Then the Fourier spectra of* f_0 *and* f_1 *satisfy that, for all* $\beta \in \mathbb{F}_2^{n-1}$,

$$\hat{f}_0(\beta) = \hat{f}(0, \beta) + \hat{f}(1, \beta), \qquad \hat{f}_1(\beta) = \hat{f}(0, \beta) - \hat{f}(1, \beta). \qquad (2)$$

Conversely, the Fourier spectrum of f *satisfies*

$$\hat{f}(0, \beta) = \frac{1}{2}(\hat{f}_0(\beta) + \hat{f}_1(\beta)), \qquad \hat{f}(1, \beta) = \frac{1}{2}(\hat{f}_0(\beta) - \hat{f}_1(\beta)). \qquad (3)$$

Proof. Clearly it suffices to prove either (2) or (3) and the other follows immediately. We prove the first part of (3), the second part can be proved analogously. By the definition of Fourier transform,

$$\hat{f}(0, \beta) = \frac{1}{2^n} \sum_{x \in \mathbb{F}_2^n} f(x)\chi_{(0,\beta)}(x)$$

$$= \frac{1}{2^n} \sum_{y \in \mathbb{F}_2^{n-1}} \left(f(0, y)\chi_{(0,\beta)}((0, y)) + f(1, y)\chi_{(0,\beta)}((1, y)) \right)$$

$$= \frac{1}{2^n} \left(\sum_{y \in \mathbb{F}_2^{n-1}} f(0, y)\chi_\beta(y) + \sum_{y \in \mathbb{F}_2^{n-1}} f(1, y)\chi_\beta(y) \right)$$

$$= \frac{1}{2^n} \sum_{y \in \mathbb{F}_2^{n-1}} f_0(y)\chi_\beta(y) + \frac{1}{2^n} \sum_{y \in \mathbb{F}_2^{n-1}} f_1(y)\chi_\beta(y)$$

$$= \frac{1}{2}(\hat{f}_0(\beta) + \hat{f}_1(\beta)).$$

□

B   The Fourier Spectrum of Disjoint Union of two Affine Subspaces

In this section we calculate the Fourier spectrum of a Boolean function whose support is the union of two disjoint affine subspaces satisfying certain properties. In particular, the two affine subspaces are of the same dimension and their Fourier spectra have minimum intersection.

Let $n \geq 1$ and $0 \leq k < n$ be integers. If V is a linear subspace in $\mathbb{F}_2^n$ of dimension $n - k$ and $a \in V^\perp$, where $V^\perp$ denotes the linear subspace that is the *orthogonal complement* of V, then it is well known that the Fourier spectrum of the indicator function of affine subspace $a + V$ is (see e.g. [25]):

$$\hat{\mathbb{1}}_{a+V}(\alpha) = \begin{cases} \frac{1}{2^k} \chi_\alpha(a) & \text{if } \alpha \in V^\perp, \\ 0 & \text{otherwise.} \end{cases}$$

Let $f : \mathbb{F}_2^n \to \{0,1\}$ be a Boolean function whose support is the union of two disjoint affine subspaces of dimension $n - k$. By a shift of the origin if necessary, we may assume that one of the two affine subspaces is a linear subspace. Therefore $f = \mathbb{1}_{a+V_1} + \mathbb{1}_{V_2}$, where V_1 and V_2 are two linear subspaces of dimension $n - k$ in $\mathbb{F}_2^n$ and $a \in V_1^\perp$. In order for $a + V_1$ and V_2 to be disjoint, a necessary condition is that their orthogonal complement subspaces have non-trivial intersection, $V_1^\perp \cap V_2^\perp \neq \{\mathbf{0}\}$. The special configuration we are interested in is when this intersection is minimal, that is when $|V_1^\perp \cap V_2^\perp| = 2$.

To this end, without loss of generality, we let $V_1^\perp = \text{span}(e_1, \ldots, e_k)$ and $V_2^\perp = \text{span}(e_k, \ldots, e_{2k-1})$ so that $V_1^\perp \cap V_2^\perp = \{\mathbf{0}, e_k\}$. Then we necessarily have[8] $\langle e_k, a \rangle = 1$. Therefore for simplicity (and also without loss of generality) we may take $a = e_k$. Therefore the Fourier spectrum of f is

$$\hat{f}(\alpha) = \hat{\mathbb{1}}_{a+V_1}(\alpha) + \hat{\mathbb{1}}_{V_2}(\alpha) = \begin{cases} \frac{1}{2^{k-1}} & \text{if } \alpha = \mathbf{0}, \\ -\frac{1}{2^k} & \text{if } \alpha \in e_k + (\text{span}(e_1, \ldots, e_{k-1}) \setminus \{\mathbf{0}\}), \\ \frac{1}{2^k} & \text{if } \alpha \in \text{span}(e_1, \ldots, e_{k-1}) \setminus \{\mathbf{0}\}, \\ \frac{1}{2^k} & \text{if } \alpha \in e_k + (\text{span}(e_{k+1}, \ldots, e_{2k-1}) \setminus \{\mathbf{0}\}), \\ \frac{1}{2^k} & \text{if } \alpha \in \text{span}(e_{k+1}, \ldots, e_{2k-1}) \setminus \{\mathbf{0}\}, \\ 0 & \text{otherwise.} \end{cases}$$

$$(11)$$

References

1. Aggarwal, D., Dodis, Y., Lovett, S.: Non-malleable codes from additive combinatorics. SIAM J. Comput. **47**(2), 524–546 (2018). Earlier version in STOC'14
2. Ben-Sasson, E., Lovett, S., Ron-Zewi, N.: An additive combinatorics approach relating rank to communication complexity. J. ACM **61**(4), 22 (2014)
3. Ben-Sasson, E., Ron-Zewi, N.: From affine to two-source extractors via approximate duality. SIAM J. Comput. **44**(6), 1670–1697 (2015). Earlier version in STOC'11

---

[8] This is because, the affine subspace $a + V_1$ can be expressed as the solutions to a system of linear equations $a + V_1 = \{x \in \mathbb{F}_2^n \mid \langle x, e_i \rangle = a_i \text{ for every } 1 \leq i \leq k\}$, where $\{e_1, \ldots, e_k\}$ is an orthonormal basis for $V_1^\perp$, and $\{a_i := \langle e_i, a \rangle\}_{i=1}^k$ are the components under this basis. Now if $|V_1^\perp \cap V_2^\perp| = 2$, and because the intersection of the two orthogonal complement subspaces is a subspace, we may take $V_1^\perp \cap V_2^\perp = \{\mathbf{0}, e_k\}$ for convenience. On the other hand, $V_2 = \{x \in \mathbb{F}_2^n \mid \langle x, e_i \rangle = 0 \text{ for every } k \leq i \leq 2k-1\}$. $a + V_1$ and V_2 are disjoint if and only if there is no solution to the two systems of linear equations combined together, which is equivalent to the condition that $\langle e_k, a \rangle = 1$.

4. Bernasconi, A., Codenotti, B.: Spectral analysis of Boolean functions as a graph eigenvalue problem. IEEE Trans. Comput. **48**(3), 345–351 (1999)
5. Bhowmick, A., Dvir, Z., Lovett, S.: New bounds for matching vector families. In: Proceedings of 45th Annual ACM Symposium on the Theory of Computing, pp. 823–832 (2013)
6. Chattopadhyay, A., Mande, N., Sherif, S.: The Log-approximate-rank conjecture is false. In: Proceedings of 51st Annual ACM Symposium on the Theory of Computing (2019, to appear)
7. Chistopolskaya, A., Podolskii, V.: Parity decision tree complexity is greater than granularity, October 2018. http://arxiv.org/abs/1810.08668
8. Chow, C.: On the characterization of threshold functions. In: Proceedings of 2nd Annual IEEE Symposium on Foundations of Computer Science, pp. 34–38. IEEE (1961)
9. De, A., Diakonikolas, I., Feldman, V., Servedio, R.: Nearly optimal solutions for the Chow parameters problem and low-weight approximation of halfspaces. J. ACM **61**(2), 1–36 (2014). Earlier version in STOC'12
10. Even-Zohar, C.: On sums of generating sets in $\mathbb{Z}_2^n$. Comb. Probab. Comput. **21**(6), 916–941 (2012)
11. Freiman, G.: Foundations of a structural theory of set addition. American Mathematical Society, Providence, RI (1973). Translated from the Russian. Translations of Mathematical Monographs, Vol 37
12. Friedgut, E.: Boolean functions with low average sensitivity depend on few coordinates. Combinatorica **18**(1), 27–35 (1998)
13. Friedgut, E., Kalai, G., Naor, A.: Boolean functions whose Fourier transform is concentrated on the first two levels. Adv. Appl. Math. **29**(3), 427–437 (2002)
14. Gopalan, P., O'Donnell, R., Servedio, R., Shpilka, A., Wimmer, K.: Testing Fourier dimensionality and sparsity. SIAM J. Comput. **40**(4), 1075–1100 (2011). Earlier version in ICALP'09
15. Green, B., Ruzsa, I.: Sets with small sumset and rectification. Bull. Lond. Math. Soc. **38**(1), 43–52 (2006)
16. Green, B., Tao, T.: Freiman's theorem in finite fields via extremal set theory. Comb. Probab. Comput. **18**(3), 335–355 (2009)
17. Hatami, H., Hosseini, K., Lovett, S.: Structure of protocols for XOR functions. SIAM J. Comput. **47**(1), 208–217 (2018)
18. Konyagin, S.: On the Freiman theorem in finite fields. Math. Notes **84**(3–4), 435–438 (2008)
19. Łaba, I.: Fuglede's conjecture for a union of two intervals. Proc. Am. Math. Soc. **129**(10), 2965–2972 (2001)
20. Lin, C., Zhang, S.: Sensitivity conjecture and log-rank conjecture for functions with small alternating numbers. In: Proceedings of 44th Annual International Conference on Automata, Languages, and Programming, vol. 80, pp. 51:1–51:13 (2017)
21. Lovász, L., Saks, M.: Lattices, Möbius functions and communication complexity. In: Proceedings of 29th Annual IEEE Symposium on Foundations of Computer Science, pp. 330–337 (1988)
22. Lovett, S.: Communication is bounded by root of rank. In: Proceedings of the 46th Annual ACM Symposium on Theory of Computing, pp. 842–846 (2014)
23. Lovett, S.: Additive combinatorics and its applications in theoretical computer science. Theory Comput. 1–55 (2017)
24. MacWilliams, F.J., Sloane, N.J.A.: The Theory of Error-Correction Codes. North Holland, Amsterdam (1977)

25. O'Donnell, R.: Analysis of Boolean Functions. Cambridge University Press, Cambridge (2014)
26. O'Donnell, R., Servedio, R.: The Chow parameters problem. SIAM J. Comput. **40**(1), 165–199 (2011). Earlier version in STOC'08
27. O'Donnell, R., Wright, J., Zhao, Y., Sun, X., Tan, L.Y.: A composition theorem for parity kill number. In: Proceedings of 29th Annual IEEE Conference on Computational Complexity, pp. 144–154 (2014)
28. Rothschild, B.L., van Lint, J.: Characterizing finite subspaces. J. Comb. Theory Ser. A **16**(1), 97–110 (1974)
29. Ruzsa, I.: An analog of Freiman's theorem in groups. Astérisque **258**(199), 323–326 (1999)
30. Samorodnitsky, A.: Low-degree tests at large distances. In: Proceedings of 39th Annual ACM Symposium on the Theory of Computing, pp. 506–515 (2007)
31. Sanders, T.: A note on Freiman's theorem in vector spaces. Comb. Probab. Comput. **17**(2), 297–305 (2008)
32. Shpilka, A., Tal, A., lee Volk, B.: On the structure of boolean functions with small spectral norm. Comput. Complex. **26**(1), 229–273 (2017)
33. Tao, T., Vu, V.: Additive Combinatorics. Cambridge University Press, Cambridge (2006)
34. Tsang, H., Wong, C., Xie, N., Zhang, S.: Fourier sparsity, spectral norm, and the Log-rank conjecture. In: Proceedings of 54th Annual IEEE Symposium on Foundations of Computer Science, pp. 658–667 (2013)
35. Tsang, H., Xie, N., Zhang, S.: Fourier sparsity of GF(2) polynomials. In: Proceedings of the International Computer Science Symposium in Russia, pp. 409–424 (2016)
36. Zhang, Z., Shi, Y.: On the parity complexity measures of Boolean functions. Theoret. Comput. Sci. **411**(26–28), 2612–2618 (2010)

Author Index

Printed in the United States
by Baker & Taylor Publisher Services

Printed in the United States
by Baker & Taylor Publisher Services